Patty's Industrial Hygiene and Toxicology

Volume I
GENERAL PRINCIPLES

Volume II
TOXICOLOGY

Volume III
THEORY AND RATIONALE OF INDUSTRIAL HYGIENE PRACTICE

Patty's Industrial Hygiene and Toxicology

Volume III
THEORY AND RATIONALE OF
INDUSTRIAL HYGIENE PRACTICE

LESTER V. CRALLEY, Ph.D.
LEWIS J. CRALLEY, Ph.D.
Editors

Contributors

E. W. Arp, Jr.
K. A. Busch
K. J. Caplan
J. S. Chapman
W. C. Cooper
J. V. Crable
L. V. Cralley
L. J. Cralley
J. C. Davis

P. M. Eller
R. L. Fischoff
F. G. Freiberger
P. J. Gehring
J. C. Guignard
R. L. Harris, Jr.
B. J. Held
V. H. Hill
S. M. Horvath
N. A. Leidel

J. R. Lynch
C. R. McHenry
M. H. Munsch
R. L. Potter
C. H. Powell
K. S. Rao
R. G. Thomas
R. L. Thomas
R. S. Waritz

A WILEY-INTERSCIENCE PUBLICATION

JOHN WILEY & SONS, New York • Chichester • Brisbane • Toronto

Copyright © 1979 by John Wiley & Sons, Inc.

All rights reserved. Published simultaneously in Canada.

Reproduction or translation of any part of this work beyond that permitted by Sections 107 or 108 of the 1976 United States Copyright Act without the permission of the copyright owner is unlawful. Requests for permission or further information should be addressed to the Permissions Department, John Wiley & Sons, Inc.

Library of Congress Cataloging in Publication Data (Revised)

Patty, Frank Arthur, 1897–
  Patty's Industrial hygiene and toxicology.

  Vol. 3 edited by L. J. Cralley and L. V. Cralley.
  "A Wiley-Interscience publication."
  Includes index.
  CONTENTS: v. 1. General principles.—
v. 3. Theory and rationale of industrial hygiene practice.
    1. Industrial hygiene. 2. Industrial toxicology.
I. Clayton, George D. II. Clayton, Florence E.
III. Battigelli, M. C. IV. Title. V. Title:
Industrial hygiene and toxicology. [DNLM: WA400.3 P322 1978]

RC967.P37 1977   613.6'2   77-17515
ISBN 0-471-02698-0

Printed in the United States of America

10 9 8 7 6 5 4 3 2 1

Contributors

EARL W. ARP, Jr., Ph.D., Associate Professor, Department of Industrial Management, Clemson University, Clemson, South Carolina.

KENNETH A. BUSCH, Chief, Statistical Services Branch, National Institute for Occupational Safety and Health, Department of Health, Education and Welfare, Cincinnati, Ohio

KNOWLTON J. CAPLAN, Industrial Health Engineering Associates, Inc., Hopkins, Minnesota

JOHN S. CHAPMAN, M.D., Professor of Internal Medicine, University of Texas Health Science Center at Dallas, Dallas, Texas.

W. CLARK COOPER, M.D., Occupational Medical Consultant, Berkeley, California

JOHN V. CRABLE, Chief, Measurements Research Branch, National Institute for Occupational Safety and Health, Department of Health, Education and Welfare, Cincinnati, Ohio

LESTER V. CRALLEY, Ph.D., Formerly Manager of Environmental Health Services, Aluminum Company of America, Pittsburgh, Pennsylvania

LEWIS J. CRALLEY, Ph.D., Formerly Director, Division of Epidemiology and Special Services, National Institute for Occupational Safety and Health, Department of Health, Education and Welfare, Cincinnati, Ohio.

JEFFERSON C. DAVIS, M.D., Colonel, USAFMC, Aerospace Medical Division, Brooks Air Force Base, San Antonio, Texas

PETER M. ELLER, Ph.D., Measurements Research Branch, National Institute for Occupational Safety and Health, Department of Health, Education and Welfare, Cincinnati, Ohio

ROBERT L. FISCHOFF, Division Industrial Hygienist, International Business Machines Corporation, Bethesda, Maryland

FRED G. FREIBERGER, Safety Engineering/Emergency Control Program Administrator, International Business Machines Corporation, Bethesda, Maryland

PERRY J. GEHRING, D.V.M., Ph.D., Director, Toxicology Research Laboratory, Health and Environmental Research, DOW Chemical, U.S.A., Midland, Michigan

JOHN C. GUIGNARD, M.B., Ch.B., Research Medical Officer, Naval Medical Research Laboratory, Michaud Station, New Orleans, Lousiana

ROBERT L. HARRIS, Jr., Ph.D., Director, Occupational Health Studies Group, University of North Carolina, Chapel Hill, North Carolina

BRUCE J. HELD, Leader, Safety Science Group, Lawrence Livermore Laboratory, University of California, Livermore, California

VAUGHN H. HILL, Consultant—Acoustics, Formerly, Applied Technology Division, E. I. duPont de Nemours and Company, Inc., Wilmington, Delaware

STEVEN M. HORVATH, Ph.D., Director and Professor, Institute of Environmental Stress, University of California, Santa Barbara, California

NELSON A. LEIDEL, Chief, Technical Evaluation and Review Branch, National Institute for Occupational Safety and Health, Department of Health, Education and Welfare, Rockville, Maryland

JEREMIAH R. LYNCH, Research and Environmental Health Division, Medical Department, EXXON Corporation, Linden, New Jersey

CHARLES R. McHENRY, Manager, Environmental Health and Safety, XEROX Corporation, East Rochester, New York

MARTHA HARTLE MUNSCH, J.D., Reed Smith Shaw and McClay, Pittsburgh, Pennsylvania

ROBERT L. POTTER, J.D., Professor of Law, School of Law, University of Pittsburgh, Pittsburgh, Pennsylvania

CHARLES H. POWELL, Sc.D., Manager, Health Services Administration, PPG Industries Inc., Pittsburgh, Pennsylvania

K. S. RAO, Ph.D., Toxicology Research Laboratory, Health and Environmental Research, DOW Chemical, U.S.A., Midland, Michigan

ROBERT G. THOMAS, Ph.D., H-4 Group Leader, Mammalian Biology, University of California, Los Alamos Scientific Laboratory, Los Alamos, New Mexico

RANDI L. THOMAS, Mammalian Biology, University of California, Los Alamos Scientific Laboratory, Los Alamos, New Mexico

RICHARD S. WARITZ, Ph.D., Senior Toxicologist, Hercules Incorporated, Wilmington, Delaware

Preface

The quantitative aspects of industrial hygiene practice have been developed primarily during the past four decades, which have seen an ever-accelerating advancement in investigative techniques. Likewise, greater emphasis has been given to making environmental controls an integral part of the industrial process.

Concurrently, the trend has been away from the use of threshold limit values (TLVs) as guidelines and toward the incorporation of these values into legally binding limits. The Occupational Safety and Health Act of 1970 firmly established the promulgation of permissible exposure limits into legally binding governmental standards. The resultant momentum in establishing such values has created a great responsibility in assuring that the values are sound and rational. It is extremely important that the data on which the values are based are both valid and reproducible.

An exceedingly high level of professionalism has always existed in the field of industrial hygiene. It is important that this aspect not be weakened. Thus it is timely that the theoretical basis and rationale of industrial hygiene practice be again examined thoroughly to restate fundamental facts and to direct attention to areas of weakness before they become incorporated into acceptable practice by virtue of precedent.

LESTER V. CRALLEY, PH.D.
LEWIS J. CRALLEY, PH.D.

March 1979
Fallbrook, California

Notation

The subject areas covered in this volume are based on information and legal interpretations of regulations available in 1978. The practice of industrial hygiene necessitates a continuing updating in these areas.

Contents

1	**Rationale**	1
	Lester V. Cralley, Ph.D., and Lewis J. Cralley, Ph.D.	
2	**The Emission Inventory**	11
	Robert L. Harris, Jr., Ph.D., and Earl W. Arp, Jr., Ph.D.	
3	**Statistical Design and Data Analysis Requirements**	43
	Kenneth A. Busch and Nelson A. Leidel	
4	**Data Automation**	99
	Robert L. Fischoff and Fred G. Freiberger	
5	**Analytical Measurements**	191
	Peter M. Eller, Ph.D., and John V. Crable	
6	**Measurement of Worker Exposure**	217
	Jeremiah R. Lynch	
7	**Biological Indicators of Chemical Dosage and Burden**	257
	Richard S. Waritz, Ph.D.	
8	**Evaluation of Exposure to Chemical Agents**	319
	Charles H. Powell, Sc.D.	
9	**Evaluation of Exposure to Ionizing Radiation**	359
	Robert G. Thomas, Ph.D., and Randi L. Thomas	

10	**Evaluation of Exposure to Nonionizing Radiation** Charles R. McHenry	405
11	**Evaluation of Exposure to Noise** Vaughn H. Hill	425
12	**Evaluation of Exposures to Hot and Cold Environments** Steven M. Horvath, Ph.D.	447
13	**Evaluation of Exposure to Vibrations** John C. Guignard, M.B., Ch.B.	465
14	**Evaluation of Exposure to Abnormal Pressures** Jefferson C. Davis, M.D.	525
15	**Evaluation of Exposure to Biologic Agents** John S. Chapman, M.D.	543
16	**Toxicologic Data Extrapolation** Perry J. Gehring, D.V.M., Ph.D., and K. S. Rao, Ph.D.	567
17	**Health Surveillance Programs in Industry** W. Clark Cooper, M.D.	595
18	**Philosophy and Management of Engineering Controls** Knowlton J. Caplan	611
19	**Personal Protection** Bruce J. Held	647
20	**Job Safety and Health Law** Martha Hartle Munsch, J.D., and Robert L. Potter, J.D.	681
21	**Compliance and Projection** Martha Hartle Munsch, J.D., and Robert L. Potter, J.D.	719
	Index	737

Patty's Industrial Hygiene and Toxicology

Volume I
GENERAL PRINCIPLES

Volume II
TOXICOLOGY

Volume III
THEORY AND RATIONALE
OF INDUSTRIAL HYGIENE
PRACTICE

CHAPTER ONE

Rationale

LESTER V. CRALLEY, Ph.D., and
LEWIS J. CRALLEY, Ph.D.

1 GENERAL

Our technological society has been built on an orderly progression wherein each stage successively led into another at a rapidly advancing pace and scope. Future technological trends will depend on how well this technology can be harnessed to serve society.

The problems of our earliest ancestors and their approaches to the challenge of providing food, clothing, shelter, and protection have been reconstructed to some extent from archaeological information. Some insight of our ancestors' philosophy and approach to living was evident in ancient biblical writings, which assigned to mankind "dominion over the fish of the sea and over the birds of the air and over every living thing that moves upon the earth." In practice this philosophy has come to include all the natural resources of the air, land, and sea.

It is thus only natural that even in the earliest times people used their inquisitiveness, initiative, and talents to find increasingly better ways of living and protection. Very early on, however, it was undoubtedly recognized that a misuse of a useful tool or activity could have a deleterious effect. It must have been evident during the period that fire became a desirable adjunct of living that smoke was associated with eye irritation and headache whenever fire was built in a confined space such as a cave without proper air movement. Thus our ancestors must have been aware that adaptations were required to live in harmony with the environment and its constituency without feeling its adverse effect.

In the advancement of civilization it was found that increased production could be realized if people specialized in their work pursuits and each shared the fruits of work with others. This led to a continuing search for better ways of producing commodities at an increased volume with minimal manpower. It was soon learned, however, that increasing production through specialization whereby industrial activities were concentrated and conducted in confined space, created health problems among workers that may not have existed when production was scattered and done by small groups of laborers. Thus along with the earlier professionals who promoted the use of an ever-expanding technology, another group began to note the deleterious effects of an uncontrolled work environment on the health of the workers. Among these earlier observers were Hippocrates, Pliny the Elder, Galen, Celsus, Ellenborg, Paracelsus, and Ramazzini.

In this earlier period it became evident that the goal for production and commerce and the need for the workers to maintain a livelihood took precedence over the concern for the welfare of the workers. This foreshadowed the laws and regulations later promulgated to secure for the worker the right to work in an environment that does not cause injury to health.

Even though history is documented with descriptions of unsatisfactory environmental working conditions, earlier attention was directed primarily to the recognition of occupational diseases and their treatment rather than their prevention. Only during the past five or so decades have the basic disciplines of chemistry, physics, mathematics, engineering, biology, and medicine united to comprehensively study occupational diseases on a quantitative basis with the intent of controlling their causes and maintaining exposures to harmful substances below predetermined levels.

Out of the application of this approach to preventing the occurrence of occupational disease, the profession of industrial hygiene was established. Today industrial hygiene is widely acknowledged and accepted as the science relating the industrial environment to worker health.

Industrial hygiene is defined by the American Industrial Hygiene Association as "that science and art devoted to the recognition, evaluation and control of those environmental factors or stresses, arising in or from the workplace, which may cause sickness, impared health and well-being, or significant discomfort and efficiency among workers or among citizens of the community." As such, the practice of industrial hygiene encompasses the recognition, evaluation, and control of potential occupational health hazards. This requires a comprehensive program designed around the nature of the industrial operation, documented to preserve a sound retrospective record, and executed in a professional manner.

The basic components of a comprehensive industrial hygiene program include (1) an ongoing data collection schedule that provides the essentials necessary for assessing the current level of health hazards in the workplace, (2) participation in research including epidemiological studies designed to generate data useful in establishing threshold limit values (TLVs) and standards, (3) a data storage system that permits appropriate retrieval to study the long-term effects of occupational exposure, and (4) an integrated

RATIONALE

program capable of responding to the need for appropriate controls resulting from changing environmental influences.

2 HEALTH HAZARD DETECTION AND MEASUREMENT

To better understand the significance of on-the-job environmental health hazards, it is first desirable to view briefly the total 24 hr stress picture of the worker. This will permit a perspective in which the overall component stresses are related to the whole of the workers' health.

Our habitat, the earth and its flora and fauna, is in reality a chemical one, that is, an entity that can be described in terms of an almost infinite number of related elements and compounds, the habitat in which the species was derived and in which a sort of symbiosis exists that permits and supports the survival of the individual and the species.

The intricacy of this relationship is illustrated in the presently recognized essential dietary trace elements (copper, chromium, fluorine, iodine, molybdenum, manganese, nickel, selenium, silicon, vanadium, and zinc). All are toxic when ingested in excess, and all are listed in the TLVs that have become incorporated into standards. Some forms of several of these trace elements are classified as carcinogens. It is most revealing that certain trace elements essential for survival are, under certain circumstances, capable of destruction. Thus it is a case of how much.

The environment is both friendly and hostile. The friendly milieu provides the components necessary for survival (air, food, water, heat, etc.). On the other hand, the hostile environment constitutes a stress in which survival is constantly challenged. Therefore the body must possess a built-in mechanism to meet the continuous hostile insults of temperature variations, radiation, atmosphere pollution, food and water contamination, viral and bacterial invasion, and so on. This is accomplished in a remarkable manner by the ectodermal and endodermal barriers, supported by the backup mesodermal protective and biotransformation mechanisms in instances of invasion. However inasmuch as these mechanisms are also viable entities, they in turn are subject to the stress of the hostile environment and are not absolute. Therefore some kind of additional accommodation is generally reached with the environmental insults that are not properly handled by the protective mechanisms above. In the case of exposure to ultraviolet radiation, for example, it is obvious that avoidance of all such exposure is impracticable. Thus in addition to the body's own protective mechanisms, such as skin pigmentation, further accommodation is reached through the use of skin barriers, eye protection by means of sun glasses, and a managed limitation to exposure. In the case of allergens in food, cosmetics, air, fabrics, drugs, and so on, discovery of the offending agents and avoidance of them constitute the prime approach to accommodation, though medical management is also highly important.

The body is also subjected to stresses directly related to the nature and pattern of living. Though such stresses may be insidious and inflicted on the body without our realizing it, their effects on health are evident. Examples of such stresses are annoyance and personal habits. It is now realized that annoyance may be reflected in biochemical changes, physiological responses, and nervous system reactions. Though it is difficult to express annoyance in quantitative terms of dosage and response, it can constitute both a high source of stress for the individual and a baseline to which other environmental stresses are additive. Accommodation is reached through knowledge and understanding of the cause, with appropriate counseling, and so on.

The foremost health stress in the area of personal habits is generally recognized to be smoking. Not only have the adverse health effects of smoking been documented, but also an additive or synergistic effect has been shown when excessive smoking is combined with exposure to a vast number of other substances including asbestos, cement, coal, cotton, grain, iron oxide, moldy hay, silica, uranium, and wool. Other personal habits that lead to stress are over use of stimulants (alcohol, caffeine, etc.), self-medication (drugs), and improper diet. These are only a small representation of causes that will be shown as environmental health studies are expanded. Again accommodation is reached largely through understanding and counseling.

Thus the off-the-job gamut of environmental health stresses is wide and formidable. It is also one to which a degree of accommodation must be reached based on human judgments, feasibility, personal option, objectives, cost-effectiveness, and other factors. It is evident that although the components of the total environmental health stresses must be considered on an individual basis, each component cannot stand alone and apart from the others.

2.1 Detection

One of the basic concepts of industrial hygiene is that the environmental health hazards of the workplace can be quantitatively measured and recorded in terms that relate to the degree of the stress. Therefore a major effort has been devoted to the implementation of this concept.

The recognition of potential health hazards is dependent on such relevant basic information as (1) detailed knowledge of the industrial process and any resultant emissions that may be harmful, (2) the toxicological, chemical, and physical properties of these emissions, (3) an awareness of the sites in the process that may involve worker exposure, (4) job work patterns with energy requirements, and (5) other coexisting environmental stresses that may be important.

This information may be expressed in many ways, but the most effective form is the material-process flow chart that lists each step in the process along with the appropriate information just noted. This allows us to pinpoint areas of special concern. The effort, however, remains only a tool for the use of the industrial hygienist in the actual assess-

RATIONALE

ment of the environmental health hazards of the workplace. In the quantitation itself, many approaches may be taken depending on the information sought, the level of effort available, and the practicality of the sampling procedures shown below:

Direct	Indirect
Body dosage	Environment
Tissues	Ambient air
Fluids	Interface of body and stress
Blood	Physiological response
Serum	Sensory
Excreta	Pulse rate and recovery pattern
Urine	Heart rate and recovery pattern
Feces	Body temperature and recovery pattern
Sweat	Voice masking, etc.
Saliva[a]	
Hair[a]	
Nails[a]	
Mother's milk[a]	
Alveolar air	

[a] Not usually considered to be excreta; see, however, Sections 10 to 12, Chapter 7.

2.2 Direct Measurement

To measure directly the quantity of the environmental stress agent actually received by the body, fluids, tissues, expired air, excreta, and so on, must be analyzed to determine the agent per se or a biotransformation product. Such procedures may be quite involved, since the evaluation of the data at times depends on previous information gathered through epidemiological studies and animal research. Studies made on animals, moreover, may have used indirect methods for measuring exposure to the stress agent, necessitating appropriate extrapolation in the use of such values. Examples are blood levels now in use for the evaluation of environmental lead exposures and urinary fluoride levels for environmental fluoride exposures.

One decided advantage of biological monitoring is that a time-weighted factor is integrated that is difficult to estimate through ambient air sampling when the exposure is highly intermittent or involves peak exposures of varying duration. Conversely, it may fail to reflect adequately peak concentration per se that may have special meaning. Urine analysis may also provide valuable data on body burden in addition to current exposures when the samples are collected at specific time intervals after the exposure, such as at the end of the work shift and before returning to the job, to incorporate a suitable time lapse.

Sampling the alveolar air may be an appropriate procedure for monitoring exposures to organic vapors and gases. An acceleration of research in this area can be anticipated because of the ease with which the sample can be collected.

2.3 Indirect Measurement

The most widely used technique for the evaluation of occupational health hazards is indirect in that the measurement is made at the interface of the body and the stress agent (the breathing zone, skin surface, etc.). In this approach the stress level actually measured may differ appreciably from the actual body dose. For example, all the particulates of an inhaled dust are not deposited in the lower respiratory tract. Some are exhaled and others are entrapped in the mucous lining of the upper respiratory tract and eventually are either expectorated or swallowed. The same is true of gases and vapors of low water solubility. Thus the target site for inhaled chemicals is scattered along the entire respiratory tract, depending on their chemical and physical properties. Another example is skin absorption of a toxic material. Many factors, such as the source and concentration of the contaminant (e.g., airborne or direct contact) and its characteristics, body skin location, and skin physiology, relate to the amount of the contaminant that may react with or be absorbed through the skin.

This indirect method of health hazard assessment is nevertheless, a valid one when techniques used are the same or equivalent to those relied on in the studies that established the TLVs or standards.

The sampling and analytical procedures must relate appropriately to the chemical and physical properties of the agents (particle size, solubility, limit of sensitivity for analytical procedures, etc.). Other factors of importance are weighted average values, peak exposures, and the job demands, with the energy requirement, which is directly related to respiratory volume and retention characteristics.

2.4 Purpose of Data Collection

The scope of data collection depends largely on the purpose of the evaluation. In an epidemiological study in which the relationship between the environmental stress and the body response is sought, the stress factors must be characterized in great detail. This may require massive volumes of data suitable for statistical analysis and a comprehensive data procuring procedure so that a complete exposure picture may be accurately constructed. This is extremely important, since the data may be used later for a purpose not known at the time of the collection.

The collection of valid retrospective exposure data may be extremely difficult. If available at all, the data may be scanty, the sample collection and analysis procedures may have been less sensitive than current procedures, and the job activities of the workers may have changed drastically during the interim period. Other factors that may have produced changes in levels of exposure to hazardous materials in an industry

RATIONALE

include changing technology, effectiveness of control procedures, and housekeeping and maintenance practices.

The effect of national emergencies may significantly change the nature and extent of worker exposure to associated environmental agents. The experience of World War II is an example. The work week hours were increased in many industries, and priority was given to production. Substitute materials had to be used in many instances. Less attention was given to maintenance and housekeeping practices. Local exhaust ventilation equipment sometimes became ineffective or completely inoperative. Though the major impact of World War II on levels of exposure of workers to harmful agents probably occurred from about 1940 into the early 1950s, the effects of these exposures may show up in the older age work force as late as the middle 1980s.

In contrast to the collection of data for epidemiological studies, data collected for the purpose of standards compliance may require relatively few samples if the values are clearly above or below the standard values. if the values are borderline, the situation may call for a more comprehensive sampling strategy and may be a matter for legal interpretation. Scientifically, though, the data should be adequate to establish a clear pattern, with no single value being given undue weight.

It is absolutely essential that the data be valid, regardless of the purpose for which they were collected. This is a key factor in establishing TLVs and standards. It is also equally important in fact finding related to compliance with standards. Since judgment and action will in some way be passed on the data, validity is paramount if there is to be a bona fide basis for action.

2.5 Data Evaluation

The evaluation of environmental data that serve as a basis for determining whether a health hazard exists requires a denominator that characterizes a satisfactory workplace. In this respect the practice of industrial hygiene is founded on the premise that there is a direct relationship between dosage and response. Often there is disagreement on the nature of the curve at the lower end of the dose-response relationship. This is primarily due to the lack of precise data at the lower levels of dosage over long periods of time and the necessity to predict relationships through extrapolations. Also, at the lower end of the dose-response curve there may be a point at which the effects on health from ever-present extraneous stresses cannot be distinguished from that of a specific source. Thus it is a question of what levels of exposure can be designated for which there can be an acceptable risk.

We greatly need more information on how specific etiologic agents are related to such occurrences as hypersensitivity, hypersusceptibility, and carcinogenicity. Such data are requisite to the proper evaluation of exposures to these agents.

During the past three decades the American Conference of Governmental Industrial Hygienists has systematically and continuously reviewed available data for the purpose of recommending guidelines now known as threshold limit values. Frequently the guidelines were based on information so limited that some degree of data extrapolation was necessary. Inevitably standards based on such values are subject to legal challenge.

3 ENVIRONMENTAL CONTROL

The cornerstones of an acceptable industrial hygiene program can be described as follows: (1) proper identification of on-the-job environmental health hazards, (2) the measurement of such hazards, (3) data evaluation, and (4) environmental control. In essence, the success of the entire program depends on the method of implementation of this last phase, that is, the control strategy. The technical aspects of the program must encompass sound practices and must be related both to the worker and to the medical preventive program. Appropriate attention must be given to the cost-effectiveness of the strategy. This constitutes a challenge to the professionalism and to the ultimate contribution of the industrial hygienist. The effort and cost of the control program must be commensurate with its effectiveness when examined in context with the other off-the-job environmental health stresses.

The heart of the control program must rest with process and/or engineering controls properly designed to protect the workers' health. The most effective and economic control is that which has been incorporated at the stage of production planning and made an integral part of the process. With new processes this can be accomplished by bringing input at the bench design, pilot, and final stages of process development. It is neither good industrial hygiene practice nor sound economics to design minimal control into a process with the intention of adding supplemental control hardware piecemeal as indicated by future production or to comply with regulations. On the other hand, the need may exist for the judicious use of personal protective equipment under unique circumstances—for example, breakdowns, spills or releases, and certain repair, maintenance, and housekeeping jobs. The adequate control program must embrace a proper mix of process and/or engineering hardware, personal protective equipment, and administrative control. No single design can be made to fit all circumstances. Rather, each program must be tailored to fit the individual situation without violating the basic tenets of industrial hygiene practice.

4 EDUCATIONAL INVOLVEMENT

The industrial hygienist is obligated to become involved in education by making professional information and opinion available to other groups having a responsibility for the health of workers and improving their environment.

Professional organizations such as the American Industrial Hygiene Association, the American Conference of Governmental Industrial Hygienists, and the American Academy of Industrial Hygiene offer an excellent opportunity for the interchange of professional knowledge and the continuing education of the industrial hygienist. These professional organizations invite participation through technical publications, lectures, committee activities, seminars, and refresher courses. This participation by experts in the many facets of the profession will enhance the overall performance of the profession and will permit members to keep abreast of newer industrial technology in the recognition, measurement, and control of associated work stresses.

The industrial hygienist should have an active role in educating management concerning environmental stresses in the plant and the programs for their control. An alert management can bring pending situations to the industrial hygienist for study and follow-up, and thus prevent the inadvertent occurrence of health problems.

The educational involvement of the worker is extremely important. The worker has a right to know the status of his or her job environment, the factors that may be deleterious to health if excessive exposures occur, and the control programs that have been instituted. Knowledgeable workers are in a position to enhance their own protection through the proper use of control equipment such as local exhaust ventilation equipment and through the proper use of personal protective devices. A worker is often the first to observe that a control system is not operating properly and can inform management of this situation. In cases of emergency such as spills and leaks or equipment breakdown, the worker who is informed of the hazardous nature of the materials involved can better follow plans prescribed for such situations.

5 SUMMARY

Gigantic strides have been made during the past four decades in characterizing and controlling environmental industrial health hazards. Nevertheless, we are not able to evaluate many commonly occurring situations. For examples, we still are not able to adequately measure or understand such phenomena as gases adsorbed on inhaled particles or the presence of simultaneous multiple stresses.

It is vital that the techniques used in measuring occupational stresses give valid and reproducible values. The practice of industrial hygiene rests on proper judgment in evaluating valid data, combined with effective follow-through. The following chapters examine comprehensively the theoretical basis and rationale of industrial hygiene.

CHAPTER TWO

The Emission Inventory

ROBERT L. HARRIS, JR. Ph.D., and
EARL W. ARP, JR., Ph.D.

1 INTRODUCTION

An emission inventory for an industrial or commercial enterprise is a compilation of information from which one can calculate or estimate the rates (quantity per unit time) at which pollutants are released to the environment. For purposes of this chapter, only the emissions that contaminate workroom or community air are considered; emissions to surface or ground water, to soil, or to other environmental receptors are treated only as they may, in turn, result directly in emissions to workplace air or community air.

An emission inventory may be simple or complex. A rudimentary inventory may consist of source location, date, identification of process, a qualitative listing of materials used, and an index of size (e.g., annual production rate) for the subject enterprise. Such an inventory, along with emission factors generated by studies of other similar processes, will permit the making of an estimate of annual emissions. A comprehensive emission inventory, on the other hand, may contain sufficient detail to permit quantitation of emissions, including temporal variations, for a number of specific materials from each point of release in a complex industrial process.

An inventory of emissions, along with various other kinds of companion information, discussed later in this chapter, permits the making of estimates of the nature, and sometimes the intensity, of exposures to airborne agents in workplaces or in the community. The level of detail needed and achievable for the inventory depends both on the purposes for which it is to be used and the data sources, or data generating efforts, that can be utilized.

2 ELEMENTS OF AN EMISSION INVENTORY

The compilation of emission inventories is a well-established, specifically identifiable activity in the field of air pollution control. Practices and procedures have been highly developed and descriptions of the technology are available (1, 2). Emission inventory has been practiced in industrial hygiene for many years but has not been identified as a categorical work area in this field to the extent that it has been in the field of air pollution control. Although the types of data used and the techniques for obtaining them vary somewhat, the same basic elements appear in emission inventories in both fields of work.

2.1 Identification of Agents

Identification, evaluation, and control of hazards are the three basic steps in the practice of industrial hygiene and community air pollution control. An emission inventory, regardless of whether it is specifically identified as such, is necessary in all three areas and is particularly important in the first two, identification and evaluation. It is clear that if a hazard is to be dealt with, it must first be identified; this identification, and the recognition of an emission, is a rudimentary emission inventory. Evaluation requires more than identification. In situations involving release of chemical agents to the air, evaluation may include obtaining additional information such as quantity, character, and temporal variations of emissions, all of which are part of the emission inventory. The design and implementation of emission control requires detailed information about the emission source that goes beyond the level of detail usually required for an emission inventory.

The federal Toxic Substances Control Act of 1976 (3), among other things, provides for the collection of information regarding commercially produced chemicals. Such information will permit preliminary assessment of potential exposures and possible effects on health and the environment. Implementation of the act will require identification, by process and location, of many chemical agents in industry and commerce. The notification, reporting, and recordkeeping provisions of the act will facilitate the agent identification component of a comprehensive, plantwide, emission inventory.

Not all materials handled in industry and commerce are hazardous. More than 4 million distinct chemical compounds have been identified, and the number is increasing at the rate of about 6000 per week (4). Of the millions of compounds that exist, some 63,000 are thought to be in common use (4). Some toxic dose information, based on experimental animal work or other observations, is available on about 16,500 compounds (5). Some of these are relatively nontoxic, others are not in common use, and for many the toxicity information is fragmentary. Probably fewer than 1000 compounds have been identified with occupational health or community air pollution problems sufficiently to permit development of workplace or air quality standards. The identification, by means of an emission inventory, of materials that are released from a process or operation, however, is a fundamental step in the recognition of those that represent a potential hazard in the workplace or the community.

For manufacturing processes preliminary identification of potentially hazardous agents often can be based upon the identification of process raw materials, intermediate and by-product materials, and process end products. In commercial enterprises the identification of materials that are handled permits the singling out of those that may represent potential hazards. For combustion sources information on the composition of the fuel used, and the type of combustion equipment, permits qualitative identification of pollutant components.

The evaluation phase of an industrial hygiene or community air pollution problem requires, in addition to identification of the agents of concern, a number of other kinds of information that can be obtained in an emission inventory. Among the most fundamental of these is the identification and description of the site or location at which the contaminant is released to the air.

2.2 Identification of Emission Sites

The most cursory emission inventory may identify the site or location of an emission source only as a particular plant or commercial establishment. Such location information is generally useful only for preliminary surveys of community air pollution or for indicating the need for more thorough workplace exposure evaluation. Any emission inventory use other than agent identification alone requires more specific emissions location information. For example, in air pollution control the identification of specific stacks, vents, and other points of emission is necessary for diffusion modeling and for most impact evaluations and emission regulatory activities. For industrial hygiene purposes the specific workplace, process point, and perhaps even a particular process equipment opening (e.g., a mixer charging port) or work paractice (e.g., the handling of shipping bags after use) may be needed. Such location information is vital to the hazard assessment process. It is necessary for identifying the workers subject to exposure from the particular source and for identifying alternatives from which to select the means for control of any hazard caused by the emission.

An emission inventory that identifies both the materials emitted from a process and the specific locations at which these materials are released can serve as the first step, and perhaps the only step necessary, in the evaluation of a hazard and initiation of a control effort.

2.3 Time Factors in Emissions

Time resolution in emission inventories may be yearly, seasonal, monthly, weekly, daily, hourly, or even less than hourly, depending on specific needs and the availability of data. For initial surveys of community air pollution, yearly average emissions by plant site may be satisfactory. At the other extreme, the assessment of emissions to workplaces that involve cyclic or intermittent operations may require use of time intervals shorter even than one hour. For short time or intermittent operations—for example, the taking of materials samples at process sampling ports—the actual time interval of emission and the frequency with which the operation occurs should be recorded in the inventory.

In some cases the interval of record for the quantity of material released is relatively long—for example, a monthly record of solvent use—even though actual release may be cyclic or may occur over short intervals. In such cases the emission inventory record should contain a sufficiently detailed description of the process or operation to permit estimation of actual emission intervals and the quantities of materials released during these intervals.

It is important that the emission inventory record include both the date on which the inventory was done and the calendar interval for which it applies. When a change in process or operation occurs that materially affects the composition, quantity, or condition of an emission, the emission inventory record should be updated to reflect the change. In the absence of any substantial change, the inventory should be revalidated at convenient intervals, perhaps annually, or as may be required by governmental regulation. When an emission inventory record is updated or revalidated, the old record should be retained; a sequential inventory file over a long period may be invaluable in future retrospective environmental epidemiologic studies.

3 QUANTITATING EMISSIONS

Emissions can be quantitated either by direct measurement, such as source testing, or by indirect means. Indirect means include techniques such as process materials balance or the determination of an index parameter—for example, a production rate—that can be related empirically to emissions.

3.1 Source Sampling

Source sampling is ordinarily associated with measurement of air pollutant emissions. Under some circumstances the techniques can be applied to industrial hygiene investigations as well. The techniques for air pollutant source testing have recently been described in detail by Paulus and Thron (6); their chapter "Stack Sampling" lists 72 references. The Environmental Protection Agency (EPA) has published stepwise procedures on source sampling for particulates (7); this publication contains a number of data recording forms that are useful in sampling not only for particulates but for other agents as well.

Source sampling ordinarily consists of withdrawing a representative sample from a contaminant-bearing gas stream in a duct or stack. Analysis of the sample yields data on concentration of the contaminant in the gas stream. Concentration data, combined with companion data on gas flowrates in the ducts or stacks, yield values for contaminant emission rates for gas streams released to the atmosphere.

The critical concern in source sampling is the representativeness of the sample. Both composition and flowrate of a contaminated gas stream may vary as the processes and operations that generate it vary. Thus representativeness of a source sample depends very much on the representativeness of processes and operations at the time of sampling.

When the contaminant is particulate, the collection of a representative sample requires isokinetic sampling and use of an unbiased sampling traverse pattern. Isokinetic sampling is performed by taking the sample at such a flowrate that the sampled gas stream enters the inlet nozzle of the sampling probe with velocity equal to that which prevails at the specific point in the cross section of the stack or duct from which the sample is being withdrawn. When the velocity of gas at the sampling point in the duct or stack is greater than that in the sampling nozzle, part of the approaching gas stream is deflected around the nozzle. Smaller particles tend to follow the deflected gas stream while larger ones, by virtue of their momentum, tend to continue their trajectories and enter the nozzle; this results in a nonrepresentative overabundance of larger particles in the sample. When the velocity of the gas stream at the sampling point in the duct or stack is lower than the velocity entering the sample nozzle, the gas stream converges into the nozzle inlet, carrying with it the smaller particles but losing some of the larger ones, which are carried past the nozzle by their momentum; the sample then is nonrepresentative because of a deficiency in larger particles.

The velocity of the gas stream in a duct or stack is not uniform throughout its cross section. For this reason, and because particles are not necessarily uniformly distributed within a duct or stack, a specific traverse pattern is ordinarily used in source sampling and the measurement of velocity for determination of flowrates. For purposes of a sampling traverse, the cross-sectional area of the duct or stack is divided into equal sized subareas; sample increments and velocity readings are taken at the centers of these subareas. Sampling time should be the same for each traverse point in a duct or stack. Isokinetic sampling and equal area traverses are discussed in detail, including descriptions of apparatus and calculation methods, in the source sampling references cited earlier (6, 7).

The sampling of gases and vapors differs from sampling of particulates in that isokinetic sampling is not required unless concentrations differ from place to place in the duct cross section, and the collection apparatus and reagents used in a gas sampling train differ from those for particulates. Filtration, inertial size classification, and impingement with capture in liquid media are the collection mechanisms used for particulates source sampling; liquid absorption, adsorption on solids, and freeze-out are the methods usually employed for gas and vapor sampling. Sampling methods and analytic procedures for gases and vapors are described in Chapter 17, Volume I. Sampling apparatus, collecting media, and analytic methods for a number of gases and vapors have been tabulated by Paulus and Thron (6).

As mentioned earlier, the techniques of source sampling are most often applied in air pollution emission measurements. They can, in some circumstances, be applied for in-plant industrial hygiene purposes as well. When a workplace is served by general dilution ventilation in the exhaust mode, the techniques of source sampling can be applied to the exhausted airstreams to determine the rate at which contaminants are released to the workplace air. The calculation methods described in Section 5.3.1 of this chapter can be used to estimate emissions (the generation rate, G) when the concentration of contaminant in the exhaust air, the ventilation rate, and the room dimensions have been determined. In such applications of the equations mixing is expected to be such that $K = 1$. The procedure is most applicable when general dilution ventilation in the exhaust mode is the sole and controlling ventilation regime for the workplace, that is, when there is no mechanical local exhaust ventilation and when there is no local exhaust component to natural ventilation. If local exhaust ventilation is used in the workplace, its influence as general dilution ventilation must be taken into account when using this technique to estimate workplace emissions.

Source sampling techniques may also be used to quantitate emissions from individual points of release in a workplace. The emissions may be captured using a temporary exhaust ventilation setup, and the amount of material released may be determined by sampling from that exhaust stream. Application of this technique to a single source or emission point in a space that contains several sources of the air contaminant requires either elimination of the influence of the other sources or correction for them. A mechanical arrangement can be provided to supply contaminant-free outside air to the test source. The exhaust stream from the test source will then contain only contaminant from that source and will be unbiased by contaminant from other sources. Alternatively, monitoring of the concentration of contaminant in room air that supplies the source test exhaust system permits correction for other sources; emissions from the test source can be determined by the difference in contaminant concentration in supply and exhaust air of the test system.

3.2 Materials Balance

In some cases knowledge of processes and operations permits determination of the amount of material released to the air of a workplace without emission measurements. If it is known that a gas or vapor is generated by chemical reaction or otherwise, and is released to workplace air in proportion to the use of a raw material or a production rate, that index of generation can be used to determine the release rate. Examples include the generation of products of combustion by unvented open flames, as is the case with direct fired unit heaters. Here fuel composition and use rate are indices of contaminant emissions. Uses of volatile solvents in which the solvents do not become part of a product, but evaporate completely into the workplace air, are common in industry. Solvent use rate, in such cases, is also an emission rate. When exhaust ventilation is applied to some operations in a workplace and not to others, distinction must be made between that portion of the material which is captured by exhaust ventilation and that which is released in the occupied workplace; only the portion that is released directly to the workplace air is used to estimate workplace exposures. For estimating community air pollution emission rates, the total amount of volatile material that evaporates into the atmosphere is taken as the emission rate regardless of whether the material is released to workroom air or through exhaust ventilation systems.

The American Petroleum Institute (API) has reported mathematical relationships that describe evaporation losses of petroleum products from tanks during loading and unloading (8). The materials balance concepts of the API procedure can be applied to estimating vapor emissions from the loading of volatile liquids into vessels that are vented to workroom air. The mass of vapor expelled by displacement when a volatile liquid is transferred into a vessel is

$$M = 1.37 \ VSP_v \ \frac{mw}{T} \qquad (1)$$

where M = mass of vapor expelled (lb)
V = volume of liquid transferred to the vessel (ft^3)
S = fraction of vapor saturation of expelled air
P_v = true vapor pressure of the liquid (atm)
mw = molecular weight of the vapor
T = temperature of the tank vapor space (°R)

Except for S, the fraction of vapor saturation, the various parameters of Equation 1 are ordinarily known or can be measured easily. For splash filling of a vessel that was initially vapor free or for the refilling of a vessel from which the same liquid has just been withdrawn, the value of S can ordinarily be taken as 1 (8).

Equation 1 may overestimate the mass of vapor emission if the vessel walls are substantially colder than the volatile liquid. The true vapor pressure depends on the temperature of the liquid; in the case of the cold vessel, however, some vapor may condense on the inner wall surfaces and fail to escape through the vent into the workroom air.

When complete evaporation of a volatile material does not take place, some index other than total use is needed for quantitating emissions. In the simple case, when the amount used and the amount remaining can be determined, the amount released as vapor can be obtained by difference. When this is not possible, more sophisticated means such as exhaust air sampling must be employed.

With complete evaporation of a mixture of volatile materials, the quantity of each component that vaporizes is simply the quantity of that material in the mixture. Partial evaporation of a mixture, however, does not necessarily yield vapor quantity of each component in proportion to the quantity of that component in the liquid mixture. According to Raoult's law, the equililbrum partial pressure of each component of a perfect solution is the product of the vapor pressure of the pure liquid and its mole fraction in the solution:

$$p_n = P_n x_n \qquad (2)$$

where p_n = partial pressure of component n
P_n = vapor pressure of pure liquid n
x_n = mole fraction of component n in the liquid mixture

Thus vapor yielded by partial evaporation from a mixture of volatile materials is richer in the more volatile and leaner in the less volatile components than is the original liquid solution. The use of Raoult's law permits estimation of emission rates of components of a solution when partial evaporation takes place. The composition of the parent solution in each case must be known; values for vapor pressures of pure liquids can be found in chemical handbooks. When a substantial fraction of a liquid mixture evaporates, the change in its composition as the fractions of more volatile components decrease should be taken into account in applying Raoult's law.

Raoult's law should be used with caution in estimating emissions from partial evaporation of mixtures; not all mixtures behave as perfect solutions. Elkins, Comproni, and Pagnotto measured benzene vapor yielded by partial evaporation of mixtures of benzene with various aliphatic hydrocarbons, chlorinated hydrocarbons, and common esters, as well as partial evaporation of naphthas containing benzene (9). Most measurements for all four types of mixture showed greater concentrations of benzene vapor in air than were predicted by Raoult's law. Of five tests with naphtha-based rubber cements, one yielded measured values of benzene concentration in air in agreement with calculated values, the other four showed measured benzene concentrations in air to be 3 to 10 times greater than those calculated using Raoult's law.

Substantial deviation from Raoult's law is not always the case, however, even with benzene. Runion compared measured and calculated concentrations in air of benzene in vapor mixtures yielded by evaporation from a number of motor gasolines and found excellent agreement (10).

In the absence of other more certain means, Raoult's law can be used to estimate emissions generated by partial evaporation of mixtures that approximate ideal solutions or mixtures in which the solution is nearly pure in one component. The applicability of Raoult's law to the mixtures being assessed should be validated, or quantitative measurements of emissions should be done, if accurate emission values are needed.

For dilute solutions the partial pressure of the component present in lower concentration is given by Henry's law, expressed as follows:

$$p_n = H_n x_n \qquad (3)$$

where p_n = partial exposure of component n
H_n = Henry's law constant
x_n = mole fraction of component n in the liquid mixture

THE EMISSION INVENTORY

Henry's law is also applicable to the solubility of a gas in dilute liquid solution, and solubilities of gases in liquid may be expressed in terms of Henry's law constants. These constants and the applicable concentration range for valid use of Henry's law can be determined only empirically.

Application of materials balance concepts for determination of emissions other than those for combustion, chemical reaction, or evaporation of volatile materials, ordinarily requires engineering analysis on a case-by-case basis.

3.3 Emission Factors

The need exists for emissions estimates for large numbers of sources in community air pollution studies, but the impracticability of source-by-source emissions tests has led to the development of emission factors. An emission factor is a pollutant emission rate for a particular type of emission source expressed as a quantity of pollutant released per unit of activity of that type source. The unit of activity chosen in each case is one that can be determined and can be related quantitatively to emissions; it may be ton of product, million Btu of heat produced, mile of vehicle travel, or other such index unit. Emission factors represent typical emissions from a class of sources and ordinarily cannot be applied with confidence to individual sources. In the absence of other information, however, emission factors for a particular source type can give useful insights into the character and general levels of emissions from individual sources of that type.

The most reliable emission factors are those based on a combination of emission measurements, process data, and engineering analysis for a large number of sources. Those that do not have a theoretical basis and are derived from only one type of data, or from data from only a few sources, should be used with caution. In some cases sufficient knowledge or information is available to permit development of empirical or analytic relationships between emission rate and some process parameter such as material composition or stream temperature. Such factors are generally the most reliable of all and can even be applied to individual sources with reasonable confidence.

Several thousands of individual air pollutant emission factors for a large number of source types have been tabulated and reported by the EPA's Office of Air and Waste Management (11). Process descriptions and emission control practices, along with typical collection performance for various types of control, are presented for most of the source types covered. Table 2.1 lists major source types for which emission factors appear in the current publication. In a separate document the EPA has published emission factors for arsenic, asbestos, beryllium, cadmium, manganese, mercury, nickel, and vanadium for processes involving these materials (12).

Emission factors in the EPA tabulation generally apply to identifiable point sources in processes or operations from which pollutants are released to the atmosphere through vents or stacks. As such they have limited applicability to in-plant industrial hygiene assessments. They do, however, identify some of the air contaminants that are

Table 2.1 Major Source Types for Which Air Pollutant Emissions Factors Have Been Adopted (11)

1. External combustion sources
    Bituminous coal combustion
    Anthracite coal combustion
    Fuel oil combustion
    Natural gas combustion
    Liquefied petroleum gas comsumption
    Wood waste combustion in boilers
    Lignite combustion
    Bagasse combustion in sugar mills
    Residential fireplaces
2. Solid waste disposal
    Refuse incineration
    Automobile body incineration
    Conical burners
    Open burning
    Sewage sludge incineration
3. Internal combustion engine sources
    Average emission factors for highway vehicles
    Light-duty, gasoline-powered vehicles
    Light-duty, diesel-powered vehicles
    Light-duty, gasoline-powered trucks and heavy-duty, gasoline powered vehicles
    Heavy-duty, diesel-powered vehicles
    Gaseous-fueled vehicles
    Motorcycles
    Off-highway, mobile sources
        Aircraft
        Locomotives
        Inboard-powered vessels
        Outboard-powered vessels
        Small, general utility engines
        Agricultural equipment
        Heavy-duty construction equipment
        Snowmobiles
    Off-highway stationary sources
        Stationary gas turbines for electric utility power plants
        Heavy-duty, natural-gas-fired pipeline compressor engines
        Gasoline and diesel industrial engines
4. Evaporation loss sources
    Dry cleaning
    Surface coating
    Storage of petroleum liquids
        Fixed roof tanks
        Floating roof tanks

             Variable vapor space tanks
             Pressure tanks
         Transportation and marketing of petroleum liquid
             Fixed roof tanks
             Floating roof tanks
             Variable vapor space tanks
             Pressure tanks
5.   Chemical process industry
         Adipic acid
         Ammonia
         Carbon black
             Furnace process
             Thermal process
             Channel process
         Charcoal
         Chloralkali
         Explosives
         Hydrochloric acid
         Hydrofluoric acid
         Nitric acid
         Paint and varnish
         Phosphoric acid
         Phthalic anhydride
         Plastics
         Printing ink
         Soap and detergents
         Sodium carbonate
         Sulfuric acid
             Elemental sulfur-burning plants
             Spent-acid and hydrogen-sulfide-burning plants
             Sulfide ores and smelter gas plants
         Sulfur
         Synthetic fibers
         Synthetic rubber
         Terephthalic acid
6.   Food and agricultural industry
         Alfalfa dehydrating
         Coffee roasting
         Cotton ginning
         Feed and grain mills and elevators
         Fermentation
         Fish processing
         Meat smokehouses
         Nitrate fertilizers

Table 2.1 (Continued)

      Orchard heaters
      Phosphate fertilizers
          Normal superphosphate
          Triple superphosphate
          Ammonium phosphate
      Starch manufacturing
      Sugarcane processing
7. Metallurgical industry
      Primary aluminum production
      Metallurgical coke manufacturing
      Copper smelters
      Ferroalloy production
      Iron and steel mills
          Pig iron manufacture
          Steel-making processes
          Scarfing
      Lead smelting
      Zinc smelting
      Secondary aluminum operations
      Brass and bronze ingots
      Grey iron foundry
      Secondary lead smelting
      Secondary magnesium smelting
      Steel foundries
      Secondary zinc processing
8. Mineral products industry
      Asphaltic concrete plants
      Asphalt roofing
      Bricks and related clay products
      Calcium carbide manufacturing
      Castable refractories
      Portland cement manufacturing
      Ceramic clay manufacturing
      Clay and fly-ash sintering
      Coal cleaning
      Concrete batching
      Fiber glass manufacturing
          Textile products
          Wool products
      Frit manufacturing
      Glass manufacturing
      Gypsum manufacturing
      Lime manufacturing
      Mineral wool manufacturing

THE EMISSION INVENTORY

       Perlite manufacturing
       Phosphate rock processing
       Sand and gravel processing
       Stone quarrying and processing
9. Petroleum industry
       Petroleum refining
          Crude oil distillation
          Converting
             Catalytic cracking
             Hydrocracking
             Catalytic re-forming
             Polymerization, alkylation, and isomerization
          Treating
             Hydrogen treating
             Chemical treating
             Physical treating
          Blending
          Miscellaneous operations
       Natural gas processing
10. Wood processing
       Chemical wood pulping
          Kraft pulping
          Acid sulfite pulping
          Neutral sulfite semichemical (NSSC) pulping
       Pulpboard
       Plywood veneer and layout operations
       Woodworking operations
11. Miscellaneous sources
       Forest wildfires
       Fugitive dust sources
          Unpaved roads (dirt and gravel)
          Agricultural tilling
          Aggregate storage piles
          Heavy construction operations

---

generated by these processes; the same contaminants represent potential in-plant exposures.

3.4 Fugitive Sources

Contaminant emissions from point sources such as tank vents or transfer points, and from processes and operations that clearly involve release of a process material—for example, release of volatile components in cementing or painting operations—are ordinarily capable of identification and quantitation. There are other types of source not so easily accommodated and sometimes neglected in emission inventories. Such sources, often called fugitive sources, may be intermittent, temporary, or unpredictable; many are unrecognized or ignored in emission inventories. Fugitive sources and the generation of secondary pollutants deserve attention in emisson inventories, however, even though all of them may not be capable of quantitation or prediction.

Fugitive sources that are of consequence primarily in the field of air pollution, and for which emission factors have been developed, include forest wildfires, unpaved roads, agricultural tilling, aggregate storage piles, and heavy construction operations (11). Other sources that are consequential from the standpoints of both air pollution and industrial hygiene include the following:

- Urban fires.
- Industrial process fires.
- Materials spills (accidents, equipment failure or malfunction).
- Sample collection and analysis.
- Process leaks (flanges and piping, valves, packing glands, conveyors, pumps, compressors, tanks and bins, etc.).
- Relief valves and control device bypasses.
- Maintenance activities (tank and vessel cleaning, cleaning a filter, replacing piping, pumps, etc.).
- Emissions from waste streams and reemissions of collected materials.
- Secondary reactions (nonproduct process reactions, extraprocess reactions).

Urban structural fires are intermittent phenomena predictable only in the aggregate and not as individual events; they are considered to be emission sources primarily in the air pollution sense. An industrial hygiene consequence of urban fires, however, is the exposure to toxic materials of firefighters and others who may be involved in rescue or control activities. A number of toxic atmospheric contaminants are generated by structural fires; these tend to vary from one fire to another. Of concern in all structural fires, however, is emission of carbon monoxide. Burgess et al. (13) have described exposures of emergency personnel to carbon monoxide emissions in real fire situations; the maximum sustained air concentrations of carbon monoxide to which these persons were exposed was about 2 percent. Materials balance using pyrolysis and oxidation processes offer one means of estimating the quantities of pollutants that may be generated in any particular case.

Estimates of emissions from industrial fires require individual analysis. The quantities and characteristics of combustion products depend on the nature of the materials that burn and the circumstances of combustion (e.g., whether open or confined). Again, consideration of pyrolysis and oxidation phenomena may permit estimation of the nature and quantities of air contaminants generated by any particular event. The likelihood and consequences of accidental fires is an appropriate consideration in industrial emission inventories.

Materials spills in manufacturing, transporting, and uses of industrial materials are not infrequent occurrences. Studies of processes and operations with specific attention to spills can reveal the frequencies and magnitudes of spills if they are usual occurrences (14). Such spills can then become part of an emissions inventory. When spills are not a usual occurrence, an emission inventory can do little more than trigger consideration of the possibilities and consequences of such events.

Process leaks are sometimes of major consequence from the standpoint of industrial hygiene. Control of process leaks, for example, has been a major factor in achieving acceptable working conditions in vinyl chloride polymerization plants. Process leaks can be identified by inspection or instrumental methods; timely repair may obviate quantitation for an emission inventory. Should quantitation be necessary, the techniques of source testing or materials balance may suffice.

As acceptable exposure levels become lower, attention to detail in emission evaluation takes on added importance. Routine tasks such as process sample collection may present a potential for excessive exposure under the traditional procedure of sampling at open manhole covers, open sample containers, and unconfined sample streams. Bell addressed this problem for vinyl chloride sampling and suggested techniques that may find application in other industries, particularly petrochemical operations (15).

Release of materials to community air or to a workplace through process relief valves occurs from time to time. The frequency of operation of these devices and magnitudes of releases, obtained from plant records or other sources, can be used for emission inventory purposes. Recognition of the existence of relief values or control device bypasses in a process is important in an emission inventory.

Kletz has stated that many of the foregoing sources of fugitive emissions can be controlled with emergency isolation valves (16). In addition to suggesting a method of control, this author also presents a leakage profile for both an olefin plant and an aromatic plant that could serve as the basis for developing a routine checklist at similar operations.

Volatile air contaminants may be released to the atmosphere from process sewers, drainage ditches, and/or collecting ponds. Such releases are of concern from the standpoint of both air pollution and industrial hygiene. Two examples are offered. Consider a case in which a gravity separator is used in an enclosed benzene recovery system to separate the organic and aqueous streams. Water from this separation may be reused for other purposes elsewhere in the plant. Even though benzene is only slightly soluble in water, such water reuse can result in measurable concentrations of benzene in the air at workplaces in the plant where benzene is not used. Consider another case in which alkaline scrubbing water is used to remove fluorides from a process waste gas stream. Discharge of the scrubbing water to a waste pond in which the pH is low permits release to the atmosphere of the collected fluorides. The use of the air pollution control scrubber in such a case only relocates the site of fluoride emission from the process stack to the waste pond. Air sampling and materials balance techniques are means for assessing emissions from waste streams and reemissions of collected materials.

Descriptions of processes and products do not always reveal whether by-products are formed or whether any by-products and their parent chemicals are stable throughout the manufacturing process and in the environment. This issue has been discussed, and an approach to assessing the significance of by-products and secondary reactions has been presented (17). The likelihood and consequences of hydrolysis, pyrolysis, oxidation, and other reactions are factors in the assessment. The identification of processes and principal materials is the first step in exploring for secondary pollutant emissions in an emission inventory. Literature review and perhaps laboratory exercises may yield indicators of the presence of secondary contaminants; air sampling may be required for validation.

Although it is not always possible to make quantitative, or perhaps even qualitative, assessment of fugitive sources and secondary pollutants, the possibilities of their occurrence and the opportunities for intervention to protect workers and the community merit scrutiny in emission inventories.

4 IN-PLANT EMISSION ESTIMATES FROM RECORDS AND REPORTS

In the absence of in-plant measurements of emissions, estimates derived from secondary sources of information can be used for inventory purposes. In some cases data are available from similar processes and estimates by analogy may prove sufficient for a given purpose. In other cases unique features of a process or operation may make estimates by analogy inappropriate. An alternative to measurement data or analogy makes use of records and reports whose basic purposes were other than environmental but whose content may be extracted, combined, or otherwise manipulated into a usable estimate of emissions and potential exposure. Vital information may be recorded in a variety of business documents including engineering, accounting, production, quality control, personnel, and governmental records. Although these records represent a rich source of information for emission inventory purposes, they can serve other purposes in the practice of industrial hygiene and occupational health research as well. Some of these purposes are mentioned from time to time in the following discussions of specific types of records.

4.1 Research and Development Records

Research and development units often issue a variety of reports to producing units as aids to bringing a new product on-stream, to ensure a required level of uniformity in a given item produced by different plants, or to make changes or improvement in operating equipment, materials, or techniques. Valuable insight with respect to materials and work practices employed in the past may be gleaned from several of these records. Among such records are the following:

1. *Product specification.* Final product and intermediate component data included in a product specification often give dimensions, weight, materials of construction, processing aids, processing conditions, tools and fabricating equipment, and special notes.

2. *Standard operating procedure.* This record may exist under a variety of names; SOP, Standard Practice, Uniform Methods, and (Company Name) Manual of Operations appear among the myriad of titles. All, however, share fairly specific operational instructions, usually in a step-by-step approach by processing sequence. Vital data from this record can include:
    a. Fabrication equipment.
    b. Authorized materials of construction.
    c. Acceptable processing aids.
    d. Processing conditions.
    e. Alternate materials for cyclic operations (e.g., summer vs. winter stocks).
    f. Change notices affecting materials or techniques.
    g. Instructions regarding protective equipment.

3. *Formulations and raw materials.* Detailed data concerning the qualitative aspects of potential exposure can be derived from study of information describing the raw materials employed in the process.
    a. Raw material specifications may contain composition data, restrictions on contaminants, and acceptance testing schedules.
    b. A listing of authorized vendors can offer clues to potential contaminants based on knowledge of a vendor's source of supply or processing methods.
    c. Formulation records frequently contain the composition of mixtures as well as the methods, equipment, quantity, and conditions of processing.

4.2 Analytical Laboratory Records

Quality control considerations often dictate that incoming raw materials be monitored for selected chemical and physical properties, that supplies from prospective vendors be subjected to acceptance testing, and that production quality be assayed by sampling at intermediate stages of the process. A single journal entry of analytical results may include a wealth of intelligence applicable to an emissions inventory. Analytical laboratory records include the following:

1. *Raw material test results.* Recorded in either the analyst's journal or on prepared forms will be entries such as:
    a. Analytical results vis-à-vis specifications.
    b. Date of analysis, and perhaps date of receipt.

c. Quantity of the shipment.
   d. Vendor.
   e. Method of analysis.
   f. Analyst

2. *Authorized vendors* Quality control laboratories often maintain vendor-supplied information regarding purchased materials. Material data sheets, product specifications, production methods, and quality control statements all offer bits of useful data.

4.3 Process and Production Records

Records generated by production units relate to such matters as scheduling, quality control, cost control, and to a lesser extent, training exercises. Depending on the product and process, production-type records can be of considerable value in generating an emissions inventory.

1. *Flow chart.* A detailed diagram of a process is particularly useful in identifying points with potential for release of air contaminants.

2. *Operating conditions.* Time, temperature, and pressure data are useful in assessing possible release of reaction products, by-products, degradation products, and unreacted raw materials.

3. *Scheduling.* Shift, cyclic, and seasonal variations may influence the emissions of contaminants. Such variations can be ascertained through production scheduling records. Included under this heading are records of batch versus continuous production modes. Gantt chart records offer a particularly attractive systematic source of scheduled activities normally both precise and detailed.

4.4 Plant Engineering Records

Virtually all industrial organizations include an engineering component responsible for the physical plant and supporting facilities. This responsibility results in a plant archive that is particularly useful in emissions inventory activity and in research that requires reconstruction of past environmental conditions.

1. *Plant layout.* The plant floor plan, coupled with elevation drawings, affords a visualization at the process, materials flow, and occupied areas, and possibly insights relating to contaminant generation and control. Furthermore many engineering drawings are cross-referenced to other drawings of equipment, emissions control features, and adjacent work areas. Useful characteristics of plant engineering files are:
   a. Processing equipment type, extent, and location appear on layouts.
   b. Department floor space showing geographic boundaries and physical barriers appear on some plates.
   c. Engineering controls, particularly ventilation systems, are depicted in mechanical equipment layouts, which normally show the location, site, type,

and rating of air-moving equipment, as well as a reference to the detailed plate covering that system.
   d. Dates of change or modification are often included on drawings, either by direct entry or by reference to a new set of plates.
2. *Equipment specifications.* Files on processing equipment installed by or under the supervision of plant engineers often include specifications that supply clues to potential problems such as process emissions, sound power rating and directivity data, or emission control features incorporated in the equipment.
3. *Project records.* Each major project of construction, modification, or installation normally proceeds as an integral project with one individual assigned as coordinator. Usually this project officer accumulates records that include data on:
   a. Equipment specifications.
   b. Performance checks and acceptance test results.
   c. Prime contractors and subcontractors.
4. *Material balance.* A detailed accounting of materials, products, and side streams, either measured or theoretical, serves as a basis for a quantitative estimate of emissions.
5. *Emission estimates for regulatory agencies.* For a number of years regulatory agencies have required an accounting of process losses for air and water pollution surveys, for permit applications, and for other regulatory purposes. These documents provide a record of:
   a. Process description.
   b. Effluent discharge.
   c. Contaminant(s) identification.
   d. Concentration parameters.
   e. Controls in effect.

4.5   Industrial Engineering Records

The practice of industrial engineering includes the analysis of work procedures to improve either work conditions or productivity or both. Especially for the period since World War II, industrial engineering records contain information on work methods and standards that are useful for current emission inventories and for estimating past emissions and exposures that may have resulted from them. Such records include:

1. *Job descriptions.*
   a. Description numbers, department listings, and simple descriptive job titles can provide the link between the tasks performed and the work history recorded in an individual's personnel file.
   b. Job location within a plant, thus the potential for various exposures, may be ascertained from the geographic area, processing equipment, or department listing contained in a job description.

c. The listing in a job description of tools, equipment, and processing aids can help establish the nature of emissions and characteristics of potential exposures (e.g., inhalation, absorption).
d. Task evaluations often include judgments concerning such health-related items as work conditions, safety requirements, hazards, skill, effort, and responsibility.

2. *Time and motion studies.* Task breakdowns are useful to establish contact with materials of interest, and the fraction of time spent on certain tasks coupled with cycle time can afford an estimate of the extent of exposure.

3. *Process analysis.* Evaluations of an entire process focus on tasks performed by all the members of a crew rather than those of individual workers. Particularly useful in process analysis reports are listings of fractions of a workday spent at different work stations by members of the operating crew, including break time and sequencing of tasks.

4.6 Accounting Records

Accounting records cover virtually all phases of an operation for both outside reporting and internal control purposes. Major areas of interest for emission inventory purposes include records on incoming materials, internal use of these raw materials, and plant production factors (18).

1. *Purchasing.* Raw materials for both production and support operations, as well as processing equipment, are ordinarily obtained from outside the company, and an acquisition record is generated upon receipt. Even when raw materials are drawn from captive sources, records of resource depletion may be available.
   a. *Purchasing.* Materials, quantities, and supplier may be available through the ledger, often in the accounts payable portion.
   b. *Fixed assets.* Capital equipment is normally amortized by a recognized accounting procedure, the evaluation of which can be employed to determine the date of acquisition. Alternately many organizations maintain a property book that contains the same information.
2. *In-plant issues.*
   a. *Raw material consumption.* Often these data can be developed through accumulation of issues to departments, cost centers, or plants through collected expense or work-in-process accounts.
   b. *Cost accounting.* Material variance accounts may serve as a surrogate for material balance data if the latter are unavailable. The absolute variance value between actual and standard material usage may be of marginal value, but changes in this quantity can be important in establishing trends.
   c. *Stores' accounts.* Nonproduction supplies are often issued to maintenance and other support departments through a "general store" account, with the receipt listing the item and quantities issued.

3. *Inventory.* Inventory control may be maintained through stock records that list material, vendor, date and quantity received, date and quantity of issue, department to which issued, and location and quantity on hand.

4. *Production.* Finished goods and work-in-process production accounts normally include an identification of the work schedule, days operated, and average production level.

4.7 Personnel Records

Personnel records, per se, provide some insights into the nature and locations of contaminant releases applicable to emission inventories. In addition they are vital to other industrial hygiene occupational health research activities. In any study of causal association between conditions of work and health experience, the work experience of each individual is an integral component. Thus accurate and reliable personnel-type records are vital. Organizational units normally developing useful records include the personnel, payroll, industrial relations, safety, and medical departments.

1. *Personnel records.* The file on a particular employee usually lists department and job assignments chronologically and gives a brief description of tasks, limitations, and periods of absence. Earnings records sometimes provide a clue to potential exposure through a listing of special pay or rates for hazardous tasks, and exposure duration or regimen by shift differentials or overtime pay.

2. *Industrial relations.* Seniority listings and work force distribution charts are useful in establishing the locations of jobs and personnel currently or at various times in the past.

3. *Safety and medical.* Insurance carrier survey reports, fire and explosion inspection reports, sickness/accident/compensation records, admininstrative control documents that limit time at a particular task or require worker rotation, along with written SOPs for activities such as entering tanks and vessels and other hazardous tasks, represent a record of emission sites and past working conditions, however subjective, suitable to at least rank-order jobs by exposure potential to certain agents.

4. *Union records.* Seniority listings and union membership rolls can help to identify the working populations of past years, and records of negotiated agreements between labor and management often contain provisions for work schedules, protective clothing and equipment, provisions for personal hygiene, clothing changes, and other health-related clauses that help to describe conditions of work.

4.8 Government Records and Reports

Data so broad as to characterize an entire industry, or so specific as to deal with a single chemical compound, are routinely and systematically collected, assembled, and published by various government agencies. Most of these records are available to the enterprising investigator, usually at little or no cost. A few of these record types are mentioned here as illustrations.

1. *Nationwide statistics.* The U.S. Tariff Commission publishes an annual listing of Production and Sales of Synthetic Organic Chemicals (19). It includes listings of raw materials, the production chemicals by use category (surface-active agents, etc.), and a directory of manufacturers of each material. Time series analyses of the entries can indicate the dates of introduction, extent of use, and dates of decline of a given material.

2. *Industrywide data.* Government directed or controlled operations such as government owned–contractor operated plants compile data for required reports, many of which find their way into the public domain. For example, information on virtually all facets of the synthetic rubber industry during the 1940–1945 period are available through the various reports of the Rubber Reserve Company (20, 21). Included are flow charts, raw materials, formulations, plant capacities, and locations.

3. *Specific products.* On June 15, 1844, Charles Goodyear received U.S. Patent 3633 for a compound consisting of gum elastic, sulfur, and white lead (22). Thus the approximate period of time at which rubber compounders could be considered as potentially exposed to white lead as a rubber accelerator is established. Roughly 4 million U.S. patent numbers have accrued since the mid-nineteenth century, doubtless including many materials, processes, and devices of interest to health investigators.

4.9 Record Location

Within most corporate industrial organizations, records of major additions or changes, as well as events affecting the entire operating units, are likely to be available at a central location, such as the division or corporate offices. Examples of these types of records are overall production levels, descriptions of uniform practices, and product specifications. More specific records, or those dealing with minor plant alterations, are ordinarily available only at the local plant level. Current records are ordinarily found in the departments identified in Sections 4.1 through 4.7 of this chapter; historical records may be found in those departments or in a central plant records storage facility. At the plant level project officers often retain personal copies of papers related to their projects; if a plant record has been discarded, such personal records may provide valuable historical data.

4.10 Output from Records

Measures derivable from secondary sources of data are limited by the nature and detail of the record content. In general, the more distant the time of interest, the less complete will be the record, and the lower will be the confidence in its accuracy. When data are sparse and only plantwide data are available, estimated measures such as gallons per cubic foot of plant building volume per month, or pounds used per employee per day, may be the limit of detail, but this may serve for plant-to-plant contrasts. More detailed data such as quantities of materials used in specific departments and descriptions of controls for specific work areas permit more detailed estimates. Material generation rates coupled with general area ventilation rates may enable one to estimate an area average concentration. Such data may be particularly useful for within-plant contrasts or for plant-to-plant studies of particular areas or processes.

Records pertaining to the detailed nature of tasks, such as standard operating procedures or job descriptions, make possible a ranking of jobs—thus individuals—by potential for exposure. In this fashion personnel with similar exposure histories can be grouped for comparison of health experience with those sharing different common exposure profiles.

As an example of the use of such information, let us offer an illustration of the type of emissions output that has been accomplished for retrospective research purposes through the use of secondary sources of information.

In Figure 2.1 the standard operating procedure is the pivotal document that identifies both tasks and materials for a step in manufacturing. A *formulation* record, or series of records, for each material used in that particular manufacturing step identifies each raw material *constituent* of each material. Examination of the raw material *specification* for each constituent then permits identification of *agents* (e.g., coal tar naphtha) that may

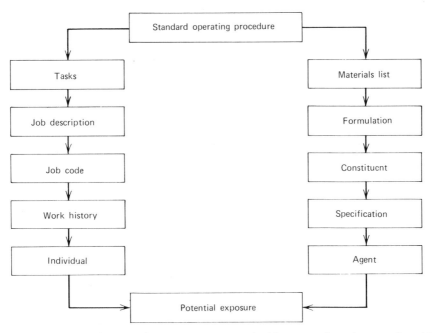

Figure 2.1 Synthesis of potential exposures associated with a manufacturing step for which a standard operating procedure is available.

be emitted in the workplace, often with additional compositional detail (e.g., aromatic content or benzene content). Because formulations, constituents, and specifications change from time to time, the listing of potential emissions thus identified must show inclusive dates on an agent by agent basis.

The *tasks* portion of the SOP includes a *job description*, which in turn has a specific *job code*. The job code is listed in the personnel record of each person who has held that particular job. Examination of the personnel records of the plant's current and past workers identifies complete *work history*, including dates and duration of each job code assignment for each *individual* who has held the specific job code of interest.

In this way the potential for contact with a particular chemical agent by an identified individual, including dates and duration of the contact, can be developed, even for events that occurred in the distant past.

5 USES OF EMISSION INVENTORIES

Emission inventories are fundamental to the identification, evaluation, and control of industrial hygiene and community air pollution hazards associated with chemical agents. Not all uses for inventory information can be mentioned here, but a few are discussed.

5.1 Regulatory Requirements

Various governmental regulatory programs in occupational safety and health, air pollution, and toxic substances control have reporting requirements that involve emission inventory information. Regulations regarding the nature and content of these reports, the agents covered, and the types of establishment with reporting obligations change periodically. Persons responsible for industrial or commercial enterprises that involve emissions to community air, or the handling or production of agents that may be covered in occupational health or toxic substances regulatory programs, should keep abreast of the specific requirements imposed by state and federal regulations. Emission inventory programs can then be so designed and operated that the emissions data necessary to satisfy reporting requirements are obtained.

5.2 Community Air Pollution Dispersion Estimates

Data on emissions of a pollutant to community air and companion meteorological data permit estimation and prediction of ground level concentrations of that pollutant at any desired location in the community. Methods and procedures for manual calculations to estimate the impact of individual sources (23), and a basic computer program for calculating concentrations resulting from emissions from large numbers of point and area sources (24) have been reported. The mathematical relationships and computational techniques are too extensive to be repeated here. Application is best undertaken with the workbook or guide in hand. Suffice it to say that both methods can be used by engineers and industrial hygienists when the necessary data on emissions, site locations, and meteorology are available.

5.3 Workplace Exposure Estimates

Qualitative estimates of the potential for emissions and the consequent exposures of workers can be made when agents have been identified and conditions of their use are known. Even the most rudimentary emissions inventory—the identification of agent, process, and location of use—permits this kind of assessment. When a very large number of materials are used at a single facility, and perhaps some of them are used only infrequently or in small quantity, the maintenance of a file on potential emissions and exposures may be a major undertaking. An operating program in which the industrial hygiene department of a large research facility systematically obtains descriptions of all uses of the many hundreds of chemical agents available for use at the facility has been described (25). The system includes a hazard rating for each agent based on its toxicity, flammability, reactivity, and special properties. Use information does not identify emissions per se, but retrieval of use information for a particular agent permits rapid subjective assessment of its emissions potential throughout the facility.

Qualitative information on emissions, either measured, calculated, or estimated, can be combined with data on workplace ventilation to estimate workroom concentrations of an agent. Such estimates may occasionally be of use in validating sampling data, but they are more likely to have value in such applications as assessing growth or decay of concentrations in nonequilibrium situations, in predicting the consequences of new processes or changes in operations, or in estimating retrospectively the levels of a contaminant to which particular groups of workers may have been exposed. This latter use is of particular interest in retrospective epidemiological studies of worker populations.

5.3.1 Gases and Vapors

Emission rates of gases and vapors are used in the mathematical relationships that yield values for concentration of contaminants in workroom air. In addition to emission rates, one must know physical dimensions or volume of the workplace and the ventilation rate. The mathematical relationships were derived using the concept of materials balance (26). The rate of change in the quantity of contaminant in the air of a workplace is described by Equation 4. All emissions of a contaminant to workroom air comprise a contaminant generation rate, G.

$$V \frac{dx}{dt} = G + QX_s - QX \tag{4}$$

Integration of Equation 4 yields:

$$X_t = X_i \exp\left(\frac{-Qt}{v}\right) + \left(X_s + \frac{G}{Q}\right)\left[1 - \exp\left(\frac{-Qt}{v}\right)\right] \quad (5)$$

where V = volume of the ventilated workplace (m³)
G = rate of contaminant generation as volume per unit time of contaminant gas or vapor (m³/min)
Q = ventilation rate, volume/unit time of air plus contaminant (for practical purposes the ventilation rate of air may be used in dilution ventilation of occupied space) (m³/min)
t = interval of time elapsed since conditions represented by X_i and $t = 0$ (min)
X = fraction of contaminant in workplace air (10^{-6} ppm)
X_i = fraction of contaminant present in workplace air at time $t = 0$
X_t = fraction of contaminant present in workplace air at time t
X_s = fraction of contaminant in incoming dilution air

Equation 5 is applicable only over intervals in which G and Q are constant. Should either G or Q change, a new value of X_i and a new t_o must be established.

At equilibrium, when t is great, Equation 5 simplifies to:

$$X = X_s + \frac{G}{Q} \quad (6)$$

or with equilibrium and clean supply air:

$$X = \frac{G}{Q} \quad (7)$$

When the concentration of the contaminant in the workplace air increases with time, the maximum concentration for the time interval t is X_t; for decay situations the maximum concentration for the interval t is X_i. Exposure guides and standards may list both maximum allowable short-term concentrations and allowable average concentrations. The average concentration over time t, also derivable from materials balance, is:

$$X_{av} = X_s + \frac{G}{Q} + \left(\frac{V}{Qt}\right)\left(X_i + X_s - \frac{G}{Q}\right)\left[1 - \exp\left(\frac{-Qt}{v}\right)\right] \quad (8)$$

As is the case with Equation 5, at equilibrium, when t is great, Equation 8 simplifies to Equation 6.

Equation 5 may be used to calculate the concentration of gas or vapor remaining in a vessel or tank at any time after purging ventilation or contaminant introduction has begun. It may also be used in the following form to estimate the interval of time required for a particular ventilation rate to cause a particular concentration X_t to be

THE EMISSION INVENTORY

reached:

$$t = \frac{V}{Q} \ln \left(\frac{G + QX_s - QX_i}{G + QX_s - QX_t} \right) \quad (9)$$

In some repetitive operations the generation rate for a contaminant may vary in a cyclical manner, being at one rate for one interval of time and at a different rate for another interval. If, in such a case, the greater generation rate is identified as G, the lower rate as G' and the time intervals t and t' are companion to G and G', respectively, when the cycle has equilibrated so that the workroom concentration pattern has become repetitive, the maximum concentration that occurs during the cycle is represented by:

$$X_{max} = \frac{\left(X_s + \frac{G}{Q}\right)\left[1 - \exp\left(\frac{Qt}{v}\right)\right] - \left(X_s + \frac{G'}{Q}\right)\left[1 - \exp\left(\frac{-Qt'}{v}\right)\right]}{\exp\left(\frac{-Qt'}{v}\right) - \exp\left(\frac{Qt}{v}\right)} \quad (10)$$

For the same cyclic operation after equilibrium has been reached, the average concentration over a large number of cycles is:

$$X_{av} = X_s + \left(\frac{Gt + G't'}{Q(t + t')}\right) \quad (11)$$

In the simple case when there is no generation during the time interval t' the term G'/Q becomes zero in Equations 10 and 11.

In an excellent analysis of dilution ventilation relationships Roach has shown that the variance of emission σ_G^2 is inversely proportional to the length of the time interval t over which emissions take place, and that the variance in concentration σ_x^2 is then inversely proportional to the product of ventilation rate Q and the square of the room volume V^2 (27). Thus for a given average emission rate, the smaller the workspace and the smaller the ventilation rate, the greater will be the fluctuations in concentration, while the greater the volume of the space and the greater the ventilation rate, the smaller will be the fluctuations.

In real situations instantaneous perfect mixing does not take place. Usually when dilution ventilation is applied, the contaminant is emitted at locations that are in the occupied portion of a workroom; not all the ventilation air mixes with the contaminant while it is in this occupied space. An empirical factor K may be used in calculations of estimated exposures to account for departures from perfect mixing. The definition of K is:

$$K = \frac{Q_{actual}}{Q_{effective}} \quad (12)$$

where Q_{actual} = ventilation rate, volume/unit time, as measured or obtained from engineering records (m³/min)

$Q_{effective}$ = ventilation rate, volume/unit time, that is effective in determining the concentration of contaminant to which people are exposed in a workplace (m³/min)

The value of $Q_{effective}$ is the ventilation rate Q to be used in Equations 5 through 11 for calculation of exposure estimates. The choice of a value for K is a matter of judgment for an investigator; it is ordinarily in the range 3 to 10 (28). Calculations using engineering records for ventilation rates, actual emission rates, and a number of concentration measurements of three different contaminants in each of two actual dilution ventilation situations, have yielded K values ranging from 2.2 to 5.0 with a mean value of 3.25 (29). In these two dilution ventilation situations, solvent vapors were being emitted in a number of locations in the workrooms at elevations that were approximately the same as breathing zone and sampling elevations. They may be considered to represent generally good mixing of emissions and ventilation air in actual application of dilution ventilation in industrial situations. The observed values are consistent with the lower values of the range of typical K values; the greater the departure from efficient mixing of emissions and ventilation air, the greater should be the value of K.

5.3.2 Particulates

Caution must be exercised in the application of dilution ventilation relationships, as described for gases and vapors, to particulate contaminants that are released to workplaces. Gravitational settling may be an appreciable factor in the decay of air concentrations of particulates. The smaller the particles, the greater is the likelihood that they will remain airborne and the more applicable are dilution ventilation relationships. When emissions of submicrometer sized particles—for example, most components of smoke and metal oxide fumes—are known, concentrations can be approximated by use of the dilution ventilation equations for gases and vapors.

Dusts with a wide range of particle sizes do not necessarily behave in air in the same way as fumes. Estimates of the relative concentrations of respirable dusts may be made with dilution ventilation equations when particulate emissions data are sufficient for quantitation of the respirable fraction. Size characteristics that define respirable particulates are described in Chapters 6 and 7 of Volume I.

In assessing exposures to particulates, particles larger than those considered respirable should not be ignored. Of these larger particles, those that are inhaled and deposited in the nasopharynx and tracheobronchial system before they can reach the pulmonary spaces of the lungs, represent ingestion exposures. Airborne concentrations of such particles generally decay at appreciable rates because of gravitational settling; dilution ventilation relationships do not apply to them.

As particles size increases, the influence of ventilation on concentration, therefore on the magnitude of exposure, decreases. Thus for large particles, those well in the nonrespirable range, an emission rate per se is likely to be a better relative index of potential exposures than is a calculated concentration based on both emission and ventilation rates.

6 RECORDS RETENTION PROGRAM

Emissions records may be generated for a variety of purposes. Some records are for purposes that can be served in a relatively short time; then the record may be considered a candidate for discarding. Other emission records may be generated specifically for purposes such as health research, which require long retention. Emission records made for various short- or long-term purposes may have secondary use in long-term community air pollution or occupational health research. Whether they represent typical operations or unusual events, such records may be invaluable in the future for reconstructing conditions of work to compare with long-term health experience of persons who engaged in that work. Since not all specific agents of future interest are identifiable in advance, it is prudent to retain emission records for as many agents as possible.

Emission records, or records from which emission estimates can be made, such as those identified in Section 4 of this chapter, should be retained for a long time. A single specific age at which a record no longer has value cannot be stated. Guidance on retention duration may be found in federal regulations, which require that records of medical examinations of workers exposed to asbestos be kept for at least 20 years (30). The same asbestos regulation requires that exposure records need be kept only 3 years. For research purposes exposure records have value equal to or greater than that of medical examination records. Because long latent periods are associated with health effects of some agents, early exposure records may be invaluable in the search for causal associations. Exposure records, or emission records that identify contact with agents or from which exposure estimates can be made, should be retained long enough to be used in studies of the mortality experiences of the worker populations to which they apply.

The need for long-term retention of environmental records is recognized in the recordkeeping provisions of a proposed Occupational Safety and Health Administration regulation on toxic substances posing potential occupational carcinogenic risk (31). The proposed federal regulation specifies that records of environmental sampling results and records of environmental variables that could affect the measurement of exposures, be kept for 40 years or for the duration of exposure plus 20 years, whichever is longer. Emission data are pertinent to both these records categories. Such retention is appropriate for all records of emissions of air pollutants or of chemical agents released in places of work.

The responsibility for retention of emission records for any establishment should be clearly defined. This responsibility may well be placed with the same organizational unit that has responsibility for retention of air monitoring records and/or health experience records. Records that serve as secondary sources of health related information, such as those identified in Section 4 of this chapter, may be kept in different locations. A checklist of all such records that are needed for emission records purposes should be assembled and kept up to date for each establishment and retained by the unit having responsibility for emissions records. Responsibility for retention of each of these secondary records should be clearly defined and this responsibility, along with location of each record set, should appear on the records checklist.

REFERENCES

1. U.S. Environmental Protection Agency, *A Guide for Compiling a Comprehensive Emission Inventory,* APTD 1135, EPA, Research Triangle Park, N.C., 1973.
2. J. R. Hammerle, in: *Air Pollution,* Vol. 3, 3rd ed., A. C. Stern, Ed., Academic Press, New York, 1976, ch. 17, pp. 717–784.
3. Toxic Substances Control Act, PL 94-469, 15 USC 2607, 1976.
4. T. H. Maugh, *Science,* **199,** 162 (1978).
5. U.S. Department of Health, Education and Welfare *Registry of Toxic Effects of Chemical Substances, 1975 Edition,* DHEW (NIOSH), Government Printing Office, Washington, D.C., 1975.
6. H. J. Paulus and R. W. Thron, in: *Air Pollution,* Vol. 3, 3rd ed., A. C. Stern, Ed., Academic Press, New York, 1976, Ch. 14, pp. 525–587.
7. U.S. Environmental Protection Agency, *Administrative and Technical Aspects of Source Sampling for Particulates,* EPA 450/3-74-047, EPA, Research Triangle Park, N.C., 1974.
8. American Petroleum Institute, *Evaporative Loss from Tank Cars, Tank Trucks, and Marine Vessels,* API Bulletin 2514, API, Washington, D.C., 1959.
9. H. B. Elkins, E. M. Comproni, and L. D. Pagnotto, *AIHA J.* **24,** 99 (1973).
10. H. E. Runion, *AIHA J.,* **36,** 338 (1975).
11. U.S. Environmental Protection Agency, *Compilation of Air Pollutant Emission Factors,* A.P.42, 3rd ed., EPA, Research Triangle Park, N.C., 1977.
12. D. Anderson, *Emission Factors for Trace Substances,* EPA-450/2-73-001, U.S. Environmental Protection Agency, Research Triangle Park, N.C., 1973.
13. W. A. Burgess et al., *AIHA J.,* **38,** 18 (1977).
14. M. Smith, *Investigation of Passenger Car Refueling Losses,* Scott Research Laboratories, Inc., San Bernardino, Calif., September 1972, National Technical Information Service, PB-212 592.
15. Z. Bell, J. Laflenn, R. Lynch, and G. Work, *Chem. Eng. Prog.,* **71:** 9, 45 (1975).
16. T. Kletz, *Chem. Eng. Prog.,* **71:** 9 (1975).
17. E. Sowinski and I. H. Suffett, *AIHA J.,* **38,** 353 (1977).
18. R. N. Anthony, *Management Accounting, Text and Cases,* 4th ed., Irwin, Homewood, Ill., 1970.
19. U.S. Tariff Commission (1958), *Synthetic Organic Chemicals, United States Production and Sales, 1958,* Report 205, TC1.9-205, 2nd Series, Government Printing Office, Washington, D.C.
20. Rubber Reserve Company, *Report on the Rubber Program 1940–1945,* February 24, 1975.
21. G. S. Whitby, Editor-in-Chief, *Synthetic Rubber,* Wiley New York, 1954.
22. G. D. Babcock, *History of the United States Rubber Company, A Case Study in Corporate Management,* Indiana Business Report 39, Graduate School of Business, Indiana University, 1966.

23. D. B. Turner, *Workbook of Atmospheric Dispersion Estimates,* U.S. Public Health Service Publication 999-AP-26, National Air Pollution Control Administration, Cincinnati, Ohio, 1969.
24. A. D. Busse and J. R. Zimmerman, *User's Guide for the Climatological Dispersion Model,* EPA-R4-024, U.S. Environmental Protection Agency, Research Triangle Park, N.C., 1973.
25. W. E. Porter, C. L. Hunt, and N. E. Bolton, *AIHA J.,* **38,** 51 (1977).
26. R. L. Harris, Jr., *Industrial Hygiene Engineering Training Course,* U.S. Public Health Service Occupational Health Field Headquarters, Cincinnati, Ohio, 1960.
27. S. A. Roach, *Ann. Occup. Hyg.,* **20,** 65 (1977).
28. *Industrial Ventilation, A Manual of Recommended Practice,* 15th ed., Committee on Industrial Ventilation, Lansing, Mich., 1978.
29. J. C. Baker, Jr., *Testing Model for Predicting Solvent Vapor concentrations in an Industrial Environment,* M.S.E.E. Technical Report, Department of E.S.E., University of North Carolina, Chapel Hill, N.C., 1977.
30. *General Industry Safety and Health Standards,* OSHA 2206 (29 CFR 1910), part 1910.1001, Department of Labor, Washington, D.C., 1976.
31. Identification, Classification, and Regulation of Toxic Substances Posing a Potential Occupational Carcinogenic Risk, 29 CFR 1990, *Fed. Reg.,* **42;** 192, 54148 (October 4, 1977).

CHAPTER THREE

Statistical Design and Data Analysis Requirements

KENNETH A. BUSCH and
NELSON A. LEIDEL

1 INTRODUCTION

Industrial hygienists can derive important professional benefits from use of sampling strategies, experimental designs, and data analysis methodologies that are based in the theory of mathematical statistics and probability. Section 2 of this chapter gives a comprehensive discussion of general areas of industrial hygiene practice where statistical methods are useful. Sections 3, 4, and 5 discuss in detail two specific areas for application of statistics. Section 3 presents some basic statistical theory in the context of its use in Section 4 as a model for random errors in measured employee exposure data, to make controlled risk decisions of compliance or noncompliance with exposure control limits. Section 5 applies the versatile statistical models given in Section 3 (for random errors in employee exposure measurements) to biological data from animal chronic exposure experiments. Section 5 discusses basic principles of statistical experimental design in the context of an animal chronic exposure experiment. The sample size tables given in Section 5 have enough generality to be applicable to experiments in the physical sciences as well as the biological sciences.

2 GENERAL AREAS OF APPLICATION FOR STATISTICS IN INDUSTRIAL HYGIENE

Statistical theory is applicable in controlled laboratory research studies (in both the physical and biological sciences) as well as in field observational studies. With proper statistical design, both laboratory experiments and field studies can identify causes of occupational health problems, evaluate effectiveness of engineering technology for source control of emissions, and evaluate effectiveness of personal protective equipment. Of course, depending on the relative cost, availability of relevant laboratory models, and interpretability of available field data, one or the other of these two general types of studies is usually a clear choice for any given research objective.

A primary problem for industrial hygienists is that safe exposure levels for some substances are unknown and must be determined. To do this, pertinent health effects data are collected and statistically analyzed. One approach is to expose suitable animal species and observe biological effects that may occur. The statistical design of animal chronic exposure studies and the statistical analyses of resulting data are discussed in detail in other sections of this chapter. Another important approach to occupational health research, the epidemiological study, is discussed only broadly in this section but references to more comprehensive treatments are supplied.

2.1 Epidemiological Studies

In the epidemiological study an attempt is made to associate the observed incidence or severity of an occupational disease in groups of human workers with their job type or with their work exposures to potentially toxic materials. Along with the important advantage of having human workers as subjects, observational studies have the frustrating disadvantage of lack of control of the exposure conditions, uncontrolled effects of the extraexposure environment, and interactions between exposure effects and demographic and socioeconomic factors.

Better control over and measurement of short-term exposures is obtained in controlled clinical studies of human volunteer subjects. Of course an overriding necessity is safety of the subjects, and this usually precludes use of exposure levels high enough or exposures long enough to produce the chronic toxicity of the workplace.

General statisticial methodology for epidemiological studies is given in books by Mausner and Bahn (1), Friedman (2), and McMahon and Pugh (3) among others. No textbook in epidemiology deals specifically with studies of occupational groups. However many specific studies have been reported in professional journals such as *Archives of Environmental Health, Journal of Occupational Medicine, American Journal of Epidemiology,* and *British Journal of Industrial Medicine.* In addition, state-of-the-art general methodology for occupational health epidemiological studies is now available in published articles. Both theoretical and practical problems of performing occupational health field studies are being solved, thanks to the extensive field experience that investigators have now accrued in occupational observational studies. The entire March 1976 issue of *Journal of Occupational Medicine* deals with epidemiology applied in this field, including a survey paper of the asbestos literature by Enterline (3) entitled "Pitfalls in Epidemiological Research." The limitations of these studies include inaccurate or uninterpretable cause of death statements on death certifi-

STATISTICAL DESIGN AND DATA ANALYSIS REQUIREMENTS

cates, improper control groups, lack of quantitative exposure data, overlapping exposure and follow-up periods, and competing (but sometimes unknown) causes of death.

The research needs in epidemiology have been summarized recently by an authoritative Second Task Force appointed by Department of Health, Education and Welfare. In Chapter 15 of the 1977 Second Task Force report (4), a subtask force chaired by Dr. Brian MacMahon makes an assessment of the state of the art of epidemiology and other statistical methods used in environmental health studies. Among the specific recommendations are the following:

- For clinical environmental research, guidelines are needed for the protection of human subjects and special attention should be given to statistical design and analysis of such studies.
- For epidemiological studies, better exposure data are needed. Health professionals should help make decisions about what environmental data are collected by governmental agencies. Also, routine surveillance is needed of disease incidence in occupational groups along with surveillance of exposure levels. Animal studies should be used to identify biochemical or physiological early indicator effects of serious chronic disease in worker groups believed to be exposed to potentially hazardous agents. In all areas of environmental health research, more powerful statistical techniques are needed in areas of multivariate analysis, time series and sequential analysis, and nonparametric methods. Concerning dose-response studies, we need better models for cost-benefit analysis related to the setting of exposure limits. Better dose-response models for mixtures of toxic agents are needed as are better models for animal-to-man extrapolation (particularly for carcinogenesis).

Another useful reference to more specific methodological techniques is the "Steelworker Series" of 10 epidemiological reports by J. W. Lloyd and Carol K. Redmond, published between 1969 (5) and 1978 (6) in *Journal of Occupational Medicine*. These reports are considered by some to constitute the evolution to the present state of the art of modern methodology for epidemiological studies of a particular occupation. In a 1975 paper Kupper et al. (7) also discuss methods for selecting suitable samples of industrial worker groups and valid control groups.

2.2 Threshold Levels and Low Risk Levels

Ideally, the industrial hygienist would like to be able to assess the net risk of an adverse health effect as a function of a worker's past, present, and future exposures in relation to age, race, sex, and other personal susceptibility factors. If such comprehensive toxicological knowledge were available, appropriate limits on future exposures of an individual, in relation to past exposures, could be recommended so as to control the chances of experiencing adverse health effects. However achieving such a high level of toxicological understanding is not realistic—one would have to know all the "dose-time-response" relationships for the toxic material of interest (i.e., the relationships between level of exposure, length of exposure, and the incidence and severity of health effects that occur to some or all of an exposed group). Rather, the industrial hygienist usually must be satisfied to merely know a "low risk" exposure level at which the incidence of severe health effects is very low. Better yet, if possible, one would like to know a "safe" level of exposure below which health effects are absent (i.e., a threshold level). The existence of thresholds is a controversial question that must be left to others to answer.

Statistical models have been developed that can aid us in extrapolating to low risk dose levels from higher level dose-response data. Since a very low risk dose level always exists (regardless of whether a true threshold exists), a low risk level of exposure can usually be selected which is "sufficiently safe." Hartley and Sielken (8) have reviewed the technical statistical aspects of estimating "safe doses" in carcinogenesis experiments. They leave the definition of an "acceptable" increase in the risk of carcinogenesis to regulatory agencies. Figures that have been suggested are 10^{-8} (1 in 100 million) or 10^{-6} (1 in 1 million). Either of these low risk exposure levels is effectively impossible to determine by direct experimentation with animals because of the large sample sizes that would be required to distinguish such miniscule tumor incidence increments. The problem becomes particularly acute when there is a "normal" (control) incidence (see Section 5 for further discussion of sample size requirements). Therefore various mathematical models have been proposed for extrapolation to low risk dose levels using a curve fitted to higher level dose-response data. Among these are the Mantel-Bryan procedure (8, 9) based on extrapolation using a conservative slope on log dose-probability graph paper. Other models are also reviewed in Hartley and Sielken's papers (10): some are more flexible—for example, Hartley and Sielken (11, 12) use a polynomial instead of a straight line to extrapolate—and some are derived from assumed biological mechanisms—Crump et al. (13) assume a multistage biological mechanism for development of cancer. The Crump model can be fitted to incidence data, whereas fitting the Hartley-Sielken model requires time-to-tumor data.

It is difficult to determine dose-time-response relationships from limited and unstructured epidemiological data such as are usually available for humans reflecting a disease state that can be attributed to their real work exposures. However permissible exposure levels (PELs) may be estimated from epidemiological studies of human workers. For example, the low range of average personal exposure levels for a large group of workers in a given plant could be taken as a conservative PEL if the "health profile" of these workers is found to be not inferior to that of a cohort group of unexposed but otherwise similar individuals. The specialized statistics of epidemiological studies and survey sampling using questionnaires is not discussed in detail in this chapter. However Section 5 presents the general statistical methods for the analysis of data from laboratory studies, and these can be employed for epidemiological studies as long as probability-based random sampling has been used. A good discussion of statistical aspects of industrial hygiene surveys is given by Frank A. Patty in Volume I (14).

STATISTICAL DESIGN AND DATA ANALYSIS REQUIREMENTS

2.3 Need for Statistical Experimental Design

For either type of occupational health study, controlled laboratory or field observational, the data have stochastic components (i.e., random or chance variations) that cannot be ignored when evaluating the data. Different types of biological data may have basically different statistical properties. Some population health effects are measurable in terms of (1) the average amount of change in a quantitative biological parameter measured on a continuous scale. Examples of continuous variable measurements are lung volume, heart rate, and body weight. Effects of other types can be recorded only as (2) the presence or absence of a qualitative biological abnormality (e.g., a pathological condition such as a tumor). Still another type of biological measurement has (3) an ordered scale, but values exist only at a limited number of discontinuous (i.e., discrete) points. For example, severity of lung histopathology has been graded on a 6-point rating scale. The first and second types of data are discussed in relation to principles of study design and data analysis in Section 5.

The objectives of an occupational health study should be the primary determinant of its statistical design, not expedient considerations of "availability" of specialized experimental facilities in a given laboratory or of specialized expertise on a given staff. Availability can be a powerful incentive for inappropriate inclusion of these in an experimental program.

It is the responsibility of the scientist subject matter specialist, not the statistician, to be sure that appropriate animal models, valid exposure techniques, and relevant, sensitive biological parameters that can be measured accurately are used in the experiment protocol. The statistician then addresses the tactical problems of numbers of subjects, assignment to exposure groups and caging plans, schedules for exposure, biological sampling, sample assay, and sacrifice, and so on. Proper selection of such design parameters can assure the experimenter that the study will have adequate (but not excessive) precision to permit the measurement of the anticipated effects that would be of interest.

Sometimes appropriate experimental facilities are available but the sample sizes that can be used are of marginal value insofar as the production of definitive results is concerned. Nevertheless, an expedient decision may be made to proceed with a study under the rationale that "some information is better than none." This practice often results in studies being performed that are unknowingly predestined to be inconclusive. The resulting waste of time and resources could have been prevented by securing a statistician's evaluation of the study plan. The statistician can usually warn the experimenter in advance if the planned experiment has low power to detect the size effects that need to be detected if indeed they exist.

2.4 Statistical Decision Theory for Compliance or Noncompliance

In Section 4, which gives statistical decision theory for deciding compliance or noncompliance with safe exposure limits, appropriately different statistical decision rules are presented for use by compliance officers and employers. In the responsible practice of their respective professions, the two parties must limit different types of "risk." The "risk" will be interpreted as the probability (i.e., the long-term relative frequency) of making incorrect decisions. The compliance officer must control at a low level the risk of falsely charging a company with noncompliance. Conversely, the employer must have high confidence that the employees are not subject to overexposure—that is, when employee exposure data are taken, they must conclusively show the employees' safety.

The basis for the statistical analyses of personal exposure data are assumed mathematical models for the statistical variabilities occurring in the sequence of physical steps used to collect and analyze the samples. For example, the normal distribution is usually assumed for chemical analysis errors. The statistical models can also be used to identify in advance a sampling strategy that will yield data having a desired level of precision. Leidel et al. (15) have written a manual published by the National Institute for Occupational Safety and Health (NIOSH) that recommends suitable statistical personal exposure sampling strategies.

In summary, the general theme of all statistical material in this chapter is the deliberate control of random measurement errors (including sampling variations) through statistical experimental design, and the rational assessment of the precision of resulting data through statistical analysis. The remainder of the chapter gives some statistical theory and related statistical methods for application in two specific areas of industrial hygiene, as mentioned in the introduction.

3 DATA MODELS AND BASIC STATISTICAL DISTRIBUTIONS FOR INDUSTRIAL HYGIENE DATA

The basic statistical distribution theory presented in this section is needed background for understanding the applied statistical methods given in Sections 4 and 5. The theory is presented in relation to random errors in employee exposure measurements. Section 4 uses the theory to deal with statistical sampling strategies for employee exposure measurements and as a basis for related data analysis methods for making controlled-risk decisions of compliance or noncompliance with federal standards for employee exposure. Section 5 uses variations on the same basic statistical theory as a foundation for the statistical design of a chronic animal exposure study and for the analyses of resulting biological data.

3.1 Model for Net Error in Measurements Data

Suppose an employee is to be sampled for the purpose of estimating the inhalation exposure to a contaminant. The questions of selection of the sampling location and choice of industrial hygiene sampling and chemical analysis methods are not addressed here. It is irrelevant to this statistical discussion whether a personal sample is required or an area sample will suffice. We start with the assumption that an appropriate sam-

STATISTICAL DESIGN AND DATA ANALYSIS REQUIREMENTS

pling/analysis method is available, which can give representative, accurate results *on the average*. The sampling equipment and laboratory instruments must be properly calibrated to reduce the risk of systematic errors. This does not imply that each and every sample gives the correct answer. To the contrary, every sample measurement will differ from the respective true average exposure concentration that existed during the period sampled. The discrepancies, which are termed random "errors," will vary in magnitude and direction in a random manner from sample to sample. Random errors, within limits, are inherent to any method and equipment. The presence of random error does not imply that the method has been improperly used (i.e., that a mistake has been made). Of course, a discrepancy outside the usual range of precision for the method could indicate that a mistake has been made and such data might be discarded, especially if the suspect result can be associated with a known irregularity that occurred during the sampling/analytical process.

To systematically approach the statistical treatment of random errors, we use a mathematical statistical model. The true average concentration is denoted by the symbol μ and a particular sample concentration measurement by the symbol X. Thus the total error of a single measurement ϵ_T is given by

$$\epsilon_T = X - \mu \tag{1}$$

The total (net) error ϵ_T is the algebraic sum of independent measurement errors due to the component sampling and analysis steps:

$$\epsilon_T = \epsilon_s + \epsilon_a \tag{2}$$

where ϵ_s is a positive or negative random sampling error and ϵ_a is a positive or negative random analytical error. For "independent" errors the size and sign of the analytical error does not depend on the size or sign of the sampling error. All ϵ's have the same units of concentration as X.

Thus any concentration measurement X that is both an industrial hygiene and statistical "sample" can be represented as the algebraic sum of the true concentration μ and the net sampling/analysis error for the particular sample measurement.

$$X = \mu + \epsilon_s + \epsilon_a \tag{3}$$

If multiple samples could be taken at the same point in time and space, they would be referred to as "replicate samples" if the true value μ were the same for all samples. Figure 3.1 presents a horizontal concentration scale on which the point μ represents the true concentration. Concentration measurements X for replicate samples occur within intervals above and below μ with predictable relative frequencies (or probabilities) that are proportional to corresponding areas under the curve (the sample "distribution curve") appearing above the concentration scale. Ordinates of the sample distribution curve do *not* give probabilities of corresponding sample concentrations. Rather, it is the *area* under the sample distribution curve between two values of X that is proportional to the relative frequency (proportion) of replicate samples that would occur in that interval. For this area to represent a probability, which is a proportion between 0

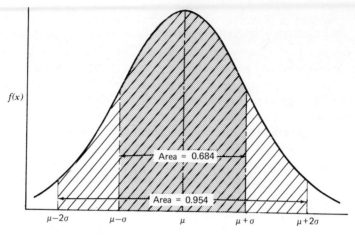

Figure 3.1. Normal distribution curve.

(impossibility) and 1 (certainty), a distribution curve is standardized such that the total area under the curve is unity (1.0).

3.2 The Normal Distribution

For many types of data a mathematical formula can be used to describe and compute the distribution curve. Such a formula, which relates ordinates $f(X)$ of the curve to values of the variables X, is called a "distribution function" or simply "distribution." [*Note.* This terminology is not universally used. Our definition of the term "distribution function" is according to Hald (16), but others refer to $f(X)$ as the "frequency function" or "probability density function."] The $f(X)$ illustrated in Figure 3.1 is of a particular type known as a normal curve; its distribution function is represented by the special notation $N(X; \mu, \sigma^2)$. That is, for the normal curve:

$$f(X) = N(X; \mu, \sigma^2) = \frac{1}{\sigma\sqrt{2\pi}} \exp\left(\frac{-\frac{1}{2}(X-\mu)^2}{\sigma^2}\right) \qquad (4)$$

Two constants, or parameters, appear in the distribution fucntion for the normal curve: μ, its mean, and σ, its standard deviation. Thus the notation $N(X; \mu, \sigma^2)$ is shorthand for "the variable X is normally distributed about the true mean μ with variance σ^2." Any normal curve has the same general appearance—a bell-shaped curve that is symmetrical about its mean. The true mean μ, also called the "expected value" of the random variable X, denoted $E(X)$, is the average value of all values of the distribution. Mathematically, the mean $E(X)$ of any distribution function, say $f(X)$, is the center of gravity of the corresponding distribution curve, defined by:

$$E(X) = \int_{-\infty}^{\infty} Xf(X)\, dX \qquad (5)$$

For the mean of the normal distribution the general $f(X)$ in Equation 4 is replaced by $N(X; \mu, \sigma^2)$ (see Equation 3) so that

$$E(X) = \int_{-\infty}^{\infty} \frac{X}{\sigma\sqrt{2\pi}} \exp\left(\frac{-\frac{1}{2}(X-\mu)^2}{\sigma^2}\right) dX = \mu \qquad (6)$$

The integration in Equation 6 is not obvious and the details of its evaluation are not presented here. The point to note is that for the normal distribution, the parameter μ in its formula is the mean of the distribution.

Similarly, it can be shown that for the normal distribution, the average value of squared deviations $(X - \mu)^2$ is σ^2, that is,

$$E(X-\mu)^2 = \int_{-\infty}^{\infty} \frac{(X-\mu)^2}{\sigma\sqrt{2\pi}} \exp\left(\frac{-\frac{1}{2}(X-\mu)^2}{\sigma^2}\right) dX = \sigma^2 \qquad (7)$$

For any distribution the mean square of deviations from the mean, denoted by $E(X - E(X))^2$, is known as the "variance" of X. The variance is the square of the standard deviation. For the normal distribution, the variance is equal to its second parameter σ^2, so that the standard deviation is σ. The "mode" of any distribution is the point on the X scale at which the maximum of its distribution function occurs. The "median" is the middle X-value, that is, the value exceeded 50 percent of the time. For the normal distribution the mode and median are both equal to the mean.

Since errors of sampling and chemical analysis tend to be proportional to the concentration analyzed, it is useful to express the standard deviation as a proportion of the mean. The ratio $\sqrt{E(X-E(X))^2}/E(X)$ is called the coefficient of variation (CV) or, by chemists, the relative standard deviation. For the normal distribution, $CV = \sigma/\mu$. It can be shown that the net error (ϵ_T), composed of independent sampling and analysis errors (see Equation 2), has the following "total coefficient of variation":

$$CV_T = \sqrt{CV_s^2 + CV_a^2} = \frac{\sigma_T}{\mu} \qquad (8)$$

The value CV_T is usually nearly constant within the normal range of application of the sampling/analysis method. At any given concentration within this normal range, the standard deviation of the net error is given by

$$\sigma_T = \mu CV_T \qquad (9)$$

where

$$\sigma_T = \sqrt{\sigma_s^2 + \sigma_a^2} \qquad (10)$$

Note that a normally distributed random variable theoretically can have negative values, whereas chemical concentrations are strictly nonnegative. Nevertheless the normal curve is usually an excellent approximation to the distribution curve for replicate samples (17). This is true because most sampling/analysis methods have errors whose standard deviation is small enough relative to the true mean concentration that the area under the corresponding normal curve to the left of zero is negligible.

Table 3.1 Areas Under the Normal Curve

X-Interval	Data Within Interval (%)
$\mu - \sigma$ to $\mu + \sigma$	68.4
$\mu - 1.645\sigma$ to $\mu + 1.645\sigma$	90.0
$\mu - 1.96\sigma$ to $\mu + 1.96\sigma$	95.0
$-\infty$ to $\mu + 1.645\sigma$	95.0
$\mu - 2\sigma$ to $\mu + 2\sigma$	95.4
$\mu - 1.96\sigma$ to ∞	97.5
$\mu - 2.576$ to $\mu + 2.576\sigma$	99.0
$\mu - 3\sigma$ to $\mu + 3\sigma$	99.7

The two parameters of a normal curve imply everything about its location and shape. The location parameter is the mean μ, which is the center point of the curve. The variability parameter is the standard deviation σ, which measures the dispersion of the X-values about their mean. Table 3.1 gives some examples of relationships between the mean, standard deviation, and proportions of data within various intervals. The first and fifth of these proportions are depicted graphically in Figure 3.1 as shaded and cross-hatched areas, respectively, under a normal curve.

It is important to note that normality (or at least approximate normality) of concentration measurements made on a set of replicate samples is to be expected from theoretical considerations. The total error in any particular sample determination is the net error resulting from many random incremental positive and negative physical influences during the various steps of a multistep sampling/analytical process. These include unavoidable technician variations, small environmental variations (humidity, temperature, relative humidity, etc.), and functional variations in component parts of the sampling and analytical apparatus (voltage, flowrate of pump, etc.). Insofar as these various random error sources operate independently, their combined influences tend to make the net error follow a normal probability density curve. The proof that this is true would be similar to the proof of the central limit theorem from the theory of mathematical statistics, which states (18):

If a population has a finite variance σ^2 and mean μ, then the distribution of the sample mean approaches the normal distribution with variance σ^2/n and mean μ as the sample size n increases.

3.3 Confidence Limits for the Mean of a Normal Distribution Whose Standard Deviation Is Known

If n independent consecutive sample measurements $X_1, X_2, \ldots, X_n$ are made that have normally distributed sampling/analysis errors with the same total coefficient of varia-

STATISTICAL DESIGN AND DATA ANALYSIS REQUIREMENTS

tion CV_T, the mean $\bar{X}$ of the measurements is also normally distributed. The normal distribution of $\bar{X}$ values has the following mean and variance:

$$\mu_{\bar{X}} = E(\bar{X}) = \left(\frac{1}{n}\right)(E(X_1) + E(X_2) + \cdots + E(X_n))$$

$$= \left(\frac{1}{n}\right)(\mu_1 + \mu_2 + \cdots + \mu_n)$$

$$\sigma_{\bar{X}}^2 = \left(\frac{1}{n^2}\right)(CV_T^2)(\mu_1^2 + \mu_2^2 + \cdots + \mu_n^2)$$

If a worker's exposure were uniform during the workday, there would be $\mu_i = \mu$ for $i = 1, 2, \ldots, n$, and the sample mean exposure $\bar{X}$ could then be treated as a random sample from a normal distribution with mean μ and variance $\sigma^2/n = (1/n)(CV_T^2)(\mu^2)$. In 95 percent of such samples, the average $\bar{X}$ would be within an interval $\mu \pm 1.96(CV_T)(\mu)/\sqrt{n}$. Equivalently, intervals $\bar{X} \pm 1.96(CV_T)(\mu)/\sqrt{n}$ would contain μ 95 percent of the time. Note that the latter intervals are centered by the randomly varying sample means $\bar{X}$.

If we now assume that the true average exposure level μ is somewhere below the standard (denoted STD), intervals surrounding the *sample means* can be computed that would contain the true mean in *at least* 95 percent of cases. Such intervals would be of the form

$$\bar{X} \pm \frac{1.96(CV_T)(STD)}{\sqrt{n}}$$

Similar lower-bounded open intervals (i.e., "one-sided" intervals) can be computed from sample means that would contain the true mean at least 95 percent of the time, given that $\mu \leq STD$. The lower bounds of such right-sided open intervals, $\mu \geq \bar{X} - 1.645(CV_T)(STD)/\sqrt{n}$, will be denoted as "LCL's," since they may be looked on as conservative (too low) 95 percent "lower confidence limits" for μ; [If σ were known, exact 95 percent lower confidence limits would be $\bar{X} - 1.645\,\sigma/\sqrt{n} = \bar{X} - 1.645(CV_T)(\mu)/\sqrt{n}$. The quantity $(CV_T)(STD)$ overestimates σ if $\mu < STD$.]

Section 4 applies the LCL (and similar UCL) concepts to the problem of decision making regarding compliance or noncompliance with federal employee exposure standards.

3.4 The Log Normal Distribution

Another distribution that is useful for approximating many industrial hygiene data distributions is the logarithmic normal distribution. It has been used by Larsen (19) to describe temporal variations in community air concentrations. Leidel and Busch (15) give numerous references to the use of the log normal distribution as a model for variability of exposure results from short-term "grab" samples (i.e., 15 min or less), day-to-day variations in 8 hr time-weighted average (TWA) exposures, and variability of exposures within job classifications. The log normal curve illustrated in Figure 3.2 is one of a family of such curves that have the general formula:

$$f(X) = \frac{1}{X(\ln \sigma_g)\sqrt{2\pi}} \exp\left(\frac{-\frac{1}{2}(\ln X - \ln \mu_g)^2}{\ln^2 \sigma_g}\right) \quad (11)$$

where $0 < X < \infty$.

This formula is called the log normal distribution function. Its general structure is similar to Equation 4 for the normal probability density function. The relationship between these two distributions is that if a random variable X is log normally distributed, it follows that the tranformed values, ln X are normally distributed. As for the normal distribution, the log normal distribution is described fully by only two parameters; however the parameters have different names for the log normal, namely, the geometric mean and geometric standard deviation. The geometric mean μ_g is defined by

$$\begin{aligned} E(\ln X) &= \int_{-\infty}^{\infty} \ln X (N(\ln X; \ln \mu_g, \ln^2 \sigma_g) \, d \ln X \\ &= \ln \mu_g \end{aligned} \quad (12)$$

The geometric standard deviation σ_g is defined by

$$\begin{aligned} E(\ln X - \ln \mu_g)^2 &= \int_{-\infty}^{\infty} (\ln X - \ln \mu_g)^2 \, N(\ln X; \ln \mu_g, \ln^2 \sigma_g) \, d \ln X \\ &= \ln^2 \sigma_g \end{aligned} \quad (13)$$

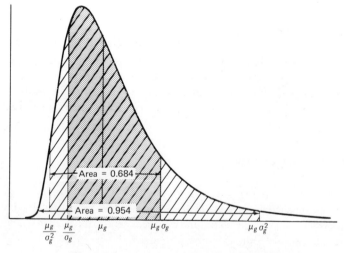

Figure 3.2. Log-normal distribution curve.

Table 3.2 Areas Under the Log Normal Curve

X-Interval	Data Within Interval (%)
μ_g/σ_g to $\mu_g\sigma_g$	68.4
$\mu_g/\sigma_g^{1.645}$ to $\mu_g\sigma_g^{1.645}$	90.0
$\mu_g/\sigma_g^{1.96}$ to $\mu_g\sigma_g^{1.96}$	95.0
0 to $\mu_g\sigma_g^{1.645}$	95.0
μ_g/σ_g^2 to $\mu_g\sigma_g^2$	95.4
$\mu_g/\sigma_g^{1.96}$ to ∞	97.5
$\mu_g/\sigma_g^{2.576}$ to $\mu_g\sigma_g^{2.576}$	99.0
μ_g/σ_g^3 to $\mu_g\sigma_g^3$	99.7

Thus μ_g and σ_g are antilogs to the base e of the mean and standard deviation, respectively, of the natural logarithmic transform of X. The interpretation of μ_g and σ_g parameters for a log normal distribution differs from the interpretation of μ and σ for a normal distribution. In Table 3.1, for data following a normal distribution, multiples of σ were added to and subtracted from μ to obtain intervals of X that contain specified proportions of the data. The factors that multiply σ are known as "Z-values"; these are listed in tables of the standard normal distribution published in statistical textbooks and many other scientific reference books. A given Z-value, say Z_p, corresponds to a probability P that a randomly chosen value of X will be within the interval $\mu - Z_p\sigma$ to $\mu + Z_p\sigma$. For example, Table 3.1 shows that $Z_{.684} = 1.000$, $Z_{.90} = 1.645$, and $Z_{.95} = 1.960$. Corresponding intervals for a log normal distribution are of the form $\mu_g/(\sigma_g)^{Z_p}$ to $\mu_g \cdot (\sigma_g)^{Z_p}$. Table 3.2 gives examples of the form of log normal intervals corresponding to the intervals for a normal distribution in Table 3.1. The first and fifth of these intervals are depicted graphically in Figure 3.2 as shaded and cross-hatched areas under a log normal curve.

Examples: Calculation of Log Normal and Normal Intervals. As a numerical example, suppose that the day-to-day variability of a worker's 8 hr TWA exposures is approximately log normally distributed with geometric mean (μ_g) of 20 ppm and geometric standard deviation (σ_g) of 1.20. One could then make the following types of implied interval prediction:

1. 95 percent of daily TWAs would be below 27.0 ppm (i.e., $20 \cdot 1.20^{1.645} = 20 \cdot 1.350$).
2. 95 percent of daily TWAs would be within 14.0 to 28.6 ppm (i.e., $20/1.20^{1.96}$ to $20 \cdot 1.20^{1.96}$ or $20/1.430$ to $20 \cdot 1.430$).

3. 90 percent of daily TWAs would be within 14.8 to 27.0 ppm (i.e., $20/1.20^{1.645}$ to $20 \cdot 1.20^{1.645}$ or $20/1.350$ to $20 \cdot 1.350$).

The log normal example just cited relates to temporal variability of the true daily TWA exposures. The following extension of the example relates to normally distributed errors of sampling and analysis. Suppose that on a particular day, the worker's true TWA exposure is 25 ppm. On that day, if the worker's exposure were measured using a sampling/analysis method that has a coefficient of variation of 10 percent (i.e., $CV_T = .10$), we would expect the value of the measurement to follow a normal distribution with mean $\mu = 25$ ppm and standard deviation $\sigma = \mu\, CV_T = 2.5$ ppm. Thus the probability is 95 percent that a single measurement would be within 20.1 to 29.9 ppm [i.e., $25 \pm 1.96(2.5)$ or 25 ± 4.9].

Figure 3.3 compares the shapes of normal and log normal distributional curves. Both curves were constructed to have the same mean (9.0) and the same standard deviation (2.5). The corresponding parameters of the log normal distribution are:

$\mu_g = 8.67$ (geometric mean)   and $\sigma_g = 1.31$ (geometric standard deviation)

For this example the arithmetic mean of the log normal distribution is 3.8 percent larger than its geometric mean. In general, for the log normal distribution, the arithmetic mean $E(X)$ is always greater than the geometric mean μ_g. The ratio $E(x)/\mu_g$ is larger when σ_g is larger, as Table 3.3 indicates.

To compute any additional ratios, use the formula:

$$\frac{E(X)}{\mu_g} = \exp(\tfrac{1}{2} \ln^2 \sigma_g) \qquad (14)$$

There are many practical industrial hygiene applications for the statistical models given

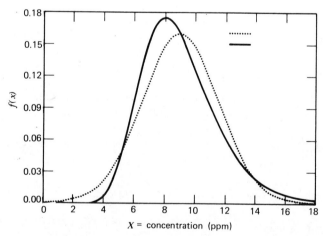

Figure 3.3. Comparison of normal (dotted) and log normal (solid) distribution curves.

STATISTICAL DESIGN AND DATA ANALYSIS REQUIREMENTS

Table 3.3 Ratios Between Geometric Mean μ_g and Arithmetic Mean $E(X)$ for Log Normal Distributions

Geometric Standard Deviation, σ_g	$E(X)/\mu_g$	$\mu_g/E(X)$
1.01	1.00005	0.99995
1.05	1.0012	0.9988
1.10	1.0046	0.9955
1.20	1.0168	0.9835
1.50	1.0857	0.9211
2.00	1.2715	0.7864
3.00	1.8285	0.5469

in this section as distributions of random errors in measurements data. Sections 4 and 5 cover two different types of application: in the physical sciences and in the biological sciences, respectively.

4 COMPLIANCE SAMPLING STRATEGIES AND DECISION THEORY

4.1 Nomenclature for Types of Statistical Sample

The decision procedures given in this section regarding compliance and noncompliance based on exposure measurements differ depending on how the samples were obtained in relation to the period of the standard, duration of the samples, and number of samples. We adopt the nomenclature used by Leidel et al. (15) with respect to four types of sampling strategy. The word "period" refers to the period of the standard. For an 8 hr TWA standard, the period is 8 hr, and for a ceiling standard it is generally 15 min. An exposure "measurement" consists of one or more samples (personal or breathing zone) taken during the measurement period. The four types of measurement are identified below and are shown diagrammatically in Figure 3.4.

4.1.1 Full Period, Single Sample Measurement

The sample is taken for the full period of the standard. This would be 8 hr for an 8 hr TWA standard and 15 min for a ceiling standard.

4.1.2 Full Period, Consecutive Samples Measurement

Several samples (equal or unequal time duration) are obtained during the entire period appropriate to the standard. The total time covered by the samples must be 8 hr for an 8 hr TWA standard and 15 min for a ceiling standard.

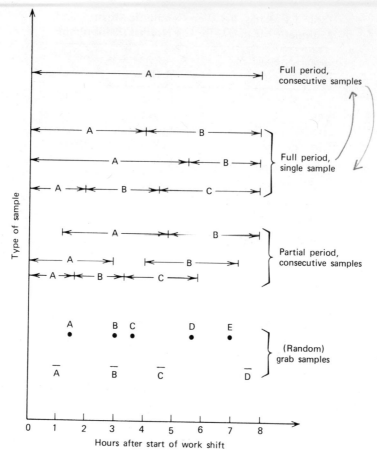

Figure 3.4. Types of exposure measurement that could be taken for an 8 hr average exposure standard. (From Leidel et al. (15).

4.1.3 Partial Period, Consecutive Samples Measurement

One or several samples (equal or unequal time duration) are obtained for only a portion of the period appropriate to the standard. For an 8 hr TWA standard this would mean that the sample or samples cover about 4 to less than 8 hr. Several samples totaling less than 4 hr (as eight 30 min samples) would probably be best described as grab (short-term) samples for the purposes of statistical analysis.

4.1.4 Grab Sample Measurements

In some cases it is impossible, because of limitations in measurement methods as with direct reading meters or colorimetric detector tubes, to collect either a single sample or a

STATISTICAL DESIGN AND DATA ANALYSIS REQUIREMENTS

series of consecutive samples whose total duration approximates the period for which the standard is defined. Under such conditions, grab samples are taken over some number of short periods of time (less than 1 hr each; generally only minutes or seconds). Grab samples are taken at random intervals selected from the total period of time for which the standard is defined.

4.2 Statistical Analysis of Exposure Measurement Sample Results

This section discusses the application of standard statistical methods to the analysis of exposure sampling results. Because of the presence of random measurement errors, any exposure average for an employee calculated from exposure measurement is only an estimate of the unknown true exposure average. The compliance and noncompliance decision procedures presented take into account the statistical distribution of measurement errors of sampling and chemical analysis. The exact difference between the measured exposure average and the true exposure average for the sampled period is unknown. However the true error in the measurement at hand can be considered to be a random sample from a statistical distribution of all such errors that could have occured. On this basis a decision can be made regarding the value of the true exposure average relative to an occupational health standard. The decision statement will be formulated to have a predetermined risk level (or equivalent confidence level) associated with it. We discuss the effect of choosing different risk levels on the probabilities of declaring compliance or noncompliance. Such considerations are among several general concepts of confidence interval limits, hypothesis testing, type I and II errors, and power function curves that are treated before any specific methodology is presented.

There are several questions that can be addressed in statistical analyses.

1. Was an employee exposure average in compliance with the health standard (either ceiling or 8 hr TWA) on a particular day?
2. What is an employee's long-term exposure estimate based on several exposure measurement daily averages?
3. What is the percentage of days an employee can be expected to be exposed to above-standard levels, based on several exposure measurement daily averages?
4. Should engineering controls be installed to reduce excessive exposures?

Only the first question, dealing with single-day sampling, is discussed here, since our purpose is to illustrate general principles of statistical decision theory, not to deal comprehensively with all the available statistical methods for analysis of exposure measurements data. The required statistical methods for question 1 are given in Leidel et al. (15), who also address questions 2, 3, and 4 (which concern multiday sampling).

4.3 Test of Significance or Hypothesis Testing

The decision tests for question 1 of Section 4.2 are based on confidence intervals that could be shown to be algebraically equivalent to appropriate statistical tests of significance. It is useful to discuss the concepts and terminology of significance tests of hypotheses and compare them with decision rules based on confidence intervals.

The industrial hygienist is interested in whether the exposure measurements are sufficiently accurate to permit a confident decision of compliance or noncompliance, as the case may be. In this context a pertinent hypothesis is an assumption about the state of the true exposure average μ relative to a threshold limit value (TLV) or standard. Statistical significance tests involve two hypotheses. Before the exposure measurement is taken, a tentative assumption about the value of the total exposure average relative to the standard is made. This tentative assumption is then accepted unless it is proved wrong by the statistical test. By "proved wrong" we mean that the sampling measurements actually obtained would have had low probability (e.g., less than .05) of occurring before the samples were taken if the tentative assumption were true. This tentative negative hypothesis is called the null hypothesis. Correspondingly, an alternative assumption, referred to as the alternative hypothesis, is made. This alternative hypothesis must be accepted whenever the null hypothesis is rejected. The null hypothesis and the alternative hypothesis are chosen in relation to the professional job responsibilities of the industrial hygienist as they relate to the risks he or she is willing to run of making an incorrect decision of one type or the other. The philosophies of an employer and a governmental compliance officer tend to differ, and the appropriate points of view are discussed below.

4.3.1 Hypotheses for the Employer

Each employer is required to furnish to each of the employees a place of employment free from recognized hazards that are likely to cause death or serious injury. To do this, the employer must keep true employee exposures at levels below the appropriate TLVs or standards. Thus the employer must make decisions regarding exposure measurements in a manner permitting confidence that there is no employee whose average exposure exceeds the average exposure standards and that no employee will at any time be exposed to levels above the ceiling exposure standards. In statistical terms, the employer must formulate the null hypothesis that the true exposure *exceeds* the standard and put the "burden of proof" on the data, which must indicate compliance even after a "limit of error" allowance has been added to the exposure measurement. The "limit of error" is chosen such that the probability would be small (e.g., .05) that a larger random measurement error would have occurred in the negative direction. For the employer's test of compliance:

Null hypothesis is H_0 $\mu >$ standard (i.e., noncompliance)
Alternative hypothesis is H_A $\mu \leq$ standard (i.e., compliance)

STATISTICAL DESIGN AND DATA ANALYSIS REQUIREMENTS 61

Sections 4.4 and 4.5 give step by step procedures for making the employer's test of compliance.

4.3.2 Hypothesis for Compliance Officer

The governmental agency must meet the substantial evidence test and has the burden of proving that a health standard has been exceeded on a particular day. This is because the Occupational Safety and Health Administration (OSHA) health standards are either average exposure standards defined for an 8 hr averaging period or ceiling exposure standards that at no time shall be exceeded*. Therefore the compliance officer should state the null and alternative hypotheses such that the data must indicate noncompliance after allowing for random measurement variability that may have occurred in the positive direction. For the compliance officer's test for noncompliance:

Null hypothesis is H_0 $\mu \leq$ standard (i.e., compliance)
Alternative hypothesis is H_A $\mu >$ standard (i.e., noncompliance)

4.3.3 Errors in Hypothesis Testing

If we use the confidence interval as a test criterion for the measured exposure average, we realize that the risk exists that the confidence interval does not include the true exposure average. Hypothesis testing uses the terms "type I" and "type II" errors to describe the two types of wrong decision we might make based on the results of our tests. If we reject the null hypothesis (accept the alternative hypothesis) when the null hypothesis is really true, we commit a type I error. On the other hand, if we fail to reject the null hypothesis when it is truly false, we commit a type II error.

In the context of the compliance officer's test, the two types of error can be shown schematically as follows:

Compliance Officer's Test of Noncompliance

Test Result	True State	
	Compliance with Standard	Noncompliance with Standard
Decide H_0: compliance	No error	Type II error
Decide H_A: noncompliance	Type I error	No error

* Code of Federal Regulations: 29 CFR 1910.1000.

A similar schematic arrangement for the employer's test is:

Employer's Test of Compliance

Test Result	True State	
	Compliance with Standard	Noncompliance with Standard
Decide H_A: compliance	No error	Type I error
Decide H_0: noncompliance	Type II error	No error

To clarify the interpretation of the statistical decision procedure, we discuss the decision table used by compliance officers. The decision criterion for use by compliance officers is:

Reject H_0 $\mu \leq$ standard, and
Accept H_A $\mu >$ standard, whenever a lower confidence limit for the true mean at the $100(1 - \alpha)$ percent confidence level exceeds the standard.

The risk (probability) of making a type I error is designated α. The maximum value of α is the test's level of significance, also referred to as the size of the test. Note that the confidence level $(1 - \alpha)$ is the complement of the probability α of a type I error. This is true because our decision rule is based on a one-sided confidence interval but was formulated to be algebraically equivalent to an α-level one-sided significance test of the null hypothesis H_0. Thus a decision rule based on a 95 percent confidence interval is the same as a significance test with a 5 percent maximum risk of committing a type I error.

The risk of making a type II error is designated by β. The value of β varies with magnitude of the real difference between the standard and the true exposure average. The relation between these two types of risks can be summarized on either an operating characteristic (OC) curve for the test or on its complement, the power function (PF) curve, which is discussed below.

4.3.4 Power Function Curves

The power of the test is the probability of accepting the alternative hypothesis when the alternative hypothesis is true. The power is designated by $(1 - \beta)$, the complement of the probability of a type II error.

Earlier the term "95 percent confidence level" was introduced in reference to statistical hypothesis testing. The term arose from the choice of a 5 percent type I error risk level for the equivalent statistical significance test to be used. The clear advantage of using statistical tests for the decision process regarding exposure standards is that the maximum desired α and β risk levels can be selected in advance and power function

STATISTICAL DESIGN AND DATA ANALYSIS REQUIREMENTS

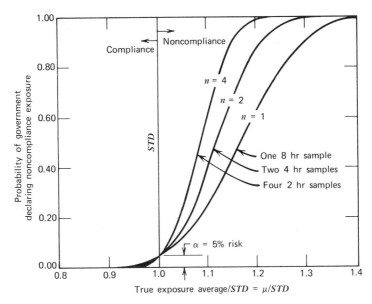

Figure 3.5. Power function (PF) curve for one-sided compliance officer's test (5 percent risk level) to detect noncompliance. Calculated for sampling analytical method with $CV_T = 0.10$ (about ±20 percent accuracy at 95 percent confidence level).

probability curves can be calculated. The PF curve gives the power $(1 - \beta)$ of the test as a function of the true mean μ. Figure 3.5 is the set of three PF curves for the compliance officer's test in the case of full period, consecutive samples with $\alpha = .05$ and $n = 1, 2,$ and 4. When the standard is for an 8 hr TWA exposure, $n = 1, 2,$ and 4 correspond to one 8 hr sample, two 4 hr samples, and four 2 hr samples, respectively.

4.4 Decision Rules for the Case of Full Period, Consecutive Samples and Uniform Exposure Over the 8 Hr Period of a TWA Standard

The case to be considered is of full period, consecutive samples when the expected values of all measurements are equal and n samples are taken for unequal durations $T_1, T_2, \ldots, T_n$. The total of sampling durations is $T = T_1 + T_2 + \cdots + T_n$, which is assumed to be equal to the period for the TWA permissible exposure limit (usually 8 hr). A one-sided confidence limit (LCL or UCL) can be used to classify average exposures into one of the three possible exposure categories. The use of the LCL (by the compliance officer) would result in a decision of either "noncompliance exposure" or "possible overexposure." The use of the UCL (by the employer) would result in a decision of either "compliance exposure" or "possible overexposure." Figure 3.6 shows the three-way classification relative to the standard. The circle in each vertical line

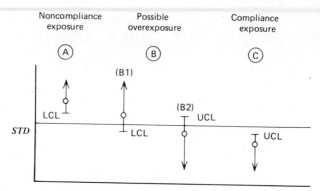

Figure 3.6. Classification system for employee exposure according to one-sided confidence limits.

represents the average exposure estimate calculated from the measurement sample results; the letter designations are explained in the following tabulation:

Classification	Definition	Statistical Criterion
A. Noncompliance exposure	There is 95% confidence (based on measurements) that a worker's exposure is above the standard	LCL (at 95%) > STD
B. Possible overexposure	Any individual who cannot be classified in A or C	
C. Compliance exposure	There is a 95% confidence (based on measurements) that a worker's exposure is below the standard	UCL (at 95%) ≶ STD

The steps of the classification procedure are as follows. (Note that other procedures are given in Reference 15 for use when the n sampled periods have unequal true concentrations.)

1. Obtain $X_1, X_2, \ldots, X_n$, the n consecutive exposure measurement values on one work shift and their durations $T_1, T_2, \ldots, T_n$. Also obtain CV_T, the sampling/analytical total coefficient of variation, from published tables of CV_T values (see Reference 15 or 25).

2. Compute the estimated TWA exposure.

$$TWA = \left(\frac{1}{T}\right)(T_1X_1 + T_2X_2 + \cdots + T_nX_n)$$

where $T = T_1 + T_2 + \cdots + T_n$ is the total sampling time.

STATISTICAL DESIGN AND DATA ANALYSIS REQUIREMENTS

3. Compute the standard deviation of the estimated TWA.

The derived standard deviation of TWA is designated by the term "standard error" (*SE*) to distinguish it from σ, which is the standard deviation of the basic measurements (i.e., of the X values).

$$SE = CV_T \cdot STD \left(\frac{1}{T}\right) \sqrt{T_1^2 + T_2^2 + \cdots + T_n^2}$$

where *STD* denotes the applicable exposure standard.

NOTE. If the sample durations are approximately equal, this short equation can be used:

$$SE = \left(\frac{1}{\sqrt{n}}\right) CV_T \cdot STD$$

4. Compute the 95 percent LCL or 95 percent UCL as follows:

$$LCL = TWA - 1.645(SE)$$

$$UCL = TWA + 1.645(SE)$$

5. Classify the TWA exposure average for the n uniform samples as follows:
   a. Compliance officer's test for noncompliance:
      If $LCL > STD$, classify as noncompliance exposure.
      If $TWA > STD$ and $LCL \leq STD$, classify as possible overexposure.
      If $TWA \leq STD$, no statistical test for noncompliance would be made.
   b. Employer's test for compliance:
      If $UCL \leq STD$, classsify as compliance exposure.
      If $UCL > STD$, classify as possible overexposure.
      If $TWA > STD$, no statistical test for compliance would be made.

4.5 Decision Rules for the Case of Short Period Samples Taken at Random Intervals within an 8 Hr Work Shift

Leidel, Busch, and Crouse have documented data showing that short-period air sample concentrations taken at random intervals tend to follow a log normal distribution (20). Therefore confidence limits for the true mean *of logs* could be calculated by making a log transformation of the concentrations and applying standard methods and tables that are available for normally distributed data. In particular, the Student t table can be employed for such calculations. The detransform (antilogs) of these log limits would then provide confidence interval limits for the true *geometric mean* concentration. However federal health standards and/or TLVs are not stated in terms of geometric means. Rather, exposure standards are stated in terms of TWA concentrations over the full period of the standard. A TWA exposure standard can be interpreted to be the arithmetic mean $E(X)$ of the set of all possible short-period grab samples which could be taken during the full period of the standard. Table 3.3 lists the ratio between the geometric mean and arithmetic mean of a log normally distributed variable for various geometric standard deviations. Table 3.3 shows that the geometric mean of grab samples could be used for direct comparison to a TWA exposure standard only if the geometric standard deviation of the grab samples was less than about $\sigma_g = 1.20$. (For $\sigma_g = 1.223$, the geometric mean of grab samples underestimates the TWA exposure by less than 2 percent.)

Since the geometric mean of grab samples is a biased (under) estimate of the TWA exposure, a special statistical test of significance was derived by NIOSH under contract (21) for making controlled-risk decisions of compliance or noncompliance with federal TWA exposure standards. The NIOSH statistical test is based on an assumed log normal distribution of grab samples, but the inference made concerns the arithmetic mean (i.e., TWA).

1. Take n grab samples at random intervals during the period of the standard; for example, $n = 5$ if 15 min grab samples were taken during five intervals selected at random from the 32 (15 min) intervals in an 8 hr workday.

2. Designate the sample concentrations as $X_1, X_2, \ldots, X_n$. Compute common logs of each concentration: $Y_1 = \log X_1, Y_2 = \log X_2, \ldots, Y_n = \log X_n$.

3. Compute the sample mean Y and sample standard deviation s_Y of the logs.

$$\overline{Y} = \frac{1}{n}(Y_1 + Y_2 + \cdots + Y_n)$$

$$s_Y = \sqrt{[1/(n-1)][(Y_1 - \overline{Y})^2 + (Y_2 - \overline{Y})^2 + \cdots + (Y_n - \overline{Y})^2]}$$

4. Calculate the mean $\bar{y}$ and standard deviation s_y of common logs of the relative concentrations, $y = \log(X/STD)$, where STD denotes the federal standard for the contaminant being investigated. The individual y-values need not be computed; rather, merely transform the mean and standard deviation computed in step 3 as follows:

$$\bar{y} = \overline{Y} - \log(STD)$$

$$s_y = s_Y$$

5. Plot a decision point on the decision chart of Figure 3.7. The abscissa (horizontal coordinate) is s_y and the ordinate (vertical coordinate) is $\bar{y}$.

6. The decision is made according to the position of the plotted point in relation to two boundary lines corresponding to the sample size n. The decision rules for use by compliance officers and employers are given below. Risks of making wrong decisions are also discussed.

Case a. Compliance Officer's Test

Null hypothesis, H_0 TWA $\leq$ standard
Alternative hypothesis, H_A TWA $>$ standard

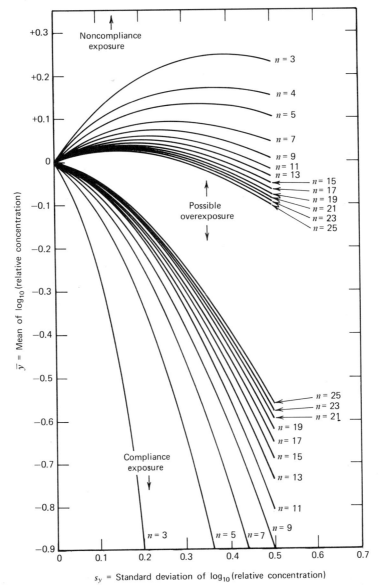

Figure 3.7. Grab sample measurement average classification chart.

Decision Rule. Decide for noncompliance (H_A) if the decision point lies on or above the upper boundary curve. Otherwise, decide for (possible) compliance (H_0).

The probability α that decision H_A is incorrect is $\alpha \leq .05$. The probability β that decision H_0 is incorrect ranges from about $\beta = .95$ when the true TWA is barely over the standard to near zero when TWA is so small relative to the standard that the range of grab sample concentrations does not encompass the standard. Because the probability β can be large, the decision for H_0 is high risk; hence the uncertainty attached to the phrase "possible compliance" is intended.

Case b. Employer's Test.

Null hypothesis, H_0 TWA > standard
Alternative hypothesis, H_A TWA $\leq$ standard

Decision Rule. Decide for compliance (H_A) if the decision point lies on or below the lower boundary curve. Otherwise, decide for (possible) noncompliance (H_0).

The probability α that decision H_A is incorrect is $\alpha \leq .05$. The probability β that decision H_0 is incorrect ranges from about $\beta = .95$ when the true TWA is barely under the standard to near zero when the TWA is so large relative to the standard that the range of grab sample concentrations does not encompass the standard. Because the probability β can be large, the decision for H_0 is high risk; hence the uncertainty attached to the phrase "possible noncompliance" is intended.

Case c. Double-Sided Test.

Null hypothesis, H_o TWA $\leq$ standard
Alternative hypothesis, H_A TWA > standard

Decision Rule. Decide for noncompliance (H_A) if the decision point lies above the upper boundary curve. Decide for compliance (H_0) if the decision point lies on or below the lower boundary curve. Reserve making a decision if the decision point lies on or below the upper curve and above the lower boundary curve.

The probability α that decision H_A is incorrect is $\leq .05$. That is, the probability of deciding for noncompliance (H_A: TWA > standard) is at most .05 when H_0 is true (TWA $\leq$ standard). Also, the probability that decision H_0 (TWA $\leq$ standard) is incorrect is less than .05.

When the decision point is between the boundary curves, the probability of making the wrong decision would be greater than .05 for *either* decision (H_0 or H_1). Therefore neither decision can be made without running a high risk of being wrong.

4.6 Sample Size Requirements for Demonstration of Compliance and Noncompliance with an 8 Hr TWA Standard

Leidel and Busch (22) have introduced a terminology for various sampling time strategies. The terminology appears in Section 4.1 and is illustrated in Figure 3.4. Any of the strategies shown in Figure 3.4 could be used to determine an estimate of the 8 hr TWA exposure for the given day on which sampling is performed. However this section discusses factors that make a particular exposure measurement strategy preferable for a particular day's measurement. There is no one "best" strategy for all situations. However some strategies are clearly better than others, and guidelines can be given for comparing alternative strategies. The following are broad considerations:

- Availability and cost of sampling equipment (pumps, filter, detector tubes, direct reading meters, etc.).
- Availability and cost of sample analytical facilities (for filters, charcoal tubes, etc.).
- Availability and cost of personnel to take samples.
- Location of employees and work operations.
- Occupational exposure variation (intraday and interday).
- Precision and accuracy of sampling and analytical methods.
- Number of samples needed to attain the required accuracy of the exposure measurement.

The subject of intraday and interday occupational exposure variation has been discussed by Ayer and Burg (25) and Leidel et al. (20). The exposure variation of specific operations is almost impossible to predict. The only generalization that can be made is that intraday and interday variation, as measured by the geometric standard deviation (GSD), typically lie between 1.25 and 2.5, as shown by data in References 25 and 20. The size of the GSD for intraday variations is the primary determinant of the number of grab samples needed.

On the other hand, for full period sampling the intraday variations do *not* affect the precision of the resulting TWA exposure estimate. This is true because ups and downs in the true concentration during the sampling period are "integrated out" by the physical sampling device. That is, at any point in time the sampled chemical is being deposited in the cumulative sample at a mass rate proportional to its air concentration. Thus only the sizes of sampling (i.e., flowrate) variations and of errors in chemical analyses of the cumulative sample(s) are related to the precision of full period samples. Again to generalize, most NIOSH sampling and analytical procedures have total coefficients of variation of .05 to .10 (5 to 10 percent).

There are other decisions that must be made about the sampling strategy which do not relate to the statistical distributions of sampling errors discussed previously. The appropriate location for sampling is primarily a nonstatistical question in the field of industrial hygiene and therefore is not discussed here. The choice of workers to be sampled does have statistical aspects, but this matter is not covered here. The choice of workers is discussed in Leidel et al. (15).

Once an appropriate place for sampling and a subject have been selected, there remain three questions: how may samples should be taken, at what times, and for how long? These questions are addressed below.

After considering both exposure variation and the precision and accuracy of sampling/analytical methods, the following general guidelines can be given:

1. The full period, consecutive samples measurement is "best" in that it yields the narrowest confidence limits on the exposure estimate. There are statistical benefits to be gained from larger sample sizes (as eight 1 hr samples instead of four 2 hr samples), but with the disproportionately large additional costs incurred (especially analytical), the benefits are usually negligible. That is, the gains from additional (shorter) samples on the same work shift in "decision-making power" are small compared with the significantly greater costs. Figure 3.8 plots the effects of increased sample size. Considering presently available sampling/analytical techniques, we can state that two consecutive full period samples (about 4 hr each for an 8 hr TWA standard) usually provide sufficient precision and are recommended as the "best" measurement to make.

2. The full period, single sample measurement (one 8 hr sample) is next to best if an appropriate sampling/analytical method is available. In this case, one 8 hr sample is essentially as good (all factors considered) as two 4 hr samples.

3. The partial period, consecutive samples measurement is the next choice. The major problem encountered with this type of measurement is handling the unsampled portion of the period. Strictly speaking, the measurement results are valid only for the duration of the period that the measurements cover (as 6 out of 8 hr). However professional judgment may allow inferences to be made concerning exposure concentrations during the unsampled portion of the period. Reliable knowledge concerning the operation is required to make this judgment. The sampled portion of the period should cover at least 70 to 80 percent of the full period.

For exposure measurements made by the employer or a representative, it is probably sufficient to assign the exposure average for the partial period to the whole period. It is assumed that the unsampled period had the same exposure average as the sampled portion. However the statistical decision tests in Section 4.4 are not fully valid in this situation. One can put confidence limits on a 6 hr exposure average, but it would not be proper to compare them with an 8 hr TWA standard since the work habits of the employee and the work operation must be identical during the sampled and unsampled portions of the work shift. This type of measurement should be avoided if possible.

For exposure measurements made by a governmental compliance officer, it is best to assume zero exposure for the unsampled period. Figure 3.9 shows the low "power" of such a partial period, consecutive samples procedure. The effect of sample size and total time covered by all samples on requirements for demonstrating noncompliance is shown by the family of four curves. The bottom curve (8 hr total sample time) is the same curve

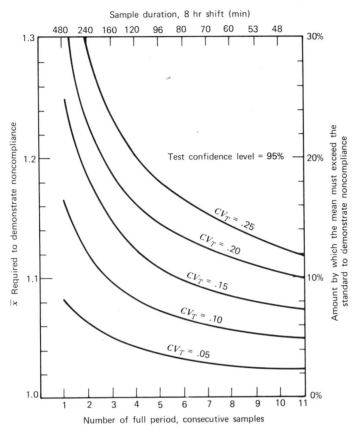

Figure 3.8. Effect of full period, consecutive sample size on noncompliance demonstration when test power is 50 percent; CV_T is coefficient of variation of sampling/analytical methods.

as the $CV_T = .10$ curve of Figure 3.8. The taking of partial period, consecutive samples is a compromise between the preferred full period sample(s) and the least desirable grab samples. Refer to Leidel and Busch (22) for analysis of these types of data when zero exposure is assumed for the unsampled period.

4. A grab sample measurement is the least desirable way of estimating an 8 hr TWA exposure. This is because the confidence limits on the exposure estimate are very wide and a low exposure average is necessary to statistically demonstrate compliance by the methods of Section 4.5. Figure 3.10 shows that the optimum number of grab samples to take for an exposure measurement is between 8 and 11. However this applies to the 8 hr TWA exposure only if the employee's operation and work exposure are relatively constant during the day. If the worker is at several work locations or operations during the 8 hr shift, at least 8 to 11 grab samples should be taken during each period of expected differing exposure that significantly contributes to the 8 hr TWA exposure. If one is

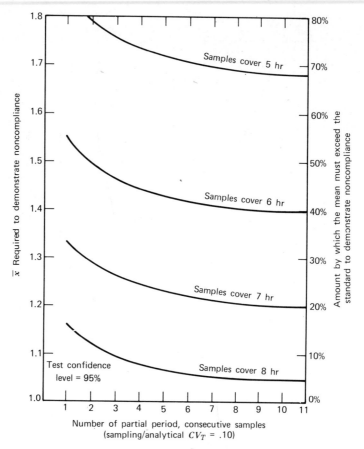

Figure 3.9. Effect of partial period, consecutive sample size and total time covered by all samples on noncompliance demonstration when test power is 50 percent.

limited to taking fewer than 8 to 11 samples at each location (or operation), choose the number of samples at each location in rough proportion to the time spent at each location. That is, *take more samples in areas where more time is spent.*

A comparison between the sampling strategies 4.1.3 and 4.1.4 is as follows. If a GSD of 2.5 is assumed in Figure 3.10, a curve of about 5½ hr on Figure 3.9 would have approximately the same $\bar{X}/STD$ ratios for moderate numbers of samples, say n = 8 to 11. Therefore it if it is not possible to sample for at least 70 percent of the time period appropriate to the standard (5½ hr for an 8 hr standard), it is better to go to a grab sampling strategy.

If grab samples are taken, their duration is important only in that enough samples must be collected for the analytical method. That is, any increase in sampling duration past the minimum time required to collect an adequate amount of material is unnecessary and unproductive. A 40 min grab sample is little better than a 10 min one. This matter is discussed by Leidel and Busch (22).

The last question to be answered concerns when to take the grab samples during the period of exposure. The accuracy of the probability level for the test depends on implied assumptions of the log normality and independence of the sample results that are averaged. These assumptions are not highly restrictive if precautions are taken to avoid bias when selecting the sampling times over the period for which the standard is defined. To this end, it is desirable to choose the sampling periods in a statistically random fashion.

For a standard defined as a TWA concentration over a period longer than the sampling interval, an unbiased estimate of the true average can be ensured by taking samples at random intervals. It is valid to sample at equal intervals if the series is known to be stationary, with contaminant levels varying randomly about a constant mean and fluctuations of short duration relative to length of the sampling interval. If means and their confidence limits were to be calculated from samples taken at equally spaced intervals, however, biased results could occur if cycles in the operation happened to be in phase with the sampling periods. Results from random sampling are unbiased even when cycles and trends occur during the period of the standard.

The word "random" refers to the manner of selecting the sample. Any particular

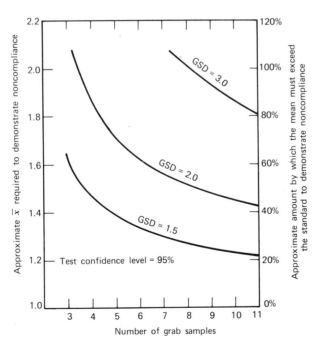

Figure 3.10. Effect of grab sample size on noncompliance demonstration. The three data geometric standard deviations (GSD) reflect the amount of intraday variation in the environment.

sample could be the outcome of a random sampling procedure. A practical way of defining random sampling is that any portion of the work shift has the same chance of being sampled as any other.

4.7 Sampling Strategies for Determination of Compliance and Noncompliance with a Ceiling Standard

Samples taken for determination of compliance with ceiling standards are treated in a manner similar to those taken for comparison with TWA standards. Two important differences should be noted.

The first is that the samples taken for comparison with ceiling standards are best taken in a *nonrandom* fashion. That is, all available knowledge relating to the area, individual, and process being sampled should be utilized to obtain samples during periods of maximum expected concentrations of the substance.

The second point is that samples obtained for comparison with ceiling standards are normally taken for a much shorter period than those taken for calculating TWAs. Each measurement usually consists of a 15 min sample (or series of consecutive samples totaling 15 min) taken in the employee's breathing zone. Leidel et al. (15) recommend that a minimum of three measurements be taken on one work shift and the highest of all measurements be used as an estimate of the employee's upper exposure for that shift.

Taking at least three measurements on a shift makes it easier to spot gross errors or mistakes. In most cases, however, only the highest value is statistically tested for compliance by the full period, single sample measurement procedure (in Section 4.4, use $n = 1$). If the samples are taken for comparison to the "maximum peak" ceiling standard, the sampling period should equal the "maximum duration" period for that particular standard. Thus in the case of detector tubes, it might be necessary to take several consecutive samples and average the results. Then the full period, consecutive samples measurement procedure (Section 4.4) would be used to analyze the results.

Even though samples taken for comparison with ceiling standards are best taken in a nonrandom fashion, the process may appear to be constant during the work shift. In this case the number of periods that should be sampled can be estimated so that representation (one or more) is assured (i.e., has a high probability) from the periods with the highest exposures (top 15 percent or top 10 percent).

The sample size recommendations given below are based on combinatorial probability formulas given in Appendix A of Reference 15. The probability theory is not presented here, but an example of the results obtained is as follows. With a ceiling standard defined for a 15 min period, there are 32 discrete nonoverlapping periods in an 8 hr work shift. A random sample of 9 of the 32 periods will have 90 percent probability of containing one or more of the six periods that have the highest exposures. (The number 6 is the nearest integer representing 20 percent of 32.) This result and other sample sizes that may be appropriate for the 15 min sampling period are as follows:

STATISTICAL DESIGN AND DATA ANALYSIS REQUIREMENTS 75

15 min Period

At Least One Period From	Confidence Level	Sample at Least
Top 20%	.90	9 periods
Top 20%	.95	11 periods
Top 10%	.90	16 periods
Top 10%	.95	19 periods

Where the ceiling standard is defined for a 10 min period, there would be 48 periods and the following sample sizes are appropriate:

10 min Period

At Least One Period From	Confidence Level	Sample at Least
Top 20%	.90	9 periods
Top 20%	.95	12 periods
Top 10%	.90	17 periods
Top 10%	.95	21 periods

Very short time samples may sometimes be taken, as with a 3 min detector tube or spot readings with a direct reading meter. Then the appropriate number of samples to take can be selected from the following values:

Less Than a 5 min Period

At Least One Period From	Confidence Level	Sample at Least
Top 20%	.90	10 periods
Top 20%	.95	13 periods
Top 10%	.90	22 periods
Top 10%	.95	28 periods

Once the appropriate number of periods has been chosen, the particular time periods to be sampled should be selected at random.

5 STATISTICAL DESIGN OF ANIMAL EXPOSURE CHAMBER STUDIES

This section discusses general statistical considerations necessary for the design of animal exposure chamber studies, including the enumeration of relevant types of error source, methods for controlling these errors through statistical experimental design, the philosophy underlying inferential methods of statistical analysis of resulting biological data, and some biological and statistical data that may help experimenters to make logistical decisions concerning the experimental protocol.

5.1 Philosophy of Experimental Design

Animal exposure experiments are performed to detect biological effects caused by the exposure. Either of two possible general plans could be followed.

Plan 1. A group of animals is exposed to the test atmosphere, and another group is exposed to control air. One or more biological parameters are then measured for both groups, and the distribution for the test group is compared with that for the control group. (In rare instances a known range of normal values determined from previous experimentation could be used instead of a concomitant group of control animals.) It is assumed that if the test group of animals were maintained under control air conditions, it would yield biological measurements similar to those of the control group. Under this assumption, the difference between average responses by the test and control groups can be taken as an estimate of the average effect of the exposure on the test group or, rather, on a larger population of similar animals from which the test group was selected as a sample. Under the condition that the control and test distributions have geometrically similar shapes, a finding that the two averages were not different would be taken to imply that no effects of the exposure on individual subjects existed.

Plan 2. Each member of a group of subjects is measured before and after exposure, and the average of paired differences is a measure of the average effect of exposure. Here each animal is used as its own control.

For short-term exposures plan 2 would usually produce more accurate estimates of average biological effects of the exposure from a given number of animals than plan 1, which employs a separate control group. This is true because the "spread" (measured as variance) in the distribution of paired differences would be less than the interanimal spread of the measurements themselves (either before or after exposure). Plan 2, however, cannot be used for chronic exposures because of possible confusion of any effects of exposure with normal changes over time. Therefore only the usual case of a separate control group is discussed. A refinement is to measure both the control and exposed groups before and after exposure (no exposure for controls) and compare averages of the two distributions of paired differences.

Since most biological responses vary widely among supposedly identically treated subjects, or even among successive measurements on a given animal, a difference between average responses by exposed and control groups of animals cannot necessarily be attributed to an effect of the exposure. Therefore the statistical philosophy underlying the analysis of the experimental data is that to be statistically significant, the observed difference must exceed any difference attributable to the random variability of the experiment. More specifically, the observed average response difference is first assumed to be merely a manifestation of random experimental variations (or "errors") and this "null hypothesis" is rejected only if the occurrence of random difference as large or larger is "improbable" under the null hypothesis. The alternative to the null hypothesis is that an average effect of the exposure exists.

5.2 Sources of Variability in Biological Data

In animal exposure chamber studies, the total random variability in biological responses by animals in a given exposure group (referred to as the total error in the response) is the algebraic sum of variations attributable to five general types of error source. These are discussed below.

1. *Interanimal normal biological variability* under identical exposure conditions is usually the largest source of variability; it is impossible to predict one animal's exact response from the response of a fellow animal. A portion of the unexplained interanimal biological variability can sometimes be removed by adjusting final biological responses for initial (preexposure) physiological nonhomogeneity of the subjects. One method for making such an adjustment has already been mentioned—the plan 2 approach of paired differences between preexposure and postexposure measurements. Covariance analysis is the name of another statistical technique often used for making similar adjustments. (Covariance analysis is discussed further in Section 5.5.)

2. *Intraanimal sampling error* is a second type of variability in biological response. For example, since it is not possible to assay an animal's total blood supply, the concentrations of biochemical or cellular constituents in an animal's total blood supply must be estimated from the concentrations in one or more blood samples. Different blood samples would yield somewhat different concentrations even if there were no measurement errors. (Such differences between the true concentration in the entire blood supply and the true concentrations in blood samples are called "blood sampling errors.") Many other types of intraanimal sampling errors exist that are analogous to blood sampling errors—for example, biochemical or cellular concentrations in samples of tissue or of bone, and repeated measurements of the tidal volume of the lungs taken during successive respirations.

Duplicate samples are usually not taken from the same animal when sampling errors are small compared to interanimal errors because only the net error from these two sources need be known to make valid statistical inferences about differences between mean responses of control and exposed *populations* of animals. To make valid statistical inferences about differences between *individual* animals, however, it would be necessary to take more than one sample from each animal. In this regard, remember that differences between assays of two aliquots taken from the same sample do not usually reveal sampling errors, but rather reveal only assay errors.

3. *Assay errors* also contribute to the total error in animal response data. Any given sample of blood contains definite red and white blood cell concentrations, hemoglobin,

and so on, but, in general the laboratory technician's assay of this blood sample does not yield results that are identical to the true concentrations in the sample because of reading errors, instrument errors, and other unavoidable random variations in analytical procedures.

Standard laboratory analytical procedures are calibrated to be unbiased (i.e., on the average, reported results are assumed to be equal to true results), but it is still a worthwhile precaution to randomize the order of assay of samples from the various exposure groups. When samples are assayed in random order, any occasional sources of systematic errors (similar errors for an entire batch of samples) such as reagent impurities, dilution errors in preparing a batch of a reagent, incorrect technique by a substitute technician, and variations in laboratory temperature, humidity, and so on, would affect results from all exposure groups equally in the long run.

Interanimal errors induced by nonhomogeneous experimental conditions represent another general type of error in biological responses. This variability may be caused by imperfectly controlled environmental factors such as animal housing and animal care, or by dose-response variations attributable to differences in gas concentrations in parts of a chamber or in separate chambers that were intended to contain identical atmospheres. A distinction is made here between intrachamber-induced errors and interchamber-induced errors.

4. *Intrachamber-induced errors* are induced interanimal response variations within the same chamber.

5. *Interchamber-induced errors* are induced response variations that are constant for all animals within a given chamber but different between replicate chambers (chambers containing the same type of atmosphere).

Obvious features of engineering design in the exposure system that are necessary for homogeneous animal exposures are seldom overlooked—for example; the possibility that chemically reactive pollutant gases may interact somewhat more before reaching chambers at the end of the gas feed pipe than for chambers at the beginning of the run. In spite of careful engineering design, however, some degree of nonhomogeneity of exposure concentrations between replicate chambers, as well as within a single chamber, will exist because of the presence of animals in the chamber. The animals themselves act as nonstationary baffles, and their hair and secretions absorb or react with gases, notably ozone and other oxidants.

5.3 Estimation of Experimental Errors from Data

In actual animal exposure experiments, the recommended practice is to take one sample from each animal and do duplicate assays.

The resulting data do not permit separate determination of all the five basic types of error that contribute to the data. Two of the types of basic error can be determined separately, namely, assay errors and interchamber-induced errors. Only the total (i.e., net) error caused by the other three types can be determined. This total error (for types

1, 2, and 4) is referred to here as the "biological" error, since interanimal biological error is usually the largest of these three component errors in animal exposure experiments. Table 3.4 shows the composition of the three types of estimable error in terms of the five basic error types. (An × in a column indicates inclusion of the basic error for that row in the error sum for the column.)

5.4 Methods for Controlling Experimental Errors

The general objective in the statistical design of an animal exposure experiment is to reduce, eliminate, or at least provide for the estimation of, the errors of the types listed above. Some general ways to accomplish this have already been referred to in the discussion of errors. Methods for overcoming errors are listed below.

Method 1. *Physical* elimination of the source of error, if feasible.

Method 2. *Mathematical* corrections for the error can be made through known theoretical relationships or through empirically determined correlations between the response variable and other measured factors. (See Section 5.5, Covariance Analysis.)

Method 3. *Blocking.* Set groups of the various treatments into "blocks" of similar experimental units. "Block" is a statistical term for a group of experimental units within which the same or similar contributions of a common error source occur for every treatment exposure in those experimental units. An example of a block is a group of exposure chambers (experimental units) located adjacent to the windows (error source) in the exposure room. Another example is furnished by the grouping of animals according to body weight (error source) with each treatment represented within each range of body weights; here the blocks are the narrow ranges of body weights that pertain to the separate groups of animals.

Method 4. *Randomization.* Randomly assign available animals and chambers to the various treatments and randomize the order of taking samples and of assaying samples. Randomization does not eliminate or reduce individual errors but it does

Table 3.4 Composition of Estimable Errors in Animal Exposure Experiments

Basic Error Type	Estimable Error Type		
	Assay	Biological	Chamber
1. Interanimal biological		X	
2. Intraanimal sampling		X	
3. Assay	X		
4. Intrachamber induced		X	
5. Interchamber induced			X

prevent bias in the experiment by distributing the errors randomly among treatments in a way that prevents confounding errors with biological effects. Randomization also permits the use of probability theory for assessment of errors using tests of statistical significance and/or confidence limits.

Method 5. Replication of experimental factors. Replicating assays, animals and chambers (in combination with randomization, method 4), reduces the error in average biological effects by "averaging out the error."

Table 3.5 indicates which methods for reduction of errors are applicable to each of the five types of error. As an example, errors of type 5 may exist for animal behavioral data related to spontaneous activity. The activity of animals is affected by their feeding and sleeping patterns, which in turn may be affected unequally in the various chambers in an exposure room by human traffic disturbances, by light from windows present along only one wall of the exposure room, by noises from motors or audible conversations outside the exposure room, by opening and closing of doors, and so on, by varying degrees of awareness animals in one chamber have of animals in other chambers, and by a myriad of other variations in the exposure environment to which animals might respond in small and unpredictable degrees.

For all these type 5 (interchamber) disturbances, physical elimination of the causes is the best cure: black drapes over windows, noise reduction, screens or walls between chambers to provide visual privacy, and so on. Mathematical corrections for such disturbances would not be possible, but blocking of treatments into groups of chambers with homogeneous environments is appropriate. If large interchamber behavioral response variations occur, blocking would be a more effective method for error reduction than randomization and replication, since too few chambers would be available to permit "averaging out" these errors. Also, plumbing and other physical restrictions would probably not permit random allocation of exposure atmospheres to chambers.

Thus because of high cost and physical restrictions on equipment, it is usually not feasible to use "extensive" replication (say three or more chambers per treatment) merely for the purpose of estimating a potential chamber-to-chamber component of the total variance. After all, by engineering design chamber differences are not intended to exist. On the other hand, the validity of an effect determined from an experiment with only one control chamber and one test chamber must be questioned, since an important potential error source (chambers) has not been replicated. In fact, any effect of the exposure factor (control vs. test) would be logically "confounded" (i.e., confused) with the chambers factor (chamber 1 vs. chamber 2). A recommended compromise is to use two chambers per treatment—not enough to estimate a chamber-to-chamber variance, but useful as a validity check of the assumption that there is no chamber-to-chamber variance. Assuming that the validity check will be satisfied, the error variance would then be taken to be the intersubject variance (after pooling subjects from duplicate chambers), not the interchamber variance. Both the control and test chambers should be duplicated for a validity check because homogeneity of *both* pairs is a fundamental

STATISTICAL DESIGN AND DATA ANALYSIS REQUIREMENTS

assumption on which the validity of the experimental estimate of the exposure effect depends.

5.5 Covariance Analysis

In animal exposure studies an apparent biological effect of the exposure could be attributable to the extraneous effects of any physiological nonhomogeneity (not due to the exposure) that exists between the respective groups, as indicated by preexposure group differences in the same response parameter or in other correlated physiological parameters. The response averages for the various exposure groups are usually adjusted for these extraneous effects by means of a statistical method called analysis of variance. This is an example of the use of a mathematical method to reduce interanimal biological errors.

The covariance adjustments are made using interanimal, intratreatment least-squares regression lines (or planes) fitted to plots of the final biological response against corresponding values for the covariate.

The effects of the covariance adjustments are (1) to reduce the amount of unexplained variability in the experiment, thereby providing more precise estimates of average biological effects of the exposure, and (2) to remove bias from estimates of the effects of the exposure caused by known initial physiological nonhomogeneity between control and exposed groups of animals. These mathematical corrections for bias are not intended to permit relaxation of the requirement for random assignment of animals to exposure groups, however, since some physiological factors that affect the biological response may be unknown or unmeasurable. It is important to avoid undetectable biases in the experiment by selecting subjects for the different exposure groups randomly from the available pool of stock animals. Then response variability attributable to randomly dispersed physiological differences will merely add to the total interanimal variability (i.e., to the experimental error) and will not be mistaken for an effect of the exposure.

Covariance adjustments to the primary biological responses are not always appropriate. The meaning of such adjustments should be considered carefully in a biological context, taking account of whatever is already known about cause and effect relationships between the primary response and the potential covariates. It is possible to unintentionally and unknowingly "adjust away" real exposure effects on the response of interest by improperly using as a covariate a concomitantly measured biological parameter that is correlated with the response because it, too, has been affected by the exposure.

5.6 Example of Statistical Analysis of Biological Data

To distinguish the effects of exposure from the variability of the experiment, comparison of the average responses of groups of control and exposed animals by means of formal methods of statistical analysis is usually required. The methods most often used are the analysis of variance for continuous-variable measurements, and the chi-square test for discrete variable measurements such as incidence of tumors or percentage mortality.

Table 3.6 gives results from a typical analysis of variance of red blood cell (RBC) data on dogs from an exposure experiment with four dogs in each of 24 chambers: three chambers for each of eight different exposure atmospheres. In this analysis the total variability in results from duplicate assays on single samples from the 96 dogs has been split into four types of variability. Three of the four sources of variability are errors attributable to chambers, interanimal normal biological variability, and assay errors (all three have been discussed). The fourth type of variability is nonrandom; it is attributable to differences in average responses for the exposure groups due to effects of the exposures. If the exposures had no effect, the mean square for the "treatment" source of variation would have the same average value as the mean square for "chambers," and probability limits for their F-ratio could be obtained by referring to standard tables of the F-distribution published in most handbooks of mathematical tables.

In this experiment, the F-ratio for "atmospheres" ($F = 4.12$) was much larger than could reasonably be attributed to mere random experimental variability—in fact, random values of F larger than 4.12 would occur in less than 1 percent of such experiments. Therefore the effect of the exposures is said to be statistically significant at the .01 probability level.

Since the four beagle bitches exposed in each chamber had been allowed to wander freely within the confines of the bottom chamber shelf, it was not possible to examine the data for response variations induced by the various possible exposure positions within a chamber. However in another experiment rats were exposed in the same type of chambers, and such an analysis was possible. The three rats of each sex present in each chamber were caged continuously in the same relative positions: adjacent to left and right walls of the middle shelf, and in the center of the top shelf. Table 3.7 gives results of the analysis of variance of hemoglobin data for the rats. Data again show a statistically significant effect of "atmospheres" at the .05 probability level. That is, results indicate nonhomogeneity of treatment means (one or more differences).

Table 3.5 Methods for Reducing Errors in Animal Exposure Experiments

	Methods for Reducing Errors			
Basic Error Type	1 Physical	2 Mathematical	3 Blocking	5 and 4 Replication (with randomization)
1. Interanimal biological	—	×	×	×
2. Intraanimal sampling	—	—	—	×
3. Assay	×	—	×	×
4. Intrachamber induced	—	—	×	×
5. Interchamber induced	—	—	×	×

STATISTICAL DESIGN AND DATA ANALYSIS REQUIREMENTS

Table 3.6 Analysis of Variance of Red Blood Cell Counts (RBC) on 96 Young Female Beagles

Source of Variation	Degrees of Freedom	Mean Square ($\times 10^{12}$)	F	Probability	Expected Mean Square
Atmospheres (treatments, T)	7	4.02	4.12	.009	$\sigma_A^2 + 2\sigma_B^2 + 8\sigma_C^2 + 24\theta(T)$
Chambers within atmospheres (C)	16	0.98	1.07	.40	$\sigma_A^2 + 2\sigma_B^2 + 8\sigma_C^2$
Animals within chambers (biological variation, B)	72	0.91	25.06	.000	$\sigma_A^2 + 2\sigma_B^2$
Duplicate assays (A)	96	0.03	—	—	σ_A^2
Experimental design 8 atmospheres: 7 contaminated, 1 clean air control 3 chambers per atmosphere 4 female dogs per chamber					

5.7 Homogeneity of Responses Within Chambers and Between Chambers

Further interpretation of the results given in Tables 3.6 and 3.7 follows:

1. No chamber-to-chamber variability of response existed beyond what is attributable to normal interanimal variability.
2. Rats responded the same, on the average, in each of the three cage positions.
3. Trivially (since the result was presumed), replicate animals did not respond identically.

5.8 Mathematical Formulas Relating Total Error of a Treatment Mean to Basic Error Components

In Tables 3.6 and 3.7 the last columns ("Expected Mean Square") give formulas for average values of the "mean square" column if the experiment were to be repeated a very large number of times. The expected mean square (EMS) formulas are functions of the variances of the estimable types of error discussed earlier and of the numbers of animals, chambers, and assays in the experimental design. The variance of a distribution of errors was defined (Equation 7) to be the second moment of the distribution about its mean value; the mean error is zero for unbiased sampling and analytical procedures. Unbiasedness is usually implied when calibration is checked periodically under a systematic laboratory quality control procedure such as has been recommended by NIOSH (24). The variance σ^2 is the square of a standard deviation σ.

For example, in Table 3.6, σ_A^2 is the variance of assay errors, σ_B^2 is the variance of biological variations, σ_C^2 is the variance of chamber (induced) errors, and $\theta(T) = \frac{1}{7} \sum_{i=1}^{8} T_i^2$ is a mean square of the average differential effects of the eight exposures.

Table 3.7 Analysis of Variance of Hemoglobin Data on Rats

Source of Variation	Degrees of Freedom	Mean Square	F	Probability	Expected Mean Square
Fixed Effects					
Atmospheres (treatments, T)	7	7.65	2.76	.04	$\sigma_A^2 + 2\sigma_B^2 + 12\sigma_C^2 + 36\theta(T)$
Cage positions (P)	2	2.00	0.49	.62	$\sigma_A^2 + 2\sigma_B^2 + 4\sigma_{CP}^2 + 96\theta(P)$
Sexes (S)	1	66.75	26.84	.000	$\sigma_A^2 + 2\sigma_B^2 + 6\sigma_{CS}^2 + 144\theta(S)$
$T \times P$ interaction	14	2.69	0.66	.79	$\sigma_A^2 + 2\sigma_B^2 + 4\sigma_{CP}^2 + 12\theta(TP)$
$T \times S$ interaction	7	4.11	1.65	.19	$\sigma_A^2 + 2\sigma_B^2 + 6\sigma_{CS}^2 + 18\theta(TS)$
$P \times S$ interaction	2	4.05	1.38	.27	$\sigma_A^2 + 2\sigma_B^2 + 2\sigma_{CPS}^2 + 48\theta(PS)$
$T \times P \times S$ interaction	14	4.34	1.48	.18	$\sigma_A^2 + 2\sigma_B^2 + 2\sigma_{CPS}^2 + 6\theta(TPS)$
Random variations ("error")					
Chambers within atmospheres (C)	16	2.77	0.94	.54	$\sigma_A^2 + 2\sigma_B^2 + 12\sigma_C^2$
$C \times P$ interaction	32	4.06	1.39	.18	$\sigma_A^2 + 2\sigma_B^2 + 4\sigma_{CP}^2$
$C \times S$ interaction	16	2.49	0.85	.62	$\sigma_A^2 + 2\sigma_B^2 + 6\sigma_{CS}^2$
$C \times P \times S$ interaction	32	2.93	218.53	.000	$\sigma_A^2 + 2\sigma_B^2 + 2\sigma_{CPS}^2$
Animals (biological variation, B)	—	—	—	—	$\sigma_A^2 + 2\sigma_B^2$
Duplicate assays (A)	144	0.01	—	—	σ_A^2

Experimental design
8 atmospheres: 7 contaminated, 1 clean air control
3 chambers per atmosphere
1 rat of each sex in each of 3 cage positions in each chamber (6 rats per chamber, caged individually)

STATISTICAL DESIGN AND DATA ANALYSIS REQUIREMENTS

The symbols in the EMS column of Table 3.7 have analogous meanings that are not discussed here because of the complexity of the statistical analysis for this multifactor experiment. Many excellent statistical textbooks are available that treat the intricacies of the analysis of variance, but the casual statistical practitioner is advised to consult with an experienced statistician before undertaking complex statistical analyses of this type.

The EMS column is discussed here because it gives formulas for the composition of the total experiment error of a treatment mean in terms of the basic error types. This information is needed to develop the theory and methodology for choosing requisite numbers of animals, assays, and chambers for another experiment. The following rule will be useful:

The variance of a treatment mean is equal to the mean square used in the denominator of the F-ratio for "treatments" divided by the number of individual assays included in each treatment mean.

In Table 3.6 the mean square for "chambers within atmospheres" is the denominator of the F-ratio for "treatments." Applying this rule to the data of Table 3.6, we see that the sample variance of the total error in a treatment mean response is $(.98 \times 10^{12})/24$ The average value of this total variance expressed in terms of variances of the component errors is denoted by $\sigma_{\bar{x}}^2$:

$$\sigma_{\bar{x}}^2 = \frac{1}{24} \text{EMS (chambers)} = \frac{1}{24}(\sigma_A^2 + 2\sigma_B^2 + 8\sigma_C^2)$$

$$= \frac{\sigma_A^2}{24} + \frac{\sigma_B^2}{12} + \frac{\sigma_C^2}{3}$$

This formula shows that the variance of the total error of a treatment mean is the sum of the ratios of the variances of each estimable error type divided by the respective numbers of *different* occurrences of that type of error within a treatment. Thus for each treatment, there were 24 assays performed, 12 animals were exposed, and three chambers were used, which are the divisors for the corresponding variances in the above formula for $\sigma_{\bar{x}}^2$. A general formula for $\sigma_{\bar{x}}^2$, for arbitrary numbers of assays, animals, and chambers, is:

$$\sigma_{\bar{x}}^2 = \frac{\sigma_A^2}{n_A n_B n_C} + \frac{\sigma_B^2}{n_B n_C} + \frac{\sigma_C^2}{n_C}$$

where n_A = number of assays per animal (of a single sample)
n_B = number of animals per chamber
n_C = number of chambers per exposure atmosphere

The forms of the $\sigma_{\bar{x}}^2$ formula suggests that to reduce errors in the treatment means efficiently, the experimental design should be so chosen as to expend available time and money to replicate primarily the operations or factors of the experiment that are associated with error sources having large variances. Usually this means that as many animals as possible should be exposed and relatively little effort should be expended on multiple assays of samples, since σ_B^2 (biological error) is very large compared to σ_A^2 (assay error). In the example σ_C^2 is near zero, but in some experiments chamber-to-chamber variability could well dominate the total error. Under the latter conditions, extensive replication of exposure chambers for each test atmosphere would theoretically assure definitive conclusions about existence of moderate-sized average exposure effects. However, replication of exposure chambers would be prohibitively expensive, so that the importance of careful engineering design for homogeneity of the exposure system and related animal environment cannot be overemphasized.

5.9 Sample Sizes for Studies Involving Quantitative Measurements

The standard deviation $\sigma_{\bar{x}}$ of the total error in a treatment mean $\bar{x}$ is the square root of the variance $\sigma_{\bar{x}}^2$ whose formula was given earlier. The "smallness" of $\sigma_{\bar{x}}^2$ for control animals is a useful measure of the overall precision of the experiment. To make this index of precison independent of the units of measurement, $\sigma_{\bar{x}}$ can be expressed as a coefficient of variation on a unit-animal basis by means of the following formula:

$$CV_T = \frac{\sqrt{n_B n_C \sigma_{\bar{x}}^2}}{\bar{x}} \quad \text{Std.D.}$$

The quantity CV_T is referred to as the "total coefficient of variation" (our term), by which we mean the coefficient of variation of a treatment mean expressed on a unit-animal basis. Combining the formulas for $\sigma_{\bar{x}}^2$ and CV_T given above, we get:

$$CV_T = \frac{1}{\bar{x}} \sqrt{\frac{\sigma_A^2}{n_A} + \sigma_B^2 + n_B \sigma_C^2}$$

Note that the last formula could also be written as follows

$$CV_T = \frac{1}{\bar{x}} \sqrt{\frac{\text{EMS (Chambers)}}{n_A}}$$

For the experimental design of Table 3.6, the formula would be:

$$CV_T = \frac{1}{\bar{x}} \sqrt{\frac{\sigma_A^2}{2} + \sigma_B^2 + 4\sigma_C^2}$$

since $n_A = 2$, $n_B = 4$, and $n_C = 3$.

A systematic methodology for objective determination of the required number of test animals (n) to be exposed to each atmosphere is outlined below for the case when $\sigma_C^2 =$

0 (i.e., for experiments in which replicate chambers yield homogeneous biological responses).*

The required sample size (n) depends on the following criteria and parameters of the experimental situation:

1. The "total coefficient of variation" of the response (measurement) for each exposure atmosphere.
2. The minimum size of the (hypothetical) true average biological effect to be detected by the experiment.
3. The "power" of the experiment as measured by the probability that the experiments will detect a real effect of the hypothetical size defined in item 2.
4. The "size" of the statistical significance test used in the statistical analysis. Figuratively speaking, the size of the test, also termed the significance probability (or "risk") is the probability that an unlikely combination of random variations will occur and will be mistaken for an effect when no average effect exists.

To determine the required number of animals (n) for each exposure group, the following procedure takes account of all four of the items above. In general, larger sample sizes are required with a larger value of the total coefficient of variation, larger "power," smaller biological effect, and smaller significance probability.

Step 1. The total coefficient of variation (CV_T), item 1, is estimated either from similar data collected in the past or from special preliminary data collected for the purpose of evaluating σ_A^2 and σ_B^2. ($\sigma_C^2 = 0$ is assumed.)

Step 2. The size of the biological effect to be detected, item 2, is measured by expressing the (hypothetical) difference between average responses, $\bar{c}$ and $\bar{x}$, for control and test populations, respectively, as a proportion of the average control response ($\Delta = |\bar{x} - \bar{c}|/\bar{c}$). The experimenter could have selected a Δ-value for the study either by choosing a "percentage" effect (i.e., 100Δ) or by choosing an $\bar{x}$ that should be detectable. In the latter case, the Δ-value would be calculated from $\bar{x}$ and the normal average response level $\bar{c}$. [Note. Symbols x and c are used in Section 5 to represent biological data for test and control groups. Symbols X (and C) were not used in Section 5 (although X was used in Section 4 to represent air sample concentrations) in order to prevent confusion between c (for control data) and C (designating "chambers").]

Step 3. The required sample sizes (n) have been calculated for a wide range of CV_T and Δ-values and the results are tabulated in Table 3.8. (For this table, "power" was arbitrarily chosen to be .95 and significance probability was conventionally chosen to

---

* *Note.* If $\sigma_C^2 > 0$, various alternative combinations of n_B and n_C would yield experiments of identical precision (i.e., identical "power" to detect an effect of a given size). The total number of animals $n = n_B n_C$ would not be the same for these various experimental designs—in fact, n increases as n_B increases. A design with smaller n_C would usually be preferred, even though the total number of animals was larger (since chambers are expensive).

be .05.) Since $n = n_B n_C$, and $\sigma_C^2 = 0$ is assumed, n_B and n_C are not determined uniquely, but their product is fixed by n from Table 3.8. Thus although this theory implies that $n_C = 1$ is permissible, in any actual experiment chambers should be replicated ($n_C \geq 2$) if possible, to validate the assumption that $\sigma_C^2 = 0$.

As an example of the use of Table 3.8, suppose we desire to have 95 percent power (i.e., to be "95 percent sure") for our planned experiment to detect a true average 25 percent change in a biological measurement that has a total coefficient of variation of .20. To determine the required group numbers of animals, in Table 3.8 find the value of n that corresponds to $CV_c = .20$ and $\Delta = .25$; the required n is found to be 17 (for each of the control and experimental groups).

The table for n (Table 3.8) was derived on the usual assumption that standard deviations of control and test groups are the same. This assumption cannot be disproved for most biological data that are analyzed. For example, EPA experiments on chronic exposure to auto exhaust generally have not shown significant differences in variances of control and exposed groups. Of course such low level chronic exposures usually have small or zero effects (i.e., low values of Δ). When a large change in a mean biological response does occur in an exposed group, for some types of responses the standard deviation has been observed to increase in proportion. If this is expected to happen in a planned experiment, the n given in Table 3.8 would be less then the required number. (Rarely, the variance is smaller in an exposed group with larger mean value.) Therefore the n from Table 3.8 should be considered as a minimum number, and it would not necessarily be wasteful to double (or even triple) n to ensure a "robust" experiment whose interpretation does not depend heavily on restrictive, statistical assumptions such as normality of distributions and equality of variances.

5.10 Sample Sizes for Other Types of Two-Group Fully Randomized Studies

Table 3.8 can be made to apply to types of experiment other than those involving animal exposure chambers. The table can be used for any two-group study (physical sciences as well as biological) that has equal numbers of control and test subjects chosen at random from respective populations whose means are to be compared. For this simplest type of experiment, n would represent the number of subjects in each of the two groups and CV_T would represent the coefficient of variation for intersubject variability in the control group. For purposes of economy in tabulating our recommended sample sizes, we have defined the control group as the one with the smaller mean.

5.11 Variability of Different Types of Biological Measurement

To give some idea of the coefficients of variation to be expected in animal exposure studies, some representative results are listed in Table 3.9. The CV_T's were estimated from data on control groups in chronic exposure studies. These data were assembled years ago (23), but there is no reason to believe that biological variability has changed. Table 3.9 shows three general ranges of CV_T's.

1. A coefficient of variation of .11 would be approximately correct for body weights of adult rodents, with higher variability for younger, growing animals (e.g., $CV_c = .17$ for 8 week old rats).

2. A CV_c of .15 to .20 is approximately correct for average body weights of infant mice in a litter, for wet lung weights and other organ weights of rodents, for lung lipid concentrations, for most hematological measurements (except white blood count), and for oxygen consumption and respiratory rate measurements.

3. Larger coefficients of variation, between .35 and .45, exist for white blood count, for life-spans, for measures of fertility such as litter size and number of litters, and for dry lung weights. Since these coefficients of variation differ among the several biological measurements that may be of interest in the same chronic exposure study, one approach to the design of such an experiment would be to base sample size recommendations on a compromise coefficient of variation of .20. This value is about right for lung physiology measurements, for most hematological and biochemical factors, and probably for most organ weight measurements (after correction for correlations with total body weights).

Having chosen a compromise value for CV_T, a final criterion is necessary for choosing n for a proposed animal exposure study, namely, the magnitude of effect we want the experiment to be able to detect. By "effect" we mean here the percentage difference between true average responses of groups of animals exposed to the test atmosphere and to the clean air. Biological effects caused by chronic exposure to contaminated workplace air are likely to be very small. For effects of the order of 10 percent ($\Delta = .10$) and a coefficient of variation of 20 percent ($CV_c = .20$), Table 3.8 provides a sample size recommendation of 95 animals per group. A sample size of 95 per exposure group would usually involve too much expense and laboratory space to be feasible for experiments with dogs or other large animals. If only 10 animals were exposed per group, effects would have to be 35 percent or larger for us to have "confidence" that they would be detected—that is, with "power" (probability) .95. If 17 animals per group were exposed, 25 percent effects could be detected. Studies of fertility and lifespan would require larger numbers of animals, since their coefficients of variation are usually larger than .20.

5.12 Sample Sizes for Incidence Data

The sample size recommendations given above are for studies of quantitative measurements that can be treated statistically as continuous variables—for example, biochemical concentrations, pulmonary function, and (approximately) hematological cell counts. However some of the most likely biological effects of workplace exposures may be measurable only in terms of qualitative incidence data (e.g., the difference between

Table 3.8 Number of Animals (per Group) to Detect a Difference ($\Delta \bar{c}$) Between Values $\bar{x}$ and $\bar{c}$ by Means of a Student's t Test[a]

CV_c = Coefficient of Variation on a Unit-Animal Basis Within a Treatment Group

Δ	.01	.02	.05	.075	.10	.15	.20	.25	.35	.40	.50	.60	.70	.80	.90	1.00	2.00
.01	26	95	561	1246	2205	4928	8736	13620	26650	34790	54310	78170	106400	138800	175900	216500	865900
.02	8	26	146	319	561	1246	2205	3430	6696	8736	13620	19600	26650	34790	44010	54310	216500
.05	3	6	26	55	95	208	362	561	1087	1419	2205	3164	4294	5603	7084	8736	34790
.075	2	5	13	26	44	95	165	254	490	637	986	1419	1924	2505	3164	3899	15500
.10		3	8	15	26	55	95	146	279	362	561	803	1087	1419	1787	2205	8736
.15		2	4	8	13	26	44	67	128	165	254	362	490	637	803	986	3899
.20			3	5	8	15	26	39	74	95	146	208	279	362	456	561	2205
.25			3	4	6	10	17	26	48	62	95	135	182	234	295	362	1419
.35			2	3	4	6	10	14	26	33	50	71	95	122	154	189	730
.40			2	3	3	5	8	11	20	26	39	55	74	95	119	146	561
.50					3	3	4	6	8	14	17	26	36	48	62	78	362
.60					2	3	3	4	6	10	13	19	26	34	44	55	254
.70					2	3	3	5		8	10	14	20	26	33	41	189
.80						3	3	4		6	8	11	15	20	26	32	146
.90						3	3	4		5	6	9	13	16	21	26	116

1.00	3	3	3	5	6	8	10	14	17	20	26	95
1.50	2	2	3	3	3	4	6	7	9	10	13	44
2.00			2	3	3	3	4	5	6	7	8	26
3.00			2	2	3	3	3	3	3	4	4	13
4.00				2	2	3	3	3	3	3	3	8
5.00							2	2	3	3	3	6
6.00								2	2	3	3	4
7.00										2	2	4
8.00											2	3
9.00												3
10.00												3

[a] Symbols as follows:

$\bar{x}$ = true mean for test population

$\bar{c}$ = true mean for control population

$\Delta = (\bar{x} - \bar{c})/\bar{c}$ = fractional effect

Power = .95 (probability of detecting a significant effect).

Size (significance probability) = .05 (1-tail test).

Table 3.9 "Total Coefficients of Variation" of Biological Measurements[a] from Animal-to-Animal of the Same Species[b]

Factor, Measurement, and Species	Sex	DF	Biological	Assay	Total (CV_T)
Fertility					
Litter size: mice	♀	114			.31
Number of litters per female: mice	♀	170			.49
Body weight					
LAF mice (32 g)	♂	14	.11	.00	.11
Rats					
8 weeks old	♂, ♀	96	.07	.00	.07
"Young adult" (approximately)					
8 months old: set 1	♂, ♀	29	.07–.09	.00	.07–.09
8 months old: set 2	♂, ♀	14	.17	.00	.17
Guinea pigs (NIH, Hart.)	♂, ♀	55			.11–.13
Average infant mice per litter	♂, ♀	114			.16
Beagles (12–20 weeks, uncorrected for age)	♀	87			.23
New Zealand rabbits, 17–18 weeks[c]	♂	14			.12
	♀	9			.09
Organ weight					
Rats (8 months old)					
Wet lung (young rats)	♂, ♀	12			.08
Wet lung (set 1)	♂, ♀	14			.15
Wet lung (set 2) adjusted for body weight	♂, ♀	29			.16
Dry lung (set 2) adjusted for body weight	♂, ♀	29			.42
Kidney	♂, ♀	14			.14
Adrenal	♂, ♀				.14
Liver	♂, ♀				.20
Heart	♂, ♀				.20
Spleen	♂, ♀				.25
New Zealand rabbits[c]					
Pituitary	♂	14			.24
	♀	9			.16
Thyroid	♂	14			.37
	♀	9			.11
Lifespan					
Rats[d]	♂	32			.16
	♀	30			.16

Table 3.9 (Continued)

Factor, Measurement, and Species	Sex	DF	Type of Variability		
			Biological	Assay	Total (CV_T)
Hematology					
MCH: young adult beagles	♀	87	.06	.07	.08
MCV: young adult beagles	♀	87	.08	.05	.09
Hemoglobin (g = %)					
LAF mice	♂	24			.07
Rats (8 weeks)	♂, ♀	96	.08	.01	.08
Young adult beagles	♀	87	.09	.03	.09
Hematocrit					
Rats (8 weeks)	♂, ♀	96	.07	.01	.07
Young adult beagles	♀	87	.09	.03	.09
RBC					
LAF mice	♂	40			.15
Rats (8 weeks)	♂, ♀	96	.08	.02	.08
Young adult beagles	♀	87	.10	.04	.11
WBC					
LAF mice	♂	40			.36
Rats (8 weeks)	♂, ♀	96	.22	.09	.24
Young adult beagles	♀	87	.24	.05	.24
Carbon dioxide (vol.-%): LAF mice	♂	20			.18
Oxygen (vol.-%)					
LAF mice	♂	20			.23
Young adult beagles	♀	87	.28	.04	.27
Oxyhemoglobin: young adult beagles	♀	87	.27	.04	.27
Biochemistry: rats					
Total lung lipids					
micromoles per gram dry weight	♂, ♀	30			.14
micromoles per gram wet weight	♂, ♀	30			.14
G-6-P dehydrogenase in RBC	♂, ♀	12	.14	.08	.15
Lactic dehydrogenase in RBC	♂, ♀	12	.11	.05	.12
Physiology					
Oxygen consumption					
Guinea pigs	♂, ♀	55			.25
LAF mice	♂	40			.14
Oxygen consumption adjusted for body weight					
Guinea pigs	♂, ♀	54			.18
LAF mice	♂	39			.12

Table 3.9 (Continued)

Factor, Measurement, and Species	Sex	DF	Type of Variability		
			Biological	Assay	Total (CV_T)
Respiratory rate					
Guinea pigs	♂, ♀	55			.19
Expiratory resistance					
Guinea pigs[e]	♂	Many			.18

[a] Unless otherwise indicated, all data are for animals used as pure air controls in chronic animal exposure studies performed by the Health Effects Research Program, Environmental Protection Agency, 1965–7.

[b] Symbols as follows:

CV_T = (pooled) "total coefficient of variation" on a unit animal basis (assuming duplicate assays unless otherwise noted and no chamber-to-chamber variability, i.e., $\sigma_c^2 = 0$)

F = degrees of freedom for the mean square used in the numerator of CV_T

[c] Krueger, Kaufmes, Price, Mason, and Bogart, "Thyroid, Pituitary, and Body Size in Rabbits," *Proc. West. Sect. Am. Soc. Am. Prod.*, **6** (July 1955).

[d] Davis, Stevenson, and Busch, "The Effects of Small Amounts of Polonium on Rats," Mound Laboratory Report to the Atomic Energy Commission, 1955.

[e] Contractor's progress report, October–December 1962.

proportions of subjects who develop byssinosis in a control group and in a group exposed to cotton dust). Detection of a given difference in incidence data usually requires many more animals than are needed to detect an equal percentage change in an average value of a continuous variable measurement, particularly when the incidence is not small in controls.

Table 3.10 gives minimum sample sizes required to be 95 percent sure to detect statistically significant differences ($P = .05$) between incidences in control and test groups. The sample size is a function of the incidence in controls (θ_1) and the size of the difference (Δ). As in the case of continuous data, one must first decide what size difference is to be detected to determine the required sample size.

For example, tumor-resistant C57 black mice have a very low incidence of lung tumors when exposed to pure air—say on the order of 1 occurence per 10^5 mice ($\theta_1 = .00001$). Suppose that preliminary evidence suggests that lung tumor incidence of C57 black mice exposed to Los Angeles type ambient air may be larger—on the order of 1.5 percent ($\theta_2 = .015$). To design an experiment to confirm such an increase in lung tumor incidence, according to Table 3.10 about 379 C57 black mice would have to be exposed to detect such a small increase in tumor incidence reliability (with .95 probability). Even if zero incidence of lung tumors in controls could be assumed ($\theta_1 = 0$), so that the occurence of even a single tumor in test animals could be taken as a statistically significant result, 199 test animals would have to be exposed to guarantee (with probability .95) that a significant result (one or more tumors) would occur.

In summary, exposures of animals to low level concentrations of workplace air contaminants usually would not cause large enough increases in tumor incidence to be statistically significant, if test animals were used in the same numbers that are sufficient to detect biological effects for continuous variable measurements. Inasmuch as exposure of large numbers of animals in the laboratory for tumor incidence studies would tie up expensive facilities for long periods of time, it may be practical to do tumor incidence studies only in the field, outside the laboratory, by exposing animals to actual contaminated workplace air or to ambient air pollution. The obvious disadvantage of field studies is the changing concentrations of the materials of primary interest, which are likely to be components of unstable, sometimes chemically and/or toxically interactive, mixtures.

Table 3.10 Sample Size (per Group) to Detect a Difference Between Binomial Incidences θ_1 and θ_2[a]

θ_1	Δ = Increase in Incidence								
	.01	.015	.02	.05	.10	.20	.25	.30	.50
0[b]		299	199	149	59	29	14	11	9	5
.00001	576	379	282	110	54	26	20	17	9
.00005	623	403	297	114	55	26	21	17	9
.0001	658	423	310	117	56	27	21	17	9
.0005	840	517	368	130	61	28	22	18	10
.001	1030	605	423	142	64	29	23	18	10
.005	2002	1071	699	197	81	34	26	21	11
.01	3113	1577	992	250	96	39	31	23	12
.02	5220	2525	1529	342	121	46	34	26	13
.05	11182	5156	3025	585	183	61	44	29	15
.10	20123	9166	5286	937	269	82	56	41	17
.15	28012	12754	7262	1243	342	98	66	48	19
.20	35199	15986	9091	1509	403	111	74	53	20
.25	49023	18294	10411	1726	479	121	79	56	20
.30	45550	20370	11494	1898	490	128	83	58	20
.40	54121	23425	13134	2114	535	134	85	58	17
.50	55220	24378	13531	2165	534	128	79	53	5

[a] Symbols as Follows:

θ_1 = lower incidence
θ_2 = higher incidence
$\Delta = \theta_2 - \theta_1$
size (significance probability) = .05 (1-tail test)
power = .95 (probability of detecting a significant effect)

[b] Of course no controls need be tested if incidence is known to be impossible (i.e., when $\theta_1 = 0$).

For typical continuous variable biological measurements, recommended sample sizes would have to approach 100 to ensure detection of biological effects as small as 10 percent. If smaller samples are used, of the order of 10 to 20 animals per group, one must be satisfied with an experiment of lower precision, that is, one that would not detect small effects reliably but would detect larger effects of size 25 to 35 percent.

REFERENCES

1. J. S. Mausner and A. K. Bahn, *Epidemiology, An Introductory Text*, Saunders, Philadelphia, 1974.
2. G. D. Friedman, *Primer of Epidemiology*, McGraw-Hill, New York, 1974.
3. B. McMahon and T. F. Pugh, *Epidemiology Principles and Methods*, Little Brown, Boston, 1970.
4. Leo G. Reeder, Overall Conference Chairman, *Advances in Health Survey Research Methods: Proceedings of a National Invitational Conference* sponsored by the National Center for Health Services Research, Health Resources Administration, U.S. Public Health Service, Department of Health, Education and Welfare Publication (HRA) 77-3154, 1977.
5. J. W. Lloyd and A. Ciocco, "Long-Term Mortality Study of Steelworker: I. Methodology," *J. Occup. Med.*, **11**, 299-310 (June 1969).
6. J. F. Collins and C. K. Redmond, "The Use of Retirees to Evaluate Occupational Hazards. II. Comparison of Cause Specific Mortality by Work Area," *J. Occup. Med.* **20**:4, 260-266 (1978).
7. L. L. Kupper, A. J. McMichael, and R. Spirtas, "A Hybrid Epidemiolgic Study Design Useful in Estimating Relative Risk," *J. Am. Stat. Assoc.*, **70**, 351: 524-528 (1975).
8. N. Mantel and W. R. Bryan, "'Safety' and Testing of Carcinogenic Agents," *J. Nat. Cancer Inst.*, **27**, 455-470 (1961).
9. N. Mantel and N. R. Bohidar, C. C. Brown, J. L. Ciminera, and J. W. Tukey, "An Improved 'Mantel-Bryan' Procedure for 'Safety Testing' of Carcinogens," *Cancer Res.*, **35**, 865-872 (1975).
10. H. O. Hartley and R. L. Sielken, Jr., "Estimation of 'Safe Doses' and Carcinogenic Experiments," *Biometrics*, **33**: 1, 1-30 (1977).
11. H. O. Hartley and R. L. Sielken, Jr., "A Parametric Model for 'Safety' Testing of Carcinogenic Agents," Food and Drug Administration Technical Report 1, Institute of Statistics, Texas A & M University, College Station, 1975.
12. H. O. Hartley, and R. L. Sielken, Jr., "A Non-parametric for 'Safety' Testing of Carcinogenic Agents," Food and Drug Administration Technical Report 2, Institute of Statistics, Texas A & M University, College Station 1975.
13. K. S. Crump, H. A. Guess, and K. L. Deal, "Confidence Intervals and Tests of Hypotheses Concerning Dose-Response Relations Inferred from Animal Carcinogenicity Data," *Biometrics*, **33**, 437-451 (1977).
14. F. A. Patty, *Industrial Hygiene and Toxicology*, Vol. I, 3rd ed., Wiley, New York, 1978.
15. N. A. Leidel, K. A. Busch, and J. R. Lynch, National Institute for Occupational Safety and Health *Occupational Exposure Sampling Strategy Manual*, U.S. Department of Health, Education and Welfare (NIOSH) Publication 77-173, 1977.
16. A. Hald, *Statistical Theory with Engineering Applications*, New York, 1952, p. 91.
17. Final Report, National Institute for Occupational Safety and Health Contract 210-76-0188, "Statistical Properties of Industrial Air Sampling Strategies," Project Leaders: Kenneth A. Busch and Nelson A. Leidel, published as a NIOSH Technical Report, 1978.
18. A. M. Mood, *Introduction to the Theory of Statistics*, McGraw-Hill, New York, 1950, p. 136.
19. R. Larsen, *United States Air Quality*, Vol. 8, American Medical Association, 1964, pp. 325-333.

20. N. A. Leidel, K. A. Busch, and W. E. Crouse, "Exposure Measurement Action Level and Occupational Environmental Variability," National Institute for Occupational Safety and Health Technical Report 76-131, December 1975.
21. Y. Bar-Shalom, D. Budenaers, R. Schainker, and A. Segall, "Handbook of Statistical Tests for Evaluating Employee Exposure to Air Contaminants," National Institute for Occupational Safety and Health Technical Report 75-147, April 1975.
22. N. A. Leidel and K. A. Busch, "Statistical Methods for the Determination of Noncompliance with Occupational Health Standards," National Institute for Occupational Safety and Health Technical Report, U.S. Department of Health, Education and Welfare Publication 75-159, Cincinnati, Ohio 1975.
23. K. A. Busch and W. F. Ludmann, "Statistical Considerations for the Conduct of Animal Exposure Experiments," paper presented at the 60th Annual Meeting of the Air Pollution Control Association, Cleveland, June 11-16, 1967.
24. David G. Taylor, Richard E. Kupel, and John M. Bryant, "Documentation of the NIOSH Validation Tests," U.S. Department of Health, Education and Welfare (NIOSH) Publication 77-185, 1977.
25. H. Ayer and J. Burg, "Time-Weighted Average vs. Maximum Personal Sample," paper presented at the American Industrial Hygiene Conference, Boston, 1973.

CHAPTER FOUR

Data Automation

ROBERT L. FISCHOFF and
FRED G. FREIBERGER

1 THE NEED FOR DATA AUTOMATION

Over the years the emphasis of industrial hygiene has gradually changed from discovering job-related causes of ill health to monitoring and controlling potentially harmful work environment situations before they result in injury to workers or the public. Associated with this modification in concept has been a significant change in industrial hygiene methodology, namely, an increasing requirement for data collection, recordkeeping, statistical analysis, and reporting. These activities are essential in determining what controls should be implemented to ensure employee health. Added to this professional responsibility for data management are the requirements of the Occupational Safety and Health Administration (OSHA) for recordkeeping and reporting on a growing list of harmful or suspected substances used in modern industrial processes.

The purpose of OSHA is to ensure, as far as possible, safe and healthful working conditions for every industrial employee in the nation by providing mandatory occupational safety and health standards. Basically the Secretary of Labor is responsible for issuing regulations requiring the employer to protect the employee. These standards also oblige the employer to maintain records of employee exposures, to give employees access to the records, to allow employees the opportunity to observe monitoring or measuring being conducted, and to notify the employees of excessive exposures, as well as to inform them of corrective action being taken. The government is also allowed access to all of the foregoing records. In some cases record retention requirements may be 20 or more years.

In addition to recording and reporting requirements, OSHA standards prescribe, as necessary, the training of the employee, suitable protective equipment, control procedures, type and frequency of medical exams, and use of warnings to ensure employee awareness of hazards, symptoms, emergency treatment, and safe use conditions. At the time of writing there are 17 specific OSHA standards, and the 1976 Registry of Toxic Effects lists 218 chemicals as recommended standards from National Institute for Occupational Safety and Health (NIOSH) criteria documents. This publication lists 100,000 chemicals, 4,000 of which are carcinogens, mutagens, or teratogens. Thus the responsibility of industrial hygiene management can become exceedingly complex. In protecting the health of employees it must recognize potential health hazards, have them evaluated, assure that controls are in place, initiate exposure monitoring procedures, enter and delete employees from the recordkeeping system, and comply with changing government requirements. Industrial hygiene management is not the only area that is in need of the information just outlined. Other persons or functional areas that are affected by the OSHA regulations and need hazard exposure evaluations and reports are:

1. Line management.
2. Employees.
3. Safety.
4. Transportation.
5. Medical.
6. Development engineering.
7. Manufacturing engineering.
8. Facilities engineering.
9. Chemical control and disposal.
10. Purchasing.
11. Shipping and receiving.
12. Personnel.
13. Laboratory.

Because of the rapidly increasing workload of recordkeeping, data analysis, and reporting, many industrial hygiene managers are turning to electronic data processing (EDP). An EDP system can greatly increase the productivity and quality of most information collecting, storage, and retrieval operations, while providing more timely and economical data analyses and reports. However automating a large industrial hygiene program is a complex task requiring careful planning and evaluation if costly mistakes are to be avoided.

This chapter presents the basic concepts of EDP systems and describes the steps necessary to plan, design, and implement an effective computer application. Two specific examples of industrial health computer applications are given; one is a relatively simple, single data base system to control plant ventilation, and the other is a more complex data base involving an environmental health information and control system. Concepts of other industrial health applications are also discussed and references are provided to assist those interested in pursuing additional aspects of data automation. A data processing glossary is included for reference.

DATA AUTOMATION

2 ELECTRONIC DATA PROCESSING CONCEPTS

2.1 Computer Equipment (Hardware)

EDP is the handling of data by an electronic computer and associated (peripheral) devices in a planned sequence of operations to produce a desired result. The many types of EDP system range in size from desk-top units to systems that fill several large rooms with interconnected devices. But regardless of the information to be processed or the complexity of equipment used, all EDP involves four basic functions:

1. Entering the source data into the system (input).
2. Storing the data in addressable locations (storage).
3. Processing the data in an orderly manner within the system (processing).
4. Providing the resulting information in a usable form (output).

These functions are performed by an input device, a storage unit, a central processing unit (CPU), and an output device (Figure 4.1).

2.1.1 Input Devices

Input devices read or sense coded data that are recorded on a prescribed medium and make this information available to the computer. Data for input are recorded on cards and paper tape as punched holes, on magnetic tape, disks, or drums as magnetized spots, or on paper documents as characters or line drawings. Section 2.3 discusses the method of recording data for machine use and the characteristics of each type of recording medium.

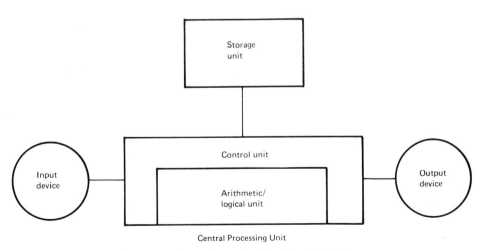

Figure 4.1. General organization of an EDP system.

2.1.2 Storage

Storage is somewhat like an electronic filing cabinet, completely indexed and instantaneously accessible to the computer. All data must be placed in storage by an input device before they can be processed by the computer. Each position of storage has a specific location, called an address, so that the stored data can be located by the computer as needed.

The computer can rearrange data in storage by sorting or combining different types of information received from a number of input devices. The computer also can take the original data from storage, calculate new information, and place the result back in storage.

The size or capacity of storage determines the amount of information that can be held in the system at any one time. In some computers, storage capacity is measured in millions of digits or characters (bytes) that provide space to retain entire files of information. In other systems storage is smaller, and data are held only while being processed. Consequently the capacity and design of storage affect the method in which data are handled by the system.

Storage capacity built into a computer is called main storage. This usually consists of main data storage, for programs or other data; control storage, which often contains special built-in "microprograms" to assist the computer in carrying out its own operations; local storage, consisting of high speed working areas (registers) for performing calculations and other processing; and large capacity storage. In addition; much more storage can be provided by auxiliary magnetic tape drives or disk storage drives connected to the computer. Examples of auxiliary storage units appear in Figure 4.2.

2.1.3 Central Processing Unit

The CPU is the controlling center of the entire EDP system. It is usually divided into two parts as shown in Figure 4.1: the control section, and the arithmetic/logical unit. The control section directs and coordinates all computer system functions. It is like a traffic cop that schedules and initiates the operation of input and output devices, arithmetic/logical unit tasks, and the movement of data from and to storage.

The arithemetic/logical unit performs such operations as addition, subtraction, multiplication, division, shifting, moving, comparing, and storing. It also has a capability to test various conditions encountered during processing and to take action accordingly.

2.1.4 Output Devices

Output devices record or write information from the computer onto cards, paper tape, magnetic tape, disks, or drums. They may print information on paper, generate signals from transmissions over teleprocessing networks, produce graphic displays or microfilm images, or take other specialized forms.

DATA AUTOMATION

Figure 4.2. Auxiliary storage units. (a) IBM 3420 magnetic tape unit. (b) IBM 3350 direct access storage.

Frequently the same physical device, such as a card reader/punch or a tape drive, is used for both input and output operations. Thus input and output (I/O) functions are generally treated together. The number and type of I/O devices that may be connected directly to a CPU depends on the design of the system and its application. Note that the functions of I/O devices and auxiliary storage units may overlap—thus a tape drive or disk file may be used both for I/O operations and for data storage. Figure 4.3 presents some examples of I/O devices.

The console (Figure 4.4) is an I/O device that provides external control of a data processing system. Keys or switches allow the computer operator to turn power on or off, start or stop operation, and control various devices in the system. There are usually lights, permitting data in the system to be displayed visually.

On some systems a console printer and keyboard provide limited output or input capability. The I/O device may print messages, signaling the end of processing or an error condition. It may also print totals or other information that enables the operator to monitor and supervise operation, or it may give instructions to the operator. On the other hand, the console keyboard may be used to key in meaningful information (such as altering instructions) to a data processing system that is programmed to respond to such messages.

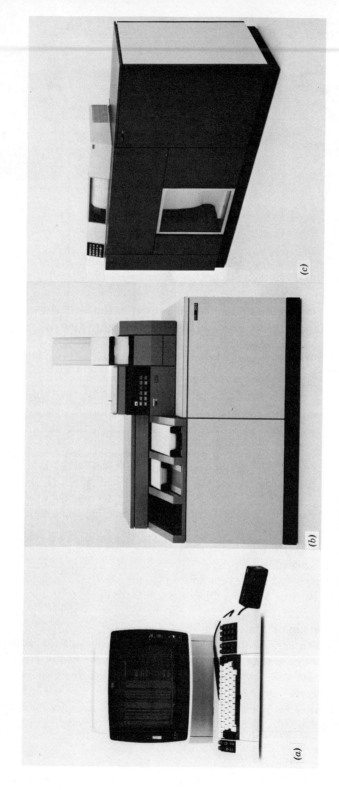

Figure 4.3. Example of I/O devices. (a) IBM 3278 display station. (b) IBM 3505 card reader. (c) IBM 3211 printer.

DATA AUTOMATION

Figure 4.4. CPU and operator console.

A remote console may offer increased efficiency and flexibility by providing duplicate operator controls at a station removed from the CPU.

2.2 Computer Programs (Software)

2.2.1 Application Programs

Each EDP system is designed to perform a specific number and type of operations. It is directed to perform each operation by an instruction. The instruction defines a basic operation to be performed and identifies the data, device, or mechanism needed to carry out the operation. The entire series of instructions required to complete a given procedure is known as an application (or problem) program.

For example, the computer may have the operation of multiplication built into its circuits in much the same way that the ability to add is built into a simple desk-top adding machine. But there must be some means of directing the computer to perform multiplication, just as the adding machine is directed by depressing keys. There must also be a way to instruct the computer where in storage it can find the factors to multiply.

Furthermore, the comparatively simple operation of multiplication implies other activity that must precede and follow the calculation. Assume that the multiplicand and the multiplier are read into storage by an input device. Once the calculation has been performed, the product must be returned to storage at a specified location, from which it may be written out by an output device.

Any calculation, therefore, involves reading, locating the factors in storage, performing the required computation, returning the result to storage, and writing out the completed result. Even the simplest portion of a procedure involves a number of planned steps that must be spelled out to the computer if the procedure is to be accomplished.

An entire application program is composed of these individual steps grouped in a sequence that directs the computer to produce a desired result. Thus a complex problem must be reduced to a series of basic machine operations before it can be solved. Each of these operations is coded as one instruction or as a series of instructions, in a form that can be interpreted by a computer, and is placed in the main storage unit as a portion of a stored program.

The possible variations of a stored program afford the EDP system almost unlimited flexibility. A computer can be applied to a great number of different procedures simply be reading in, or loading, the proper program into storage. Any of the standard input devices can be used for this purpose, because instructions can be coded into machine language just as data can.

The stored program is accessible to the computer, giving it the ability to alter the program in response to conditions encountered during an operation. Consequently the program selects alternatives within the framework of the anticipated conditions.

2.2.2 Control Programs

To make possible the teleprocessing networks and the orderly operation of many types of I/O devices that may be on-line with a computer, control programs have been developed. Control programs, also known as monitor programs or supervisory programs, act as traffic directors for all the application programs (which solve a problem or carry out a particular operation or process on a set of data), then relinquish control of the computer to the control program. The control program may be constructed to allow the computer to handle random inquiries from remote terminals, to switch from one problem program within the computer to another, to control external equipment, or do to whatever the application requests.

The concept of maintaining optimum computer usage by interleaving and interspersing application programs under the direction of control programs gives rise to the use of two terms—time sharing and multiprogramming.

Briefly, time sharing may be thought of as the cooperative use of a central computer by more than one user (company, division, branch of a company, institution, or government agency). Each user receives a share of the time available, with the result that many jobs are being performed within a congruent time (either simultaneously or seemingly simultaneously). This service may be achieved by interspersing programs rapidly on one computer system, by multiprogramming, or by using two computers that are connected. Multiprogramming is usually thought of as a system of control programs and computer equipment that permits many application programs to go on concurrently. This is accomplished by interleaving the programs with each other in their use of the CPU, storage, and I/O devices.

2.3 Data Representation

2.3.1 Recording Media

Symbols convey information; the symbol itself is not the information but merely represents it. Presenting data to a computer is similar in many ways to communicating with another person by letter. The intelligence to be conveyed must be reduced to a set of symbols. In the English language, these are the familiar letters of the alphabet, numbers, and punctuation marks.

Similarly, communication with a computer system requires that data be reduced to a set of symbols that can be read and interpreted by data processing machines. The symbols differ from those commonly used by people because the information to be represented must conform to the design and operation of the machine. The choice of these symbols (and their meaning) is a matter of convention on the part of the designers. Just as there are rules of grammar that dictate the proper use of a language for clear communication among people, so there are rules of syntax that prescribe how a set of symbols must be used for communication between people and machines. Use of the assigned set of symbols in accordance with the prescribed rules constitutes a programming language.

When a computer program is written, it must be recorded in a medium that can be read by a machine. It may be put in the form of punched cards or paper tape, magnetic tape, direct access storage devices (DASD) such as magnetic disks or drums, magnetic ink characters, optically recognizable characters, microfilm, display screen images, communication network signals, and so on. Figure 4.5 illustrates some of these data recording forms.

Data are represented on the punched card by small rectangular holes in specific locations of the card. In a similar manner, small circular holes along a paper tape represent data. On magnetic tape, or DASD, the symbols are small magnetized areas, called spots or bits, arranged in specific patterns. Magnetic ink characters are printed on paper. The shape of the characters and the magnetic properties of the ink permit the printed data to be read by both man and machine. <u>Each medium requires a code or specific arrangement of symbols to represent data.</u>

An input device of the computer system is a machine designed to sense or read information from one of the recording media. In the reading process, recorded data are converted to, or symbolized in, <u>electronic form;</u> then the data can be used by the machine for data processing operations.

An output device is a machine that receives electronic information from the computer system and records it on the designated output medium.

All I/O devices cannot be used directly with all computer systems. However data recorded on one medium can be transcribed to another medium for use with a different system. For example, data on cards or paper tape can be transcribed onto magnetic tape. Conversely, data on magnetic tape can be converted to cards, paper tape, printed reports, or plotted graphs.

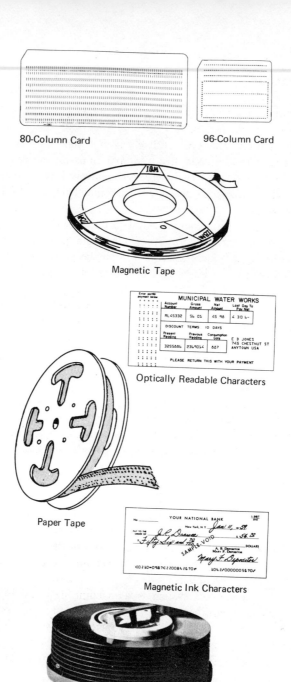

Figure 4.5. Data recording media.

DATA AUTOMATION

2.3.2 Machine Data

Not only must there be a method of representing data on physical media such as cards or magnetic tape, there must also be a method of representing data within a machine. In the computer, data are represented by many electronic components: transistors, magnetic cores, wires, and so on. The storage and flow of data through these devices are represented as electronic signals or indications. The presence or absence of these signals in specific circuitry is the method of representing data in the machine, much as the presence of holes in a card represents data.

Binary States. Digital computers function in binary states; this means that the computer components can indicate only two possible states or conditions. For example, the ordinary light bulb operates in a binary mode: it is either on or off. Likewise, within the computer, transistors are maintained either conducting or nonconducting; magnetic materials are magnetized in one direction or in the opposite direction; and specific voltage potentials are present or absent (Figure 4.6). The binary states of operation of the components are signals to the computer, as the presence or absence of light from an electric light bulb can be a signal to a person.

Representing data within the computer is accomplished by assigning a specific value to a binary indication or group of binary indications. For example, a device to represent decimal values could be designed with four electric light bulbs and switches to turn each bulb on or off as illustrated in Figure 4.7. The bulbs are assigned decimal values of 1, 2, 4, and 8. When a light is on, it represents the decimal value associated with it. When a light is off, the decimal value is not considered. With such an arrangement, the single decimal value represented by the four bulbs will be the numeric sum indicated by the lighted bulbs.

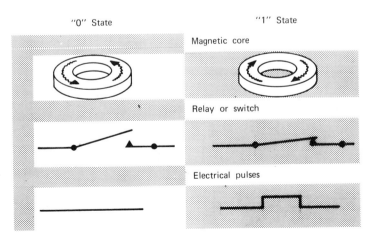

Figure 4.6. Binary components.

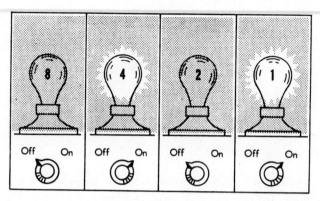

Figure 4.7. Representing decimal data with binary components.

Decimal values 0 through 15 can be represented. The numeric value 0 is represented by all lights off; the value 15, by all lights on; 9, by having the 8 and 1 lights on and the 4 and 2 lights off; 5, by the 1 and 4 lights on and the 8 and 2 lights off; and so on.

The value assigned to each bulb in the example could have differed from the respective values used. This change would involve assigning new values and determining a new scheme of operation. In a computer the values assigned to a specific number of binary indications become the code or language for representing data.

Because binary indications represent data within a computer, a binary method of notation is used to illustrate these indications. The binary system of notation uses only two symbols, zero (0) and one (1), to represent specific values. In any one position of binary notation, the 0 represents the absence of a related or assigned value, and the 1 represents the presence of a related or assigned value. For example, to illustrate the indications of the light bulbs in Figure 4.7, the following binary notation would be used: 0101.

The binary notations 0 and 1 are commonly called bits. For computers using the binary system of data representation (typified by the IBM System/370), the basic unit of information is contained in eight consecutive bit positions, called a byte. Four bytes (32 consecutive bits) constitute a word.

Computer Codes. The method used to represent (symbolize) data is called a code. In a computer the code relates data to a fixed number of binary indications (symbols). For example, a code used to represent alphabetic and numeric (alphameric) characters may use eight positions of binary indication. By the proper arrangement of the binary indications (0 bit, 1 bit), all numbers and characters can be represented with these eight binary positions.

Some computer codes in use are six-bit alphameric code, eight-bit alphameric code, two-out-of-five-count code, and six-bit (packed) numeric code. Most computer codes are self-checking; that is, they have a built-in method of checking the validity of the coded

DATA AUTOMATION

information. Code checking occurs automatically within the machine as the data processing operations are carried out. The method of validity checking is part of the design of the code.

3 STEPS IN EDP APPLICATION DEVELOPMENT

In designing and implementing any new computer application there are logical steps of analysis, planning, development, testing, and installation that must be carried out. Usually these steps are accomplished in the sequence described below.

3.1 Defining and Sizing

When a new computer application is desired, the first task is to define the problem. One should be able to state clearly and concisely just what is to be performed by the proposed system, and the scope (size) of the effort. A short written report should be prepared to describe the problem in terms of subject, scope, objectives, and recommendations. This report will establish the basis for communicating needs to the system analyst and to management.

3.2 Analysis

The analysis phase of a design effort is not really an isolated step. The system analyst, who views a new computer application in terms of its scope and objectives, must work closely with the users to determine their needs, the information in use in the present system, and the information needed in the new system. The analyst must also consider what equipment would be the most cost-effective for the necessary functions of the new system. The basic specification for the new system may then be modified or expanded as necessary during the development (programming) phase. Often, as the result of the preliminary analysis and discussions with the users and programmer(s), the original problem statement is redefined.

3.3 Charting of Tasks

System design is a creative process in which the analyst must identify each activity or procedure required to arrive at the desired result. A flow chart is used to help thinking through the entire process and keeping track of each step, (Figure 4.8). Once the new system has been fully designed, the analyst may use a PERT (program evaluation review technique) chart to define and schedule the key system development tasks. A PERT chart is a graphic representation of the interrelationships and chronological dependencies of all activities required to complete a project.

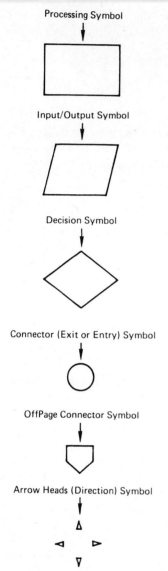

Figure 4.8. Flow chart symbols.

3.4 Data Base

A data base is a collection of related, nonredundant data files that can be accessed by more than one user. It is usually shared by several application areas such as personnel, marketing, and medical, but it may be dedicated to a single application. When the data base is shared by various users, it is necessary to restrict access of each type of user to

DATA AUTOMATION

the part of the data needed for the respective jobs, so that personal or company confidential information can be controlled.

The process of structuring a data base to accommodate the needs of different users can become quite complex. The system analyst usually defines the data base requirements in terms of the input data sources available and the type of output (reports) needed by the users.

3.5 Programming

Traditionally, the system analyst will prepare a system, or program, specification document, which is a detailed blueprint for the development programmer to follow in writing the computer program needed. The specification explains the overall function of the system and includes detailed descriptions of the input data sources, data base structure, processing requirements, security considerations, and output requirements (reports and new or updated data files). The programmer then prepares the detailed computer instructions to accomplish the specified tasks. Often, however, experienced programmers do both jobs—analysis and programming; this is especially likely to happen on smaller applications.

The programmer develops the logical sequence of computer instructions and functions using a flow chart; or a form of shorthand called "pidgin" may be used with the newly developed technique of structured programming. Then the detailed computer instructions are written in program segments and each section is tested as it is completed.

3.6 Testing and Debugging

The programmer works closely with the system analyst, or with the ultimate users of the new system, to be sure that the new programs are processing the unit data correctly and producing the required data on reports or other output desired. The user's help may be requested in developing adequate test data to cover every conceivable condition or error in the input data. Finding syntax or logic errors in a new program and eliminating them is called "debugging." Each program segment is tested individually as it is developed; then all the pieces are put together for final testing.

When all required programs have been written and tested and the necessary hardware has been installed, all components of the new system are put together for an integration test. The hardware, software, operating procedures, and security controls are tested for validity and reliability using test and actual or "live" data. It is very unusual for all aspects of a new system to run smoothly the first time.

The system must be debugged, and procedures and controls may need to be modified. Backup procedures (to enable the system to recover in case of a system failure or accidental loss of data files) must also be tested and any potential problems identified and corrected. Audit procedures and security controls should be checked out by persons other than those who designed or developed the system, to avoid the possibility of intentional fraud.

3.7 Pilot Test

The pilot test is the last phase of implementing a new system. This is also called conversion, because the old system or procedure is converted to the new system. There are two commonly used types of pilot tests: parallel and phase-in.

In the parallel method the old system or procedure is continued while the new system is installed and checked out. This method may be more expensive because two systems are being run, but it is usually considered the best approach. The old system is a backup to guarantee that the required reports and processes will be carried on even if the new system does not run smoothly, and the output of the two systems can be compared for completeness and accuracy. When the reliability of the new system has been demonstrated, all files and procedures are converted to the new system and the old system is eliminated.

In the phase-in method of conversion the new system is installed in segments. As each new segment or function is put in place, the corresponding function of the old system or procedure is dropped, until, eventually, the new system is in full operation and the old system has been eliminated. This method is well suited to the installation of a new computerized system to replace a manual system.

3.8 Modify/Extend System

Once the new system has been installed, users may see ways to improve or expand it. Also, computer operations personnel, programmers, or system analysts may have ideas for cutting costs or improving performance that were not obvious during the development of the system. Any new system should be formally evaluated after installation to determine whether it is performing to expectations and within anticipated costs. Any modifications or additional functions should be thoroughly analyzed to determine the potential benefits and impact on the system, the users, or related systems. If approved, the changes should be implemented in the same process of planning, development, testing, and installation used for the introduction of the original system.

Actually, any operating system should be constantly monitored to determine whether it is performing properly. New ideas for restructuring the data base, eliminating unnecessary files or reports, or adding additional capabilities should be encouraged, to maintain the system in an efficient manner and to ensure that it stays current with the needs of the users.

4 APPLICATION CONSIDERATIONS

As discussed in Section 3, the computer system analyst is usually the first person contacted when one wants to utilize a computer. This person determines whether the program is feasible and ascertains whether computer time is justifiable and available. Thus in many organizations it is necessary to justify the use of the computer with both

DATA AUTOMATION

direct management and computer operations management, since resources and time may be limited. However, certain basic questions should first be answered about the potential application to determine whether data computerization is practical and economically feasible. The "user" questions are:

1. What do I want to accomplish?
2. How should it be accomplished?
3. How do I obtain the information?
4. How will the data be used?

Once these questions have been answered satisfactorily, the proposed application must be examined by management, and these additional questions answered:

1. What other methods are available to accomplish the program?
2. Is the computer the best method?
3. What are the advantages and disadvantages of computerization?
4. And, most important, is there a cost savings by using the computer?

One must fully understand the purpose and functions to be accomplished before they can be explained to a system analyst who is not familiar with the industrial hygiene field. Often the scope of a new computer application is unnecessarily limited because insufficient thought was given to the functions that could be designed into the system. Some typical industrial hygiene data system responsibilities are:

1. Making available and maintaining material safety data.
2. Maintaining medical/industrial hygiene data on employees and their exposures.
3. Providing selective data to medical systems and the personnel department.
4. Establishing medical/industrial hygiene requirements.
5. Maintaining company records with legally required documentation, under OSHA and Toxic Substance Regulation, at both federal and state levels.

Early system planning should also be concerned with future growth and flexibility of the system. Some considerations for future development are:

1. The system should have the ability to process the anticipated work load effectively and economically within current limitations imposed by technology.
2. It should possess the flexibility to accommodate changing requirements to meet business needs.
3. It should allow for capacity growth at little increase in cost.
4. Long-range expansion should be accommodated readily.

Appendix 4A lists typical industrial hygiene system requirements.
Answers to the following questions about the application should be helpful.

1. What do I really want to accomplish (scope and purpose)?
2. How is the concept defined, including sizing and complexity?
3. What information is required?
4. Who should obtain information?
5. What is the best method to input data to the computer?
6. Who is responsible for information?
7. What methods of checking validity are feasible?
8. Who needs the data?
9. How will data be used?
10. How should data be presented?

Although such information is necessary for the thought process, the system analyst will be concerned with data base design, how the data will be acquired, how managed, and finally, how presented. These aspects of a potential application are discussed below.

4.1 Data Base Design

The essence of any information system is the data base. But one of the most difficult exercises faced by the industrial hygienist and management is determining what data are to be included in the data base. Some considerations are:

1. How will data be used?
2. What format is needed?
3. How will access be obtained and protected?
4. When are data needed (is instant retrieval a requirement, or will there be lead time)?

Design aids in the form of design standards and programs can be used to improve data base quality. Data base design aids (DBDA) is a productivity tool that allows interim data design information to be catalogued and stored in a dictionarylike system for quick recall during the design phase. Such an automated analysis can reduce quite significantly the time required for generating a working, structural model. Better design quality will result, since the DBDA program performs a more thorough analysis of the data requirements than is normally possible with manual methods. This increases the likelihood of attaining a consistent and effective design.

Standardization of the new data base is achieved by using requirements from various information source applications in the same company or classification. If the personnel department records employees at work in days, and industrial hygiene documents exposures in hours, needless complexity in cross-referencing exists. DBDA greatly reduces this possibility. Careful data design is critical to future effective use. The design aid program is a software package created for use with a particular system. This package can be purchased commercially or prepared by a programmer. Use of the

DATA AUTOMATION

design aid will provide a sound data base by detecting omissions, inconsistencies, and redundancies in the data requirements.

In addition, the design aid helps to create new data bases, redesign and integrate existing data, add new applications to existing data, and add new elements or associations to existing information. During this process a structural model is created to identify human decision points. The process of designing a data base can be generally divided into the following six steps:

1. Gathering requirements.
2. Generating a structural model (organization).
3. Constructing a physical model (hardware).
4. Design evaluation.
5. Physical implementation.
6. Performance evaluation.

In summary, generating a structural model combines analysis of the data requirements of the applications and synthesis of these needs into a single network. Such a network will help the designer address the questions that arise during construction of a physical or working model.

Data base design activities normally are performed in the following order:

1. Gathering the data requirements for the data base.
2. Identifying and correcting inconsistencies, omissions, and duplications of data elements.
3. Selecting the elements and associations that must be included in the data base and those that can be derived from them.
4. Grouping elements and keys into segments.
5. Determining physical and logical relationships.
6. Arranging segments into hierarchical structures.
7. Selecting the access method.
8. Providing data elements for retrieval, insertion, update, and deletion.
9. Updating, reorganizing, and backup requirements.

To illustrate data base design considerations, the outline of a typical application could be as follows:

1. *Goal.* System will provide and assemble chemical health and safety information for all company locations. Chem-Safe.
2. *Basis.* Corporate chemical material data bank.
3. *Responsibility.* Corporate data center and procurement office.
4. *Concerns.*
    a. Standarized material identification.

  b. Hazardous ingredients.
  c. Labeling.
  d. Medical examination requirements.
  e. Workplace measurement rules.
  f. Compatibility.
  g. Ventilation.
  h. Handling and storage.
  i. Emergency control.
  j. Protective equipment.
  k. Overexposure effects.
  l. Disposal.
  m. Toxicology.
5. *Access by.*
  a. All locations—management/medical.
  b. Corporate technical review center.
6. *Data sources.*
  a. Personnel records.
  b. Medical files.
  c. Purchasing.
  d. Manufacture engineering.
  e. Facility engineering.
  f. EPA and OSHA.
7. *System configuration.*
  a. Magnetic storage.
  b. Keyboard input.
  c. Printer/cassette output (1) punch card option; (2) phone patch.

Starting with development of a master chemical numbering system, from uniform labels to format design, an information system of this magnitude requires a lot of basic data gathering before any electronic data processing is inolved. These steps must be in place before the system analyst and other EDP professionals are approached.

Note. Business justification for an industrial hygiene data system can be based on the premise that many pieces of data already exist in computer format. Personnel records include job description and assigned workplace—add materials each employee worked with and air sampling data and an essential part of an industrial hygiene record system is present.

4.2 Data Acquisition Methods

For the purpose of this discussion data acquisition methods have separated into three categories: manual, automated, and sensors. Actually, sensors would fall into the

DATA AUTOMATION

automated category but because of their value to the industrial health field, a separate section is devoted to these devices.

4.2.1 Manual Methods

The cost of data entry has increased primarily because manual methods of entry involved labor. Producing faster computers and printers could not completely offset this cost. Manual transcriptive data entry involves discrete steps, such as source document handling, recording and transcribing, and transporting, which are slow. If manual methods must be used, graphic display units are in most cases the best method, because they offer the most direct entry into the computing system by use of a keyboard. This aspect is especially relevant for entry of scientific data. Because these display units can be connected remotely to the computer by means of telephone lines, the data can be entered at the point of origin, which is of value to the scientist. Another advantage is that data can be entered directly from the source document, by the person who generated the data, if necessary.

If errors are detected in the data entry, the operator can correct them dynamically. Certain display consoles allow entry also by selector pen that enables the operator to enter several commands with one action. Once the transaction has been processed, the data in the working field are updated immediately. The computer is always current. Thus if manual methods must be used, greater system efficiency results from video input because of higher input rates, lower error rates, ease of update, and data handling.

4.2.2 Automated Methods

The automated methods have the advantage of putting the data directly into the computer. As such they are considered "real time" in computer terminology, with obvious advantages as compared to manual input. For example, a chemical operator presses a button to start a motor. The computer searches for safety conditions that must be met if the motor is to be turned on. If the conditions are met, it turns the motor on; otherwise it informs the operator by signal or printout why the motor cannot be turned on, and corrective action can be taken. A more responsive computer system might be able to correct the situation before the motor is turned on, bypassing the chemical operator.

In industrial hygiene time studies are important, especially where there is a need to calculate a time-weighted exposure. This could be done by the individual recording the time spent in an exposure area. However this is not a dependable method because the person may not keep an accurate record. It would also be expensive to have someone assigned to this recording task. One method that has been very helpful with collection of data of this type is the badge reader. Into the door of the location is inserted a card that includes the workers identifying characteristics. The door does not open unless a card is inserted into the system. The time of entry is then recorded. When the worker leaves the area the same process is repeated and the computer calculates the time in the area.

This method is very useful for keeping track of exposure times to maintenance personnel who can be found working in any part of the plant. Once the times in a specific area are known, the computer can calculate a time-weighted average (TWA) for the person based on the air sampling data that have also been programmed.

The same type of system can also be used to exclude persons from an area. That is, the door will open only when an approved badge is placed in it. Such a system is also of value in work with carcinogens, where the law requires a logging of persons entering the area. It can also be used to exclude from an area persons not having proper chemical training. Examples of sensors-based applications appear in Figure 4.9.

Another automated form of data collection makes use of machine-readable forms. It seems redundant to have to transcribe data from one form to another, or to a card to be placed in the computer. The government, as well as the medical and legal aspects of industrial health, requires significant recordkeeping. Machine-readable systems have been of value with air sampling data, analytical results, medical histories, and questionnaires of all types. The machine-readable form takes information from the source directly to the computer and eliminates much of the intermediate handling.

4.2.3 Sensors

Sensors receive data from the real world. Since the sensor-based computer receives input from these sensors, it has distinct advantages over the non-sensor-based system, which receives input prepared manually for entry into the system, and the output must be interpreted manually before action can be taken. The sensor-based computer on the other hand can also send output directly through a sensing device to control a physical action. The most obvious advantage to a sensor-based system is time, since the interval between inputting the data and using the output is reduced; another advantage is accuracy. With the sensor-based computer, human intervention is eliminated at several points, thus decreasing the possibility of error. Figure 4.10 indicates the differences between sensor-based and non-sensor based computers.

Sensors in use today are able to measure temperature, pressure, force, flow, acceleration, velocity, and sound. A sensor is often called a transducer because it converts one form of energy to another. For computer usage we need to convert mechanical or thermal energy to digital electrical signals.

There are three types of sensor signal: contact, voltage, and process interrupt. For contact sense, if a switch is activated the contact closure would be sensed and interpreted by a digital input device as being a bit or no-bit condition (1 or 0). This is useful information for the computer, as described for input/output devices in Section 2.3. The same is true for voltage sense (a voltage is presented to the digital input circuits) and for process interrupt (a specific features is being measured). From a programming standpoint, one must decide when and how the digital input device should read the data.

When transferring data in the opposite direction, that is, from the CPU, the sensor-related hardware is a digital output device that can convert digital data into a voltage

DATA AUTOMATION

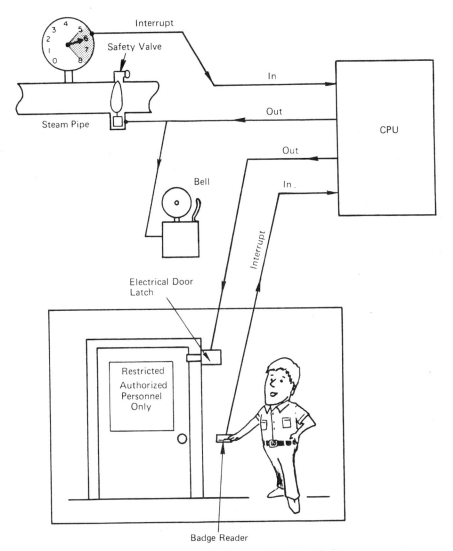

Figure 4.9. Examples of sensor-based applications.

capable of operating various pieces of equipment. These output devices fall into two general categories: contact output and voltage power. The contact output device can perform functions by activating a circuit to close or open a switch. Thus it provides the switching power to turn a given piece of equipment "on" or "off." Voltage power, as the word implies, has the ability to supply voltage to a given piece of equipment (e.g., to turn on a light or operate a motor). Depending on the equipment to be operated, voltage is supplied or removed by the digital output device to operate or stop the equipment.

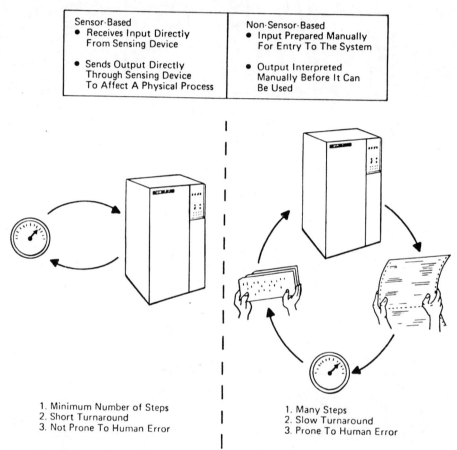

Figure 4.10. Differences between sensor-based and non-sensor-based computers.

On the other hand, analogue signals from sensors must be converted to digital input for computer usage. This is accomplished by an analogue-to-digital converter (ADC). If the output signal needs to be analogue, then a digital-to-analogue converter (DAC) is needed for the equipment to function. In the latter process the digital data are converted to a voltage, which could operate a motor or other piece of equipment as previously described.

DATA AUTOMATION

Thermocouples, resistance thermometers, thermistors, strain gauges, and flow meters are commonly used analogue input/output sensors. The first three measure temperature and the last two measure respectively, force and flow. All these devices have been used accurately, successfully, and reliably with the computer. The thermocouple, for example, is a sensing device capable of measuring temperatures as high as 4000°F. Since it is also able to detect temperature changes, it could be coupled with other logic and parameters being measured to raise or lower the temperature as required.

Each type of sensor has advantages and disadvantages that make each one more suited for certain applications. For example, resistance thermometers are more accurate than thermocouples, but the measuring temperature range is less. Thermistors are very small, relatively inexpensive, and can be made into various shapes and sizes. However they are not suitable for all operations because they are less stable than the other devices.

There are many applications in industrial health for sensor-based computers. Section 5 has information on both sensor-based and non-sensor based applications. Acquisition of data from instruments, processes, physiological functions, and other sources is of importance in industrial health because with the sensor-based computer, the data acquisition is direct and produces real-time information. The response to the information can also be direct if output sensors are used. The sensor-based applications can be separated into three general categories: data acquisition, control, and automation.

A sensor-based data acquisition type of system, as the name implies, is primarily concerned with the collection of information from the various sensors attached to the system. The sensor-based computer can apply some degree of control over the data acquisition system, but this control function would be secondary to the application of data acquisition. Quite often in data acquisition systems data are only collected, or only partially processed, with complete processing done at a later date. This may be necessary because most sensor-based computers are small. In the case of the data collected from an electrocardiogram, for example, the data could be interpreted by a physician, interpreted by another computer, or stored for comparison with another electrocardiogram taken from the patient at a later date.

Another example applying to a sensor-based computer data acquisition system could be water and air pollution monitoring (see also Section 5.2.4). A computer could be attached to a wind speed guage or an overfall weir. When the sensor measured a certain pressure or flow, the water or air pollution sampling equipment would be turned on to collect data. This simple case could be controlled manually. However the computer could monitor from 1 to 1000 sensors, predict impending pollution alerts, or operate a large number of sampling devices at very remote locations. Thus preprogrammed sampling procedures, with predetermined limits, are placed in the sensor-based computer. When the limits are met, the data acquisition process starts.

If the sensor-based computer application is for control of an operation or process, data acquisition is still important. The control application involves the acquisition of data, either analogue or digital, from a given operation. The computer processing of the data could make control correction calculations to ensure the proper functioning of the process or operation. Depending on the application for the computer and the parameters of the computer program, the computer would produce output control signals, either analogue or digital, for control of the process. In this case the task of the sensor-based computer would be to receive input from the attached sensors, analyze the data, and effect a change in the process or operation by activating one of its sensor-based output devices. The sensor-based computer has advantages over other types of systems that can accomplish the same control because it can collect data, make calculations, and initiate control in real time (i.e., during the actual time that the operation or process is transpiring).

A sensor-based control system could be employed in many industrial applications. One application receiving much attention today is the use of a sensor-based computer control system for energy management. The computer monitors sensors attached to chillers, standby boilers, air-handling systems, and cooling louvers. The computer regulates air conditioning and heat, and controls humidity and temperature by zones, using data from inside and outside weather stations. The computer can also automatically turn off lights at specific times. Although the computer has the logic for the application desired, its use should be carefully monitored where industrial health aspects are evident. For example, one would not want to use an energy management system that automatically turned off the lights in an industrial building if safety or health hazards could occur, if the lights went off while operations were still being conducted. Likewise one would not want to automatically turn fans off in laboratory hoods, or cycle fans in areas that require makeup air.

Another use for a sensor-based control system could be in controlling air and water pollution from a process. In this case direct reading water and air pollution equipment would be continuously operated. Data from the sensors in the equipment would be directed to the computer for analysis. If concentrations reached a certain predetermined level, the computer would take control over the process to ensure that concentrations did not exceed legal limits. With this type of control the process could be operated closer to the allowable pollution level, obtaining, in most cases, maximum production and also providing the capability to shut the process down during a malfunction or accidental release.

Computer-based sensor control systems are often termed "closed loop" or "open loop" systems. The air and water pollution control example just cited is a closed loop system because the computer obtains all the data, makes calculations, and produces control of the operation. It is faster than the open loop, which requires manual intervention. If delays of several minutes can be tolerated by the desired application, or if direct sensor-based control of the operation is not practical, the open loop method would have to be used. In this case the computer acquires the data, performs complex calculations, and gives the operator an audible or visual indication that something has happened. It is then up to the operator to make adjustments to the process.

Given rising costs, it is important to bear in mind that use of the computer should result in a manpower savings, since one computer could be used to control any number of operations throughout a plant. From the industrial hygiene standpoint, a chemical operator in a computer control center will not receive a significant exposure as compared to the

DATA AUTOMATION

number of persons that would have to work near the operation, with the resultant high exposure in a manually controlled situation. Thus sensor-based computer applications result in more effective use of personnel, quicker diagnosis of operation changes, and concise presentation of process status, and exposure of personnel to hazardous materials is reduced, as well.

The final type of sensor-based computer application, automation, is a combination of data acquisition and process control. However in automation the need for data acquisition perhaps exceeds the need for process or operational control. Automation is concerned with monitoring individual items; determining how they operate and what is happening as they move from one step to another, testing each step to ensure proper operation in an expeditious manner, and collecting the resulting data. Not only is the information obtained of value to the operator of the equipment in regard to timely corrective measures, periodic exception reports, and summary reports, but the information supplied to management on performance of personnel and equipment allows management to make corrective decisions.

For example, an automated system would be of value in an industrial health application involving the attachment of sensors to laboratory equipment. Some of the incentives for the laboratory to use a sensor-based automation computer system are precise timing and control of equipment, buffering of large quantities of data, validity checking, simulation, decreased response time, and better utilization of personnel. More detailed aspects of laboratory automation, as well as expanding the computer system for information reporting, are discussed in Sections 5.2.3 and 5.2.4.

4.3 Data Management

This section instructs potential computer users in the concepts of data management as related to information processing. Data management is related to the control, retrieval, and storage of information to be processed by a computer.

4.3.1 Control

To make the process of "data management control" understandable, the relationship between user and information system must be defined in computer terminology. A computer information system provides information to a user, on request, by processing the data through the computer. Therefore a user of an information can obtain data from the computer. Control is the authorization and supervision of the data management process. Authorization is the validation of a user's right to access (read) the data. A higher level of user authorization may permit the user to alter (update) the data. User rights are established by the person responsible for the program. A user is never allowed to alter the system without permission of the person responsible for the program. The data manager for the program is responsible for monitoring the location of the information ensuring against data loss (data integrity), and ensuring that the information in the system is current. On the other hand, the person responsible for the program must ensure that the data placed into the computer are valid.

4.3.2 Data Retrieval

Before we can discuss data retrieval, or for that matter data storage as related to management, it is important to examine information and how it is represented and used in an information processing system. Information is ideas and facts about things, people, places, chemicals, operations, machines, equipment, and so on. Each of these ideas or facts is referred to as an entity.

The user of an information system wishes to know about certain entities. Some facts may be off-limits to certain users. This is very important in the field of industrial health, since some facts, such as personal medical data, should be retrieved only by authorized persons. To retrieve the facts needed, an authorized user must define the entities of interest and supply for each entity a list of information attributes.

For example, an industrial physician is concerned about an embryotoxin being used in the plant; the entities of interest are female employees exposed to the chemical "DMAC." The information attributes would be defined as:

1. Name.
2. Sex.
3. Department.
4. Employee number.
5. Chemical.
6. Exposure level.

The list of information attributes for each entity is a logical record. A logical record may have meaning only to the person using the system to obtain information. For the information to be obtained by the user, it must be programmed so that the entities, as well as the information attributes, can be retrieved and only that information. In this case, for example, the physician might not be authorized to obtain salary data for the female employees.

Thus it is the function of data management of build meaningful information by bringing together the proper context, data, and data representation as well as user control. The retrieval function is the process of locating, structuring, and ordering the information in the system for the user. Locating information involves determining what data are needed and where they may be found. If the information is not in a form suitable to the user, it must be structured to meet the user's needs and perhaps ordered in a different sequence.

4.3.3 Storage

The last function of data management is storage, which is the technique for representing the information both logically and physically on a storage device such as a disk,

DATA AUTOMATION

tape, or punched card. It also includes the order in which the information is stored. Consideration must also be given to the way it may be physically accessed or addressed as well as the physical method of representing the data. Depending on usage, certain data may be placed in the computer main storage for quick recall rather than being stored outside the computer. Decisions concerning control, retrieval, and storage of data are made by the computer programmer and data manager and may be modified, as the need arises, to provide the best possible data management for the program.

4.4 Data Presentation Methods

Any computerized system produces a significant amount of data. For example, the monitoring of radioactivity may involve the collection of hundreds of samples weekly, which could be checked for the alpha, beta, and gamma activity by a scintillation counter. The results, such as sample number and total counts, are placed into the computer, which sorts and edits the data, calculates counts per minutes, arranges the data accordingly, and sends the data to the laboratory for further editing. It is essential that the computer do as much work as possible rather than furnish an unorganized mass of data that must be worked manually to obtain desirable results.

Data screening, reduction, and analysis by the computer are factors that must be considered in the presentation of information. An unorganized array of data, which is possible because of the mass of information that can be retrieved from the computer, may mean little or nothing unless it can be put into a compact and reviewable form. Methods of screening the data must be employed so that useful information can be obtained to convey some notion of the nature and dimensions of the entire aggregate of data. For example, controls in industrial hygiene are very expensive. Management responsible for approving the expenditures for controls does not want to review a mass of unorganized data. Management is usually interested in the "bottom line" or net results. Presenting the data properly is important and can augment the point to be made. This can be done effectively with the computer, thus saving time in analysis of the data.

The most frequently used method of screening is to arrange the data into a statistical distribution. A person desiring to set up a computer system should be thoroughly versed in statistics because such methods of analysis can be programmed into the computer. In a well-defined data screening operation the computer can not only compute mean, median, standard deviation, and so on, but also can point out items with missing data. Statistical analysis can also be performed on data entered in the computer in a completely random design (randomized block design) or in a Latin square design.

Screening is very important when collecting physiological data. Even a preliminary screening often can reduce the number of computing problems attributable to poor or missing data.

The averaging technique is another method of screening data. For example, a large number of bioelectric responses from a repetitive stimulus can be averaged together to reduce or eliminate random errors. A filtering technique can be used also to eliminate frequencies outside a specified bandwidth. By applying these techniques to data files before sophisticated analysis is undertaken, time and resources and be saved and meaningful information produced.

4.5 Responsibilities of an Industrial Health Data System

What should an industrial hygiene data system be capable of doing? Let us first review some typical responsibilities:

1. Making available and maintaining material safety data.
2. Maintaining medical/industrial hygiene data on employees and their exposures.
3. Providing selective data to medical systems and personnel.
4. Establishing medical/industrial hygiene requirements.
5. Keeping corporate records with required legal documentation, under OSHA and Toxic Substance Regulation at both federal and state levels.

Requirements for future development include the following:

1. The system should have the ability to process the known and to anticipate current work load effectively and economically within limitations imposed by technology.
2. It should possess the flexibility to accommodate changing requirements to meet business needs.
3. Small increases in cost should suffice to secure necessary capacity growth.
4. Long-range expansion should be accommodated readily.

These responsibilities are user oriented and quite obvious to the experienced system analyst. The subtle needs of a system are more difficult to ensure.

System security is a great concern. Aside from the conflicts that arise when personal data are divulged to an unauthorized source, think for a moment what would happen if a competitor were to have access to supplier contracts. Data security is complex and should be part of the basic programming effort. Only those with a valid "need to know" should be able to access the system.

Both the programmer and user have a severe responsibility to keep the confidence of those whose data are recorded in the system by ensuring that only authorized persons can retrieve such data. Anything less would be a breach of the trust and confidence placed in the occupational health specialists and the supporting data processing team. For example, OSHA records for annual lost time, injuries, and sickness do not need to include details of employee name (as long as management can, if necessary, obtain the information).

Accuracy and timeliness are probably the most important systems functions. Why get data fast by automation if they are not up to date and accurate? Simplicity should also be a goal of an effective system. The more complex an information system becomes, the more restrictive it is to its users.

DATA AUTOMATION **129**

Documented programming is essential to the longevity of an information data base. People come and go, but <u>the effectiveness of the system should not be dependent on the skills of specific individuals.</u> Standard note taking and the programming instruction should be understandable by any successive series of programmers. A test of this is the ability to duplicate the program by those who have never seen the original package. An additional safeguard is the locked-in aspect of critical data. A power failure, user mistake, or programming error should not be able to cause a major loss to the system. Depending on the sophistication of the data, dedicated phone lines, codes, and ciphers may be necessary to assure data integrity.

5 DATA BASE APPLICATIONS AND CONCEPTS

5.1 Example of a Single Data Base System

Ventilation is a primary method in industrial hygiene for controlling exposures to chemicals. The following example of a computer application involves a single program for the testing of ventilation systems. The same principles and adaptations, however, could be used for other control systems to prevent excessive exposures to employees.

5.1.1 Establishing a Ventilation Testing Program

The example data base system, designed and installed at an IBM Corporation manufacturing facility in Rochester, Minnesota, has been found to be very effective in conducting an exhaust ventilation testing program. Testing and inspection of the ventilation systems are performed manually with ventilation measuring instrumentation. The data are recorded in the computer for recall as required. This simple method of recordkeeping is used to document, in compliance with government regulations, that the system was tested quarterly and within 5 working days of any process or ventilation change that might affect workplace employee exposure. Data placards are placed on the ventilation system to indicate that the system was tested and meets specifications. The ventilation readings and the next inspection date are also recorded. This gives the user of the system and line management a visual record of the information being placed in the computer system.

Although this relatively simple recordkeeping function of the computer could have been performed manually, the use of the computer has advantages. It can produce, in a short interval of time, a clear and concise report of all the exhaust systems at the location. It can also produce historical facts about any single system. Most important, the manager can be assured that a system will receive its required testing. The computer notifies Maintenance when testing and inspection are required, and it prints delinquency notifications concerning any system not inspected until the work is completed.

A more advantageous type of testing would be a system that continually monitors the operation and function of the environmental controls. This can best be accomplished by use of the sensor-based computer. Gauges can be placed in the ventilation system and coupled to an on-line computer for sensor-based monitoring or control. If the computer can not correct a fault condition by an output sensor response, it notifies Maintenance that abnormal conditions or functions exist, to permit corrective action to be taken. As with the less complex system just described, the computer records data for legal purposes and information regarding system efficiency.

In automating the recordkeeping and reporting functions for the manual measurement and control system described above, Ronald Long and William Karoly (at IBM's Rochester, Minnesota location) had to devote much time and energy to achieve the results they wanted. Liaison with the system analyst, the computer programmer, the engineering and maintenance departments, and line management was necessary to ensure that each understood its responsibilities. Then a procedure was written. Appendix 4B gives portions of this procedure to demonstrate the complexity of putting together the computer and noncomputer procedures for a simple ventilation testing program.

5.1.2 Workings of the System

In this program the computer is utilized to notify Maintenance when a particular ventilation unit requires inspection and measurements. The measurements are made manually. For the system to work, Environmental Safety and Health, Facility Engineering, Facility Maintenance, and the using line management at the location must know and complete their responsibilities as required in the program. The computer is programmed to carry out its responsibilities and functions. For example, Facility Maintenance must be trained in proper air measurement technique and in the evaluation of each system. If this department fails to measure or send the data to the computer for recording, the computer report received by Environmental Safety and Health will so indicate, and corrective action can be initiated.

The action sequence for the entire program appears in Figure 4.11, that is, the action to be taken and what is to be done when the system does not meet requirements. These actions are not computer-required responses. The computer printout of the ventilation data work completed is sent to the industrial hygienist on a monthly basis for review. The computer printout (Figure 4.12) is interpreted in Appendix 4B. The industrial hygienist also receives a weekly computer printout indicating the ventilation systems that did not meet requirements, the reasons for the discrepancies, and the action that is being taken. With this system the industrial hygienist has an up-to-date record of all plant ventilation systems and can do whatever is deemed necessary to control conditions.

5.2 Other Industrial Health Applications and Concepts

5.2.1 Information Search and Recovery

A very important facet in conducting any type of health-related work is reliance on information. Information and the communication of the information to required areas is

DATA AUTOMATION

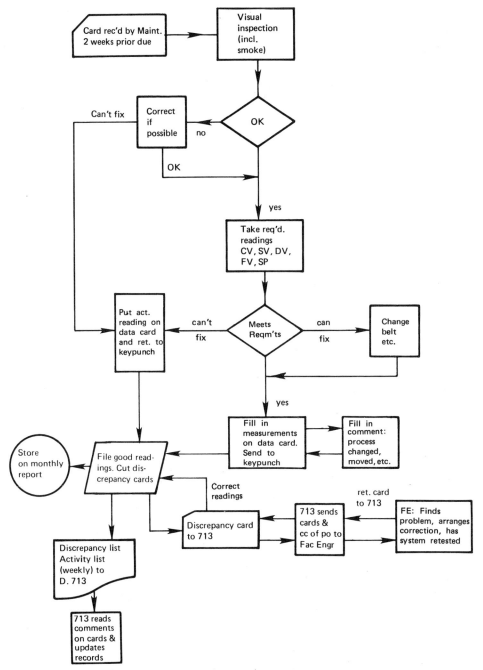

Figure 4.11. Ventilation testing action sequence.

```
D.P.JOB 70805                              MONTHLY EXHAUST VENTILATION SYSTEMS       DATE/TIME   04/22/78  02.49              34

 VENT   STK FAN    OPERATION      MATERIALS AND  OSHA CLASS TYPE----DIMENSIONS FACE/SLOT TANK DUCT---- -------AIR FLOW DATA-------
 HOOD                              CONDITIONS     HAZ/LEVEL       L1    L2   W1   W2    D     A1   A2    A3   MIN    AVE    VOL  ST   FQ/
                                                                                              .00  .00  .000  VEL    VEL         PS  DATE
106036  26 000 HOT WATER RINSE H20  130F                     SV   30    27    3   24 0000    63 0000 0000
                                                                                  REQUIRED MEASUREMENTS        900   1000    630  0      3
                                                                                  LATEST   MEASUREMENTS       1500   1660   1045  0 780309
                                                                                  PREVIOUS MEASUREMENTS       1150   1200    756  0 771212
106037  26 000 CUPOSIT CU 9F   928502514 RT   CORR           SV   31    27    3   24 0000    65  454 0000
                                                                                  REQUIRED MEASUREMENTS        400    500    325  0      3
                                                                                  LATEST   MEASUREMENTS       1800   1900   1235  0 780309
                                                                                  PREVIOUS MEASUREMENTS       1150   1200    780  0 771212
106038  26 000 IMMERSION TIN   915101204      C3 CORR        SV   30    27    3   24 0000    63  454 0000
                                                                                  REQUIRED MEASUREMENTS        400    500    315  0      3
                                                                                  LATEST   MEASUREMENTS       1600   1750   1102  0 780309
                                                                                  PREVIOUS MEASUREMENTS       1150   1200    756  0 771212
106039  26 000 CUPOSIT 19      93440044     RT C4 CORR       SV   29    27    3   24 0000    57  454 0000
                                                                                  REQUIRED MEASUREMENTS        400    500    285  0      3
                                                                                  LATEST   MEASUREMENTS       1700   1800   1026  0 780309
                                                                                  PREVIOUS MEASUREMENTS       1150   1200    684  0 771212
106040  26 000 HCL TANK        922012535 DRT  C4 CORR        SV   32    27    3   24 0000    67  454 0000
                                                                                  REQUIRED MEASUREMENTS        900   1000    670  0      3
                                                                                  LATEST   MEASUREMENTS       1800   2000   1340  0 780309
                                                                                  PREVIOUS MEASUREMENTS       1150   1200    804  0 771212
106041  26 000 ELESS CU        910010013 RT   TOXC           SV   60 0000     3 0000 0000   126 0000 0000
                                                                                  REQUIRED MEASUREMENTS        900   1000   1260  0      3
                                                                                  LATEST   MEASUREMENTS       1500   1625   2047  0 780309
                                                                                  PREVIOUS MEASUREMENTS       1150   1200   1512  0 771212
106042  26 000 NEUTRA CLEAN 7  928118105      D4 CORR        SV   30 0000     3 0000 0000    63 0000 0000
                                                                                  REQUIRED MEASUREMENTS        400    500    315  0      3
                                                                                  LATEST   MEASUREMENTS       2000   2300   1449  0 780309
                                                                                  PREVIOUS MEASUREMENTS       1150   1200    756  0 771212
106043  26 000 SULFURIC ACID   922017734 DRT  B4 CORR        SV   30 0000     3 0000 0000    63 0000 0000
                                                                                  REQUIRED MEASUREMENTS        400    500    315  0      3
                                                                                  LATEST   MEASUREMENTS       1700   1850   1165  0 780309
                                                                                  PREVIOUS MEASUREMENTS       1150   1200    756  0 771212
106044  26 000 CU PYROPHOSPHAT 924296806 120F C3 CORR        FV   72 0000    36 0000 0000  1800 0000 0000
                                                                                  REQUIRED MEASUREMENTS        250    300   5400  0      3
                                                                                  LATEST   MEASUREMENTS         50     78   1404  0 780309
                                                                                  PREVIOUS MEASUREMENTS        250    300   6042  0 771212
```

Figure 4.12. Ventilation system inspection computer printout.

the backbone of any industrial health program. For the program to be effective, the information must be obtained in an expedient manner.

As industrial hygienists with responsibility for multiple locations, we receive calls daily from industrial hygienist colleagues, physicians, engineers, and others, asking such questions as: What is the compatibility of acetone and formaldehyde? Is the chemical "MDS" a carcinogen? Should pregnant women to be exposed to lead? We have switched from using the chemical "TDI" to "MDI," the toxicology is similar—should the medical surveillance be the same? What are the analysis and sampling procedures for chlorobenzene?

In some cases the answer is available in a standard text and the necessary facts can be recovered quickly. With more complex questions textbooks are of little help and may be outdated. Perhaps the best method for information search and recovery for "routines" is a computer, to provide a viable response time. For example, why not have a computer list of all chemicals that are carcinogenic, mutagenic, or embryotoxic? Likewise, why not have the medical surveillance information for all chemicals computerized? The sampling and analysis procedures that may be needed by chemists or industrial hygienists also can be reduced to program form. The use of computer information system also assists in maintaining uniformity between locations.

The computer files just discussed are rather simple and could be undertaken by the responsible person programming the data on a portable computer. Updates to the system could also be made by the coordinator in addition to sending the hard copy to the requestor of information. If the requests are too frequent, this information system would occupy a significant portion of the person's time, and it becomes advisable to have the responsible persons separated from the request for information. This could be accomplished by providing the ancillary locations with cathode ray terminals for on-line retrieval of information.

An advantage of computerized information is that the information is always current. With the computer, because the obsolete data are removed, an update is available to everyone immediately.

It is also obvious that when such electronic files of information become very large, the number of users or frequency of use must increase for the computer to be competitive with other methods. In industrial health practice also, significant data are stored for medical/legal reasons. This reduces storage cost when compared with bulk storage of paper. The retrieval of computerized information is of course more efficient. The factors that make electronic files of information worthwhile are implicit in the following questions: Is real-time response needed? Are labor costs with other methods prohibitive? Are other users available to reduce the cost of the computer usage?

An example of a computerized information data base is the computer-derived exposure list of common contact dermatitis antigens (1). By use of this system, information can be obtained concerning product ingredients. Although its original concept was to supply information to dermatologists, the system would be of value to industrial physicians and hygienists.

When programming information systems, ease of user interaction for the data must be addressed. Therefore it is important to select the "best" key words and synonyms for retrieval purpose. Information systems that involve chemicals can become quite complex. Programming systems of single ingredients are not too complex provided all generic and trade names are programmed for the user. Mixtures with chemicals in various proportions increase the programming complexity as well as the ability to retrieve the information the user wants. Another factor to be considered is the loss or inadequacy of data. This results because many manufacturers will not give complete chemical ingredients to protect proprietary information. New legislation may resolve some of these problems.

Group usage definitely brings the cost of computerization to an acceptable level in relationship to the benefit derived. Bibliographic research is an area where computerization is of extreme value. Expanding technology produces myriads of information for publication. Related to any field of interest are many journals and government, industry, and academic publications, as well as proceedings from technical symposia. It is physically impossible to keep current manually without a large budget, time, and manpower. Even after retrieval of the articles, time must be spent to read as well as classify information. Many are of little use and may be outdated. Thus it is advantageous to have access to a system that has computerized bibliographies of interest.

The computer can deal effectively with the foregoing type of information and data handling if it has been adequately formalized for this special usage. The advantages of computerization are better collection, storage, retrieval, and tabulation. Once the data have been programmed, the computer is able to search through abstracts of vast electronic libraries in a matter of minutes to select all articles or abstracts containing a key word—for example, "epoxies." One computer may have more information than most large libraries and can retrieve information faster than the most efficient noncomputerized library. The computer relieves much of the administrative and clerical burden and also reduces errors.

Because of the need for quick and efficient bibliographical research, many organizations are programming such information on computers, and technical information retrieval centers are becoming quite commonplace in very large organizations. For example, chemical information from the widely used chemical abstracts of the American Chemical Society are now on a computer data base. The National Library of Medicine has medical and toxicology information in the computer data base called "Toxicon." IBM and many other companies have computer files and abstracted bibliographies for their own internal technical activity. In this way technical and professional personnel can keep up with state of the art more quickly and economically. To recover the cost of programming, some companies that have established large electronic libraries are offering access to these facilities on a fee basis.

Computerized technical information retrieval centers usually operate under the following format. Descriptions of documents are keyed into machine-readable form. The summaries include title, author, and other bibliographic information. In most cases the abstract text is the same as written by the author. A word text searching technique is usually used for retrieval and dissemination of information. The computer searches every word of input from title through bibliographic data, including index terms in addition to the complex text of the abstract. As a result, searching strategy is extremely flexible and accurate. The "answer" to an information search may be a complete bibliography on the effects of lead poisoning, or the effects on the central nervous system alone. Likewise, one can search for only a single reference.

One of the principal advantages of electronic libraries is that a standing profile can be left in the system by a user. Each time new data on a subject (e.g., benzene) are placed in the data files, the subscriber automatically receives the reports. Figure 4.13 is an example of a computer printout from the IBM Technical Information Retrieval Center. The subscriber's key indicators, listed at the bottom of the abstract, are chemical analysis, gas analysis, gas sampling, industrial hygiene, and occupational safety and health. Any new abstracts of articles containing those key words are forwarded to the subscriber monthly.

Table 4.1 lists data that were collected from two computerized technical libraries. The main purpose of the data gathering was to discover industrial health applications using the computer. The key word in the information search was "computers." Since this would bring information from many sources, modifier words had to be projected for retrieval purposes, to reduce the number of publications that the computer would print and the user would have to review. The computer was instructed by the library specialist to only print abstracts that also contained the words "health and safety," "air pollution," "chemicals," and so on.

The computer "run" produced more than 2000 abstracts. Since Table 4.1 contains approximately 200 references, it is seen that only 10 percent of the abstracts were of value. Because of programming, the computer was not able to break the data into the categories listed in the table. This was accomplished manually by reading each abstract and in many cases going to the library to obtain the article for further information. If

```
DOCUMENT NUMBER: HHH76H002292

TITLE       PB-245 851/1SL NIOSH ANALYTICAL METHODS FOR SET B STANDARDS
            COMPLETION PROGRAM.  OCT 75.
SEQNO       76H 02292
AUTHOR           STANFORD RESEARCH INST., MENLO PARK, CALIF.*NATIONAL INST. FOR
            OCCUPATIONAL SAFETY AND HEALTH, CINCINNATI, OHIO.
SOURCE      PB-245 851/1SL
            NIOSH-SCP-B
ABSTRACT         53P.  INDUSTRIAL HYGIENE SAMPLING AND ANALYTICAL MONITORING
            METHODS VALIDATED UNDER THE JOINT NIOSH/OSHA STANDARDS COMPLETION
            PROGRAM FOR SET B ARE CONTAINED HEREIN.  MONITORING METHODS FOR THE
            FOLLOWING COMPOUNDS ARE INCLUDED: CAMPHOR; MESITYL OXIDE;
            5-METHYL-3-HEPTANONE;  ETHYL BUTYL KETONE;  METHYL (N-AMYL) KETONE;
            AND OZONE.
SUBJECT     CHEMICAL ANALYSIS       GAS ANALYSIS       GAS SAMPLING
            INDUSTRIAL HYGIENE
            OCCUPATIONAL SAFETY AND HEALTH
COSATI      FIELD 07D, 06J, 99A, 57U, 68G
```

Figure 4.13. IBM Technical Information Retrieval Center abstract.

Table 4.1 A Bibliography of Computer Applications Arranged by Various Subjects

General Category	Subject	Reference
Industrial hygiene	Information systems	1, 4, 5, 13–20, 208, 210–213
	Sampling	2–12, 88, 208, 209
	Noise	21–28
	Ventilation/energy	29–36
	Heat	36–40
Medical	Information systems	1, 21, 61–69, 110, 111, 189, 208, 210–213
	Diagnosis	41–60, 70, 75, 86
	Radiology	50, 53, 55, 70–76, 103
	Heart	72, 77–86, 104
	Lung	47, 51–53, 55, 56, 69, 79, 87–108, 193, 194
	Electroencephalogram	42, 109
	Audiometry	118–122
	Multiphasic health screening	63, 64, 66, 112–117
Analytical	Spectrograph	123–131, 134, 135, 137
	Microscopy	9, 10, 132, 154, 158
	Gas chromatograms	11, 69, 133–145
	Other methods	9, 67, 73, 95, 126, 146–157
Miscellaneous	Safety	159–166
	Environmental	3, 37, 128, 155, 167–176
	Statistics	48, 68, 176, 177
	Toxicology/research	1, 38, 42, 87, 95, 178–186
	Epidemiology	18, 65, 68, 187–194
	Bioengineering	38, 47, 49, 202–204
	Training	51, 174, 205–207
	Health physics	20, 55, 73, 88, 91, 92, 175, 195–201

the interest in computer applications involved only industrial hygiene sampling, the references as listed in the table would provide such information. The same would be true for any of the other categories.

If the computer had been programmed specifically to retrieve the word headings in the table, the task of producing the table would have been less time-consuming. This points out the need for proper programming and search methods. For example, in the case of audiometry it would have been necessary to review 2000 abstracts to find the five references. Because some of the reference searching is complex, many of the computer libraries have cathode ray terminals for direct user interaction with the information.

DATA AUTOMATION 137

With this method the user can directly modify the key search words to obtain the information desired.

5.2.2 Physiological Monitoring

Conducting intensive physiological monitoring for patient care or research can be tedious and time-consuming. The physiological monitoring also requires highly trained personnel, which also results in significant cost. Many of the physiological data being collected are redundant or irrelevant. On the other hand it may be difficult to tell which variables are significant without measuring and studying all the information. This results in the collection of a significant amount of data to find useful information. Likewise, to produce effective research or to make valid patient diagnosis, all the short- and long-term trends of the data collected must be studied and compared. In most cases this requires complex analysis and calculations involving several related or secondary factors. Physiological monitoring, therefore, is amenable to computerization, especially, if real-time information or analysis is needed.

Since physiological systems usually emit bioelectric impulses of an analogue or continuous nature that can be measured and quantified into digital or discrete form for computer processing, the computer, which offers the advantages of accuracy, unlimited storage, mathematical computation, and logic capabilities, is suitable for physiological monitoring. The data can be stored for long periods so that trends can be studied. The data can also be compressed, ensuring that only relevant information is stored. When data acquisition is on-line, by positioned sensors on the subject, real-time or extended-period measurements of physiological parameters can be made. In addition, alarms or notification can be based on the immediate situation (e.g., ventricular fibrillation, increased blood pressure, decreased respiration). Another use is for the analysis of special tests such as electroencephalogram or electrocardiograms.

The rapidly advancing state of sensor development now makes available a wide selection of accurate and reliable devices. These can be transducers, thermistors, or an array of electrodes and other components, which can present the variable being measured in the form of an electrical signal that is proportioned to the value of the variable. The choice of the sensor is important. As with any detection system, the device should not adversely affect the signal or variable being measured. In addition, it should introduce little or no distortion, be relatively insensitive to movement, and offer minimum restriction to activity and comfort.

The nervous, cardiovascular, respiratory, and digestive systems are areas in which data acquisition by computers can be accomplished. Table 4.2 breaks down by system the typical physiological measurements that can be made. For example, in cardiovascular research blood pressure can be measured by placing a needle in a subject's artery and recording the data on analogue tape. Selected portions of the tape can be placed in an analogue computer to convert the data to digital form for transmittal to a digital computer. Upon receipt, calculations are made, as well as comparisons with

Table 4.2 Typical Physiological Measurements for Computerization

Cardiovascular system
  Electrocardiogram
  Phonocardiogram
  Arterial pulse
  Venous pulse
  Blood counts
  Peripheral blood flow

Digestive system
  ph
  Motility
  Electrogastrogram

Nervous system
  Electroencephalogram
  Electromyogram
  Sensory signals

Respiratory system
  Rate and/or depth of breathing
  Oxygen, carbon dioxide, and pH levels in the blood
  Lung volumes

stored mathematical models. Once normal conditions have been projected, experimental conditions can be undertaken. An electroencephalogram or other physiological system could be tied into the same program if these parameters are related and must also be measured and analyzed.

Complex mathematical models of physiologic systems have also been developed using the computer. The general procedure is to develop a set of differential equations that describe the proposed model. By using the computer these equations can be solved, and least-squares techniques (to determine the unknown variables and constraints in the equations) allow testing of the validity of the model. For example, a model of lung clearance could be developed and tested for a dust or radioisotope. The area in which the most progress has been made with computers is radionuclide scanning of organs. Scanning programs using isotopes and computers are available for almost all organs.

5.2.3 Laboratory Analysis, Control, and Information Systems

Automated analytical instruments, virtually nonexistent a few years ago, are becoming commonplace in the laboratory. Since the output of this type of equipment is an electrical signal, it can be accepted in either analogue and digital form by a sensor-based computer. Sensors are discussed in Section 4.

DATA AUTOMATION

The continuous flow and multiple channel types of laboratory analyzer are very suitable for computer analysis. For reasons already given, the computer has demonstrated its capability to automatically acquire data from a variety of instruments, to perform analysis, and to make calculations without error. In addition, since the computer can present data in myriad ways, it lends itself to a total laboratory information system. Data presentation methods, which include screening, reduction, and analysis, are given in Section 4.4. Laboratory instrument analysis data that are not available for direct computer input by way of sensors can be processed by batch methods of data handling. Specifically, many instruments, such as colorimeters, chromatographs, and spectrophotometers can be computerized by the on-line sensor method. A bibliography of application examples and uses can be found in Table 4.1. The gas chromatograph and its subsequent advantages for computerization to the analyst are briefly discussed here. The reader can make the same analogies to other laboratory instrument and equipment types. For example, since the introduction of gas chromatographs to the laboratory more than 20 years ago, these instruments have continually undergone redesign and improvements (new and more sensitive detectors, columns, materials, and liquid phases, etc.).

Perhaps one of the more important reasons for computerization is that analysis of chromatographic data, because of the new technology, can become increasingly complex and time-consuming. Consider the drifting of the baseline or the occurrence of several peaks. Although the new developments have resulted in better information, laboratory managers find that output has decreased. Thus in the face of increased cost per sample, computerization is the best solution, especially in a laboratory with a high volume of sample analysis.

The major benefits of computerizing gas chromatographic analysis are faster results, increased reproducibility, greater accuracy, increased production, increased utilization of equipment, and better utilization of analyst's time as well as complete control over all functions. Some of the computer equipment available today has the ability to handle up to 20 chromatographs simultaneously. As previously stated most of the advantages above can also be related to other laboratory equipment, for example, clinical laboratory instruments.

Once the laboratory equipment has been computerized, the output of data from the laboratory will increase. Even in the most advanced noncomputerized laboratory the analyst is still bogged down with paperwork for a significant portion of time. The most efficient utilization of both equipment and laboratory personnel is an on-line data acquisition and analysis system with information reporting. Where complete automation cannot be accomplished, data from other instruments can be introduced into the computer system by way of batch processing (Section 2). This enables all laboratory instrumentation to take advantage of the information reporting portion of the computer program. Input into the information system for non-on-line instruments can be accomplished semidirectly by way of keyboard input or other devices, also mentioned in Section 2.

For programming purposes and to obtain a better view of the entire process, complete laboratory computerization can be separated into the following steps or tasks:

1. Instruments to be completely automated.
2. Instruments requiring batch processing.
3. Entry of analytical request.
4. Development of master log.
5. Generation of work list.
6. Automatic instrument reading and data logging.
7. Entry of test results—automatic and manual.
8. Mathematical computation and analysis.
9. Verification of instrument performance and quality control.
10. Verification of test results.
11. Reporting of test results.
12. Production of statistical information used for laboratory management control.
13. Storage of data for future recall.

The overall computer system of analysis and control as well as information can be viewed diagrammatively (Figure 4.14); its overall advantages are increased analyst production, improved control, and improved laboratory responsiveness.

5.2.4 Environmental Protection/Sampling and Energy Management Programs

Environmental protection is of primary consideration in the design and operation of industrial facilities, especially those using chemicals in their processes. Environmental protection includes air and water sampling and analysis and operational control procedures. Today there are stringent laws regarding air and water effluents from the plant and hygiene standards for in-plant air quality.

Another management problem is the continual downward revision of "acceptable" levels for effluent and in-plant air quality. Thus managements of industrial facilities have the dilemma of complying with increasing stringent control standards with as little as possible impact on production. Therefore many industrial facilities are utilizing the computer for environmental protection, sampling, and energy management programs. This provides real-time control over pollutants with less impact on the day-to-day operation of the plant. As the laws change, the computer program can be altered to meet the new parameters.

Important elements in an integrated environmental protection strategy include information obtained from the monitoring of stacks, outfalls, and pollution abatement facilities, as well as the actual production processes. By use of the computer, the facility is able to collect and analyze continuously plant emissions and effluent data. A computer system can also be used to calibrate automatically the pollution instruments, to maintain overall integrity of the data.

The computer relates the foregoing information to the environmental protection control parameter established for the plant. The computer can also initiate action to ensure that pollution does not occur. In Figure 4.15, for example, shutdown of operation is one possible result of activation of the pressure sensor. A computer that could act in this

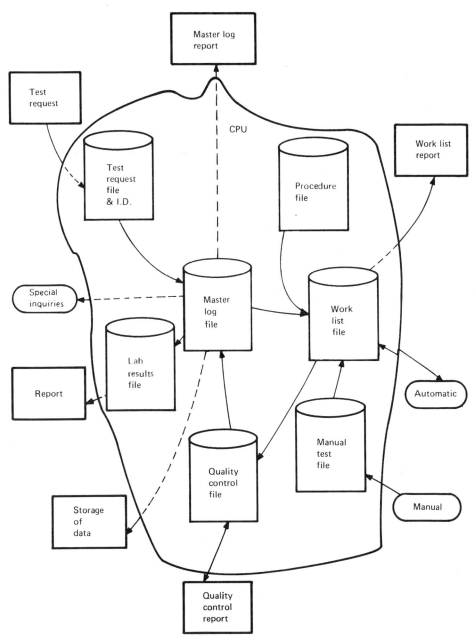

Figure 4.14. Computerized laboratory system.

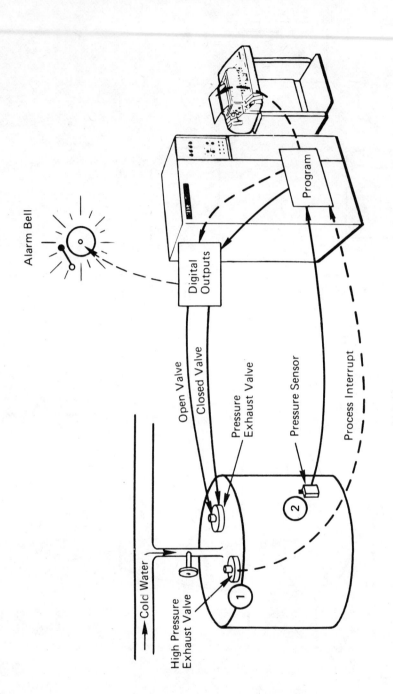

Figure 4.15. Sensor-based computer process control.

DATA AUTOMATION

manner would have to be sensor based: a non-sensor-based computer can do no more than give a signal for manual interruption. For the compliance reporting portion, the computer automatically generates required detail and summary reports and maintains a complete historical record of the environmental data.

In setting up environmental protection programs the control of effluents is usually more easily defined because the sources are easily identified. In particular, needed sampling locations of water effluents can be identified without difficulty because of the "static" type of source. Air emission control represents a slightly more difficult problem, except for stack monitors, which can also be considered to be static. For other air emission control information, multiple monitoring stations must be positioned over a wide area. In the simplest case these locations are at the boundary of the plant. More sophisticated placement can be undertaken by use of the computer. Utilizing air pollution theory and formulas, the location of the maximum downstream concentration from a single or multiple source can be predicted under average climatic conditions. The computer has also been used successfully for in-plant monitoring and control. A recent article gave useful information regarding the use of the computer for automatic monitoring of vinyl chloride in working atmospheres (209). The use of the computer was justified on the need to comply with the OSHA "Emergency Temporary Standard for Exposure to Vinyl Chloride."

The sampling and analysis portion of the vinyl chloride monitoring system consisted of Union Carbide Model 1200 chromatographs, with flame ionization detectors located throughout the plant in 19 areas that would be representative of employee exposures. The locations were also selected on the basis of greatest potential for process leaks into the worker environment. The sampling and analysis regime was under the control of a Union Carbide Model 2800 Microcomputer used in conjunction with a General Electric TermiNet 300 Printer.

For result viewing at the monitoring location for employee or management, a strip-chart recorder was used. The microcomputer accepted the same raw data from the analyzer and performed validations to ensure the reliability of the data. A computer program was established to calculate probable employee exposures as well as to provide work shift, daily, and monthly reports of information. In addition, the computer was also used to signal undesirable conditions for corrective action to reduce exposures. The operational and analytical instrumentation personnel associated with industrial hygiene collaborated to define the data requirements for the system. This is an important facet of any computer and sampling program. However even though considerable effort was made to define all situations, one parameter was overlooked. In most cases once programming has been established, it is difficult to correct such errors without reprogramming. This often results in additional cost. Sections 3 and 4 indicate procedures in data base design to minimize such omissions.

On the positive side, the authors felt that the most important function of the computer system they designed was preventing employees from exposure to high concentrations of vinyl chloride monomer. If the concentration of the chemical was above either of the two target conditions, an alarm notified employees. Another important contribution of the computer system was that by locating sources and timing occurrences of leaks, the company was able to reduce emissions. This enabled production engineers to redesign problem areas. This also resulted in cost savings, since valuable product loss to the atmosphere was reduced. In addition, the system provided a permanent record of the estimated exposure for each job for industrial hygiene and medical evaluations and for compliance purposes.

In most industrial plant facilities engineering is responsible for environmental protection as well as energy conservation. If further justification is needed for the purchase of a computer or usage, energy conservation management should be considered. Facility computer-controlled power management systems have been found to save energy consumed by preprogrammed criteria. Such energy management programs have resulted in cost savings. These programs could also be placed in the same computer, thus reducing its overall cost to use.

In some cases ready-to-use programs can be obtained from computer software vendors. The computer is programmed to turn on or shut down operations at prescribed times. It can also interrupt for prescheduled short periods selected energy devices that do not have to operate continuously. Also by continuously monitoring the rate of electrical consumption in the facility and comparing the rate to variable and/or preestablished targets, it can inform management of problem areas. Management can then take action regarding personnel or equipment requiring correction action.

5.2.5 Medical Examination and Surveillance

Periodic health examination to aid in early detection of disease, as well as medical surveillance techniques such as bioassay, are being administered with increasing frequency. Many industries have inaugurated programs in which employees are periodically given medical exams because of exposures to hazardous materials or processes, or to comply with legal requirements. In the latter case positive tracking is needed to ensure the that appropriate exams are conducted. This poses increased administrative and clerical demand on the medical department.

To provide the best possible preventive medicine for the employees and to ensure that legal and management requirements are met, the programming of medical examinations and surveillance has been found to be very helpful. It offers an easy method of tracking to ensure that physicals are conducted as required. Manual methods of tracking are very prone to error, especially in a busy medical department. When programmed accordingly, the computer can also act as an audit to ensure that the physicals are indeed conducted after notification. The computer can be programmed to print delinquency reports until the physical is entered into the computer as being completed. Computerization also aids in scheduling and enables the medical department to be more efficient. The medical department administrator can see at a glance what type of exams are scheduled for next period and can allot proper time for their completion. The manufacturing managers are extremely happy with this type of efficiency because it reduces

DATA AUTOMATION

unproductive work time caused when employees are waiting needlessly for appointments.

Figure 4.16 is a printout of a very basic type of computerized scheduling arrangements. It is used at a small facility and does not require all the additional programming as indicated earlier. It is described to give an idea of a computerized medical tracking program. Such a system could be easily modified by reprogramming to meet all the special needs previously mentioned. This system is effective because the medical department manager knows who is scheduled for what exams and where each person works, so that the proper manager can be notified to have the employee make an appointment with the medical department. The printout across the top line gives all that information in addition to the birthdate of the employee. The birthdate appears because certain

EMPLOYEE PHYSICALSBY DOB/M -

SERIAL	EMPLOYEE NAME		DEPT-S	BIRTHDATE	PHY/SCH
047564	JAMES	O	092-1	01/11/43	AHY
478152	HAROLD		706-1	01/21/31	ART
301754	ROBERT	E	716-1	01/11/41	ATH
373433	DALLAS	L	641-1	01/21/25	ATY
679602	ROBERT	W	707-2	01/04/35	ATY
682363	CLAYTON	M	707-2	01/08/25	ATY
235337	WILLIAM	J	406-1	01/14/53	AZY
753162	EDWARD	E	716-1	01/08/37	AZY
173494	GORDON	H	920-1	01/15/24	AZY
783552	DONALD	P	41D-1	01/11/33	AZY
403709	JAMES	E	528-1	01/05/51	EBE
902641	THOMAS	J	49F-1	01/12/31	EBO
841771	MICHAEL	J	42A-1	01/17/53	EBO
086541	FREDERICK	J	49E-1	01/22/39	EBO
987770	RAYMOND	G	948-1	01/06/23	EBO
788862	LAWRENCE	C	945-1	01/08/50	EBO
403704	EDWARD	L	421-1	01/12/35	ECO
152512	LAVERN	F	446-1	01/08/22	ECO
379645	CHARLES	D	424-1	01/24/45	ECO
895866	WILLIAM	S	429-1	01/19/45	ECO
341825	JAMES M		959-1	01/01/42	ENY
446832	MICHAEL	L	946-1	01/13/48	ENY
956663	JOHN	D	948-1	01/15/35	ENY
723932	JANICE	M	867-1	01/18/42	FZY
359284	SHARON	K	867-1	01/11/41	FZY
268522	CAROLE	E	867-1	01/25/39	FZY
088415	DONNA	G	867-1	01/08/29	FZY
103413	GAIL	H	707-2	01/03/38	HRT
216961	THOMAS	J	572-1	01/18/27	LID

Figure 4.16. Computer printout for scheduling of medical exams.

additional health measurements such as electrocardiograms are administered on the basis of the age of the employee.

The computer printout is rather simple to interpret, except for the last column across the top line (PHY/SCH). For example, the first employee listed is James O. (last name was deleted to declassify the information). His employee serial number is 047564 and he was born on January 11, 1943. He also works in Department 092, where he is notified to come to the medical department for his exam. The last number in the "department" column indicates which shift the individual works. In this case "1" corresponds to the first shift and "2" is the second shift. [The printout indicates that three people to be scheduled for exams work on the second shift. In most medical departments special plans must be made because exams cannot be conducted on the second shift. Simple items like this often become important in computerization of data. If shift designation were needed but not supplied, there would be two alternatives: to go back into the data base and revise all the information, or to get the information from each chart. To save embarrassment regarding reprogramming, it is wise ensure that all the information needed can be obtained from the items included prior to programming the data.]

The PHY/SCH column indicates to the person scheduling the exams the type of physical James is to have. To reduce "space," which may be needed for future programming, the physical type and the scheduled frequency has been letter coded. (Table 4.3) Thus "AHY" for James means an audiogram and a "heat treat" physical yearly. When the physical is completed an input card is forwarded to the computer center requesting a recall next year for a repeat physical and indicating that the physical this year was completed as required.

Some employees who work in environmental chambers or with lasers may require different types of physical depending on the degree of exposure. This is covered in the classification block (2nd) of the PHY/SCH column. For example, Lavern F. (152512) is required to enter an environmental chamber. She should receive the C class physical and she is scheduled for reexamination every other year. For persons requiring annual physicals, the third block can also be used to indicate the type of physical. One example of this coding is Gail H. (103413), who works in heat treating, is a trucker, and may be required to wear respirators. She therefore must receive all three physical examination annually. Thus a simple computer program and system can be expanded to indicate when bioassay tests for chemicals (lead, mercury, etc.) should be conducted.

Although the actual physical exams in the example above were conducted manually, many of the portions of the examination processes, including the history and summary report, can be computerized. For example, the person requiring the examination is given prepunched computer cards with the employee number indicating the tests required. The history can be taken by presenting questions to the employee on a cathode ray tube or an optical image unit. Using a special electrical probe, the subject can respond in "yes" or "no" to the questions. Depending on the response, the unit is advanced to another question. For example, if a positive response is given to the question "Have you ever been exposed to chemicals?" the next question will elicit the dura-

Table 4.3 Computer Coding Information for Medical Exams

Type of physical (1st, 2nd or 3rd block of exam code)
   A = Audio
   F = Cafeteria
   M = Chemical materials
   E = Environmental chambers
   H = Heat treat
   X = Radiation (CBC—no fasting)
   S = Security
   T = Trucker
   R = Respirators
   Z = Filler
   L = Laser
   W = Three wheel scooter (regular trucker nurse assessment) sit down scooter (regular trucker nurse assessment)

Classification (2nd block)
   N = A chamber
   B = B chamber
   C = C chamber
   I = A laser
   J = B laser
   K = C laser

Schedule dates (3rd block)
   Y = Yearly
   E = Even years
   O = Odd years
   U = Three year recheck
   D = No recheck

Instructions

One periodic physical	Fill in 1st block with appropriate symbol Fill in 2nd block with Z Fill in 3rd block with Y, E, or O
Two periodic physicals	Fill in 1st and 2nd blocks with symbol Fill in 3rd block with Y, E, or O
Three periodic physicals	Fill all 3 blocks with symbols
Chamber physical	Fill in 1st block with physical symbol Fill in 2nd block with classification Fill in 3rd block with Y, E, or O
Laser exams	Fill in 1st block with physical symbol Fill in 2nd block with classification Fill in 3rd block with Y, U, or D

tion of exposure. A negative response to the first question will cause the program to skip to the next category. With this type of branching logic, an almost complete medical history could be taken on-line with the computer.

Other on-line testing with the computer can also be conducted as part of the medical exam. For example, the technician prepares the patient for an electrocardiogram. The test card with the employee information is entered into the terminal, notifying the computer that the subject is ready for the test. The computer then activates the device, collects and analyzes the data, and records the results in the employee record. Other physical examination tests that are amenable to on-line testing are audiometry, blood pressure, clinical laboratory chemistry, hematology (red blood count, white blood count, hemoglobin, hematocrit), and spirometry.

Off-line tests involving the input response to the computer of a physician or paraprofessional can also be obtained. Such tests include chest X-rays, differential blood cell counts, measurements of height and weight, vision checks, and urinalysis. The results are entered into the test booklet on a mark-sense card, which is placed in a reader connected to the computer. The computer reads the data and records the results of the employee examination.

When all the tests have been completed, the physician is given a patient summary report, which is standardized by the computer. The physician can then conduct any additional measurements deemed necessary or ask for the additional information based on the test results or the history. This information can also be computerized if necessary or simply placed in the employee file along with the computerized data. Figure 4.17 diagrams the entire process.

When computerized, the medical examination will afford greater accuracy and efficiency in history taking, laboratory testing, and the physical measurement portion of the exam. Advantages include better utilization of professional and paraprofessional personnel as compared to traditional examination methods. Since the clerical functions are drastically reduced, personnel can devote more time to the actual medical aspects of the examination.

5.2.6 Portable Computers

To persons in the industrial health field, perhaps one of the most important computer developments is the availability of the small scientific and engineering computers that are truly portable. These computers, which can be hand carried to a location, have all the abilities of the large computers: significantly less storage is their only limitation. Yet when the equipment is used in a remote-type location for a problem-solving need, great storage of data or voluminous program libraries are usually not required. The most valuable attribute of the portable computer is that it can be taken to a work site for data entry or problem solving with little prior planning or facility installation cost. Thus the small computer with the ability of a larger computer can be placed at the location on a surface no larger than a desk top; it does not require any special communication lines or environmental control or electrical facilities, as are needed with the larger computer.

DATA AUTOMATION

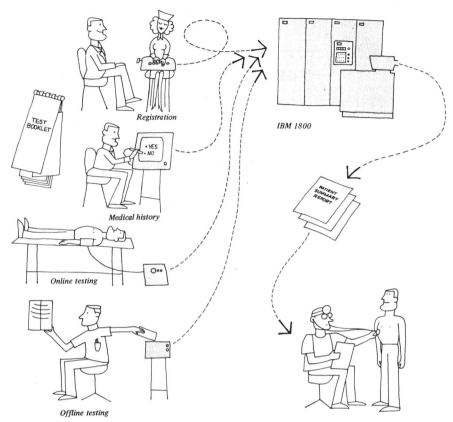

Figure 4.17. Sequence of a computer-assisted medical examination.

The benefits of using the small portable computer are somewhat analogous to the advantages that are evident in industrial hygiene by having direct reading instrumentation for field use. By the collection of data at the site either directly or indirectly through the computer, on-the-spot evaluation of information can be made. A personal computer, readily available for use with the full-function capabilities of high level computers for problem-solving, offers many uses. The ability to collect accurate data at the source for immediate answers is an advantage to any problem-solving professional. The main limitation in using the computer in an industrial environment is that there are "hostile" environments for computers, just as exposure to certain substances is hazardous to humans. In some locations, depending on the environmental conditions, it may be impossible to install some of the portable computers if environmental controls cannot be provided. Where this is a problem, remote sensor-based relays or other data acquisition methods, as discussed in Section 4, should be used. Another solution is to use the slightly larger sensor-based computer equipment, which can operate in a more hostile environment without damage to the electronic components.

Figure 4.18. Small portable computer.

The small portable computer illustrated in Figure 4.18 is about the size and weight of a large typewriter and has main storage capacity of up to 64,000 storage positions. This computer comes equipped with a familiar typewriterlike keyboard, which is easy to use for data entry or programming. A small visual display is usually provided to show the data being placed in the program or the information being retrieved. When a permanent copy of the data is needed, a printer can be attached to produce such "hard" copy. The data, information, or program can also be placed on a magnetic tape cartridge for storage and retrieval later. To reduce weight, the visual display provided with the system is small and really usable only by the operator and perhaps one or two other persons. Where there is a need for demonstrations or computer-assisted instruction for larger groups, the unit should be attached to a larger television monitor. The usefulness of demonstrations and the advantages of computer-assisted instruction in the industrial health field are discussed in Section 5.2.7.

The removable magnetic tape cartridge enhances the small computer by providing a source of increased storage of data and programs beyond the main capacity. As very sophisticated programs are developed, future use without time-consuming reprogramming becomes possible. Having data or programs on magnetic tape cartridges also is advantageous insofar as the tape can easily be transported to a remote location where a compatible computer is available for use. Also available from the vendors of portable

DATA AUTOMATION

computers are problem-solving programs already on magnetic tape. Thus programmed routines of mathematical or statistical equations such as complex differential equations, analyses of variance or regression, and design analyses may be readily available for use. Cartridges are also available to show how to use the computer as well as its programs. Presentation techniques used for plotting line graphs, bar charts, histograms, and curve fits have also been preprogrammed. Here it is necessary to attach the computer to a matrix printer. Data presentation methods are discussed in Section 4.4.

Enhancing the uniqueness and advantages of the small computer is the capability of connecting it to a teleprocessing facility, offering communications to larger data bases and voluminous program libraries in other computers. Thus information and answers are available from the portable computer or other machines where the need is greatest. Directly, this can be accomplished by attaching the computer to teleprocessing connections or to an industrial process or location by a sensor, or with a research device coupler, allowing it to be tied into laboratory or a variety of instrumentation devices. With direct data entry, immediate solutions at the source of concern are available with unmatched speed and precision. Real-time control, not as fast, can also be obtained by manual input of data directly at the site to provide most scientific and engineering problem-solving information or to indicate needed control methods. Of extreme value in signaling when real-time controls are needed is the ability to use the computer to tie into remote areas that are interrelated to the solution. This can be accomplished by using two or more computers tied into each other by connecting transmission lines or multiple sensors from the location to a single computer.

5.2.7 Computer Based Training

Computer-based training (CBT) is an attempt to resolve two major industrial training problems, outlined as follows.

1. Rapidly rising training requirements to:
   a. New products and technological changes requiring a significant amount of employee retraining.
   b. New systems or procedures.
   c. Lack of experienced, qualified new employees.
   d. Changing, more demanding regulations.
2. Continually increasing training costs.

With this background in mind, it is understandable that in every technical endeavor, the search continues for more effective approaches that will enable instructors to increase their productivity. New technologies are being employed, and interactive–self-study methods are being adapted for training.

Industrial health and safety training requirements require a significant amount of time. CBT, when properly conceived and utilized, can provide the protective knowledge and documentation required by federal regulations and sound business sense. Figures

4.19 through 4.21 describe of chemical safety, electrical safety and laser safety courses for which IBM has utilized CBT.

Two types of CBT can be structured.

1. Computer assisted instruction (CAI), in which the majority of instructions are presented through the terminal, possibly with some off-terminal assignments.

2. Computer-managed instruction (CMI), in which the majority of instruction is off-terminal through audiovisual (A-V) materials (films, audio or video tapes, slides, etc.), reading assignments, lectures, or perhaps practical laboratory assignments. The student may be tested through the terminal and directed to appropriate off-terminal material. Special, individualized instruction can be provided by the terminal to fit each student's needs, based on ability to master the subject and required repetition of course information.

There are no rigid distinctions between these two approaches. Both can be used in the same course. However the course developer usually is in a better position to select the technique and materials than the training group or department.

One of the most significant benefits of CBT is its potential contribution to cost-effec-

Course title	Chemical Safety
Course number	6006 (CAI)
	5007 (Hazardous Material—Department 713)
Length of course	11 hr (CAI)
	1 hr (Hazardous Material—Department 713)
	As required for on-the-job training
Method of instruction	CAI (CHEMSAFE)
	Classroom
	On-the-job training
Courses appropriate for	Recommend CAI course for department technicians and personnel with prolonged contact with chemicals. Recommend the "Hazardous Materials" presentation or on-the-job training for production people and those with limited exposure to chemicals.
Abstract	CAI course includes instruction on the terminal, slide/tape presentation regarding chemicals in the eye, and a film on lab safety. The content of CAI and classroom courses covers protective equipment, chemical hazards, safety practices, and laboratory safety.
Responsible for implementation	Department manager monitors need and initiates request for training.
Instruction given by	Education Department (CAI)
	Department manager and/or Department 713 (Hazardous Materials)
Documentation	On formal classes held, forward the completed rosters to the Education Department for processing into the PSD system. Department manager maintains on-the-job training records.
Refresher training	Annually

Figure 4.19 Outline of a CBT course in chemical safety.

DATA AUTOMATION

Course title	Electrical Safety
Course number	6010 (CAI)
	4021 (708-Electrical Safety)
Length of course	3 hr (CAI)
	1 hr (Department 708 Safety Presentation)
	As required for on-the-job training
Method of instruction	CAI Course (ELSAFE)
	Classroom or department meetings
	On-the-job training
Course appropriate for	Recommend CAI course for test equipment, builders, electrical engineers, manufacturing engineers, and Facility engineers, Recommend 1 hour presentation or on-the-job training for production people and those with limited exposure to electrical hazards.
Abstract	CAI contains instruction on terminal, audio tape/slide presentation. Describes basic safety precautions, effects of electricity on the body, and action to be taken in case of an electrical accident. Also includes instruction for high potential operations and insulating testing. Classroom discussion of UL-approved equipment, grounding, and potential electrical hazards and corrective action.
Responsible for implementation	Department manager monitors need and initiates requests for training.
Instruction given by	Education Department (CAI)
	Department 708 (1 hr presentation)
	Department manager (on-the-job)
Documentation	On formal classes held, forward the completed rosters to the Education Department for processing into the PDS system. Department manager maintains on-the-job training records.
Refresher training	Every 18 months

Figure 4.20 Outline for CBT course in electrical safety.

tive training programs. It can help increase productivity by reducing such costs as instructor time, travel-time and expense, and non-productive waiting for traning. In the business environment, CBT can be integrated with existing computer hardware on terminal-based systems. CBT can be effectively piggybacked into the main data processing operation of a firm.

In multilocation businesses the employee "to be trained" can take CBT courses at work location. To end a training session, a simple course sign-off returns the same terminal to production applications for the same or another employee. While one terminal (or several) is engaged in training, others in the system are independent to provide business support. Terminals may be as close or remote as business needs dictate. Normally telephone lines provide the necessary "remote" link to the base operating unit.

Depending on the technology for which the CBT is based, the course context can be formulated and implemented through the combined efforts of the course developer, data processing staff, and the training function. Though several CBT programs can coexist, a central coordinating responsibility should be assigned. This helps to reduce duplication and inconsistencies, and contributes to efficient use of the system facilities.

Course title	Laser Safety
Course number	6018 (CAI)
	4019 (classroom)
Length of course	3 hr (CAI)
	1 hr (classroom)
Method of instruction	CAI (LASAFE)
	Classroom
	On-the-job training
Course appropriate for	Recommend the CAI course for Development personnel, Test Equipment Building personnel, and those who are closely associated with an in prolonged contact with lasers.
	For production personnel and those with minimal exposure, the one hour classroom program will be adequate.
Abstract	CAI course includes two motion pictures, an audio tape/slide presentation, plus questions and instructions from the terminal. This training will provide the ability to:
	1. Identify laser terminology.
	2. Identify laser hazards.
	3. Identify control measures to minimize hazards.
	The classroom program will simplify and condense the major points and highlights of the CAI course for those with minimum exposure.
Responsible for implementation	Department manager monitors needs and initiates requests for training. Ref. Safety Manual Index 36-8-02
Instruction given by	Education Department (CAI)
	Department 713 for classroom program.
	Department manager for on-the-job training.
Documentation	On formal classes held, forward the completed rosters to the Education Department for processing into the PDS system. Department manager maintains on-the-job training records.
Refresher training	18 months

Figure 4.21 Outline for CBT course in laser safety.

For the sake of our discussion, the CBT control function is addressed as the "training group." Figure 4.22 shows the communications and functional responsibilities required for a CBT system. CBT is usually part of training, along with classroom instruction, audiovisual aids, and other functions.

Applications for CBT. The applications for computers in occupational training are almost unlimited. For example, CBT can be used to:

1. Orient and test new employees to identify each one's training needs (The computer could be used to input unique occupational health hazards when keyed by each employee's department number at the time of signing on).

2. Provide a common background to a group of students before they enter a class (or undertake a new process).

DATA AUTOMATION

3. Teach operators (almost any process) how to track their productivity, equipment, or material use through a terminal.
4. Provide basic skill training to new (or reassigned) employees as necessary before technical training (e.g., in the use of protective equipment prior to learning required skills).

In addition to the instructional uses, the same computer data processing system can have some other applications, such as:

1. Maintaining training records of each employee (an OSHA requirement).
2. Providing performance certification testing.
3. Recording employee survey or questionnaire responses.
4. Recording employee hazard awareness.

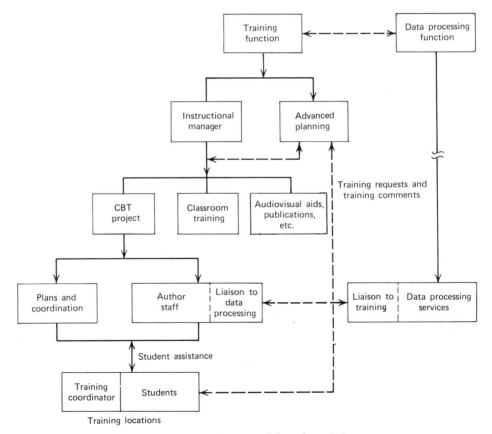

Figure 4.22. Functional responsibilities for a CBT system.

5. Testing to identify the general training needs of a given group, to help trainers decide what new training must be developed.

6. Comparing legal training requirements with employee hazard recognition.

Getting Started with CBT. The adoption of CBT does not mean that existing training materials must be discarded or must undergo major revision. Instead CBT should be viewed as just one more training medium to be combined with other appropriate techniques. For example; CMI could be used to optimize material for a given subject, consisting of programmed instruction (PI) text, audio tape, laboratory, slides, and lectures. Figure 4.23 illustrates the incorporation of several instructional modes into a CMI training program. In each given subject, the results of a terminal-administered test are used to determine what additional instruction is needed. The pretest (Figure 4.23) is a useful first step in sizing the additional instruction. Figure 4.24 reflects the course development process as it applies to several training media.

Benefits of CBT. A benefit must be measureable if it is to be assessed accurately and balanced against a cost. To do this, the benefit must be translated into a tangible cost saving. For example, a reduction in training time, can be measured in terms of the student and instructor salaries saved.

Compared to non-self-study media, CBT should reduce (perhaps eliminate) instructor involvement once the course has been developed (except for occasional updates). This is because CBT can:

1. Eliminate repetitive preparation and presentation for each class.
2. Eliminate instructor time and expense to travel to and from the teaching location.
3. Reduce class time by placing introductory and common basics of existing courses on CBT.

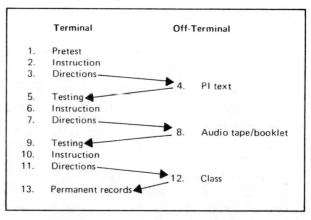

Figure 4.23. Sample CMI course sequence.

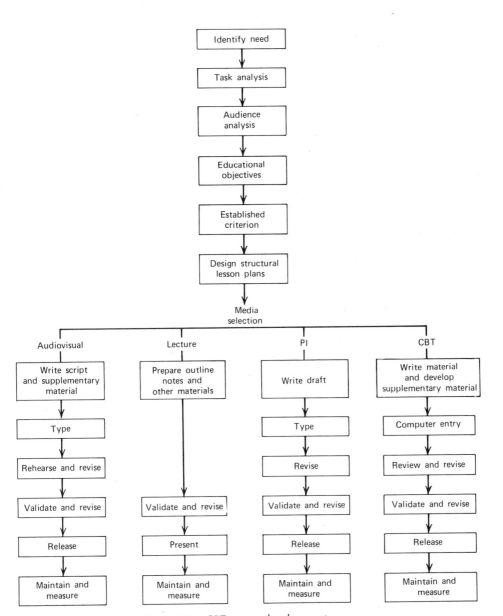

Figure 4.24. CBT course development process.

Benefit	Dollar Calculation
• Reduced on-job-training, one-on-one training (supervisor) Supervisor salary rate × hours =	_____
• Reduced travel time (student and instructor) Number of hours × salary rate =	_____
• Reduced per diem (student and instructor) Dollars × number of days =	_____
• Reduced instructor preparation time per class Number of hours × salary rate =	_____
• Reduced classroom time (student and instructor) Number of hours × salary rate =	_____
• Reduced time waiting for training (student) Waiting time × salary rate × fraction (ineffectiveness) =	_____
• Reduced publication (printing) costs Printing costs × number of copies =	_____
• Reduced publication (distribution) Costs × unit wrapping and postage cost =	_____
• Reduced cost of changes (additions/deletions) Cost of changes × printing and distribution =	_____

Figure 4.25 Factors in the calculation of CBT benefits.

The test or measure is not only savings in travel and instructor time but also the new courses (CBT and others) that can be achieved in the time gained by use of CBT. Figure 4.25 shows how some of the major tangible dollar savings can be calculated. Many items apply equally to instructor- and student-based cost reduction.

5.3 Example of a Complex/Shared Data Base System

5.3.1 Establishing a Computerized Environmental Health and Control Information System

Thus far we have discussed the fundamentals of electronic data processing, the ability of EDP to accomplish required tasks, and its use, requirements, advantages, and disadvantages. The fundamentals of industrial hygiene are the ability to recognize, evaluate, and provide controls, as needed, to prevent health hazards. This is the basis of solving any single industrial health problem, but an overall industrial health program is more complex. A complete industrial health program consists of problem solving, inspections, control procedures, routine exposure monitoring, and other functions. These programs produce the basic goal of industrial health, which is to ensure that employees, the general public, and the environment are not adversely affected by company operations.

The same analogy can be made to the sections of this chapter. Up to this point, specific applications and possible uses for the computer have been given. Other uses for

DATA AUTOMATION

the computer as related to specific expertise, operations, and needs are evident. To couple the best possible utilization of the computer and benefits with the most effective control, a complete industrial health computer program for these applications should be combined as much as possible, or at least coordinated under one management system. This section puts the parts together to indicate the concept of a complete package, termed a computerized environmental health information and control system. First, to develop the necessary computer systems, we discuss the noncomputer programs. Although some of the parts of the concept for the overall system may not be needed for a particular operation or problem, our main goal is to provide background information with the concept, to start the reader toward the design and development of a computerized environmental heatlh information and control system to answer individual requirements.

Noncomputer Program Aspects. Before discussing the computerization aspects of an environmental health information and control system it is necessary to list all the individual applications or programs in the total package that could be computerized. This section treats the programs from the industrial health standpoint. The next section reviews them from the computer processing point of establishment. Both sections reflect the general philosophy that individual computer applications are necessary and advantageous, but it is the interrelationship or tie-in coordination between applications that provides the superior total program and the best operational results.

In viewing any program, total or individual, each of the steps or requirements must be indicated and reviewed from creation to realization. Table 4.4 lists the individual industrial health programs required for an environmental health information and control system aimed at protecting the employees, the general public, and the environment. Once operational it should be able to provide information to the company areas affected by the program as listed in Section 1 as well as complying with government requirements and maintaining adequate recordkeeping.

Some of the other programs named in Table 4.4 were discussed earlier. For example, Section 5.2.5 (Medical Examinations and Surveillance) is related to programs 7 and 15, and Section 5.2.7 (Computer-Based Instruction) is especially related to programs 5 and 8 as well as most of the other programs. In the same manner it may not be necessary or possible to computerize all the programs listed in Table 4.4.

However for the best operational and computerization results it is necessary to review and separate each of the individual programs into three elemental parts. Once these parts of each program have been defined and established, the programs can be blended into the overall computer system because the analyst has a better view of their interrelationships. The first element is the administrative aspects of each program. In this portion of the review we must define what is necessary and assign to management and support functions their responsibilities and instructions for implementing and maintaining each program.

The second element is to provide specific operating procedures for each program. For example, programs 9, 10, and 11 would call for well-defined procedures for approval,

Table 4.4 Noncomputer Industrial Health Programs Required

1. Complete chemical data information.
2. Advance evaluation before approval for use.
3. Control mechanisms to limit exposure.
4. Regulated area (carcinogens, etc.).
5. Protective equipment authorization and control.
6. Emergency plans and programs.
7. Physician certification and selective job placement.
8. Management and employee training programs.
9. Procurement and inventory control, and labeling and storage.
10. Shipping controls.
11. Authorization, distribution, and quantity control.
12. Monitoring and maintenance of control mechanisms.
13. Exposure measurements and determination.
14. Laboratory methods and analysis.
15. Medical surveillance.
16. Employee exposure data bank.
17. Employee notifications.
18. Exception reporting system.
19. Epidemiology studies.
20. Emergency incidents and releases.
21. Waste disposal programs.
22. Compliance audits.
23. Complete recordkeeping.

---

ordering, handling, storage, and use of chemicals, as well as hazardous materials. There would also be an interrelationship between these procedures and other procedures (e.g., programs 2, 3, and 8). In this context ensure that there is not conflict in procedures.

To be an effective program monitoring and auditing must be provided; this is the last element as well as the most important part of each program. To obtain compliance with corporate as well as governmental regulations, a mechanism to ensure the effectiveness of the administration and operating procedures is needed. This could be called the check and balance of the individual as well as overall program. It provides us an indication of where corrections, additions, deletions, and revisions should be made. Thus all the programs given in Table 4.4 are related to programs 22 and 23. Putting together the three elements above is by no means easy. The foregoing elements were utilized to establish the computer program application indicated in Section 5.1.1 for establishing a ventilation testing program. This can also be used as a guide for establishing each of the three elements for an individual program. But once each of the elements has been established and where possible computerized, a very effective program should be in place.

Computerization Concepts and Requirements. Now we turn to concepts and requirements for a computerized environmental health information and control system.

DATA AUTOMATION

Although we discuss the computer aspects of the system in this section, our objectives remain the same as before. Simply restated, these objectives are:

1. To protect the employees from heatlh hazards.
2. To protect the general public from health hazards.
3. To protect the corporation and management from undue citations and litigation.

Likewise the programs remain the same as indicated in Table 4.4, but for computerization the purpose changes. Now the purpose becomes the utilization of the programs in Table 4.4 to develop an environmental health program with management system to be responsive to:

- The individual operational needs.
- Management and government requirements.
- Cost effectiveness, where possible.

To accomplish this purpose the 23 programs of Table 4.4 would be reduced seven program or systems (Table 4.5). The information that was developed for the 23 programs would be utilized to establish the seven required systems. All the aspects of each system must be developed utilizing the 23 "noncomputer" programs established to put them in a "computerized form." Each task for every system must be determined, reviewed, and established before computerization can begin. As was evident with the 23 programs, interrelationships exist and coordiantion is necessary among the seven systems. For example, all the first five systems are related to systems 6 and 7.

The task involved in system 1 for chemicals is to relate the person to the chemicals to the job and department. If the personnel computer system has an employee numbering

Table 4.5 Computer Programs

1. System to identify chemical, physical, and biological hazards; their location, use, and personnel exposed.
2. System that establishes baseline data measurements and monitors to assure that they are evaluated periodically.
3. System that assures that hazard information is generated and kept current for management, support operations, and employees.
4. System that assures that necessary controls are monitored and alerted to process changes.
5. System that assures that necessary and appropriate medical surveillance and training are conducted.
6. System that assures that review of results of items 1–5 and where indicated proper and expeditious action will take place.
7. System of recordkeeping in compliance with federal regulations that is functional and allows ready retrieval of data.

and job coding system, part of the work is already accomplished. If there are no computer descriptive terms for operation and code as shown in Table 4.6, these must be developed. The next task is putting the data into the computer. A form similar to Figure 4.26, which is machine readable, must be developed for line management use. Special cases may also occur—for example, an OSHA-regulated area, necessitating indication by the computer system of persons authorized to enter the area. The computer system, by use of a sensor badge reader, could permit access to the area and at the same time maintain the daily company roster as required by the regulation (Figure 4.19). Thus in this special case system 1 is also related to system 4.

In the same manner as in the Table 4.4 programs, the noncomputer tasks of how, when, and where to obtain exposure measurements are developed which are related to system 2 in Table 4.5. Now some of the computerization task for system 2 are:

1. To relate persons to their exposure data.
2. To calculate time-weighted averages for multiple exposure conditions.

Table 4.6 Descriptive Terms for Operations

Code		Code	
101	Abrading	124	Handling
102	Assembling	125	Heating
103	Blasting	126	Impregnating
104	Burning	127	Irradiating
105	Casting	128	Laminating
106	Cementing	129	Machining
107	Cleaning	130	Melting
108	Coating	131	Milling
109	Condensing	132	Mixing
110	Crystal growing	133	Molding
111	Crystal slicing	134	Packaging
112	Cutting	135	Paint removing
113	Deburring	136	Painting
114	Degreasing	137	Photolithographing
115	Developing	138	Plating
116	Diffusing	139	Polishing
117	Dipping	140	Pouring
118	Doping	141	Soldering
119	Encapsulating	142	Spraying
120	Etching	143	Stripping
121	Gassing	144	Trucking
122	Glass blowing	145	Welding
123	Grinding	146	Wiping

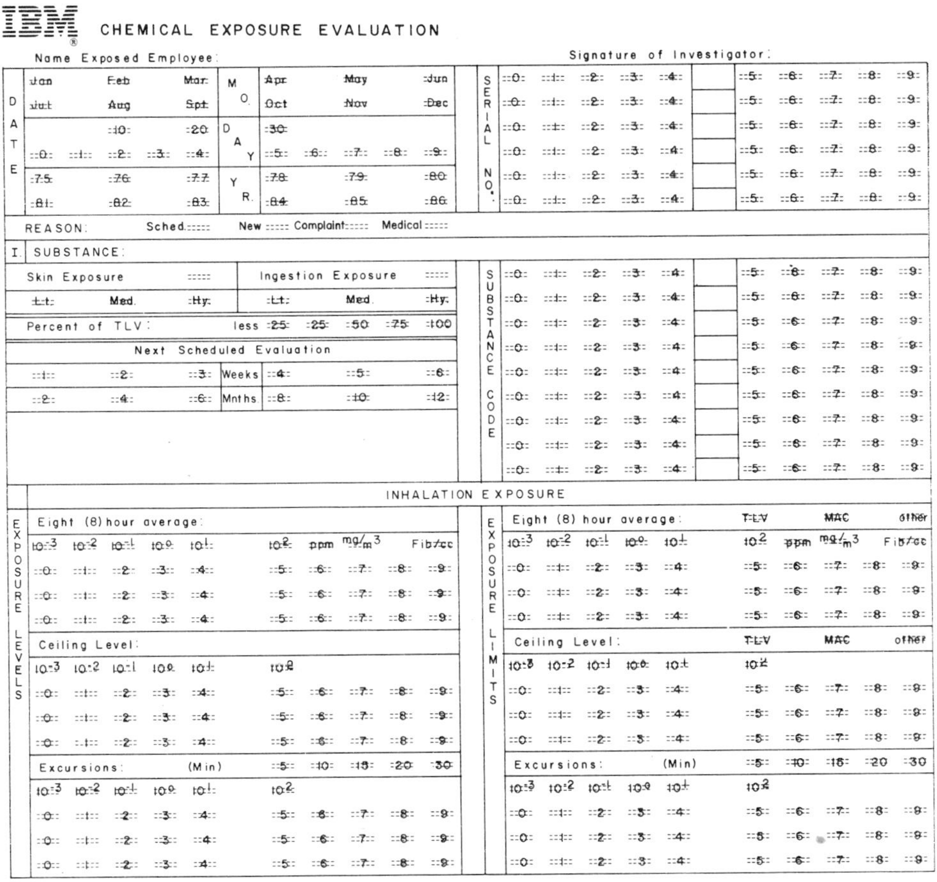

DATA AUTOMATION 163

Figure 4.26. Computer input form for employee exposure recordkeeping.

3. To record all exposures, calculate the mean, and indicate the maximum for the year.
4. To print the entire work exposure history for a person.
5. To indicate persons exceeding legal requirements and whether they were notified of exposure results within 5 working days.

The last two tasks are of course related also to systems 6 and 7. See also Appendix 4A for additional tasks. Likewise the data obtained from system 1 and system 2 must be meshed with the noncomputer information obtained from system 5. Thus flow charting similar to that in Figure 4.27 must be developed for particular needs. Additional tasks include the development of medical forms similar to those in Figures 4.28 and 4.29.

For establishing the computer programs for system 3, the information in Section

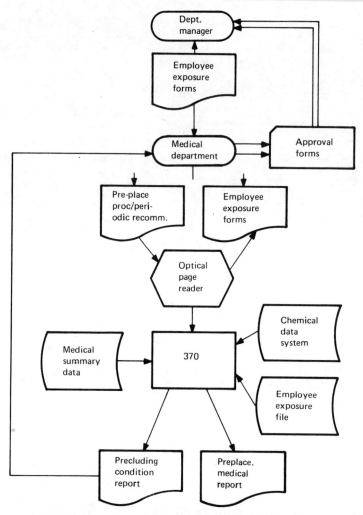

Figure 4.27. Flow chart of medical surveillance program.

5.2.7 on CBT will be helpful regarding training of management and employees. As well as training, a computer program to supply health and safety information to other areas of the company (chemical control, shipping, purchasing, etc.) needs to be established. For example, suppose a physician wants to know what chemicals being used at a given location may be carcinogenic. Input of the data as in Figure 4.30 into the computer terminal is all that is necessary to obtain the information immediately. Once the data have been inputted into the computer, a search of the files will produce the answer to the question (Figure 4.31).

DATA AUTOMATION

```
              Name              Internal Zip    City & State

    To:
                                                                  IBM
```

```
         Date:
Name & Tie/Ext.:
Title/Dept. Name:   Medical Department
    Zip/City, State: 400/9M3/044
 U.S. mail address: FSD Manassas

      Subject:    Periodic Physical

    Reference:   Name:                        Serial:
```

Your employee has been scheduled for a _____ physical.

In order to meet the requirements for this examination, he/she must have a preliminary work-up by the nurse and a review and examination by Dr. Nichols.

Please notify your employee that he/she is scheduled for their preliminary work-up on _____ at _____.

The appointment for the review and examination by Dr. Nichols is on _____ at _____.

Since it is a job requirement, it is your responsibility to see that your employee keeps this appointment.

Special Instructions: ( )   None

                                           ( )   12 hour fast - nothing to eat or drink except sips of water.

Figure 4.28. Medical appointment card.

As previously mentioned, all the programs above are related to system 6. For example, in system 6 some of the related tasks are:

1. Indicate control program documentation.
2. Indicate situations (emergencies or incidents) requiring reports to OSHA.
3. Identify exposure situations requiring special action.
4. Indicate delinquency reports of work not completed for any of the systems.
5. Conduct epidemiology studies.

In the last task, the method of programming the data is extremely important. Often the data are programmed to be retrievable only in a certain format. For epidemiology studies the hierarchy of programmed data should be accomplished to permit retrieval of any segment (chemical, job, bioassay, exposures greater than certain level, etc.).

```
┌─────────────────────────────────────────────────────────────────┐
│ ╱IBM                        MEDICAL DEPARTMENT EXAMINATION REPORT│
│                                                                  │
│   ─────────────  ──────────────  ──────────────  ──────────────  │
│      INITIALS        NAME             DEPT.          DATE        │
│                                                                  │
│   YOU RECENTLY PARTICIPATED IN A MEDICAL DEPARTMENT EXAMINATION  │
│   CONCERNING ─────────────────────────                           │
│                                                                  │
│   RESULTS INDICATE THAT:                                         │
│                                                                  │
│       1. ( ) THE REPORT REVEALED NO SIGNIFICANT ABNORMALITY.     │
│                                                                  │
│       2. ( ) PLEASE CONTACT ─────────────────────────────────    │
│              IN THE MEDICAL DEPARTMENT.                          │
│                                                                  │
│   M04-0011-0                                         IBM C67743  │
└─────────────────────────────────────────────────────────────────┘
```

Figure 4.29. Medical department examination report.

When the tasks for the seven systems are completed, one must decide whether to have a completely separate or combined computer system. Traditionally data files are designed to serve individual applications, such as inventory control, payroll, engineering drawing release, and manufacturing planning. Each data file was specifically designed with its own storage space in the computer, on tape, or in direct access devices. Most companies already have personnel and manufacturing data bases, as Figure 4.32 indicates.

It is evident from Figure 4.32 that some of the information regarding health and safety is already in the personnel and manufacturing data bases. Thus if an environ-

"SUPPOSE A PHYSICIAN WANTS TO KNOW WHAT CHEMICALS MAY BE CANCER CAUSING."

INPUT		EXPLANATION
SELECT		
CODES:	9.	
SELECTION, VALUES, OR FUNCTION ? S		
PROPERTIES:	DESC CANCER.	DISPLAY CHEMICAL NAME.
EXPRESSION:	CANCER HAS 'YES'	CANCER IS PROPERTY NAME CONTAINING 'YES/NO' CARCINOGEN -- INDICATOR.

Figure 4.30. Input to computer necessary to obtain information.

```
CODES:     9.
SELECTION, VALUES, OR FUNCTION? S
PROPERTIES:    DESC CANCER.
EXPRESSION:
        CANCER HAS 'YES'

IBM CODE,      : 924336332
DESCRIPTION,   : ARSENIC TRIOXIDE
               : ARSENOLITE, WHITE ARSENIC, CRUDE ARSENIC
               : CLAUDETITE, ARSENIOUS ACID, ARSENOUS ANHYDRIDE
CARCINOGEN,    : YES

IBM CODE,      : 943010610
DESCRIPTION,   : BETA NAPHTHALAMINE
CARCINOGEN,    : YES

IBM CODE,      : 951835296
DESCRIPTION,   : DICHLORO BENZIDINE
CARCINOGEN,    : YES

IBM CODE,      : 970160103
DESCRIPTION,   : BETA PROPRIOLACTONE
CARCINOGEN,    : YES

EXPRESSION:
```

Figure 4.31. Output from computer answering question.

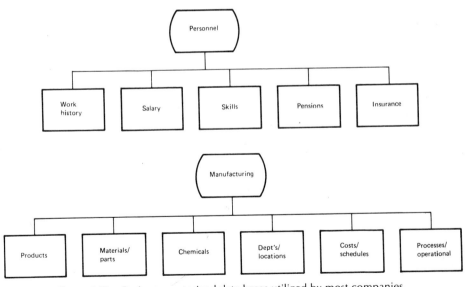

Figure 4.32. Basic computerized data bases utilized by most companies.

mental computer system is established the same data will reside in different application files. What is needed is a data base that provides for the integration or sharing of common data. This concept is depicted in Figure 4.33. The undesirable attributes of the separate data files have been eliminated by the sharing or overlap concept. This type of data base offers flexibility of data organization. Independence is achieved by removing the direct association between the application program and the physical storage of data. This facilitates the addition of data to the existing data base without modification of existing application programs.

The advantages of this common data base concept are:

1. Elimination of redundant data and implied redundant maintenance.
2. Consistency through the use of the same data by all parts of the company.
3. Application program independence from physical storage and sequence of data.
4. Reduction in application costs, storage costs, and processing costs.

If we use the concept of the common data base for the personnel, manufacturing, and environmental areas, a computerized environmental health information and control system results (Figure 4.34). This concept also had the effect of combining the 23 programs and seven systems as given in Table 4.4 and 4.5.

The other alternative is to have a completely separate system, such as the computer network developed by (210). The entire system is operated in the health department domain and provides all members of the team with access to vital data. Figure 4.35 diagrams Amoco's basic system health data input into the computer. The computer system designs for the four modules appear in Figure 4.36.

The literature contains few descriptions of computerized environmental health information and control systems. Westinghouse (211), DoW (212), and IBM (213) have published information concerning their computer systems. Other companies, such as DuPont, Western Electric, and CIBA-Geigy are in the process of developing com-

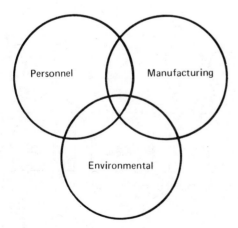

Figure 4.33. Integration of company data base systems.

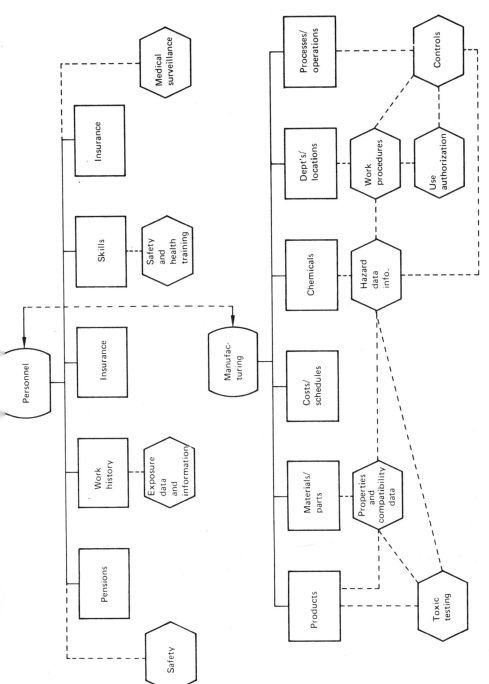

Figure 4.34. Computerized environmental health information control system.

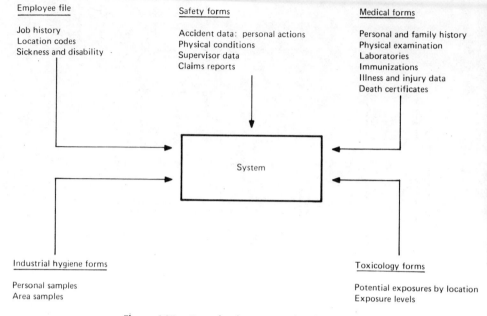

Figure 4.35. Completely separate data base system.

puterized recordkeeping and information systems. As we stated in Section 1 and throughout the chapter, there is a definite need for data automation industrial health management systems. Computerization does have its advantages as related to "manual" recordkeeping and control systems.

APPENDIX 4A

4A.1 Typical System Requirements: Industrial Hygiene

Typical systems requirements are as follows:

1. Relate person to chemical and exposure and job/or department.
2. Calcualte time-weighted averages for multiple exposure.
3. Record all exposures, calculate mean, and indicate maximum for year.
4. Print entire work history while working for this firm as related to items 1, 2, and 3.
5. Relate person to any bioassay performed.
6. Repeat 3 and 4 for bioassay.
7. Indicate quarterly bioassay or exposure evaluations not conducted.
8. Indicate routes of exposure (e.g., skin, lungs, etc.).

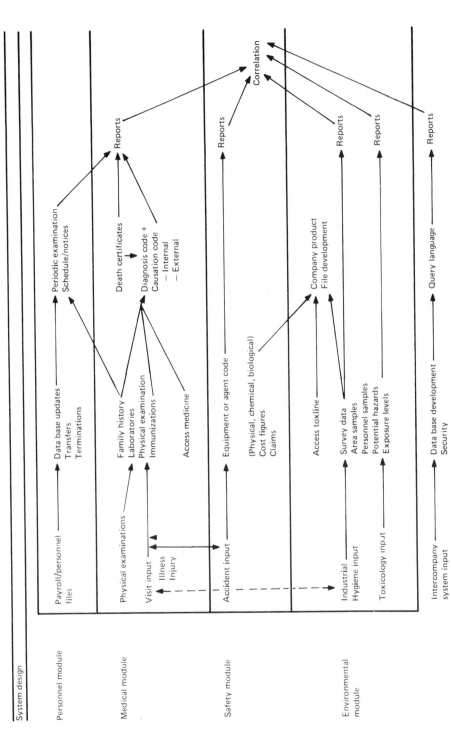

Figure 4.36. Modules designed for a separate data base system.

9. Hierarchy of programmed data should permit retrieval of any segment of the above for epidemiology studies (e.g., a certain chemical, job exposures greater than certain level, action level, etc.).
10. Indicate persons exceeding legal requirements and whether they were notified of exposure results within 5 working days.
11. Regulated areas:
    a. Indicate persons authorized to enter.
    b. Indicate daily roster—tie badge reader into system.
12. Indicate for each person the physicians's work certification for chemicals and physical exposures (laser, etc.).
13. Indicate person received training annually.
14. Indicate quarterly delinquency regarding training.
15. System should be amenable to including physical and biological exposures immediately or at future date as indicted in items above.
16. Compliance program
    a. Indicate control program documentation (engineering controls investigated) for exposures exceeding limits—OSHA requirement for documentation of wearing personal protection.
    b. Indicate situations (emergencies or incidents) requiring reports to OSHA.
    c. Identify exposure situations requiring special action or correction.
    d. Indicate quarterly delinquency regarding above.
17. Emergency program
    a. Indicate that above has been accomplished for areas requiring OSHA and company requirements.
    b. Indicate delinquency quarterly.
18. Determine persons wearing respirators.
19. Indicate if medical surveillance performed.
20. Indicate if training conducted.
21. Indicate quarterly delinquencies regarding items 19 and 20.
22. Record and track ventilation measurements quarterly.
23. Indicate that measurements meet control values and/or the corrections being made.
24. Measurment made within 5 days of changes to meet new control values and OSHA ventilation requirements.
25. Indicate delinquencies to ventilation measurements quarterly (items 22–24).

DATA AUTOMATION   **173**

APPENDIX 4B

4B.1   Ventilation Testing Program

4B1.1   Responsibilities

The responsibilities are as follows:

1. Environmental safety and health (D/713–D/708).
   a. Determine the need for installation of local exhaust ventilation.
   b. Determine required velocity readings for exhaust systems in accordance with OSHA and/or recommended practice (joint effort between D/705 and D/713).
   c. Review installation of new exhausts, alteration and/or relocation of existing exhausts (normally included in project folder sign-off).
   d. Assign testing frequency schedule to all systems based on regulations and/or recommended practice.
   e. Develop and maintain a computerized ventilation program for the periodic maintenance of exhaust systems.
   f. Assign identification numbers to new systems and enter on computer.
   g. Apply for and obtain permits.
2. Facility engineering.
   a. Design, write specifications for, and classify all new or altered exhaust systems for inclusion in project folders.
   b. Balance exhaust systems in accordance with OSHA and/or recommended practice (joint effort between D/705 and D/713).
   c. Supervise comprehensive initial testing for each system. Refer to recommended sequence of testing for initial test.
   d. Create field data form and file in central location accessible to both Facilities and Environmental Safety and Health.
   e. Notify Environmental Safety and Health of changes initiated by Facility Engineering that would not appear on a project.
3. Facility maintenance.
   a. Perform initial and periodic testing on all systems. Refer to measurement sequence flow chart.
   b. Determine the need for maintenance and make repairs.
   c. Notify Environmental Safety and Health of failures or under specified conditions in existing exhaust systems.
4. Using department line management.
   a. Contact Environmental Safety and Health if a need for local exhaust appears.
   b. Advise Maintenance Department 706 of any deterioration of exhaust systems in their area or systems that have not been tested when due.

    c. Advise Environmental Safety and Health of any process changes where materials may be substituted, temperatures changed, or any adjustment to exhaust requirements might be needed.

4B.1.2 Measurements

All exhaust systems should be thoroughly tested upon installation and periodically tested throughout the life of the system. A valid comparison between the design basis and optimum system performance can be made only at the time of the initial survey of the system. Experience has shown that periodic surveys are required to assure that the system is performing adequately.

Basis for Evaluation of Ventilation Systems.

1. To assure adequacy and performance.
2. To assure that system performance is maintained.
3. To determine the feasibility for expanding the system.
4. To establish improved design parameters for new systems.
5. To assure compliance with federal, state, and/or local regulations.

Initial Tests.

1. Review system specifications and drawings.
2. Inspect systems (fan rotation, belts, dampers, etc.).
3. Select and identify test locations; make drawing on back of data sheet.
4. Measure air volume, fan static pressure, motor rpm, and pressure drops across all components.
5. Record test data and design specifications on test data sheet.
6. Compare test data with design specifications, make alterations to meet specifications, codes, and so on.
7. Retest system after adjustments have been made, record final test data; note on sketch the changes made.
8. Retain test data for life of system.
9. Use data sheet as base for inputting system on computer and updating master map.

Periodic Test for all Systems.

1. Computer will issue maintenance cards based on frequency required. Refer to master map for location.
2. Visual inspection.
   a. Physical damage (corroded ducts, etc.).
   b. Proper operation of components.
   c. Smoke test.

DATA AUTOMATION

3. Take measurements (method specified on data card).
    a. Same locations as initial tests or
    b. Static pressure measurement (only when system is operating adequately).*
4. If system is performing as required (per data card):
    a. Fill in blanks on card.
    b. Comment if process has changed at all.
    c. Send card to Keypunch.
5. If system is not performing adequately:
    a. See if corrections can be made quickly (putting on new fan belt, removing hood blockage, etc.).
    b. If corrected, reread, fill out card, and send to Keypunch.
    c. If engineering help is reqired to find and arrange correction of problem, put actual readings on the card along with any comments and return to Keypunch.
    d. If hood performance significantly changed (≥ 50 percent), place "do not operate" tag on hood face and alert using department manager and Environmental Safety and Health.
6. Facility Engineering will receive discrepancy cards indicating systems that are not performing adequately.
    a. System should be corrected as soon as possible.
    b. Once corrected, remeasurement is essential.
    c. The card is then filled in and sent to Department 713.

Computer Functions.

1. Information available from computer:
    a. Vent/hood.
    b. Stack.
    c. Fan.
    d. Operation.
    e. Materials and conditions (temperature, etc.).
    f. OSHA class (open surface tank).
    g. Principal hazard.
    h. Toxicity level (TLV or measured).
    i. Dimensions face/slot, tank, duct.
    j. Frequency of measurements required.
    k. Data of next measurement.
    l. Required readings: average velocity.
    m. Latest readings: minimum velocity.
    n. Previous readings: volume, static pressure, type (SP, DV, FV, SV, CV).
2. Measurement designations (see Appendix 4C).
    a. Capture or control velocity (CV).
    b. Slot velocity (SV).

*Whenever alterations are made to a system, new initial testing is necessary as outlined in initial tests.

c. Duct velocity (DV).
  d. Face velocity (FV).
  e. Static pressure (SP).
  f. Velocity pressure (VP).
3. Maintenance cards sent to Maintenance every week identifying systems that need measuring in 2 weeks.
4. Weekly list of systems measured and discrepancies found sent to Department 713.
5. Discrepancies sent to Facility Engineering with master list weekly.
6. Monthly update listing sent to Department 713.

APPENDIX 4C

4C.1 Data Processing Glossary

The following definitions will assist the person interested in data processing in obtaining insight into the specialized language of data processing.

ADDRESS: An identification for a register, location in storage, or other data source or destination; the identification may be a name, a label, or a number.

ALGOL: Algorithmic language. A data processing language utilizing algebraic symbols to express problem-solving formulas for machine solution.

ALGORITHM: A prescribed set of well-defined rules or processes for the solution of a problem in a finite number of steps.

ANALOGUE DATA: Data represented in a continuous form, as contrasted with digital data in which are represented in a discrete (discontinuous) form. Analogue data are usually represented by means of physical variables, such as voltage, resistance, and rotation.

APL: A programming language. A problem-solving language designed for use at remote terminals; it offers special capabilities for handling arrays and for performing mathematical functions.

APPLICATION PROGRAM: A program written for or by a user that applies to the individual's own work.

AUXILIARY (PERIPHERAL) EQUIPMENT: Equipment not actively involved during the processing of data, such as input/output equipment and auxiliary storage utilizing punched cards, magnetic tapes, disks, or drums.

BASIC: An algebralike language used for problem solving by engineers, scientists, and others who may not be professional programmers.

BATCH PROCESSING: A system approach to processing: similar input items are grouped for processing during the same machine run.

DATA AUTOMATION

BINARY: (1) The number representation system with a base 2. (2) A characteristic or property involving a selection, choice, or condition in which there are only two possibilities.

BINARY CODED DECIMAL (BDC): A type of notation in which each decimal digit is identified by a group of binary ones and zeros.

BIT: A binary digit.

BYTE: A continguous set of binary digits operated on as a unit.

CENTRAL PROCESSING UNIT (CPU): The unit of a computing system that contains the circuits that calculate and perform logic decisions based on a man-made program of operating instructions.

CHARACTER: One of a set of elementary symbols acceptable to a data processing system for reading, writing, or storing.

CHARACTER RECOGNITION: The technique of reading, identifying, and encoding a printed character by optical means.

CLEAR: To put a storage or memory device into a state denoting zero or blank.

COBOL: Common business-oriented language. A data processing language that resembles business English.

COMPILE: To convert a source-language program such as COBOL to a machine-language program.

COMPUTER PROGRAM: A series of instructions or statements, in a form acceptable to a computer, prepared to achieve a certain result.

CONSOLE: The unit of equipment used for communication between the operator or service engineer and the computer.

CONTROL PROGRAM: A program that is designed to schedule and supervise the performance of data processing work by a computing system.

CORE STORAGE: A form of magnetic storage that permits high speed access to information within the computer. See *Magnetic core.*

CPU: Central processing unit.

CRT: A display device on which images are produced: cathode ray tube.

DATA PROCESSING SYSTEM: A network of machine components capable of accepting information, processing it according to man-made instructions, and producing the computed results.

DEBUG: To detect, locate, and remove errors from a programming routine or malfunctions from a computer.

DIGITAL DATA: Information expressed in discrete symbols.

DIRECT ACCESS: See *Random access.*

DISK STORAGE: A method of storing information in code, magnetically, in quickly accessible segments on flat rotating disks.

DOWNTIME: The elasped time when a computer is not operating correctly because of machine or program malfunction.

DRUM STORAGE: A method of storing information in code, magnetically, on the surface of a rotating cylinder.

EDP: Electronic data processing.

EXECUTE: To perform a data processing routine or program, based on machine-language instructions.

FILE: A collection of related records—for example, in inventory control, one line of an invoice forms an item, a complete invoice forms a record, and the complete set of such records forms a file.

FILE MAINTENANCE: The processing of information in a file to keep it up to date.

FLOW CHART: A graphic representation for the definition, analysis, or solution of problem in which operations, data flow, and equipment are depicted symbolically.

FORTRAN: Formula translating system. A data processing language that closely resembles algebraic notation.

HARD COPY: A printed copy of machine output (printed reports, listings, documents, etc.).

HARDWARE: The mechanical, magnetic, electrical, and electronic devices of a computer.

INPUT/OUTPUT: Commonly called I/O. A general term for the equipment used to communicate with a computer; the data involved in such communication.

INQUIRY: A request for information from storage (e.g., a request for the number of available airline seats on a given flight).

INSTRUCTION: A statement that calls for a specific computer operation.

INTEGRITY: Preservation of data or programs for their intended purpose.

LANGUAGE: A defined set of symbols that can be converted to machine language (ALGOL, FORTRAN, COBOL, etc.).

LIBRARY ROUTINE: A special-purpose program that can be maintained in storage for use when needed.

MACHINE LANGUAGE: A code used directly to operate a computer.

MACHINE READABLE: A medium that can convey data to a given sensing device.

MAGNETIC CORE (MAIN CORE): A configuration of tiny doughnut-shaped magnetic elements in which information can be stored for use at extremely high speed by the CPU.

MAGNETIC INK CHARACTER RECOGNITION (MICR): A method of storing information in characters printed with ink containing particles of magnetic material. The information can be detected or read at high speed by automatic devices.

DATA AUTOMATION

MAGNETIC TAPE: A plastic tape with a magnetic surface on which data can be stored in a code of magnetized spots.

MARK-SENSE: To mark a position on a card or paper form with a pencil. The marks are interpreted electrically for machine processing.

MATHEMATICAL MODEL: A set of mathematical expressions that describes symbolically the operation of a process, device, or concept.

OBJECT PROGRAM: The machine-language program that is converted to electrical pulses that actually guide operation of a computer.

ON-LINE SYSTEM: In teleprocessing, a system in which the imput data enter the computer directly from point of origin or in which output data are transmitted directly to where used.

OPERATING SYSTEM: An integrated collection of computer instructions that handle selection, movement, and processing of programs and data needed to solve problems.

OPTICAL READER: A device used for machine recognition of characters by identification of their shapes.

OUTPUT: (1) The final results after data have been processed in a computer. (2) The device or set of devices used for taking data out of a computer system and presenting them to the user in the form desired.

PAPER-TAPE READER: A device that senses and translates the holes in a roll of perforated paper tape into machine-processable form.

PL/1: A high-level programming language, designed for use in a wide range of commercial and scientific computer applications.

PRINTER: A device that prints results from a computer on paper.

PROGRAM: (1) (n) The plan and operating instructions needed to produce results from a computer. (2) (v) To plan the method of attack for a defined problem.

PUNCHED CARD: A card punched with a pattern of holes to represent data.

RANDOM ACCESS: A technique for storing and retrieving data: it does not require a strict sequential storage of the data nor a sequential search of an entire file to find a specific record. A record can be addressed and accessed directly at its location in the file.

RAW DATA: Data that have not been processed or reduced.

REAL TIME: Computing that occurs while a process takes place so that results can be used to guide operation of the process.

ROUTINE: A sequence of machine instructions that carry out a specific processing function.

SIMULATE: To represent the functioning of a system or process by a symbolic symbolic (usually mathematical) analogous representation of it.

SOFTWARE: (1) The collection of man-written solutions and specific instructions needed to solve problems with a computer. (2) All documents needed to guide the operation of a computer (manuals, programs, flow charts, etc.).

SOLID STATE COMPONENT: A component whose operation depends on electrical activity in solids (e.g., performance of a transistor or cystal diode).

SORT: To arrange data in an ordered sequence.

SOURCE LANGUAGE: A language nearest to the user's usual business or professional language, which enables the user to instruct a computer more easily. FORTRAN, COBOL, ALGOL, BASIC, PL/I are a few examples.

STORAGE: Pertaining to a device in which data can be entered and stored and from which they can be retrieved at a later time.

STORED-PROGRAM COMPUTER: A digital computer that stores instructions in main core and can be programmed to alter its own instructions as though they were and can subsequently execute these altered instructions.

TELEPROCESSING: The use of telecommunications equipment to transmit data between two computers in different locations, or between input/output devices and a centralized computer when I/O is at a location remote from the computer.

TERMINAL UNIT: A device, such as a key-driven or visual display terminal, which can be connected to a computer over a communications circuit and can be used for either input or output from a location either near or far removed from the computer.

REFERENCES

1. J. H. MacEachran, W. E. Clendenning, and R. E. Gosselin, "Computer-Derived Exposure Tests for Common Contact Dermatitis Antigens," *Contact Dermatitis*, **2**, 239 (1976).
2. R. O. Moss and H. J. Ettinger, "Respirable Dust Characteristics of Polydisperse Aerosols," *Arch. Env. Health* **31**, 546 (1970).
3. J. R. Goldsmith, J. Terzaghi, and J. D. Hackney, "Evaluation of Fluctuating Carbon Monoxide Exposures," *Arch. Environ. Health*, **7**, 33–49, 647–663 (1963).
4. J. E. Peterson, H. R. Hoyle, and E. J. Schneider, "The Application of Computer Science to Industrial Hygiene," *Am. Ind. Hyg. Assoc. J.*, **27**, 180 (1966).
5. "Computers Turn on for Industrial Hygiene," *Occup. Haz.* **30**, 37 (1968).
6. R. G. Edwards, Jr., C. H. Powell, and M. A. Kendrick, "Dust Counting Variability," *Am. Ind. Hyg. Assoc. J.*, **27**, 546 (1966).
7. D. C. Stevens, W. L. Churchill, D. Fox, and N. R. Large, "A Data Processing System for Radioactivity Measurements on Air Samples," *Ann. Occup. Hyg.*, **13**, 177 (1970).
8. F. J. Haughey and R. M. Maganelli, "An Experimental System for Aerosol Research," *Am. Ind. Hyg. Assoc. J.*, **29**, 268 (1968).
9. M. E. Jacobson, E. W. White, P. B. Denee, K. M. Morse, et al., "Dust Measurement and Control, Coal Workers' Pneumoconiosis," *Ann. N. Y. Acad. Sci.*, **200**, 661 (1972).

10. E. J. Jones, "Practical Aspects of Counting Asbestos on the Millipore, PIMC," *Microscope*, **23**, 93 (1975).
11. J. F. Remark, "Computerized Three-Dimensional Illustrations of Gas Equations," *J. Chem. Educ.*, **52**, 61 (1975).
12. J. E. Evans, and J. T. Arnold, "Monitoring Organic Vapors," *Environ. Sci. Technol.*, **9**, 1134 (1975).
13. D. P. Schlik and R. G. Peluso, "Respirable Dust Sampling Requirements under the Federal Coalmine Health and Safety Act of 1969," U. S. Bureau of Mines, Information Circular 8484, Publication Distribution Branch, Pittsburgh, Information Circular July 1970.
14. R. Barry, "Computer Control," *Safety*, **40**, 13 (1968).
15. M. Robert, "The International Occupational Safety and Health Center," *Ann. Occup. Hyg.*, **16**, 267 (1973).
16. A. S. Reid, "Classification and Sources in Environmental Safety and Health," *Ann. Occup. Hyg.*, **16**, 257 (1973).
17. E. B. Duncan, B. McGovern, and S. A. Hall, "BOHS Symposium on Cooperation in Information Handling and Retrieval in Environmental Hygiene," *Ann. Occup. Hyg.*, **16**, 305 (1973).
18. B. Herman, "Computer Analysis in the Evaluation of Occupational Medical Records," *Ann. Occup. Hyg.* **16** (1973).
19. R. F. Cumberland and M. D. Hebden, "A Scheme for Recognizing Chemicals and Their Hazards in an Emergency," *J. Hazardous Mat.*, **1**, 35 (1975).
20. W. C. McArthur and B. G. Kneazewycz, "A Nuclear Power Plant Radiation Monitoring System," *Health Phys.*, **29**, 427 (1975).
21. E. N. Corlett, V. J. Morcombe, and B. Chanda, "Shielding Factory Noise by Work-in-Progress Storage," *Appl. Ergonomics*, **1**, 73 (1970).
22. R. C. Miller and A. Sagar, "Analysis of Machining Noise and Its Potential Application to Adaptive Control," *Mach. Prod. Eng.*, **121**, 840 (1972).
23. F. G. Hagg, "Using Computer Programs as Noise Control Tools," *Sound Vib*, **5**, 22 (1971).
24. J. S. Bendat, "Modern Methods for Random Data Analysis," *Sound Vib*, **3**, 14 (1969).
25. D. W. Merritt and R. R. James, "Isograms Show Sound-Level Distribution in Industrial Noise Studies," *Sound Vib.*, **7**, 12 (1973).
26. K. Drechsler, "Studies on Vibration Transmission to the Driver's Cab of Automotive Agricultural Machinery," *Agratechnik*, **24**, 553 (1974).
27. J. Donovan, "Community Noise Criteria," *Am. Ind. Hyg. Assoc. J.*, **36**, 849 (1975).
28. J. F. Bell, A. C. Johnson, and J. C. K. Sharp, "Pulse-Echo Method of Investigating the Properties of Mechanial Resonators," *J. Acoust. Soc. Am.*, **57**, 1085 (1975).
29. B. Stanton, "The Computer in Development Testing and Evaluation," *ASHRAE J.*, October 1973, p. 35.
30. G. W. Dunn, "Central Plant Design for Utilization of Computer Controls," *ASHRAE J.* July 1973, p. 55.
31. E. S. Rubin, "A New Application of Building Programs," *ASHRAE J.*, February 1973, p. 46.
32. L. W. Nelson, "Reducing Fuel Consumption with Night Setback," *ASHRAE J.*, August 1973, p. 41.
33. R. S. Bycraft, "Ventilation Energy System Analysis by Computer," *ASHRAE J.*, June 1973, p. 46.
34. K. N. Feinberg, "Use of Computer Programs to Evaluate Energy Consumption of Large Office Buildings," *ASHRAE J.*, January 1974, p. 73.
35. H. Rentsch, "The Use of Computers for Calculating the Soundproofing Requirements of Ventilation Systems," *Luft-Und Kalletechnik*, **7**, 11 (1971).

36. B. I. Medvedev and V. A. Pavlovskii, "Computer Simulation of Emergency Ventilation Conditions in Miners with Intense Heat Transfer," *Izv. Vyssh. Uchebn. Zavode. Gorn. Zh.*, **15**, 81 (1972).
37. K. E. Hicks, "Computer Calculation and Analysis of the P4SR Heat Stress Index," *Environ. Res.*, **4**, 253 (1971).
38. P. E. Smith, and E. W. James II, "Human Responses to Heat Stress," *Occup. Med.*, **9**, 332 (1964).
39. R. L. Harris, Jr., "Computer Simulation and Radiant Heat Load and Control Alternatives," *Am. Ind. Hyg. Assoc. J.*, **35**, 75 (1974).
40. H. M. Berlin, L. Stroschein, and R. F. Goldman, "A Computer Program to Predict Energy Cost, Rectal Temperature, and Heart Rate Response to Work, Clothing, and Environment," Edgewood Arsenal Special Publication ED-SP-74011, Aberdeen Proving Ground, Aberdeen, Md.
41. "A Study of Hypertension During Work and Computerized Analysis of Data Obtained," *Cah. Med. Interprof.*, **43**, 25 (1971).
42. W. S. Neisel and D. C. Collins, "Structural Languages and Biomedical Signal Analysis Using Interactive Graphics," Air Force Office of Scientific Research, Report AFOSR-TR-72-0616, Arlington, Va., March 1972.
43. C. D. Jenkins, R. H. Rosenman, and S. J. Zyzanski, "Prediction of Clinical Coronary Heart Disease by a Test for Coronary-Prone Behavior Pattern," *New Eng. J. Med.*, **290**, 1271 (1974).
44. R. E. Birk et al., "Approach to a Reliable Program for Computer-Aided Medical Diagnosis," *Aerosp. Med.*, **45**, 659 (1974).
45. M. Helberman et al., "Complex Man-Machine System for Delivery of Outpatient Medical Care," *Aerosp. Med.*, **45**, 975 (1974).
46. G. N. Bycroft and L. Seaman, "Mathematical Models of Head Injuries," Stanford Research Institute, SRI Project 1633, Menlo Park, Calif., July 1973.
47. D. P. Discher, F. J. Massey, and W. Y. Halleti, "Quality Evaluation and Control Methods in Computer-Assisted Screening," *Arch. Environ. Health*, **19**, 323 (1969).
48. H. R. Newman and M. L. Schulman, "Renal Cortical Tumors: A 40-Year Statistical Study," *Urol. Surv.*, **15**, 2 (1969).
49. E. Edwards and F. P. Lees, "The Influence of the Process Characteristics on the Role of the Human Operator in Process Control," *App. Ergonomics*, **5**, 21 (1974).
50. C. H. Suh, "The Fundamentals of Computer Aided X-Ray Analysis of the Spine," *J. of Biomech.*, **7**, 161 (1974).
51. J. A. Crocco et al., "A Computer-Assisted Instruction Course in the Diagnosis and Treatment of Respiratory Diseases," *Am. Rev. Respir. Dis.*, **111**, 299 (1975).
52. W. R. Ayers et al., "Description of a Computer Program for Analysis of the Forced Expiratory Soirogram. II. Validation." *Comput. Biochem. Res.*, **2**, 220 (1969).
53. M. V. Merrick, "The Role of Radioisotopes in the Diagnosis of Bronchial Cancer, *Scand. J. Respir. Dis.*, **85**, 106 (1974).
54. A. L. Rector and E. Ackerman, "Rules for Sequential Diagnosis," *Comput. Biomed. Res.*, **8**, 143 (1975).
55. N. Konietzko, W. E. Adam, and H. Matthys, "Use of Radioisotopes in Modern Diagnosis of Pulmonary Function Impairment," *Munch. Med. Wochensch.*, **116**, 159 (1974).
56. F. Wiener, "Computer Simulation of the Diagnostic Process in Medicine," *Comput. Biomed. Res.*, **8**, 129 (1975).
57. B. N. Feinberg and J. D. Schoeffler, "Computer Optimization Methods Applied to Medical Diagnosis," *Comput. Biol. Med.*, **5**, 3 (1975).

58. R. G. Mancellas and A. Ward, "Machine Recognition in Pathology," *Comput. Biol. Med.*, **5,** 39 (1975).
59. A. W. Sills, V. Honrubia, and W. E. Kumley, "Algorithm for the Multi-Parameter Analysis of Nystagmus Using a Digital Computer," *Aviat., Space Environ. Med.*, **46,** 934 (1975).
60. E. Anzaldi and E. Mira, "An Interactive Program for the Analysis of ENG Tracings," *Acta Otolaryngol.*, **80,** 120 (1975).
61. J. L. Craig and C. M. Derryberry, "Applied Concepts of Automation in an Occupational Medical Program," *Ind. Med. Surg.*, **40,** 9 (1971).
62. J. Ortega, "The Application of Data Processing Techniques to Occupational Health Records," *Cah. Notes Doc. Secur. Hyg. Trav.*, **62,** 61 (1971).
63. M. R. Scott and W. S. Frederik, "Electronic Data Processing and Multiphasic Health Screening," *J. Occup. Med.*, **14,** 457 (1972).
64. W. S. Frederik and M. R. Scott, "Medical Statistics, System Monitoring and Provisional Normals," *J. Occup. Med.*, **14,** 466 (1972).
65. S. Pell, "Epidemiological Studies in a Large Company Based on Health and Personnel Records," *Public Health Rep.*, **83,** 399 (1968).
66. J. Planques, "Report of the Commission for Medical Records in Occupational Medicine," *Arch. Mal. Prof.*, **34,** 25 (1973).
67. I. Cavill et al., "A System for Data Processing in Haematology," *J. Clin. Pathol.*, **27,** 330 (1974).
68. J. Planques et al., "Statistical Evaluation on Medical Records for Occupational Health Purposes," *Arch. Mal. Prof.*, **32,** 129 (1971).
69. J. P. Horwitz et al., "Adjunct Hospital Emergency Toxicological Service," *JAMA*, **235,** 1708 (1976).
70. II. LaRocca and I. Macnab, "Value of Pre-Employment Radiographic Assessment of the Lumbar Spine," *Ind. Med. Surg.*, **39,** 31–36, 253–258 (1970).
71. A. Beck and J. Killus, "Normal Posture and Spine Determined by Mathematical and Statistical Methods," *Aerosp. Med.*, **44,** 1277 (1973).
72. R. S. Sherman, C. A. Bertrand, and J. C. Duffy, "Roentgenographic Detection of Cardiomegaly in Employees with Normal Electrocardiograms," *Am. J. Roentgenol. Radium Ther. Nuclear Med.*, **119,** 493 (1973).
73. F. Bockat and S. N. Wiener, "An Electrostatic Printer Display for Computerized Scintiscans," *Amer. J. Roentgenol.*, **117,** 146 (1973).
74. R. S. Sherman, "An Automated System for Recording Reports of Chest Roentgenograms," *Amer. J. of Roentgenol. Radium Ther. Nuclear Med.*, **117,** 848 (1973).
75. J. R. Jagoe and K. A Paton, "Reading Chest Radiographs for Pneumoconiosis by Computer," *Br. J. Ind. Med.*, **32,** 267 (1975).
76. Digital Film Library NIOSH 54638, National Institute for Occupational Safety and Health, Rockville, Md., August 1975.
77. F. Yanowitz et al., "Quantitative Exercise Electrocardiography in the Evaluation of Patients with Early Coronary Artery Disease," *Aerosp. Med.*, **45,** 443 (1974).
78. W. H. Walter et al., "Dynamic Electorcardiography and Computer Analysis," **44,** 414 (1973).
79. R. A. Taha et al., "The Electrocardiogram in Chronic Obstructive Pulmonary Disease," *Am. Rev. Respir. Dis.*, **107,** 1067 (1973).
80. C. A. Bertrand et al., "How the Computer Helps the Cardiologist Interpret ECG's," *Ind. Med. Sur.*, August 1973, pp. 14–19.
81. S. Talbot et al., "Normal Measurements of Modified Frank Corrected Orthogonal Electrocardiograms

and Their Importance in an On-Line Computer-Aided Electrocardiographic System," *Br Heart J.*, **3**, 475 (1974).

82. M. L. Simoons et al., "On-Line Processing of Orthogonal Exercise Electrocardigorams," *Comput Biomed. Res.*, **8**, 105 (1975).
83. R. W. Morris, "The Value of the Effort Electrocardiogram," *South Afr. Med. J.*, **49**, 1553 (1975).
84. H. Karlsson, B. Lindberg, and D. Linnarsson, "Time Courses of Pulmonary Gas Exchange and Heart Rate Changes in Supine Exercise," *Acta Physiolog. Scand.*, **95**, 329 (1975).
85. P. Franket et al., "A Computerized System for ECG Monitoring," *Comput. Biomed. Res.*, **8**, 56 (1975).
86. A. A. Sarkady, R. R. Clark, and R. Williams, "Computer Analysis Techniques for Phonocardiogram Diagnosis," *Comput. Biomed. Res.*, **9**, 349 (1976).
87. K. Horsfield et al., "Models of the Human Bronchial Tree," *J. Appl. Physiol.*, **31**, 207 (1971).
88. P. G. Voilleque, "Computer Calculation of Bone Doses Following Acute Exposure to Strontium-90 Aerosols," Health Services Laboratory, U.S. Atomic Energy Commission, National Reactor Testing Station, Idaho Falls, Idaho, 1970.
89. J. M. Beeckmans, "The Deposition of Aerosols in the Respiratory Tract—Mathematical Analysis and Comparison with Experimental Data," *Can. J. Physiol. Pharmacol.*, **43**, 157 (1965).
90. D. P. Discher and A. H. Palmer, "Development of a New Motivational Spirometer—Rationale for Hardware and Software," *J. Occup. Med.*, **14**, 679 (1972).
91. J. D. Brain and P. A. Valberg, "Models of Lung Retention Based on ICRP Task Group Report," *Arch. Environ. Health*, **28**, 1 (1974).
92. D. B. Yeates et al., "Regional Clearance of Ions from the Airways of the Lung," *Am. Rev. Respir. Dis.*, **107**, 602 (1973).
93. A. W. Brodey et al., "The Residual Volume—A Graphic Solution to the Functional Residual Capacity Equation," *Am. Rev. Respir. Dis.*, **109**, 87 (1974).
94. G. J. Trezek, "Predictions of the Dynamic Response of the Lung," *Aerosp. Med.*, **44**, 8 (1973).
95. M. K. Loken et al., "Dual Camera Studies of Pulmonary Function with Computer Processing of Data," *Am. J. Roentgenol. Radium Ther. Nuclear Med.*, **121**, 761 (1974).
96. W. R. Beaver, "On-Line Computer Analysis and Breath-by-Breath Graphical Display of Exercise Function Tests," *J. Appl. Physiol.*, **34**, 128 (1973).
97. R. S. Sherman, "An Automated System for Recording Reports of Chest Roentgenograms," *Am. J. Roentgenol., Radium Ther. Nuclear Med.*, **117**, 848 (1973).
98. W. R. Ayers et al., "Description of a Computer Program for Analysis of the Forced Expiratory Spirogram. I. Instrumentation and Programming," *Comput. Biochem. Res.*, **2**, 207 (1969).
99. L. Jansson and B. Jonson, "A Method for Studies of Airway Closure in Relation to Lung Volume and Transpulmonary Pressure at a Regulated Flow Rate," *Scand. J. Respir. Dis.*, **85**, 228 (1974).
100. M. Paiva, "Gaseous Diffusion in an Alveolar Duct Simulated by a Digital Computer," *Comput. Biomed. Res.*, **7**, 533 (1974).
101. A. A. Smith and E. A. Gaensler, "Timing of Forced Expiratory Volume in One Second, "*Am. Rev. Respir. Dis.*, **112**, 882 (1975).
102. J. H. Ellis et al., "A Computer Program for Calculation and Interpretation of Pulmonary Function Studies," *Chest*, **68**, 209 (1975).
103. W. A. Barrett, "Computerized Roentgenographic Determination of Total Lung Capacity," *Am. Rev. Respir. Dis.*, **113**, 239 (1976).
104. H. Pessenhofer and T. Kennes, "Method for the Continuous Measurements of the Phase Relation between Heart and Respiration," *Pfluegers Arch.*, **335**, 77 (1975).

105. C. Simecek, "Formulae for Calculation of Membrane Diffusion Component and Pulmonary Capillary Blood Volume," *Bull. Physiol-Pathol. Respir.*, **11**, 349 (1975).
106. J. N. Davis and D. Stagg, "Interrelationship of the Volume and the Time Components of Individual Breaths in Resting Man," *J. Physiol.*, **245**, 481 (1975).
107. A. Crockett and R. L. Smith, "Use of On-Line Computer Facilities in a Respiratory Function Laboratory," *Med. J. Aust.*, **2**, 486 (1975).
108. H. Guy et al., Computerized, Noninvasive Tests of Lung Function," *Am. Rev. Respir. Dis.*, **113**, 737 (1976).
109. C. Xintaras et al., "Brain Potentials Studied by Computer Analysis," *Arch. Environ. Health*, **13**, 223 (1966).
110. J. C. G. Pearson and D. Radwanski, "Principles of Design of Occupational Health Records," *J. Soc. Occup. Med.*, **24**, 17 (1974).
111. L. Nottbohn, "Collection and Electronic Processing of Data from Routine Medical Examinations," *Arbeitsmedizin-Solialmedizin-Arbeitshygiene*, **5**, 127 (1970).
112. G. H. Collings et al., "Follow-up of MHS," *J. Occup. Med.*, **14**, 462 (1972).
113. A. Yedidia, "California Cannery Workers Program, Multiphasic Testing as an Introduction to Orderly Health Care," *Arch. Environ. Health*, **27**, 259 (1973).
114. "Multiphasic Health Screening . . . A Tool for Industrial Hygiene?" *Nat. Saf. News*, **102**, 72, (1970).
115. H. A. Haessler, "Industrial Experience with Automated Multiphasic Health Examinations," *Ind. Med. Surg.*, **39**, 24-26, 335-337 (1970).
116. S. Bangs, "Multiphasic Screening: Wave of the Future or Droplet?" *Occup. Hazards*, **31**, 51 (1969).
117. D. F. Davies, "Progress Toward the Assessment of Health Status," *Prev. Med.*, **4**, 282 (1975).
118. N. Righthand et al., "A Computer-Oriented Hearing Conservation Program," *J. Occup. Med.*, **16**, 654 (1974).
119. R. A. Campbell, "Computer Audiometry," *J. Speech Hear. Res.*, **17**, 134 (1974).
120. H. M. Sussman, P. F. MacNeilage, and J. Lumbley, "Sensorimotor Dominance and the Right Ear Advantage in Mandibular—Auditory Tracking," *J. Acoust. Soc. Am.*, **56**, 214 (1974).
121. I. Klockhoff et al., "A Method for Computerized Classification of Pure Tone Screening Audiometry Results in Noise-Exposed Groups, *Acta Otolaryngol.*, **75**, 339 (1973).
122. V. Mellert, K. F. Siebrasse, and S. Mehrgardt, "Determination of the Transfer Function of the External Ear by an Impulse Response Measurment," *Anal. Spectrogr.*, **134**, 135, 137 (1974).
123. W. Niedermeier, J. H. Griggs, and R. S. Johnson, "Emission Spectrometric Determination of Trace Elements in Biological Fluids," *Applied Spectrosc.*, **25**, 53 (1971).
124. D. W. Lander et al., "Spectrographic Determination of Elements in Airborne Dirt," *Appl. Spectrosc.*, **25**, 270 (1971).
125. A. G. Schoning, "A Computer-Based Storage and Retrieval System for Electronic Absorption Spectra," *Anal. Chem. Acta*, **71** 17 (1974).
126. H. G. Langer, T. P. Brady, and P. R. Briggs, "Formation of Dibenzodoxins and Other Condensation Products from Chlorinated Phenols and Derivatives," *Environ. Health Perspect.*, **5**, 3 (1973).
127. R. J. Gelirke and R. C. Davies, "Spectrum Fitting Technique for Energy Dispersive X-ray Analysis of Oxides and Silicates with Electron Microbeam Excitation," *Anal. Chem.*, **47**, 1537 (1975).
128. B. M. Golden and E. S. Yeung, "Analytical Lines for Long-Path Infrared Absorption, Spectrometry of Air Pollutants," *Analy. Chem.*, **47**, 2132 (1975).
129. D. Schuetzle, "Analysis of Complex Mixtures by Computer Controlled High Resolution Mass Spectrometry. I. Application to Atmospheric Aerosol Composition," *Biomed. Mass Spectrom.*, **2**, 288 (1975).

130. H. W. Dickson et al., "Environmental Gamma Ray Measurements Using in Situ and Core Sampling Techniques," *Health Phys.*, **30**, 221 (1976).
131. B. Versino et al., "Organic Micropollutants in Air and Water," *J. Chromatogr.*, **122**, 373 (1976).
132. I. Harness, "Airborne Asbestos Dust Evaluation," *Ann. Occup. Hyg.*, **16**, 397 (1973).
133. M. V. Sussman, K. N. Astill, and R. N. S. Rathore, "Continuous Gas Chromatography," *J. Chromatogr. Sci.*, **12**, 91 (1974).
134. M. Axelson, G. Schumacher, and J. Sjovall, "Analysis of Tissue Steroids by Liquid-Gel Chromatography and Computerized Gas Chromatography–Mass Spectrometry," *J. Chromatrogr. Sci.*, **12**, 535 (1974).
135. P. J. Arpino, B. G. Dawkins, and F. W. McLafferty, "A Liquid Chromatography/Mass Spectrometry System Providing Continuous Monitoring with Nanogram Sensitivity," *J. Chromatogr. Sci.*, **12**, 574 (1974).
136. B. S. Finkle, R. L. Foltz, and D. M. Taylor, "A Comprehensive GC-MS Reference Data System for Toxicological and Biomedical Purposes," *J. Chromatogr. Sci.*, **12**, 304 (1974).
137. E. Jellum, O. Stokke, and L. Eldjarn, "Application of Gas Chromatography, Mass Spectrometry, and Computer Methods in Clinical Biochemistry," *Anal. Chem.*, **45**, 1099 (1973).
138. H. Nau and K. Biemann, "Computer-Assisted Assignment of Retention Indices in Gas Chromatography–Mass Spectrometry and Its Application to Mixtures of Biological Origin," *Anal. Chem.*, **46**, 426 (1974).
139. A. Bye and G. Land, "Determination of 3-(5-Tetrazolyl) Thioxanthone 1010-Dioxide in Human Plasma, Urine, and Feces," *J. Chromatogr.*, **115**, 93 (1975).
140. R. C. Lao, R. S. Thomas, and J. L. Monkman, "Computerized Gas Chromatographic–Mass Spectrometric Analysis of Polycyclic Aromatic Hydrocarbons in Environmental Samples," *J. Chromatogr.*, **112**, 681 (1975).
141. N. Buchan, "Computer Analysis of Amino Acid Chromatograms," *J. Chromatogr.*, **103**, 33 (1975).
142. J. Einhorn et al., "Computerized Analytical System for the Analysis of the Thermal Decomposition Products of Flexible Urethane Foam," Flammability Research Center, University of Utah, Salt Lake City, December 1973.
143. G. F. Gostecnik and A. Zlatkis, "Computer Evaluation of Gas Chromatographic Profiles for the Correlation of Quality Differences in Cold Pressed Orange Oils," *J. Chromatogr.*, **106**, 73 (1975).
144. W. Cautreels and K. Van Cauwenberghe, "Determination of Organic Compounds in Airborne Particulate Matter by Gas Chromatography–Mass Spectrometry," *Atmost. Environ.*, **10**, 447 (1976).
145. S. R. Heller, J. M. McGuire, and W. L. Budde, "Trace Organics by GC/MS," *Environ. Sci. Technol.*, **9**, 210 (1975).
146. T. Mamuro et al., "Activation Analysis of Polluted River Water," *Radioisotopes*, **20**, 111 (1971).
147. A. H. Qazi et al., "Identification of Carcinogenic and Noncarcinogenic Polycyclic Aromatic Hydrocarbons Through Computer Programming," *Am. Ind. Hyg. Assoc. J.*, **34**, 554 (1973).
148. K. D. Hapner and K. R. Hamilton, "Basic Computer Program for Amino Acid Analysis Data," *J. Chromatogr.*, **93**, 99 (1974).
149. M. A. Fox., "Programs for Use with Automatic Amino Acid Analyser to Identify, Compute, and Correlate Amino Acid Concentrations in Biological Samples," *J. Chromatogr.*, **89**, 61 (1974).
150. I. Cavill and C. Ricketts, "Automated Quality Control for Haematology Laboratory," *J. Clin. Pathol.*, **27**, 757 (1974).
151. D. M. Linekin, "Multielement Instrumental Neutron Activation Analysis of Biological Tissue Using A Single Comparator Standard and Data Processing by Computer," *Int. J. Appl. Radiat. Isot.*, **24**, 343 (1973).

152. J. S. Ploem et al., "A Microspectrofluorometer with Epi-Illumination Operated Under Computer Control," *J. Histochem. Cytochem.*, **22**, 668 (1974).
153. B. Sarkar and T. P. A. Kruck, "Theoretical Considerations and Equilibrium Conditions in Analytical Potentiometry," *Canadian Journal of Chem.*, **51**, 3541–3548 (1973).
154. B. T. Dew, T. King, and D. Mighdoll, "An Automatic Microscope System for Differential Leukocyte Counting," *J. Histochem. Cytochem.*, **22**, 685 (1974).
155. A. L. Linch et al., "Nondestructive Neutron Activation Analysis of Air Pollution Particulates," *Health Lab. Sci.*, **10**, 251 (1973).
156. L. Kryger, D. Jagner, and H. J. Skov, "Computerized Electroanalysis. Part I. Instrumentation and Programming," *Anal. Chem. Acta*, **78**, 241 (1975).
158. E. W. White and P. B. Denee, "Characterization of Coal Mine Dust by Computer Processing of Scanning Electron Microscope Information," *Ann. N. Y. Acad. Sci.*, **200**, 666 (1972).
159. E. A. Curth, "Causes and Prevention of Transportation Accidents in Bituminous Coal Mines," Bureau of Mines, Information Circular 8506, Pittsburgh, 1971.
160. W. R. Miller, "System Analysis Approach to Safety," *Nat. Saf. Congr. Trans.*, **11**, 19 (1972).
161. "How Safety/Security Directors Use EDP," *Occup. Hazards*, **30**, 41 (1968).
162. S. E. Hall, "Procedure for Army Safety Sampling (PASS)," National Technical Information Service, Springfield, Va., December 1970.
163. V. Steinecke, "Electronic Data Processing for Safety," *Sicherheitsingenieur*, **4**, 364 (1973).
164. "How a Computer System Cuts Accident Costs," *Occup. Hazards*, **35**, 45 (1973).
165. G. W. Radl, "Can Visibility Be Improved on a Fork-Lift Truck?" *Mod. Unfallverhüt.*, Vulkan-Verlag Haus, Der Techniq Essen Germany), **17**, 41 (1973).
166. J. Wanat, "A System for Computer Analysis of Work Accidents Using a Modern Form of Information Bank Utilization in the Area of Industrial Safety," *Ochr. Pr.*, **26**, 4 (1972).
167. R. V. O'Neil and O. W. Burke, "A Simple Systems Model for DDT and DDE Movement in the Human Food Chain," Ecological Sciences Division, Publication 415, Oak Ridge National Laboratory, Oak Ridge, Tenn., October 1971.
168. W. J. Stanley and D. D. Cranshaw, "The Use of a Computer-Based Total Management Information System to Support an Air Resource Management Program," *J. Air Pollu. Control Assoc.*, **18**, 158 (1968).
169. E. B. Cook and J. M. Singer, "Predicted Air Entrainment by Subsonic Free Round Jets," *J. Spacecr. Rockets*, **6**, 1066 (1969).
170. H. A. James and H. Currie, "Punched Card Information Retrieval System for Air Pollution Control Data," *J. Air Pollu. Control Assoc.*, **14**, 118 (1964).
171. N. M. Rochkind, "Infrared Analysis of Gases: A New Method," *Environ. Sci. Technol.*, **1**, 434 (1967).
172. B. J. Huebert, "Computer Modelling of Photochemical Smog Formation," *J. Chem. Educ.*, **51**, 644 (1974).
173. M. I. Hoffert et al., "Laboratory Simulation of Photochemically Reacting Atmospheric Boundary Layers: A Feasibility Study," *Atmos. Environ.* **9**, 33 (1975).
174. F. C. Hamburg and F. L. Cross, Jr., "A Training Exercise on Cost-Effectiveness Evaluation of Air Pollution Control Strategies," *J. Air Pollu Control Assoc.*, **21**, 66 (1971).
175. H. A. Hawthorne et al., "Cesium-137 Cycling in a Utah Dairy Farm," *Health Phys.*, **30**, 447 (1976).
176. E. W. Crampton, "Husbandry Versus Fluoride Ingestion as Factor in Unsatisfactory Dairy Cow Performance," *J. Air Pollut. Control Assoc.*, **18**, 229 (1968).
177. R. G. Edwards, Jr., C. H. Powell, and M. A. Kendrick," Dust Counting Variability," *Am. Ind. Hyg. Assoc. J.*, **27**, 546 (1966).

178. J. L. Spratt, "Computer Program for Prohibit Analyses, " *Toxicol. Appl. Pharmacol.*, **8**, 110 (1966).
179. C. Xintaras et al., "Brain Potentials Studied by Computer Analysis," *Arch. Environ. Health, 13,* 223 (1966).
180. J. K. Raines, M. Y. Jaffrin, and A. H. Shapiro, "A Computer Simulation of Arterial Dynamics in the Human Leg," *J. Biomech.*, **7**, 77 (1974).
181. D. A. Hobson and L. E. Torfason, "Optimization of Four-Bar Knee Mechanisms—A Computerized Approach," *J. Biomech.*, **7**, 371 (1974).
182. R. Penn, R, Walser, and L. Ackerman, "Cerebral Blood Volume in Man," *JAMA,* **234**, 1154 (1975).
183. P. J. Lewi and R. P. H. M. Marshoom, "Automated Weighing Procedure for Toxicological Studies on Small Animals Using a Minicomputer," *Lab Anim. Sci.*, **25**, 487 (1975).
184. M. Salzer, "Model for Describing Tremor," *Eur. J. Appl. Physiol. Occup. Physiol.*, **34**, 19 (1975).
185. S. Schottenfeld and S. Rothenberg, "An Automated Laboratory Control System: Collection and Analysis of Behavior and Electro-Physiological Data," *Comput. Programs, Biomed.* **5**, 296 (1976).
186. N. H. Sabah, "A Presettable Multichannel Digital Timer," *J. Appl. Physiol.*, **38**, 757 (1975).
187. A. L. Henschel et al., "An Analysis of Heat Deaths in St. Louis During July 1966," *Am. J. Public Health*, **59**, 2232 (1969).
188. F. Burbank, "A Sequential Space-Time Cluster Analysis of Cancer Mortality in the United States: Etiologic Implications," *Am. J. Epidemiol.*, **95**, 393 (1972).
189. S. Pell, "Epidemiological Studies in a Large Company Based on Health and Personnel Records," *Public Health Rep.*, **83**, 399 (1968).
190. W. E. McConnell, "The Wavr File," *Aerosp. Med.*, **44**, 210 (1973).
191. S. Milham, Jr., *Occupational Mortality in Washington State 1950-1971*, Vol. 3, National Institute for Occupational Safety and Health, Department of Health Education and Welfare Publication (NIOSH) 76-175-C, Cincinnati, Ohio, April 1976.
192. D. J. Kilian, D. J. Picciano, and C. B. Jacobson, "Industrial Monitoring: A Cytogenetic Approach," *Ann. N.Y. Acad. Sci.*, **269**, 4 (1975).
193. Q. T. Pham et al., "Methodology of an Epidemiological Survey in the Iron Ore Mines of Lorraine—Research into the Long-term Effects of Potentially Irritant Gases on the Pulmonary System," *Ann. Occup. Hyg.*, **19**, 33 (1976).
194. C. A. Mitchell, R. S. F. Schilling and A. Bouhuys, "Community Studies of Lung Disease in Connecticut," *Am. J. Epidemiol.*, **103**, 212 (1976).
195. P. G. Voilleque, "Computer Calculation of Bone Dose Following Acute Exposure to Strontium-90 Aerosols," U.S. Atomic Energy Commissions, Idaho Falls, Idaho, 1970.
196. J. R. Mallard and T. A. Whittingham, "Dielectric Absorption of Microwaves in Human Tissues," *Nature*, **218**, 366 (1968).
197. S. A. Beach, "A Digital Computer Program for the Estimation of Body Content of Plutonium from Urine Data," *Health Phy.*, **24**, 9 (1973).
198. R. E. Ellis et al., "A System for Estimation of Mean Active Bone Marrow Dose," Food & Drug Administration, Rockville, Md., August 1975.
199. A. I. Burhanov, "Combined Effects of the Basic Constituents of Mixed Metal Dusts," *Gigi. Tr. Prof. Zabol.*, **3**, 30 (1975).
200. R. E. Goans, "Two Approaches to Determining Pu-239 and Am-241 Levels in Phoswich Spectra," *Health Phys.*, **29**, 421 (1975).
201. W. R. Wood, Jr., and W. E. Sheehan, "Evaluation of the Pugfua Method of Calculating Systemic Burdens," *Am. Ind. Hyg. Assoc. J.*, **32**, 58 (1971).

202. B. Bergstrom et al., "Use of a Digital Computer for Studying Velocity Judgments of Radar Targets," *Ergonomics,* **16,** 417 (1973).
203. D. M. S. Peace and R. S. Easterby, "The Evalaution of User Interaction with Computer-Based Management Information Systems," *Hum. Factors,* **15,** 163 (1973).
204. A. B. Trump-Thorton and R. Daher, "The Prediction of Reaction Forces from Gait Data," *J. Biomech.,* **8,** 173 (1975).
205. H. R. Warner, F. R. Woolley, and R. L. Kane, "Computer Assisted Instructions for Teaching Clinical Decision-Making," *Comput. Biomed. Res.,* **7,** 564 (1974).
206. D. N. Ostrow, N. Craven, and R. M. Cherniack, "Learning Pulmonary Function Interpretation: Deductive Versus Inductive Methods," *Am. Rev. Respir. Dis.,* **112,** 89 (1975).
207. A. U. Valish and N. J. Boyd, "The Role of Computer Assisted Instruction in Continuing Education of Registered Nurses: An Experimental Study," *J. Contin. Educ. Nurs.,* **6,** 13 (1975).
208. D. P. Schlick and K. R. Werner, "Computerized Programming of Respirable Dust Sampling Data," U. S. Department of the Interior, Bureau of Mines, Report 8504, March 1971.
209. G. L. Baker and R. E. Reiter, "Automatic Systems for Monitoring Vinyl Chloride in Working Atmospheres," *Am. Ind. Hyg. Assoc. J.,* **38,** 24 (1977).
210. P. S. Kerr, "Standard Oil's Computer Network Makes Health Management Possible," *Occup. Health Saf,* November–December 1977, pp. 44–46.
211. H. R. Jennings and K. L. Rohrer, "A Computerized Industrial Hygiene Program," *Plant Eng.,* October 14, 1976, pp. 149–151.
212. M. G. Ott et al., "Linking Industrial Hygiene and Health Records," *Am. Ind. Hyg. Assoc. J.,***10,** 760 (1975).
213. N. J. Gilson, W. I. Bitter, and H. G. Barrett, "Automated Monitoring System for Exposure to Toxic Materials," IBM Research Center, Yorktown Heights, N.Y., February 8, 1972.

CHAPTER FIVE

Analytical Measurements

PETER M. ELLER, Ph.D., and
JOHN V. CRABLE

1 INTRODUCTION

Analytical chemistry is important to the industrial hygienist in several areas: personal monitoring, source sampling and analysis, and area measurements. In most cases the chemical properties of the analyte must be considered to make the proper choice of sampling device and to recover the analyte quantitatively. Thus collection efficiency, stability of the sample, possible interferences, and desorption efficiency may be functions of the chemical interactions between analyte and sampling device. This chapter discusses selectively, but not exhaustively, the analytical techniques employed in industrial hygiene and their relationships to sampling systems.

2 VALIDITY OF THE MEASUREMENT

2.1 Accuracy and Precision

The value of the analytical result to the industrial hygiene program and the relevance to the individual worker are ultimately dependent on the accuracy of the result. This chapter deals with contributions of the sampling and analytical method to accuracy; other considerations such as sampling strategies and personal versus area sampling are discussed elsewhere in this volume.

The accuracy of an individual measurement is determined by the occurrence of two types of error. These are determinate errors, also called bias, and indeterminate errors, or random errors. Examples of determinate errors are improperly calibrated sampling pumps and desorption efficiencies that are less than quantitative. Indeterminate errors are typified by instrumental noise or interanalyst variations and are frequently the limiting factor in defining detection limits.

Contamination, both of the sampling device in the field and of the sample by reagents during analysis, is one of the most common sources of error. The positive biases introduced by the various sources of contamination are best eliminated by efforts to reduce the sources of contamination and the analysis of appropriate blanks. Another source of random error is the analysis of samples that are outside the optimum working range of the analytical method. For most methods the coefficient of variation (CV) increases rapidly with decreasing sample size near the lower end of the working range. For many methods, such as spectrophotometry, an increase of CV with increasing sample size is also seen above the upper limit of the working range. Thus the best precision is obtained with sample sizes that are within an optimum range. Losses, through chemical degradation on the sample device, or in the analytical process, lead to negative biases that may be reduced by prompt analysis and the practice of subjecting known amounts of analyte to the same processing steps as the samples. Table 5.1 summarizes these and other sources of error.

2.2 Reference Materials

The use of reference materials of known composition to calibrate analytical procedures is an important step in methods development or quality control. When evaluating the results obtained by a candidate procedure using a reference material, both precision and accuracy are important. The critical question is, "Does the candidate procedure give a result that falls within the range certified?" Thus the procedure will be categorized as one of three types: inaccurate (poor precision, possibly with bias), biased (good precision, but with bias), and accurate (good precision, with bias absent). When precision is poor, it is not correct to classify the procedure as accurate (1). These definitions are illustrated by three hypothetical candidate methods using a reference material with certified value 100 ± 5 (Table 5.2 and Figure 5.1). Method A is inaccurate because the estimate it provides extends outside the certified range of 95 to 105. It makes no difference that the mean value falls within the certified range; the entire set of values expressed by

$$\frac{\bar{X}_A \pm t s_A}{(N)^{1/2}} \qquad (1)$$

must fall within the certified range. The t-value, 2.23 in this example, is selected to include a desired proportion of the values obtained by method A (95 percent of the values in this example). Method B is inaccurate but precise; systematic errors are present and must be identified. The amount of bias that must be removed is the difference between the mean obtained $\bar{X}_B$ and the nearest boundary of the certified range. In this case the difference is 107.59 minus 105, or 2.59. Method C qualifies as a

ANALYTICAL MEASUREMENTS

Table 5.1 Some Common Errors in Sampling and Analysis

Source of Error	Direction	Remedy
Sampling		
Contamination	±	Analyze blank sampling device
Flowrate uncertainty	±	Calibrate sampling train
Sample too small for precise analysis	±	Take larger sample
Interfering substances	±	Take also bulk, area, or rafter samples
Loss of analyte	−	Avoid high temperatures, long storage of sample
Low collection efficiency	−	Decrease sampling time, temperature, flowrate; use fresh collection reagents
Analysis		
Contamination	+	Use reagent blank
High recovery with standard additions	+	Correct for nonlinear calibration curve
Biased analytical method	±	Calibrate with reference material or method
Matrix effects	±	Match sample and standard matrices
Interferences	±	Analyze bulk, area, rafter samples; apply correction for matrix
Too little or too much sample	±	Work in linear portion of calibration curve
Low desorption efficiency	−	Determine recovery with spiked samples
Loss during sample processing	−	Carry standards through same processing

reference method because at the 95 percent confidence level it gives a mean value that falls entirely within the certified range. In practice, a larger number of determinations than 10 would be obtained. A detailed explanation of the steps needed to establish reference methods is given elsewhere (2).

The U.S. National Bureau of Standards offers a large number of reference materials (3), some of which are directly applicable to industrial hygiene. Table 5.3 lists these; several gases, organic solvents, metals, and quartz are included. A need exists for additional reference materials, particularly biological standards [e.g., lead in blood (4)]. The application of reference materials to proficiency testing of laboratories is covered in Section 2.5.

2.3 Validation of Sampling and Analytical Methods

After selection of candidate sampling and analytical methods, further validation under conditions appropriate to industrial hygiene use is usually desirable.

Table 5.2 Evaluation of Hypothetical Methods Using a Reference Material with Certified Value 100 ± 5

	Method		
	A	B	C
Results	99.0	103.9	93.9
	97.3	111.7	101.7
	84.8	106.0	96.0
	94.7	108.1	98.1
	100.6	107.9	97.9
	106.6	109.8	99.8
	111.3	107.3	97.3
	91.0	106.5	96.5
	103.2	109.5	99.5
	117.8	105.2	95.2
Mean, $\overline{X}$	100.63	107.59	97.59
Standard deviation, s	9.68	2.33	2.33
$ts/(N)^{1/2}$	6.83	1.64	1.64
95% confidence limits of mean	93.80–107.43	105.95–109.23	95.95–99.23

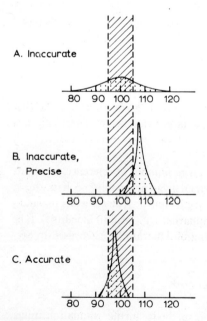

Figure 5.1. Evaluation of three methods using a reference material with certified value of 100 ± 5.

ANALYTICAL MEASUREMENTS

Table 5.3 Reference Materials Useful for Industrial Hygiene (3)

Standard Reference Material No.	Name	Substance Certified
1625-27	Sulfur dioxide permeation tube	Sulfur dioxide
1629	Nitrogen dioxide permeation device	Nitrogen dioxide
1665-69	Propane in air	Propane
1673-75	Carbon dioxide in nitrogen	Carbon dioxide
1677-81	Carbon monoxide in nitrogen	Carbon monoxide
1683-87	Nitric oxide in nitrogen	Nitric oxide
2203	Potassium fluoride	Fluoride ion
2661-67	Organic solvents on charcoal	Benzene, m-xylene, p-dioxane, 1,2-dichloroethane, chloroform, trichloroethylene, carbon tetrachloride
2671	Freeze-dried urine	Fluoride ion (2 levels)
2672	Freeze-dried urine	Mercury (2 levels)
2675	Beryllium on filter media	Beryllium (3 levels)
2676	Metals on filter media	Lead, cadmium, zinc, manganese (3 levels)
2679	Quartz on filter media	Quartz (4 levels)

One set of validation criteria has been applied successfully to more than 200 personal sampling and analytical methods using laboratory-generated atmospheres (5-7). The basic accuracy requirement for validation was that the overall sampling and analytical method must be capable of giving a value within ± 25 percent of the true air concentration at the standard set by the Occupational Safety and Health Administration (OSHA), at the 95 percent confidence level. This requires that the overall coefficient of variation be 12.8 percent or less for a method shown to contain no systematic errors. If the method contains a bias, precision requirements are more stringent. The maximum allowable bias correction, aside from a correction for desorption efficiency, was set at 10 percent. The minimum acceptable desorption efficiency was set at 75 percent. The criterion for sample stability was that no more than a 10 percent change in analyte concentration should occur after 7 days storage at room temperature. The methods were to be suitable for sampling periods of one hour or longer, except where shorter term standards apply. Detailed instructions for sampling and analysis are given, as are data relating to accuracy, precision, working range, and possible interferences, in a standard format (Table 5.4).

These validation criteria were met with the following protocol (criteria added later in the program are marked with an asterisk):

Table 5.4 Outline of Sampling and Analytical Method and Sampling Data Sheet (7)

Sampling and Analytical Method
Dimethylamine

Analyte:	Dimethylamine
Matrix:	Air
OSHA standard:	10 ppm (18 mg/m^3)
Procedure:	Adsorption on silica gel, desorption in 0.2N H$_2$SO$_4$ in methanol, GC
Method no.:	S142
Range:	7.0–29.5 mg/m^3
Precision (CV_T):	.062
Validation date:	6/6/75

1. Principle of the method
2. Range and sensitivity
3. Interferences
4. Precision and accuracy
5. Advantages and disadvantages of the method
6. Apparatus
7. Reagents
8. Procedure
    Cleaning of equipment
    Calibration of personal pumps
    Collection and shipping of samples
    Analysis of samples
    Determination of desorption efficiency
9. Calibration and standards
10. Calculations
11. References

Sampling Data Sheet
Sampling Data Sheet No. S142

Substance
OSHA standard
Analytical method
Sampling equipment
Sample size
Sampling procedure
Special considerations
Bulk samples
Shipping instructions
References

ANALYTICAL MEASUREMENTS

1. Analysis of six replicate standards each at 2, 1, and ½ the amount equivalent to the OSHA standard.
2. Analysis of six samples collected from laboratory-generated atmospheres at 2, 1, and ½ times the OSHA standard, with verification of the generated air concentrations by an independent method.
3. For solid sorbents, measurement of desorption efficiency for six replicates each at 2, 1, and ½ times the OSHA standard.
4. For solid sorbents, measurement of breakthrough capacity at twice the OSHA standard. For moisture-sensitive sorbents, the measurement is made with at least 80 percent relative humidity.*
5. Storage of six samples collected at the OSHA standard for 7 days at room temperature before analysis.*

2.4 Classification of Sampling and Analytical Methods

To describe the degree of confidence that can be expected in a method, it is desirable to develop a descriptive classification scheme. Of several that have been proposed, one denotes a "definitive" method as one that is directly related to fundamental (e.g., SI) units (1). An example of such a method is isotope dilution mass spectrometry. This scheme further defines a "reference" method as one that has been extensively collaboratively tested by the scientific community.

Another classification system (8) consists of five categories or classes: E (proposed), D (operational), C (tentative), B (recommended), and A (accepted). In this system the determining factors are the degrees of intra- and interlaboratory evaluation of the method. To move from E to D requires the successful use of the method on at least 15 field samples. Class C is reserved for methods in general use by other laboratories, but not evaluated for a particular industrial hygiene application. Validation of a method using standards and generated samples to characterize any biases and to obtain separate estimates of sampling and analytical error is required for a class B method. Successful field and collaborative testing upgrades the class B method to class A.

2.5 Proficiency Testing

The goal of proficiency testing is to measure the relative abilities of laboratories, using methods of their choice, to obtain accurate results. Because proficiency samples may receive special treatment, and because they are prepared in relatively uncomplicated matrices, proficiency testing is not necessarily a measure of the accuracy of the day-to-day work of the laboratory, however. Also, it is not well suited for methods development.

For a successful test, several requirements must be met: (1) samples must be uniform and prepared reproducibly, (2) samples must have adequate shelf life, and (3) reliable analytical methods must be available. If separate estimates of intra- and interlaboratory variation are desired, each laboratory must analyze two or more samples of a given analyte (although not necessarily duplicate samples). Because the resulting data appear to be log normally distributed (9), the use of geometric, rather than arithmetic, means has been found to be useful in their interpretation. Two statistical measures of performance may be defined: the mean ratio and the error ratio. The mean ratio, a measure of the accuracy of each laboratory, is defined by

$$M = \frac{(X_1 X_2 X_3 \cdots X_n)^{1/n}}{G} \qquad (2)$$

where $X_1, X_2, \ldots, X_n$ are the results obtained by the laboratory in question on its n samples of a given analyte and G is the grand geometric mean, excluding outliers, of the results obtained on this analyte by all laboratories. The value of M that signifies perfect accuracy is 1.00, and limits can be established, based on the data of a given testing session, which signify excessive deviation (bias) relative to the other participating laboratories.

The error ratio, or residual, gives a measure of intralaboratory variation. That is, it is a measure of the ability of a given laboratory to obtain consistent results. A residual is calculated for each analytical result submitted by each laboratory and is defined by

$$R_i = \frac{\dfrac{X_i}{(X_1 X_2 X_3 \cdots X_n)^{1/n}} \Big/ F_i}{(F_1 F_2 F_3 \cdots F_n)^{1/n}} \qquad (3)$$

where X_i is one of the values $X_1, X_2, X_3, \ldots, X_n$, and F_i is the geometric mean result obtained on the sample (excluding residual outliers) by all laboratories. Perfect precision is indicated if the values of R_i are 1.00. As with the mean ratio, statistical limits can be calculated to determine whether a given R value is abnormal. For a given testing session, the product $R_1 R_2 R_3 \cdots R_n$ is 1.0 for each analyte.

The performance of a laboratory is indicated by both its mean ratio and its residuals. Four general cases exist:

1. *Mean ratio and all residuals within limits.* Both accuracy and precision are acceptable. The closer the ratios to unity, the better the performance.

2. *Mean ratio within limits, and one or more residuals out of limits.* In this case interlaboratory variation is acceptable but precision of the intralaboratory results is not. If one of the residuals is grossly different from the others, a calculation error, contamination, or other mistake may be the cause. For example, failure to correct for a nonlinear working curve whose slope decreases at high concentrations may lead to abnormally low residuals for the high concentration samples.

3. *Mean ratio out of limits, all residuals within limits.* A bias common to all the analyses is a possible cause. Calibrations, standards, and calculations should be checked.

4. *Mean ratio and one or more residuals out of limits.* This unacceptable inter- and intralaboratory variation may be due to any of the above causes.

ANALYTICAL MEASUREMENTS

As an example, the Proficiency Analytical Testing (PAT) program of the National Institute for Occupational Safety and Health (NIOSH), begun in 1972, sends bimonthly samples to more than 200 participating laboratories (9). The samples include spiked filters containing cadmium, lead, and zinc (as the nitrates), chrysotile asbestos (in alumina matrix), quartz (with sodium silicate), and organic solvents (toluene, benzene, carbon tetrachloride, chloroform, ethylene dichloride, p-dioxane, trichloroethylene, and xylene) on charcoal tubes. Each set contains four samples, at different concentrations, of each analyte. Two limits are calculated and used for judging the performance. The outer limit defines a region above and below $M = 1$ or $R = 1$ that contains, at the 99 percent confidence level, at least 98 percent of the analytical results. The inner limits, slightly closer to $M = 1$ or $R = 1$, are the limits beyond which it can be stated with 99 percent confidence that a result belongs to the outer two percent of the distribution. An investigation into possible sources of error is recommended whenever (1) a mean ratio or residual exceeds an outer limit, (2) results on two consecutive rounds exceed an inner limit, or (3) results in six consecutive rounds give mean ratios on the same side of $M = 1$.

Because a variety of analytical methods may be used for a given analyte in a given PAT round, interlaboratory variance is higher than in a collaborative test. Typical values of CV are 5 to 10 percent for the metals, 40 to 60 percent for silica, 70 to 80 percent for asbestos, and 10 to 25 percent for the solvents. Some dependence of CV on concentration is noted (9).

3 SAMPLE COLLECTION AND PROCESSING

3.1 Personal Sampling Devices

The ideal personal sampling device is small and lightweight, in addition to having the sample collection characteristics discussed in Section 2.3. Thus it can be attached to the worker's clothing in the breathing zone and used to sample for extended periods of an hour or more to determine the time-weighted average concentration to which the worker is exposed. Flowrates through the sampler are set to assure efficient collection, and the minimum and maximum sample sizes are dictated by analytical sensitivity and sampler capacity, respectively.

For sampling particulates, 37 mm diameter filters in closed-face cassettes to avoid contamination and at flowrates of 1.5 to 3 liters/min are commonly used. A number of different filter materials are used, as shown by the applications summarized in Table 5.5. Considerations in the selection of a filter for a particular application are usually related to the processing involved in the recovery and determination of the analyte. Glass fiber filters are inert to all but the most vigorous chemical treatment (involving hydrofluoric and phosphoric acids) and provide relatively small pressure drops during sampling, however they may contain sufficient trace metal contamination to produce unacceptably high blank values in some cases (10). Teflon filters also have a high degree of chemical inertness and provide a hydrophobic substrate as well.

Table 5.5 Some Personal Sample Collection Methods

Collection Device	Example	Flowrate (liter/min)	Reference[a]
Filters			
Cellulose membrane	Arsenic compounds	1.5	P&CAM 268
	Beryllium compounds	1.5	S 339
	Chromium compounds	1.5	S 323
	Lead compounds	1.5	S 341
	Oil mist	1.5	S 272
	Phthalic anhydride	1.5	S 179
	Selenium compounds	1.5	P&CAM 181
Polyvinyl chloride (PVC)	Chromates	1.5	S 317
PVC-Acrylonitrile	Zinc oxide	1.5	P&CAM 222
Silver membrane + cyclone	Crystalline silica	1.7	S 315
Solid sorbents			
Molecular sieve 5A	Sulfur dioxide	0.2	P&CAM 204
Silica gel	Aniline	≤ 0.2	S 310
	Dimethylamine	≤ 1	S 142
	o-Toluidine	≤ 1	S 168
Activated charcoal	Chloroform	1	S 351
	Stoddard solvent	≤ 0.2	S 382
Tenax GC	Nitroglycerine	1	S 216
	White phosphorus	≤ 0.4	P&CAM 257
Silica gel + $HgCl_2$	Stibine	0.2	S 243
Molecular sieve + tri-ethanolamine	Nitrogen dioxide	≤ 0.05	P&CAM 231
Chromasorb P + silver	Mercury	≤ 0.05	P&CAM 175
Liquids			
Bubbler (0.01 N NaOH)	Hydrogen bromide	1	S 175
Impinger (1% KI, 1 N NaOH)	Ozone	1	S 8
Impinger (0.5% H_2O_2)	Sulfur dioxide	1	P&CAM 163
Combination			
Filter (glass fiber) + bubbler (isooctane)	Tetrachloronaphthalene	1.3	S 130

[a] Numbers following "P&CAM" or "S" refer to methods found in Reference 7.

For applications in which the sample is wet or dry ashed, or in which it is desired to dissolve the filter in concentrated nitric acid, acetic acid, sodium hydroxide, acetone, or dioxane, cellulose acetate or mixed cellulose ester membrane filters may be used. Polycarbonate filters provide a microscopically smooth surface that is compatible with photomicrography, and they are soluble in strong bases and dioxane, while being resistant to acids. Silver membrane filters have found application for the sampling and analysis by X-ray diffraction, without further sample processing, of crystalline species. Polyvinyl chloride (PVC) and PVC-acrylonitrile filters are relatively unaffected by relative humidity of the air sampled and thus find application where filters must be weighed; in addition, the latter type of filter is transparent in much of the infrared region; however this property has not yet been utilized in industrial hygiene analysis.

Personal sampling devices for gases may take one of four forms: whole-air ("bag") samplers, reactive solid sorbents, reactive liquids, and continuous monitors. Most whole-air samplers are useful for relatively short periods only and are effectively "grab" samples. Such samples may be useful for some applications (e.g., determination of peak, or ceiling exposures) but are less accurate for estimation of time-weighted average concentrations than are integrated samples of longer duration (Chapter 6). A recent development, a pocket-size whole-gas sampler, has been shown to eliminate this disadvantage for several gases; furthermore the requirement for a pump is eliminated (11). The device consists of an evacuated chamber with internal volume approximately 100 cm^3 connected to a sampling port containing a flow-limiting orifice. By choice of the critical orifice size, sample times of 8 hr or longer may be obtained. Connection of the sampler to a gas chromatograph by means of a simple gas handling system with provision for measurement of pressure allows for relatively easy analysis.

The second category of samplers for gases is reactive solid sorbents, including tubes filled with granular materials such as activated charcoal, silica gel, porous polymers, or other materials. The attraction between solid sorbent and gas may arise from relatively weak physical forces (e.g., collection of organic solvents on activated charcoal) or from chemical reaction involving electron transfer between gas and sorbent (e.g., sulfur dioxide on cuprous oxide) (12). Analysis may be direct in some cases such as the direct reading indicator tubes; in others desorption and analysis are performed in a laboratory. Figure 5.2 shows a three-section silica gel tube that has been applied to the sampling of aromatic amines (7, 13). The analyst can use a relatively small (150 mg) collection section, or a larger (850 mg) section, which allows flexibility for various combinations of air concentration, relative humidity, and sampling time. As solid sorbents are studied for collection of an increasing number of substances, more sorbents are found to be useful, including a variety of porous polymers that previously were used as

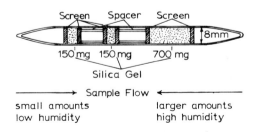

Figure 5.2. Sampling tube for aromatic amines (7).

chromatographic columns (14). An example is the use of Tenax GC, Porapak Q, or the Century Chromasorb series for the sampling of elemental phosphorus vapor in air (7, 15). Another approach that is simple and elegant is the use of a solid sorbent device in a passive mode. Palmes et al. have developed a triethanolamine-coated sorbent for use in a tube open at one end, into which nitrogen dioxide diffuses (16). The stability of both unexposed and exposed samplers is good, and analysis is by a colorimetric method.

Reactive liquids constitute a less favorable type of sampling medium from the point of view of the industrial hygienist because the handling and use of liquids in bubblers or impingers during sampling and shipping may lead to spillage, with resultant hazards and loss of sample. The mode of action of the liquid collection medium may be one of simple solvation (e.g., collection of tetrachloronaphthalene in isooctane), or it may involve a chemical reaction that fixes analyte (e.g., ozone + alkaline potassium iodide = molecular iodine).

Finally, sampling of gases is accomplished in some cases by continuous monitoring systems based on detection of a property such as thermal conductivity, electrical conductivity of liquid solution, or infrared absorption. These systems are usually more suited to area or source sampling than to personal sampling because of restricted portability and large physical size.

3.2 Air Sampling Parameters

The ideal sample collection device has high collection efficiency, high breakthrough capacity, and high desorption efficiency. Collection efficiency is defined as the ratio of quantity of analyte collected to quantity sampled. Breakthrough capacity is the maximum quantity of analyte collected before 5 percent of the influent appears on a backup sampling stage; it should correspond to an air volume that is at least 50 percent larger than the maximum air volume to be sampled (5). Breakthrough capacity may be a function of sampling rate, air concentration, and relative humidity. For collection of toluene on activated coconut charcoal, breakthrough capacity was seen to vary from 30 mg of toluene/100 mg of charcoal at 2040 μg of toluene/liter of air (15 liters sampled) to 19 mg of toluene/100 mg of charcoal at 45 μg of toluene/liter of air (420 liters sampled) at room temperature and less than 7 percent relative humidity (17). When the relative humidity was increased to 80 percent, breakthrough volume and capacity decreased by about 50 percent. Capacity is also a function of sampling rate, as was demonstrated for vinyl chloride on activated coconut charcoal (18). Air concentrations of 500 μg of vinyl chloride per liter were sampled; the breakthrough volumes for 100 mg of sorbent were 0.9 liter at 1.0 liter/min, 2.4 liters at 0.2 liter/min, and 5.2 liters at 0.05 liter/min.

Desorption efficiency, a term usually applied only to solid sorbents, is a measure of the quantity of analyte that can be recovered from the collection device during analysis. Measurement of this parameter is done by spiking unexposed sorbent with known quantities of analyte, allowing sufficient time for equilibration, and determining the percentage of recovery. Desorption efficiency may be a function of age of the sample and should be determined for realistic storage conditions. Some typical desorption efficiencies measured for coconut charcoal are: cyclohexanone, 47 percent; methyl ethyl ketone, 62 percent; 1,1,2,2-tetrachloroethane, 68 percent; styrene, 80 percent; methylisobutyl ketone, 89 percent; toluene, 98 percent; and benzene, cyclohexane, and trichloroethylene, 100 percent (12). In a collaborative test of seven organic solvents collected on charcoal tubes, an overall average desorption efficiency of 96 percent was observed for each of the following: benzene, carbon tetrachloride, chloroform, ethylene dichloride, trichloroethylene, and m-xylene. For dioxane the average desorption efficiency was 91 percent. The variation among laboratories was significant. For example, the range of values for benzene was 0.87 to 1.01 (19).

3.3 Sample Processing

Preparation of air filters, solid sorbents, or biological samples for analysis is an important part of the overall sampling and analysis procedure. Where appropriate, it is desirable to carry standards through all or part of the processing. Table 5.6 illustrates some sample treatment applications, including the use of various solvents, wet ashing, and thermal desorption. Not shown is the technique of low temperature, oxygen plasma ashing, which has found application for the determination of volatile elements in biological and other samples (20). The conditions under which a particular solvent is effective depend on the solute, the particle size, the temperature, and the degree of agitation. For some analytes, the use of an ultrasonic bath during desorption is required (e.g., amines on silica gel).

Preparation of air filters or biological samples for determination of metals by elemental analysis usually requires destruction of part or all of the organic matter present. Selection of an appropriate ashing technique depends on the risk of losing the analyte through volatilization, retention in the ashing vessel, or conversion to a chemical form that is unreactive or unavailable to the analytical method (21). Mercury, for example, is volatile under many wet-ashing and dry-ashing conditions. In the absence of chloride ions, however, samples containing mercury may be ashed with minimal losses using mixtures of nitric, sulfuric, and perchloric acids (21).

Volatilization losses of other elements may occur under oxidizing conditions (Rh, Os), reducing conditions (Se, Te), or in the presence of chloride ion in acidic solutions. The last category is the largest, including losses of antimony, arsenic, chromium, germanium, lead, and zinc. The chlorides of these elements are the volatile species, and the source of chloride may be the sample matrix or reduction of perchloric acid used for ashing. Retention of the analyte may occur by precipitation (e.g., loss of lead as the sulfate when sulfuric acid is used) or by reaction with the ashing vessel (e.g., formation of glasses with silica containers, or reduction of the element and diffusion as occurs with copper in silica). Finally, losses may be due to conversion of the element to a chemical form that is inert toward the analytical method. An example is the antimony(IV) compound formed in the absence of a strong oxidizing agent; the element in this form does not form the desired complex with Rhodamine B (22). Table 5.7 compares the common ashing techniques.

Table 5.6 Sample Treatment Methods Used in Industrial Hygiene

Method	Example	Reference[a]
Desorption with solvent		
Dilute sulfuric acid (aqueous)	Chromates	S 317
Dilute sulfuric acid (H_2O + CH_3OH); ultrasonic	Dimethylamine	S 142
Dilute ammonia (aqueous)	Phthalic anhydride	S 179
Dilute triethanolamine (aqueous)	Nitrogen dioxide	P&CAM 231
Ethanol	Aniline	S 310
	o-Toluidine	S 168
	Nitroglycerine	S 216
Carbon disulfide	Chloroform	S 351
	Stoddard solvent	S 382
Chloroform	Oil mist	S 272
Isooctane	Tetrachloronaphthalene	S 130
Xylene	White phosphorus	P&CAM 257
Thermal desorption	Sulfur dioxide	P&CAM 204
	Mercury	P&CAM 175
Wet ashing		
Nitric acid	Arsenic compounds	P&CAM 139
	Chromium compounds	S 323
	Lead compounds	S 341
Nitric + sulfuric acids	Selenium compounds	P&CAM 181
Nitric, sulfuric, and perchloric acids	Arsenic compounds	40
	Beryllium compounds	S 339
Dry ashing, redeposition	Crystalline silica	P&CAM 259
pH Adjustment	Hydrogen bromide	S 175
	Sulfur dioxide	P&CAM 163
No treatment	Chrysotile asbestos	P&CAM 245
	Crystalline silica	S 315
	Ozone	S 8
	Zinc oxide	P&CAM 222

[a] Numbers following "P&CAM" or "S" refer to methods found in Reference 7.

4 SAMPLE ANALYSIS

4.1 General Considerations

Analytical methods have undergone rather dramatic changes in the past decade. The classical "wet" techniques such as acid-base or oxidation-reduction titrations, or gravi-

ANALYTICAL MEASUREMENTS

Table 5.7 Methods for the Destruction of Organic Matter

Method	Reagents Used	Temperature (°C)	Comments	Reference
Dry ashing	None	500–700	Slow, less supervision, for larger samples, loss of volatile elements	21
Wet ashing	Nitric, sulfuric, and perchloric acids	100–140	May also use hydrogen peroxide or potassium permanganate	21
	$(CH_3)_4NOH$	25–60	For tissue	23
Low temperature ashing	Oxygen plasma	25–100	Small samples only, little volatilization	20

metric determinations have given way to instrumental methods for many analyses. Characteristics of the instrumental methods are less dependence on the skill of the analyst, complex electronic equipment, and dramatically lowered detection limits. One of the major dividends of these methods is that specific compounds, rather than only elemental composition, can be determined. For example, chromatography in its various forms, X-ray diffraction, molecular spectrophotometry, and mass spectrometry all provide a great deal of information about the molecular species present. The importance of this to industrial hygiene analysis can be appreciated by examining a list of toxic substances (24). Approximately 95 percent of the substances are specific compounds rather than classes of compounds based on elemental analysis (Table 5.8). It is not meant to imply that the burden of speciation falls entirely on the analytical method, however. In many cases (e.g., separation of gases from particulates) the sampling method is an integral part of the speciation process.

Table 5.8 Specificity Required in Industrial Hygiene Measurements[a]

Type of Analysis	Number	Examples
Specific compound	503	Ammonia, benzene, lead arsenate
Elemental analysis only	18	Beryllium, cadmium, fluorine
Elemental analysis, soluble/insoluble forms	10	Tungsten, zinc
Elemental analysis, different oxidation states	1	Chromium

[a] Adapted from Reference 24.

In certain situations it is desirable to determine more than one substance in a given sample (e.g., several metals in welding fume; mixed organic solvents). The (usually) small samples obtained in personal sampling limit the total amount of information obtainable from any one sample, however. In the development of future sampling and analytical procedures, two avenues appear promising: increased analytical sensitivity, and the use of methods with ability to determine more than one substance at a time. Examples of the first kind are provided by the development of atomic absorption spectrophotometry and gas chromatography. The early atomic absorption instruments, in the 1960s, used flame atomization. In more recent developments detection limits have been lowered by two or more orders of magnitude by various microsampling systems, including hydride generation and nonflame atomization. The development of new and more sensitive detectors for gas chromatography has also led to lower detection limits (25). Thus samples can be diluted to larger volumes and the capability for multiple determinations on a single sample is enhanced. Some techniques with capability to determine multiple substances simultaneously are emission spectroscopy, polarography, anodic stripping voltammetry, neutron activation analysis, X-ray fluorescence, and the various forms of chromatography.

4.2 Analytical Methods

The choice of which analytical method to use is frequently influenced by the sensitivities of the approaches available. For example, a number of methods have been applied to the determination of trace metals. As Figure 5.3 depicts, the range of sensitivities is wide,

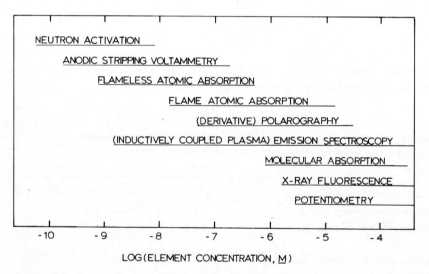

Figure 5.3. Approximate working ranges of some elemental analysis techniques.

covering more than 5 orders of magnitude between neutron activation analysis, one of the most sensitive methods for some elements, to potentiometry with specific ion electrodes. Improvements to several classical methods have made them considerably more sensitive. One of the most dramatic examples is the development of inductively coupled plasma techniques; improvements of 3 or more orders of magnitude in detection limits have resulted compared to older emission techniques (26).

In the field of industrial hygiene heavy reliance has been placed on two techniques: gas chromatography and atomic absorption spectrophotometry. An estimate of their importance is obtained from recent methods development work in the NIOSH-OSHA Standards Completion Program (5). As summarized in Table 5.9, 191 of 215 methods use one of the two techniques, with gas chromatography accounting for 165 of these. This does not constitute a claim that the two methods dwarf others in importance; it does indicate that they are extremely versatile however. Table 5.10 gives some applications of analytical methods to approximately 20 substances. The working ranges listed in the table refer to the total amount of·substance in the air sample. Because the aliquot size depends on the method, varying from the entire sample for methods such as anodic stripping voltammetry to less than 1 percent of the sample in gas chromatography, instrumental detection limits must be adjusted for total sample size. For example, desorption of solid sorbents for gas chromatographic analysis requires a minimum of 1 to 2 ml of solvent even though the aliquot to the chromatograph is a few microliters. Similarly the minimum volume to which an ashed filter can be diluted accurately is 5 to 10 ml, while flame atomic absorption requires 1 to 3 ml per determination, and graphite furnace atomic absorption uses only 5 to 50 μl per aliquot analyzed. In the determination of arsenic by flame atomic absorption spectrophotometry, the instrumental detection limit of 0.1 μg/ml translates to 1 μg of arsenic per 10 ml of sample. For the more sensitive hydride generation mode (see Figure 5.4) the instrumental detection limit is approximately 0.05 μg of arsenic (7) and the entire sample can be used for a single determination. The instrumental sensitivity of graphite furnace atomic absorption spectrophotometry is 100 pg of arsenic per determination; with the limitation of sample preparation, this equals 0.05 μg of arsenic per 10 ml of sample. Figure 5.5 illustrates similar interpretation of analytical methods for the determination of lead. The regions of applicability are shown for the lead specific ion electrode, anodic stripping voltammetry (ASV), flame atomic absorption spectrophotometry, and other AAS variations, including tantalum sampling boat (27), Delves sampling cup (28), and two varieties of graphite atomizer (29, 30).

Interaction of the sample treatment procedures with variations in analytical methods has resulted in a large number of published methods for the determination of lead in air or blood samples (Figure 5.6). Other methods, including ASV, are not shown but are nonetheless important and in general use. The determination of lead in blood is complicated by matrix effects in most of the methods shown, which makes differences in results between methods a perplexing problem and the need for a standard reference material more urgent.

Table 5.9 Analytical Methods in NIOSH–OSHA Standards Completion (5)

Analytical Method		Number of Methods
Gas chromatography	Flame ionization	144
	Electrolytic conductivity	8
	Flame photometric	6
	Electron capture	5
	Alkali flame ionization	2
Atomic absorption spectrophotometry		
Flame		23
Flameless		3
UV/visible spectrophotometry		13
Specific ion electrode		4
Combustible gas meter		2
Gravimetric		2
X-Ray diffraction		2
Fluorescence		1
High pressure liquid chromatography		1

4.3 Some Newer Techniques

4.3.1 Inductively Coupled Plasma–Optical Emission Spectroscopy (ICP–OES)

The capability for simultaneous, multielement determinations over large dynamic ranges has made ICP–OES a promising technqiue (26). Minimal matrix effects are seen, and the determination of elements in 25 to 50 μl volumes of biological fluids has been demonstrated (31). Some possible applications include the determination of more than 45 elements with good detection limits (< 0.03 $\mu g/ml$), including elements difficult to determine by other methods such as boron, phosphorus, hafnium, uranium, tungsten, yttrium, zirconium, and the rare earths.

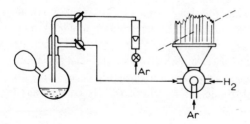

Figure 5.4. Apparatus for hydride generation atomic absorption spectrophotometry.

ANALYTICAL MEASUREMENTS

Table 5.10 Analytical Methods Used in Industrial Hygiene

Method	Example	Sample Working Range (μg)	Reference[a]
Chromatography			
Gas/flame ionization	Aniline	100–1200	S 310
	Chloroform	375–11,000	S 351
	Dimethylamine	180–2700	S 142
	Stoddard solvent	900–27,000	S 382
	o-Toluidine	250–3000	S 168
Gas/electron capture	Nitroglycerine	3–45	S 216
Gas/electrical conductivity	Tetrachloronaphthalene	20–600	S 130
Gas/flame photometric	White phosphorus	0.2–20	P&CAM 257
High pressure liquid/UV	Phthalic anhydride	100–3600	S 179
Spectrophotometry			
Molecular UV/visible	Chromates (as CrO_3)	0.6–14	S 317
	Nitrogen dioxide	0.5–18	P&CAM 231
	Ozone	5–20	S 8
	Selenium compounds	0.06–1	P&CAM 181
Molecular fluorescence	Oil mist	5–1500	S 272
Atomic absorption/flame	Chromium compounds	5–250	S 323
	Lead compounds	5–180	S 341
Atomic absorption/graphite	Beryllium compounds	0.06–0.6	S 339
Atomic absorption/hydride	Arsenic compounds	0.02–3	40
Atomic absorption/vapor	Mercury	0.001–2.5	P&CAM 175
X-Ray diffraction	Crystalline silica	14–300	S 315
	Crystalline silica	20–2000	P&CAM 259
	Zinc oxide	25–2000	P&CAM 222
Electrochemistry			
Anodic stripping voltammetry	Lead compounds	0.05–3	P&CAM 195
Specific ion electrode	Hydrogen bromide	100–3000	S 175
Titration (colorimetric)	Sulfur dioxide	50–1000	P&CAM 204
Mass spectrometry	Sulfur dioxide	200–60,000	P&CAM 204

[a] Numbers following "P&CAM" or "S" refer to methods found in Reference 7.

4.3.2 Electron Spectroscopy for Chemical Analysis (ESCA)

ESCA is a technique for the examination of surfaces to a depth of only a few Ångsroms. Since the energies of photoelectrons from the target atoms are measured, their binding energies can be calculated. This gives the analyst information about the valence

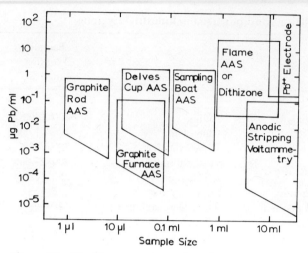

Figure 5.5. Working ranges for analytical methods for lead.

states of the target atoms as well as a semiquantitative, multielement analysis. Through the use of ESCA, ambient air pollution particulates were studied and found to contain sulfur tentatively identified as sulfate, sulfite, and sulfide ions, in addition to neutral sulfur and surface-bonded sulfur dioxide and sulfur trioxide (32, 33). Figure 5.7 presents an ESCA spectrum obtained on an air particulate sample (Nuclepore filter)

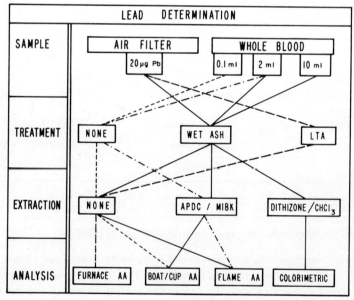

Figure 5.6. Sample treatment procedures used in some analytical methods for lead.

ANALYTICAL MEASUREMENTS

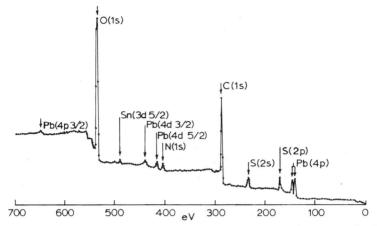

Figure 5.7. ESCA spectrum for air particulates from a copper smelter. From Eller (34).

from a smelter. The presence of lead, tin, and sulfur is indicated; the sulfur was present as sulfate in this sample (34).

4.3.3 X-Ray Fluorescence (XRF)

Advantages of XRF for environmental samples are simultaneous multielement analyses, with no sample treatment necessary in most cases (35, 36). The analysis is nondestructive, and individual particles of diameter approximately 1 μm or larger can be analyzed. Figure 5.8 is a scanning electron photomicrograph of air particulates sampled on a

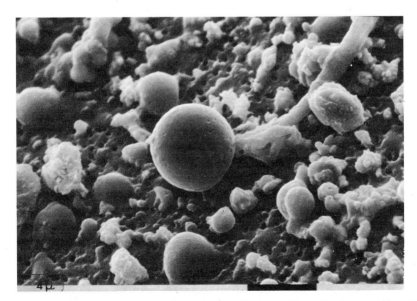

Figure 5.8. Scanning electron micrograph of air particulates on 0.4 μm Nucleopore filter (4600 ×).

Nuclepore filter at a sulfite pulp mill employing calcium bisulfite. Microanalysis of several of the particles by XRF determined that the major elements present were sulfur, calcium, and iron. Smaller amounts of phosphorus, silicon, aluminum, and magnesium were also found.

4.3.4 Ion Chromatography (IC)

A gap in inorganic and organic analysis, the rapid quantitation of anions and cations, has some promise of aid from the recent development of ion chromatography. Essentially a liquid chromatograph using pellicular ion exchange analytical columns and an electrical conductivity detector, the equipment elutes aqueous samples with carbonate-bicarbonate or other buffers for anion analysis, or acidic eluents for cation analysis (37). Applications have been demonstrated for the determination of sulfates and nitrates in ambient aerosols (38), and for electrolytes, including ammonium ion, in serum, urine, and other biological fluids (39). Conditions for the determination of a number of amines and low molecular weight organic acids have also been established (37). A chromatogram showing separation of seven anions in less than 15 min appears in Figure 5.9 (34).

4.3.5 Derivative Spectroscopy

Using wavelength modulation, derivative ultraviolet or visible spectra are obtained, providing a sensitive means of detection of gases (40). Detection limits for some gases are

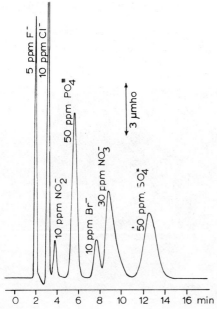

Figure 5.9. Chromatogram of anions; eluent: 0.003 M HCO_3^-, 0.0024 M CO_3^{2-}. From Eller (34).

nitric oxide, 5 ppb; nitrogen dioxide, 40 ppb; ozone, 40 ppb; benzene, 25 ppb; formaldehyde, 200 ppb; and mercury vapor, 0.5 ppb.

5 SAMPLING AND ANALYTICAL GAPS

A number of factors relating to sampling, analysis, or the analyte itself may hinder the development of validated sampling and analytical methods. This section discusses some of these factors, along with examples (41), with the hope that future research will fill these gaps.

5.1 Sampling-Related Problems

Three problems related to sampling are poor collection efficiency (low capacity), poor sample storage characteristics, and low desorption efficiencies. An example of the first type is the low capacity of conventional charcoal tubes for some of the chlorofluoromethanes, ethyl chloride, methyl acetylene, and stibine. Alternate solid sorbents should be investigated. Substances that tend to deteriorate over a period of several days when sorbed on charcoal tubes are methyl formate, nitromethane, nitroethane, 2-nitropropane, crotonaldehyde, and diglycidyl ether. Desorption efficiencies below 75 percent are characteristic of ethanolamine, n-butylamine, diisopropylamine, and other amines on silica gel for freshly spiked samples. Also, chloroacetaldehyde, diphenyl, furfural, and 1-nitropropane show low desorption efficiencies from activated charcoal.

5.2 Analysis-Related Problems

In some cases the analytical method is the limiting factor because it is lacking in specificity, sensitivity, or precision. Adaptation of a fluoride-specific electrode to the determination of perchloryl fluoride and chlorine trifluoride is unsuccessful because of interference by other fluorine compounds such as oxygen difluoride, and chlorine monofluoride. Also, oxidizing agents such as chlorine, bromine, nitrites, and chloramine interfere with colorimetric determinations for iodine (42) and chlorine dioxide (43).

For personal samples, gas chromatography is not sensitive enough in the cases of acrylamide, dinitrotoluene, nitrogen trifluoride, and perchloromethylmercaptan, while atomic absorption spectrophotometry lacks sensitivity for hafnium, osmium, and tantalum.

Poor analytical precision is characteristic of the particle-count methods that have been used for graphite, mica, portland cement, soapstone, and talc. Also showing poor reproducibility are gas chromatography for dibutyl phosphate, maleic anhydride, acetic acid, methylamine, and Arachlor 1242, and spectrophotometric methods for anisidine (44), p-phenylenediamine (45), and oxalic acid (46).

5.3 Problems Related to the Analyte

Some analytes, such as boron trifluoride, *tert*-butyl chromate, decaborane, diborane, nickel carbonyl, and sulfur monochloride, are very reactive in the presence of oxygen or water vapor, therefore very difficult to collect. Possible solutions include direct measurement in air by spectral means or specific, reactive sorbents.

REFERENCES

1. U.S. National Bureau of Standards, *Accuracy in Trace Analysis: Sampling, Sample Handling, Analysis. Proceedings of the Seventh Materials Research Symposium,* Superintendent of Documents (No. C13.10:422), Washington, D.C., 1976.
2. J. P. Cali, G. N. Bowers, Jr., and D. S. Young, *Clin. Chem.*, **19**, 1208 (1973).
3. U.S. National Bureau of Standards, *NBS Standard Reference Materials,* NBS Special Publication 260, Washington, D.C., 1977.
4. P. M. Eller and J. C. Haartz, *Am. Ind. Hyg. Assoc. J.*, **38**, 116 (1977).
5. D. G. Taylor, R. E. Kupel, and J. M. Bryant, "Documentation of the National Institute for Occupational Safety and Health Validation Tests," Department of Health, Education and Welfare Publication (NIOSH) 77-185, Washington, D.C., 1977.
6. *National Institute for Occupational Safety and Health Manual of Sampling Data Sheets,* 1977 ed., Department of Health, Education and Welfare Publication (NIOSH) 77-159, Washington, D.C., 1977.
7. *National Institute for Occupational Safety and Health Manual of Analytical Methods,* 2nd ed., Department of Health, Education, and Welfare Publications (NIOSH) 77-157-A,B,C, and 78-175, Washington, D.C., 1977, 1978.
8. J. V. Crable and R. G. Smith, *Am. Ind. Hyg. Assoc. J.*, **36**, 149 (1975).
9. P. Schlecht, National Institute for Occupational Safety and Health, personal communication.
10. P. M. Eller, J. C. Haartz, R. D. Hull, and B. Frumer, *Am. Ind. Hyg. Assoc. J.* (submitted for publication).
11. F. W. Williams, J. P. Stone, and H. G. Eaton, *Anal. Chem.*, **48**, 442 (1976).
12. E. V. Ballou, Ed., *Second National Institute for Occupational Safety and Health Solid Sorbents Roundtable,* Department of Health, Education and Welfare Publication (NIOSH) 76-193, Washington, D.C., 1976, p. 3.
13. G. O. Wood and R. G. Anderson, *Am. Ind. Hyg. Assoc. J.*, **36**, 538 (1975).
14. L. D. Butler and M. F. Burke, *J. Chromatogr. Sci.*, **14**, 117 (1976).
15. H. K. Dillon, W. J. Barrett, and P. M. Eller, *Am. Ind. Hyg. Assoc. J.*, **39**:608 (1978).
16. E. D. Palmes, A. F. Gunnison, J. DiMattio, and C. Tomczyk, *Am. Ind. Hyg. Assoc. J.*, **37**, 570 (1976).
17. E. V. Ballou, Ed., *Second National Institute for Occupational Safety and Health Solid Sorbents Roundtable,* Department of Health, Education and Welfare Publication (NIOSH) 76-193, Washington, D.C., 1976, p. 55.
18. R. H. Hill, Jr., C. S. McCammon, A. T. Saalwaechter, A. W. Teass, and W. J. Woodfin, *Anal. Chem.*, **48**, 1395 (1976).
19. R. L. Larkin, J. V. Crable, L. R. Catlett, and M. J. Seymour, *Am. Ind. Hyg. Assoc. J.*, **38**, 543 (1977).
20. T. H. Lockwood and L. P. Limtiaco, *Am. Ind. Hyg. Assoc. J.*, **36**, 57 (1975).
21. T. T. Gorsuch, in Reference 1, p. 491.

22. A. A. Al-Sibbai and A. G. Fogg, *Analyst,* **98,** 732 (1973).
23. L. Murthy, E. E. Menden, P. M. Eller, and H. G. Petering, *Anal. Biochem.,* **53,** 365 (1973).
24. American Conference of Governmental Industrial Hygienists, *Threshold Limit Values,* ACGIH, Cincinnati, Ohio, 1977.
25. C. H. Hartman, *Anal. Chem.,* **43,** 113A (1971).
26. V. A. Fassel and R. N. Kniseley, *Anal. Chem.,* **46,** 1110A (1974).
27. H. L. Kahn and J. S. Sebestyen, *At. Absorpt. Newsl.,* **9,** 33 (1970).
28. M. M. Joselow and J. D. Bogden, *At. Absorpt. Newsl.,* **11,** 99 (1972).
29. J. F. Lech, D. Siemer, and R. Woodriff, *Environ. Sci. Technol.,* **8,** 840 (1974).
30. *National Institute for Occupational Safety and Health Manual of Analytical Methods,* 2nd ed., Department of Health, Education and Welfare Publication (NIOSH) 77-157-A,B,C, Washington, D.C., 1977; P&CAM 214.
31. R. N. Kniseley, V. A. Fassel, and C. C. Butler, *Clin. Chem.,* **19,** 807 (1973).
32. N. L. Craig, A. B. Harker, and T. Novakov, *Atmos. Environ.,* **8,** 15 (1974).
33. T. Novakov, P. K. Mueller, A. E. Alcocer, and J. W. Otvos, *J. Colloid Interface Sci.,* **39,** 225 (1972).
34. P. M. Eller, unpublished results.
35. J. R. Rhodes, *Am. Lab.,* July 1973.
36. T. G. Dzubay, Ed., *X-Ray Fluorescence Analysis of Environmental Samples,* Ann Arbor Science Publishers, Ann Arbor, Mich., 1977.
37. H. Small, T. S. Stevens, and W. C. Bauman, *Anal. Chem.,* **47,** 1801 (1975).
38. J. Mulik, R. Puckett, D. Williams, and E. Sawicki, *Anal. Lett.,* **9,** 653 (1976).
39. C. Anderson, *Clin. Chem.,* **22,** 1424 (1976).
40. R. N. Hager, Jr., *Anal. Chem.,* **45,** 1131A (1973).
41. D. G. Taylor, National Institute for Occupational Safety and Health, personal communication.
42. J. K. Johannesson, *Anal. Chem.,* **28,** 1475 (1956).
43. American Public Health Association, *Standard Methods for the Examination of Water and Wastewater,* 13th ed., APHA, New York, 1971.
44. J. T. Steward, T. D. Shaw, and A. B. Ray, *Anal. Chem.,* **41,** 360 (1969).
45. *National Institute for Occupational Safety and Health Manual of Analytical Methods,* 2nd ed., Department of Health, Education and Welfare Publication (NIOSH) 77-157-A,B,C, Washington, D.C., 1977, P&CAM 142.
46. J. Bergerman and J. S. Elliot, *Anal. Chem.,* **27,** 1014 (1955).

CHAPTER SIX

Measurement of Worker Exposure

JEREMIAH R. LYNCH

1 INTRODUCTION

This chapter explains why measurements of air contaminants are made, discusses the options available in terms of number, time, and location, and relates these options to the criteria that govern their selection and the consequences of various choices.

A person at work may be exposed to a certain number of potentially harmful agents for as long as a working lifetime, upward of 40 years in some cases. These agents occur in mixtures, and the concentration varies with time. Exposure may occur continuously or at regular intervals or in altogether irregular spurts. As a result of exposure to these agents, the worker is being dosed, and depending on the magnitude of the dose, some harmful effect may occur. All measurements in industrial hygiene ultimately relate to the dose received by the worker and the harm it might do.

Early investigators of the exposure of workers to toxic chemicals encountered conditions so obviously unhealthy as evidenced by the existence of frank disease that quantitative measurements of the work environment to estimate the dose received by the afflicted were not needed to establish cause-and-effect relationships. At the same time the ability of these early industrial hygienists to make measurements was severely limited because suitable sampling equipment did not exist and existing analytical methods were insensitive. Pumps were driven by hand, equipment was large and heavy, filters changed weight at random, gases were collected in fragile glass vessels, absorbing solutions spilled or were sucked into pumps, and laboratory instrument sophistication was bounded by an optical spectrometer. To collect and analyze only a few short period samples required several days of work and the probability of failure due to one of many possible equipment defects or other mishaps was high. Consequently few measurements were made and much judgment was applied to maximize the representativeness of the measurements, or even as a substitute for measurement.

Changes in working conditions, in technology, and in society have caused the old methods of measurement to be reexamined.

- With few exceptions workplace exposure to toxic chemicals is not grossly excessive but is close to what is commonly accepted as a safe level.
- As a consequence of the reduction of exposure, frank occupational disease is rarely seen and much of the disease now present results from multiple factors of which occupation is only one.
- Workers are demanding to know how much toxic chemical exposure they are receiving, and this results in a need even to document the negative.
- Technology has provided enormously improved sampling equipment that is rugged and flexible. This equipment, used with analytical instruments of great selectivity and sensitivity, has largely replaced the old wet chemical methods.

As a consequence of these changes in the workplace and advances in technology, it is now both necessary and possible to examine in far more detail the way in which workers are exposed to harmful chemicals. Personal sampling pumps permit collection of contaminants in the breathing zone of a mobile worker. Pump-collector combinations are available for long and short sampling periods. Systems that do not require the continual attention of the sample taker permit the simultaneous collection of multiple samples. Automated sampling and analytical systems collect data day after day. Sorbent–gas chromatograph techniques permit the simultaneous sampling and analyses of mixtures and, when coupled with mass spectrometers, identify obscure unknowns. Sensitivities have improved to the degree that tens and hundreds of ubiquitous trace materials begin to be noticeable.

At the same time the demands placed on our information gathering systems are more acute. Now we must answer not only the question, "Is exposure to this agent likely to harm anyone?" we are being called on to provide data for a great many other purposes. Worker exposure must be documented to comply with the law (1). Employees are demanding to be told what they are breathing, even in the absence of hazard (2). Epidemiologists need data on substances not thought to be hazardous, to relate to possible future outbreaks of disease. Design engineers need contaminant release data to relate to control options. Process operators want continuous assurance that contaminant levels are within normal bounds. Management information systems that issue status reports when queried require monitoring data inputs. Data needs are so pervasive that a tendency to monitor for the sake of monitoring develops.

2 OBJECTIVES OF EXPOSURE MEASUREMENT

The central question that must be asked before any sampling strategy is selected is, "What use is going to be made of the data?" That is, what questions will the data

MEASUREMENT OF WORKER EXPOSURE

answer, or what external information need will be satisfied? At the same time it is useful to consider other questions that will need answers or other information that will be needed so that an efficient scheme to satisfy several requirements can be devised.

2.1 Legal Requirements

Section 6b7 of the Occupational Safety and Health Act provides for ". . . monitoring or measuring employee exposure at such locations and intervals, and in such a manner as may be necessary for the protection of the employees" (3). Responsible and effective implementation of this congressional intent requires that regulatory language be devised both to discriminate between employers who may have a hazard present and those who almost certainly do not, and to do so by a scheme that is easy to understand and implement. The ideal regulation should very clearly not apply to the vast majority of establishments, which have no conceivable hazard resulting from the substance being regulated, and proceed to apply requirements of increasing strictness and burden as the significance of the hazard in an establishment increases, ultimately calling for measurement of sufficient frequency to ensure that the potential for harm is fully assessed in the few establishments where exposures are great enough to create a real danger.

This sorting of workplaces by level of hazard can be done in the regulatory context, by prescribing a series of thresholds or triggers that lead to increasingly stringent requirements for a decreasing number of employers. First, all employers who do not have the substance present in the establishment are not required to monitor or to train or inform employees about that substance or to offer medical examinations. Although the "presence" of a material seems a simple enough criterion that everyone would interpret in the same way, the extreme bounds of interpretation, which are of concern in legal arguments, include the presence of as little as a few molecules of a gas or a single asbestos fiber. As analytical techniques become more sensitive, it is being found that almost everything is almost everywhere, at least at the level of a few molecules. What is needed in a regulation is an exclusion, such as a percentage in a liquid, below which the substance is not considered to be present, to prevent ridiculous interpretations.

For employers who have the substance present in the workplace above the excluded level, it must be determined whether there is any possibility that the substance is released into the workplace such that workers may be exposed. The setting of this threshold must reflect consideration of the conditions under which the substance is present and the consequences of release. Thus nuclear reactor decay products are continuously monitored against the possibility of leaks even when they are hermetically sealed; they are not released into the workplace except under very rare emergency circumstances. Likewise cadmium released into the workplace as a result of silver soldering should be monitored, but it is not necessary to measure exposure to cadmium resulting from coatings on parts in an auto store stockroom. No simple "potential for release" trigger has yet been found; thus there is some regulatory error (employers included who should not have been, and vice versa) at this decision point.

A further step is needed, therefore, before a full monitoring program with its consequent expense is mandatory. One such step is to require a single measurement of the exposure of the maximum risk employee under conditions when the exposure is likely to be the greatest. If the result of this measurement is sufficiently below the occupational exposure limit (OEL) to permit confidence about the result (4), we could say that no significant exposure is occurring and no more would be required of this employer than if the substance were not released in the first place. This single initial measurement scheme, however, is not appropriate where there is such a massive presence of a highly toxic substance that continuous vigilance must be maintained against the possibility of leaks or other inadvertent releases.

In cases where it has been established, by means of an initial measurement or from data from other sources such as prior measurements of worker illness, that significant exposure is occurring, possibly over the OEL on occasion, a regular program of periodic monitoring should be required. The frequency of monitoring should relate to the level of exposure and should consider the probability of trends between measurements that might lead to conditions with unacceptable consequences. Figure 6.1 presents an example of a scheme proposed by the National Institute for Occupational Safety and Health (NIOSH) that incorporates some of the foregoing factors.

2.2 Exposure Evaluation

The usual reason for measuring worker exposure to a toxic chemical is to evaluate the health significance of that exposure. To do this, it is necessary to have some reference level against which to compare the result. Traditionally, the Threshold Limit Values (TLVs) for Airborne Contaminants of the American Conference of Governmental Industrial Hygienists (ACGIH) (5) have been used to represent safe levels or "conditions under which it is believed that nearly all workers may be repeatedly exposed day-after-day without adverse effect." Unfortunately, published TLV's cover only a small fraction of the variety of chemicals that occur in industrial workplaces, albeit the most common ones. Where there is exposure to a substance for which there is no TLV or comparable reference level such as a legal standard, it is necessary to develop a local standard for use in a particular plant or company. These standards for substances whose toxicology is not well known are generally set to avoid acute effects in man or animals and, often by analogy to other better-documented substances, at a level low enough to make chronic effects unlikely. When very little is known about a substance, it may be possible only to estimate a lower level at which it is reasonably certain that no adverse effects occur and an upper level at which adverse effects are likely. The width of the gap between these two levels is a measure of the uncertainty that needs to be considered in evaluating the results of exposure measurements.

2.3 Control Effectiveness

When changes are made that affect the release of substances that are contributing to worker exposure, measurements of the magnitude of that change may be needed. In the

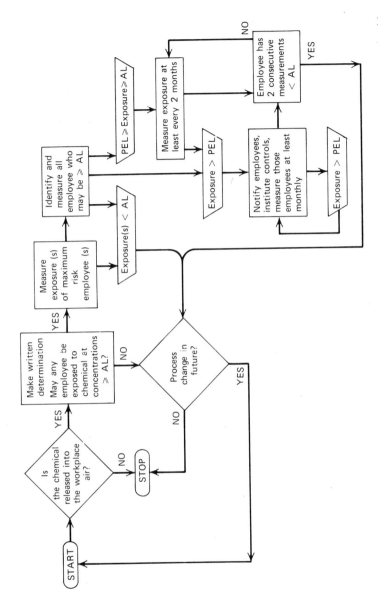

Figure 6.1. NIOSH-recommended employee exposure determination and measurement strategy. Each individual substance health standard should be consulted for detailed requirements: AL = action level; PEL = permissible exposure limit.

simplest case, before-and-after measurements are made when a new control, such as a local exhaust hood, is installed on a contaminant release point. From the results of these measurements, it is possible to predict the reduction in worker exposure, which can be confirmed by subsequent measurement.

Unfortunately the situation is rarely that simple. Most worker exposures are caused by multiple release points, creating a work environment of complex spatial and temporal concentration variations through which the worker moves in a not-altogether-predictable manner. Furthermore, interaction between several release points or other factors in the environment may confound the results. A needed improvement such as a new exhaust hood may be seen to be without effect because the building is air starved, or a poor hood may seem to function well because of exceptional general ventilation. The time of contaminant release may depend on obscure and uncontrollable process operation factors. As with measurements made for other purposes, control evaluation studies must be carefully designed and are likely to consist of a series of factorial measurements analyzed by statistical methods.

2.4 Methods Research

Industrial hygiene research hypotheses often take the form "method A is the same as method B." If we are unable to reject this hypothesis in a carefully designed experiment, then we accept that methods A and B are equivalent within our limits of error and given the bounds of the experimental conditions (6–9).

Method equivalence research usually requires extensive laboratory work, but in most cases field sampling is necessary because completely realistic samples usually cannot be generated in laboratory chambers, and the difficulties of making field measurements introduce errors that may affect one method more than another. For these reasons most practicing industrial hygienists tend to distrust assertions of equivalence that are not backed up by field data. To be credible, experiments of this kind should clearly define the range of conditions over which the equivalence has been tested. Personal versus area equivalence of coal mine dust measurements made in long-wall mines should not be assumed to hold in room-and-pillar mines. Manual versus automated asbestos counting relationships based on chrysotile are of little value when counting amosite fibers.

Enough data should be collected not only to determine whether the methods are correlated, but also the ability of a measurement by one method to define the confidence limits on a prediction of the result that would have been obtained by the other method (Figure 6.2). Often a high correlation coefficient is obtained when many pairs of measurements have been made, indicating that the two methods are certainly related; yet the scatter is such that one method may be used as a predictor of the other method only to within an order of magnitude. The design of experiments for the purpose of measuring method equivalence is beyond the scope of this chapter. However such considerations as environmental variability as related to location, time, and numbers,

MEASUREMENT OF WORKER EXPOSURE

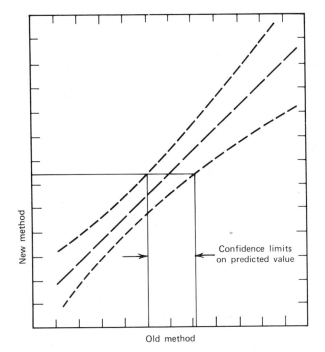

Figure 6.2. Methods comparison.

which are described below, should be taken into account, to maximize the range of conditions over which the equivalence is rated without introducing so much error that the relationship has no predictive ability.

2.5 Leak Detection

Quite apart from measurements for the purpose of evaluating intentional changes in control systems are measurements to detect leaks or other inadvertent loss of control. In addition to leaks and other "fugitive" emissions, which are a constant danger in chemical plants, local exhaust systems can fail because of plugged ducts or fan failure. Patterns of contaminant generation may change because of loss of temperature control in a vessel or tank, dullness of a chisel in a mine or quarry, or bacterial contamination of a cutting oil. These events cannot always be detected by changes in process parameters, and it may be unacceptable to wait until they show up in exposure measurements because the margin for error in a control system designed for a very strict standard, such as vinyl chloride, may be too small, or as in the case of hydrogen sulfide, because the consequences of overexposure, even for a short period, are too serious.

2.6 Illness Investigations

When an employee presents frank occupational disease, such as lead poisoning, confirmed by both physical findings and analysis of biological materials, or is known to have been overexposed based on analysis of biological materials, the industrial hygienist must determine the cause of the overexposure. Although the conditions that led to the overexposure may still exist and so may be evaluated after the event, it is also possible that the overexposure resulted from past episodes that were not observed and evaluated when they occurred. Under such conditions it is necessary to construct a history of exposure opportunities and to estimate how significant they may have been in the etiology of the present case. Exposure, of course, need not always be by inhalation and may include off-the-job activities. In some cases it may be necessary to reenact or simulate an event to measure what may have happened—being careful, of course, to ensure that all participants are protected.

A much more difficult investigation is the search in an occupational setting for the cause of an outbreak of illness or complaints of illness that may or may not be of occupational origin, or even if related to the job may result from factors other than exposure to toxic substances. Investigation of possible chemical exposure, however unlikely, can be extremely complex, since a release from unusual sources (off-gassing of plastics, air-conditioning system contamination) that may have occurred in the past must be considered.

3 SOURCES OF WORKER EXPOSURE

The core concern of industrial hygiene is the prevention of disease arising out of the workplace. Toxic substances cause disease when some amount, or dose, enters on or in the body. Workers have contact with, or are exposed to, toxic substances by inhaling them with the air they breathe and by several other routes. To accurately appraise the total dose being received by a worker, thus to predict biological effect, it is necessary to understand how exposure occurs.

3.1 Production Operations

Industry is generally thought of in terms of continuous repetitive operations that generate air contaminants to which workers are more or less continuously exposed. Paint is sprayed on parts passing continuously in front of a worker. Dust is generated by a foundry shakeout on a continuous casting line every few seconds. Fumes seep steadily from cracks in aluminum smelting pot enclosures. Welders join structural members on a production basis, with only short breaks between welds. Operators watch controls in the midst of a chemical plant that allows the steady release of fugitive streams. In some case, such as a grinder cleaning sand from a casting, the concentration of the contaminant is closely related to the work performed and is probably high in the workers breathing zone than it would be several feet away. Other workers, such as dorfers and creelers in a cotton spinning mill, are exposed to dust released from hundreds of bobbins, and their own activities, short of leaving the workplace, have little effect on their exposure. All these continuous exposures are actual rather than potential and present the least difficulty in evaluation.

3.2 Episodes

Much of the exposure workers receive occurs as a result of events or episodes that occur intermittently. Glue is mixed on Wednesday. The shaker mechanism breaks and a mechanic must enter the baghouse. Coke strainers preceding a pump need to be dumped when the pressure drop becomes excessive. A drum falls off a pallet and ruptures on the floor. Samples of product are taken every two hours. A pressure relief valve opens. A packing gland bursts. A reactor vessel cover is driven from its hinges by overpressure in the vessel.

These exposure events can be periodic or they can occur at irregular intervals. They may be planned and predictable or altogether unanticipated. Some events are frequent and result in small exposures, whereas others may be catastrophic events causing massive exposures and even fatalities. As a class, episodic exposure events result in a significant fraction of the exposure burden of many workers and may not be ignored. Their evaluation, however, is extremely complex, and often only broad estimates of the probability of an unlikely event and the consequent risk may be available.

3.3 Noninhalation Exposure

When relating exposure to biological effect, either in the case of an epidemiologic study or because of an outbreak of illness that might be related to occupation, the total dose is the relevant quantity, not merely the dose received by inhalation. Amounts of a substance entering the body by any route may contribute to the total dose.

Many substances, especially fat-soluble hydrocarbons and other solvents, can enter the body and cause systemic damage directly through the skin when the skin has become wet with the substance by splashing, immersion of hands or limbs, or exposure to a mist or liquid aerosol. Some substances, such as amines and nitriles, pass through the skin so rapidly that the rate at which they enter the body is like that of substances inhaled or injested. The prevention of skin contact to phenol is as important as preventing inhalation of airborne concentrations. A few drops of dimethyl formamide on the skin can contribute a body burden similar to inhaling air at the TLV.

For some substances with low vapor pressure, like benzidine, skin absorption is the most important risk. Benzene, on the other hand, though absorbed through the skin, is absorbed at such a low rate that skin contact probably contributes little to the body burden. To judge the degree to which skin absorption is contributing to exposure, it is necessary to consider <u>both the rate of absorption and the degree of contact</u>. Clothing wet with a substance that remains on the worker for prolonged periods provides the maximum contact short of immersion. On the other hand, the poultice effect does not occur on wet unclothed skin; thus evaporation can take place and the result is less severe. Contact with mist that does not fully wet the clothes or body but merely dampens them is not as severe as being splashed with the bulk liquid. Theoretically gases and vapors may be absorbed through the skin, but the rate would be extremely small except under extreme conditions. Protective clothing, which is impervious to the substance, will reduce adsorption to nil on protected areas (10). However a leaky glove that has become filled with a solvent is providing contact with the hand equivalent to immersion. Barrier creams are often used to prevent dermatitis but may not always prevent skin absorption (11). Although skin contact and absorption must be considered as contributing to the dose for many materials, few quantitative data on rate of absorption are available, and these are in a form that is difficult to apply in an industrial setting (12, 13). Furthermore, such factors as part of the skin exposed, sweating, and the presence of abrasions or cuts can cause order of magnitude differences (14).

While on the subject of skin absorption, it should be noted that toxic substances may pass through the skin by intentional or unintentional injection. Opportunities for all kinds of materials to enter the body by injection connected with drug abuse or therapeutic accident are obvious. Bulk liquids may also break through the skin and enter the bloodstream without the aid of a needle when driven into the body as high velocity projectiles released from high pressure sources. Airless paint spray and hydraulic systems (15) often use pressures in this range, and such pressures often occur inside pipes and vessels in chemical plants. Inadvertent cracking of a flange under pressure can cause the traumatic introduction of a toxic substance into the body. Solid particles may also enter this way and if soluble or radioactive, they may cause damage beyond the initial injury.

3.4 Ingestion

Although ingestion is an uncommon mode of exposure for most gases and vapors (16), it cannot be ignored in the case of certain metals such as lead. Indeed, for workers exposed in pigment manufacture and use, ingested lead, either coughed up and swallowed or taken on food, though less well adsorbed than inhaled lead, constitutes a significant fraction of the total burden. Spot tests (17) and tests with tracers have shown that materials present in a workplace tend to be widely dispersed over surfaces, thus support the essential rule prohibiting eating and smoking where toxic substances are present.

3.5 Nonoccupational Exposure

In addition to the exposure to toxic substances a worker receives at work, an increment of dose may also be delivered during nonworking hours. Air (18), water (19), and food all contain small amounts of toxic substances that may also be found in the occupational

MEASUREMENT OF WORKER EXPOSURE

environment. As a rule the nonoccupational dose is at least an order of magnitude below the occupational dose, just as community air standards are much lower than TLVs, and many substances present in industry are very rare in the community. Yet in certain cases these pollutants have an impact. The consequence of arsenic exposure resulting from copper smelting in northern Chile is difficult to assess, since arsenic poisoning there is endemic as a result of naturally occurring drinking water contamination. Chronic bronchitis resulting from air pollution in the industrial midlands of England is confounded with the lung diseases of coal miners. The cardiovascular consequences of urban carbon monoxide exposure may be not unlike those of marginal industrial exposures.

In addition to these somewhat involuntary exposures to toxic substances, many workers have hobbies or other leisure activities. Acoustic traumata from loud music and target practice are well known. Lead exposure on police and presumably private firing ranges is significant. Epoxies are used in home workshops, and garden chemicals contain a variety of economic poisons. Although no data come to hand, leisure activity exposure to toxic substances rivaling high but permissible work exposure must be rare. As a result of toxic substance legislation and product liability questions, many toxic substances are no longer sold as consumer products. Benzene, carbon tetrachloride, lead paint, and asbestos have been disappearing from retail stores. Soon it will be a rare event indeed to find a known carcinogen or a serious toxin in any consumer product. Furthermore although a worker may apply far more diligence to a hobby than to the job, hobby activity tends to be intermittent, with long lapses. Serious acute exposures are not unlikely, but chronic disease is rare if not unknown. Thus although a supplementary increment of dose may be inferred from an off-the-job activity known to involve the toxic substance in question, it should not be assumed to be the primary source unless the assumption is supported by measurements.

Intentional exposure to solvents for their narcotic effects is well documented (20–25). Glue sniffing, gasoline sniffing, ingestion of denatured alcohol, methanol, and even turpentine have been reported. This category of chemical abuse results in exposure and often damage far beyond that normally encountered in industry. The industrial hygienist must be alert to the possibility that a case of disease may be related to intentional addictive exposure.

3.6 Environmental Variability

An important factor in the design of any measurement scheme is the degree of variability in the system being observed. This variability has a primary effect on the number of samples to be taken and the accuracy of the results that can be expected. When the system being observed is the exposure of a worker to a toxic substance in a workplace, variability tends to be quite high. During the course of a day there are minute-to-minute variations and daily averages vary from day to day.

A typical recording of actual intraday environmental fluctuations appears in Figure 6.3. Highly variable environmental data of this kind, which are truncated at zero,

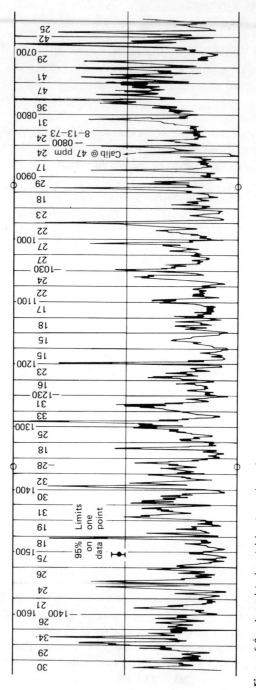

Figure 6.3. Actual industrial hygiene data showing intraday environmental fluctuations. Range of carbon monoxide data on chart is 0 to 100 ppm.

generally have been found to be best described by the log normal rather than the normal distribution (Figure 6.4). This two-parameter distribution (26) is described by the geometric mean (GM) and the geometric standard deviation (GSD), which is the antilog of the standard deviation of the log-transformed data (27). A rough equivalence between GSDs and the more familiar coefficient of variation (Table 6.1) is valid up to a GSD of about 1.4. Figure 6.5 illustrates the consequence of various values in terms of spread of data. Models and data derived from community air pollution measurements (28–31) have been useful in studying the in-plant micro environment. In general, studies of occupational environmental variability (32, 33) have confirmed the usefulness of the log normal description of data and have yielded GSDs in the range of 1.5 to 1.7, with few less than 1.1 and as many as 10 percent exceeding 2.3.

One can speculate on the probable causes for this variability in worker exposure. Fugitive emissions, which are like frequent small accidents rather than main consequences of the production process, occur almost randomly. Production rates change. Patterns of overlapping multiple operations shift irregularly. Distribution of contaminants by bulk flow, random turbulence, and diffusion is uneven in both time and space. Through all this our target system, the worker, moves in a manner that is not altogether predictable. These and other uncertainties are the probable causes of the

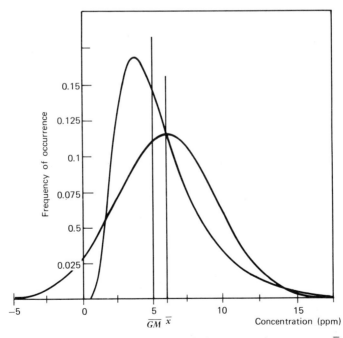

Figure 6.4. Log normal and normal distributions with the same arithmetic maan ($\bar{x}$) and standard deviation: $\overline{GM}$ = geometric mean.

Table 6.1 Log and Arithmetic Standard Deviation Equivalence

Geometric Standard Deviation (GSD)	Coefficient of Variation (CV)
1.05	.049
1.10	.096
1.20	.18
1.30	.27
1.40	.35

variability typically observed, but the situation is far too complex to pinpoint each individual source of variation and its consequences.

Sampling schemes devised to fit this situation may deal with the variability from all sources as a single pool and derive whatever accuracy is required by increasing the number of samples. Alternatively one can postulate that a large part of the variability is due to some observable factor or factors and by means of a factorial design account for

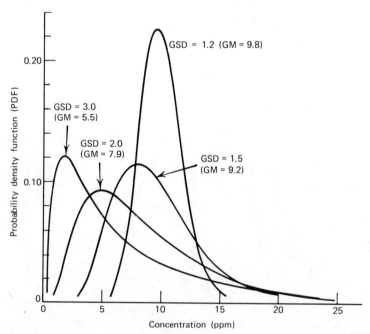

Figure 6.5. Log normal distributions for arithmetic mean concentration of 10 ppm.

MEASUREMENT OF WORKER EXPOSURE

this portion of the variance, leaving only the residual to be dealt with as error. No hard and fast rules can be made regarding the choice except that it seems logical to expect the factors identified to account for a statistically significant fraction of the variance (F test) if it is to be worthwhile to sample and analyze the data in this manner. Shift, season of the year, wind velocity, and rate of production are some of the factors affecting worker exposure that may be worth singling out to ensure that part of the gross variance can be assigned to an accountable cause. Even if successful, it is likely that the residual or error variance will still be large, much larger than the variance due to measurement method inaccuracy, and will exert a major influence on our decisions.

4 CHARACTERISTICS OF EXPOSURE AGENTS

The physical and chemical properties of a substance are important because of the effect they have on exposure measurement quite apart from their effect on the magnitude of the exposure and its consequences.

The most important properties of any substance that affect sampling are those that determine whether it can be collected on a sampling medium, and treated or removed in a manner that permits analysis. Many vapors are readily adsorbed and desorbed from one or another of a variety of solid sorbents, but some high boiling materials are very difficult to desorb and some gases do not adsorb well. Similarly, most dust particles are collected efficiently by membrane filters with 0.8 μm and even larger pore sizes, however fresh fumes may pass through and hot particles can burn holes. Beyond these apparent physical and chemical dependencies of exposure measurement methods, there are some less apparent complications that frequently occur.

4.1 Vapor Pressure

Since liquids with high vapor pressure tend to evaporate completely when they are aerosolized, it is rarely necessary to measure them as mists. Liquids with very low vapor pressure may be present only as mists and must be measured with methods appropriate for aerosols. The situation can be very complicated for a liquid of intermediate volatility, high enough to produce a saturated vapor 10 percent or more of the TLV, yet low enough that mists will not quickly evaporate. If samples are collected on filters, only the mist will be caught and part of this liquid may be lost by evaporation into the air flowing through the filter during sampling, particularly if samples are collected over long periods. This effect undoubtedly results in a significant loss of the lower molecular weight two- and three-ring aromatic compounds present in coke oven effluents, when sampled by the usual filter methods.

When mixed mist-vapor atmospheres are sampled with a solid sorbent, the vapor is likely to be caught and retained efficiently, but the charcoal granules and associated support plugs do not constitute an efficient particulate filter (34). As a consequence, compound devices with a filter preceding a sorbent tube have been developed. Obviously the filter must come first, since what escapes the filter will be caught by the charcoal, but not vice versa. Despite severe sampling rate limitations, these systems are satisfactory when the sum of aerosol and vapor concentrations is to be related to a standard or effect. If, however, the aerosol and vapor mist are to be considered separately, perhaps because of different deposition sites or rates of adsorption, the single sum of the two concentrations is not enough. Furthermore, it cannot be assumed that the material collected on the filter is all the aerosol, since part of the liquid evaporates and is recollected on the charcoal. Since the reverse cannot be true, in cases when the vapor is more hazardous, this assumption is "safe" (i.e., protective). Size selective presamplers that collect part of the aerosol and none of the vapor may approximate the respiratory deposition-adsorption differential and so deliver a sample weighted toward the vapor to a biologically appropriate degree. However even here the possibility of evaporation from the cyclone, elutriator, impactor, or other component must be acknowledged. Additional research is needed in this area.

4.2 Reactivity

Most toxic substances are stable in air at the concentrations of interest under the usual ranges of temperature and pressure encountered in inhabited places to the degree that they may be sampled, transported to the laboratory, and analyzed without significant change or loss. Some few substances are in a transient state of reaction during the critical period when the worker is exposed. In the spraying of polyurethane foam, isocyanates present in aerosol droplets are still reacting with other components of the polymer while the aerosol is being inhaled, thus must be collected in a reagent that halts the reaction and yields a product that can be related to the amount of isocyanate that was there. Similarly solid sorbents such as charcoal can be coated with a reagent to react with the substance collected to yield a new compound; which will be retained and can be analyzed.

Unwanted reactions can also occur on the collecting media. For example, a substance that will hydrolyze may do so if brought in contact with water on the solid sorbent, particularly if a sorbent like silica gel, which takes up water well, is used. Substances that may coexist in air and not react or react only slowly because of dilution may react rapidly when concentrated on the surface of a collection medium.

4.3 Particle Size

The route of entry, site of deposition, and mode of action of an aerosol all depend on some measure of the size of particle, usually but not always the aerodynamic equivalent diameter. Exposure measurement methods must discriminate among different size particles to sort out those of greater or lesser biologic effect. The most common instance of such size selection is the exclusion of particles not capable of penetrating into the terminal alseoli when measuring exposure to pneumoconiosis-producing dust. Such "respirable mass sampling" by cyclone, horizontal elutriator, or various impactors is discussed at length elsewhere in this series and is not covered in detail here. However penetration into the smallest parts of the lung is not the only division point in size selective sampling. Cotton dust particles probably do not produce the chest tightness response by penetration into the deep lung. Rather, histamine may be released in large airways (bronchioles), which can be reached by larger particles; thus a different size selective criterion (50 percent at 15 μm) and device (vertical elutriator) are used (35). Even larger particles may enter the body by ingestion if they are caught in the ciliated portion of the bronchus, coughed up, and swallowed.

At some size particles are so large they cease to be "inhalable," thus should be excluded from exposure samples. It is difficult to select a criterion on which to base size selectors for "inhalable dust" samplers. Particles having falling (setting) speeds greater than the upflow velocity into the nose could be said to be "noninhalable" except that some individuals breathe through their mouths. These particles turn out to be quite large, thus have falling speeds that cause them to be removed from all but the most turbulent or recently generated dust clouds. Because of their size, of course, they represent a mass far out of proportion to their number. By using the general tendency of open face filter samples to undersample large particles when pointed down (36), it is possible to use this vertical elutriation effect to discriminate against noninhalable particles in most cases.

4.4 Exposure Indices

Most toxic substance encountered in the workplace are clearly defined chemical compounds that can be measured with as much specificity as we please. Benzene need not be confused with other compounds, and any other closely related compounds (e.g., ethyl benzene) have different toxicology. For some toxic substance we tend to think of an element (lead) in any one of a number of possible compounds. Thus specificity is defined in terms of the element rather than the compound. Often, however, the situation becomes more complex because the compound containing the element of interest has a significant effect on its uptake, metabolism, toxicity, or excretion (37). Although the influence of the chemical structure containing the element probably varies even among similar compounds, for simplicity the compounds are usually grouped as organic/inorganic (lead, mercury), soluble/insoluble (nickel, silver), or by valence (chromium).

The problem of differentiating the several classes of compounds of a toxic element in a mixed atmosphere adds complexity to sampling method selection, and it is sometimes necessary to make, and clearly state alongside the results, certain simplifying assumptions. It is commonly assumed when measuring lead exposure in gasoline blending, for example, that all the lead measured is organic. Similarly when measuring the more toxic soluble form of an element, the "safe" assumption may be made that all the element present was soluble.

The greatest complexity occurs when toxicity is based on the effects of a class of compounds or of a material of a certain physical description. Some polynuclear aromatic hydrocarbons (PNA) are carcinogens of varying potency, and they usually exist in mixtures with other PNAs and with compounds (activators, promotors, inhibitors) that modify their activity. Analysis of each individual compound is very difficult and when done does not yield a clear answer, since given the complexity of the mixture of biologically active agents and their interactions, a calculated equivalent dose would have little accuracy. In these instances it is common to measure some quantity related to the active agents and to base the TLV on that index. For PNAs a TLV has been based on the total weight of benzene- or hexane-soluble airborne material. Alternatives include the single carcinogenic PNA benzo[a]pyrene, the sum of a subset of six carcinogenic PNAs (Table 6.2) or 14 or more individual PNAs (38).

Asbestos is another toxic substance for which the parameter of greatest biological relevance is difficult to define (39). In the early studies when measurements were made with an impinger, few fibers were seen and consequently a count of all particles present was used as an index of overall dustiness. More recently the TLV has been based on counts of fibers longer than 5 μm as seen with a light microscope. Since most fibers present are usually shorter than 5 μm and too thin to be visible under a light microscope, it has been suggested that counts of all fibers seen by an electron microscope would provide the most meaningful estimate of risk. Long fibers may be more dangerous, however, short fibers will dominate the count; thus some adjustment may be necessary.

Byssinosis appears to be caused not by cotton itself but by inhalation of cotton plant debris dust baled with the cotton. The total dust airborne in a cotton textile mill is mostly lint (35), and indices have aimed at excluding these "inert" cellulose fibers by collecting a "lintfree" fraction by use of a screen or vertical elutriator. More relevant indices could include plant debris only or the specific biologically active agents if known.

Indices are used where a group of compounds interact to produce a biological effect, or where the active agent is unknown or unmeasurable. In choosing an index we try to maximize biological relevance with a method of measurement that is practical to use. On the one hand very simple parameters like gross dust or total count are easy to use but include much irrelevant material. On the other hand counting fibers by scanning electron microscope or detailed analyses of individual PNAs are difficult, expensive, and not likely to be undertaken frequently. In making the choice it is important to remember that the primary objective is the protection of the worker. Very exact and highly relevant methods, though scientifically satisfying, may be so tedious that very few samples

Table 6.2 Carcinogenic PNA Subset

Benz[a]anthracene
Benzo[b]fluoranthene
Benzo[j]fluoranthene
Benzo[a]pyrene
Benzo[e]pyrene
Benzo[k]fluoranthene

are taken, and because of the larger variability of worker exposure, the true accuracy of the exposure estimate is lower than it would have been if many samples had been taken by a less specific method.

When the overall contaminant level has been reduced and with it the level of the biologically active agent, the level of all correlated indices will be lower. The danger of less relevant, more indirect indices is that serious systematic bias may occur, particularly when an index from the workplace where the health effect relationship was estimated is used in other quite different workplaces. The carcinogenic risk of roofers using asphalt is far lower than for coke oven workers at the same level of exposure as measured by benzene solubles. Byssinosis patterns may be different in mills that garner old rags or process linters and in raw cotton card rooms. In using a measurement that is not perfectly specific, it must be remembered that the result obtained has a less than direct connection with the biological process, and the stronger the effect of extraneous factors, the more care is needed in interpreting the result.

4.5 Mixtures

Industrial workplaces rarely contain only one airborne contaminant, although it is uncommon for there to be several toxic substances each at or near its TLV. The measurement problems caused by the presence of gases, vapors, or dust—some in even higher concentrations than those of primary interest—are discussed below. The question of what to measure when a worker is simultaneously exposed to several agents involves biology and can be answered only by considering the mode of action of the substance in the body. Although possible interactions of substances are extremely complex, it has been the custom to accept the simplifying assumption that "In the absence of information to the contrary, the effects of the different hazards should be considered as additive" (5), and to sum the concentrations C_n of each substance as a fraction of its limit T_n.

$$\frac{C_1}{T_1} + \frac{C_2}{T_2} + \cdots + \frac{C_n}{T_n}$$

When there is "good reason to believe that the chief effects of the different harmful substances are not in fact additive, but independent, as when purely local effects on different organs of the body are produced by the various components of the mixture," exposure is judged by comparing the concentrations to each TLV. Strictly interpreted, the words "purely local effects" would place very few substances in the "independent" class, but it is common practice not to adjust TLVs for mixtures of substances that affect very different organs, even though systemic rather than local. Thus the TLV for dust in a mine is not usually reduced when carbon monoxide is present. Other interpretations are possible, and in general it should be recognized that disease from any cause lessens resistance to all other insults.

Hydrocarbons used as fuels or solvents are usually a mixture of a large number of individual aliphatic compounds and their isomers, often so numerous that analyzing for each individual compound and comparing the result with an individual limit is impractical. Indeed, since TLVs exist for only a few of the compounds present, limits must be stated for the mixture by considering the compounds as a class. Gasoline, for example, may contain aliphatic and aromatic hydrocarbons and additives such as tetraethyl lead. In view of the large differences in toxicity of the several substances and variations in content, calculation of a "TLV" for gasoline is a complex matter (40, 41). One difficulty in expressing a TLV for any vapor mixture is that it has been customary to state TLVs for gases or vapors in parts per million. Analytical results, typically from gas chromatographs, emerge initially as the weight in milligrams of each fraction present, from which the concentration, in milligrams per cubic meter, can be calculated. To convert this to parts per million, it is necessary to know the mole weight, which will not be known for unidentified homologues or for the mixture as a whole. For this reason it is preferable to state TLVs for vapor mixtures as a weight concentration (mg/m^3) rather than as a volume fraction (ppm).

Certain combinations of substances present a far more complex situation than can be described by either independent or simple interaction. Benzo[a]pyrene and particulates, with and without sulfur dioxide, carbon monoxide and hydrogen cyanide, ozone, and oxides of nitrogen, and some other mixtures result in complex interactions that may cause effects beyond those predicted by the merely additive case.

4.6 Period of Standard

The concentration of industrial air contaminants varies with time, and recordings such as that in Figure 6.3 are typical. In general, where the concentration is varying with time, the height of the maxima and depths of the minimum are greater as the period of the measurement is shortened. If it were possible to make truly "instantaneous" or zero-time measurements, the peaks and troughs would be very great indeed. Real measurements using continuous reading instruments do not show quite such wide extremes, because the response time of the instrument causes some averaging, thus prevents true zero-time measurements. Even so, instruments with short response times, such as those with solid state sensors, show wide variation from the average, and even those with relatively long response times such as a beta adsorption type of dust monitor, still reveal peaks and valleys that are more than double or less than half the average. As longer and longer period measurements are made, the extremes regress toward the average and obviously, if we define our average over an 8 hr period, a single integrated sample over that period would show no extremes above or below the average. However even daily averages have highs and lows compared to monthly or yearly averages in all but perfectly nonvarying environments, which do not occur in the real world.

Given the variance of the universe of instantaneous concentrations and the probability distribution of the concentration over one averaging time, it should be possible to determine the probability distribution of the concentration over any other averaging time. The mathematics of this conversion have not yet been developed, but the relationship of averaging time and environmental variance has the consequence that measurements made over different integrating times have a relationship to each other that is a function of the variance.

For example, if a given value of the concentration had a 50 percent (or 90 percent) chance of occurring over one averaging time, there would be a unique higher and lower pair of values that had an equal chance of occurring over a shorter averaging time. Since not all values of environmental variance are equally likely but rather tend to be in the range of GSDs of 1.5 to 3 or 4, it is possible to say in some cases that certain values for different averaging times are inconsistent with each other. Thus it is very unlikely that any environment would be so variable that the probability of exceeding the 8 hr average of 2 μg of beryllium per cubic meter would be the same as exceeding the 25 μg limit for a 15 min averaging time. To state this another way, if the probability of the 2 μg, 8 hr limit being exceeded is 50 percent, the probability of exceeding the 25 μg, 15 min limit is very much less than 50 percent.

In another example, it is very unlikely that an industrial environment would be so constant that the 100 ppm, 8 hr limit for styrene could have any chance of being exceeded without exceeding the 125 ppm, 15 min short-term exposure limit (STEL). The effect of setting short-term limits close to long-term limits is to force the effective long-term limit down. Thus to avoid exceeding a 125 ppm, 15 min STEL, it will probably be necessary to achieve an 8 hr average value of much less than 100 ppm of styrene, perhaps even lower than 50 ppm. There may be valid toxicological reasons (42, 43) for setting a short-term limit based on, for example, acute irritation, and a TWA that is aimed at preventing some chronic effect; however it should be recognized that the two are not independent, and when they are set outside the range of approximately equal likelihood, holding concentrations below the limit for one averaging time means holding them far below the limit for the other averaging time: thus one limit is in effect forcing the other.

In terms of sampling strategy, the significance of limits for different averaging times that are statistically inconsistent is that since one has a relatively greater likelihood of being exceeded than the other, regardless of the absolute likelihood of either, there is an opportunity to devise schemes that emphasize measurements to detect the likely event and use these measurements and knowledge of variance derived from them to draw inferences about the less likely event.

5 MEASUREMENT OPTIONS

The armory of techniques available to the industrial hygienist for measuring worker exposure to toxic substances is much fuller now than even a few years ago, although there are still many gaps. This section describes the range of choices available with location at which the measurement is made, time period or averaging time of ...urement, ability to select certain size aerosols and reject others, and degree to ...uman involvement can be lessened by automation and computer analysis.

5.1 Personal, Breathing Zone, and General Air

The several sampling options available with respect to locations are usually termed personal, breathing zone, and area or general air. A personal sample is one that is collected by a sampling device worn on the person of the worker, which travels with the worker. The device usually consists of a pump on the worker's belt or pocket and a sampling head containing the sorbent tube, filter, or other collection medium, clipped to the lapel close to the nose. Alternatively, some personal samplers do not pump the air to the collection medium but passively allow the toxic gas or vapor to diffuse (44) onto the solid sorbent. These dosimeters or "gas badges" (45) are also worn close to the nose. Since personal samplers are worn by the worker, they must be lightweight, portable, and not affected by motion or position. These restrictions tend to limit the size of sample that can be collected, and the use of wet collectors, except in "spillproof" vessels. Gravitatorial size selectors such as horizontal elutriators are also affected by position, although inertial devices such as cyclones or impactors are satisfactory.

When limitations of weight, complexity, wet collecting media, and so on prevent successful personal sampling, it is still possible to make an approximate measurement of worker exposure by collecting the sample in the "breathing zone" of the worker. This vaguely defined zone is the envelope of air surrounding the worker's head, which is thought, based on observation and the nature of the operation, to have approximately the same concentration of the contaminant being measured as the air breathed by the worker. This "breathing zone" may be sampled by a fixed sampler with the inlet near the nose of a stationary worker or, in the case of a mobile worker, by carrying the sample collecting equipment and holding the sample inlet near the worker's head, while moving around the work site with the worker. The obviously awkward and time-consuming nature of this kind of sampling limits its usefulness.

When measurements of worker exposure are not needed or indirect estimates are adequate, the concentration of a contaminant in the general air of the workplace as measured at some fixed station may be useful. Many of the equipment limitations imposed on personal and breathing zone sampling systems do not apply to general air samplers. Portability is not critical, so electrical components may be either battery or line powered. When line power is available, powerful pumps may be used to provide enough vacuum for critical orifices to obtain precise flow control or to operate high volume vacuum sources capable of collecting very large samples. Wet devices and both horizontal and vertical elutriators are practical. Very large samples, which may be needed to obtain sufficient sensitivity for trace analyses, may be collected on heavy or bulky collecting media. Multiple samplers of different types may be arrayed close to each other to provide sampler comparison data. Most new methods of measurement of

5.2 Sampling: Grab, Short, and Long Period

Available sampling methods have limited flexibility in the period of time over which the sample can or must be collected. Some methods are inherently grab samplers, although the increased interest in long period samplers has lead to their adaptation. Most detector tubes (46) were originally designed as direct reading colorimetric indicators intended to produce a result after a few pump strokes of up to several hundred milliliters. The interval between strokes could be lengthened but instead of increasing the sampling period, this produces an average of several short samples taken over a longer time. Automatic systems have been developed (47) as fixed station samplers that can extend the low range of some tubes by repeated pump strokes spread out over a long period. In the continuous flow mode, short period detector tubes have been recalibrated for use at very low flowrates over long periods and special tubes have been developed (48) as long-term samples for up to 8 hr at flowrates of 5 to 50 cm^3/min. Other inherently short period methods are those that use a liquid, particularly a volatile liquid in a bubbler or impinger. As the sample air passes through the sampler, the collecting medium will evaporate, eventually to dryness with loss of sample, unless terminated in time. Usually, without additions to the liquid, sampling periods of in excess of 30 min are impractical with wet collection media.

Vessels that collect a whole air sample are convenient grab samplers (49), but some can be adapted for longer period sampling. Many low flowrate personal samplers have an air outlet fitting that can be used to fill a bag. Allowing for possible contamination due to necessity for the air to pass through the pump, long period personal samples may be collected when the bag is carried in a sling on the worker's back.

Much greater time flexibility is available with solid sorbents. Even though there is a fixed volume of air that can be sampled at a concentration before breakthrough occurs, the freedom to use a wide range of low flowrates permits personal and fixed sampling over periods of 8 hr or more. Such influences as other air contaminants competing for active sorbent sites and particulates adding to the resistance to flow limit the maximum volume of air that can be sampled in much the same way as breakthrough, but within the new limit the time/volume/flowrate relationships apply.

Systems that tend to be suitable for long-period samples only are usually those where the sensitivity of the analytical method requires a minimum amount of analyte. Thus a personal sample for respirable quartz using a 10 mm cyclone size selective presampler operated at 2 liters/min or less will sample less than a cubic meter of air in 8 hr and thus less than 100 μg of quartz at the TLV. Shorter samples will confront the serious sensitivity limitations of most methods for quartz (50). The same general class of problem occurs with personal samples for beryllium and for detailed analyses for multiple compounds such as PNAs.

Of course higher volume fixed station, or even personal samplers are possible using large size selectors and filters where necessary to allow shorter sampling periods that are still longer than "grab samples." Even when analysis is not necessary and only gross or respirable weight is being measured, analytical balance limitations prevent very short samples for particulates. There are, however, various "instantaneous" mass monitors based on beta absorption (51) or the piezoelectric effect (52) that are capable of making a measurement in as little as one minute.

5.3 Size Selection

When it is necessary in sampling for particulates to include or exclude certain size particles from the sample, limitations are created by the nature of the devices available. The unsuitability of elutriators as personal samplers due to their orientation requirements has already been mentioned. All size selectors make their stated cut only at a predetermined flowrate, which must be held over the period of the sample against changing filter resistance and battery condition. Whereas cyclones and impactors tend to compensate for flowrate changes and elutriators compound the error, all need pumps that not only sample a known volume over the sampling period but do so at a known and constant flowrate. Such pumps are usually larger and heavier than those low flow pumps, which need only sample a reliably known volume, and approach the limit of practicality as personal samplers. In addition to the requirement for constant flow, pump pulsation must be damped out if it upsets the size selector.

Isokinetic conditions usually must be established when sampling for particulates in high velocity streams (53). In stacks, particles with high kinetic energy due to their weight and velocity can be improperly included or excluded from the sample under nonisokinetic conditions. Generally sampling in the workplace is done at low ambient air velocities (less than 300 ft/min), and large particles have settled out or are excluded from the sample by vertical elutriation, since they are not inhalable. Thus isokinetic sampling is required only under exceptional circumstances.

5.4 Sequential, Continuous, and Automatic Monitors

Leak detection by means of fixed sensors in critical locations connected to alarms and remote indicators is commonly used for carbon monoxide, hydrogen sulfide, hydrocarbons, and other acutely toxic or explosive gases. These systems may use passive sensors, they may pump the air through a detection cell located at the point of collection, or they may pump contaminated air to a remote analyzer. Sequential valving arrangements allow one measuring device to be coupled to many sample lines. Provision for automatic calibration and zeroing may be included.

Adaptation to the estimation of worker exposure of these systems, which make measurements automatically at a number of locations and gather the data at a central readout point, has lead to the development of elaborate computer-based monitoring systems (54) (Figure 6.6). Since the sensors do not detect the presence of a worker, some data on worker location must be added to estimate exposure. Time and motion studies that yield percentage of time in various measured locations could be used with the daily average fixed station measurement for that location to calculate a weighted average exposure. The drawbacks are that time/activity distributions exhibit considerable variation even under routine conditions, whereas the most significant exposures occur during nonroutine periods. Also an assumption is made that the concentration at a time is independent of the worker's activity.

Alternative schemes provide a device that reads a card carried by each worker to signal the computer of each entry and departure from a monitored area. Time in the area can be multiplied by the general or weighted average concentration from the sensors in that area as measured while the worker was present. This situation is analogous to estimating exposure from fixed station sampler measurements. An even more elaborate system places on each worker small transmitters that are tracked automatically by a sensing network, and these detailed worker location data are combined with fixed station measurements for various locations to estimate exposure.

5.5 Mixtures

As was pointed out previously, toxic substances seldom occur by themselves. In addition to causing difficulties in estimating the biological consequences of exposure, mixtures also complicate exposure measurement. Even when only one component of an airborne mixture of gases and vapors is being measured, the other substances present need to be

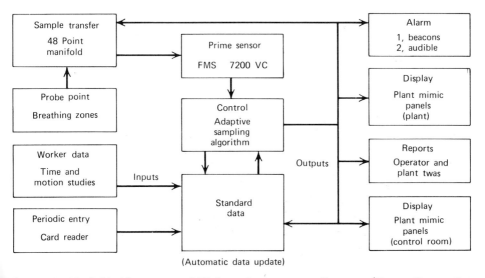

Figure 6.6. Vinyl chloride monomer (VCM) monitoring system. Courtesy of Eocom Corporation.

considered, often even when they are far below toxic effect levels, because of their effect on the sampling system. Charcoal is a very useful sampling medium because it will adsorb and retain so many substances; but because of this property it is possible for all the active sorption sites to be occupied by other materials, causing the substance being measured to break through long before the recommended sample volume, based on collection of the pure substance, has been passed through the tube.

An analogous case is the measurement of a low concentration of asbestos fibers in an environment containing a great deal of other, nonfibrous dust. To collect enough fibers to give the fiber density necessary for an adequate count without counting on a reasonable number of fields would result in the collection of so many grains that the fibers would become obscured. Thus the sample volume is limited by the total dust present, and the low fiber density is compensated for by counting a large number of fields. Overload from other airborne substances can also result in plugging of filter samplers, particularly when liquids accumulate on the surface of membrane filters and "blind" the pores. The use of thick depth filters of glass or cellulose fibers provides greater capacity, although at the possible sacrifice of some efficiency.

6 SAMPLING STRATEGIES

Thus far this chapter has discussed the reasons for making measurements of air contaminants in the industrial environment, the physical and biological characteristics of toxic substances as these relate to their measurement, and the attributes of the instruments and systems for carrying out the measurement (Table 6.3). The next subject is a plan of action or strategy for making some set of measurements in the industrial environment that will acomplish the desired purpose effectively and efficiently. A starting point in this problem-solving process must be a clear understanding of the purpose of the measurement. Rarely are data collected purely for their own sake. Even when data are collected because of a demand by others, such as the government or employees, the use of the data should be considered.

A convenient way of analyzing the data is to ask, What questions will be answered by these data? What decisions could depend on those answers? For example, is it desired to answer the question, Are these workers overexposed? And if so, will a decision to take control action follow? Is the control likely to be a minor change in a work practice or an expensive engineering modification? Or is the reason to combine the data with health effects data to answer the question, What level of exposure is safe? (55). Will it then be decided to modify the TLV or establish a new permissible exposure limit (PEL) by regulation? Or are the numbers to be assembled to answer the question, What level of control is presently being achieved in industry? And may this answer lead to new decision regarding what control is feasible? Last, will workers use the result to find out whether their health is at risk and, as a consequence, decide to change jobs or seek compensation. It frequently happens that there are many questions that need answers, thus data are being collected for multiple purposes. As the discussion that follows indicates, the purpose of the data has significant impact on the design of the measurement scheme. All too often data intended for multiple purposes turn out to be not suitable for any purpose. Thus it is usually necessary to focus on the prime need to be sure the strategy will meet this requirement; then, if possible, minor adjustment or additions can be made to meet other needs, if this can be done without losing the main purpose.

The optimum sampling strategy is that which so advantageously combines the choice of method and sampling scheme with respect to location, time, and number that we are confident that the answer to the objective question is adequate for the decisions which follow.

6.1 Location

When our purpose requires the measurement of worker exposure, it is preferable to take the samples as close to the mouth as possible. Even 5 in. difference in the placement of the filter head of a personal sampler may make a significant difference in the concentration measured (56).

Personal sampling is not possible, even today, for all air contaminants and was even less attainable years ago. Furthermore there is a fear that workers will react negatively to wearing personal samplers, and the modern sophistication of computer-based automated monitoring systems has great appeal. Many attempts (57–61; 67) have been made to estimate worker exposure from fixed station air sampling schemes, including computer-based systems and those that also incorporated time and motion study data. Leidel (27) analyzed these studies and concluded that general air (area) monitoring was inadequate for measuring employee exposures. In the words of Linch (62) "only by personal monitoring could a true exposure be determined."

If it is decided that personal sampling cannot be used and some other means of estimating exposure must be accepted, breathing zone measurements made by a sample collector who follows the worker come closer to measuring exposure. However this intrusive measurement may influence worker behavior, and the inconvenience of the measurement will limit the number of measurements, therefore reduce accuracy, as discussed below. When fixed station samplers are used, knowledge of the quality of the relation between their measurements and the exposure of the workers is necessary. An experimental design that collects large numbers of pairs of measurements of quantities that are in any way related will yield a significant correlation coefficient. The important question in the use of general air measurements is, What confidence can be placed in the estimate of worker exposure? Thus not only the regression line is important, but also the width of the bounds on the confidence limits of a predicted exposure value from some set of fixed station measurement as shown in Figure 6.2.

Although worker exposure measurements are most closely realted to health hazards, not all measurements made for the protection of health should be measurements of

Table 6.3 Sampling and Analysis Method Attributes of Importance in Selecting Sampling Strategy

Method	Sampling Period	Ability to Concentrate Contaminate	Ability to Measure Mixture	Time to Result	Intrusiveness[a]	Proximity to Breathing Zone
Personal sampler/solid sorbent						
Sorption only	Medium to long	Yes	Yes—gases	After analysis	Medium	Very close
Sorption plus reaction	Medium to long	Yes	No	After analysis	Medium	Very close
Personal sampler/filter						
Gross gravimetric	Medium to long	Yes	Yes—particulate	After weighing or analysis	Medium	Very close
Respirable gravimetric	Long	Yes	Yes—particulate	After weighing or analysis	Medium	Very close
Count	Medium to long	Yes	Yes—particulate	After counting	Medium	Very close
Combination filter and sorbent	Medium to long limited	Yes	Yes	After analysis	Medium	Very close
Passive dosimeter	Long	Yes	Yes—gases	After analysis	Low	Very close
Breathing zone impinger/bubbler						
Analysis	Medium-limited	Yes	Yes	After analysis	High	Close
Count	Medium-limited	Yes	Yes—particulate	After counting	High	Close
Detector tubes						
Grab	Short	NA	No	Immediate	High	Close
Long period	Medium to long	NA	No	Immediate	Medium	Very close
Gas vessels						
Rigid vessel	Short	No	Yes—gases	After analysis	High	Close
Gas bag	Short to long	No	Yes—gases	After analysis	High	Close
Evacuated/critical orifice	Medium to long	No	Yes—gases	After analysis	Medium	Very close
Direct reading portable meters						
Nonspecific (flame ion, combination gases)	Instantaneous or recorder	NA	Yes	Immediate	High	Slightly distant
Specific (carbon monoxide, hydrogen sulfide, ozone, sulfur dioxide, etc.)	Instantaneous or recorder	NA	No	Immediate	High	Slightly distant
Multiple compound (infrared, gas chromatography, etc.)	Instantaneous or recorder	Some	Yes	Almost immediate	High	Slightly distant
Mass monitor (beta absorber, piezoelectric)	Short	Yes	No	Almost immediate	High	Slightly distant
Particle counters (optical, charge)	Short	No	No	Almost immediate	High	Slightly distant

Method	Duration	Personal	Collects	Accuracy	Stability of sample	Sensitivity	Cost	Specificity	Operation
High volume	Medium to long	Yes	Yes—particulate		After analysis		Low		Remote
Horizontal or vertical elutriator	Long to short	Yes	Yes—particulate		After analysis		Low		Remote
Installed monitor	Short to long	Some	No		Almost immediate		Low		Remote
Freeze trap	Medium	Yes	Yes—vapors		After analysis		Low		Remote
Personal sampler/solid sorbent									
Sorption only	High by analysis			High	Good			Elution—Yes; thermal des. no	Good
Sorption plus reaction	High by analysis			High	Good			Yes	Good
Personal sampler/filter									
Gross gravimetric	None for weight only—high by analysis			High	Fair			Yes	Good
Respirable gravimetric	High by analysis			Medium	Fair			Yes	Fair
Count	Fair—depends on particle identification			High	Good			Yes	Poor
Combination filter/sorbent	High by analysis			Medium	Good			Yes	Fair
Passive dosimeter	High by analysis			Very high	Good			Yes	Fair
Breathing zone impinger/bubbler									
Analysis	High by analysis			Low	Poor			Yes	Fair
Count	Fair—depends on particle analysis			Low	Poor			Yes	Poor
Detector tubes									
Grab	Medium—some interference			High	No sample			No	Fair
Long period	Medium—some interference			High	No sample			No	Fair
Gas vessels									
Rigid vessel	High by analysis			Low	Fair			Yes	Good
Gas bag	High by analysis			Low	Fair			Yes	Good
Evacuated/critical orifice	High by analysis			High	Good			Yes	Good
Direct reading portable meters									
Nonspecific (flame ion, combination gases)	None—total of measured class			High	No sample			No	Good
Specific (carbon monoxide, hydrogen sulfide, ozone, sulfur dioxide, etc.)	Medium—some interference			High	No sample			No	Good
Multiple compound (infrared, gas chromatography, etc.)	Medium—frequent overlap			Medium	No sample			No	Fair
Mass monitor (beta absorber, piezoelectric)	Mass only			High	No sample			Not usually	Fair
Particle counters (optical, charge)	Count/size only			High	No sample			No	Fair

(footnotes on following page)

Table 6.3 (Continued)

Method	Specificity	Convenience Rating[b]	Sample transport-ability	Recheck of Analysis Possible	Accuracy[c]
Fixed station[d]					
High volume	High by analysis	Low	Fair	Yes	Good
Horizontal or vertical elutriator	High by analysis	Low	Fair	Yes	Good
Installed monitor	Medium—may be interferences	High	No sample	No	Good
Freeze trap	High by analysis	Very low	Poor	Yes	Fair

[a] Degree to which irtrusion of sampling system into work situation may affect worker behavior.
[b] Rating of the amount of work or difficulty involved in collecting a sample.
[c] Estimated usual overall sampling and analytical accuracy—may be better or worse for specific substances.
[d] All methods above can be used as fixed station samplers.

exposure per se. When it has been established that an industrial operation does not produce unsafe conditions when it is operating within specified control limits, fixed station measurements that can detect loss of control may be the most appropriate monitoring system for worker protection. Local increases in contaminant concentration caused by leaks, loss of cooling in a degreaser, or fan failure in a local exhaust system can be detected before important worker exposure occurs.

6.2 Time

Free of all other constraints, the most biologically relevant time period over which to measure or average worker exposure should be derived from the time constants of the body burden of the toxic substance and its elimination (63, 64). These periods range from minutes in the case of fast-acting poisons such as chlorine or hydrogen sulfide, to days for slow systemic poisons such as lead or quartz. In the adoption of guides and standards, this broad range has been narrowed and the periods have not always been selected based on speed of effect or half-life. For most substances a time-weighted average over the usual work shift of 8 hr was accepted as convenient, since it was long enough to average out extremes and short enough to be measured in one work day. Several systems have been proposed for adjusting limits to novel work shifts (65, 66).

Once the time period over which exposure is to be averaged, has been decided for either biological or other reasons (67, 68), there are available several alternate sampling schemes to yield an estimate of the exposure over the average time. A single sample could be taken for the full period over which exposure is to be averaged (Figure 6.7). If such a long sample is not practical, several shorter samples can be strung together to make up a set of full period, consecutive samples. In both cases, since the full period is being measured, the only error in the estimate of the exposure for that period is the error of the sampling and analytical method itself. However when these full period measurements are used to estimate exposure over other periods not measured, the interperiod variance will contribute to the total error.

It is often difficult to begin sample collection at the beginning of a work shift, or an interruption may be nessary during the period to change samples. Several assumptions may be made with respect to the unsampled period. It may be assumed that exposure was zero during this period, in which case the estimate for the full period could be regarded as a minimum. Alternatively it could be assumed that the exposure during the unmeasured period was the same as the average over the measured period. Although this is the most likely assumption, in the absence of information that the unsampled period was different it is difficult to calculate confidence limits on the overall exposure estimate; since the validity of the assumption is a factor, there is no internal estimate of environmental variance, and the statistical situation is complex.

When only very short period or grab samples can be collected, a set of such samples can be used to calculate an exposure estimate for the full period. Such samples are usually collected at random; thus each interval in the period has the same chance of

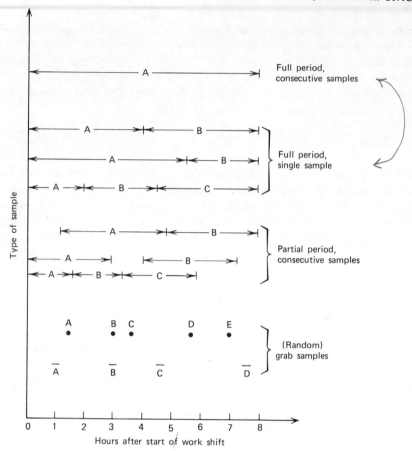

Figure 6.7. Types of exposure measurement that could be taken for an 8 hr average exposure standard.

being included as any other and the samples are independent. This sampling scheme of discrete measurements within a day is analogous to a set of full period samples used to draw inferences about what is happening over a large number of days. In both cases the environmental variance, which is usually large, has a major influence on the accuracy of the results.

Short period sampling schemes can be useful against dual standards. For example, if a toxic substance has both a short period, say 15 min, limit and an 8 hr limit, 15 min samples taken during the 8 hr period could be used to evaluate exposure against both standards. This involves some compromise, however, since samples taken to evaluate short period exposure should be taken when exposure is likely to be at a maximum rather than at random. Statistical techniques for evaluating exposure with respect to dual standards are also available (69). As discussed earlier, when a dual standard is

MEASUREMENT OF WORKER EXPOSURE

inconsistent, so that one limit is more likely to be exceeded than the other, sampli.
schemes that evaluate exposure with respect to the limit more likely to be exceeded can
be used to provide some confidence about the other limit.

A traditional method of estimating full period exposure has been by calculation of the
"time-weighted-average." In this method the work day is divided into phases based on
observable changes in process or worker location, which are assumed to be homogeneous with respect to exposure. A measurement or measurements, usually shorter
than the length of the phase, are made in each phase, and the exposure estimate is calculated according to the formula:

$$E = \frac{C_1 T_1 + C_2 T_2 + \cdots + C_n T_n}{8}$$

where C_n = concentration measured in phase n
T_n = duration (hr) of phase n ($\sum T = 8$)

Figure 6.8 represents this procedure schematically. Although the exposure estimate
itself is simple to make, calculation of the confidence limits on this estimate can be very
complex. Each phase must be treated as a period, and some set of samples must be
collected to determine the mean and standard deviation for the phase. These data then
must be combined in a manner that weights the variance to obtain an error estimate for
the whole. This complex calculation does not include, however, any consideration of the
imprecision in selection of phase boundaries. Given the number of samples required in

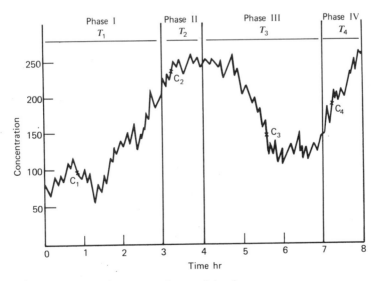

Figure 6.8. Time-weighted average.

each phase to provide adequate error estimates and the lack of confidence in the end results due to the several layers of assumptions, an equal number of random grab samples may well yield more reliable data.

The recommendation that samples should be taken over a full shift to determine employee exposure to toxic substances has been questioned (70). It is maintained that it is possible in some instances to characterize a worker's exposure with a few short period samples. There are arguments to support both sides; ultimately, however, each case can be decided based on the answer to two questions: Can the risk of error in the decision to be based on these measurements be calculated from the data? Is this risk acceptable, given the consequences of the decision?

When workplace measurements are made for purposes other than the estimation of worker exposure, different considerations apply. A single 8 hr sample may give a good estimate of a worker's exposure during that period, but the exposure undoubtedly was not uniform, and the single sample gives no information on the time history of contaminant concentration. To find out when and where peaks occur, with the aim of knowing what to control, short period samples or even continuous recordings are needed. Similarly, when a control system is evaluated or sampling methods are compared, measurements need be only long enough to average out system fluctuations and provide an adequate sample for accurate analysis. As in the case of the decision on location, purpose of the measurement has a primary impact.

6.3 Number

By increasing the number of measurements, the magnitude of the confidence limits on the mean result can be reduced; thus it becomes possible to arrive at a decision at a given level of confidence or to become more confident that a decision is correct. In Figure 6.9 decisions are possible in cases A and C, but not in case B. By collecting more samples it might be possible to tighten the lower confidence limit (LCL) in case B_1, for

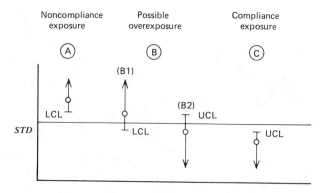

Figure 6.9. One-sided confidence limits.

MEASUREMENT OF WORKER EXPOSURE 251

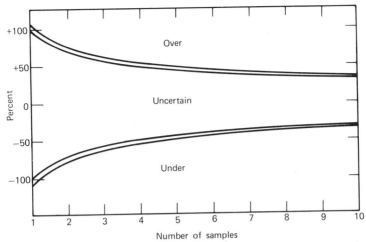

Figure 6.10. Difference between mean measured concentration and TLV required for a decision at the 95 percent confidence level versus number of samples averaged.

example, permitting the conclusion that these data do in fact represent an overexposure. The choice of the number of samples to be collected rests on three factors: the magnitude of the error variance associated with the measurement, the size of a difference in results that would be considered important, and the consequence of the decision based on the result.

The error variance associated with the measurement depends to a great extent on the environmental variance. In the rather limited instance of evaluating the exposure of a worker over a single day by means of a full period measurement, only the sampling and analytical error need be considered and confidence limits tend to be quite narrow. Usually, however, our concern is with the totality of a worker's exposure, and we wish to use the data collected to make inferences about other times not sampled. There is little choice; unless the universe of all exposure occasions is measured, we must "sample," that is, make statements about the whole based on measurements of some parts.

As discussed earlier, the universe has a large variance, quite apart from the error of the sampling and analytical method. In terms of our decision-making ability, the combined error of the sampling and analytical method may have very little impact. In Figure 6.10 the inner pair of curves define the decision zone for an environmental coefficient of variance equal to .60 with no sampling or analytical error; the outer zone includes a typical detector tube error having a coefficient of variance of .25. Obviously, even the relatively large error of one of the less accurate methods results in only a slight increase in the no-decision zone (71). The similiar curves in Figure 6.11 also illustrate the effect of sample size or our ability to conclude, in this case based on a number of grab samples, that we are confident that an overexposure did not occur. As can be seen,

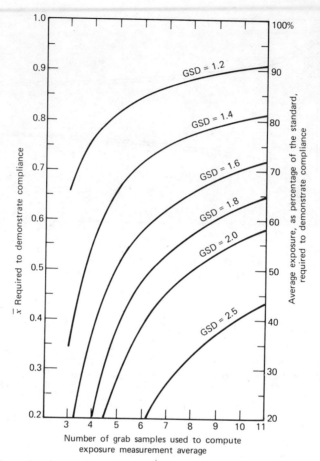

Figure 6.11 Effect of grab sample size on compliance demonstration: GSD = variation of grab samples.

the difference between the mean and the standard necessary to achieve confidence in the conclusion decreases sharply as the number of samples used increases from 3 to 11.

In selecting a sample size it must be kept in mind that it is possible to make a difference significant by increasing the number of samples even though the difference may be of small importance. Thus although it may be possible to show that a mean concentration of 1.12 ppm is significantly different from a TLV of 1.0 ppm, the difference has no importance in terms of biological consequence and the statistical significance is not useful. We should consider, therefore, how small a difference would be considered to be important in terms of our use of the data, selecting a sampling scheme that would prove this difference significant, but not lesser differences.

The consequences of the decision made on the basis of the data collected should be the deciding factor in selecting the level of confidence at which the results will be tested.

Although 95 percent (1 in 20) is common, usually arbitrarily chosen because its bounds are 2 standard deviations from the mean, other levels of confidence may often be appropriate. When measurements are made in a screening study to decide on the design of a larger study, it may be appropriate to be only 50 percent confident that an exposure is over some low trigger level. On the other hand, when a threat to life or a large amount of money may hang on the decision, confidence levels even beyond the 3 standard deviations usually used in quality control may be appropriate. No firm rule may be given except to carefully consider the consequence of being wrong and decide on an acceptable level of risks.

Since sampling and analysis can be expensive, some thought should be given to ways of improving efficiency. Sequential sampling schemes in which the collection of a second or later group of samples is dependent on the results of some earlier set are a possibility. This common quality control approach results in infrequent sampling when far from decision points, but increases as a critical region is neared. Another means of economizing is to use a nonspecific, direct reading screening method, such as a total hydrocarbon meter, to obtain information on limiting maximum concentrations that will help to reduce the field of concern of exposure to a specific agent.

Few firm rules can be provided to aid in the selection of a sampling strategy because the data can be put to such a wide variety of uses; the sequence of decisions to be made can be listed, however.

- Decide on the purpose of the measurements in terms of what decisions are to be made. When it seems that there are multiple purposes, select the one or ones most important for design.
- Consider the ways in which the environmental exposure and the nature of the agent relate to measurement options.
- Identify the methods available to measure the toxic substance as it occurs in the workplace.
- Select an interrelated combination of method, location, time, and number that will allow a confident decision in the event of an important difference with a minimum of effort.

REFERENCES

1. M. Corn, *J. Am. Ind. Hyg. Assoc.*, **37**, 353–356 (1976).
2. U.S. Public Health Service–National Institute for Occupational Health and Safety, "The Right to Know: Practical Problems and Policy Issues Arising from Exposures to Hazardous Chemical and Physical Agents in the Workplace," Department of Health, Education and Welfare, PHS–NIOSH, Cincinnati, Ohio, 1977.
3. Occupational Safety and Health Act of 1970, PL 91-596.
4. J. R. Lynch, *Nat. Saf. News*, **113**: 5, 67–72 (1976).
5. American Conference of Governmental Industrial Hygienists, "Threshold Limit Values for 1977," ACGIH, Post Office Box 1937, Cincinnati, Ohio, 45201.

6. E. M. Thompson et al., *J. Am. Ind. Hyg. Assoc.*, **38,** 523–535 (1977).
7. H. M. Donaldson and W. T. Stringer, "Beryllium Sampling Methods", Department of Health, Education and Welfare Publication (NIOSH) 76-201, Cincinnati, Ohio, 1976, p. 21.
8. C. S. McCammon, Jr., and J. W. Woodfin, *J. Am. Ind. Hyg. Assoc.*, **38,** 378–386 (1977).
9. L. D. Horowitz, *J. Am. Ind. Hyg. Assoc.*, **37,** 227–233 (1976).
10. G. R. Oxley, *Ann. Occup. Hyg.*, **19,** 163–167 (1976).
11. R. R. Lauwerys et al., *J. Occup. Med.*, **20,** 17–20 (1978).
12. S. Fukabori et al., *J. Sci. Labour*, **53,** 89–95 (1977); abstracted in *Ind. Hyg. Dig.*, **41,** 16 (1977).
13. S. Fukabori et al., *J. Sci. Labour*, **52,** 67–81 (1976); abstracted in *Ind. Hyg. Dig.*, **40,** 19–20 (1976).
14. E. Cronen and R. B. Stoughton, *Arch. Dermatol.*, **36,** 265 (1962).
15. J. V. LeBlanc, *J. Occup. Med.*, **19,** 276–277 (1977).
16. M. M. Key, Ed., *Occupational Diseases, A Guide to Their Recognition,* Government Printing Office, Washington, D.C., 1977.
17. R. W. Weeks, Jr. et al., *Occup. Health Saf.* **46,** 19–23 (1977).
18. T. D. Sterling and D. M. Kobayashi, *Environ, Res.*, **3,** 1–35 (1977).
19. D. T. Wigle, *Arch. Environ. Health*, **32,** 185–190 (1977).
20. A. Poklis and C. D. Burkett, *Clin. Toxicol.*, **11,** 35–41 (1977).
21. J. S. Oliver, *Lancet*, **1,** 84–86 (1977).
22. R. Korobkin et al., *Arch. Neurol.*, **32,** 158–162 (1975).
23. J. W. Hayden et al., *Clin. Toxicol.* **9** (2), 169–184 (1976).
24. R. A. Warriner III et al., *Arch. Environ. Health*, **32,** 203–205 (1977).
25. B. L. Weisenberger, *J. Occup. Med.*, **19,** 569–570 (1977).
26. J. Aitchinson and J. A. C. Brown, *The Lognormal Distribution,* University Press, Cambridge, England, 1963.
27. N. A. Leidel, K. A. Busch, and J. R. Lynch, *Occupational Exposure Sampling Strategy Manual,* Department of Health, Education and Welfare, Publication (NIOSH) 77-173, Cincinnati, Ohio, 1977.
28. K. E. Bencala and J. H. Seinfeld, *Atmos. Environ.*, **10:** 11, 941–950 (1976).
29. R. I. Larsen, "A Mathematical Model for Relating Air Quality Measurements to Air Quality Standards," U.S. Environmental Protection Agency, Publication AP-89, Research Triangle Park, N.C., 1971.
30. Y. Kalpasanor and G. Kurchatora, *J. Air. Pollut. Control Assoc.*, **26,** 981 (1976).
31. A. C. Stern, Ed., *Air Pollution,* 3rd ed., Vol. 3, *Measuring, Monitoring, and Surveillance of Air Pollution,* Academic Press, New York, 1976, p. 799.
32. H. Ayer and J. Burg, "Time-Weighted Average Versus Maximum Personal Sample," paper presented at the American Industrial Hygiene Conference, Boston, 1973.
33. N. A. Leidel, K. A. Busch, and W. E. Crouse, *Exposure Measurement Action Level and Occupational Environmental Variability,* Government Printing Office, Washington, D.C., 1975.
34. C. I. Fairchild and M. I. Tillery, *J. Am. Ind. Hyg. Assoc.*, **38,** 277–283 (1977).
35. J. R. Lynch, "Air Sampling for Cotton Dust," *Transactions of the National Conference on Cotton Dust and Health,* University of North Carolina, Chapel Hill, 1970, p. 33.
36. C. N. Davies, *J. Appl. Phys.*, Ser. 2, **1,** 921–932 (1968).
37. B. R. Roy, *J. Am. Ind. Hyg. Assoc.*, **38,** 327–332 (1977).
38. K. A. Schute, D. J. Larsen, R. W. Hornung, and J. V. Crable, "Report on Analytical Methods used in a Coke Oven Effluent Study," National Institute for Occupational Safety and Health, 1974.

39. R. K. Zumwalde and J. M. Dement, "Review and Evaluation of Analytical Methods for Environmental Studies of Fibrous Particulate Exposures," Department of Health, Education and Welfare, Publication (NIOSH) 77-204, NIOSH, Cincinnati, Ohio, 1977, p. 66.
40. H. E. Runion, *J. Am. Ind. Hyg. Assoc.*, **38**, 391–393 (1977).
41. H. J. McDermott and S. E. Killiany, *J. Am. Ind. Hyg. Assoc.*, **39**, 110–117 (1978).
42. D. Turner, *Ann. Occup. Hyg.*, **19**, 147–152 (1976).
43. D. M. Ferguson, *Ann. Occup. Hyg.*, **19**, 275–284 (1976).
44. E. D. Palmes et al., *J. Am. Ind. Hyg. Assoc.*, **37**, 570–577 (1976).
45. F. C. Tompkins, Jr., and R. L. Goldsmith, *J. Am. Ind. Hyg. Assoc.*, **38**, 371–377 (1977).
46. American Industrial Hygiene Association, *Direct Reading Colorimetric Indicator Tubes Manual*, AIHA, Akron, Ohio, 1976.
47. K. Leichnitz, *Detector Tube Handbook*, 3rd ed., Drägerwerk, Lübeck, Germany, 1976.
48. K. Leichnitz, *Ann. Occup. Hyg.*, **19**, 159–161 (1976).
49. R. W. Miller et al., *J. Am. Ind. Hyg. Assoc.*, **37**, 315–319 (1976).
50. National Institute for Occupational Safety and Health Manual of Analytical Methods, 2nd ed., Department of Health, Education and Welfare (NIOSH) Publication 77-157, U.S. Public Health Service–NIOSH, Cincinnati, Ohio, 1977 (3 volumes).
51. P. Lilienfeld and J. Dulchunos, *J. Am. Ind. Hyg. Assoc.*, **33**, 136 (1972).
52. G. J. Sem et al., *J. Am. Ind. Hyg. Assoc.*, **38**, 580–588 (1977).
53. N. A. Fuchs, *Atmos. Environ.*, **9**, 697–707 (1975). Isokinetic
54. G. L. Baker and R. E. Reiter, *J. Am. Ind. Hyg. Assoc.*, **38**, 24–34 (1977).
55. E. E. Campbell, *J. Am. Ind. Hyg. Assoc.*, **37**, A-4 (1976).
56. B. B. Chatterjee, M. K. Williams, J. Walford, and E. King, *J. Am. Ind. Hyg. Assoc.*, **30**, 643 (1969).
57. A. J. Breslin, L. Ong, H. Glauberman, A. C. George, and P. LeClare, *J. Am. Ind. Hyg. Assoc.* **28**, 56–61 (1967).
58. R. J. Sherwood, *J. Am. Ind. Hyg. Assoc.*, **27**, 98–109 (1966).
59. H. F. Shulte, "Personal Sampling and Multiple Stage Sampling," peper presented at ENEA Symposium on Radiation Dose Measurements, Stockholm, Sweden, June 12–16, 1967.
60. A. L. Linch, E. G. Wiest, and M. D. Carter, *J. Am. Ind. Hyg. Assoc.*, **31**, 170–179 (1970).
61. B. D. Baretta, R. D. Stewart, and J. E. Mutcheler, *J. Am. Ind. Hyg. Assoc.*, **30**, 537–544 (1969).
62. A. L. Linch and H. V. Pfaff, *J. Am. Ind. Hyg. Assoc.*, **32**, 745–752 (1971).
63. S. A. Roach, *J. Am. Ind. Hyg. Assoc.*, **27**, 1 (1966).
64. S. A. Roach, *Ann. Occup. Hyg.*, **20**, 65–84 (1977).
65. R. S. Brief and R. A. Scala, *J. Am. Ind. Hyg. Assoc.*, **36**, 6 (1975).
66. J. W. Mason and H. Dershin, *J. Occup. Med.*, **18**, 603–606 (1976).
67. E. J. Calabrese, *J. Am. Ind. Hyg. Assoc.*, **38**, 443–446 (1977).
68. J. L. S. Hickey and P. C. Reist, *J. Am. Ind. Hyg. Assoc.*, **38**, 613–621 (1977).
69. R. S. Brief and A. R. Jones, *J. Am. Ind. Hyg. Assoc.*, **37**, 474–478 (1976).
70. D. D. Douglas, *J. Am. Ind. Hyg. Assoc.*, **38**, A-6 (1977).
71. J. R. Lynch, "Uses and Misuses of Detector Tubes," *Transactions of the 32nd Meeting of the American Conference of Governmental and Industrial Hygienists*, ACGIH, Cincinnati, Ohio, 1970.

CHAPTER SEVEN

Biological Indicators of Chemical Dosage and Burden

RICHARD S. WARITZ, Ph.D.

1 INTRODUCTION AND DEFINITIONS OF TERMS

"In analysis of biological specimens for solvents, however, it appears that each solvent represents a different problem, and some knowledge of the peculiarities of the metabolism and excretion of each is necessary, else one may be led astray in one's interpretation of the results" (1). Elkin's words of caution are as valid today as they were when he wrote them in 1954. Elkins defined "solvent" to include any liquid industrial chemical. The advice applies equally well to solids, gases, vapors, organic chemicals, and inorganic chemicals. It also is important to know whether the exposure chemical or a metabolite is the toxicologically determining material.

This chapter is not an in-depth review of particular biologic evaluations such as lead or mercury in blood or urine. In most cases an extensive literature is devoted to the individual biological analyses, and this material should be consulted by those interested in a particular analysis. Nor are analytical methodologies discussed in detail; they also are readily located in specialized texts and scientific articles. Biochemistry, enzyme kinetics, pharmacokinetics, and blood-protein binding of chemicals must be considered in the development of any biological assay, and the reader is referred to the relevant specialized texts and scientific papers for detailed guidance in these areas (see, e.g, References 2–5).

This chapter discusses the general concept of biological analysis as a tool to measure workplace exposure to exogeneous chemicals with some biological activity (xenobiotics), possible utilities of the concept in the workplace, the problems to be expected in developing a biological monitoring method for a particular chemical or element, the problems in applying a method, and the problems and uncertainties in interpreting the results. Examples from the published literature are used to illustrate.

The terms "biological analysis" or "biological monitoring" are used in this chapter to indicate analysis of exhaled air, analysis of some biological fluid, such as urine, blood, tears, or perspiration, or analysis of some body component, such as hair or nails, to evaluate past exposure to a chemical. The chemical of analytical interest is called the "index" chemical.

2 RATIONALE FOR BIOLOGICAL EVALUATION OF EXPOSURE

Blood and urine analyses for certain components have long been used by the medical professions to help differentiate between the normal and diseased states (see, e.g., Reference 6). Since most industrial chemicals that might cause systemic effects are transported by the blood system, metabolized by enzymes, and excreted by one of the excretory systems, it seems a logical extension of the medical groundwork in these areas to attempt to measure exposure to industrial chemicals by biological analysis for the industrial chemical(s) of interest.

The objective of industrial hygiene is to prevent an effect level of an agent from reaching that agent's target organs or tissues in the worker. The agent may be physical or chemical. Physical agents such as sound or electromagnetic radiation or particulate radiation occur in one form only, and controlling that one form will control the worker's total exposure to that agent. However workers may be exposed to chemicals in the form of gases, vapors, liquids, or solids, singly or combined. Workers may be exposed to chemicals through inhaling, absorbing the chemical through the skin, or ingesting. Exposure may be by several routes concurrently.

Within limits, the effect of a chemical on an organ or tissue is directly proportional to the amount of the chemical reaching or reacting with that organ or tissue. Ideally, then, to relate toxicity to dose, the amount of chemical in the target organs or tissues should be correlated with the effect seen. This would first be done in animal experiments, and these results would be extrapolated to man (see Chapter 16). However it is even more impractical to continually and routinely make such direct measurements on an organ in a worker than it is to make them in a laboratory test animal. Therefore it would be desirable to have secondary or tertiary measurements that could be made easily and routinely on a potentially exposed person and would reliably bear a known relationship to the appearance or absence of toxic effects and would give some idea of what additional burden of the chemical or its metabolites could be assumed before reaching the threshold of a toxic effect.

Within limits, the amount of the chemical reaching the target organs or tissues usually varies directly with the amount unbound (see Section 6.1.4) in the blood. Therefore the amount reaching these target organs or tissues can be reasonably approximated by measuring blood levels. As with organ analysis, however, the taking of blood samples is an invasive technique and is not suited to routine daily monitoring of workers.

In most cases, again within limits that vary for each chemical, the amount reaching the target organs or tissues varies directly with the dose received by any particular route. Historically, the major exposure route of concern for the industrial worker has been inhalation. Consequently, industrial hygienists have concentrated on keeping atmospheric levels of chemicals below effect levels. The Threshold Limit Value (TLV) Committee of the American Conference of Governmental Industrial Hygienists (ACGIH) has set atmospheric levels for several hundred industrial chemicals to which they believe "nearly all workers may be repeatedly exposed day after day without effects" (7). The TLVs for some of these chemicals also have been incorporated in the Occupational Safety and Health Act of 1970 (PL 91-596).

However examination of the TLV list shows that classes of organic compounds listed have at least one member of the class that has the notation "Skin," indicating that toxicologically significant amounts of that chemical in the liquid form can be absorbed through the skin. If a worker spills the liquid form of one of these chemicals on the skin, puts the hands in the liquid, or spills the liquid on the shoes or clothing, the target organs or tissues will receive the chemical both through inhalation and through the skin. Obviously in this case atmospheric levels of the chemical do not accurately indicate the amount of chemical that might reach target organs or tissues. In fact, the amount reaching the target organs or tissues from inhaling the chemical may be only a small part of the total reaching them from both routes of entry (8–17). Since the amount of the liquid that might contact the skin in an industrial situation is variable, this concurrent route of exposure cannot be reliably assessed as an adjunct to an inhalation exposure of test animals and cannot be factored into the determination of safe inhalation levels of the chemical.

Multiple routes of exposure also are possible when workers are exposed to particulate material. For example, inhaled dust particles that are cleared from the lungs by way of the trachea will be ingested; thus exposure of workers by inhalation also is an exposure by more than one route. If the absorption coefficient for gastrointestinal absorption is greater than that for pulmonary absorption, the second route of exposure again may be more important toxicologically than inhalation. The ratio of absorption by the two routes also may vary with particle size. However in this situation, in contrast to the inhalation–skin absorption situation, inhalation exposure of test animals will give some measure of concurrent absorption by ingestion. However if particle sizes differ in the industrial situation and in the animal exposures, the amounts concurrently absorbed by the two routes may be more or less than expected from the inhalation studies in animals. Consequently the amount of the chemical reaching the target organs or tissues may vary, and the animal exposures may not be quantitatively predictive of the effect on exposed humans.

Obviously, then, measuring atmospheric concentrations of a chemical in a plant does not always give a reliable measure of the amount of the chemical that may reach that chemical's target organs or tissues in exposed workers.

As already mentioned, the ideal measure would be the measure of the concentration in the target organ. This is impractical on a continuing basis as mentioned, although it may be practical for single, isolated analyses. Blood sampling, the next more direct measure of the amount in the target organs or tissues also may be impractical on a continuing basis, because of hazards introduced in taking samples continuously and also because of worker resistance to the procedure. However this technique is suited for occasional isolated or repeated analyses.

The ideal method for measuring all the xenobiotic absorbed, thus the total risk to the employee, is a noninvasive technique that accurately reflects the systemic burden of the chemical and is not disruptive or objectionable to the employee.

3 HISTORY OF BIOLOGICAL MONITORING

As mentioned earlier, the concept of biological monitoring has been used by the medical professions for decades. The formal application of this technique on a broad scale to industrial hygiene is probably most rightfully attributed to Hervey Elkins (1), although individual correlations of exposure levels and blood or excretory levels of industrial chemicals precede his broad-scale advocacy by many years (16, 18–22). Since his initial advocacy, Elkins has published several papers on the subject (23–25) and Dutkiewicz (26) has proposed an integrated index of absorption.

Elkins initially tried to correlate exposure levels, urinary levels, and toxicity. Since then, the concept of biological monitoring for xenobiotics has been extended to include monitoring of the biliary-fecal route, exhaled air, perspiration, tears, fingernails and toenails, hair, milk, and saliva.

Although the analysis of feces, perspiration, tears, nails, hair, milk, or saliva has been used in special situations to give an estimation of body burden of a chemical, these substances are not suited for routine, frequent monitoring of workers to determine immediate past exposure. Analysis of exhaled air for volatile xenobiotics or their metabolites and analysis of urine for water-soluble xenobiotics and metabolites have received the greatest attention as measures of industrial exposure.

4 USEFULNESS OF BIOLOGICAL MONITORING

Since biological monitoring usually occurs after exposure, it cannot replace good industrial hygiene practices. It does, however, supplement a good industrial hygiene program and provides additional information of value in the overall worker protection program. Some specific utilities of biological monitoring are detailed in Sections 4.1 to 4.6.

4.1 Indication of Unsuspected Employee Exposure

The unsuspected exposure may arise from skin absorption, as mentioned, or from unsuspected equipment leaks.

4.2 Guidance to Physician in Deciding Whether to Administer Therapy That Also Carries Some Risk

Some antidotes, such as atropine or 2-PAM (prescribed for overexposure to certain organophosphates), are used at dosages that would themselves be toxic if the patient had not received a prior excessive dosage of the hazardous substance.

4.3 Documentation of Overexposure or Acceptable Exposure

If a worker has received an excessive dose of an agent, good industrial hygiene practice dictates that the worker be removed from exposure. Biological monitoring can document the acceptable exposure or overexposure. In the case of overexposure, it can give the degree of overexposure and indicate how long the worker must remain free of additional exposure.

4.4 Preemployment Screen or Metabolic Abnormalities

Some people are unusually sensitive to the adverse effects of chemicals (7, p. 2). For their own protection, they should not be exposed to levels of a chemical that will affect them, but not the rest of the exposed population. In some cases this unusual sensitivity is due to atypical metabolic routes or to atypical enzyme kinetics. This could be evaluated in a preassignment physical by administering controlled, low doses of chemicals to which the worker would be exposed and determining if the individual biological half-life values for these chemicals are in the normal range, and if the usual metabolites are excreted.

A subnormal half-life would be acceptable, but a greater than normal half-life could predict bioaccumulation and possible hazard under exposure conditions that would not be harmful to most people (27). Abnormal metabolic products might or might not indicate greater hazard.

4.5 Validating TLVs Calculated for Concurrent Exposure to Chemicals Affecting the Same Target Organ(s)

When a worker is concurrently exposed to more than one chemical affecting the same target organ(s), exposure levels that were safe for the individual chemicals of exposure may not be safe for the combined exposure. The TLV handbook contains a formula for calculating reduced atmospheric exposure levels for this situation (7, p. 45). Direct biological monitoring for the exposure chemicals or their metabolites will not document the safety of the calculated safe exposure levels. However if the organ damage caused by overexposure results in elevated levels of some usual blood or excretory fluid component, normal postexposure levels of the component(s) would confirm the validity of the calculation. Conversely, abnormal values would indicate the inapplicability of the correction formula in this situation (see Section 5).

4.6 Validation of Area Monitoring

Personal monitoring, manual area monitoring, and automated area monitoring are the most common techniques now used to monitor worker exposure to atmospheric chemicals. All three offer opportunities for error. Regular biological monitoring may be useful to validate the atmospheric monitoring. However biological monitoring contains its own innate possibilities for error, as discussed below.

5 TYPES OF BIOLOGICAL MONITORING

Most monitoring of biological fluids and in some cases hair, nails, and expired air can utilize one of two approaches: direct or indirect. Direct analysis is analysis for the xenobiotic of interest or its metabolites per se.

Indirect analysis is the quantitation of second-generation effects that result from the action of the agent on some body system, organ, or tissue. For example, blood-urea-nitrogen or urinary glutamic pyruvic transaminase levels could be utilized as measures of effect on kidneys (6, 28). Changes in blood cell number per unit volume and distribution between types could be chosen to measure effect on the hematopoietic system (6). Changes in isoenzyme patterns could serve as a measure of effects for example, on liver or heart (28) and changes in cholinesterase levels in whole blood or plasma could be employed as a measure of organophosphate exposure (29).

Indirect analysis usually is not a desirable approach for routine worker monitoring for at least the following reasons:

1. The measured effect usually appears only after damage to the worker has occurred, and the thrust of industrial hygiene should be to prevent deleterious effects on workers from agents in the workplace.
2. Appearance of the measured effect may not occur for some time following exposure to the agent, thus may be difficult to attribute to a specific cause. Also, in this situation, there could have been protracted overexposure before the secondary effect was seen, and this is undesirable.
3. If the effect measured does not vary directly with amount of exposure, it will be difficult to determine the degree of exposure.

4. The effect may not be specific and may have several possible etiologies, not all of which occur in the workplace. This should not be interpreted to mitigate the health significance of the change for the individual worker, however.

5. If blood sampling is required, it may not be suited for frequent routine monitoring in the workplace.

6 BIOCHEMICAL PROBLEMS ASSOCIATED WITH BIOLOGICAL MONITORING

As with most techniques, increased study has illuminated many possible sources of error, other than those associated with methodology, that apply not only to analyses carried out on most of the fluids or body components mentioned above but also apply to the interpretation of the results.

General factors that must be considered in developing, applying, and interpreting a biological analysis include:

1. Metabolism.
2. The ratio of bound to free chemical in the blood.
3. Special situations in which excreted levels of the index chemicals do not indicate current exposure levels.
4. Concentration changes due to volume changes in the bioassay material.
5. Natural occurrence of the index chemical in the body and its resulting natural variations in concentration in the bioassay material.
6. Age of the worker.
7. Disease.
8. Sex of the worker.
9. Normal range of the index chemical to be expected in the bioassay material.
10. Time required for the index chemical to appear in the bioassay material.
11. Analytical methodology.
12. Route of exposure.

6.1 Metabolism

Xenobiotics can be handled by the body in many ways, including:

1. Metabolism to a component of a naturally occurring series of reactions and then oxidization to carbon dioxide and water or anabolism (biochemical synthesis) to a body component.
2. Excretion of the unchanged chemical.
3. Metabolism to a chemical that is more readily excreted.
4. Storage in an organ, bone, or body structure or organized element of some sort such as fat.

5. Accumulation without storage (e.g., in the blood or the hepatobiliary-intestinal loop).

Many xenobiotics follow a combination of routes 2 and 3. For these, all the factors that can affect metabolism became important in determining how much of the original exposure chemical is excreted unchanged and by what route and how much of each possible metabolite is excreted and by what route.

Obviously factors that can affect either or both the rate of excretion and the ratios of the xenobiotic and the various possible metabolities excreted can affect the estimate of exposure based on analysis of excreted chemicals.

Anything that can increase or decrease the normal accumulation or the storage of xenobiotics handled by routes 4 or 5, will have the inverse effect on excreted materials. Thus biological analyses relied on to indicate level of exposure will indicate correspondingly high or low results in these situations.

Most of these factors apply to organic compounds. There are not as many opportunities for individual variation in biological monitoring of metal or inorganic metal salt exposures since, unlike organic compounds, these materials usually are not metabolized. Nevertheless, one should expect individual variations in route and rate of excretion of these materials. Also, to the extent that storage or excretion of such salts may be hormonally controlled, concurrent exposure to medicinal substances or industrial chemicals that affect hormonal levels should be considered when evaluating the body burden of metals or metal salts through analysis of excreted materials.

Inorganic metal salts usually are not absorbed through the skin. Thus monitoring of the workroom air usually gives a good estimation of worker exposure to such salts in contrast to organic chemicals that may be absorbed through the skin as well as inhaled. Organic salts and complexes of metal ions or metals (e.g., butyl chromate, tetraethyl lead, tetramethyl lead, alkyl mercury, methylcyclopentadienyl manganese tricarbonyl, and organic tins) may be absorbed through the skin. The metabolism of the organic portion of these compounds, consequently the storage and excretion of the metal component, may be affected by the factors just mentioned, which are discussed more fully below.

Some organic industrial chemicals such as polychlorinated or polybrominated biphenyls (30) and some metals such as lead and mercury are stored by the body. However just as with foodstuffs, most industrial chemicals are metabolized and excreted (see, e.g., References 31 and 32). Chemicals that are totally catabolized (biochemical breakdown) usually do not yield characteristic end products and are not suitable for analysis other than in blood shortly after exposure. Chemicals that are excreted either without modification, or, after modification, as metabolites not occurring naturally, offer excellent opportunities for measuring worker exposure by measuring the amount excreted.

Most excretory systems of the body are aqueous, and if the xenobiotic is not catabolized to a component of one of the body's many metabolic sequences, the thrust of its metabolism will be toward increased water solubility. This means, in general, that a

BIOLOGICAL INDICATORS OF CHEMICAL DOSAGE AND BURDEN

hydrocarbon xenobiotic will be hydroxylated or at least one carbon will be oxidized to a keto group. Frequently alkyl side chains on aromatic compounds are oxidized to a carboxylic acid moiety or all the way to benzoid acid. An aliphatic alkane may have a terminal carbon oxidized to a carboxylic acid moiety. Thus the chemical(s) excreted following exposure may not be the same as the exposure chemical. Figure 7.1 presents examples of some of the metabolic routes that have led to metabolites in blood or to excretion products in liquid excretion media.

As can be seen, several metabolites of one chemical may be excreted in the urine, as well as glucuronates, sulfates, mercapturic acids, or amides of metabolites. These derivatives are called conjugates. Therefore analyses of biological excretory liquids should measure conjugated, bound, and free forms of the index chemical.

Since the direction of metabolism usually is toward increased water solubility, metabolites generally are less volatile than the chemical of exposure. Hence metabolites generally are not excreted in exhaled air.* Accordingly, some of the complications introduced by metabolism are correspondingly less important when biological monitoring is based on analysis for the workplace chemical in exhaled air. For this monitoring, generally the exposure chemical only need be considered, regardless of the route by which it was absorbed.

Neither the kinetics of a particular metabolic pathway nor the relative kinetics of alternate metabolic pathways are constant in the same person at all times, or between individuals. Since a biological monitoring program relies on the rate of appearance of and/or the amount of the index chemical in the bioassay material, anything that affects this rate or amount will affect the estimate of the amount of workplace exposure.

Factors that can affect metabolic rates, thus excretion, include:

1. Individual variations in enzyme complement.
2. Diet.
3. Stimulation or inhibition of enzymes in the metabolic sequences.
4. Ratio of bound to free chemical in the blood.
5. Dose of the exposure chemical.
6. Competition for a necessary enzyme.

6.1.1 Individual Variations in Enzyme Complement

Viable mutations resulting in altered enzyme profiles occur spontaneously in man. These have been reviewed extensively (52) and are not covered here. If a worker is missing an enzyme or has an enzyme with specific activity different from that of most other workers and that enzyme is critical in the metabolism and excretion of a workplace chemical, analysis of that person's excretory products for the expected

---

* There are some notable exceptions. For example, some metals, such as selenium and tellurium are methylated by the body. The methylated compound is more volatile than the metal and is excreted in exhaled air (33).

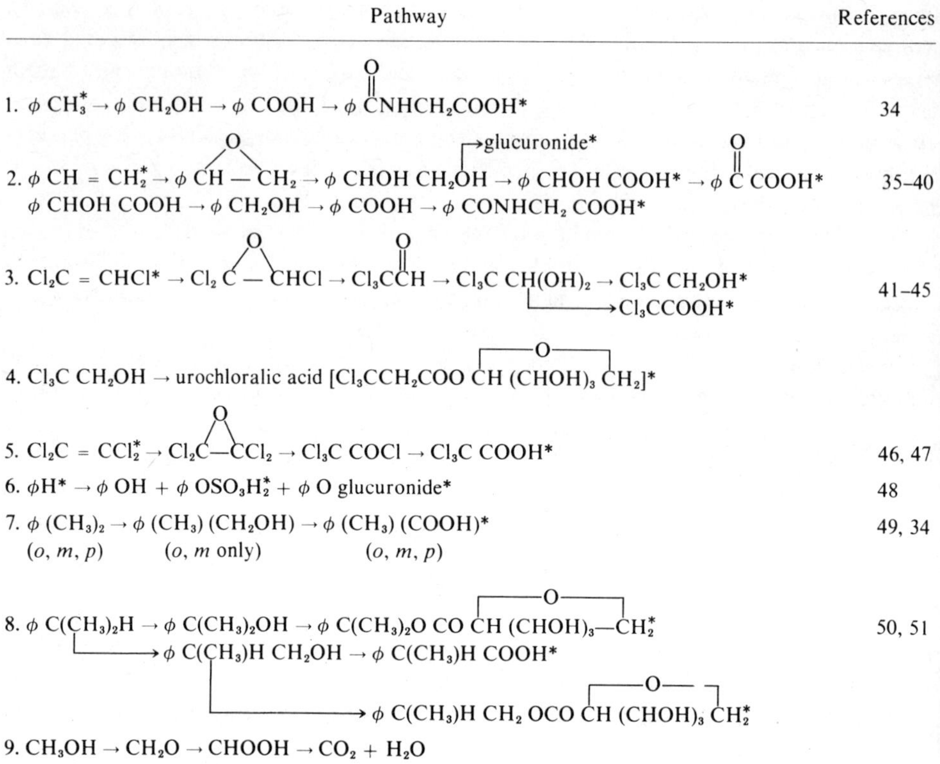

Pathway	References

1. $\phi\,CH_3^* \to \phi\,CH_2OH \to \phi\,COOH \to \phi\,\overset{O}{\overset{\|}{C}}NHCH_2COOH^*$ 34

2. $\phi\,CH = CH_2^* \to \phi\,\overset{O}{\overset{\triangle}{CH-CH_2}} \to \phi\,CHOH\,CH_2OH \overset{\to glucuronide^*}{\to} \phi\,CHOH\,COOH^* \to \phi\,\overset{O}{\overset{\|}{C}}\,COOH^*$ 35–40
$\phi\,CHOH\,COOH \to \phi\,CH_2OH \to \phi\,COOH \to \phi\,CONHCH_2\,COOH^*$

3. $Cl_2C = CHCl^* \to Cl_2\,\overset{O}{\overset{\triangle}{C-CHCl}} \to Cl_3C\overset{O}{\overset{\|}{C}}H \to Cl_3C\,CH(OH)_2 \to Cl_3C\,CH_2OH^*$ 41–45
                                                              $\to Cl_3CCOOH^*$

4. $Cl_3C\,CH_2OH \to$ urochloralic acid $[Cl_3CCH_2COO\,\overset{\overset{O}{\frown}}{CH\,(CHOH)_3\,CH_2}]^*$

5. $Cl_2C = CCl_2^* \to Cl_2\overset{O}{\overset{\triangle}{C-CCl_2}} \to Cl_3C\,COCl \to Cl_3C\,COOH^*$ 46, 47

6. $\phi H^* \to \phi\,OH + \phi\,OSO_3H_2^* + \phi\,O$ glucuronide*   48

7. $\phi\,(CH_3)_2 \to \phi\,(CH_3)\,(CH_2OH) \to \phi\,(CH_3)\,(COOH)^*$ 49, 34
  (o, m, p)   (o, m only)   (o, m, p)

8. $\phi\,C(CH_3)_2H \to \phi\,C(CH_3)_2OH \to \phi\,C(CH_3)_2O\,CO\,\overset{\overset{O}{\frown}}{CH\,(CHOH)_3-CH_2^*}$ 50, 51
        $\to \phi\,C(CH_3)H\,CH_2OH \to \phi\,C(CH_3)H\,COOH^*$
        $\to \phi\,C(CH_3)H\,CH_2\,OCO\,\overset{\overset{O}{\frown}}{CH\,(CHOH)_3\,CH_2^*}$

9. $CH_3OH \to CH_2O \to CHOOH \to CO_2 + H_2O$

Figure 7.1 Principal metabolic pathways of some industrial chemicals: ϕ = phenyl; * = identified in the urine of man or animals. Other minor metabolites may be formed (see, e.g., Reference 34) and excreted, but they are not of value for biological monitoring.

metabolite(s) will give a false picture of the individual's exposure. If, as a result of this aberrent enzyme, the elimination half-time ($t_{1/2}$) of a workplace chemical is increased, the error will be of the worst type; that is, it will indicate less exposure than the worker has received. As discussed earlier, increased $t_{1/2}$ for a chemical to which a worker will be exposed could lead to unexpected accumulation in the body, with possible toxic consequences.

Vessell and Passananti (53), for example, found that the $t_{1/2}$ in plasma for dicumarol varied almost tenfold in fraternal twins. The $t_{1/2}$ for the medicinals antipyrine and phenylbutazone varied threefold and sixfold, respectively, in the same twins. Variations at least as large are to be expected in the general population, and similar variations in the metabolism and excretion of industrial chemicals are to be expected.

Tang and Friedman (54) examined the activity of liver microsomal oxidase* from 10

---

* If a xenobiotic does not fit a specific enzyme, its metabolism generally is affected by the nonspecific microsomal enzymes. These enzymes do not seem to require the enzymatic "fit" that characterizes most enzymatically catalyzed reactions. They are active in varying degrees on a variety of structures. They are known to carry out epoxidations, demethylations, reductions, and hydrolyses.

humans varying in age from 3 days to 92 years. Six were between the ages of 22 and 57, inclusive. Both sexes were represented. Some had died violent deaths, some had died from disease, and one had died following surgery for an acute condition. Thus not all had died of a disease and not all were aged, so the findings should be representative of what could be expected for a worker population. Activity varied from "undetectable" to "highly active" as measured by mutations induced in *S. typhimurium* TA-100 incubated with various aromatic amines known to require microsomal oxidase activation for mutagenic activity.

6.1.2 Diet

Many studies have been carried out showing the effect of diet on enzymatic activity, consequently on the metabolism of chemicals. For example, excessive (3 to 9 g/day) levels of niacinamide have been reported to cause elevation of SGOT (see Abbreviations section preceding Bibliography) and APase (55). Protein deficiency has been reported to cause reduced hepatic metabolism in rats (56) and could be expected to have the same effect in man. High protein diet, on the other hand, has been shown to enhance the metabolism of chemicals in man (57). A charcoal-broiled beef diet has been shown to increase microsomal oxidase [benzo(a)pyrene hydroxylase] activity in the liver of pregnant rats (58), suggesting a stimulation of the nonspecific microsomal oxidase systems. However when humans were placed on a diet containing charcoal-broiled beef, there was no change in the plasma half-life of the test chemical used to monitor changes in microsomal oxidase activity (59).

Brussels sprouts, cabbage, and cauliflower have been shown to contain chemicals that, when ingested by rats, increase liver microsomal oxidase activity as measured by oxidation of benzo(a)pyrene. Furthermore, specific activity was found to vary between cultivars of Brussels sprouts and cabbage (60).

It is thus probable that diet, particularly a diet unbalanced with regard to any component, could lead to changes in enzymatic activity followed by changes in metabolism and in excretion of workplace chemicals.

6.1.3 Stimulation or Inhibition of Enzymes That Metabolize the Chemical

As mentioned earlier, many xenobiotics are metabolized through the nonspecific microsomal enzymes, which can catalyze at least four types of reactions: oxidations (epoxidations), demethylations, reduction, and hydrolysis. More than 200 chemicals (61) are known to stimulate some or all of the known microsomal enzymes, thus increasing their metabolism of many other xenobiotics. This could result in an elevated concentration of the metabolites in excreted fluids. If the evaluation of worker exposure is being based on the normal concentration of excreted metabolites formed by way of the microsomal enzymes, the estimate of worker exposure could be erroneously high. Conversely, if the analysis is based on unmetabolized xenobiotic, the estimate of worker exposure could be erroneously low.

Some of the medicinal chemicals, pesticides, and industrial chemicals that have been shown to stimulate the microsomal enzyme systems include phenobarbital [one of the most potent stimulators (61)] and spironolactone (62). Pesticides that are known to stimulate microsomal enzymes include chlordane (63), DDT (64, 65), and methoxychlor (66). Industrial chemicals stimulating microsomal enzymes include condensed polynuclear aromatic hydrocarbons and polychlorinated or polybrominated biphenyls.

Cigarette smoke also has been found to stimulate the activity of liver microsomal enzymes in pregnant rats (59).

Pretreatment with phenobarbital has been shown to stimulate the oxidation of styrene in rats (39) and the oxidative step in the metabolism of T_3CE and T_4CE in rats and hamsters (27). In both cases the oxidation rate was almost doubled.

Other chemicals [e.g., morphine (62)] are known to inhibit the microsomal enzyme systems. Styrene and benzene epoxidation by microsomal enzymes also has been reported to be hindered in the presence of toluene (67). Obviously reduced activity could lead to decreased appearance of metabolites in the excretory fluid analyzed and could lead to a low estimate of workplace exposure if metabolite concentration in the excretory fluid were the basis of estimation of workplace exposure.

Regardless of whether the effect is stimulation or inhibition of the microsomal enzymes, the effect could alter the rate of appearance of metabolites in the excretory fluids only if the microsomal enzymes catalyzed one or more rate-limiting steps in the metabolism of the xenobiotic.

It also has been known for many years that continued exposure to a chemical can stimulate its own metabolism (3, 68). Ikeda (27) has shown that pretreatment with T_3CE can either stimulate or depress metabolism of subsequently administered T_3CE, depending on the pretreatment time and dose. Pretreatment with T_4CE had only a metabolic stimulatory effect on subsequently administered T_4CE.

Ikeda (69) also reported on a subject who was addicted to sniffing T_3CE and whose $t_{1/2}$ for urinary excretion of metabolites was twice the usual $t_{1/2}$. It was not known whether the prolonged $t_{1/2}$ preceded or followed the addiction. Thus repeated exposure to a chemical may not always stimulate its metabolism.

6.1.4 Ratio of Bound to Free Chemical in the Blood

Many medicinal chemicals have been found to be transported in the blood in the unbound (free) state as well as bound to proteins. Generally the binding is to albumin in the plasma, but it also may be to other plasma proteins and/or to proteins that are structural parts of formed elements in the blood (see e.g., Reference 62). Since industrial chemicals basically differ from medicinal chemicals only in their application and relative magnitude of biological effect, there is no a priori reason for them to differ from medicinal chemicals in binding. Therefore it is to be expected that industrial chemicals also would be transported in the blood in both the bound and free forms. Also, as with medicinal chemicals, the ratio of bound to unbound would be expected to vary from chemical to chemical.

The binding phenomenon probably is a simple complex formation, subject to all the mass action mathematics and dynamics of inorganic or other organic complexes. The stability constants also probably vary over several orders of magnitude, just as do those for classical complexes. Thus bound and unbound fractions would be in dynamic equilibrium. This has important ramifications in biological monitoring, as discussed below.

Total blood concentration, bound plus free, should reflect workplace exposure of chemicals that are absorbed. However amount excreted per unit time generally reflects unbound concentration in the blood. Again, as already mentioned, systemic toxicity generally varies with the concentration of unbound chemical. Situations can occur in which the normally bound fraction of an industrial chemical, the fraction that was bound when the correlations between dose, toxicity, and amount excreted were developed, is decreased. In these situations the amount of index chemical excreted still represents the unbound concentration in the blood, thus the hazard to the worker (see, e.g., Reference 70). However it will not be as closely related to workplace exposure and will suggest higher recent exposures than occurred.

As a rule changes in the ratio of bound to unbound chemical in the blood are caused by a decreased number of available binding sites. This usually is due either to a deficiency of binding protein or to concurrent administration of a chemical that is more tightly bound. However disease also has been shown to reduce binding of chemicals (70).

Decreased amount of binding protein may be due to genetic factors. However it most frequently is due to disease. Chronic conditions accompanied by lowered plasma levels of protein include chronic liver disease, rheumatoid arthritis, diabetes, and essential hypertension. Nephritis and gastrointestinal diseases such as peptic ulcer and colitis also can be accompanied by lowered plasma protein. Section 6.7 discusses other effects of disease on the ratio of bound to unbound chemical.

A decrease in the number of available binding sites accompanying a normal level of transport protein usually is caused by the presence of other chemicals that are more tightly bound. Phenylbutazone, a drug occasionally prescribed for arthritic and other aches and pains, is very tightly bound to plasma proteins. It is so tightly bound that it will displace most other bound chemicals from the blood transport proteins. If present in sufficient concentration in the blood, it could displace significant amounts of a bound workplace chemical. This would result in elevated blood levels of the free form of the latter, thus in an increased rate of metabolism and excretion. If workplace exposure were being estimated from excreted levels of the exposure chemicals and/or metabolites, this could produce an erroneously high estimate of exposure. However the excreted level would reflect the *hazard* to the worker, since that would usually vary with the concentration of unbound chemical in the blood.

Another common pharmaceutical that also is strongly bound is salicylic acid (70). Less common pharmaceuticals that could displace bound workplace chemicals include clofibrate, a cholesterol-lowering drug (70).

Since it is likely that workplace chemicals also are bound in varying degrees to blood proteins, as already mentioned, concurrent high exposure to either or both of two or more industrial chemicals could alter the binding ratio of the weaker bound of the two.

Among the diseases, cirrhosis has been shown to result in decreased binding of drugs with accompanying increased blood levels of the unbound form (70). There undoubtedly are other diseases that also reduce binding.

6.1.5 Effect of Dose of Exposure Chemical on Metabolism

It has been known for many years that the metabolism of chemicals can change with the dose administered. This obviously could affect the dose-excretion curves for metabolites used to measure exposure. Quick (71), for example, found that when high doses of benzoic acid were administered in humans, the glucuronide became an important excretory product, as well as hippuric acid. Von Oettingen (21) found that after about 8 hr of exposure to approximately 300 ppm of toluene, benzoic acid glucuronide also was excreted in addition to hippuric acid.

Tanaka and Ikeda (72) found the usual metabolic pathway for TCA formation in humans from T_3CE became saturated after about 8 hr inhalation exposure at 150 ppm. The excretion rate of TCA excretion reached a plateau at that concentration.

More recently, Ikeda et al. (73) reported that the usual metabolic pathway for T_4CE was saturated after exposure at 50 ppm for several hours. They also discovered that in the rat, formation of phenylglyoxylic and mandelic acids plateaus at styrene exposure levels $\geq$ 100 ppm for 8 hr (74).

Götell (75) reported that in man, the urinary concentration of phenylglyoxylic acid not only reached a peak after exposure for approximately 8 hr at 150 ppm, but actually decreased at higher atmospheric levels.

Watanabe et al. (76) also have reported that the usual metabolic pathway for VCM is saturable in the rat. At low VCM levels it is principally metabolized, and the metabolites are excreted in the urine. Only a small percentage is excreted unchanged in exhaled air. At high VCM levels the reverse is true, and most of the VCM is excreted unchanged in exhaled air.

These studies all indicate the importance of not extrapolating biological monitoring equations or dose-response curves beyond the conditions used to derive the equations or reference curves.

6.1.6 Competition for a Necessary Enzyme

If a worker is concurrently exposed to two or more chemicals that require the same enzyme(s) for metabolism, the resulting pattern of metabolites and parent compound in blood and excreted fluids can be dramatically altered. The chemical that is preferentially metabolized by the enzyme(s) will have the most nearly normal pattern. This is elegantly illustrated by concurrent exposure to ethanol and methanol.

The first step in the metabolism of ethanol and methanol is a dehydrogenation catalyzed by the liver enzyme alcohol dehydrogenase (ADH) as shown in reactions 1 and 2:

$$CH_3CH_2OH \xrightarrow{ADH} CH_3\overset{\overset{O}{\|}}{C}H \qquad (1)$$

$$CH_3OH \xrightarrow{ADH} H\overset{\overset{O}{\|}}{C}H \qquad (2)$$

ADH acts preferentially on ethanol. Therefore in the presence of ethanol, only a small fraction of the methanol is dehydrogenated, and most of the methanol circulates in the blood as unbound and unmetabolized methanol. Consequently blood, urinary, and exhaled methanol levels are elevated over what they would be if the same dose of methanol had not been competing with ethanol for the available ADH (77). Under these conditions, biological monitoring would suggest higher workplace exposure to methanol than occured. [In this situation the elevated methanol levels reflect not increased hazard of eye damage, but rather decreased hazard, since the metabolite causing the eye damage appears to be formic acid, a metabolite (78).]

Enzyme competition also may explain the observation of Stewart et al. (79) that the urinary excretion of TCE was greater than expected and that of TCA was less than expected after T_3CE exposure, if the subject consumed alcohol. If T_3CE exposure was being measured by urinary TCE excretion, this also yielded higher estimates of exposure than occurred.

From a toxicological standpoint, the importance of all these possible metabolism variations depends on whether the most important toxicant is a metabolite of the exposure chemical or the exposure chemical itself. Obviously anything that decreases the concentration of a toxicant decreases the hazard, and anything that increases the concentration increases the hazard.

From a regulatory or a monitoring standpoint, these causes of variation indicate that the normal range of concentration of a workplace chemical or its metabolites in blood, exhaled air, or excretory fluids following a given exposure will be broad rather than narrow. This also would be expected if based on an examination of the range of blood and urine concentrations reported for normally occurring chemicals in these fluids in humans (6). Two- to threefold variations between the upper and lower "normal" values are usual. Furthermore, recognizing this normal range, physicians tend not to rely solely on urine or blood levels of a chemical in diagnosing a disease condition. From a worker monitoring standpoint, one must be particularly careful about interpreting blood or excretory fluid levels of a chemical that is always present in this fluid and also may arise from workplace exposure. In this situation the concentration range to be expected from a particular exposure level may vary by four- to sixfold.

6.2 Changes in the Ratio of Bound to Free Chemical in the Blood

Section 6.1.4. treated changes in the ratio of bound to free chemical in the blood as one of the factors that can alter metabolic rates. As indicated there, this also can affect the rate of appearance of the index chemical in the bioassay material and consequently the estimate of workplace exposure.

6.3 Special Situations in Which Excreted Levels of the Index Chemical would not Indicate Current Exposure Levels

Situations also exist when all of an index chemical in the bioassay material may not arise from recent exposure or, conversely, the index chemical is not excreted in the bioassay material at the rate expected. Obviously, one should be sure that such factors are not operating in any particular individual assay.

It should be stressed here that to the extent the assayed index chemical reflects the hazard to the worker, steps should be taken to reduce the hazard, regardless of the etiology of the toxicant. However if the toxicant did not largely arise from the workplace, trying to reduce workplace exposure will not reduce the hazard to the worker (see also Section 6.5).

Competition for the same enzyme between concurrently administered methanol and ethanol is an example that has already been mentioned. Leaf and Zatman (77) found that the urinary methanol levels after methanol exposure were up to 100 percent higher in the presence of ethanol than in its absence. Obviously in this situation it would not be valid to utilize correlations of urinary methanol with exposure, made in the absence of concurrently administered ethanol. Use of such correlations would lead to greatly overestimating the workplace exposure.

Drug therapy to mobilize stored or recycling chemicals can be responsible for misleading levels of a chemical in biological monitoring materials. For example, the use of penicillamine to treat Wilson's disease (80) or schizophrenia (81) or the use of tetraethylthiuram disulfide to treat alcoholism would lead to elevated excretory levels of copper that would be totally unrelated to current workplace exposure.

Any medicinal that could release stored, recycling, or protein-bound index chemical could cause such erroneous results, regardless of the reason for administering the medicinal.

In the case of urinary measurement of the index chemical, decreased renal clearance, active lactation, or excessive sweating all could cause decreases in urinary level that would result in decreased estimates of workplace exposure.

Absorption of chemicals through the skin or by the lungs also can vary among individuals. If a person absorbed significantly more or less of the administered dose than did the subjects used to make the correlations of exposure and excretion, the individual analysis would indicate correspondingly greater or lower exposure, respectively, than occurred. The hazard estimate, however, would be correct.

Individual absorption may vary as much as 100 percent. In a study of oral absorption of ampicillin (82), the amount absorbed varied from 32 to 64 percent in nine healthy men aged 20 to 40 years. The absorption of benzene by inhalation has been reported also to vary by 100 percent between males and females, with females absorbing 41.6 percent of the dose and males 20.3 percent (83).

Assuming no metabolic differences, absorption differences between individuals thus could lead to a 100 percent difference between two individuals in the level of index chemical in the bioassay material following the same exposure.

6.4 Concentration Changes due to Volume Changes in Biological Assay Liquids

The rate of appearance of the index chemical at the excretory organ generally is independent of the rate of excretion of the biological assay material. Thus the concentration of the index chemical in the biological assay material will change with the volume of excretion of the latter. This is not a serious problem for hair or nail analyses. In the case of exhaled air analyses, the subject generally is resting when samples are taken and is not doing anything to greatly increase respiratory volume per unit time. Therefore this also is not an important factor in exhaled air analyses. However the excretion rate of liquids can vary greatly. The rates of perspiration excretion during and closely following exertion versus resting, or on a hot day versus a cool day obviously will differ. Similarly, the rate of excretion of tears while crying is different from the rate when not crying. Salivation rate also can be altered by emotional, physical, and chemical factors. Since these biological fluids are seldom used for industrial biological assay, the rate changes for these fluids are not usually significant for biological monitoring. However excretion rate is a variable that must be considered in any assay utilizing these fluids.

Urine volume normally varies by as much as fourfold (600 to 2500 ml/24 hr) between individuals and in the same individual at different times (6). Factors that can cause volume variations include fluid intake, emotional state, medicines, pregnancy, disease, menstruation, ambient temperature, and body temperature.

Another dilution factor to be considered for urine samples is dilution by urine reaching the bladder before the index chemical reaches the kidneys. This is most important when end-of-shift "spot" urine samples (see Section 7.3.4) are analyzed, the index chemical has a relatively long excretion half-life, and several hours elapse between the penultimate urine sample and the sample taken for analysis. Voiding at a set time period before taking the urine sample for analysis might reduce the spot urine analysis variability due to this factor. This technique has been tried by several investigators (15, 73, 84, 85).

If the content of index chemical in the total 24 hr volume of urine is being measured, the dilution effect will be minimized. However if only spot samples of urine are being analyzed, dilution factors may be extremely important.

Recognizing that the same kinetics of presentation at the excretion organ apply to normal constituents of urine as well as to xenobiotics and their metabolites, workers have used various properties or components of urine to normalize analytical values to compensate for concentration changes due to volume differences.

As mentioned earlier, all other things being equal, the same ultimate insertion factors that control the concentration of xenobiotic or metabolite in urine generally control the concentration of normal constituents in the urine. Therefore if one could measure a colligative property of urine, or the concentration of a usual urinary constituent that has a constant rate of presentation to the urine and rarely evolves from a xenobiotic, it should be possible to normalize urinary concentration, thus correct for volume differences.

Specific gravity and osmolality are colligative properties of urine that are easy to measure. Specific gravity normally varies by only about tenfold, whereas osmolality varies by more than twentyfold. The concentration of xenobiotic or its metabolites makes some contribution to specific gravity, but it is generally small compared to the effect due to normal constituents. The xenobiotic or its metabolites would probably make a greater contribution to changes in osmolality. For whatever reasons, specific gravity has been used widely to normalize urine values, and osmolality has not. Of the possible reference chemicals for normalizing urine analyses, creatinine has been used most extensively. Creatinine excretion normally varies less than twofold [1.0 to 1.6 g/24 hr or 15 to 25 mg/kg of body weight in 24 hr; (6)].

Instead of specific gravity or creatinine content, Elkins normalized on the basis of sulfate excretion, using the same rationale discussed previously (1). However this normalization has not been used extensively.

Ogata (86) has found that the "rate" of excretion of hippuric acid from 23 young adult human males following inhalation exposure to toluene is a more consistent measure than normalizing by the use of specific gravity. To determine rate, he collected urine samples every 3 or 4 hr during a 3 or 7 hr exposure. The total amount of hippuric acid in the sample was divided by the minutes of exposure to give milligrams per minute. The values for the 1 to 4 hr urine samples or the 4 to 8 hr samples were then plotted against the toluene concentration. The standard deviations for the specific gravity and "rate" normalizing methods were not greatly different for the 1 to 4 or 4 to 8 hr values at a toluene concentration of 100 ppm. The standard deviations in the "rates" were less than the standard deviations of concentrations corrected for specific gravity at toluene concentrations of 200 ppm. Table 7.1 summarizes Ogata's data.

As can be seen, a direct dose-response relationship appeared with both methods of normalizing, but one standard deviation from the value at 100 ppm overlapped one standard deviation at 200 ppm by either method of normalizing. Shorter sampling periods would be more desirable to validate the concept. The approach would not be too practical in an industrial situation, since it would require that the worker void urine a specific number of hours prior to taking the urine sample to be analyzed. Levine and Fahy (87) found that calculating as "rates" did not reduce the variability of lead analyses in the urine of exposed workers.

Medicinals and diseases also can affect specific gravity and creatinine levels of urine (6), possibly invalidating their use for normalizing urine analyses in these two situations.

Table 7.1 Effect of Two Normalizing Treatments on Urine Analyses for Hippuric Acid Following Toluene Exposure (86)

Exposure sampling period (hr):	0–3		4–8	
Toluene exposure (ppm):	100	200	100	200
Hippuric acid concentration in urine, uncorrected (mg/liter)				
Mean	2.95	3.74	3.09	8.19
S.D.[a]	0.83	0.59	0.70	2.62
Specific gravity, corrected (mg/liter)[b]				
Mean	2.58	3.71	2.81	5.85
S.D.	0.40	0.67	0.66	1.24
Rate (mg/min)				
Mean	2.09	3.13	3.10	4.61
S.D.	0.35	0.36	0.84	0.80

[a] S.D. = standard deviation.
[b] Corrected to 1.024.

Normalizing urine analyses in industrial hygiene on the basis of specific gravity was first reported by Levine and Fahy (87). They used a specific garvity of 1.024 as their common denominator. Other workers, particularly the Japanese, have used 1.016. Engström (88) used 1.018 in his studies on Finnish workers. There has been some controversy whether 1.016 or 1.024 more nearly represents the average urinary specific gravity (89–91).

Normalizing will have the same effect on variability regardless of what number is used, because of the nature of the correcting equation, as shown below. Levine and Fahy (87) used Equation 3 to correct to specific gravity 1.024. Equation 4 is used by the National Institute for Occupational Safety and Health (NIOSH) (92) to correct to specific gravity 1.024.

$$\text{Corrected value} = \text{observed value} \times \frac{24}{(\text{specific gravity} - 1) \times 10^3} \quad (3)$$

$$\text{Corrected value} = \text{observed value} \times \frac{24}{\text{last 2 digits of specific gravity}} \quad (4)$$

It can be seen that Equation 3 reduces to Equation 4 in practice, since (specific gravity − 1) for a specific gravity of, say, 1.021, = 0.021. When this number is multiplied by 10^3, it becomes 21. These are the last two digits of the specific gravity. Obviously Equation 3 or 4 could be used to normalize to any other specific gravity by substituting the last two digits of that specific gravity for "24" in the numerator of Equation 3 or 4.

The question of which specific gravity to use can be avoided by normalizing the urinary analyses on the basis of creatinine content, for the reasons already discussed. In the field of industrial hygiene, this approach was initially used by Hill (11) to normalize analyses of aniline in urine. Its use as a basis for normalizing has increased since then.

There have been mixed results reported from normalizing with either specific gravity or creatinine. As the work of Ikeda or Ogata reported in Table 7.2 demonstrates, neither creatinine nor specific gravity corrections consistently gave significantly smaller standard deviations with respect to each other or compared with uncorrected values. This also has been the experience of other invesigators. Seki (93) found that specific gravity correction slightly reduced the scatter, but creatinine correction did not. Conversely, Pagnatto (94) discovered that creatinine correction reduced scatter over that found in uncorrected values. Ellis (95) reported that neither correction significantly reduced scatter over that found in uncorrected samples.

In practice, since both specific gravity and creatinine are usual urinary determinations, many authors now report values corrected by both methods as well as the uncorrected value. The data presently available do not indicate that either method of normalizing analyzes reduces the standard deviation from that found in uncorrected values.

6.5 Nonworkplace Progenitors of the Index Chemical and Resulting Baseline Variation in its Concentration in the Bioassay Material

If the index chemical also has a background normal concentration range in the bioassay material being analyzed, any biological monitoring procedure must determine (a) the normal range, (b) nonoccupational factors that affect the normal range, and (c) that a small fraction of the maximum safe workplace exposure to the index chemical or its workplace progenitor causes a significant elevation in the normal concentration of the chemical. If one has a series of analyses on a worker over a period of time, interindividual variations become less important. However if there is an elevation in one analysis, possible nonworkplace causes should be investigated.

For example, extremely varied normal levels of various metals have been reported for human scalp hair. The cause may be topical contamination of the hair (e.g., from cosmetics or hair dyes), which does not represent workplace exposure. Because of this great variation, hair levels that might be assumed to represent industrial exposure must be set quite high, unless one is regularly monitoring indiviudal workers.

Examples of a similar situation with organic compound include the use of urinary hippuric acid levels for monitoring styrene, toluene, or certain n-alklybenzene exposure and the use of urinary phenol levels as a measure of benzene of phenol exposure.

As mentioned earlier, styrene, toluene, and certain n-alkylbenzene compounds are metabolized to benzoic acid which usually is condensed with glycine to form hippuric acid, the excretory product. Hippuric acid occurs normally in the urine, generally as a result of ingestion of benzoic acid, sodium benzoate, or quinic acid, a metabolic precursor of benzoic acid. Sodium benzoate is used as a food preservative. Coffee beans, prunes, plums, and cranberries are known to contain benzoic acid or quinic acid.

Table 7.2 Effect of Corrections on Variations in Urine Analyses[a]

Exposure Chemical		Index Chemical					
			Concentration in Urine[b]				
Name	Concentration in Atmosphere (ppm)	Name	Uncorrected	Corrected to Specific Gravity 1.016	Corrected for Creatinine (g/g)	Number of Subjects	Reference
None	0	Phenylglyoxylic acid	0.017 (59–159)	0.011 (64–155)	0.013 (69–138)	35	74
None	0	Mandelic acid	0.057 (61–163)	0.036 (64–158)	0.043 (63–160)	35	74
None	0	Hippuric acid	0.35 (57–177)	0.29 (55–175)	0.24 (58–171)	31	85
Toluene	20	Hippuric acid	1.84 (66–152)[c]	1.18 (72–139)[c]	1.06 (76–132)[c]	10	85
	60	Hippuric acid	2.27 (59–171)[c]	1.21 (58–174)[c]	1.14 (66–151)[c]	10	85
None	0	Phenol	0.30 (25–420)	0.29 (21–480)	0.23 (20–512)	36	84
None	0	Total trichloro compounds	0.026 (32–312)	0.023 (31–316)	0.019 (32–318)	36	84
T_3CE	3	Total trichloro compounds	0.039 (67–149)	0.035 (75–133)	0.041 (74–135)	9	73
T_3CE	45	Total trichloro compounds	0.339 (72–138)	0.253 (77–130)	0.338 (79–126)	5	73
T_3CE	120	Total trichloro compounds	0.915 (85–118)	0.481 (86–116)	0.519 (68–146)	4	73

[a] Figures in parentheses are standard deviation ranges as a percentage of the geometric mean.
[b] Spot samples (see Section 7.3.4).
[c] Exposure values uncorrected for background level.

With widespread distribution of benzoic acid and its progenitors in foodstuffs, it is not surprising that its background level in urine should vary. As Table 7.3 reveals, the normal urinary level fluctuated threefold in United States subjects, up to thirteenfold in Japanese subjects, and almost fortyfold in Finnish subjects.

In rats, approximately 10 percent of an intraperitoneal dose of styrene is excreted in the urine as hippuric acid (74). If an equivalent conversion is assumed for inhaled styrene in man, an inhalation absorption coefficient in man of 0.8 (97), a 24 hr urine volume of 1.21, and an 8 hr exposure during which 10 m^3 of air were breathed, an 8 hr exposure to 200 ppm of styrene would be required to raise the average level of urinary hippuric acid as reported by Stewart et al. (38) to the upper normal level they reported. Stewart et al., in fact, found that exposure to 100 ppm of styrene for 7 hr did not significantly elevate urinary hippuric acid up to 48 hr postexposure, suggesting the conversion in man is even less than 10 percent at an exposure level of 100 ppm.

In partial agreement with Stewart's findings, Ikeda et al. (74) found significantly elevated urinary hippuric acid levels in most subjects exposed to styrene for up to 160 min at concentrations up to 200 ppm, but no significant increase at concentrations of ≤ 60 ppm for 2 hr or less. Twenty-four hour urine samples were collected and specific gravity was corrected to 1.016. However spot urine samples taken by the same investigators after approximately 6 hr exposure of workers at levels up to 30 ppm styrene did not show significant elevations of hippuric acid (98).

The insignificant urinary elevation of hippuric acid following these latter exposures might be explained by the delayed appearance of hippuric acid in the urine following styrene exposure. Ikeda et al. (74) found that hippuric acid did not appear in workers until about 24 hr postexposure. This would not explain Stewart's results, since he collected total urine samples for more than 24 hr postexposure.

Thus the combination of small percentage conversion and high, variable, natural levels must be assumed to argue against the use of urinary hippuric acid as an index for styrene exposure.

Hippuric acid in urine has been used successfully as a marker for toluene exposure. Since toluene is almost 70 percent metabolized to hippuric acid (86), the high background levels of hippuric acid become less important. Any exposure in excess of 50 ppm for 8 hr would be expected to elevate an average hippuric acid level beyond Stewart's upper normal (38), assuming 79 percent absorption (86), 70 percent metabolism to hippuric acid, 10 m^3 of air respired, and a 24 hr urine volume of 1.2 liter.

In agreement with this, Engström (88) noted very poor correlation between blood levels of free toluene and urinary hippuric acid (creatinine corrected) in painters concurrently exposed for 8 hr to airborne levels of toluene less than 50 ppm and unspecified low concentrations of m- and p-xylene.

The use of urinary phenol levels as a measure of benzene exposure also could be frustrated at low benzene levels by the presence of phenol in urine from nonbenzene sources. For many of these sources urinary phenol levels would correspond to much lower worker hazard than would exist if the phenol had arisen from benzene.

Phenol, as well as metabolic precursors of urinary phenol such as the amino acid tyrosine and the essential amino acid phenylalanine, occur in the diet. Though the

Table 7.3 Normal Urinary Levels and Ranges (mg/liter) of Hippuric Acid as Reported by Various Authors[a]

Uncorrected	Specific Gravity, Corrected	Creatinine, Corrected	Subjects: Number, Sex, Country	Analytical Procedure	Reference
1583[b] (833–2583)	ND	ND	9M, U.S.	UV absorption	38
350[c] (199–616)	290[c,d] (160–510)	240[c]	31M, Japanese	Paper chromatography (96)	85
301[c]	290[c,d]	229[c]	36M, Japanese		
398[c]	570[c,d]	449[c]	30F, Japanese		
NR	1037[e] (126–4844)	739 (86–2340)	39M, Finnish	Gas chromatography	88
800 (400–1400)	ND	ND	NR, M, U.S.	UV absorption	94
NR	184[f] (35–444)	ND	NR, M, Japanese	Paper chromatography	96
335[b]	ND	ND	6M, U.S.	Titration	21

[a] Symbols as follows: ND = not determined; NR = not reported; M = male; F = female.
[b] Calculated by RSW from data in reference. Assumed 1.2 liter urine volume for 24 hr.
[c] Geometric mean.
[d] Corrected to specific gravity of 1.016.
[e] Corrected to specific gravity of 1.018.
[f] Corrected to specific gravity of 1.024.

normal urinary output usually is considered to be about 10 mg/day (32), the normal range can be much greater than this (Table 7.4). In three groups of Japanese subjects (in Japan) totaling 97 subjects, the geometric means of the groups ranged from 18.2 to 34.8 mg/liter. The 95 percent fiduciary limits varied from 7.3 to 123.8 mg/liter.

These values were for urine corrected to specific gravity 1.016. If the values were corrected to specific gravity 1.024 as recommended by NIOSH (92), they would be 50 percent higher. The use of geometric means by Ikeda et al. (84, 85) also suggests a distribution skewed toward the high values. Roush and Ott (100) also found preexposure urinary phenol values of up to 80 mg/liter (corrected to specific gravity 1.024). Fishbeck et al. (91) reported that the urinary phenol concentration of one person could vary by 100 percent (from 5.0 to 11.0 mg/liter) in a 6 week period. Because of such variations, Roush and Ott (100) have suggested that urinary phenol measurements do not reliably indicate benzene exposure if the exposure was less than 8 hr at 5 ppm. VanHaaften and Sie's data (89) support this contention. Thus urinary phenol measurements would not be suitable for routine monitoring of employee exposure to benzene under the proposed standard (102) for atmospheric benzene levels in the workplace. [An 8 hr time-weighted average concentration of 1 ppm with no excursions exceeding 5 ppm averaged over 15 mins.]

Another factor that argues against the unqualified and sole use of some arbitrary urinary phenol level as an indicator of excessive benzene or phenol exposure is the presence of urinary phenol progenitors in many medicinals. Phenylsalicylate, phenol, and sodium phenate are active ingredients in common over-the-counter medicinals and prescription medicinals. Phenylsalicylate is used as an enteric coating in other common medicinals. Fishbeck et al. (91), for example, found that Pepto Bismol® (a registered trademark of Morton-Norwich Products, Inc., Chicago) or Chloraseptic® (a registered trademark of R.T.M. Eaton Laboratories, Norwich, N.Y.) cold lozenges, both of which contain phenylsalicylate, when taken as directed led to peak urinary phenol levels of 480 and 498 mg/liters, respectively (corrected to a specific gravity of 1.024). Common medicinals containing phenol include "P&S" liquid or ointment, a shampoo suggested for use in cases of psoriasis and seborrheic dermatitis; Oraderm® (a registered trademark of the R. Schattner Co., Washington, D.C.), a lip balm; and Campho-Phenique® (a registered trademark of Sterling Drug, Inc., New York), a formulation suggested for cuts, burns, cold sores, and fever blisters.

Thus before concluding that particular urinary phenol levels indicate particular benzene or phenol exposures, it should be determined that the subject is not taking any over-the-counter or prescription pharmaceuticals that contain metabolic precursors of urinary phenol. It also would be desirable to have background levels of urinary phenol for the subject in addition to determining that there have been no changes in dietary habits that would increase urinary phenol levels.

6.6 Age of the Worker

Blood level and excretion of medicinals following a given dose have been found to change with age. Since metabolism and excretion of industrial chemicals utilize enzymes

Table 7.4 Normal Urinary Levels and Ranges (mg/liter) of Phenol as Reported by Various Authors[a]

Uncorrected	Specific Gravity, Corrected	Creatinine, Corrected	Subjects: Number, Sex, Country	Analytical Procedure	Reference
NR	5.5[c]	NR	20 M/F, U.S.	Gas chromatography	89
26.1[d] (8.3–81.5)[e]	23.3[c,d] (7.3–73.7)[e]	18.9[d] (6.0–60.1)[e]	36M, Japanese	Colorimetric (99)	84
25.2[d] (8.4–75.5)[e]	34.8[c,d] (9.8–123.8)[e]	28.5[d] (11.3–71.7)[e]	30F, Japanese	Colorimetric (99)	84
22.8[d] (8.9–58.3)[e]	8.2[c,d] (8.0–41.5)[e]	14.5[d] (7.1–29.9)[e]	31M, Japanese	Colorimetric (99)	84
(1–80)			52, U.S.	Gas chromatography	100
(4–5.5)	(5–11.0)[f]	NR	1, U.S.	Gas chromatography	91
NR	(12–144)[c]	NR	109, Great Britain	Colorimetric (99)	90

[a] Only values obtained using the Gibbs colorimetic method (99) or a gas chromatographic method are reported, since the other common colorimetric method [Theiss and Benedict (101)] is known to also give positive results with *p*- cresol (89). The *p*- cresol content in control urine may be 10 times greater than the phenol content (32, 89). Both Gibbs and the Theis and Benedict methods give reactions with *o*- and *m*-cresol, but these are usually present in urine in much lower concentrations than *p*-cresol.
[b] Symbols as follows: NR = not reported, M = male, F = female.
[c] Corrected to specific gravity of 1.016.
[d] Geometric mean.
[e] 95 per cent fiduciary limit.
[f] Corrected to specific gravity of 1.024.

and the same excretion mechanisms as medicinals, similar changes could be expected for industrial chemicals. For example, in disease-free and pathology-free elderly subjects, Triggs et al. (103) found a 50 to 100 percent increase in the elimination half-life of two medicinals when compared with disease-free and pathology-free young subjects. Absorption of the medicinal by these groups did not differ significantly.

In a biological assay, an increase in the half-life of a chemical usually would result in lowered concentrations of the index chemical in the biological assay material, if sampled at the usual time. This would lead to the conclusion of an exposure lower than that which actually occurred.

At the other end of the age spectrum, the enzymatic complement of the circumnate is not known for all enzymatically catalyzed reactions. For some reactions it seems to be as complete and as active in the immediate prepartum and postpartum stages as it is in the adult. For other reactions the activity may be totally missing or the specific activity of the enzyme may be severely decreased. Postpartum enzymatic activity of the circumnatal subject usually is of little concern industrially (see Section 11 for exceptions), but prepartum activity is. Too few investigations have been carried out at this time to permit specific statements about various groups of enzymes.

Also, with few exceptions, we do not know which industrial chemicals and which of their metabolites cross the human placenta. Therefore biological monitoring has limited utility at present to estimate fetal exposure or burden.

6.7 Disease

In addition to the effect of some diseases on the amount of binding proteins, diseases can affect the ratio of bound chemical to free chemical, thus the rate of appearance of the index chemical in the biological assay material.

Any disease that affects metabolism and/or excretion also could be expected to alter the rate of appearance of the index chemical in the bioassay material. Decreased renal clearance, for example, could result in higher blood levels and lower urinary levels of the chemical than would be expected from studies on subjects with normal renal clearance (70). A biological monitoring program based on urinary levels of the index chemical thus would suggest exposure levels lower than those that occurred in this situation. If the index chemical values were normalized by calculating the creatinine ratio and creatinine also were correspondingly reduced, the error should be self-correcting. If the biological monitoring was based on blood, tears, perspiration, or saliva analysis, this situation would result in a higher than normal excretion by way of these fluids. This would result in higher estimates of workplace exposures than actually occurred.

6.8 Sex of the Worker

Chemicals are known to have different lethality in the two sexes. For example, many of the organophosphate insecticides are 3 to 4 times more acutely lethal to one sex than to the other (104); the more sensitive sex varying with the chemical. This difference may be due to differences in absorption, metabolic route, or metabolic kinetics. Since the same factors would affect the appearance of other toxic signs or symptoms, it should not be surprising to find differences in the latter effects between sexes. In these situations a given level of the index chemical in the biological assay material could indicate different exposures for males and for females, and different correlation factors would have to be used for each sex to estimate exposure based on the biological assay.

Daniel and Gage (105) found that after dosing for up to 50 days, female rats had stored about 50 percent more orally administered BHT in the fat than males, at the dietary levels used. After single oral doses, females excreted about 50 percent more in the urine than did the males.

Ikeda and Imamura (27) found the urinary half-life of total trichloro compounds from T_3CE or T_4CE exposure to be almost twice as great in female workers as in male workers. They reported similar findings for other index chemicals used to assess exposure to these two compounds. Conversely, Nomiyama and Nomiyama (106) reported a significantly greater ($p < .05$) urinary excretion of TCA in females than males in the first 24 hr following inhalation exposure to T_3CE. For the first 12 hr following exposure, males excreted twice as much TCE as females. In accordance with these findings, Nomiyama (107) found that it was necessary to use a larger factor for females than males when correlating exhaled air concentration of T_3CE with the exposure concentration.

Nomiyama and Nomiyama (83) also found that female subjects absorbed approximately twice as much of a given inhalation dose of benzene as did males (41.6 vs 20.3 percent) as mentioned above.

Ikeda and Ohtsuji (84) found that the normal concentration of hippuric acid in about 30 female Japanese students was approximately twice that of about 36 male Japanese students (specific gravity and creatinine corrected). Conversely, the normal urinary level of TCA was twice as high in males as in females.

6.9 Defining Normal Range of the Index Chemical to be Expected in the Bioassay Material

Section 6.5 discussed the wide concentration range of index chemical that occurs in the biological assay material if the index chemical also can arise in the assay material from nonwork sources. All the factors discussed in the preceding eight subsections suggest that when the index chemical is a metabolite, a broad range of concentrations of the index chemical in the biological assay materials is to be expected, even if the index chemical does not occur naturally. This range also would be expected from medical experience with urinary analyses. If measurements are made repeatedly on the same group of workers over a period of time, the ranges should be narrower and an excursion more readily seen and quantitated.

In breath analyses the index chemical usually is the exposure chemical, not a metabolite. Therefore the concentration range of exhaled chemical for a particular exposure concentration would be expected to be narrower. Nevertheless, even here a range is to be expected.

6.10 Time Required for the Index Chemical to Appear in the Bioassay Material

The excretion half-life of chemicals varies from a few hours to days (Table 7.7: Section 7.3.4). This means that some xenobiotics will appear in biological assay materials in a few minutes and others will not appear in significant concentrations for days. Obviously in the latter situation it would be useless to monitor end-of-shift urine, for example, as a measure of that day's exposure to the chemical. There also may be differences in rate of appearance of the various possible biological assay materials. Thus although many inorganic cations appear reasonably promptly in urine, tears, perspiration, or saliva, they do not appear in the external part of the hair shaft or nails for weeks, because of the slow generation of these assay materials.

Hippuric acid from styrene exposure, for example, does not appear in urine until about 24 hr after exposure (74). Similarly, TCA does not appear in significant amounts in the urine of exposed subjects until 24 to 36 hr after exposure to T_3CE (108).

6.11 Analytical Methodology

Another factor that must be considered in biological monitoring is the analytical procedure used. The analytical procedure potentially introduces two principal sources of variation:

1. Conjugated index chemical versus free index chemical.
2. Analytical procedure for index chemical measurement.

As Figure 7.1 revealed, many index chemicals are excreted, in urine at least, in both the conjugated and the free form. Also, as discussed in Section 6.1.4, chemicals may be transported in the blood in either the free or protein-bound form. Obviously analyses of the free chemical only should not be compared with analyses of the total chemical: free + (conjugated or bound). The data of Slob (40) illustrate this very well (Table 7.5). The three subjects were exposed to various unspecified concentrations of styrene in the workplace and the exposure increased from subject 1 to subject 3. The analyses are for mandelic acid. As can be seen, the apparent urinary concentration of mandelic acid almost doubled following hydrolysis of the conjugate.

The measuring methodology also can be an important source of variation. In general, two types of instrument methodology have been used: visible or ultraviolet spectrophotometry, and other instruments.

Many chemicals present in biological assay materials may have visible or ultraviolet absorption spectra similar to those of the index chemical, thus potentially introducing large and uncontrolled error. Light absorption values of colored derivatives measured at

Table 7.5 Effect of Conjugate Hydrolysis on Apparent Urinary Mandelic Acid Concentration in Man (40)

Subject	Urinary Mandelic Acid Concentration (mg/liter)		Increase (%)
	Without Hydrolysis	With Hydrolysis	
1	12	16	33
2	61	118	93
3	349	598	71

a particular wavelength also can be misleading, since most chromophoric reagents are nonspecific and may react with other chemicals having structures similar to the index chemical. For example, the Theis-Benedict reagent for phenol (101) also detects *p*-cresol, which normally may be present in urine from unexposed subjects at 10 times the level of phenol (89). The analytical procedure of Ikeda and Ohtsuji (109) for urinary TTC also detects dichloro compounds. This lack of specificity can be overcome to some extent by removing interfering materials or isolating the index chemical. However the isolation procedure may not be complete.

For example, Pagnatto et al (94) and Walkley et al. (110) attempted to purify their urinary phenol by steam distillation, not knowing what interfering materials they were trying to remove. Unfortunately, *p*-cresol, the major analytical interference, should steam distill almost as well as phenol (111, 112). Therefore their purification procedure did not remove the principal impurity that would give false positive results with their colorimetric procedure. For this reason the data this group obtained with this method have qualitative value but their quantitative value is uncertain. Either removal of interfering materials or isolation of index chemical frequently leads to losses of the index chemical, which creates further error if the losses are not regular. Even if the losses are regular, discrepancies will be introduced between the method requiring cleanup and other methods that do not. In addition, the removal of interfering substances may be too time-consuming to be practical for routine analysis of biological materials from workers.

Gas chromatography, high pressure liquid chromatography, and atomic absorption spectrophotometry are three instrumental techniques that usually are readily amenable to separation of the index chemical without loss and give results that are characteristic and unique for the index chemical.

6.12 Route of Exposure

One of the hopes for biological monitoring is that it will measure total exposure to a chemical regardless of the route by which it is abosrbed. This assumption must be shown to be true for each method developed. Yet it appears in at least two reported cases that this assumption may not be true.

Dutkiewicz and Tyras (13) found that about 5 percent of ethylbenzene absorbed through the skin is converted to urinary mandelic acid, whereas Bardodej and Bardodejova (113) found 60 percent of inhaled ethylbenzene excreted as urinary mandelic acid. Dutkiewicz and Tyras therefore concluded that urinary monitoring of mandelic acid as a measure of total ethylbenzene exposure would not be acceptable.

Yant and Schrenk (10) also reported different relative methanol levels in several organs and tissues when the methanol was administered by inhalation versus subcutaneous administration. This suggests that blood levels of unbound methanol varied for the two routes. If that is the case, excretion levels would be expected to vary for the two routes of administration.

7 URINE ANALYSIS TO MEASURE INDUSTRIAL EXPOSURE

7.1 Correlation of Exposure with Index Chemical Concentrations in Urine

Urine analysis to measure exposure usually relies on analysis for a metabolite. As already discussed, a great number of factors affect the concentration of the unbound exposure chemical and its metabolites in the blood, thus the amount excreted per unit of urine per unit time. It should not be surprising, therefore, that great variations are reported by authors for the translation of urinary levels of the index chemical to exposure levels of the workplace chemical.

Despite the large uncertainty factor in translating urinary levels to body burden or exposure, there is no question but that for many, and probably all, chemicals, a relationship does exist. Also, since we do not yet know all the factors that contribute to variability, we do not know how to correct for them all.

At the present stage of development urine analyses for most organic compounds probably are best suited to (a) preemployment monitoring to detect obvious metabolic abnormalities that could increase individual hazard, and (b) regular monitoring of an employee to develop a continuing baseline against which an overexposure would be obvious and could lead to remedial measures in the workplace. At the present stage of development for most industrial organic chemicals, a single urine analysis could be correlated only within a broad range of body burden or prior exposure.

The problem is compounded when the index chemical can arise from diet or other nonoccupational sources, as has been discussed for the use of urinary phenol to evaluate phenol or benzene exposure and hippuric acid to evaluate styrene or toluene exposure. In the latter three instances, natural levels are so high and so variable that they invalidate the technique for exposure levels of interest for routine worker monitoring. For example, as previously discussed, urinary levels of phenol from subjects without any workplace exposure to benzene or phenol can exceed 80 mg/liter (100). Roush and Ott (100) and also VanHaaften and Sie (89) have published data suggesting that urinary phenol levels are not statistically valid measures of individual benzene exposure at levels of 5 ppm or below, because of normal variability. For the same reasons Ikeda and Ohtsuji (84) suggested that urinary hippuric acid was not statistically valid for estimating toluene concentrations lower than 130 ppm. As already mentioned, the data of Ogata et al. (86) suggest that this chemical may not be valid at concentrations lower than 200 ppm. Several publications suggest that urinary hippuric acid is not a usable index of styrene concentration at atmospheric levels of workplace interest because of this variability and high background level (74, 98).

Except in cases when urinary phenol is elevated because of medication, it probably is a good index of workplace phenol exposure. The work of Ohtsuji and Ikeda (114) indicates that an 8 hr exposure to 5 ppm in the air would be expected to result in about 400 mg of phenol/per liter of urine in a spot sample of urine taken at the end of the workday.

7.2 Biologic Threshold Limit Values

Instead of using urinary levels of an index chemical to estimate the level of exposure, it has been proposed that excretion levels of an index chemical be used simply to indicate the presence or absence of overexposure. The term "biological threshold limit value" has been proposed by Elkins (25). The concept is not restricted to metabolites or to urine analyses, but would apply to any index chemical: the exposure chemical and/or its metabolites. It would apply to blood and any liquid, solid, or gaseous biological excretion.

Nothing presently known about metabolism and excretion and their kinetics invalidates the concept. In general, the normal variation in excretion of chemicals, particularly metabolites, would have to be considered in setting such discriminator levels. The application of this concept would be further complicated where the index chemical occurs naturally in urine, thus has a high and variable background level. In this situation, two courses are possible: (1) the discriminator number could be set high enough to eliminate all "normal" values, or (2) it could be set at a lower fiducial, or otherwise derived limit, that included many people who had not been overexposed to the workplace chemical. If the first alternative is utilized, the discriminator level might never be reached for some overexposed individuals.

The second alternative would require some other confirmation that the elevated levels really were due to workplace exposure. Urinary hippuric acid as an index of toluene or styrene exposure has been discussed in this regard. The U.S. Department of Labor (102) has proposed that urinary phenol levels exceeding 75 mg/liter, adjusted to specific gravity of 1.024, be considered indication of overexposure to benzene, leading to certain subsequent blood measurements. Yet as discussed in Section 6.5, Roush and Ott (100) have shown that urinary phenol levels in unexposed individuals may exceed this. Fishbeck et al. (91) also have shown that the recommended dosage of a common over-the-counter medicinal can lead to urinary phenol levels more than 5 times this great. Even prescribed use of a common lozenge recommended for sore throats can cause a five- to sixfold elevation of urinary phenol. As discussed earlier, other common medicinal uses of metabolic precursors of urinary phenol also could cause levels of urinary phenol that could be interpreted as industrial overexposure to benzene.

In terms of worker hazard, the second course would be the safest, but it easily could be abused in cases of an index chemical that had nonworkplace progenitors. However as mentioned before, if urinary levels of the index chemical truly reflect hazard to the worker, the source of the chemical is irrelevant. Every effort should be made to find the cause of the elevated concentration and reduce the concentration. But if the source is not the workplace, trying to reduce workplace exposure will not reduce the hazard to the worker.

For reasons to be discussed, the discriminator number for organic compounds also must be based on a urine sample collected a certain number of hours postexposure, or erroneously high or low values relevant to the discriminator number conditions will be obtained.

Table 7.6 lists some of the workplace exposure chemicals that have shown a urinary dose-response relationship. Also shown are the index chemicals, the exposure-excretion (dose-response) equations, the type of analysis, and the type of urine sample. In some cases the investigators did not calculate the dose-response equation, and this was done by me. These instances have been noted. If the equation was based on analyses corrected by any of the methods discussed in Section 7.3.4., this also is noted. Expected metabolite concentrations based on the excretion equations also are given for a particular atmospheric concentration of the workplace chemical (biological threshold limit values).

The table illustrates many of the problems of estimating total workplace exposure from the analysis of an index chemical in excretions. These are discussed in the following sections, using the data in the table for illustration.

7.3 Factors to be Considered in Using the Methodology

7.3.1 Slope of the Dose-Response Curve

Methylchloroform metabolites (e.g., in urine) follow a normal dose-response relationship, at least to 50 ppm in air, the upper atmospheric exposure level studied. However the standard deviation is approximately 40 percent of the exposure value, and 2 standard deviations would be 80 percent. When this is coupled with the shallow slope of the curve (0.073 to 0.28, depending on the metabolite measured), it is doubtful that 2 standard deviations below the measured urinary metabolite level resulting from exposure to 350 ppm would be greater than 2 standard deviations above the zero exposure level of metabolites. The problem of the importance of the slope of the dose-response curve is treated in greater detail by Imamura and Ikeda (117). In general, because of the variability of metabolism and excretion, the dose-response curve should have a fairly steep slope. This is particularly important when the index chemical has nonworkplace progenitors.

7.3.2 Nonlinear Response

The equations found by Seki (93), which relate urinary metabolite levels to inhalation exposure of methyl chloroform, are straight lines, at least up to atmospheric concentrations of 50 ppm for 8 hr. However he did not extend his observations to 350 ppm exposures, the present TLV of methyl chloroform. In some cases, as already discussed, the urinary metabolite concentration plateaus (74) or may even decrease (75) with increasing concentration. Thus one cannot extrapolate from lower to higher concentrations but must actually carry out measurements at the exposure levels of interest.

7.3.3 Different Analytical Responses for Conjugated and Unconjugated Forms of the Index Chemical

Section 6.11 dealt with the importance of hydrolysis of conjugates prior to analysis. As Table 7.6 reveals most analytical procedures include a procedure to hydrolyze conjugates.

7.3.4 Spot Urine Samples Versus 24 hr Sample

In his analyses Seki (93) used so-called spot samples of urine. These are samples taken at a particular time, in contrast to 24 hr samples in which all urine voided over a 24 hr period or longer is collected. Elkins (24) has raised the question of whether spot samples or 24 hr samples are better. I believe this is not a matter of real concern. I believe it is more important that:

1. The sample be taken after there has been sufficient time for adequate metabolism and excretion to minimize variation due to analysis of small amounts and to reach a more or less steady state of excretion versus intake. This will vary with the urinary elimination half-life, which may vary from less than 4 hr for phenol to almost a week for tetrachloroethylene. Table 7.7 gives representative elimination half-life values for some industrial chemicals.

2. The sample always be taken at the same time relative to exposure. This applies not only to the original investigator, but to others who want either to compare results or to use the original investigator's results. This is necessary because the kinetics of absorption, distribution, and excretion may give different dose-response equations for metabolite levels at different times after exposure or after single exposure versus multiple exposures. This is particularly likely if the chemical has a long elimination half-life.

7.3.5 Equivalency of Analytical Methods

Seki (93) used a colorimetric method of analysis. The problems associated with various methods of analysis already have been discussed. If comparisons are to be made between

Table 7.6 Analytical Details for Analyses of Urine from Males Exposed to Various Chemicals

Exposure Chemical	Index Chemical	Equation	Correction	Analytical Method: Hydrolyzed Conjugates
Methyl chloroform (1,1,1-trichloroethane)	TTC[a]	$Y = 0.27X + 0.54$	None	Yes
		$= 0.18X + 0.29$	Specific gravity = 1.016	
		$= 0.28X + 0.16$	Creatinine (mg/g)	
	TCE	$= 0.19X + 0.27$	None	
	TCA	$= 0.073X + 0.21$	None	
T_3CE	TTC	$Y = 8.37X + 17.12$	Specific gravity = 1.024	Yes
	TCE	$= 5.19X + 12.28$	None	
	TCA	$= 3.17X + 4.84$	None	
T_3CE	TTC	$Y = 7.25X + 5.5$	None	Yes
		$= 4.97 + 1.9$	Specific gravity = 1.016	
		$= 5.50X + 6.2$	Creatinine (mg/g)	
	TCE	$= 5.57X + 4.4$	None	
	TCA	$= 2.74X + 0.7$	None	
Benzene	Phenol	$Y = 19X + 30$[b]	Specific gravity = 1.016	Yes
Phenol	Phenol	$Y = 108X + 28$[b]	None	Yes
		$= 80X + 40$[b]	Specific gravity = 1.016	
		$= 70X + 45$[b]	Creatinine (mg/g)	
Toluene	Hippuric acid	$Y = 23X + 800$[b]	None	—
		$= 16X + 600$[b]	Specific gravity = 1.016	
		$= 18X + 350$[b]	Creatinine (mg/g)	
Toluene	Hippuric acid	$Y = 40X + 550$[b]	Specific gravity = 1.024	—
		$= 24X + 500$[b]	Creatinine (mg/g)	
Toluene	Hippuric acid	$Y = 9.8X + 400$[b]	None	—
Toluene	Hippuric acid	$Y = 31X + 400$[b]	None	—
		$= 25X + 300$[b]	Specific gravity = 1.024	
m-Xylene	m-Methylhippuric acid	$Y = 27X$[b]	None	—
		$= 26X$[b]	Specific gravity = 1.024	
p-Xylene	p-Methylhippuric acid	Not calculable	None	—
		Not calculable	Specific gravity = 1.024	
Styrene	Mandelic acid	$Y = 18.4X + 149$[d]	Specific gravity = 1.024	No
	Mandelic acid	$= -14.0X + 6125$[e]	Specific gravity = 1.024	
	Phenylglyoxylic acid	$= 2.7X + 79$[d]	Specific gravity = 1.024	
	Phenylglyoxylic acid	$= -1.0X + 540$[e]	specific gravity = 1.024	

[a] Symbols as follows: U = unknown; Y = concentration of index chemical in urine in milligrams per liter of urine or milligrams per gram of creatinine; X = concentration of exposure chemical (ppm); NA = not available. For other abbreviations, see list at end of chapter.
[b] Calculated by RSW from investigator's data.
[c] Calculated by investigators.

Maximum Atmospheric Concentration Evaluated	Urinary Concentration (mg/liter) Calculated at Given Exposure ppm	Standard Deviation (ppm) at Given Exposure ppm	Urine Sample Type	Analytical Method	Reference
50 ppm × 8 hr × 3 days	14 at 50 9 at 50 14 at 50 10 at 50 4 at 50	~20 at 50[b] ~15 at 50[b] ~15 at 50[b] ~20 at 50[b] ~25 at 50[b]	Spot; after 3 days of exposure	Colorimetric (72)	93
40 ppm × 8 hr × 2½ days	854 at 100[c] 531 at 100[c] 322 at 100[c]	NA	Spot; after 2½ days of exposure	Colorimetric (72)	108
175 ppm × 8 hr × 3 days	730 at 100 500 at 100 550 at 100 575 at 100 210 at 100	~40 at 100[b] ~20 at 100[b] ~40 at 100[b] ~40 at 100[b] —	Spot; after 3 days of exposure	Colorimetric (72)	73
150 ppm × 8 hr	120 at 50[b]	NA	Spot; at end of workday	Colorimetric (99)	90
3.5 ppm × 8hr	568 at 5[b] 440 at 5[b] 395 at 5[b]	~1 at 3.5[b] ~1 at 3.5[b] ~1 at 3.5[b]	Spot; at end of workday	Colorimetric (115)	114
240 ppm × 6 hr	3100 at 100[b] 2200 at 100[b] 2150 at 100[b]	~45 at 100[b] ~45 at 100[b] ~45 at 100[b]	Spot; near end of workday	Colorimetric (84)	85
170 ppm × 8hr	4550 at 100[b] 2900 at 100[b]	NA NA	Spot; at end of workday	Spectrophotometric[f]	94[g]
600 ppm × 8 hr 800 ppm × 6 hr	1380 at 100[b]	NA NA	24 hr and end of day	Titration with 0.5N NaOH	21
200 ppm × 7 hr	3100 at 100[b] 2500 at 100[b]	~10 at 100[b] ~15 at 100[b]	Spot; at end of exposure	Colorimetric (116)	86
200 ppm × 7 hr	2700 at 100 2600 at 100	~55 at 100[b] ~30 at 100[b]	Spot; at end of exposure	Colorimetric (116)	86
100 ppm × 7 hr	1420 at 100[h] 3090 at 100[h]	NA NA	Spot; at end of exposure	Colorimetric (116)	86
290 ppm × 8 hr with occasional excursions of 1500 ppm × 10 min	1000 at 50[c] 2000 at 100[c] NA NA	NA	Spot; at end of workday	Colorimetric (98)	75

[d] Exposures <150 ppm.
[e] Exposures >150 ppm.
[f] Toluic acid interferes.
[g] Includes female workers, but results for females not separated by investigators.
[h] Observed values.

Table 7.7 Human Urinary Excretion Half-Life Values for Some Industrial Chemicals

Exposure Chemical	Index Chemical	Excretion Half-Life (hr)	Urine Samples	Reference
Styrene	Phenylglyoxylic acid	8.5	24 hr	74
	Mandelic acid	7.8	24 hr	
Toluene	Hippuric acid	~4 (α-phase)[a]	24 hr	86
	Hippuric acid	~12 (β-phase)[a]	24 hr	
Toluene	Hippuric acid	6.3[b]	24 hr	86
T_3CE	Total trichloro compounds	41 hr	Spot	27
T_3CE	Trichloroethanol	~7 (α-phase)[a]	Spot	108
		~76 (β-phase)[a]	Spot	
Phenol	Phenol	3.4[b]	Every 2 hr	15
Xylene	Methyl hippuric acid	3.8[b]	24 hr	86
T_4CE	Total trichloro compounds	144[b]	24 hr	27
1,1,1-Trichloro-ethane	Total trichloro compounds	8.7	Spot	93

[a] Estimated by RSW from graphs in publication: α-phase is the initial, rapid decay; β-phase is the later, slower, decay.

[b] Calculated by Ikeda and Immamura (27) from data in publication.

investigators, all participants should use the same analytical procedure, or the equivalency of different procedures should be demonstrated. If a discriminator number to indicate overexposure is to be used, the same requirements should be met.

When the same analytical procedure and comparable urine samples are used, comparable results are possible as shown by the equations developed by Ogata et al. (108) and by Ikeda et al. (73) for urinary TTC, TCE, and TCA following T_3CE exposure. Accordingly, the urinary levels they predict following exposure at 100 ppm T_3CE are very close.

Despite the similarity of the equations developed by Ogata and Ikeda and their co-workers, when the urinary TCA data and TCE data reported by Stewart et al. (118) are substituted in these equations, the exposures by Stewart et al. are underestimated by approximately one-third on the basis of TCE content and by approximately one-half on the basis of TCA content. Stewart et al. analyzed 24 hr urine samples after three 7½ hr exposures to 100 ppm of T_3CE, in contrast to Ogata and Ikeda who analyzed spot urine samples after two ½ hr or three 8 hr exposures to maxima of 40 or 175 ppm, respectively, of T_3CE. This probably is not a major procedural difference. The principal difference appears to be in analytical procedures. Ogata et al. and Ikeda et al. used the same colorimetric procedures for TTC, TCE, and TCA. Stewart et al. used a gas-chromatographic procedure. The data suggest that the slope of the standard curve for the colorimetric procedure is greater than the slope of the standard curve for the gas-chromatographic procedure.

BIOLOGICAL INDICATORS OF CHEMICAL DOSAGE AND BURDEN

The deviations to be expected with different analytical procedures also are illustrated in Table 7.6 by the equations developed by various investigators for urinary hippuric acid levels following toluene exposure. In this case the situation is further complicated by the high variable background level of hippuric acid, as already mentioned. However the investigators carried out the exposures at atmospheric levels that would minimize this interference, and the differences probably also principally reflect analytical differences. Von Oettingen (21), for instance, whose equation indicates one of the lowest background levels of hippuric acid, precipitated the hippuric acid from urine and then titrated it. The precipitation almost certainly was not quantitative. Ogata (86), whose equation indicates the next lowest background level of hippuric acid, extracted the hippuric acid before forming a derivative with a characteristic light absorption spectrum. It is possible that the initial extraction was not complete. The method of Pagnotto et al. (94) also depends on an extraction, followed by determination of light absorption at a specific wavelength. Ikeda (85) spearated the hippuric acid by paper chromatography, followed by reaction to form a colored derivative, extraction of the derivative from the paper, and determination of its light absorption at a characteristic wavelength. Ogata (116) later showed that the color depended on the filter paper used.

7.4 Analysis for Inorganic Ions

Urine analyses also can be used to measure exposure to metallic ions. Some of the metal ions excreted in urine include beryllium, cadmium, chromium, copper, iron, lead, lithium, magnesium, manganese, mercury, and zinc. The factors to be considered in developing the methodology, applying it, and interpreting the results do not differ from those already discussed. Dose-response relationships exist and can be utilized. Again, it is unlikely that metabolic factors will be as important for inorganic salt excretion as they are for organic chemical excretion. Therefore the normal range of excretion accompanying a given exposure may be narrower.

Metals are unlike organic chemicals in that they cannot have precursors that are less toxic than the industrial exposure chemical. Thus metal ion excretion levels will represent the true hazard to the worker, except in the few instances discussed. This means that biologic threshold limit values for metals can be set with greater assurance that excursions above the level represent greater hazard to the worker. However it should be remembered that exposure to metals also can occur outside the workplace. Regardless of where the exposures occurred, excretion levels generally indicate the true hazard and exposure should be reduced, regardless of the source.

Elkins (25) has reviewed urine levels for several metals. His experience leads him to believe that 0.2 mg of lead per liter of urine (corrected to specific gravity of 1.024) represents significant absorption, but not necessarily a toxic level. Similarly he reported that 0.25 mg of mercury per liter would represent significant absorption of mercury, but not necessarily a toxic effect.

The 1972 NIOSH review (118a) of the data correlating urinary lead concentration with biological effects concurs with Elkins's assessment.

Urine analyses also have been used to measure exposure to inorganic anions. Extensive work has been reported, for example, on the correlation of urinary fluoride and exposure to inorganic fluoride in the aluminum industry. NIOSH (118b) has suggested that the work of Kaltrieder et al. (118c) and Derryberry et al. (118d) indicates that postshift urinary values not larger than 7 mg of fluoride per liter of urine (corrected to specific gravity of 1.024) indicate exposure to inorganic fluoride levels that would not be expected to cause osteofluorosis.

However because of the ubiquity of fluorine atoms (e.g., inorganic fluoride in drinking water, tea, cereal grains, and organic fluoride in refrigerants, degreasing solvents, aerosol formulations, etc.), urinary fluoride levels may not always reflect solely occupational exposure to inorganic fluoride. This urinary level also cannot be taken to indicate safe exposure to inorganic fluorides such as oxygen difluoride, nitrogen trifluoride, sulfur pentafluoride, sulfur tetrafluoride, tellurium hexafluoride, or any other inorganic fluoride with innate toxicity significantly greater than the inorganic fluorides used to set this standard.

Organic fluoride levels may indicate more or less hazard than corresponding inorganic fluoride levels, depending on the comparative toxicity of the organofluorine compound and any of its metabolites.

7.5 Optimum Conditions for Using the Methodology

The studies to date indicate that urine analyses for a particular organic index chemical will be most reliable if:

1. The index chemical has no nonworkplace progenitors.
2. The slope of the dose-response curve is fairly steep or at least greater than 0.5.
3. The elimination half-life is no greater than 8 hr and preferably no greater than 4 hr.
4. The analytical method is specific for the index chemical. This requirement tends to eliminate colorimetric procedures. If gas-chromatographic procedures are used, the peak identity should be confirmed with a second column or a gas chromatography–mass spectrometer combination should be used while developing the method.
5. The method is validated in humans at the highest exposure level of interest. The dose-response curves should not be extrapolated beyond the highest experimental level.
6. Urine collection times are consistent and reflect consideration of excretion half-life.
7. Urine samples are analyzed shortly after collection. If they cannot be analyzed promptly, they should be frozen or at least refrigerated (25, 90).
8. The method and equations are first validated for the group of workers of interest before routinely applying the correlation equations.
9. The worker is not on a diet, suffering from a disease, or taking a medicine that could alter the relevant kinetics of the reactions of interest or any normalizing procedures used.

10. The worker is not being exposed off-the-job to the index chemical or another progenitor of the index chemical.

11. The urinary level of index chemical is relatable to the amount of exposure chemical absorbed by all routes.

12. The dose-response equations have been shown to apply to both men and women, or separate ones are developed for, and applied to, each sex.

13. The workdays of the group of interest and of the group used to develop the equations and/or discriminator number are the same, if the half-life of elimination is much greater than 8 hr.

8 EXHALED AIR ANALYSIS TO MEASURE INDUSTRIAL EXPOSURE

8.1 Background and Advantages of Methodology

It has been known for decades that many industrial chemicals that are inhaled and enter the vascular system of the human body are later excreted to some degree in exhaled air. Table 7.8 gives a representative listing.

It has been observed in the past few decades that the concentration of many of these chemicals in the exhaled air decreases regularly with time (see, e.g., Reference 120). It further has been observed that some chemicals absorbed through the skin also are excreted, unchanged, in exhaled air, and that the concentration of the chemical in the exhaled air decreases regularly with time in this situation also (see, e.g., Reference 12). If the concentration of the exposure chemical in exhaled air varies in some regular fashion with body burden, regardless of the route of absorption, this would provide a very desirable method for measuring industrial exposure. It would have at least the following advantages over various other methods:

1. Metabolism usually would not be involved, so all the metabolic factors that can affect the rate of appearance of the index chemical would not be involved. In most cases physical or physical-chemical factors would be involved, and these should be fairly constant between individuals.

2. The index chemical appears rapidly in the exhaled air. It is not necessary to wait hours or weeks for the index chemical to appear in the bioassay material, as in many other biological assays.

3. The analysis would be amenable to gas-chromatographic techniques. These can be made quite specific, thus eliminating analytical interference from nonindex chemicals. Gas-chromatographic techniques are suitable for analysis of small amounts and can be used to analyze for several chemicals concurrently. Thus concurrent exposure to several chemicals could be quantitated fairly rapidly and inexpensively.

4. Several samples can be taken in rapid succession, so the assessment of exposure can be based on either a kinetic analysis or substitution in an already derived equation.

Table 7.8 Some Industrial Chemicals Detected in Human Exhaled Air Following Exposure

Chemical	References
Benzene	119
Carbon tetrachloride	12, 120
Diethyl ether	121
Methanol	1
Methyl acetate	1
Methylene chloride	12, 122
Styrene	38, 75
Toluene	21
1,1,1-Trichloroethane	12
T_3CE	12, 106
T_4CE	123–125
Vinyl chloride monomer	126

5. In many cases the subject could be observed while providing the sample, to assure that instructions are being followed.

6. Very few nonworkplace progenitors exist for the index chemical. Thus the excreted material more likely represents workplace exposure than in many analyses of other excretion materials.

7. The technique is noninvasive.

8. The technique measures individual exposure without the bother of a personal monitoring device.

8.2 Factors Affecting Exhaled Air Levels of Index Chemicals

Assuming that transport through the lungs into the vascular system and back into the lungs is simple diffusion rather than "active transport" (2, p. 113), several factors can be proposed that possibly will affect the postexposure concentration of index chemical in exhaled air.

1. *Concentration of index chemical in inhaled air.* If movement of the chemical into the bloodstream is a simple diffusion process, the amount entering the blood will vary with the rate of diffusion through the alveolar wall and the partial pressure (concentration) of the chemical in the workplace atmosphere. It should not vary with the time of exposure, once equilibrium between blood and the atmosphere has been established.

2. *Blood concentration.* If movement in and out of the blood is a simple diffusion process, the concentration of the index chemical in exhaled air will vary directly with the concentration of that chemical in the blood. There will, of course, be some lag due to

the rate of diffusion. The important blood concentration will be the concentration of unbound chemical. (Factors affecting binding were discussed in Section 6.1.4.) The rate of metabolism of the chemical also will affect the unbound concentrations in the blood. (Factors affecting this were discussed in Section 6.1.) In addition to the effects of disease (Section 6.7), emphysema can uniquely affect the interpretation of the results. If the effective lung surface is decreased, the total diffusion rate of the index chemical from the blood into the atmospheric side of the lungs will be decreased, resulting in a lowered concentration in the exhaled air in comparison to the undiseased person. This would decrease the calculated exposure. Since the absorption coefficient for this emphysematous person also would be decreased in comparison to the normal person, the exhaled air concentration of index chemical still would be a measure of his body burden, but the equation relating the exhaled air concentration to exposure concentration would be different.

3. *Solubility in tissues or fat and binding to these materials.* Some chemicals are more soluble in, or more strongly bound to, these materials than others. Except in unusual situations characterized by a high degree of solubility or binding, this should not have a significant effect on exhaled air concentrations of the index chemical at the concentrations and during the time frame of interest.

4. *Dilution in exhaled air.* This is the pulmonary counterpart of urinary dilution discussed in Section 6.4. Every postexposure breath dilutes the index chemical in the alveoli with air that does not contain the index chemical. This lowers the concentration of index chemical in the exhaled air, which usually decreases the precision of the analysis. This dilution can be overcome to some extent by analyzing "end tidal" or "alveolar" air. To do this, the subject inhales and exhales normally through the collecting apparatus 2 or 3 times. At the end of the last breath, the final few milliliters are either collected in the sampler for later analysis (123) or diverted directly into the analytical instrument (125). According to DiVincenzo et al. (122), the alveolar air may be 50 percent richer in index chemical than the usual exhaled air. This technique is possible because instruments such as the gas chromatograph can routinely analyze a few hundredths of a milliliter of gas with high precision and accuracy.

Factors that have been shown to affect index chemical concentration in exhaled air include:

1. *Nonworkplace progenitors of the index chemical.* Although the instrumentation usually used for exhaled air sampling usually is quite specific, it cannot tell whether the chemical it is detecting appeared in the breath sample from the bloodstream or the mouth. It cannot tell whether the chemical got into the bloodstream from prior inhalation or ingestion. For example, phenol from a lozenge for sore throats will invalidate exhaled air analyses for phenol. Acetone from severe untreated diabetes will similarly contaminate exhaled air analyses for acetone from the workplace. Depending on the chemical of interest, lozenges, candy, chewing gum, tobacco, mouthwash, or toothpaste also could be sources of interference.

2. *Respiratory rate.* Until the blood and extracellular fluid compartment are saturated, respiratory rate can affect the rate of uptake. It similarly can affect the rate of desorption, as shown by DiVincenzo et al. (122). This suggests that correlation equations that are derived from subjects at rest should not be used on subjects that have been exerting themselves, and vice versa.

3. *Sex.* It has been reported that the absorption coefficients for some vapors vary for the two sexes (15, 83). This will affect both the time to saturation and, probably, the rate of desorption.

Nomiyama (107) also has reported an absolute difference between sexes in the concentration of T_3CE in the exhaled air in the β-phase of respiratory elimination (see footnote a, Table 7.7). The female concentration is lower. However the slopes of the two β-phase curves are parallel. The difference is due to a prolonged α-phase in women. Stewart et al. (123), on the other hand, found no sexual difference in the decay curves for T_3CE in the exhaled air of men and women following concurrent exposure of both sexes to T_3CE.

If there are sexual differences, the use of equations derived for one sex on analyses of exhaled air from the other sex would give erroneous estimates of workplace exposure. Another factor that will affect the rate of saturation of the blood and extracellular fluid compartment, is the difference in blood volume and, probably, total extracellular fluid. For instance, the male blood volume is 75 ml per kilogram of body weight, and the female volume is only 90 percent of that, or 67 ml per kilogram body weight (6). Since there easily can be a 100 percent difference in body weights between the sexes, there could be large differences in the volume of this compartment, thus the amount of the index chemical in the compartment after a given exposure. Again, this could result in the use of an inappropriate correlation equation to estimate workplace exposure.

4. *Skin absorption of the exposure chemical.* In addition to the foregoing factors that could lead to erroneous correlations between exposure levels and index chemical concentration in exhaled air, Stewart and Dodd (12) have shown that the alveolar decay kinetics for some solvents may be different for skin absorption exposure and for inhalation exposure. If this is true for other solvents, exhaled air concentration may not be a good tool for evaluating total body burden when the burden arises from various routes of exposure.

Stewart and Dodd studied the skin absorption of carbon tetrachloride, methylene chloride, T_3CE, T_4CE, and methylchloroform (1,1,1-trichloroethane). They found that for thumb immersion or nonoccluded topical exposure, the postexposure alveolar air decay was linear, whereas for total hand immersion the decay was exponential. Thus the breath excretion kinetics may vary with the size of the dose for noninhalation administration. The decay curves developed to date for excretion in breath following inhalation administration have all been exponential decays.

As the technique is further studied, additional possible causes of variation undoubtedly will be found.

Obviously, just as with urine analyses, it will have to be demonstrated for each chemical that the kinetics of its excretion in breath are the same no matter what the route of exposure, or multiple routes of exposure will invalidate the methodology for measuring body burden and exposure.

8.3 Shortcomings of the Methodology

Studies to date indicate that the technique has some shortcomings that must be recognized.

1. It does not appear to be widely usable for samples taken within, variously, 0 to 2 hr postexposure. The decay curves for various concentrations of xenobiotic and times of exposure are frequently indistinguishable in this time period. If a chemical is excreted so rapidly in exhaled air that none remains the next morning, samples would have to be taken by the workers on their own time and returned for analysis the next morning. DiVincenzo et al. (122) have found that workers may not exactly follow instructions for collection of alveolar air samples.
2. Although the body tends to integrate the exposure, thus giving exhaled breath analyses representative of the average exposure, some data suggest that samples taken shortly after exposure principally reflect the latest exposure level (123).

8.4 Experimental Studies

For inhalation administration the results published to date have been very encouraging for all the compounds studied. The methodology has given good correlation between concentration decrease of the index chemical in postexposure exhaled air and prior exposure, at least for the first 3 to 5 hr postexposure.

As mentioned previously, the decay curve is exponential and generally fits Equation 5:

$$\frac{C_t}{C_0} = K_A e^{-\alpha t} + K_B e^{-\beta t} \cdots + K_N e^{-nt} \qquad (5)$$

where C_t = the concentration of the index chemical in exhaled air at time t
C_0 = the exposure concentration
α, β, n = rate constants for the respective decay periods
K_A, K_B, K_N = zero time intercepts for the α, β, and n portions of the curve (i.e., zero-time coefficient for that segment of the curve)

Except for protracted studies over many hours or chemicals for which the decay curve seems to have more than two segments, all terms beyond the first two usually are unnecessary. K_A, α, K_B, β, K_N, and n all can be obtained graphically from the semilogarithmic plot of C_t/C_0 versus t (4). The value of C_t is determined experimentally at time t. The equation can then be applied to unknown exposure situations and C_0 can be calculated.

The data of Fernandez et al. (125) suggest that the time of exposure, at least for T_4CE, determines the decay curve. They studied alveolar air concentrations up to 4 hr postexposure. They found that the same equation described the postexposure decays after 2 hr exposure to 100 or 200 ppm, if the ratio of C_t/C_0 was plotted against postexposure time. A similar result was reported for 4 hr exposures to 100, 150, or 200 ppm of T_4CE and for 8 hr exposures to the same three concentrations. The curves for the three time periods were different. Their equations appear in Table 7.9. Baretta et al. (126) found that the decay curves for VCM in human alveolar air following 8 hr exposures to 50, 100, 250, or 500 ppm also were a family for the decay period studied. Calculation of C_t/C_0 for all curves at various common reported postexposure times yielded a common curve, within experimental limits, for the 20 hr decay period reported (See footnote d, Table 7.9). Because neither the y intercept(s) for these curves nor the data points for the first hour postexposure are shown in the publication, it is not possible to tell whether the first term in the equation describing the decay curves will be the same for all. Nevertheless, the data are strongly suggestive, and the time period reported would be suitable for use to calculate exposure levels. Decay curves for shorter exposure times were not reported, so it is not known whether those curves would differ from those for the 8 hr exposures.

Within experimental limits, Stewart et al. (123) also found that human alveolar air decay curves for the first 20 hr following seven ½ hr exposures to 20, 100, or 200 ppm T_3CE also were a family. For this family, as with the T_4CE decay curves of Fernandez et al. (125), the curves expressing the ratio of C_t/C_0 versus postexposure time were coincident within experimental limits and could be represented by the one equation in Table 7.9. Likewise, the curves for alveolar decay ratios following 1 or 3 hr exposures at 100 and 200 ppm were identical for both exposures at each time period. The coefficients and exponents for the curves from all three exposure times, 1, 3, and 7½ hours, were not identical. The decays following the 1 and 3 hour exposures at 20 ppm were not followed by Stewart et al. for a long enough time to provide enough data points for comparisons.

The data of DiVincenzo et al. (122) also suggest that the equations for the decay curves of the C_t/C_0 ratio following 2 hr human exposures to either 100 or 200 ppm of methylene chloride also will be the same for both concentrations. This equation is shown in Table 7.9.

Both Stewart et al. (38) and Götell (75) studied the decay of styrene in exhaled air following 8 hr exposures of humans. Stewart's subjects were exposed in a chamber to 99 ppm of styrene. Götell's subjects were workers exposed to a time-weighted average of 89 to 139 ppm in the workplace. Stewart collected alveolar air samples for analysis Götell did not specify the samples collected.

Despite these inconsistencies, the agreement between the two decay curves was remarkable. Both decay curves, when transformed to C_t/C_0 versus time (in hours) could be expressed as the same equation (see Table 7.9.) over the period studied. This is in

Table 7.9 Postexposure Xenobiotic Decay Equations in Human Exhaled Air

Chemical	Hours of Exposure	Hours Studied Postexposure	Concentrations Evaluated (ppm)	Air Sample[b]	Number of Subjects[c]	Coefficients and Exponents[a] $(Y = K_A e^{-\alpha t} + K_B e^{-\beta t})$				Reference
						K_A	K_B	α	β	
Methylene chloride	2	3	100, 200	E	11M	1.8×10^{-2}	10^{-1}	1.1×10^{-1}	10^{-2}	122
Styrene	8	5	25, 115, 260	E	15M	1.5×10^{-2}	5×10^{-3}	1.01	1.3×10^{-1d}	75
	7	6	99	A	6M	1.5×10^{-2}	5×10^{-3}	1.01	1.3×10^{-1d}	38
T_4CE	7	110	101	A	16M	3.5×10^{-2}	3.5×10^{-2}	8.3×10^{-2}	8.8×10^{-3d}	79
	2	4	100, 150, 200	A	23M, 1F	2.9×10^{-1}	1.6×10^{-1}	1.2×10^{-2}	8.0×10^{-3}	125
	4	4				2.1×10^{-1}	2.0×10^{-1}	9.9×10^{-2}	7.5×10^{-3}	
	8	4				10^{-1}	2.2×10^{-1}	4.9×10^{-2}	4.4×10^{-3}	
T_3CE	1	22	20, 100, 20	A	3–9 (M & F)	4.5×10^{-2}	4.8×10^{-2}	4.6×10^{-1}	5.1×10^{-2d}	123
	3					5×10^{-2}	1.3×10^{-2}	1.6	8.1×10^{-2d}	
	7½					9×10^{-2}	2.6×10^{-2}	1.8	6.9×10^{-2d}	

[a] Y = ratio of concentration in exhaled air at time t (hr) to exposure concentration (i.e., C_t/C_0).
[b] E = normal exhaled air; A = alveolar air.
[c] M = male; F = female.
[d] Calculation made by R. S. Waritz, using data obtained from graphs presented in the author's original scientific paper and without access to the original raw data. Therefore they are subject to errors introduced by the original translation to graph form and by printing. They should be considered illustrative only.

contrast to the findings of Fernandez et al. (125) and Stewart et al. (79) for T_4CE breath decay. In this case the former group found the decay entered the β-decay phase about 1½ to 2 hr following an 8 hr exposure to 100 ppm. Stewart's group did not find the β-phase starting until approximately 45 hr following a 7 hr exposure to 101 ppm.

There seems to be no question that to calculate exposure from exhaled air decay curves, one must know the exposure time.

Although this aspect of the use of exhaled air decay curves to calculate exposure is very encouraging, the variability reported for some solvents is not. For example, DiVincenzo et al. (12) found that with 11 subjects, the breath decay curves for 100 and 200 ppm of methylene chloride were within 2 standard deviations of each other. Similarly, Götell's (75) data for styrene indicate his decay curves would not reliably distinguish between 25 and 115 ppm or 115 and 260 ppm.

Conversely, Baretta et al. (126) found that exhaled air decay curves for VCM could readily distinguish between 50 and 100 ppm or 100 and 250 ppm with only four to six subjects. Unfortunately, not enough studies have been published with such comparisons to indicate the probable general situation.

The technique appeared to demonstrate worker accumulation of T_4CE after four and five 7 hr exposures to about 100 ppm (79).

Overall, exhaled air analyses seem to hold good promise as a way of determining previous exposure to certain industrial chemicals. As Table 7.9 indicates, not all chemicals are excreted at the same rate, and calibration curves will have to be developed for each chemical. It is probable that curves also will have to be developed for each group of workers, but not enough data have yet been developed to judge. The data strongly suggest that to apply the technique quantitatively, the time of exposure must be known. However Stewart, et al. (123) have published data suggesting that within certain concentration and time limits, the decay curve is determined by the product of concentration and time.

For optimum utilization of the technique, it may be necessary to develop indiviudal decay data over a period of time. The technique, with or without concurrent urine analyses, should be usable to screen workers for abnormal metabolism and excretion of many workplace chemicals prior to assigning them to areas where they may be exposed to these chemicals.

9 BLOOD ANALYSIS AS A MEASURE OF INDUSTRIAL EXPOSURE

Blood analysis is an invasive technique and carries some resultant risk to the worker. Many workers may also find it objectionable and would probably object strongly to daily or even weekly samples being drawn. Generally, the order of worker preference would be expected to be: exhaled air, urine, and blood.

Blood analysis has not been used extensively in industrial hygiene for measuring exposure, and comparatively few papers have been published in this area.

Blood analyses are used extensively in pharmacology and in clinical trials for new medicinals and are a valuable tool in these fields. They have provided great insight into the metabolism and excretion of xenobiotics and the individual variation in these body processes. Pharmaceutical chemists and biochemists have been responsible for the greatest developments in these areas and the reader is referred to texts in this specialized field for in-depth information on its utility and its drawbacks (see, e.g., References 5, 62).

There is no question that blood levels of xenobiotics and metabolites reflect dosage. There also is no question that the factors discussed in Sections 6 and 7 play a great role in individual variations. Thus all the caveats presented in those sections must be considered in interpreting individual analyses for organic and inorganic industrial chemicals.

The correlations between inorganic lead exposure, biological effects, and blood levels for inorganic lead probably have been more extensively studied than for any other industrial compound. The early work by Kehoe and others has been reviewed by Kehoe (126a). This work and more recent work also have been reviewed by NIOSH (118a). The reader is referred to these reviews for excellent summaries and discussions of the research that has led to the present proposals for biological monitoring of inorganic lead exposure.

Kehoe (126a) suggested the maximum acceptable blood level for inorganic lead in adult workers is less than 80 $\mu g/100$ g of blood. NIOSH concurred (118a). OSHA (127) proposed a level of more than 60 $\mu g/100$ g as the level that would dictate worker removal from exposure areas. EPA (128), in the proposed National Ambient Air Quality Guide, has suggested that mean blood lead levels in excess of 15 $\mu g/100$ ml of blood in children aged 1 to 5 years could be accompanied by biological effects.

As discussed in Section 5, blood analyses are routinely used in the medical field to detect organ damage by measuring transferase enzyme levels, blood urea nitrogen, and so on. Damage to organs that resulted in an increase of these transferases would cause the increase whether the damage was caused by disease, by medicinals, or by industrial chemicals. Thus such measurements could be used as a nonspecific measure of organ damage. However since the goal of industrial hygiene is to prevent worker injury, this approach is deficient in two respects:

1. It is not specific to a particular industrial chemical or even to industrial chemicals.
2. It measures an effect that occurs after injury, instead of measuring a leading effect that could be used to forestall injury.

In summary, there is no question that blood analyses can be developed to measure exposure to workplace chemicals. It is equally certain that blood levels can be used to set acceptable exposure levels for industrial chemicals. However they will show the variability already seen in the medical profession for naturally occurring blood chemicals or for medicinals, and as discussed here for urine and exhaled air analyses. However, the technique probably will not be used if urine analyses or exhaled air analyses can provide equivalent reliability in measurements of worker exposure.

10 HAIR ANALYSIS TO MEASURE INDUSTRIAL EXPOSURE

It has been known for many years that hair contains metals. Indeed, it may be considered to be an excretory mechanism for these metals, since the metals appear to have no functional role in the hair protein and in at least a few cases the content of metal in hair seems to vary with exposure to that metal (129–132).

Some of the metals that have been reported in hair are aluminum, arsenic, beryllium, cadmium, calcium, chromium, copper, iron, mercury, lead, manganese, molybdenum, potassium, selenium, silicon, thallium, titanium, and zinc (see, e.g., References 121, 129, 132–142).

In addition, the nonmetals chlorine and phosphorus have been reported (132).

Hair is not suitable as a dynamic system for evaluation of immediate past exposure to metals because the individual hair shafts do not have access throughout their length to any fluid transport system. Thus the metal content reflects that available at the time any particular portion of the shaft was being synthesized. Since hair grows at the rate of about 1/cm in 30 days, clippings would be expected to reflect a historical exposure at best; How many months in history would be indicated by the length of the clipping and the length of the remaining proximal hair shaft, in centimeters.

However if the range of metal content of hair normally is sufficiently narrow in a population, and the hair concentration varies regularly with blood concentration, suitably timed postexposure analyses could be used to confirm or refute suspected overexposure or continuing acceptable exposure. Unfortunately, as with most evolving areas of science, there are conflicting reports in the literature on the utility of hair analyses. Also, in addition to some of the factors that are known to lead to aberrant exposure estimates based on urinary concentration of chemicals, hair analyses have unique factors that may produce aberrant conclusions. These must be considered in developing and applying any procedure for correlating hair levels with exposure levels.

One of these additional unique problems of hair analysis is that of suitable cleansing of the hair to remove adsorbed metal contaminant prior to analysis. Hair normally develops an oily coating, and this oil can trap exogenous metal. Since this adsorbed material does not represent body burden, it must be removed before analysis.

Several preanalysis hair cleaning procedures have been reported. The simplest is washing with detergent followed by distilled deionized water rinses and oven drying (130, 134, 143). Additional washings have included acetone (136, 137), nitric acid (129, 134), ether (133, 136), and trisodium ethylenediamine tetraacetic acid (134). Most investigators assume that all adsorbed metals are removed by detergent washes followed by (1) distilled water washes to remove detergent and (2) organic solvent washes. In some cases the initial wash has been with the organic solvent (144). Nishiyama (134) found that washing procedures that removed all adsorbed cadmium also removed the endogenously derived cadmium in the hair. Experimental results similar to those of Nishiyama on adsorbed cadmium have been reported for hair analyses of other metals (138).

Petering et al. (136) suggested that ionic detergents were more appropriate than nonionic ones because the former could complex the metal ion or form salts with it, thus aiding its removal from the exterior of the hair. Obviously if the detergent charge could be a factor in metal ion removal, anionic detergents would be more suitable than cationic detergents, unless the metal is present as a negatively charged complex or radical.

Renshaw et al. (133) reported that the concentration of lead in single hairs from one woman increased with the distance from the scalp. Since they had cleansed the hair by diethyl ether reflux in a Soxhlet extractor prior to analysis, they assumed that all adsorbed external lead had been removed and this distal increase represented lead that had deposited on the hair from external sources, then diffused into the body of the hairs. No details of the work history, residence, or cosmetics use of the woman were given. The increased lead content at the distal portion also could be surface lead that had not been completely removed by the ether wash. Certainly the mass of the evidence suggests that hair concentrations of metals, if the hair is adequately cleansed, do bear a relationship to body burden (129, 140, 145). The uncertainty regarding contamination by airborne material presumably could be removed by using body, pubic, or axillary hair instead of scalp hair. However hair from these sites may grow at different rates from scalp or vertex hair. Axillary hair grows at about two-thirds the rate of scalp hair and pubic hair at about half the rate. Body hair grows at approximately the same rate (142). Rate differences obviously would affect time correlations.

Many of the metals are transported in blood predominantly in the bound form or are stored in the body. Anything that caused their release, with a resulting increase of the free metal in liquid transport systems accessing the hair root, could result in a short or extended shaft section with elevated concentration. This could be interpreted as a short-term or long-term overexposure. Medicinals or industrial chemicals that were more tightly bound to transport or storage proteins could replace bound metals, thus causing their release. This would be followed by increased hair uptake of the released metal ion. Wasting diseases that liberated stored metals could have a similar effect.

Therefore interpretation of isolated hair analyses for judgments against workplace exposure should be coupled with a careful and complete medical history to assure that elevated local concentrations along the hair shaft due to nonwork causes are not attributed to work exposure.

Diet also can be expected to affect the level of metals found in the hair. Green plants are notorious scavengers of metals from the ground in which they are grown. Thus levels of metals in hair could be due not only to the direct ingestion of fruits, vegetables, and cereals, but also to the ingestion of meat from animals grazed or fed hay or cereal grains. Conversely, high fiber diets apparently lower uptake of metals from the intestine

(146) and would be expected to result eventually in lowered metal content of hair from nonindustrial exposure.

Cosmetics and hair dyes also may contain various metal salts or complexes and may contribute to hair levels of metals either (a) through absorption of the metal salt or complex followed by uptake by the hair root, or (b) by adsorption on the hair shaft. Various salts or complexes of metal ions were permitted in the coloring agents of hair dyes, the dye formulations, or cosmetics as of December 31, 1977. These included aluminum, arsenic, barium, chromium (III), cobalt, copper, iron, mercury, titanium, and zinc (147). Thus many of the metals found in hair also appear in cosmetics or hair dyes.

Obviously, the cosmetic and hair dye use history of the worker also must be determined before trying to correlate metal content of a worker's hair with workplace exposure to a metal.

To use metal content of hair as an index of occupational exposure, sample preparation procedures (e.g., washing) and the analytical method must be validated. In addition, a normal baseline must be established. From the medical experience with urinalyses and the industrial experience with urinalyses mentioned earlier, this would be expected to be a range, rather than a line. Variations with sex also might be expected. The literature data in Table 7.10 and 7.11 indicate that this is the case. For some metals there also were variations due to age. Therefore, before making a judgment on isolated hair levels of metals with regard to occupational exposure, the hair analyses must be judged against not only medical, dietary, cosmetic use, and hair dye use backgrounds but matched against control ranges for sex and age or prior analyses of the same employee's hair.

Table 7.10 reveals that in general, women with no known industrial exposure had higher hair levels of lead than their male counterparts in the studies. The exception was the group of middle-class urban white females from Cincinnati, Ohio, studied by Petering et al. (136). Petering also found that the lead level in male hair decreased with age from 2 to 88 years. In his study it increased rapidly in women from age 14 to 30, then decreased rapidly from age 30 to 84. For both males and females, several values of around 35 to 40 μg of lead per gram of hair were observed. Other studies tried to correlate lead content of hair with age. This would appear to be necessary from the data of Petering et al., although the slope of the line for men was not great and the upper 95 percent fiducial limit at 60 years was within the fiducial limits at 20 years. The changes with age were dramatic for the women in Petering's study, and age matching for women of working ages definitely would appear to be necessary. If individual historical controls were used, allowance would have to be made for the changes that accompany age.

Klevay (144), however, reported an age relationship with hair lead content only for males in Panama. In his population it appeared to reach a nadir between 11 to 20 years. He reported ranges of 1.1 to 52.1 μg of lead per gram of hair for males and 8.7 to 78.7 μg of lead per gram of hair for females.

It also is obvious that the normal control adult average lead concentration varies by a factor of about 2 for males and about 1.5 for females. Furthermore, values varying by factors of 10 to 50 between control individuals have been reported (143).

Table 7.10 Hair Lead Concentrations in Control Populations[a]

Arithmetic Mean Value (μg/g) and Range (bracketed)

Male	Female	Sex Unknown	Significantly Different?	Analytical Method	Reference
14.7 (A)	19.2 (A)		Yes; $p < .001$	Atomic absorption	143, 148
9.9 (A)	14.6 (A)		ND	Unknown	140
17.8 (A)	19.0 (A)		No; $p > .05$	Atomic absorption	139
	12 (U) [4–25]		—	Atomic absorption	133
		9.4 (U) [3–26]	—	Dithizone (149)	129
6.1 (T)			—	Atomic absorption	131
24.5 (AA) [1.1–52.1]	34.6 (AA) [8.7–78.7]		Yes; $p < .001$	Atomic absorption	144
22 (P)			ND	Atomic absorption	137
17 (T)	6 (T)				
14 (W)	11 (W)				
11 (R)	10 (R)				
		14.5 (P)	—	Atomic absorption	130
4.1 (W)			—	Unknown	142

[a] Symbols as follows: A = adult; AA = all ages, urban and rural; ND = not determined; P = preschool; R = >60 years; T = 6–20 years; U = age unknown; W = 20–60 years.

Suzuki et al. (140) suggested that 30 μg of lead per gram of hair be considered the upper normal level. El-Dakhakhny and El-Sadik (129), on the basis of correlation of hair lead levels and clinical signs and symptoms, suggested that hair lead content greater than 30 μg/g be considered indicative of excessive lead exposure. They found that this corresponded to approximately 90 μg of lead/100 g of blood.

However the blood and hair analyses were carried out concurrently, and no mention was made of the residual hair length or the length of the hair sample. Therefore unless the work exposure to lead had not changed over the number of months represented by the distance of the hair sample from the scalp in centimeters, the comparison is not valid.

Klevay's data (144) indicate that 30 μg/g may be too low, at least for women. He proposed an upper normal value of 35 to 40 μg/g, and some of his hair samples from presumably unexposed persons even exceeded this value. However some of these high values came from urban, nonindustrial areas and may reflect unique urban exposures. In Klevay's study (144), values exceeding 35 μg/g were seen in male hair only from subjects younger than 10 years of age. They were seen in female subjects of all ages. Petering et al. (137) also reported urban male and female values of approximately 40

Table 7.11 Hair Content of Various Metals in Control Populations[a]

Metal	Range of Concentrations (μg/g)			Significantly Different?	Analytical Method	Reference
	Male	Female	Sex Unknown			
Aluminum	1.6–7.8 (T)		2–9 (AA)	ND	Neutron activation	138
	1.2–9.2 (TP)				Unknown	142
	4.4–5.5 (W)			ND	Neutron activation	138
Antimony	0.1–1.4 (T)		0.5–4 (AA)			
	0–4.4 (T)			—	Unknown	142
	0.07–0.2 (W)			—	Atomic absorption	130
Arsenic	0.4–7.9 (T)		0.18 (P)	—	Neutron activation	138
	0.7–5.3 (T)			—		
Cadmium	1–2 (P)			ND	Atomic absorption	137
	2 (T)	1–1.3 (T)				
	1.5–2 (W, R)	1.3–2 (W)				
		2–1.5 (R)				
			30–530 (AA)	—	Atomic absorption	143
			1.06 (P)	—	Atomic absorption	130
			>1000 (U)	No ($p > .05$)	Unknown	134
	2.76 (AA)	1.77 (AA)			Atomic absorption	139
	0.47 (W)			—	Unknown	142

Element	Values			Method	Ref
Copper	13–30 (P)	20–25 (AA)	ND	Atomic absorption	136
	30 (T)				
	30–15 (W)				
	15–10 (R)				
	7–93 (T)		7.8–234 (AA)	—	Neutron activation	138
	8–150 (T)				
	16.1 (AA)	55.6 (AA)	$p < .001$	Atomic absorption	139
	15–17 (W)		—	Unknown	142
Mercury	0.3–34 (T)		0.1–33 (AA)	—	Neutron activation	138
	0.5–53 (T)				
	1.7–1.9 (W)		—	Unknown	142
Zinc	100–110 (P)	200 (P)	No	Atomic absorption	136
	110–140 (T)	200–180 (T)	No		
	140–125 (W, R)	180–150 (W)	—		
	101–186 (T)		51–602 (AA)	—	Neutron activation	138
	85–166 (T)				
	167 (AA)	172 (AA)	No ($p > .05$)	Atomic absorption	139
	150–190 (W)		—	Unknown	142

[a] Symbols as follows: A = adult; AA = all ages, urban and rural; ND = not determined; P = preschool; R = >60 years; T = 6–20 years; U = age unknown; W = 20–60 years.

μg/g from subjects with no apparent industrial exposures. Neither Petering or Klevay reported any clinical signs or symptoms of plumbism in their subjects, but there is no indication that they looked for them. Thus the data suggest that even 40 μg/g may not be the upper level for lead content in hair that will be indicative of systemic overexposure.

The data in Table 7.11 show variations reported for hair content of cadmium, copper, and zinc. In some cases they appear to be significant and in others, not. For instance, the data of Petering et al. (136) show an age-related decrease in zinc content for both males and females, but the slopes of the lines are so shallow and the equations of the lines so similar that very little allowance need be made for age or sex. Copper content of hair peaked at about 10 years for males in the study of Petering et al. but showed only a gradual increase with age for females. The 95 percent fiducial limits for the two sexes overlapped for about 20 years. Thus the data suggest that hair analyses for copper need to be evaluated against controls matched for age and sex.

The available data suggest that because of the variability of individual hair content of metals, it would be difficult to set trigger numbers or action numbers for most metals that taken by themselves, would indicate excessive exposure. Also, since appearance of the metal in the external hair shaft would follow exposure by possibly a month or so, such analyses would not provide as early a warning of overexposure as would urine analyses.

Because of the individual variations reported, hair analyses would be of most value if individual histories of metal content could be developed over a period of years. As with urine and exhaled air analyses, comparisons between populations are difficult, particularly if different washing procedures and analytical techniques are used.

Also, as with other biological assays, high metal content of the hair can come from nonworkplace sources or may not represent recent past workplace exposure for the other reasons discussed. Obviously in these cases, workplace exposure may be trivial, and trying to reduce it will not reduce the worker's hazard. If an undesirably high level of metal in the hair is found, the source should be conscientiously sought so that possible hazard to the worker may be reduced.

11 MILK ANALYSIS TO MEASURE INDUSTRIAL EXPOSURE

Historically, human breast milk analysis has not been used to monitor industrial exposure to chemicals. It has been realized for more than 90 years that milk can be used by the body as an excretory mechanism and that even a mouse mammary cancer virus could be transmitted by way of the mother mouse's milk. However it is only within the past two decades that its importance as an excretory mechanism has been appreciated. Because of the small number of lactating women in the workplace, milk analysis has very restricted utility for monitoring workplace exposure.

Nursed infants may have body weights as small as one-twentieth the body weight of the mother. Thus milk levels and corresponding blood or tissue levels that would not be an effect level for the mother could be an effect level, or could accumulate to an effect level, in the infant. In addition, the neonatal enzyme system for metabolizing the xenobiotic could be incomplete, leading to toxic accumulations.

These considerations, rather than industrial exposure monitoring, have been principally responsible for the recent interest in monitoring mother's milk. It is unlikely that such analyses will achieve much use for industrial monitorings, but nursing industrial employees might want assurance that they did not have possible effect levels of workplace chemicals in their milk.

Since the concern principally has been with nonindustrial exposure, very few studies have related human milk levels of industrial chemicals to blood or plasma levels or to storage levels.

Pesticide levels reported in mother's milk were recently summarized (150). Among the pesticides reported in mother's milk are β-benzene hexachloride "benzenehexachloride," DDT and metabolites, dieldrin, heptachlor epoxide, "hexachlorobenzene," and polychlorinated biphenyls. Strassman and Kutz (151) also reported finding oxychlordane and *trans*-nonachlor, metabolites of chlordane and heptachlor, in mother's milk. Curley et al. (152) also have announced finding α-benzene hexachloride, endrin, aldrin, and mirex in mother' milk.

Mercury and lead have been reported in mother's milk (153). Molybdenum also may be excreted in milk (141). Iron also is excreted in mother's milk (154).

Many medicinals are excreted in mother's milk. These have been recently reviewed (155, 156). Medicinals reported in mother's milk include barbiturates, sulfonamides, some hormones, lithium salts, narcotics, ergotamine, some hypoglycaemic agents, acetylsalicylic acid (aspirin), antibiotics, opiates, and caffeine.

Mandelic acid, benzoic acid, and chloral, metabolites of the industrial chemicals styrene, certain alkyl benzenes and toluene, and of trichloroethylene, respectively, have been reported in mother's milk (155).

Although few, if any, industrial chemicals may have been sought in mother's milk following industrial exposure, their appearance should be expected, since the factors governing the appearance of pesticides or pharmaceuticals would also operate for industrial chemicals. Although few studies have been carried out to elucidate the factors, there appear to be many. Thus a simple spot analysis of an individual mother's milk may be indicative of a range of possible plasma concentrations rather than a specific concentration.

Because milk has an aqueous and a lipid phase, concentrations of excreted chemicals might be expected to be related to the unbound blood concentration of the chemical (Stowe and Plaa, 1968), and also to levels stored in fatty depots. This is in contrast to urine, saliva, perspiration, and tears that have no, or a much smaller, lipid component. The pH of mother's milk is slightly lower (ca. 7.0) than that of plasma (ca. 7.4). Therefore, weak, unbound, bases might be expected to partition preferentially to milk. Polychlorinated biphenyls appear to partition into milk (150).

Other factors that appear to affect the concentration of xenobiotics in mother's milk may be age, weight, and number of previous pregnancies (150).

Increased photoperiod has been shown to increase milk output 15 percent in Holstein cattle (157). The effect on the concentrations of chemical components was not studied. It is not known whether photoperiod affects human milk output or the concentrations of xenobiotics and metabolites. In any case it is likely that the fiducial limits of the normal value are considerably greater than ± 15 percent; thus any effect would be insignificant in determining whether the measured concentration reflected a safe or unsafe exposure level for the mother.

Factors that could be expected to affect milk levels of xenobiotics include:

1. Exposure to chemicals more strongly bound to transport and storage proteins or storage sites.
2. Impairment of alternate excretory mechanisms such as kidney.

Conversely, lactation could be expected to alter the usual urinary ratios and concentrations for a nonbound, fat-soluble xenobiotic and its metabolites. This would lead to erroneous calculated exposure levels of the xenobiotic in the workplace if only the urine were analyzed and the fact of lactation were ignored. Furthermore, the error would be one of underestimation of workplace exposure. Similar considerations would hold for analyses of sweat and tears.

Xenobiotics and their metabolites would be expected to appear in milk at least as rapidly as in urine. Thus milk analysis would have an advantage over hair and nail analyses in that it would give a measure of current or immediate past exposures rather than historical exposures.

In summary, mother's milk is a possible excretory fluid for assessing workplace exposure to chemicals. However because of its limited occurrence, development of correlations with plasma levels and toxic effect levels in the mother is difficult to justify in preference to similar developments for urine and exhaled air. Nevertheless, monitoring mother's milk to assure that effect levels of workplace chemicals were not being transmitted to nursed children appears to be justified.

12 SALIVA, TEARS, PERSPIRATION, AND NAIL ANALYSIS AS A MEASURE OF INDUSTRIAL EXPOSURE

Saliva, tears, and perspiration, although not usually considered excretory fluids, do contain chemicals transferred from blood. Xenobiotics can partition between these fluids and blood. Saliva and perspiration levels of chemicals seem to reflect unbound concentrations of the chemicals in blood (158, 160). It is likely that tear levels also will reflect blood concentrations.

Very little work has been reported on the use of these excretion fluids for biological monitoring. It is probable that all the factors affecting xenobiotic levels discussed in Sections 6 and 7 will apply to these fluids. There are obvious problems collecting samples of tears and perspiration.

Saliva analysis has demonstrated some utility for following blood levels of drugs such as the antiepileptic phenytoin (158, 159). Saliva analyses might overcome one of the problems of urine analysis: variable dilution of the xenobiotic or its metabolites by urine already in the bladder. Samples could be collected easily at the end of the workday and could provide a viable alternative to urine or blood analyses for nonvolatile workplace chemicals.

Some of the chemicals of industrial importance that have been reported in perspiration include lead (159), arsenic, mercury, iron, manganese, zinc, magnesium, copper, ethanol, benzoic acid, salicylic acid, urea, and phenol (154). Arsenic appears in perspiration very quickly after administration (142).

Salicylic acid and urea are among the industrially important chemicals reported in saliva and tears, respectively (154).

Although metals have been detected in human fingernails or toenails (142, 161, 162), analysis of this material has not been used as a technique to monitor industrial exposure. Some of the metals reported in human nails are copper (162) and zinc (161). It is very likely that the same considerations discussed under hair analysis will apply to nail analysis, and they are not discussed further here.

In relating nail levels of a metal to previous exposure, it should be noted that human fingernails grow approximately 100 μm/day and that human toenails grow only approximately 25 μm/day (142).

ABBREVIATIONS

Abbreviations used in this chapter:

ADH	alcohol dehydrogenase
APase	alkaline phosphatase
BHT	butylated hydroxytoluene
SGOT	serum glutamic acid–oxalic acid–transaminase
SGPT	serum glutamic acid–pyruvic acid–transaminase
TCA	trichloroacetic acid
T_3CE	1,1,2-trichloroethylene
T_4CE	1,1,2,2-tetrachloroethylene
TCE	1,1,1-trichloroethanol
TTC	total trichloro compounds
VCM	vinyl chloride monomer
UV	ultraviolet

REFERENCES

1. H. P. Elkins, *AMA Arch. Ind. Hyg. Occup. Med.,* **9,** 212 (1954).
2. E. S. West and W. R. Todd, *Textbook of Biochemistry,* 4th ed., Macmillan, New York, 1966.
3. J. B. Neilands and P. K. Stumpf, *Outlines of Enzyme Chemistry,* 2nd ed., Wiley, New York, 1958, pp. 379–381.
4. M. Gibaldi and D. Perrier, *Pharmacokinetics,* 1st ed., Dekker, New York, 1975, pp. 284–287.
5. B. N. LaDu, H. G. Mandel, and E. L. Way, *Fundamentals of Drug Metabolism and Drug Disposition,* 1st ed., Williams & Wilkins, Baltimore, 1971.
6. J. Wallach, *Interpretation of Diagnostic Tests,* 2nd ed., Little, Brown, Boston, 1974.
7. American Conference of Governmental Industrial Hygienists, "TLV's: Threshold Limit Values for Chemical Substances and Physical Agents in the Workroom Environment and Intended Changes for 1977," ACGIH, Cincinnati, Ohio, 1978.
8. W. A. Eldridge, *Report 29, Chemical Warfare Service* (1924); through B. R. Allen, M. R. Moore, and J. A. A. Hunter, *Br. J. Dermatol.,* **92,** 715 (1975).
9. C. P. McCord, *Am. J. Public Health,* **24,** 677 (1934).
10. W. P. Yant and H. H. Schrenk, *J. Ind. Hyg. Toxicol.,* **19,** 337 (1937).
11. D. L. Hill, *AMA Arch. Ind. Hyg. Occup. Med.,* **8,** 347 (1953).
12. R. D. Stewart and H. C. Dodd, *Ind. Hyg. J.,* **25,** 439 (1964).
13. T. Dutkiewicz and H. Tyras, *Br. J. Ind. Med.,* **24,** 330 (1967).
14. T. Dutkiewicz and H. Tyras, *Br. J. Ind. Med.,* **25,** 243 (1968).
15. J. K. Piotrowski, *Br. J. Ind. Med.,* **28,** 172 (1971).
16. C. P. McCord, *Ind. Eng. Chem.,* **23,** 931 (1931).
17. O. Tada, K. Nakaaki, S. Fukabori, and J. Yonemoto, *J. Sci. Labour* Part 2 (*Rodo Kagaku*), **51,** 143 (1975); through *Chem. Abstr.* **83,** 54178w (1975).
18. E. M. P. Widmark, *Biochem. Z.,* **259,** 285 (1933).
19. M. Neymark, *Skand. Arch. Physiol.,* **73,** 227 (1936); through *Chem. Abstr.,* **30,** 4930^2 (1936).
20. H. H. Schrenk, W. P. Yant, S. J. Pearce, F. A. Patty, and R. R. Sayers, *J. Ind. Hyg. Toxicol.,* **23,** 20 (1941).
21. W. F. von Oettingen, P. A. Neal, and D. D. Donahue, *JAMA,* **118,** 579 (1942).
22. R. R. Sayers, W. P. Yant, H. H. Schrenk, J. Chornyak, S. J. Pearce, F. A. Patty, and J. G. Linn, "Methanol Poisoning: I. Exposure of Dogs to 450–500 ppm Methanol Vapor in Air," R.I. 3617, U.S. Department of the Interior, Bureau of Mines, 1942; through *Chem. Abstr.,* **36,** 4596^2 (1942).
23. H. P. Elkins, *Pure Appl. Chem.,* **3,** 269 (1961).
24. H. P. Elkins, *J. Am. Ind. Hyg. Assoc.,* **26,** 456 (1965).
25. H. P. Elkins, *J. Am. Ind. Hyg. Assoc.,* **28,** 305 (1967).
26. T. Dutkiewicz, *Bromatol. Chem. Toksykol.,* **4,** 39 (1971); through *Chem. Abstr.,* **76,** 10786u (1972).
27. M. Ikeda and T. Imamura, *Int. Arch. Arbeitsmed.,* **31,** 209 (1973).
28. H. H. Cornish, *Crit. Rev. Toxicol.,* **1,** 1 (1971).
29. K. R. Long, *Int. Arch. Occup. Environ. Health,* **36,** 75 (1975); through *Chem. Abstr.,* **85,** 67416p (1976).
30. R. S. Waritz, J. G. Aftosmis, R. Culik, O. L. Dashiell, M. M. Faunce, F. D. Griffith, C. S. Hornberger, K. P. Lee, H. Sherman, and F. O. Tayfun, *J. Am. Ind. Hyg. Assoc.,* **38,** 307 (1977).
31. R. T. Williams, *Detoxication Mechanisms,* 2nd ed. Wiley, New York, 1959.

32. D. V. Parke, *The Biochemistry of Foreign Compounds*, 1st ed., Pergamon Press, New York 1968, p. 146.
33. K. P. McConnell and O. W. Portman, *J. Biol. Chem.*, **195**, 277 (1952).
34. O. M. Bakke and R. R. Scheline, *Toxicol. Appl. Pharmacol.*, **16**, 691 (1970).
35. C. P. Carpenter, C. B. Shaffer, C. S. Weil, and H. F. Smyth, Jr., *J. Ind. Hyg. Toxicol.*, **26**, 69 (1944).
36. I. Danishefsky and M. Willhite, *J. Biol. Chem.*, **211**, 549 (1954).
37. A. M. El Masri, J. N. Smith, and R. T. Williams, *Biochem. J.*, **68**, 199 (1958).
38. R. D. Stewart, H. C. Dodd, E. D. Baretta, and E. D. Schaffer, *Arch. Environ. Health*, **16**, 656 (1968).
39. H. Ohtsuji and M. Ikeda, *Toxicol. Appl. Pharmacol.*, **18**, 321 (1971).
40. A. Slob, *Br. J. Ind. Med.*, **30**, 390 (1973).
41. T. C. Butler, *J. Pharmacol. Exp. Ther.*, **97**, 84 (1949).
42. J. R. Cooper and P. J. Friedman, *Biochem. Pharmacol.*, **1**, 76 (1958).
43. B. Souček and D. Vlachová, *Br. J. Ind. Med.*, **17**, 60 (1960).
44. K. H. Byington and K. C. Liebman, *Mol. Pharmacol.*, **1**, 247 (1965).
45. W. J. Cole, R. G. Mitchell, and R. F. Salamonsen, *J. Pharm. Pharmacol.*, **27**, 167 (1975).
46. S. Yllner, *Nature*, **191**, 820 (1961).
47. J. W. Daniel, *Biochem. Pharmacol.*, **12**, 795 (1963).
48. J. W. Porteous and R. T. Williams, *Biochem. J.*, **44**, 46 (1949).
49. H. G. Bray, B. G. Humphris, and W. V. Thorpe, *Biochem. J.*, **45**, 241 (1949).
50. D. Robinson, J. N. Smith, and R. T. Williams, *Biochem. J.*, **59**, 153 (1955).
51. W. Seńczuk and B. Litewka, *Br. J. Ind. Med.*, **33**, 100 (1976).
52. F. A. Hommes, Ed., *Inborn Errors of Metabolism*, Academic Press, New York, 1973.
53. E. S. Vessell and G. T. Passananti, *Clin. Chem.*, **17**, 851 (1971).
54. T. Tang and M. A. Friedman, *Mutat. Res.*, **46**, 387 (1977).
55. S. L. Winter and J. L. Boyer, *N. Engl. J. Med.*, **289**, 1180 (1973).
56. P. M. Newberne, *Lab. Anim.*, **4**: 7, 20 (1975).
57. A. P. Alvares, K. E. Anderson, A. H. Conney, and A. H. Kappas, *Proc. Nat. Acad. Sci. (U.S.)*, **73**, 2501 (1976).
58. Y. E. Harrison and W. L. West, *Biochem. Pharmacol.*, **20**, 2105 (1971).
59. A. H. Conney, E. J. Pantuck, K. C. Hsiao, R. Kuntzman, A. P. Alvares, and A. Kappas, *Fed. Proc. Fed. Am. Soc. Exp. Biol.*, **36**, 1647 (1977).
60. W. D. Loub, L. W. Wattenberg, and D. W. Davis, *J. Nat. Cancer Inst.*, **54**, 985 (1975).
61. A. H. Conney, *Pharmacol. Rev.*, **19**, 317 (1967).
62. J. R. Gillette, *Ann. N.Y. Acad. Sci.*, **179**, 43 (1971).
63. L. G. Hart, R. W. Shultice, and J. R. Fouts, *Toxicol. Appl. Pharmacol.*, **5**, 371 (1963).
64. L. G. Hart and J. R. Fouts, *Proc. Soc. Exp. Biol. Med.*, **114**, 388 (1963).
65. F. K. Kinoshita, J. P. Frawley, and K. P. DuBois, *Toxicol. Appl. Pharmacol.*, **9**, 505 (1966).
66. H. C. Cecil, S. J. Harris, J. Bitman, and P. Reynolds, *J. Agric. Food Chem.*, **23**, 401 (1975).
67. M. Ikeda, H. Ohtsuji, and T. Imamura, *Xenobiotica*, **2**, 101 (1972).
68. W. E. Knox and A. H. Mehler, *Science*, **113**, 237 (1951).
69. M. Ikeda, H. Ohtsuji, H. Kawai, and M. Kuniyoshi, *Br. J. Ind. Med.*, **28**, 203 (1971).
70. M. M. Reidenburg, *Med. Clin. North Am.*, **58**, 1103 (1974).
71. A. J. Quick, *J. Biol. Chem.*, **92**, 65 (1931).

72. S. Tanaka and M. Ikeda, *Br. J. Ind. Med.,* **25,** 214 (1968).
73. M. Ikeda, H. Ohtsuji, T. Imamura, and Y. Komoike, *Br. J. Ind. Med.,* **29,** 238 (1972).
74. M. Ikeda, T. Imamura, M. Hayashi, T. Tabuchi, and I. Hara, *Int. Arch. Arbeitsmed.,* **32,** 93 (1974).
75. P. Götell, O. Axelson, and B. Lindelof, *Work Environ. Health,* **9,** 76 (1972).
76. P. G. Watanabe, G. R. McGowan, and P. J. Gehring, *Toxicol. Appl. Pharmacol.,* **36,** 339 (1976).
77. G. Leaf and L. J. Zatman, *Br. J. Ind. Med.,* **9,** 19 (1952).
78. T. R. Tephly, *Fed. Proc. Fed. Am. Soc. Exp. Biol.,* **36,** 1627 (1977).
79. R. D. Stewart, E. D. Baretta, H. C. Dodd, and T. R. Torkelson, *Arch. Environ. Health,* **20,** 224 (1970).
80. J. M. Walshe, *Am. J. Med.,* **21,** 487 (1956).
81. G. A. Nicolson, A. C. Greiner, W. J. G. McFarlane, and R. A. Baker, *Lancet,* February, 1966, p. 344.
82. C. MacLeod, H. Rabin, R. Ogilvie, J. Ruedy, M. Caron, D. Zarowny, and R. O. Davies, *Can. Med. Assoc. J.,* **111,** 341 (1974).
83. K. Nomiyama and H. Nomiyama, *Ind. Health* (Kawasaki,), **7,** 86 (1969); through *Chem. Abstr.,* **72,** 82685a (1970).
84. M. Ikeda and H. Ohtsuji, *Br. J. Ind. Med.,* **26,** 162 (1969).
85. M. Ikeda and H. Ohtsuji, *Br. J. Ind. Med.,* **26,** 244 (1969).
86. M. Ogata, K. Tomokuni, and Y. Takatsuka, *Br. J. Ind. Med.,* **27,** 43 (1970).
87. L. Levine and J. P. Fahy, *J. Ind. Hyg. Toxicol.,* **27,** 217 (1945).
88. K. Engström, K. Husman, and J. Rantanen, *Int. Arch. Occup. Environ. Health,* **36,** 153 (1976).
89. A. B. Van Haaften and S. T. Sie, *J. Am. Ind. Hyg. Assoc.,* **26,** 52 (1965).
90. S. G. Rainsford and T. A. L. Davies, *Br. J. Ind. Med.,* **22,** 21 (1965).
91. W. A. Fishbeck, R. R. Langner, and R. J. Kociba, *J. Am. Ind. Hyg. Assoc.,* **36,** 820 (1975).
92. Department of Health, Education and Welfare, *Criteria for A Recommended Standard . . . Occupational Exposure to Benzene,* Cincinnati, Ohio, 1974, p. 112.
93. Y. Seki, Y. Urashima, H. Aikawa, H. Matsumura, Y. Ichikawa, F. Hiratsuka, Y. Yoshioka, S. Shimbo, and M. Ikeda, *Int. Arch. Arbeitsmed.,* **34,** 39 (1975).
94. L. D. Pagnotto and L. M. Lieberman, *J. Am. Ind. Hyg. Assoc.,* **28,** 129 (1967).
95. R. W. Ellis, *Br. J. Ind. Med.,* **23,** 263 (1963).
96. M. Ogata, K. Sugiyama, and H. Moriyasu, *Acta Med. Okayama,* **16,** 283 (1962); through References 84 and 94.
97. W. V. Lorimer, R. Lilis, W. J. Nicholson, H. Anderson, A. Fischbein, S. Daum, W. Rom, C. Rice, and I. J. Selikoff, *Environ. Health Perspect.,* **17,** 171 (1976).
98. H. Ohtsuji and M. Ikeda, *Br. J. Ind. Med.,* **27,** 150 (1970).
99. H. D. Gibbs, *J. Biol. Chem.,* **72,** 649 (1927).
100. G. J. Roush and M. G. Ott, *J. Am. Ind. Hyg. Assoc.,* **38,** 67 (1977).
101. R. C. Theis and S. R. Benedict, *J. Biol. Chem.,* **61,** 67 (1924).
102. U.S. Department of Labor, *Fed. Reg.,* **43,** 5917 (1978).
103. E. J. Triggs, R. L. Nation, A. Long, and J. J. Ashley, *Eur. J. Clin. Pharmacol.,* **8,** 55 (1975).
104. W. J. Hayes, Jr., *Clinical Handbook on Economic Poisons,* Government Printing Office, Washington, D.C., 1963, p. 13.
105. J. W. Daniel and J. C. Gage, *Food Cosmet. Toxicol.,* **3,** 405 (1965).
106. K. Nomiyama and H. Nomiyama, *Int. Arch. Arbeitsmed.,* **28,** 37 (1971).
107. K. Nomiyama, *Int. Arch. Arbeitsmed.,* **27,** 281 (1971).

108. M. Ogata, Y. Takatsuka, and K. Tomokuni, *Br. J. Ind. Med.,* **28,** 386 (1971).
109. M. Ikeda and H. Ohtsuji, *Br. J. Ind. Med.,* **29,** 99 (1972).
110. J. E. Walkley, L. D. Pagnotto, and H. B. Elkins, *Ind. Hyg. J.,* **22,** 362 (1961).
111. D. R. Stull, *Ind. Eng. Chem.,* **39,** 517 (1947).
112. S. Glasstone, *Elements of Physical Chemistry,* 1st ed., Van Nostrand, New York, 1949, pp. 362–363.
113. Z. Bardodej and E. Bardodejova, *Cesk. Hyg.,* **6,** 537 (1961); through *Chem. Abstr.,* **65,** 6086g (1966).
114. H. Ohtsuji and M. Ikeda, *Br. J. Ind. Med.,* **29,** 70 (1972).
115. M. Ikeda, *J. Biochem.* (Tokyo), **55,** 231 (1964); through *Chem. Abstr.,* **60,** 16407b (1966).
116. M. Ogata, K. Tomokuni, and Y. Takatsuka, *Br. J. Ind. Med.,* **26,** 330 (1969).
117. T. Imamura and M. Ikeda, *Br. J. Ind. Med.,* **30,** 289 (1973).
118. R. D. Stewart, C. L. Hake, A. J. Lebrun, J. E. Peterson, et al., "Biologic Standards for the Industrial Worker by Breath Analysis: Trichloroethylene," Department of Health, Education and Welfare, Cincinnati, Ohio, 1974, p. 96.
118a. Department of Health, Education and Welfare, "Criteria for a Recommended Standard . . . Occupational Exposure to Inorganic Lead," Cincinnati, Ohio, 1972.
118b. Department of Health, Education and Welfare, "Criteria for a Recommended Standard . . . Occupational Exposure to Inorganic Fluorides," Cincinnati, Ohio, 1975.
118c. N. L. Kaltreider, M. J. Elder, L. V. Cralley, and M. O. Colwell, *J. Occup. Med.,* **14,** 531 (1972).
118d. O. M. Derryberry, M. D. Bartholomew, and R. B. L. Fleming, *Arch. Environ. Health,* **6,** 503 (1963).
119. J. Teisinger, V. Bergerová-Fišerová, and J. Kudrna, *Prac. Lék,* **4,** 175 (1952); through *Chem. Abstr.,* **49,** 4181i (1955).
120. R. D. Stewart, H. H. Gay, D. S. Erley, C. L. Hake, and J. E. Peterson, *J. Occup. Med.,* **3,** 586 (1961).
121. H. W. Haggard, *J. Biol. Chem.,* **59,** 737 (1924).
122. G. D. DiVincenzo, P. F. Yanno, and B. D. Astill, *J. Am. Ind. Hyg. Assoc.,* **33,** 125 (1972).
123. R. D. Stewart, C. L. Hake, and J. E. Peterson, *Arch. Environ. Health,* **29,** 6 (1974).
124. E. Guberan and J. Fernandez, *Br. J. Ind. Med.,* **31,** 159 (1974).
125. J. Fernandez, E. Guberan, and J. Caperos, *J. Am. Ind. Hyg. Assoc.,* **37,** 143 (1976).
126. E. D. Baretta, R. D. Stewart, and J. E. Mutchler, *J. Am. Ind. Hyg. Assoc.,* **30,** 537 (1969).
126a. R. A. Kehoe, in: *Industrial Hygiene and Toxicology,* 2nd rev. ed., Vol. 2, F. A. Patty, Ed., Wiley, New York, 1963, p. 941.
127. U.S. Department of Labor, Occupational Safety and Health Administration, *Fed. Reg.,* **40,** 45934 (1975).
128. U.S. Environmental Protection Agency, *Fed. Reg.,* **42,** 63076 (1977).
129. A. El-Dakhakhny and Y. M. El-Sadik, *J. Am. Ind. Hyg. Assoc.,* **33,** 31 (1972).
130. A. M. Yoakum, National Technical Information Services Report EPA-600/1-76-029, 1976.
131. D. I. Hammer, J. F. Finklea, R. H. Hendricks, C. M. Shy, and R. J. N. Norton, Air Pollution Control Office (U.S.) Publication AP91, p. 125, 1972.
132. E. C. Henley, M. E. Kassouny, and J. W. Nelson, *Science,* **197,** 277 (1977).
133. G. D. Renshaw, C. A. Pounds, and E. F. Pearson, *Nature,* **238,** 162 (1972).
134. K. Nishiyama and G. F. Nordberg, *Arch. Environ. Health,* **25,** 92 (1972).
135. J. A. Hurlburt, National Technical Information Services Report TID-4500-R64, 1976.
136. H. G. Petering, D. W. Yeager, and S. O. Witherup, *Arch. Environ. Health,* **23,** 202 (1971).
137. H. G. Petering, D. W. Yeager, and S. O. Witherup, *Arch. Environ. Health,* **27,** 327 (1973).
138. L. C. Bate and F. F. Dyer, *Nucleonics,* **23,** 74 (1965).

139. H. A. Schroeder and A. P. Nason, *J. Invest. Dermatol.,* **53,** 71 (1969).
140. Y. Suzuki, K. Nishiyama, and Y. Matsuka, *Tokushima J. Exp. Med.,* **5,** 111 (1958); through L. M. Klevay, *Arch. Environ. Health,* **26,** 169 (1973).
141. M. Anke, A. Hennig, M. Diettrich, G. Hoffman, G. Wicke, and D. Pflug, *Arch. Tierernaehr.,* **21,** 205 (1971).
142. H. C. Hopps, *Trace Subst. Environ. Health,* **8,** 59 (1974).
143. V. G. Oleru, *J. Am. Ind. Hyg. Assoc.,* **37,** 617 (1976).
144. L. M. Klevay, *Arch. Environ. Health,* **26,** 169 (1973).
145. Z. S. Jaworowski, *Atompraxis,* **11,** 271 (1965); through L. M. Klevay, *Arch. Environ. Health,* **26,** 169 (1973).
146. Anon., *Lancet,* August 13, 1977, p. 337.
147. Commerce Clearing House, Inc., *Food Drug Cosmetic Law Reporter,* Chicago, 1978.
148. H. Draut and M. Weber, *Biochem. Z.,* **317,** 133 (1944); through L. M. Klevay, *Arch. Environ. Health,* **26,** 169 (1973).
149. R. G. Keenan, D. H. Byers, B. E. Salzman, and F. L. Hyslop, *J. Am. Ind. Hyg. Assoc.,* **24,** 481 (1963).
150. Z. W. Polishuk, M. Ron, M. Wasserman, S. Cucos, D. Wasserman, and C. Lemeson, *Pestic. Monit. J.,* **10,** 121 (1977).
151. S. S. Strassman and F. W. Kutz, *Pestic, Monit. J.,* **10,** 130 (1977).
152. A. Curley, M. F. Copeland, and R. D. Kimbrough, *Arch. Environ. Health,* **19,** 628 (1969).
153. J. A. Knowles, *Clin. Toxicol.,* **7,** 69 (1974).
154. P. Lanzkowsky, *N. Engl. J. Med.,* **298,** 343 (1978).
155. C. M. Stowe and G. L. Plaa, *Ann. Rev. Pharmacol.,* **8,** 337 (1956).
156. R. L. Savage, *Adverse Drug Reaction Bulletin,* 1976, p. 212.
157. R. R. Peters, L. T. Chapin, K. B. Leining, and H. A. Tucker, *Science,* **199,** 911 (1978).
158. D. Schmidt, *Lancet,* September 18, 1976, p. 639.
159. J. W. Paxton, B. Whiting, F. J. Rowell, J. G. Ratcliff, and K. W. Stephen, *Lancet,* September 18, 1976, p. 639.
160. B. R. Allen, M. R. Moore, and J. A. A. Hunter, *Br. J. Dermatol.,* **92,** 715 (1975).
161. R. W. Goldblum, S. Derby, and A. B. Lerner, *J. Invest. Dermatol.,* **20,** 13 (1953).
162. W. B. Barnett, *Clin. Chem.,* **18,** 923 (1972).

CHAPTER EIGHT

Evaluation of Exposure to Chemical Agents

CHARLES H. POWELL, Sc.D.

1 INTRODUCTION

The position of an industrial hygienist as the central professional in the evaluation of occupational health hazards is still almost unique to the United States. However to properly evaluate the environment, the industrial hygienist must call on members of other scientific and professional disciplines, such as occupational physicians, toxicologists, chemists, control engineers, nurses, and production personnel. The industrial hygienist may take a position of leadership, but the evaluation and control of the chemical and physical workplace environment is a team effort requiring the input and knowledge of many disciplines; thus this position is one of coordination and recognition of the significance of the contributions made by the other members of the team. In addition to the professional assistance, the worker who may be exposed can make a significant contribution to the evaluation of the workplace. It must always be remembered that the plant management may know how it should be done, but the worker knows *how it is done*. The inclusion of both management and the worker as part of the evaluation team also improves the possibility of effective communication beteween management, the workers, and the health professionals. In many cases this type of communication is necessary to fulfill both the ethical responsibilities of the industrial hygienist and the physician, and the requirements of the Occupational Safety and Health Act of 1970 (1).

Most workers are employed in industrial establishments that are small in total employment, and as a result it is not possible for the management to have a staff of health professionals with the diversity of background necessary to make a valid,

:hensive evaluation of possible exposures to chemical agents. Thus it is often ary to make such an evaluation through the use of consultants, or the individuals avilable must have some knowledge of the other disciplines and the necessary contributions of these disciplines. This can be achieved through study of professional publications, training, or exchanges of ideas at professional meetings. It is important that the individual making the evaluation recognize that the first responsibility is to the worker and that recommendations and evaluations must only extend to the limit of professional competency.

To be able to make a comprehensive, scientific evaluation of a chemical workplace hazard, the individual charged with this responsibility must have at least a working knowledge of the chemistry of the compound, its toxicologic effect on man, the routes of entry and elimination, the meaning of specific medical tests, the related signs and symptoms, the industrial process in which possible exposure could occur, and the methods to control exposure. Whatever the professional training and experience of the individual making the evaluation, some knowledge in the other professional disciplines is required.

2 OBJECTIVES OF DATA ACQUISITION

The general principles for evaluation of an occupational environment remain the same regardless of the specific objective for which the information is being collected. The final evaluation is dependent on the sampling strategy and planning, and familiarization with the process prior to the collection of any data. The basic principles of evaluation have been outlined by Hosey et al. (2). These principles include recognition, evaluation, and control. It is important to consider that the recognition of a possible health hazard, an understanding of the industrial process and how it may result in a significant employee exposure, and the biologic action that may occur are all part of the preplanning necessary to determine the extent of exposure. To plan for an effective evaluation program, the final objective must be determined prior to sampling, so that the sampling and the study can be planned in conjunction with other available information to answer the specific evaluation objective.

In the preplanning stage, consideration must be given to the following:

1. *What to sample.* The workroom environment and/or biologic samples from the worker.
2. *Where to sample.* The breathing zone of the worker, the general room air, specific samples of workroom air collected while machinery is in operation, biological specimens collected from the worker, raw material, settled particulate samples, and wipe samples of contaminated surfaces.
3. *Who to sample.* The selection of the workers exposed, the number to be tested, and the duration of sampling.

EVALUATION OF EXPOSURE TO CHEMICAL AGENTS

4. *When to sample.* On all shifts, only during the day, on the weekend, with doors and windows open or closed, for seasonal variations, during "normal" operations, or during maintenance or shutdown.

Once all the data have been collected, they must be considered in light of the specific objectives. Normally the objective of data collection falls into one of four categories, elaborated in Sections 2.1 to 2.4.

2.1 Compliance with Governmental Regulations

The procedures to be followed for compliance with government regulations may be somewhat more comprehensive and restrictive than those for almost any other type of evaluation except for research. This restriction takes the form of the limitation of scientific judgment of the industrial hygienist and substitutes for that individual scientific judgment specific government regulations. To accomplish this objective, it is necessary to be familiar with the government regulations and the intent of the regulations and the law from which they are derived. Many times the intent of the regulations and the specific action to be taken have not been included in the regulation, and it may be necessary to contact the local or higher government office to determine precisely what is expected. Many government regulations are derived from administrative laws and as a result are little help in determining requirements, except on a broad philosophical basis. As an example, most of the present health standards of the Occupational Safety and Health Administration (OSHA) (3) were derived from the 1968 American Conference of Governmental Industrial Hygienists, (ACGIH) threshold limit values (TLVs) (4). Since analytical methods are not included in the TLVs, the method being used by OSHA must be determined and either the same or similar analytical procedures selected to ensure comparability of the two methods. At a minimum, the collection of data for compliance should include the following materials, steps, and procedures.

2.1.1 Sampling Equipment

Selection of sampling equipment and procedures that are the same as, or similar to, the equipment or procedures used by the regulating agency.

2.1.2 Calibration of Sampling Equipment

Calibration is as important in sampling for government compliance as it is in the collection of data from any other industrial hygiene evaluation effort.

2.1.3 Analytical Methods

Procedures must be similar to, or the same as, those used by the governmental agency, to ensure comparability of results. In addition, it is advisable to have the analytical work accomplished by a laboratory that is accredited by the American Industrial Hygiene Association (AIHA). This accreditation will add credence to the assessment of the data.

2.1.4 Sampling Strategy

Consideration must be given to the selection of the environmental sampling procedures to be used, based on the type of standard that has been promulgated (time-weighted average, ceiling value, etc.). In some cases it is desirable to collect samples of several different types even though the regulatory agencies may not require them. For example, the industrial hygienist might decide to collect data for both respirable and total dust when the exposure substance in question is free silica and only one type of sample, the respirable, is required by the regulatory agency. The total dust samples can be used to develop specifications for engineering control. The number of samples that must be collected is seldom delineated in the standard. However the National Institute for Occupational Safety and Health (NIOSH) (5) has developed a statistical method to determine how many samples should be collected for the determination of noncompliance with occupational health standards.

2.1.5 Quality Control

It is very important in the collection of data for the evaluation of compliance with government regulations that every effort be made to assure that the sampling, analysis, and evaluation are not only of the highest quality, but the sequence of events can be well documented and a well-designed quality control system has been included. The use of internal "check samples" for analytical procedures and a cooperative cross-check of samples with other laboratories, and utilization of the Proficiency Analytical Testing (PAT) program of NIOSH and the AIHA Laboratory Accreditation Program are recommended. In addition, good quality control on preventive maintenance, repair, and calibration of sampling equipment is essential.

2.1.6 Recordkeeping

Data are only as good as the records that are maintained. The development of a "log" or record of the chain of events that have occurred in any system for the collection and analysis of data is important under any circumstances, but particularly if the data are to be used as exhibits to establish exposure levels in a legal sense, to meet the recordkeeping requirements of a regulatory agency, to save for use at some future data in epidemiologic studies, or to evaluate environmental levels where clinical symptoms have occurred. Detailed recordkeeping may be less critical if the data will be used for short-term evaluations only, such as use of general room or source sampling to determine the effectiveness of new control methods. Even under these circumstances it may be advisable to maintain records for future use, to show good faith to a regulatory agency.

Recordkeeping is also required for accreditation of laboratories by AIHA. Such a system need not be complicated, but the records must be maintained in an orderly manner and must be easily retrievable. Comprehensive records are important in trying to develop retrospectively a picture of exposure. It is interesting to note that such records have not been maintained in many industrial operations, and as a result it is frequently impossible to develop a long-term exposure record.

When a recordkeeping system is being developed, one must plan to include environmental, medical, and personnel data that can be used to establish a dose-response relationship. Many industrial concerns are presently developing computer-based systems for long-term use for clinical and epidemiologic studies as well as short-term use for regulatory requirements and for management information systems to determine the present status of the health of the worker and environmental levels, as well as trends in environmental levels over an extended period (see Chapter 4).

Quality control and maintenance of records is particularly important in any data acquisition for compliance evaluation, and a sequential chain of records is valuable to assure acceptability of data by a regulatory agency.

2.2 Surveillance

Surveillance data acquisition serves several purposes and can be acquired in such a manner as to answer several objectives at the same time. These data can be used not only to characterize the individual worker exposure but, also, at least in part, to characterize a specific work area or job as well as to establish trends and exposure levels for any given operation. It is likewise possible to use the information collected in this manner for long-term research studies and to measure effectiveness of control procedures. The specifics for this type of atmospheric measurement have been discussed in detail in Chapter 6. Several considerations must be kept in mind when collecting data for surveillance. These are outlined next.

2.2.1 Sampling and Analytical Procedures

The sampling and analytical procedures must be selected so that the data can be related, either directly or by some meaningful comparison factor, to data that have been collected previously and to similar investigations that have been reported in the published literature. It is important that the data collected be usable in establishing trends and that they be standardized sufficiently to be usable at a later date for epidemiologic studies.

2.2.2 Position Descriptions

Position descriptions to assure proper identification of duty assignments for the workers are necessary. These must include all job elements of any exposures that have been determined to be related to the specific job. It may be that the position description used by the Personnel Office does not "fit" the actual duty of the worker and his or her exposure. As a result, some companies have found it advisable to prepare an Occupational Health Position Description that more adequately describes the exposure conditions of a given job. These "occupational health" position descriptions are usually maintained in a separate health file and do not become part of the personnel file.

The position descriptions must be constantly updated and changed as the worker moves from one position to another. This information is very significant in any future use of the data for a retrospective epidemiologic study.

2.2.3 Chronological Log

During the collection of the data, a chronological log of events that have occurred is necessary for the proper evaluation of the data. This is true for both short-term "ceiling values" and other types of sampling as well as for assessment of 8 hr time-weighted average exposures. This information should also indicate whether the exposures are continuous or sporadic, and whether the equipment is operating properly; any unusual occurrences should be described. Evaluation of the data necessitates the recording of any unusual activities, the condition of housekeeping, the use of provided ventilation, any recent engineering control changes, and the size of the work force available on the day of sampling. The housekeeping may be poor on Monday morning, for example, if the maintenance crew has not worked over the weekend. This type of housekeeping could result in higher exposures on Mondays than would normally be the case throughout the rest of the work week. The condition and operation of any local exhaust ventilation system can have a similar effect. Times of vacations or high absenteeism can result in increased work demands for the individual worker. These conditions need to be noted in relation to the data being evaluated.

2.2.4 Personal Protective Devices

The use of personal protective devices by the worker can also have an effect on the evaluation of surveillance data. For example, the difference between the levels of exposure of a worker to a toxic chemical when a respirator is worn as opposed to when such protection is not used is as significant, as is the information that the respirator worn is the right type for the hazard, is effective at the level of exposure, and is approved by NIOSH/MSHA (the U.S. Mine Safety and Health Administration) for the specific hazard. The same type of information in relation to skin absorption, protective clothing, smoking, and eating habits in the work area must also be considered and recorded.

2.2.5 Weather Conditions

Weather conditions can have a major impact on the evaluation of the industrial hygiene surveillance data. The differences that closed or open doors and windows can make on the effectiveness of general room or dilution ventilation are well documented and can

EVALUATION OF EXPOSURE TO CHEMICAL AGENTS

exert a major effect on the environmental levels. This can also be true, usually to a lesser extent, for local exhaust systems. The effects of rain in reducing the dust levels in any outside operation are easily recognized. A change in humidity can also influence environmental levels of some contaminants. Although these are not items that can be controlled, they do represent information that is valuable in the evaluation of environmental surveillance data.

2.2.6 Supplemental Data

Supplemental data that do not relate directly to the environmental data obtained, but must be considered as a major modifier to any evaluation of environmental data, are the effectiveness of engineering controls and the attitude and information available to the employee.

2.2.7 Engineering Controls

The effectiveness and the availability of engineering controls have an effect on the evaluation of the environment. For example, if it is found that the local exhaust ventilation system is normally turned off to conserve energy when the exhaust system is not in use, the evaluation must consider whether the local exhaust system, when operational, is sufficient to meet the requirements for environmental control, and whether employee training is sufficient to assure that the system would be turned on when necessary for hazard control.

Efficiency of control measures must be considered in relation to the maintenance of equipment as well as design characteristics. The evaluation of surveillance data cannot proceed without knowledge of the methods used to assure that preventive maintenance programs are effective and the time from submission of a work order until the work is completed; there must also be assurance that all work orders are handled on a priority basis to meet health needs.

Housekeeping is an important yardstick of plant attitude and concern about safety and health. For example, the accumulation of dust in the general work area, on machinery, rafters, and floors not only create a secondary health hazard but in some cases may make it impossible to evaluate the effectiveness of the existing ventilation system until the dust level is reduced by cleanup of the secondary dust sources.

Any ventilation system that is installed must be evaluated to assure that the actual capacity is sufficient to control the hazard. Frequently additional exhausts incorporated into an already existing and well-designed ventilation system have served to reduce the effectiveness of the system to an unsatisfactory level; furthermore the system may have become ineffective because of changes made during maintenance, such as reversing the fan direction, or lack of maintenance leading to plugging of the ductwork may have resulted in reduced airflow.

If the general room or dilution ventilation is in use, consideration must be given to the seasonal effectiveness of this type of ventilation, since the patterns of air movement may change when windows and doors are open during warmer periods of the year. The use of dilution ventilation must also be evaluated in conjunction with the toxicity, size, and number of sources of contaminants in the work area.

Isolation of the worker from hazardous materials as a method of engineering control must also be taken into account. The isolation of man from the hazardous material by the use of respiratory protection or other personal protective devices, or the use of isolated hazard-free work areas, such as control booths, can be effective, as well as isolating the source from the general work area and the worker. If the source of hazard has been enclosed as a means of control, the enclosure should be checked to determine whether there is any leakage, perhaps because the enclosure is under positive pressure.

A respirator program is difficult to assess but must be considered to determine whether the proper respirator protection is available for the type and concentration of the contaminant and whether the respirator is approved by a federal agency such as NIOSH/MSHA. It also must be noted whether the use of respirators is required or voluntary, whether there is a record of environmental data to support the type of respirator chosen and used, the type of respirator "fit" program that is used to assure that each person who must wear a respirator has one that gives a good fit, and that there is a program for care and cleaning of the respirators, and so on.

The effectiveness of other controls is also relevant in an evaluation of worker protection. "Other controls" include length of the workday, administrative procedures to reduce the period of exposure, the substitution of less toxic substances for a more hazardous material, and the availability of adequate (for the hazard) shower room facilities, changes of work clothes, and so on.

2.2.8 Worker Training and Attitude

It is vital that the attitude of the employees and their understanding of work procedures designed to assure their ability to work safely are understood. The Occupational Safety and Health Act speaks of informing the employee of the exposure hazards; of even more importance to the evaluation of surveillance activities, however, is the question, Does the worker understand these hazards and the proper work procedures? The answer usually must be arrived at subjectively, for few companies have testing programs for such health work rules. The evaluation of worker attitude is always subjective. The attitude of the workers, many times, is similar to the attitude of management, or a reflection of normal management relationships. However the attitude of a single worker under certain circumstances may reflect a negative attitude toward the health professionals collecting the industrial hygiene data. This is usually exhibited by "salted" samples, pumps out of calibration, and similar misadventures designed to confuse the health professional making the survey.

2.3 Research on Characterization of Chemical Agents

A number of different strategies must be employed in the acquisition of data for use in occupational health research. For the most part, the approach that gives the most definitive results will be similar to the strategies for the collection and analysis of workroom environments that are used for other types of evaluation, including surveillance. The main difference between the other types of evaluation and those for research is the additional consideration given to the design of the research program to assure that the results will be statistically significant.

The research effort usually involves the determination of methods to characterize the environment for epidemiology, animal studies, or clinical cases. The basic objectives of such research could also be to evaluate government health requirements, new products or by-products, and to reevaluate old processes or exposures.

2.3.1 Characterization Methods

The characterization of exposure patterns many times involves the development of a sampling protocol for the in-depth evaluation of a specific operation or industry using presently available methods, or it may require the development of a totally new means of sampling and analysis. For example, exposures to dust containing silica have been evaluated by existing methods and through the development of new procedures. In an extensive report in 1929 on "The Health of the Worker in the Dusty Trades, II. Exposure to Silicatious Dust in the Granite Industry" (6), the impinger was used as the means of collection. The particles counted were divided into two groups; those above and those below 10 μm in size. At the same time samples were collected and weighed and reported on the basis of total dust, even though a modification of this technique was not generally accepted until many years later.

With increased knowledge in respirable diseases and sampling techniques, it was later concluded that measurement of the total mass weight of dust and the analysis of the actual percentage of free silica in airborne dust would give a better measure of silica exposure. In addition, the range of particle size that had the highest biologic activity had been narrowed to the 1 to 5 μm range. This knowledge resulted in size-selective respirable mass collection of particulates, which has been used in both research and surveillance (7), including a reevaluation of the silica exposures in the granite industry (7, 8). The analytical procedures for analysis of free silica have also been changed to take advantage of new and improved instrumentation. Although the colorimetric analytical method of Talvitie (9–11) is the most widely used procedure for the determination of free silica, the use of both X-ray diffraction and infrared instrumentation is gaining in acceptance because of the stability of the samples, the ability to select specific silica polymorphs and to differentiate from high background material, and the lowered level of operator dependence.

In any research evaluation the development of an understanding of the pattern of possible exposure throughout the normal work cycle is necessary to determine the significance of exposures and the biologic effect. The biologic effects may be best measured for a short-term high exposure, or for a time-weighted average exposure throughout the normal work shift, or in some cases a combination of several different types of exposure patterns. The evaluation of work patterns and job cycles is probably more important for this type of research characterization than any other type of evaluation. This characerization, coupled with some expectation of the route of entry into the body and the biologic action, will determine to a great extent the type of sampling and analysis procedures to be utilized.

In some cases, when the exact chemical configuration of the hazardous material is not known, it may be necessary to predict possible contaminants and to base the design of the sampling strategy on these predictions. An understanding of the chemistry and reactions in the process will improve the accuracy of the predictions, which are usually not as comprehensive and as sensitive as would be desirable. Consequently minor yet possibly highly toxic contaminants may be overlooked in the development of a prediction model. For example, predictions of the decomposition products of certain missile fuels were based on the decomposition of the fuel and the oxidizer, and consideration was given to the primary constituents only. As a consequence, hydrogen cyanide, a highly toxic material that was later found to be a decomposition product in the parts per million range, was not considered in the prediction model.

One example of an effort to develop new methods for both sampling and analysis, necessary to characterize the chemical content of an operation and to attempt to predict compounds that may be hazardous agents, is the research underway to characterize the emissions from aluminum reduction operations, to predict what chemicals may be in the mixture, and if possible to select a compound or compounds that could be used as an "index" of exposure.

In this case it is necessary to select the proper solvent to remove the combination of particulates and vapors from a collecting medium and to quantitate the compounds that may be considered to be carcinogens or cocarcinogens. Thus the possible contaminants that may be in the emissions must be predicted and measured separately. For these studies measurements are being made of chemical compounds with three to six organic rings because of the suspected carcinogenic potency of this class of organic compounds. Although the compounds have not been precisely identified, nor is the percentage of each compound in the emissions known, this information, coupled with the data on the carcinogenic action of each compound, is necessary to develop a dose-response relationship. Predicting the dose-response relationship of individual compounds is difficult. The problem is far more complex with chemical mixtures, and even worse than the case described when the composition of the mixture and the percentage of chemicals in it are subject to change with minor adjustments in the process operations.

2.3.2 Epidemiology

Evaluation of environmental data for the purposes of epidemiologic studies may present a significant challenge, since it is usually necessary to collect data over an extended period of time and/or to attempt to reevaluate data collected earlier in relation to the methods used at the present time.

Chapter 5, Volume 1, covers the epidemiologic approach and the evaluation of disease in specific populations in detail, and this material is not repeated here. It is essential, however, to discuss the importance of environmental data that must be made available for any epidemiologic study designed to produce dose-response data.

If the epidemiologic method chosen is, for example, a prospective study, it is possible to collect the necessary data for fulfilling the objective of the epidemiologic study without assembling or evaluating environmental data. The basic question in this type of study is, Is there a difference in the incidence of disease between the exposed and nonexposed groups? Many studies, such as those involving death certificates, workmen's compensation records, can be classified as this type of investigation. Even in these studies, however, there are several reasons for considering the availability of environmental data. If the results are positive (there is a difference in disease patterns between the exposed and nonexposed groups), the next and logical question is, What is the dose-response relationship between the disease and the chemical agent? In other investigations, when a casual relationship has been shown to exist between a disease and a job or industrial operation and the specific chemical agent or agents have not been identified, efforts to characterize the environment and to develop a dose-response relationship have been intensified. In some cases a prospective study is run in parallel with efforts to develop field sampling and analytical methods for the characterization of the occupational environment.

Another method of difference is the retrospective study, which measures the relative risk by difference in exposure between "disease" and "no disease" groups. This type of observational research does not necessarily require collection of environmental data. Milham (12) compared "disease" and "no disease" groups by classification of specific industrial process without measurement of exposure; his work is an example of this type of investigation.

The method of concomitant variation poses the question, Does the disease vary as exposure varies? This approach requires at least some attempt to measure or estimate the environmental exposure. When the latent period between onset of exposure and disease production is long, the best estimate may be only one of generalized exposure levels, such as high, medium, and low. This environmental evaluation is difficult and may be totally dependent on "opinions" of individuals who have been familiar with the operation over an extended time. If this is the only information available, the conclusions may not be entirely valid.

In addition, the method of intervention could be used: here the suspected causitive agent is removed or controlled, and the decline or absence of disease is measured. This type of epidemiologic study may be done as a follow-up to previous studies and after engineering controls have been instituted.

These epidemiologic methods may not require environmental data initially, but in almost all cases the need exists for a follow-up study to establish a dose-response relationship. This is of ever increasing importance, even when the results of the epidemiologic study are negative. As more emphasis is placed on mandatory health standards, there is an increasing demand to establish the lower limit of biologic response. This requires quantitating negative disease data with environmental levels of exposure.

The development of research protocol must be such that the use of available environmental data is maximized, as well as the assessment of the disease. This means that the environmental data developed delineate the exposure conditions in the greatest possible detail as outlined earlier in this chapter (i.e., duration of exposure, changes in process, housekeeping, and job requirements). This information is valuable in the evaluation and in any follow-up studies.

Changes in sampling methods and analytical procedures can have a major impact on the establishment of any dose-response information and can result in an attempt to develop a relationship between the old and the new methods. This many times presents more new questions than it answers, since it is not possible to hold static other parameters in an industrial process. The efforts to relate old analytical and sampling methods to new procedures have been addressed in epidemiologic studies in the past by Reno (8) for silica exposures and by Lynch and Ayer (13) in the investigation of asbestos exposures. These problems are primarily of concern in retrospective evaluation of the environment, but can also present difficulties if new techniques are developed during the epidemiologic studies.

Development of retrospective environmental data is the most difficult of tasks and must be addressed with skepticism about the value of any old data as well as some data developed in other countries, where it may not be possible to define with certainty the conditions under which the data were collected. Detailed information is needed to relate analytical methods, field sampling techniques, reproducibility of analytical results, calibration procedures for sampling equipment, duration of sampling, and the development of time-weighted averages. The method of sampling (e.g., general room or breathing zone), length of workday, and any change in the process must not be overlooked.

It is desirable to be able to estimate the effectiveness of control methods and to know whether they were operational during previous studies, particularly when the control is designed to isolate the worker from the hazardous condition. If it is felt that the door to the isolation booth, for example, was never closed, or that the respiratory protection provided was not generally in use, these types of control should be discounted when evaluating environmental levels and biological reaction to exposure.

The exact nature of the job as related to exposure must be ascertained: personnel job descriptions may not give a true picture of exposure. Any interferences resulting from exposure to multiple contaminants must be evaluated, particularly if they may have an adverse effect on the same organ system or if they use the same route of entry. Changes in process may result in exposure to chemical intermediates and/or by-products that could cause a biologic variation. The question of an interaction of two or more chemicals and a possible synergistic or antagonistic or carcinogenic cofactor must be evaluated. For some chemicals the possibility of two or more routes of absorbtion must also be considered. For example, the route of entry of ethylene dibromide (EDB) may be percutaneous absorption of the liquid or inhalation of the vapor (14).

The possibility must be closely scrutinized that the signs and symptoms that furnish the corroborative evidence of intoxication may be the result of or intensified by nonoccupational disease, or may emanate from other exposures.

2.3.3 Animal Studies

In recent years there has been increasing recognition that the development of research protocols and the interpretation of laboratory research information requires evaluation by physicians, industrial hygienists, and others in addition to the evaluation made by the toxicologist. This expertise and differing professional background is required in the assessment of animal studies that must be related finally to an assessment of the effect of the workroom environment on man and compared with data collected in epidemiologic investigations, clinical cases, chemical tests, and in some cases similarity of molecular structure (see Chapter 16).

The passage on October 11, 1976, of the Toxic Substances Control Act (PL 94-469) coupled with increasing demands by OSHA for definitive data for the development of a scientific base for federal regulatory standards, may result in the loss of some latitude of the individual researcher in the manner and procedures that will be used for the collection and interpretation of research data related to exposures in the workplace. There is recognition today that the interpretation of a single animal study may be the sole basis for an enforceable federal standard. This is not the objective for much of the laboratory animal research that is done, and it puts an increasing burden on the researcher. In the development of a protocol for an animal study, many more parameters of the experiment can be controlled than in epidemiologic or clinical studies, such as duration of exposure and exposures at a specific concentration. However failure to control even one of these many variables can result in a difference in response in the test animal by 100- to 1000-fold, according to OSHA in the proposed cancer policy and standard (15).

The acquisition of meaningful data from animal studies on which evaluation of exposure of the worker to chemical agents can be based, requires that careful consideration be given to the design of the study, so that the response expected or seen in the experimental animal can be extrapolated to man and to the exposure parameters that can be expected to occur either on the job or in the community environment. For many animal studies the route of entry that is representative of worker exposure is by inhalation, and this requires long-term exposure studies that can be related to chronic inhalation exposures in man. In this type of investigation one must take account of the methods for generation of atmospheric concentrations of the chemical and for evaluation of the environmental data from the animal exposure. Inhalation studies may be planned after some suggestive toxic effect has been determined by acute and/or subacute investigations. These studies may not require the generation and assessment of environmental exposures, but they do produce valuable data that can be used to give some estimate of the type of response that can be expected, the environmental concentrations that would be most appropriate, and reasonable time periods for exposure of the animal.

The generation of environmental data for the purpose of animal inhalation studies presents some interesting problems not only in the generation of known concentrations of a specific chemical, but in the ability to maintain this concentration in equilibrium with the experimental parameters for a long time. For example, the design and construction of animal exposure chambers may result in condensation of the chemical being tested on the walls of the exposure chamber. In addition, the equilibrium of the chemical concentration in the exposure chamber may be disturbed by condensation of the chemical on the skin or hair of the experimental animal. This condensation will increase as the area of the skin surface of the animal is increased in relation to the total volume of the exposure chamber and the concentration of the chemical. In some instances these difficulties with generation and maintenance of specific exposure conditions throughout the exposure chamber necessitate measurement and evaluation of the exposure levels in the breathing zone of the animals.

The choice of the sampling and analytical methods, as with other research and surveillance efforts, can have a significant impact on the evaluation of the data collected. The sampling strategy, of course, must reflect consideration of the exposure parameters of the study, both as in any other sampling procedure and with respect to sampling in a small closed system and relating these results to the occupational environment. These methods must meet the "test" of being applicable to both the laboratory and the workplace environments. The need for quality control of analytical results is, of course, critical in the evaluation of environmental data from animal inhalation studies.

2.3.4 Clinical Cases

Epidemiologic and clinical cases have the advantage that it is not necessary to extrapolate animal data to humans. It is seldom that a human population can be controlled in such a manner as to reduce the effect of outside influences, such as smoking, dietary considerations, and intake of drugs or alcohol. Even when it is possible to control such influencing factors by clinical experimentation, the number of subjects is greatly reduced, and the differences in individual susceptibility must be considered to be a major factor in any recorded variation. The controlled clinical studies can be of great value, as illustrated by the study of lead absorption and excretion pursued for a number of years at the Kettering Laboratory, University of Cincinnati (16). The clinical studies such as those carried out at the University of Cincinnati represent controlled clinical experiments where most of the parameters of the experiment are closely controlled, and are evaluated with confidence in the results. The environmental concentrations are also well controlled even over extended periods. These controlled experiments are usually developed around the use of recognized sampling and analytical procedures. The evaluations of chemical exposures in clinical experiments are, as in the lead studies, necessarily of low levels of exposure to evaluate preclinical signs or symptoms.

The retrospective reconstruction of work exposure conditions that resulted in a clinical manifestation of disease presents a difficult exposure evaluation. In these clinical cases there is no opportunity to plan "beforehand" the parameters of the study. Rather, the investigator must attempt to develop a retrospective exposure profile in much the same way as is done in epidemiologic studies when the question of dose/response must be answered. Because of the limited number of people normally exposed in the clinical cases, it is necessary to probe in more depth for other possible causes of the signs and symptoms than in the morbidity studies. Although it is possible to estimate exposures that are reasonably accurate for clinical cases of acute exposure, the longer the latent period, the more difficult it is to make a reasonable estimate of past exposures. Even in acute exposure cases, it must be recognized that any sample collection and analysis may or may not represent the same environmental conditions that resulted in the clinical response.

In investigations of recent acute clinical cases any retrospective environmental sampling and analysis can at least be accomplished with methods that are presently available and generally accepted. In any attempt to evaluate data from clinical cases that occurred in the past, it must be recognized that it may not be possible to verify the analytical results and to develop a retrospective profile of environmental conditions, let alone relate these experiences to present conditions. Under these circumstances the environmental exposure data may be grouped with confidence only into categories of high, medium, and low exposure levels.

These limitations in the use of clinical cases have been recognized for a number of years. For example, in 1928 Bloomfield and Blum (17), when reporting on a study of workers in chrome-plating operations, made an attempt to approximate the extent of exposure of 17 of 19 employees who had symptoms of chromic acid exposure. They indicated that the estimate of exposure could be considered only an approximation except for a few of the workers who had been employed during the period that the ventilation system had been in operation. The estimate of exposure had been based on the occupational histories and the results of 39 air samples taken in six plants where the 19 workers were employed. The value of the data from these grouped clinical cases may be somewhat questionable, but they do provide some information of the cause and effect.

Lacking any more definitive information, many times data collected in clinical cases can be used with caution to develop exposure guidelines to assure some degree of worker protection that otherwise would not be possible. The study by Bloomfield et al. (17) has been used as partial justification by the ACGIH to establish a TLV for chromic acid (18).

The development of a retrospective enviornmental profile necessitates the careful consideration of all factors that may influence the conclusions to be made from the clinical studies. The possibility of interference and/or synergism or antagonism from other exposurs must be investigated to assure that the cause is well defined when it is related to an effect. An example of the effect of exposure to a second chemical, particularly an off-the-job exposure, is the interaction between inhalation of trichlorethylene and the ingestion of ethyl alcohol, which can result in a marked increase in the effect of the trichloroethylene on the body (19, 20). In clinical cases the possibility of reaction from medication that was being given at the time of exposure to the industrial chemical must be examined. Exposure from a second job and from hobbies are other possibilities that must be investigated as having been either the primary cause or an aggrevation of the industrial chemical exposure. The need to develop a complete environmental profile of exposure is necessary to assure that a true cause-and-effect relationship is established when attempting to reconstruct the circumstances that led to a medical aberration in the worker.

2.4 Adequacy and Performance of Control Program

Acquisition of data on the adequacy and performance of engineering controls, and the effectiveness of administrative controls, may be considered to be integral parts of the data collection for routine surveillance activities. However under some conditions the collection of performance data for engineering design specifications and/or the determination of the effectiveness of administrative or operational controls may dictate specific data acquisition strategies.

The development of specifications for engineering control calls for the measurement of the environmental levels of the chemical to be controlled, and the measurement of the airflow in any existing ventilation systems. The assessment of the effectiveness of the new engineering controls requires repetition of the environmental measurements for comparison with previous environmental levels. If a new ventilation system is added or any existing system is modified, airflow measurements should be taken.

The selection of specific methods of control is discussed in Chapter 18 of this volume, and the design specifications of a number of hoods and local exhaust ventilation systems are developed in detail in the *Industrial Ventilation Manual* (21).

2.4.1 Testing Procedures

The requirement for environmental testing to develop engineering specifications depends on the type of engineering control under consideration and may include process sampling or sampling at the source of generation of the contaminant, general room sampling, personal monitoring, and short-term monitoring.

Environmental sampling at the source of generation or the release of a contaminant into the work area is usually done with stationary samplers. The selection of specific sampling equipment and the procedures for acquisition of these data depend on the conditions of release of the contaminant. Source sampling gives data on the amount of material being released into the work area. If this information is not available by material balance or rate of consumption calculations, this type of data acquisition is essential when total enclosure of the source of the chemical contaminant may be necessary or when the use of local exhaust ventilation to control the hazardous material is being considered.

The collection of environmental data is only one element in the design of engineering controls. Evaluation must be made of the effect of air movements caused by motion of machinery, the workers, the material being processed, general room air currents, and thermal air movements from any hot processes or the heating and cooling equipment, as well as the capture velocity of any local exhaust ventilation system in the general area. Sampling at source is usually not applicable when dilution ventilation is being considered, since the consumption or evaporation rate of the chemical contaminant in a work area normally can be calculated. General room sampling or personnel monitoring will give a more definitive assessment of the exposure and of the dilution rate necessary to reduce exposures to below the TLV if dilution ventilation is the method of control under consideration.

The method for the calculation of airflow required to maintain the levels below the TLV is covered in the *Industrial Ventilation Manual* (21). The limitations for effective use of dilution ventilation must be considered before this approach is selected as the control method. Normally it can be used to successfully control exposures only when the toxicity of the material is low (high TLV), the rate of consumption and the rate of evaporation are low, the workers do not work directly at the source of the contaminant, and the hazardous material is released at a constant rate throughout the workday. The use of general room and personnel monitoring are acceptable methods for the evaluation of the effectiveness of dilution ventilation.

Continuous monitoring can be a very effective method of evaluating the engineering control devices used for highly toxic and life-threatening acute hazards. For example, air samples have been collected on continuous tape and analyzed directly for beryllium in aerospace operations, and permanent fixed monitors for hydrogen sulfide have been used extensively in the petroleum refining industry and has been recommended by NIOSH as a means of monitoring for hydrogen sulfide under certain conditions (22).

When it is necessary to analyze any samples collected to evaluate the adequacy of the performance of controls, the choice of analytical procedure would normally be the one that best fits the requirement for selectivity, precision, and accuracy for the levels of the chemical contaminant expected to be collected. The standard method for compliance or surveillance purposes is not necessary if the results will only compare "before" and "after" the installation of controls; however if the results are used to compare worker exposure before and after installation of controls, or for compliance purposes, the same analytical method that is used for compliance and routine surveillance would be the procedure of choice.

2.4.2 Evaluation of Local Exhaust Ventilation

Airflow measurements are the best method to evaluate the effectiveness of any existing exhaust system and to determine whether any new ventilation system is meeting its design criteria. Airflow measurements have a number of applications and are necessary to assure proper evaluation of any ventilation system.

These measurements can be used to evaluate any new exhaust system to determine whether it meets the design criteria, to balance the airflow if the exhaust system is designed for the use of blast gates, to obtain additional data for similar installations that may be required in the future, and to determine whether the delivered airflow meets the requirements of any regulatory or voluntary code.

In any exhaust systems that are already installed, airflow measurements can be used to determine the system's capability to deliver the necessary airflow to meet governmental and trade association codes, to collect design data for similar installations, to determine whether the present system has sufficient capacity for additional exhaust outlets, and for repeated evaluation of the airflow to determine whether repair, maintenance, or readjustment of the airflow is necessary.

When an evaluation is made of either an existing or new installation, the results of all air measurements, the location in the exhaust system where the measurements were obtained, and the comparison made with any previous collected air measurements should be retained just as data from environmental measurements are retained in the industrial hygiene files.

Air measurements on any new system should include velocity and static pressure measurements in the main, and in all branches of the ductwork. Static pressure measures should be obtained at each exhaust opening, static and total pressure measurements should be made at the inlet and outlet fan openings, and the differential pressure should be determined at the inlet and outlet of the collection equipment.

These testing procedures need not be repeated at regular intervals, since there is usually little change in any existing exhaust system unless the process is changed, or the exhaust system is altered, or unless routine maintenance and repair is inadequate. For routine evaluation of any existing ventilation system, the measurement of hood suction is the usual and acceptable procedure. The criteria for selection of measurement devices for assessment of ventilation systems are discussed in detail in the *Industrial Ventilation Manual* (21).

2.4.3 Administrative Control

Administrative controls have been widely used where engineering controls of exposure are inadequate—for example, to reduce the hazard of working with risks through training programs, to inform the employees of the hazards, and remove an employee from exposure to allow sufficient time to recover from the effects of exposure. "Recovery time" has many uses in industry: as a cooling-off period for workers who have been exposed to hot environments or for those who need to recover from the effects of abnormally high or low air pressure (e.g., deep sea divers, caisson workers, tunnelers, and airplane crew members). Chemical exposures have also resulted in the need for recovery time while the worker metabolizes or excretes the hazardous chemical and the body burden is reduced to acceptable levels. During the recovery time the worker is placed in a nonhazardous job or excused from work. These procedures have been used extensively in response to exposures to heavy metals, particularly lead and mercury, and to a lesser extent to organophosphorus compounds, among others. These administrative procedures require the ability to monitor the biologic fluids of the worker as the basis for deciding when the individual should be removed from, and returned to, work. This approach usually is an indication of lack of sufficient engineering controls or that the chemical is extremely hazardous and employees must be monitored on a continuing basis to be assured that they are in fact being protected from the hazardous material. The use of biologic monitoring is covered later in this chapter.

The administrative controls that require a detailed evaluation of the environment are those that should be used only until the proper engineering controls can effectively protect the workers. These administrative controls require that the workers spend only part of their time during the normal work shift in a highly hazardous area (above the TLV) and that the remainder be in a low or nonhazardous area to assure that the combined exposure is below the TLV.

EVALUATION OF EXPOSURE TO CHEMICAL AGENTS 337

This type of administrative control would normally be used only for chemicals whose exposure standard is based on a time-weighted average and no adverse acute effect from high exposures, and the exposure is reasonably consistent throughout the work shift. Silica exposure in quarry operations, for example, could be controlled by this approach. Benzene exposure should not be controlled by administrative reduction of exposure time because an acute high exposure can cause systemic poisoning, can act as a primary irritant, and is absorbed through the skin. In addition, chronic exposure by inhalation can cause blood changes, and benzene may be a carcinogen.

Prior to use of administrative control, a sampling strategy that will result in a complete understanding of the exposure characteristics of the job must be selected. Any cyclicity of the exposure, of work procedures, and exposure times must be estimated. If the preliminary evaluation suggests that the exposure would be reasonably constant throughout the work shift, it may be possible to collect data over the entire work period, at least as a preliminary analysis of exposure; later, detailed time and motion studies must be made of the work process, with specific exposure values for each operation and total time-weighted average exposure for the work shift.

The assessment of these data forms the basis for determining whether reduction of exposure is possible by administratively reducing exposure time. It must be remembered that such administrative control usually means that some other worker or workers must assume part of the toxicologic burden of the employee who had worked on this specific high hazard job. If the time spent off the high hazard job has no exposure, the high hazard exposure time is the total exposure. For the worker moved to a low exposure for the remainder of the workday, however, the total exposure will be the time-weighted average exposure of both the high and the low exposure jobs. This total exposure assessment must also be made for any employee who helps "fill in" on the high hazard job.

The use of such control demands close and continuing monitoring of these exposures both in the high and low or no exposure positions to assure that no overexposure occurs. The increase in the number of workers who are exposed on the high hazard operation is not an acceptable long-term alternative to engineering control. Even for a short period the impact of the increase in the number of workers exposed may result in additional medical examinations, specific medical procedures related to the exposure, informing the employee of the hazards, possible fitting with respirators, and increased environmental monitoring.

The analytical method as well as the sampling procedures must be the acceptable method used for routine surveillance and regulatory sampling to assure that the results can be compared to other data collected.

3 IMPORTANCE OF MODES OF ENTRY

An understanding of the physical and chemical characteristics of a workroom contaminant, its route of entry, and the effect of the hazardous material in man is ential in the development of sampling strategies that will satisfy evaluation objectives. Several methods are used for classification of the modes of entry and the biologic reaction of the hazardous material in man. The specific and detailed discussion of these classification systems are covered in Chapter 6, Volume I. These classification systems present a convenient method for the development of evaluation strategies.

Three classification methods—physiological, physical, and chemical—should be considered individually and collectively: these allow for an organized method to answer the basic questions for the development of an evaluation strategy. How does the toxic agent enter and react in the body (physiological classification)? What are the physical characteristics of the contaminant (physical classification)? What is its chemical structure (chemical classification)? This section deals with these classification systems and how they are used in the development of sampling strategies.

3.1 Inhalation

The inhalation of airborne hazardous material into the respiratory system is the most common route of entry for the contaminants found in the industrial workplace. Most of the occupational health standards and the TLVs of the ACGIH are based primarily on this route of entry. A few standards are based on other modes of entry, but for the most part they are considered to be an additional and/or complicating route of entry. A group of hazardous chemicals that may use one or several of the routes of entry into the body are those that result in a carcinogenic reaction in man. The standards for this biologic effect group may not be based on an evaluation of an airborne dose-response relationship. The basic industrial hygiene approach to the evaluation of occupational hazards has been the collection of environmental data and the relating of that information to an adverse effect in man. This approach relies on the assumption that the primary, and usually only, route of entry is the inhalation of the airborne contaminant into the respiratory system from the workroom air. The effect of exposure by other routes is less well defined.

The methods of collection of airborne contaminants are based primarily on their physical characteristics, and the method of analysis is based on their chemical characteristics. The selection of a method for field collection and/or chemical analysis may have to be modified as a function of other considerations. For example, the "analysis" for asbestos has been based on its physical characteristics, not its chemical classification. In this case any fibers exhibiting a defined physical characteristic are considered to be asbestos. This evaluation technique is not able to differentiate between fibers that have different chemical identities, such as tremolite asbestos and fibrous talc dust, which cause entirely different diseases in man. This suggests the need for a more complex chemical analysis to differentiate between the several types of fibers. Asbestos is found in many commerical talcs (23).

As a group, particulates, including dust, fibers, fumes, and mist, have been difficult to assess. Generally the collection and analysis of dust and fibers have been based on physical characteristics and modified in relation to the reaction of the respiratory system

EVALUATION OF EXPOSURE TO CHEMICAL AGENTS

to the contaminant; for example, the sampling method for dust has been modified based on the deposition in the lung of sphericallike particulates of a specific aerodynamic size. The lung retention models used for aerodynamic sphericallike particulates that have been used for silica, lead, and other particulates were found to be inapplicable to asbestos because of its fiber configuration. This resulted in the development of the concept of counting only fibers with a certain aspect ratio between fiber length and diameter.

Normally fumes and mist, though having been formed by different physical conditions, are collected by the same methods employed for the other particulates, since they all exhibit similar deposition characteristics in the respiratory tract. For example, NIOSH recommends that for breathing zone sampling for lead dust and fumes (24), chromic acid dust or mist (25), asbestos fibers (26), and silica dust (27), all should be collected on some type of filter media. These four chemical particulates all require some type of evaluation after collection, since direct reading instrumentation is not normally used and three of these (silica, lead, and chromic acid) require chemical analysis. Only the asbestos assessment is based on physical characteristics alone.

The collection and analysis of environmental samples of gases and vapors can be considered together, since the need to differentiate between the two physical states is primarily in the evaluation of the hazard potential of a contaminant, not in the consideration of how the substance is collected and analyzed. In the case of gases and vapors the collection method may be based on the contaminant's physical or chemical characteristics. The basic collection methods may be physical entrapment on one hand, or chemical reaction of the contaminated air with an absorbing or adsorbing medium. Physical entrapment is usually accomplished by collecting the air to be analyzed in an evacuated container, such as a bottle or a plastic bag. The methods based on the chemical characteristics of the airborne contaminant and of the collection medium rely on the collecting medium to remove the comtaninant from the airstream as it moves through or over the medium, which may be a solvent or a sorbent, such as silica gel or activated carbon.

Under certain conditions gases or vapors are deposited on particulates while still in the air and may be inhaled as particulates. Where both particulates and gases and vapors may be contaminants, such as in aluminum reduction and coke oven operations, consideration must be given to the method of collection and analysis of both the particulates and the gases and vapors.

3.2 Ingestion

Ingestion is not considered to be a major route of entry of a toxic material in the workplace. But it is a possible route of entry and cannot be completely discounted, and at least some of the material that is inhaled may be swallowed and absorbed into the digestive tract. Ingestion is considered as presenting a significant problem when the contaminant is highly toxic or carcinogenic. The usual method of evaluation is not by environmental sampling but by evaluation of work practices. A toxic material that is deposited on food, tobacco, or in liquids may be ingested. Possible sources of contamination, and the possibility of ingestion caused by poor work habits, must be evaluated. For example, ingestion of lead dust by a worker who held his sanding disc between his teeth when not grinding on automotive bodies presented evidence of lead intoxication, although the environmental levels for lead were essentially negative.

3.3 Skin Absorption

Skin absorption, along with ingestion, is not considered a major route of entry for industrial contaminants. Not all chemicals that cause skin irritation effectively penetrate through the skin barrier and reach the bloodstream. In the 1977 list of TLVs, the ACGIH indicates that about 25 percent of the chemicals included may make a potential contribution to the overall exposure by absorption through the skin, including the mucous membranes and the eyes (28). Included on that list are such highly toxic chemicals as parathion and tetraethyl lead.

It is not possible to make environmental measurements for the effect from skin absorption, as with ingestion, and any evaluation of the contaminent must be by observation of work habits, including personal hygiene, changes of clothes, and use of personal protective equipment. The use of wipe samples of the skin area has been suggested as a method for evaluating skin absorption, but the approach has not been widely accepted. As with ingestion, the compounds that penetrate the skin usually may also be inhaled. When making any assessment of the occupational environment, it is important that the possibility of an additional exposure resulting from skin absorption and/or ingestion be considered in addition to the results of samples collected in the breathing zone of the worker.

4 WORK EXPOSURE EVALUATION

Sampling of the occupational environment, as outlined under Section 2, "Objective of Data Acquisition," is done for a number of reasons, all of which, except for sampling to measure the effectiveness of control programs, and animal experimentation, are directly related to worker exposure. In all cases the biologic reaction of the worker or former workers to the chemical insult must be measured.

The environmental evaluation for compliance and surveillance may be related to the worker only in the abstract, that is, to an existing occupational health standard not to a direct effect in a worker. Environmental assessment for clinical cases and epidemiologic purposes are directly related to the evaluation of the worker's health, even if the assessment of the worker's health is by evaluation of past medical records.

Much of the same is true of medical assessments; that is, many of the medical examinations are for medical evaluations of nonoccupational and preventive purposes not related to occupational exposure. Occupationally related medical assessment may also be done only in the abstract—not directly related to actual environmental exposure levels, but for compliance and routine medical surveillance and related directly to existing standards or "normals" for clinical tests and examinations. Medical assessment for epidemiologic studies, and for specific chemical hazard evaluations, are directly related to worker exposure and biologic response. It is in this area of epidemiology and clinical case studies that the closest scrutiny is given to the medical and environmental records that have been maintained and to the medical examinations and environmental assessments that are to be completed.

The probable cause for the poor quality of many of the records available today is that at the time of collection, it was not intended that these data be used for epidemiologic or clinical case evaluation. For this reason environmental and medical data should be collected and maintained as if they were going to be used for clinical and epidemiologic research purposes. It is in relation to epidemiologic and clinical case studies that the direct connection between the worker and exposure is most likely to exist and the data collected must satisfy not only the objectives of environmental or medical monitoring, but the interactions between the two sets of data. The basic differences between epidemiologic studies and clinical cases from the standpoint of the assessment of the data involve the number of workers covered in the study, whether it is necessary to draw a sample of the population, and whether inferences can be made to a larger universe.

The techniques and uses of epidemiology for occupational health assessment, and the effect of the routes of entry, have been discussed in detail in Chapters 5 and 6, of Volume I and are not repeated here.

Choice of specific measurement technique for worker exposure evaluation many times is limited by the information that is available and/or can be collected, and the technique chosen may not be the preferred method. For example, the lack of any or adequate past medical records or a change in sampling or analytical procedures may limit the value of any retrospective study with the objective of defining a dose-response relationship. The choice of measurement techniques must be based first on the best epidemiologic method to answer the main objective of a study and second, on the information that is available or can be collected and whether it allows for valid conclusions to be drawn. If it does not appear that the data would lead to valid conclusions, the next best approach must be considered, and so on, until data availability and evaluation goals are compatible.

The final selection of a study protocol may be further modified by a reexamination of the evaluation goals to determine whether the end result will satisfy all the questions raised by the limitation of availability of data. For example, if the objective was to study the relative risk in specific job classifications by differences in exposure between "disease" and "no disease" groups and the most detailed job classification information available was classification by industry (chemicals, textile mills, lumber, tobacco, etc.), the study would give no information by specific job classifications. In fact, it might mask serious health risk problems in a specific job classification by comparing it grouped with all other job classifications within its overall industrial classification. Under these circumstances different methods would have to be employed to compare the relative risk of disease by job classification.

The problem with the recognition of potential interference from other factors has been covered in Chapter 5, Volume I, and in this chapter (Section 2.3.1). However it is important to emphasize several of the compounding factors that have resulted in invalid interpretation of data. Unfortunately the erroneous interpretations and conclusions may not be apparent when the results of the specific study are published. These factors of special concern are the reliance on inadequate environmental data and not treating all medical data with equal scrutiny. For example, evaluating both primary and secondary causes of death for the exposed group but not the unexposed group. If the epidemiologic measurement technique is not compatible with the information that will be available and the study objectives, either consideration should be given to another approach or potential sources of error of the study should be acknowledged. It must always be considered that because of lack of data, animal experimental studies may be more appropriate than the utilization of the epidemiologic approach.

In prospective studies it is possible to decide which specific biologic parameters should be measured, the type of environmental samples that should be collected, and the intervals at which the measurements should be made. It is possible to include in such investigations specific biologic parameters that are precursors to disease and can be used to measure subclinical manifestations. This, of course, can occur only after development of the correlation between biologic test levels and environmental exposure levels.

The methods of measurement will change with the mode of entry and biologic action and must be considered in the protocol design. For example, in a protocol for the study of a pulmonary irritant, the medical examinations could be designed to measure the possibility of short-term (acute) and long-term (chronic) effects in the worker, and the environmental evaluation could be designed to detect short-term excursions (ceiling values) and long-term (time-weighted average) exposure levels. In the case of an irritant, use of detailed medical history and/or questionnaires may be necessary in addition to the medical examination procedures, to determine the incidence of cough, sputum, dyspnea, and cigarette smoking, to obtain information that may not be evident upon physical examination. In addition, careful scrutiny must be given to the action of any other chemicals that many times are found in the workplace in conjunction with irritants.

The prospective assessment of exposure to a carcinogen is usually a follow-up study, after the carcinogen has been identified by retrospective epidemiologic or animal studies. Long-term tests in experimental animals are valuable in assessing the potential carcinogenesis of a chemical used in industry. These animal studies do not necessarily confirm that a chemical is carcinogenic in man, but they are valuable corroborating evidence.

Several approaches are possible in a carcinogenic prospective study. Traditionally investigators have examined differences between exposed and nonexposed groups, with follow-up studies to determine the effect of exposure level on the occurrence of disease, and/or the latent period from first exposure to onset of disease. Further evaluation of specific tests to measure subclinical signs or symptoms of disease may be necessary, and additional animal experimentation and chemical tests may be made to confirm the results of the epidemiologic study. Specific attention must be paid to other possible exposures, particularly for any compound that may act as a cocarcinogen or a potentiator. The complexity of the chemical structure of many of the known carcinogens requires specific analytical procedures. As a result, new field sampling and analytical methods may have to be developed.

The identification of chemicals having mutagenic potential has traditionally been from retrospective studies. The more recent development of tissue culture techniques and use of animal experimentation has resulted in the use of these techniques for screening for mutagenic potential.

Clinical studies in man and epidemiologic methods are very difficult to use effectively, since the effect of mutagenic reaction may not be apparent for several generations. However workers exposed to chemical compounds that cause sterility or semisterility, depression of the bone marrow, teratogenic or carcinogenic effects, inhibition of spermatogenesis or oogenesis, inhibition of mitosis, inhibition of immune response, or chemicals whose structure is closely related to known mutagens, should be closely followed.

Systemic poisons, on the other hand, may be ideal candidates for prospective, as well as retrospective, studies. Chapter 5, Volume I, lists examples of systemic poisons and the organs that are injured by exposure to them. The latent period for extensive damage to an organ system may be long; however early signs and symptoms, and results of specific tests, indicate some biologic reaction prior to permanent damage. Thus it is logical to subject systemic poisons to prospective assessment. The standardization of medical examinations, which can be achieved; will minimize inconsistencies between examinations and can be directed for in-depth evaluation of the organ system affected. Careful recording of the data collected also allows for the determination of the significance of subjective signs and work complaints between exposed and nonexposed groups. It is also possible to correlate the results of specific and nonspecific tests with the results of environmental exposure. The epidemiologic method of intervention also can be used effectively with chemicals that cause systemic poisoning if removal from exposure causes a reversal of evidence of the disease in the worker. This requires the development of sufficient medical and environmental data for in-depth analysis and correlation between exposure and effect. In the use of these techniques the possibility of some other mode of entry, as well as biologic action, must be considered—for example, ingestion of lead and absorption through the skin of aromatic hydrocarbons, in addition to inhalation.

Unusual physical conditions and physical agents can affect the assessment of the biologic insult caused by a chemical agent. These confounding factors must be considered and adjustments made in the analysis of the data, just as adjustments are made for age, smoking, and so on. Heat, cold, altitude, increased pressure, high demands for physical activity, and work schedules are examples of some of these physical conditions that are addressed later in this chapter. Another example of the interaction between a chemical and a physical agent is the photosensitivity effect on the skin of certain chemicals when the skin is exposed to ultraviolet electromagnetic energy. If at all possible, any combined effect should be eliminated in the study population; otherwise such factor(s) must be given careful attention when conclusions are based on the data.

The establishment of an equilibrium between the amount of a chemical agent in the body tissue with environmental levels is covered in detail in Chapter 6, Volume I and Chapter 7, Volume III. It is important to recognize that in clinical cases and in morbidity studies, this equilibrium of chemicals in blood, urine, tissues, and expired air and its rate of detoxification and elimination is critical to the determination of the proper time interval between exposure and collection of biological specimens. The effect of intermittent exposures must also be considered.

5 RELATING ENVIRONMENTAL EXPOSURE AND CLINICAL DATA

Defining the relationship of environmental exposure data to clinical measurements is the basis for the practice of occupational health by the specialties of industrial hygiene and occupational medicine, as well as the contributions made by many other scientific disciplines, including chemistry, toxicology, epidemiology, and engineering.

The importance of developing the relationship between the environmental measurement and clinical reaction cannot be overemphasized, for without establishment of this relationship, the collection of data is of little value. This relationship is imperative for the evaluation of exposure regardless of whether the data are being used for research, routine plant surveillance, or engineering control. The objectives of any evaluation program must be to assess this basic relationship clearly and concisely. Unfortunately it is not possible to control all the parameters that can affect this relationship in surveillance of worker exposures. The two basic questions in the final assessment of the data collected must be: Does the environmental exposure and clinical data fit a predictable pattern? And is there a rational basis for any aberrations from the predictable relationship? Consideration must be given in depth to each of these questions before any final conclusions or recommendations for corrective action are made.

5.1 Do Environmental Exposure and Clinical Data Fit a Predictable Pattern?

The absence of any expected sign of symptom (negative data) can be as troublesome as the finding of clinical signs (positive data) that were not expected. In the evaluation of a single or a few clinical cases, the variation in human tolerance to the chemical stress must be considered. This is also true in epidemiology studies but perhaps is a less significant point than in individual clinical cases. The worker or workers are more often than not exposed to more than one contributing cause, both on and off the job, that result in a more complex evaluation of the effect of workroom exposures than may have been anticipated initially. In these cases chemical exposure may only be an additional contributor to an already existing disease or clinical symptom. For example, the consideration of food intake is particularly important when the symptoms are being evaluated in relationship to environmental exposure to inorganic arsenic and to urinary arsenic levels. The urinary arsenic levels may be excessively high when the worker has eaten seafood within 48 hr of the collection of urine samples (29).

NIOSH, in the criteria document on inorganic mercury (30), also pointed out several other pitfalls in attempting to correlate clinical data and environmental levels, including a lack of substantive data to support a relationship between mercury excretion with signs and symptoms, and environmental levels. NIOSH based these conclusions on the belief that it is impossible to establish a level at which no signs and symptoms are observed because the signs and symptoms of mercury exposure, which are also prevalent in the general population, are nonspecific. In addition, it was pointed out that the validity of sampling and analytical methods for environmental monitoring was open to question. The use of biologic sampling for mercury has resulted in a wide variance in correlation factors with environmental and clinical results. As a consequence, the use of a biologic standard for mercury would be meaningless for assessment of the general worker populations. Biologic monitoring of several biologic media (urine, blood, breath, and hair being the most common: see also Chapter 7) has been used with a variety of results for other chemicals, including organic and inorganic lead, benzene, arsenic, fluorides, toluene, carbon monoxide, tetraethyl lead, DDT, organic phosphate insecticides, halogenated hydrocarbons, hydrocarbons, aromatic nitrogen compounds, ketones, aldehydes, and alcohols. For additional details on the interpretation of these biological tests in relation to environmental exposures, see "Biological Monitoring for Industrial Chemical Exposure Control" (31).

The correlation of biologic analysis and environmental data was considered to be so significant for carbon monoxide that NIOSH (32), in its recommendation for a standard, based the proposed standard on the correlation between the carboxyhemoglobin levels and a time-weighted average exposure to carbon monoxide for an 8 hr period (33).

The assessment of the total life experience of the worker in conjunction with the clinical picture and the environmental exposure information is necessary, particularly when assessing a few clinical cases—for example, atypical clinical effects resulting from physical effort on the job may be the expected results if it is known that the employee has a cardiac problem or is working in a hot environment.

Interpretation of epidemiologic-related environmental exposure data may not follow the predicted pattern. This may be caused by only a slight excess in relative risk, use of an inappropriate control group, lack of measurement of other possible chemical, physical, or emotional stresses, and other conditions (smoking habits, nutrition, disease, family history, marital status, economic status, age, sex, race, work schedule, and off-the-job stress, etc.) that may affect the response of the workers in the cohort under investigation. The increase in the number of workers included in the sample may help reduce the effect of the variability of some of these parameters, but the increase in the number of workers studied may result in inability to investigate in depth all possible causes of the variation from the expected pattern of correlation. Inconsistent results from different epidemiologic methods are not uncommon, such as negative results between the exposed and nonexposed groups and positive results between first-time exposed and disease development. This may suggest the need for additional analysis and possibly a follow-up study to establish a dose-response relationship.

Nonenvironmental parameters can have a major impact on the results and conclusions drawn from any study. The effect of some of these nonenvironmental conditions have been measured in epidemiologic studies by using wives of the workers for controls (34). Kalačić (35), in an attempt to determine the effect of domestic and nonoccupational factors on cement workers, studied workers in nondusty trades and wives of the cement workers and the nonexposed workers, who were also used as controls. He concluded that there was no significant difference between the two groups of wives and suggested that domestic and nonoccupational factors probably were not significant in the worker exposure. He was not able to draw any conclusions on the effect of smoking habits, probably because of the lack of similarity in smoking habits between the workers and the wives. Higgins (36), in the study of coal miners in England, used wives as a control and by comparing differences between the workers and the wives was able to determine a difference in the incidence of bronchitis between those who worked and their wives.

Changes in environmental measurement techniques and diagnostic procedures, with the passage of time, have made it difficult in some cases and impossible in others to develop any meaningful retrospective studies, and these changes have a direct impact on the establishment of any predictable pattern between chronic occupational diseases and environmental exposure measurements.

The validity of an observed correlation between environmental exposure and clinical data may be open to question until the work has been replicated, if the observed relationship is not the one expected or has not been previously reported in the literature.

5.2 Is There a Rational Basis for the Aberration of the Predictable Relationships?

The search for the cause of deviations from the expected relationships between environmental exposure and clinical change should start with a reevaluation of the environmental sampling strategies discussed in Section 2 and of the clinical examination objectives and procedures. Special consideration should be given to the parameters of the evaluation strategies that were designed to hold constant as many as possible of the parameters that could affect the measurement of the occupational exposure and to determine whether the methods employed did "in fact" hold constant these interfering factors (e.g., were the analytical results correct, and was there a possibility of an additional exposure to some other chemical contaminant?). Any new information that may have become available from other studies or sources should be assimilated.

A number of specific parameters discussed in the following sections should be dealt with in the reevaluation of the results of an occupational health study if the expected correlation of environmental and clinical data is not realized. If careful consideration is not

EVALUATION OF EXPOSURE TO CHEMICAL AGENTS

given to these parameters as well as others that are specific for the hazard being evaluated, the relationship between exposure and effect cannot be well defined.

5.2.1 Unknown Past Exposures and Medical Histories

In many cases the medical histories of employees are of questionable value, particularly when medical information has been obtained from several different employers. In some large industrial concerns and in many smaller ones, no preplacement medical examinations or periodic follow-ups have been given, or such exams have been given only over the last few years. In these cases it may not be possible to make a meaningful correlation of exposure to clinical evidence of chronic occupational disease.

In other circumstances involving unknown past exposure records, unknown work histories, and retired or deceased employees, or persons who have moved to other work (i.e., cannot be followed), these employees may be lost to a study, adversely affecting the validity of the correlation of the clinical and environmental results. The presence of workers or former workers with poor work or medical histories that lack such data as age, sex, race, and family medical history, can also adversely affect the correlation of exposure and clinical data. This lack of data may be the result of poor recordkeeping and/or failure to recognize what information may be of value in the future. For example, the history of smoking habits today is a very important and necessary piece of information for many studies in which the route of entry is by inhalation and chronic results are expected. Detailed smoking histories were obtained in the past as part of only a few medical histories, and the information that may be available is usually too superficial to meet present requirements.

The collection of environmental data and the maintenance of environmental records was even less well organized in the past than was the maintenance of medical records. Environmental data usually were kept a few years and then destroyed, only occasionally were the data kept with the employee's medical file. These recordkeeping practices were justified at that time by acknowledging that the constant changes in medical procedures and environmental sampling techniques reduced the value of old records to the questionable level at best. Two examples of changes in procedures that improved the practice of industrial hygiene and occupational medicine but reduced the value of existing records are found in the change in the collection and analysis for asbestos (*a*) from counts for dust particles to (*b*) collection of fibers and counting fibers that have a certain length-to-diameter ratio. Even though there have been several attempts to correlate the two procedures, insufficient data have been collected to be convincing (13, 37). The change in the radiologic classification system for pneumoconioses and the improvement in the techniques for taking and developing X-rays have had a similar effect on the evaluation of chest films.

Records of both medical and environmental data should be retained even after changes in sampling and/or medical procedures, since these records may have at least some value in future years for research purposes and will certainly be of value for compliance and legal use.

5.2.2 Off-the-Job Stresses

Off-the-job stresses must be investigated as possible sources of deviation from the expected relationship between environmental data and clinical experience and can be considered in three basic classifications: additional exposure off the job to the same chemicals used on the job, exposure to other hazardous chemicals, which may complicate the exposure on the job, and off-the-job stresses that are not related to chemical exposure but may be the result of physical stress or infectious disease. An example of continuation of exposure while off the job is provided by a worker who upon medical examination was found to have been overexposed to uranium (radon daughters), presumably from the work of removing uranium paint from aircraft instruments. The work was done in a completely enclosed glove box. Removal from the job did not cause a change in the worker's condition, and investigation of off-the-job exposure revealed that the employee painted luminous street markers as a private enterprise from home.

In addition, a number of cases of toxicologic effects in the community or in the worker's family suggest that continued exposures may occur to workers while away from the the workplace. In one reported case (38), a mother of a beryllium worker died of beryllium disease. The only known exposure had been at home, from the beryllium dust that had been brought in on the clothes of the worker. As early as 1966 a total of 60 neighborhood cases of beryllium disease had been reported to the Beryllium Registry (39). Similar family and community cases have been reported from asbestos exposure (40).

Community exposures must also be considered. For example, high levels of lead are known to be in both community air and water supplies in some metropolitan areas.

The complicating effect of alcohol on workers exposed to trichloroethylene is a classic example of the effect of off-the-job stress in addition to workplace exposures. Exposures to home cleaning solvents, degreasing agents, photographic chemicals, paint pigments and solvents, and isocyanates are all recognized as possible sources of exposure at home or in the workshop.

Infectious and parasitic diseases, including anthrax and brucellosis, have long been associated with agricultrual workers. These conditions could also contribute to a worker's combined effect of at-home and on-the-job exposures if the worker lives on a farm. It is also possible that poor nutrition, substandard housing, and poor economic conditions can reduce resistance to workplace exposures and result in the development of other diseases. Air pollution, physical stress, heat, and altitude can have some effect on workers who may have progressive respiratory and circulatory difficulties brought about by occupational exposures. The major off-the-job, and in some cases on-the-job, stress is the smoking habits of the worker.

5.2.3 Work Schedule

Little research has been done in this country relating the effects of the work schedule or shift work to occupational disease; however the physiological effects caused by shift work can certainly influence the health of the worker. El Batawi and Noweir (41) indicated that the stress of night work adversely affected the blood pressure of garage workers. Such stress could result in peptic ulcer, fatigue, nervousness, irritation, and insomnia. In another study (42) it was reported that night workers may have a higher incidence of peptic ulcer than those working on the day shift. Peptic ulcers were also found in (night) shift workers in Egypt, as well as chronic gastritis. The authors suggested that this may be due to the stress of shift work, as well as nutritional factors (41).

5.2.4 Additive, Synergistic, and Antagonistic Effects

The TLVs for chemical substances (28) contain in Appendix C a formula for estimating the combined or additive effect of two or more chemical hazards. This formula treats the combined exposure as if the maximum effect of the combination could not be larger (no synergistic effect) than the maximum of each chemical. This reduces the exposure proportional to the concentration and the TLV of each compound. Even assuming that the mechanism of action is the same, this approach can underestimate the effect of combined exposure. In addition to the formula for mixtures, ACGIH suggests (28), (Appendix C) the use of the formula for independent biologic effects when the toxicologic response is different. In these cases the formula assumes that there is no interactive effect and that it is possible to be exposed up to the TLV for each compound without exceeding the TLV for the combined exposure.

It is unfortunate that it is necessary to use such an approach in the case of exposures to two or more chemicals when it is not known whether the action in the body is additive (as suggested by the mixture formula), synergistic (which would not give sufficient protection), or antagonistic (affords more protection than is necessary). There is a synergistic effect from the combination of cigarette smoking and asbestos inhalation in the production of bronchial carcinoma (26), and there is increasing evidence of the synergistic effect of cigarette smoking in other worker populations, such as uranium miners. There is some evidence of a synergistic reaction between arsenic and sulfur dioxide (43). Laskin (44) reported a synergistic effect to rats inhaling benzo[a]pyrene and sulfur dioxide in combination in the production of bronchial mucosal changes and tumors of bronchogenic origin. Both these chemicals are present in some industrial operations (e.g., in the production of coke from coal). Antagonistic effects result in a reduction in the biologic response over what would have been normally expected, such as that which occurs when the combined fumes of nitrogen oxides and iron oxides are found, as in some welding operations (45).

5.2.5 Abnormal Temperature or Pressure

Individuals who must work in elevated temperatures usually become acclimatized or move to other jobs that are less demanding. This natural selection reduces the effects that would be expected from increased temperature; however the effects of temperature and heavy work may produce heart failure in workers with cardiac problems. Arteriosclerosis and enlarged hearts have been found in excess of that expected in workers in steel and glass industries. Organic heart defects are reported in another hot industry, foundries.

What the combined effect of chemical exposures may have been is not known. In all probability the effect from excessive heat is a more significant contributing factor to the inability of a worker (who already has some medical complications) to resist any additional insult from a chemical or physical stress. Linemen, loggers, surveyors, and construction workers are also exposed to a variety of plant and wood hazards; the most common of these are poison oak and ivy, in addition to exposure to chemical hazards and heat stress.

Workers exposed to cold temperatures may work out of doors and not be exposed to high concentrations of chemical agents. Some workers, such as packing house, construction, and highway workers, and mail carriers and firefighters, are likely to be exposed during a workday to a combination of heat and cold that may adversely affect the ability to resist infectious diseases. Other workers, such as police officers and highway construction and tunnel workers, may be exposed in addition to chemical agents such as carbon monoxide, and exhaust gases from internal combustion engines. Liquified gas and oil field workers may be exposed to hydrogen sulfide as well as to cold. Little is known about the effect of cold in combination with chemical agents. It is known, however, that cold as well as hot environments result in increased frequency of minor industrial accidents. This suggests that under adverse temperature conditions the employee does not follow normal expected safety rules and in all probability may not be as conscientious in using personal protective devices and procedures as under more normal work conditions.

The primary concern with increased pressures has been associated with tunnelers, caisson workers, and divers. Recently compression and decompression problems have been intensified by the development of offshore drilling for crude oil, which is requiring divers to go to greater depths. There has been increased concern about exposure to compressed air because of the growing number of skin divers (off-the-job stress) and tunnel workers. In the case of compressed air, concern is not only of the consequences of breathing compressed air for long periods but also the possibility of contamination of the breathing air. The possibility of contaminated compressed air requires controls of the air similar to those required for use with air-supplied respirators. The increased adverse effects of both carbon dioxide and oxygen under pressure are well known.

5.2.6 Extreme Job Energy Requirements

Energy requirements may have a very significant effect on the expected relationship between environmental data and the clinical response observed. Extreme or excessive energy requirements can result from work periods longer than the normal 8 hr workday, and higher than normal physical activity during the regular workday. The

EVALUATION OF EXPOSURE TO CHEMICAL AGENTS

expected clinical response is normally based on a time-weighted average exposure over an 8 hr workday. In the preface to the TLVs for chemical substances (28), the ACGIH defines the conditions that are to be used for the evaluation of workroom exposures by a time-weighted average as being for a normal 8 hr workday over a 40 hr work week. This defines the conditions by which comparisons are made between environmental data and clinical response.

The concentration of the toxic agent that reaches the site of the biologic response in the body can be increased by extending the time of exposure, or by increasing the breathing rate by increasing the physical activity (energy demands) of the job within the 8 hr period. The ventilation rate of the lungs increases about twentyfold from measurements taken at rest to those taken during heavy exercise. As a consequence, it is expected that heavy work demands will increase the amount of toxic material that is inhaled and deposited in the body with a resultant increased toxic action. Although work conditions for the most part do not normally require heavy energy demands, when such demands are made even for short periods, they could result in a departure from predictable patterns of biologic response if the energy demands have not been taken into consideration in the initial evaluation.

5.2.7 Altitude

The effects of high altitude flying are minimized by the use of pressurized aircraft and supplementary oxygen. The effect of high altitude on man, when it is the living environment, is one to which the worker is subjected not only on the job, but at home as well. There are physiological differences between populations native to high altitudes and those that recently have moved to the high areas. Workers who are native to the mountainous area may have working capacity similar to those who work at sea level, whereas those who have been acclimatized may hyperventilate during exertion or exercise. This difference in reaction to increased altitude may result in some inconsistencies between the observed and expected response to a specific toxic chemical, if the results are considered together for both the native and acclimatized workers. The hyperventilation in the acclimatized worker will result in the same type of reaction that would be expected from a heavy energy demand job, with additional effects due to lowering of carbon dioxide tension.

Decreased pressure also requires calibration of field sampling equipment at the altitude at which environmental sampling is done.

5.2.8 Need for New Data Assessment

The data and information necessary to assess properly the effect of the specific parameters that can cause deviations from the predicted relationship, as well as others such as differences in sex and race, are not readily available, and much more research must be done. Many of these conditions, such as the altitude effect, will affect the

Table 8.1 Organizations Recommending Occupational Health Limits

Source	Limits		Applicable Daily Time Period	Comment
American Conference of Governmental Industrial Hygienists (ACGIH) P.O. Box 1937 Cincinnati, Ohio 45201	1.	Threshold limit value (TLV)	8-hr/day time-weighted	For both chemical and physical agents; limits are published annually
	2.	Ceiling value	15-min	
American National Standards Institute (ANSI) Z.37 Series[a] 1430 Broadway New York, New York 10018	1.	Time-weighted average	8-hr	Published irregularly
	2.	Ceiling value	Related to 8-hr average	
	3.	Maximum for peaks above ceiling	Time specified	
Pennsylvania Department of Health P.O. Box 90 Harrisburg, Pa. 17120	1.	Short-term limits	5, 15, and 30 min	Last published April 24, 1970

Organization		Type of standard	Averaging time	Remarks
National Academy of Science–National Research Council (NAS–NRC) 2101 Constitution Avenue NW Washington, D.C. 20418	1.	Short-term public limits (STPL)	10, 30, 60 min and 4–5 hr/day, 3–4 days/month	Published irregularly
	2.	Public emergency limits (PEL)	10, 30, 60 min	
American Industrial Hygiene Association (AIHA) 475 Wolf Ledges Pkwy Akron, Ohio 44313	1.	Emergency exposure limits	5, 15, 30, and 60 min	Published irregularly
	2.	Hygienic guides		
		a. Maximal atmospheric concentration	8 hr	
		b. Short exposure tolerance	Time specified	
		c. Atmospheric Concentration immediately hazardous to health	Time specified	
National Institute for Occupational Safety and Health 5600 Fishers Lane Rockville, Maryland 20852	1.	Time-weighted average	8 hr/day	Published as guidance for OSHA
	2.	Short-term	Time-specified	

Source: R. S. Brief, *Basic Industrial Hygiene—A Training Manual*, Exxon Corp., New York, 1975. Reprinted by permission.

[a] The (ANSI) Z.37 Committee is no longer developing standards.

evaluation of relatively few working environments; however others such as synergism apply to a much wider segment of the working population, and all can cause the development of erroneous information.

6 WORKROOM ENVIRONMENTAL EXPOSURE LIMIT

Any procedure or evaluation of worker exposure to noxious chemicals carries the assumption that at some level of exposure there will be some type of biologic effect, and it is necessary at some level of exposure to initiate engineering or administrative control of that exposure.

The development of such exposure guides had been based on their use by a competent industrial hygienist familiar with daily exposure levels, an understanding of the basis of the exposure limit, and a subjective evaluation of all factors to determine the appropriate control action necessary.

A number of exposure limits that do not have the status of law have been used for many years in the United States; these include the Hygienic Guide Series of the AIHA, the standards of the American National Standards Institute (ANSI), Z.37 Committee, the TLVs for Chemical Substances of the ACGIH, the standards of the National Academy of Sciences–National Research Council (NAS–NRC) and NIOSH, as well as the legal standards developed by OSHA, MSHA, and some state authorities where the basis for regulating is not derived from federal law, as in Pennsylvania. These limits include time-weighted averages, ceiling values, maximum peaks above ceiling, short-term limits, and emergency exposure limits. Richard S. Brief, in *Basic Industrial Hygiene—A Training Manual*, developed a table covering most of these standards (Table 8.1).

Both ACGIH TLVs and the ANSI Z.37 guidelines have been adopted, at least in part, as OSHA standards (3). Neither of these exposure guides will fill the requirement for standards as outlined in the Occupational Safety and Health Act, but they were promulgated as OSHA standards either as national consensus standards (ANSI) or as established federal standards (ACGIH TLVs) because they had already been adopted as federal standards under the Walsh-Haely Act. In cases where there were both federal standards and national consensus standards, the Secretary of Labor was to promulgate the standard that assured the greatest protection to the employee. The ANSI standards as consensus standards were adopted in preference to the TLVs of ACGIH when there was a choice to be made.

The largest list of environmental exposure limits has been developed by the ACGIH's Chemical Agent Threshold Limit Committee, then adopted by the membership of that organization. The ACGIH publishes a new list of TLVs each year, including revisions to already existing TLVs, and adds chemical compounds when it has determined that limits are needed because of exposures in the workplace. It has been estimated that approximately 95 percent of the exposures in industry in the United States is covered by an ACGIH TLV. Any new TLVs or any changes in the limits are publicly announced to the ACGIH membership and are placed on a tentative list of TLVs to allow interested professionals to present additional information and comment on the proposed limit.

The ACGIH TLVs have been adopted in other countries, even though the organization has strongly opposed the use of its TLVs in countries where working conditions differ from those in the United States (28). ACGIH also pointed out that the TLVs should be interpreted only by a person trained in industrial hygiene; and the TLVs should not be used as a relative index of hazard, or toxicity, for the evaluation or control of community air pollutants, for evaluating toxic effect, for continuous uninterrupted exposures or for other extended periods, or to verify or disprove the existence of a physical condition or disease (28).

This wide acceptance of ACGIH TLVs has resulted in these limits being used as the basic working standard for the measurement, evaluation, and control of workplace hazards for many years. The important differences between these suggested guides and the legal standards being developed by OSHA involve the ability of a professional committee of ACGIH to act in a reasonably short time to subjectively evaluate data and to propose TLVs, and for the industrial hygienist, who uses the TLVs, to exercise professional judgment in evaluating a work hazard.

The preface to the TLVs for Chemical Substances in workroom air should be read each year for the minor changes that occur in the preface from year to year. The following quotation from the 1977 TLV list (28) of chemical substances gives a good indication of how the TLV Committee believes that the TLVs should be used, and their intent.

Threshold Limit Values refer to airborne concentrations of substances and represent conditions under which it is believed that nearly all workers may be repeatedly exposed day after day without adverse effect. Because of wide variation in individual susceptibility, however, a small percentage of workers may experience discomfort from some substances at concentrations at or below the Threshold Limit; a small percentage may be affected more seriously by aggrevation of a preexisting condition or by development of an occupational illness.

A separate ACGIH publication, "The Documentation of Threshold Limit Values for Substances in Workroom Air," gives the rationale for the TLV and quotes the primary references used to support the TLV. This publication and supplements to the documentation are published at irregular intervals; in the interim periods the annual proceedings of ACGIH contain the recommendations and discussions of the TLV by the membership of ACGIH at the time of their adoption at the annual meeting. The basis for the TLVs may be epidemiologic studies in industrial, clinical experience, or animal studies.

REFERENCES

1. Occupational Safety & Health Act, Public Law 91-596 S2193, 91st Congress, December 29, 1970, Superintendent of Documents, Government Printing Office, Washington, D.C.

2. A. D. Hosey and H. L. Kusnetz, "General Principles in Evaluating the Occupational Environment," in: *The Industrial Environment—Its Evaluation and Control,* 2nd ed., C. H. Powell and A. D. Hosey, Eds., Public Health Service Publication 614, Superintendent of Documents, Government Printing Office, Washington, D.C., 1965, p. B-1-1.
3. Department of Labor, Occupational Safety and Health Administration, "Occupational Safety and Health Standards," *Federal Register,* Vol. 39, No. 125, Part II, Subpart G-1910.93, June 27, 1974, p. 23540.
4. (American Conference of Governmental Industrial Hygienists,) "Threshold Limit Values of Airborne Contaminants for 1968, Recommended and Intended Values," ACGIH, Cincinnati.
5. N. A. Leidel and K. A. Bush, "Statistical Methods for the Determination of Noncompliance with Occupational Health Standards," Department of Health, Education, and Welfare, Publication (NIOSH) 75-159, Superintendent of Documents, Government Printing Office, Washington, D.C., 1975.
6. A. E. Russell, R. H. Britten, L. R. Thompson, and J. J. Bloomfield, "The Health of the Workers in Dusty trades. II. Exposure to Siliceous Dust (Granite Industry)," Public Health Bulletin 187, Superintendent of Documents, Government Printing Office, Washington, D.C., 1929.
7. G. P. Theriault, W. A. Burgess, L. J. DiBerardinis, and J. M. Peters, *Arch. Environ. Health,* **28,** 12 (1974).
8. S. J. Reno, H. B. Ashe, and B. T. H. Levadie, "A Comparison of Count and Respirable Mass Dust Sampling Techniques in the Granite Industry," presented at the Annual American Industrial Health Conference, Pittsburgh, 1966.
9. N. A. Talvitie, *Anal. Chem.,* **23,** 623 (1951).
10. N. A. Talvitie and F. Hyslop, *Am. Ind. Hyg. Assoc. J.,* **19,** 54 (1958).
11. N. A. Talvitie, *Am. Ind. Hyg. Assoc. J.,* **25,** 169 (1964).
12. S. Milham, "Cancer Mortality Patterns Associated with Exposure to Metals," in *Occupational Carcinogenesis, Ann. N.Y. Acad. Sci.,* **271,** 243 (1976).
13. H. E. Ayer, J. R. Lynch, and J. H. Fanney, *Ann. N.Y. Acad. Sci.,* **132,** 274 (1965).
14. E. V. Olmstead, *AMA Arch. Indust. Health,* **21,** 525 (1960).
15. Department of Labor, Occupational Safety and Health Administration, *Federal Register,* Vol. 42, No. 192, Part VI, [29 CFR Part 1990], October 4, 1977, p. 54149.
16. R. A. Kehoe, "The Metabolism of Lead in Man in Health and Disease," *The Harben Lectures,* 1960, Reprinted from J. Royal Inst. of Public Health and Hyg., (1961), by McCorquodale and Co. Ltd., London.
17. J. J. Bloomfield and W. Blum, *Public Health Rep.,* **43,** 2330 (1928).
18. American Conference of Governmental Industrial Hygienists, "Documentation of the Threshold Limit Values for Substances in the Workroom Air," 3rd ed., ACGIH, Cincinnati, 1971.
19. R. J. Vernon and R. K. Ferguson, *Arch. Environ. Health,* **18,** 894 (1969).
20. A. Ahlmark and S. Forssman, *Arch. Ind. Hyg. Occup. Med.,* **3,** 386 (1951).
21. American Conference of Governmental Industrial Hygienists, *Industrial Ventilation—A Manual of Recommended Practice,* 14th ed., ACGIH, Cincinnati, 1971.
22. "Criteria for a Recommended Standard . . . Occupational Exposure to Hydrogen Sulfide," Department of Health, Education, and Welfare Publication (NIOSH) 77-158, Superintendent of Documents, Government Printing Office, Washington, D.C., 1977, p. 3.
23. B. Weiss, and E. A. Boettner, *Arch. Environ. Health,* **14,** 304 (1967).
24. "Criteria for a Recommended Standard . . . Occupational Exposure to Inorganic Lead," Department of Health, Education, and Welfare Publication (HSM) 73-11010, Superintendent of Documents, Government Printing Office, Washington, D.C., 1972, p. VII-1.

25. "Criteria for a Recommended Standard . . . Occupational Exposure to Chromic Acid," Department of Health, Education, and Welfare Publication (HSM) 73-11021, Superintendent of Documents, Government Printing Office, Washington, D.C., 1973, p. 64.
26. "Criteria for a Recommended Standard . . . Occupational Exposure to Asbestos," Department of Health, Education, and Welfare Publication (HSM) 72-10267, Superintendent of Documents, Government Printing Office, Washington, D.C., 1973.
27. "Criteria for a Recommended Standard . . . Occupational Exposure to Crystalline Silica", Department of Health, Education, and Welfare Publication (NIOSH) 75-120, Superintendent of Documents, Government Printing Office, Washington, D.C., 1974, p. 101.
28. American Conference of Governmental Industrial Hygienists, "TLV's: Threshold Limit Values for Chemical Substances and Physical Agents in the Workroom Environment with Intended Changes for 1977," ACGIH, Cincinnati.
29. H. H. Schrenk and L. Schreibeis, Jr., *Am. Ind. Hyg. Assoc. J.,* **19,** 225 (1958).
30. "Criteria for a Recommended Standard . . . Occupational Exposure to Inorganic Mercury," Department of Health, Education, and Welfare Publication (HSM) 73-11024, Superintendent of Documents, Government Printing Office, Washington, D.C., 1973, pp. 75, 79.
31. A. L. Linch, *Biological Monitoring for Industrial Chemical Exposure Control,* CRC Press, Cleveland, 1974.
32. "Criteria for a Recommended Standard . . . Occupational Exposure to Carbon Monoxide," Department of Health, Education, and Welfare Publication (HSM) 73-11000, Superintendent of Documents, Government Printing Office, Washington, D.C., 1973, p. V-5.
33. R. F. Coburn, R. E. Forster, and P. B. Kane, *J. Clin. Invest.,* **44,** 1899 (1965).
34. P. E. Enterline, *Public Health Rep.,* **79,** 973 (1964).
35. I. Kalačić, *Arch. Environ. Health,* **26,** 84 (1973).
36. I. T. T. Higgins, "An Approach to the Problem of Bronchitis in Industry: Studies in Agricultural, Mining, and Foundry Communities," in: *Industrial Pulmonary Diseases,* E. King and C. M. Fletcher, Eds., Little, Brown, Boston, 1960.
37. W. C. Cooper and J. L. Balzer, "Evaluation and Control of Asbestos Exposures in the Insulating Trade," *Proceedings of a Working Conference on the Biological Effects of Asbestos,* Dresden, 1968.
38. L. B. Tepper, H. L. Hardy, and R. I. Chamberlin, *Toxicity of Beryllium Compounds,* Elsevier, New York, 1961.
39. H. L. Hardy, E. W. Rabe, and S. Lorch, *J. Occup. Med.,* **9,** 271 (1967).
40. P. Champion, *Am. Rev. Respir. Dis.,* **103,** 821 (1971).
41. M. A. El Batawi and M. H. Noweir, *Indr. Health* (Kawasaki) **4,** 1 (1966).
42. A. Aanonsen, *Ind. Med. Surg.,* **28,** 422 (1959).
43. "Criteria for a Recommended Standard . . . Occupational Exposure to Inorganic Arsenic," Department of Health, Education, and Welfare Publication (NIOSH) 75-149, Superintendent of Documents, Government Printing Office, Washington, D.C., 1975.
44. S. Laskin, M. Kuschner, and R. T. Drew, in: *Inhalation Carcinogenesis,* M. G. Hanna, Jr., P. Nettesheim, and J. R. Gilbert, Eds., Atomic Energy Commission, Washington, D.C., 1970, p. 321.
45. H. E. Stokinger, in: *Occupational Diseases,* W. M. Gafafer, Ed., Public Health Service Publication 1097, Superintendent of Documents, Government Printing Office, Washington, D.C., 1964, p. 15.

CHAPTER NINE

Evaluation of Exposure to Ionizing Radiation

ROBERT G. THOMAS, Ph.D., and
RANDI L. THOMAS

1 INTRODUCTION

Radiation has been known as a toxic agent for a relatively short time (approximately 80 years). It has received considerable attention in recent decades, however, primarily as a result of military applications during World War II. Compared to chemically toxic agents, the properties of ionizing radiation generally enable its detection at much lower levels and with much greater accuracy. This condition has contributed greatly to the feeling that radiation has great toxic potential, but in many (or perhaps most) cases this is mainly because its presence can be detected at these low levels. It appears that much of the concern over health hazards due to radiation is the direct result of this.

There are two distinct possibilities for human radiation exposure: irradiation from sources external to the body, and irradiation from sources deposited internally that have entered the body by various routes, such as inhalation, and ingestion, and on or through the skin. This chapter devotes considerably more emphasis to the internal source of potential radiation health problems, primarily because accumulation through the internal routes is often less controllable. Also measurement of radiation from external sources is much more reliably related to body exposure than is for instance, the reconstituted air concentration of a radionuclide to the calculated long-term dosage that may be received by an internal organ. There are many difficulties associated with the latter estimates—determining such factors as the particle size, the amount deposited in various sections of the respiratory tract, solubility in body fluids, elemental position in the periodic table—and naturally, the differential metabolism of foreign materials by exposed individuals. The latter factor may involve such basic parameters as sex differences, and body and organ weight. Because the degree of injury is probably related to the concentration of a radioactive material in an organ, it is easy to see how organ weight could become such an important factor.

1.1 Types of Radiation Exposure

1.1.1 External Radiation

External radiation, in its less exotic forms, is generally considered to be either from an electromagnetic (wavelike) or a neutron source. There are other sources of external ionizing radiation, but these rarely lead to industrial exposure of the worker. With electromagnetic radiation (gamma rays, X-rays) ionization in a medium may be caused by one of three primary processes, namely, the photoelectric effect, Compton scattering, and pair production. These three types of ionization are illustrated in elementary textbooks on nuclear physics.

The photoelectric effect generally prevails when tissues are affected at photon energies less than 100 keV. In this process an electron is ejected from an atom in the medium, and it subsequently becomes the major source of further ionization within the medium. Ionization by electromagnetic radiation is somewhat "inefficient"; therefore pathways may be very long compared to the length of ionization path of the secondary electron released. Hence ionization events may not occur in close proximity to each other, and effects on tissues may be diffuse. As the secondary electron traverses the tissue, it promulgates additional ionizing electrons known as delta rays. These are discussed in paragraphs to follow. In the photoelectric ionization process all the incident energy is imparted to the ejected electron, which explains the effectiveness of this process at the lower energies.

Between the low (about 100 keV) and very high (about 10 MeV) energy regions, the ionizing process known as Compton scattering prevails. Under these conditions the incident photon is partially degraded by interacting with an orbital electron in an atom of the medium being traversed. The energy remaining in the photon carries it further through the medium, to continue interaction with other atoms in various molecules. The ejected electron (delta ray) continues with its imparted energy (the fraction obtained from the incident photon) to ionize other atoms near to the original event. This "clustering" of events in the immediate vicinity of the first ionization is due to the relatively heavy ionizing density of the particulate radiation (ejected electron), described in Section 1.1.2. Because the principal range of energies for Compton scattering spans most of the common external radiation sources, this ionizing process is most prevalent in radiobiological effects.

Pair production is confined to high energy photons (>1.2 MeV). This event has little place in radiobiological reactions because of the energies involved and the atomic structure required. In this process the photon energy, in the proximity of a highly charged nucleus, is transformed into mass because of the strong nuclear electromagnetic forces, thus forming one negatively charged electron mass and one that is positively charged. Each of these newly formed particles is capable of ionizing atoms to an extent that is dependent on their energies, as described.

Neutrons, uncharged particles in the simplest sense, damage the traversed medium in a different manner. The neutral nature of neutrons allows them to have large penetrating distances. Because they are neutral, they generally penetrate the outer electron clouds of atoms and in essence, strike the nucleus. This is due to the relative size of the target. If the neutron is of sufficiently high energy (fast), the nucleus is dislodged and actually is driven some distance through the surrounding tissue, causing damage as it goes. The nucleus ultimately picks up enough electrons to once again become its original "self" as it comes to rest. The elements of greatest abundance in soft body tissues are carbon, hydrogen, oxygen, and nitrogen. The abundance of hydrogen nuclei make this element the most vulnerable target for fast neutron interaction, therefore hydrogen will have a much greater absorbed proportion of the incident neutron than will the less abundant larger atoms. Slower (low energy) neutrons may be captured upon encounter with a nucleus. This can result in an unstable condition in which the nucleus will release energy in many forms, thus becoming a primary source of ionizing energy. This generally limits the radiobiological action to the proximity of the initial event.

Protons and other charged particles are also a source of external radiation and their ionization is similar to alpha and beta particles, as described below.

1.1.2 Internal Radiation

Internal radiation sources, as far as industrial exposure is concerned, are generally comprised of alpha or beta particle emitters. Beta particles, electrons by charge and weight, may interact with orbital electrons of atoms within molecules, or with the nucleus. When the beta particle approaches an orbital electron, energy is imparted, the orbital electron may be ejected as a delta ray, and the incident beta particle will continue at a lowered energy and with an altered direction. This procession by the incident electron will continue until the electron dies by capture somewhere in the medium (tissue). The delta rays (secondary electrons) will traverse the surrounding area and operate in a manner similar to that of the original primary incident electron. Because of the charge and the mass, compared to a photon (Section 1.1.1), the ionizing events associated with electrons are clustered and occur within a very small volume of tissue.

Alpha particles, being helium nuclei of 2+ charge and 4 mass units, are much more heavily ionizing than are electrons (beta particles). Thus their path is straighter because there is less deflection upon energy loss, and their ionizing events are confined to much smaller volumes of tissue. A 5 MeV alpha particle in soft tissue has a linear path length of approximately 40 μm. Since beta particles from the nucleus of radioactive atoms have a spectrum of energies, their path length is variable. However the range of the emitted beta particle is many times that of an alpha particle of the same incident energy. Emitted alpha particles are released with discrete energy(ies).

1.2 Radiation Measurement

1.2.1 Electromagnetic Radiation

Electromagnetic radiation represents the easiest form to measure where radiobiological exposures are concerned. It is usually uniform within the location of possible worker exposure, and in most cases one could probably argue that the energy characteristics from a relatively stable source, whether spectral or quite narrow, would fall within a consistent range. This may be speculated because of the more common nature of the source of such "leakage" radiation. If the electromagnetic radiation is from a source of gamma emission, such as from ^{60}Co, the energy distribution is consistent, making interpretation of any measurements, even of a gross nature, much easier and probably more accurate.

Ionization in air has been accepted as the primary measuring standard on which all other (secondary) assessments are based. The amount of ionization in a specific volume of air under precise conditions of temperature and pressure serves to determine this primary standard. When electronic equilibrium is achieved (secondary electron influx into a specific volume of air is exactly equal to secondary electron efflux) within a given measuring device, the amount of current generated by the ionization is proportional to the amount of incoming energy of the photons. It is this incident energy that is directly proportional to the defined units used in radiation studies. The roentgen, which is a measure of this ionization under primary standard conditions, is defined as the quantity of X- or γ-radiation such that the associated corpuscular emission per 0.001293 gram of air produces, in air, under standard conditions of temperature and pressure, ions carrying 1 electrostatic unit of quantity of electricity of either sign. With this primary basis for measuring incident electromagnetic radiation, it is not difficult to conceive that many instruments have been devised to measure ionization in media that may be related to the primary standard.

In recent years crystalline and solid state detectors have been most successful for measuring external radiation dosage. The former will yield spectral energy data if sophistication is desired but may be used as constantly recording, full energy monitors without this sophistication. Electronic windows may be applied to the detection device to limit detection to a rather specific range of energies. The solid state variety works extremely well in sharp spectral resolution, but the need for such equipment in routine monitoring for the worker is often questionable. This is not to imply that solid state detectors are not widely used.

1.2.2 Neutrons

Neutrons are generally separated by energy into thermal, intermediate, and fast, as indicated in Section 1.1.1. The classical scheme of energies associated with these is less than

0.5 eV, 0.5 eV to 10 keV, and 10 keV to 10 MeV, respectively. Thermal neutrons are primarily degraded in matter by capture, and the capture cross section is inversely proportional to the velocity of the neutron. For instance, when hydrogen ($_1H^1$) captures a neutron ($_0n^1$) the result is a $_1H^2$ atom, plus a gamma-ray as excess energy. The intermediate neutrons fall into a range in which there are resonant peaks in the capture cross section, leading to a slowing down of the neutron in matter. This process is generally inversely proportional to the energy of the particle. Fast neutrons interact by scattering and may be elastic or inelastic, depending on whether the energies are toward the low or high end of the range, respectively. As mentioned in Section 1.1.1, the most important interaction in tissue is with hydrogen, each particle having essentially the same mass. Relativistic neutrons, another category, are not really of importance to this chapter.

Methods of detection and measurement of neutron fluxes have been based on neutron properties, as noted earlier. Calorimetry, the measurement of temperature rise due to ionization in a medium, is always very accurate, but it is difficult to measure the small changes in temperature achieved. Ionization depends on many factors, and the medium being ionized within the instrument should have very precise and determinately known qualities. The size (dimensions) and structure of the ionization chamber cavity are very sensitive parameters and must adhere to criteria that are related to the energy of the incoming neutron. For instance, an instrument designed to work reasonably well with thermal neutrons will not perform for fast neutrons. As with electromagnetic radiation, chemical means of detection and measurement are choices that give extreme accuracy. The problem in either solid (e.g., photographic) or liquid media is that the dose-response curve is not linear, and impurities play an important role in the resulting estimation. It is difficult to obtain a liquid chemical detection system that is pure enough to give no spurious neutron interactions. Perhaps one of the most reliable methods of detecting and measuring neutron radiation is through its secondary gamma rays following an event. Depending on the energy of the neutron, hence the reaction involved, secondary gamma rays will undoubtedly be emitted. Accurate determination of time of irradiation and magnitude of the gamma rays produced in the reaction of neutrons with the absorbing medium, allows calculation back to the incident flux.

All these general principles are operable in their own right. However since some require equipment that is not suitable for field work, bulkiness becomes an important drawback with most. In recent years the solid state detection systems have become popular and probably will ultimately replace all other means for personal detection. These include the thermoluminescent detectors used routinely in many laboratories for personnel monitoring. They have many advantages and are still under development for more precise detection characteristics.

1.2.3 Particles Other than Neutrons

Assessment of the impact from particle radiation is perhaps the least gratifying of all measurements that can be made in the case of accidental exposure to the worker, although recent advances are encouraging. The use of standard filter sampling at constant volume through small pore filters is simple and relatively inexpensive. The filter may be constantly monitored, the buildup of activity recorded, and when a certain predetermined radiation level is reached, the result is announced by an alarm. This represents a steady, consistent method of detection of leakage of radioactive materials from a given operation. This type of sampling is usable for both alpha and beta emitters. The difficulty inherent in this method of monitoring arises when an inhalation accident occurs.

Until recently little time and effort have been devoted to methods of determining solubility and particle size of the sample that is collected and is presumed to be representative of the atmosphere breathed by the worker. In the case of external radiation one has some reasonable chance of mocking up the exposure conditions, as was attempted in the Lockport exposures (1), and to closely estimate the worker dose. One generally does not have the opportunity to do this when accidental inhalation exposures to radioactive aerosols are involved; this has been the case with most accidental exposures to plutonium in the atomic energy industry (2, 3).

The routine use of continuous air monitors that will allow determination of particle size must be a compromise between sophistication and economics. Ideally, five- or seven-stage cascade impactors of the Anderson (4) or the Mercer (5) variety could be used in a constant monitoring mode so that if an accident occurred, the exposure atmosphere could be carefully determined retrospectively with regard to particle size distribution. However the constant problem of sophisticated samplers becoming clogged with room dust makes routine sampling sometimes impractical. More important with the use of this type of sampler, coupled with the particle size data, is the ability to do solubility measurements on each size fraction. The combination of these two determinations would enable a reasonably accurate estimation of the quantity deposited in the worker's lower respiratory tract, and the extent to which the deposited particles would be soluble in body fluids. The type of therapy to be applied after an accident could be much more judiciously determined if this kind of information were available at an early time postexposure. If the deposited particles were completely insoluble, one would consider lung lavage (6); but if they were soluble and entering the blood, chelation therapy (see Section 4.2.1) would become an option.

Perhaps the most practical air monitors are those that attempt a relatively crude separation of "respirable" and "nonrespirable" sizes (7–11). Solubility studies from such samplers can be made, and the results perhaps are equally satisfying as those obtained using the more complicated particle sizing instruments, given the spectrum of errors inherent in the measurements.

For certain operations a more specialized system has been devised that enables the detection of various daughter products, by types of emission or energies involved, as in the radium series (12). Collectors may be continuously monitored for various energies and types of emissions so that ratios of two different daughters can be determined, or a total sample can be captured and total daughter activity at a given time determined. Many types of sampler for the uranium mining industry have been developed along these lines with various methods of analysis. In all cases any counts above "background" may be instrumented to be indicative of an accidental release or an abnormal working environment. These systems have a clear application in the uranium mining industry.

1.3 Dosimetry

1.3.1 Electromagnetic Radiation

As implied, dosimetry with electromagnetic radiation in the human exposure case may be more straightforward than for exposure to airborne radionuclide particles or vapors, primarily because of the ability to mock up exposure conditions. Historically, the determination of radiation dose has proceeded through a series of changes since the discovery of X-rays by Roentgen in 1895. For quite some time the erythema dose was used to determine proper treatment in radiation therapy. In the 1920s it was recognized that ionization in air was the best approach to estimating dosage for therapeutic purposes. The roentgen was defined in 1928, and this ultimately led to widespread usage of the R unit, followed by the rep, the rem, and finally, in current use today, the rad. The rad is a measure of the energy deposited in any medium, and one rad is equivalent to the deposition of 100 ergs per gram of that medium.

Varous kinds of instrumentation have been used to define dosimetrically the energy arising from a gamma-or X-ray source. These include air ionization chambers, semiconductors, thermoluminescent devices, and other instruments using heat, light, and chemical changes as means to quantitate the ionizing events being produced. The most practical and common personnel dosimeters in use are the film badge (optical) and the thermoluminescent dosimeter. In any event, although no dosimetry appears simple and straightforward, electromagnetic radiation lends itself to giving the most consistent results where exposure to man is concerned.

1.3.2 Neutrons

The accurate determination of neutron dosage is fraught with problems that entail many factors. In general, neutron dosimetry has actually represented a measurement of accompanying protons, through the latter's interaction with some medium that is effective in giving a proportionate relationship to the incident neutron flux. Neutron dosimetry is obviously very energy dependent because the secondary "reactant" serving as the primary detector (of protons) is highly dependent on the type and efficiency of the reaction from which it was derived. Thus, neutron detectors, per se, are very energy dependent, and awareness of the source term is mandatory when detection instrumentation is selected.

The International Commission on Radiation Units and Measurements report of 1977 (13) on neutron dosimetry in biology and medicine separates dosimetric methods and instrumentation into several general types: (1) gaseous devices, (2) calorimeters, (3) solid state devices, (4) activation and fission methods, and (5) ferrous sulfate dosimeters. The details of these devices may be obtained from the reference (13), and this chapter does not duplicate that fine summary of the problems inherent in accurate determination of dosage from incident neutrons. The types of device and instrumentation described are quite broad, however, and deserve mentioning here by specific name, for those who may wish to pursue a given type of detection. The more specific devices mentioned are (1) ionization chambers, (2) proportional counters, (3) Geiger-Müller counters, (4) photographic emulsions, (5) thermoluminoscent devices, (6) scintillation devices, (7) semiconductors, and (8) nuclear track recorders. Some of the more general categories listed above (e.g., calorimetry) represent an overall methodology and do not require the individual specificity of a given type of instrumentation.

Neutron dosimetry is not as straightforward as one would desire, and it is certain that research in this field will continue for decades.

1.3.3 Particulate Radiation

The most common dose calculation for internal emitters comprised almost solely of alpha and beta radiation utilizes the "average dose" concept. In the classical sense, if the amount of radioactivity in an organ can be estimated by any means, it is assumed, for simplicity, that the radioactivity is distributed uniformly throughout that organ. This assumption is almost always in error with soft beta emitters and alpha particles, because of their short range in tissues (35 to 40 μm for the average alpha from heavy radioelements). However when one attempts to estimate dose-effect relationships on a microdistribution or "hot spot" premise, errors are equally forthcoming. For purposes of this chapter it is assumed that average organ or tissue dose is sufficient, with full knowledge that this concept may be somewhat in error.

To arrive at an average tissue dose to an organ from the estimated air concentration breathed, many steps are required. These steps are described below.

Particle Distribution. Airborne particles in real life are generally log normally distributed by mass and can be described by a mass median diameter—the particle size above and below which lies one-half of the total mass of the sample collected. The aerodynamic diameter of a particle size distribution is becoming a more popular descriptor than it was, say 20 years ago, and it incorporates the density of the material and a correction for very small particles (14). The size spread of the log normally distributed aerosol particles is defined by the geometric standard deviation, symbolized by σ_g. Thus a distribution of particles collected on a cascade impactor in an industrial situation would be described by $\bar{\mu} \pm \sigma_g$, where $\bar{\mu}$ is the median diameter defined in mass or aerodynamic terms and is commonly expressed in micrometers.

Respiratory Tract Deposition. It has been long known that site of particle deposition and retention in the respiratory tract during and following inhalation is dependent on particle size. Very simply, the larger particles deposit in the upper areas of the tract (nasopharyngeal region, trachea) and are quickly removed. The smaller particles traverse past the tracheal bifurcation and deposit all the way to the alveolar region

Deposition characteristics were described in detail in 1966 (15), and despite some modifications, this review remains the most comprehensive.

Respiratory Tract Retention. Respiratory tract retention parameters are also elaborated very thoroughly in the 1966 reference (15). Ciliary activity obviously accounts for a large, rapid clearance of deposited material from the upper respiratory passages. This material does not present a major problem from the viewpoint of toxicity unless the material is readily absorbed during passage through the gastrointestinal tract, or unless it is present in very large quantities (high doses). It is the material that lingers in the deeper (alveolar) spaces of the lung that seems to dictate problems such as carcinogenicity. If the deposited material is very soluble, it will enter the bloodstream through the lung or the gastrointestinal tract and will proceed by way of the circulatory system to deposit in the organ dictated (primarily) by its chemistry. If the material is insoluble, three alternatives present themselves: to remain in the alveoli indefinitely (life of individual), gradually to be transferred upward by ciliary action through an initial step of phagocytosis by macrophages, or to be transported to the pulmonary lymph nodes. Numerous mechanisms and kinetics describing these alternatives have been reported in the literature, and individual subjects of interest can be found in the proceedings of some recent symposia (16–19).

Uniform Organ Retention. Once the inhaled material has entered and has been deposited in the organ of choice, its kinetics of loss (retention function) can easily be described if a few facts are known. These include rate of loss from the tissue to blood, and the rate of reentry to the same types of tissue versus rate of excretion in urine or feces. This is rather simply described in terms of a general whole-organ concept, but the next section discusses the more complicated situation of highly localized dose, using bone as an example. The problem is that the rate constants described are seldom known accurately, particularly for man. If first-order kinetics prevail for loss from the organ, and the organ concentration of the radionuclide is relatively well estimated, average dose calculations are quite straightforward. Many such data are described for animals, but the extrapolation to humans is at best, satisfactory. An example of a dose calculation is shown below, using specific units for purposes of demonstration.

$$D \text{ (rads)} = C_0 \times k_1 \times \bar{E} \times k_2 \times k_3 \times \int_0^t R_t \qquad (1)$$

where C_0 = organ concentration of the radionuclide at $t = 0$ (μCi/g of tissue)
k_1 = conversion factor (2.22×10^6 dpm/μCi)
$\bar{E}$ = average energy (MeV/disintegration)
k_2 = conversion factor (1.62×10^{-6} MeV/erg)
k_3 = conversion factor (10^{-2} for the definition, 100 ergs/gram = 1.0 rad)
R_t = retention function in the organ of deposition

This equation assumes that all the associated particulate energy is absorbed in the tissue involved. The retention function is obviously the key to accuracy in this equation, and it is the most difficult to estimate for the exposed industrial worker. Only estimations of the assumed inhaled atmosphere, the inhalation-related parameters, and the metabolism of the material involved, based on animal work or suitable related publications (20, 21), can be made.

Reversal of the calculation above, using the total dose allowable or recommended in standards will permit solution for other parameters of interest such as C_0, and working backward through the processes just described, a recommended maximum air concentration may be determined. This process is not as simple as it seems, but it does allow for computer computation for any radioelement about which some biological parameters are known.

Localized Organ Retention. Skeletal deposition may serve as a good example of localized retention with regard to internal dosimetry problems. The dosimetry of radionuclides deposited in bone is very complicated where alpha- or soft beta-emitting radionuclides are involved. The complications are related to the long-discussed "hot spot" problem. With radionuclides (as with any metallic element) the ionic chemistry of the isotope dictates the site in which the particle will locate. Durbin has written an excellent review of the effect of ionic radius on site of deposition (22). One of the chemical groups of elements associated with bone is the bivalent alkaline earths: calcium, barium, strontium, and radium. One of the most important structures in bone is the hydroxyapatite crystal into which mineral calcium is laid down. Alkaline earths are termed volume seekers because they are primarily incorporated into the mineral portion of the bone tissue, where they readily become a part of compact bone. The turnover of these elements is slow once they have become incorporated, even though remodeling of bone occurs throughout life. Thus if an area surrounding a haversian canal is not in the process of remodeling at the time one of the radioactive alkaline earth elements enters the plasma, the chances of that isotope entering that particular haversian system are reduced considerably. The unique quality of bone, however, allows a given haversian system to be in the process of remodeling while at the same time the surrounding systems are in a state of rest. In this case, with a relatively high concentration in the plasma of, say strontium or radium, there is a finite probability that the ionic form of the alkaline earth will be deposited on the remodeling sites. Based on this general description, it is obvious that "hot spots" would readily form in the bone as a result of the remodeling system's seeming inconsistency. Autoradiograms of bone observed under these circumstances indicate a very nonuniform distribution, the pattern of which depends on the microanatomic nature of the bone at the time of deposition.

Whereas the alkaline earths are termed volume seekers, the transuranic elements such as plutonium and americium are commonly referred to as surface seekers. These descriptive terms are not entirely all-inclusive because the radionuclides are not strictly confined to the described areas but are deposited to some extent throughout all portions of the bone. The surface seekers tend to attach to the membrane of the osteogenic cell or to enter that cell, attaching to proteins, collagen, or other molecular structures, and remaining in the matrix, particularly if that area is not in a remodeling stage. Thus surface seekers tend to affect the most sensitive part of the bone, that is, the epithelial lining or the areas containing osteogenic cells. Osteogenic cells are stem cells, and because they are proliferative they have a high turnover rate. With surface seekers emitting alpha radiation, damage to osteogenic cells is highly probable. Conversely, if an alpha emitter like radium is buried fairly deep in the mineral portion of bone where the path length of the alpha particle is extremely small (high density), the probability of an ionization occurring in an osteogenic cell to create radiobiological damage is small.

The hot particle problem in bone, as described, is extremely important. Uneven distribution is what leads to the difference in quality factors between some of the elements like radium and plutonium, but the biological effects are very difficult to interpret on this basis.

Some of the foregoing material is from a draft of an International Commission on Radiation Protection (ICRP) Task Group report to be published in 1979 and to which the senior author has contributed considerably on this subject (23).

2 GENERAL BIOLOGICAL EFFECTS

2.1 Electromagnetic Radiation and Neutrons

Acute and long-term biological studies with external sources of irradiation have been rather extensively defined, and this is one area in which there is also a sizable amount of human data. One of the best sources of information is from the persons exposed to the atom bomb radiation at Hiroshima and Nagasaki. These data are updated periodically by various United States and Japanese research groups. A recent publication reviews findings through the first 30 years (24). Additionally, the epidemiological studies on radiologists have given considerable firsthand information (25). A few isolated accidental exposures to gamma and X-rays and/or neutrons may also be cited. These human data are considered along with supporting or pertinent animal research data in the material to follow.

2.1.1 Acute External Radiation Effects

The acute radiation syndrome has been described by many authors, and Table 9.1 from Upton (26) provides a clear description. He lists doses of 400, 2000, and 20,000 rems and indicates that with the second two doses death has occurred by at least the third and fourth weeks. Certain symptoms (i.e., nausea, vomiting, and diarrhea) are present at all doses immediately following exposure to these higher levels. The gastrointestinal tract is one of the most sensitive organs to electromagnetic irradiation.

2.1.2 Intermediate Dosage Effects

There is a radiation dosage range, quite individually variable and quite large, at which many deleterious symptoms may appear, but at which recovery may occur and the individual may return to relatively normal health. The hematopoietic system is also one of

Table 9.1 Effects of Radiation on Man: Major Forms of Acute Radiation Syndrome

Time After Irradiation	Cerebral and Cardiovascular Form (20,000 rems)	Gastrointestinal Form (2000 rems)	Hematopoietic Form (400 rems)
First day	Nausea Vomiting Diarrhea Headache Erythema Disorientation Agitation Ataxia Weakness Somnolence Coma Convulsions Shock Death	Nausea Vomiting Diarrhea	Nausea Vomiting Diarrhea
Second week		Nausea Vomiting Diarrhea Fever Erythema Emaciation Prostration Death	
Third and fourth weeks			Weakness Fatigue Anorexia Nausea Vomiting Diarrhea Fever Hemorrhage Epilation Recovery (?)

Reproduced with permission, from *The Annual Review of Nuclear Science*, Volume 18. © 1968 by Annual Reviews, Inc.

those most sensitive to radiation, and its symptoms often indicate whether death is imminent or recovery will occur. The lymphocytes are perhaps the most sensitive blood cells and are affected within the circulatory system as well as at the precursor level. (Most standard texts dealing with radiation biology discuss in detail the effects of external irradiation on the hematopoietic system.) Other blood cells are affected, but this is primarily the result of the radiation damage to precursors in the bone marrow. The circulating lymphocytes show a decrease immediately following dosage and within limits, may serve as an indicator of the severity of the radiation exposure. This phenomenon has been considered by many as a possible dosimeter for radiation exposure. Figure 9.1, from Arthur C. Upton, Director of the National Cancer Institute, (26) shows a typical pattern of hematopoietic response with subsequent recovery at lengthy times postirradiation.

The testes and ovaries are also very radiosensitive and react to very low doses (a few rems). The testes recover to normal eventually, the time for recovery and severity of damage depending on the dose. The ovary is in a different category because its oocytes are present in entirety, never to be replaced after they have been permanently damaged.

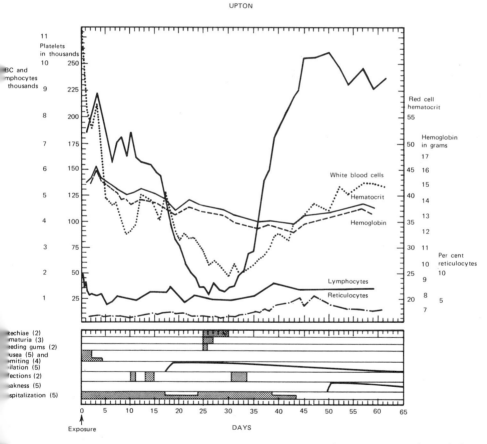

Figure 9.1 Hematological values, symptoms, and clinical signs in five men exposed to whole-body irradiation in a criticality accident. The blood counts are average values; the figures in parentheses denote the number showing the symptoms and signs indicated. From Andrews et al. (26a).

Most other tissue cells, such as those of the central nervous system, heart, and lung, are relatively radioresistant.

2.1.3 Late Effects

Low level, chronic or repeated exposure to ionizing radiation is the primary cause of industrial concern. This type of exposure is often difficult to measure accurately because of its quantitative similarity to background radiation; but more important, it is difficult to control. Some types of occupation such as those involving radiographical usage or those associated with reactor sites, inherently are accompanied by a certain low level radiation. Industry has done a remarkable job in controlling such problems, but reduction of levels from a fraction above normal background to natural background cannot be feasibly accomplished in many cases.

Late effects of radiation in mammals are generally categorized as a generalized accelerated aging process or a specific carcinogenic process. "Life shortening" is a relatively nonspecific term and is considered to mean a premature death resulting from a variety of causes, most of which are specifically unidentified in the given individual. Classical animal experiments, such as the early work of Lorenz (27), indicate that, in mice, life-span may be shortened as a "function of many factors, some of which are unknown." This type of radiation effect is not well understood and merely represents a mammalian system dying of old age before its time. This type of effect on survival would appear to be minimal compared to the more readily diagnosed causes of death such as carcinogenesis.

The types of cancer observed with low level radiation have generally been associated with skin or the hematologic system. Radiation dermatitis was common among workers who had chronic irradiation of the skin, primarily the hands. After latent periods of many years (e.g., 20) it was not uncommon to observe the formation of skin tumors primarily squamous and basal cell carcinomas (26). Similar lesions have been obtained experimentally with beta particles in rats (28) in which damage to the hair follicles was observed to be implicated in the formation of skin cancer. With modern day industrial hygiene practices, the observance of radiation-induced skin cancer in workers has essentially disappeared.

Many types of leukemia have been associated with radiation in the human population, mainly in persons who received radiation from nuclear weapons or from radiotherapy techniques (26). Many factors seem to be important in the onset of radiation-induced leukemia, including age at exposure, sex, dose, and perhaps dose rate. One of the greatest sources for study of the leukemias is the Hiroshima-Nagasaki survivors as indicated, although the patients receiving radiation therapy for ankylosing spondylitis (29–31) and from other radiation sources also add significance to the overall interpretation of induction of leukemia from radiation (32). A summary of many of the related radiobiological findings in the survivors of the bombings in Japan was compiled in 197 (24) and covers the major categories of dosimetry, biological effects, future research, and health surveillance. Ishimara and Ishimara stated in that report that the intense study of the relationship between radiation dose and incidence of leukemia provides an important link to the effects of external radiation on man. They quote from other sources (33) that "the apparent excess incidence (leukemia) of A-bomb survivors is about 1.8 cases per million person-year rads for the period 1950-1970." This reference also states that "those who were in either the youngest or oldest age brackets at the time of exposure were more sensitive to the leukemogenic effects of radiation" (0 to 10 and more than 49 years of age). These leukemias were generally of early formation, however, and the risk of chronic granulocytic leukemia among survivors appeared to be greatest at 5 to 10 years postexposure. Doses calculated to be as low as 50 rads were expressed as being associated with the onset of leukemia, with those exposed at Hiroshima having a higher incidence. This is attributed to a higher relative biological effectiveness (RBE) for neutrons than gamma rays, and a greater abundance ratio of the former in Hiroshima.

2.2 Radionuclide Radiation

As stated in Section 1, the internal exposure to radionuclides is probably the most pertinent to practical toxicity problems in the industrial world. Radionuclides are difficult to control, particularly if their emissions do not allow for detection at reasonable distances from the source. Plutonium is a good example of the latter, in that it has alpha particle emission and only a very weak X-ray component, too weak to detect through the ordinary glove box. Section 1.3.3 covered the difficulties in determining biological dosage following exposure. There has been a wealth of information collected over the past 30 years in animal studies, but extrapolation to man is not straightforward. We do have the advantage of the excellent epidemiological studies of the radium dial painters (34) and the radium-treated ankylosing spondylitic patients. (29) Other studies may also prove useful from the point of view of assessing human exposure to radionuclides [e.g., thorotrast (35) and plutonium (36, 37)]. The problem with studies such as those with the plutonium workers is that since no effects have been observed, there is no way to evaluate dose-effect relationships. Laboratory experimental data are presented in the following sections and where possible, the pertinence to the human situation is discussed.

The problems inherent with radiation dose from internally deposited radionuclides (internal emitters) are much more involved than with whole-body electromagnetic radiation. Compartmentalization of the various radioelements is responsible for the complications of dose estimation, and it is the interplay between these sites of deposition in the body that results in the kinetic modeling carried out with internal emitters. The ultimate goal is to arrive at a time-integrated value, in ergs expended per gram of a tissue, and to relate this through the proper constant(s) to the desired radiation dosage unit. The dose arrived at in this manner is an average over the entire tissue or organ and is generally, today, characterized by the rad, equivalent to 100 ergs per gram of medium. For biological samples of soft tissue, where most beta emitters are involved, this is a reasonably good estimate for comparative purposes. This type of radiation distributes rather uniformly because of the relatively long path length in soft tissue, hence average dose is perhaps a reasonable approach to assessing biological damage or effect. With alpha emitters the pathway is very small (Section 1.3.3), and average dose is generally not as suitable for comparison of effects. Except where intricacies of localized dose are of concern, however, the average is still the most acceptable (often the only) means of relating dose to effect.

In biological modeling to arrive at radiation dose and interplay between organs, one generally uses first-order kinetics to define the parameters. This approach allows one to think in terms of the quantity leaving or entering a given depot as being directly related to the amount in that repository at any one time. Through a series of rate constants representing the various compartments of deposition in the body, half-lives (0.693/rate constant) of retention may be calculated, and investigators in this area of radiation biology or biophysics have learned to think and write in these terms. This is an acceptable concept and will be in use in this field for a long time, although many researchers prefer the power function method of fitting retention kinetics (38–40).

Figure 9.2 gives an example of gross modeling to arrive at the time course of a radionuclide residing in the body. Here the simple model consists of organs A, B and C, with excretion occurring through two routes to external compartments D and E. The development of equations to fit this scheme proceeds according to the following reasoning. Because inhalation appears to be the most practical route of entry to the body, this is a simplified version of a model in which the lung (A) is one of the organs involved and B is the blood; C may represent any other internal organ (kidney, liver, spleen, etc.), and D and E are gastrointestinal tract and urinary excretion, respectively.

1. The amount of radionuclide entering compartment D, gastrointestinal excretion (feces), from the lung is straightforward and is represented by

$$\frac{dD}{dt} = k_2 A \tag{2}$$

2. Compartment B, the blood, represents a much more complicated scheme. An expression for the rate of accumulation in the blood may be derived through a series of differentials as follows:

$$\frac{dB}{dt} = k_1 A = \text{to blood from lung} \tag{3}$$

$$\frac{dB}{dt} = k_6 C - k_5 B = \text{to blood from tissue } C \tag{4}$$

$$\frac{dB}{dt} = -k_3 B - k_4 B = \text{to } D \text{ and } E \text{ from blood} \tag{5}$$

An additional rate constant and equation are applicable for absorption from gastrointestinal tract to blood, but this is assumed to be negligible here for purposes of simplicity.

EVALUATION OF EXPOSURE TO IONIZING RADIATION

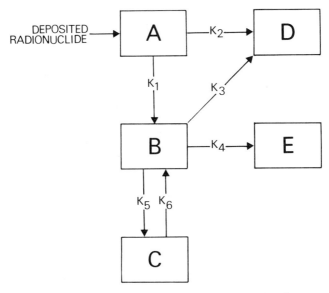

Figure 9.2 Example of first-order kinetics modeling from which the descriptive equations are derived: A = lung, B = blood, C = an internal organ, D = gastrointestinal tract excretion, E = urinary excretion.

3. The ultimate goal is to determine the time course of radiation energy released into organ C. The process is initiated by combining some of the equations above. The amount in C at any time is the integral of the rate of buildup or decline.

$$\frac{dC}{dt} = k_5 B - k_6 C \qquad (6)$$

An expression of B may be obtained from the integral of

$$\frac{dB}{dt} = k_1 A + k_6 C - k_5 C - k_3 B - k_4 B \qquad (7)$$

An expression for A at any time may be obtained from a form of equations

$$-\frac{dA}{dt} = k_1 A + k_2 A \qquad (8)$$

and

$$A = A_0 e^{-(k_1 + k_2)t} \qquad (9)$$

Equation 7 now becomes

$$\frac{dB}{dt} = k_1 A_0 e^{-(k_1 + k_2)t} + C(k_6 - k_5) - B(k_3 + k_4) \qquad (10)$$

This may be solved using linear differential equation techniques, giving an expression for B in terms of A_0 and C. The form generally recognized for integration is

$$\frac{dB}{dt} + B(k_3 + k_4) = k_1 A_0 e^{-(k_1+k_2)t} + (k_6 - k_5)C \quad (11)$$

$$B = \exp\left[-\int_t^0 (k_3 + k_4)\, dt\right] \int_t^0 \exp\left[\int_t^0 (k_3 + k_4)\right] \quad (11)$$

$$[k_1 A_0 e^{-(k_1-k_2)t} + (k_6 - k_5)C]\, dt + \text{constant} \quad (12)$$

4. By proper substitution from Equation 12, an expression for C can be obtained through integration. To quantitate the scheme, one needs to know the initial quantity of radionuclide deposited in the lung A_0, and the various rate constants k_1 through k_6. These may be obtained using the time course of measured radionuclide in blood and excreta at any time. An analogue computer is almost a necessity when the number of compartments is large.

Thus through a series of first-order kinetic manipulations, an expression for dose to an organ may be obtained with which to match biological effect. Although this modeling scheme might appear to belong under Section 1.3.3, it seemed logical to place it in a relationship to the ensuing biological effect.

2.2.1 Effects on Lung

Many radiation effects studies on the respiratory system have been carried out following intratracheal injection (IT) and inhalation (INH). There is always some question of the validity of using IT data inasmuch as this route is a nonpractical means of acquiring a radionuclide; thus the tendency is to utilize information from INH whenever possible. (IT instillation in the lung is generally nonuniform, and the material is in a solution or suspension; an aerosol, used for the INH route, is more uniform and realistic.) Major INH studies using dogs have been carried out in recent years at Battelle Pacific Northwest Laboratories (BPNWL) (41) and at the Lovelace Foundation (the Inhalation Toxicology Research Institute: ITRI) (42). Doses have been expressed as average energy released to the tissue (rads) or in terms of radioactive unit of material deposited per gram of organ (e.g., nCi/g). Either manner of presentation is acceptable.

In both laboratories the inhalation exposures have been carried out in a manner in which only the nose is subjected to the aerosol. With whole-body exposure during inhalation, the animal is covered with the particles and obtains extremely large amounts of external contamination. Even when the aerosol is essentially insoluble in body fluids, the small amount that may be absorbed through the gastrointestinal tract because of the animal's licking (cleaning) itself can result in a deposit in internal organs. If the half-time of residence in that organ system is extremely long compared to that in lung, the possibility for appreciable radiation dose to that organ exists. Thus nose-only exposures are desirable. At BPNWL the primary radioactive inhalation exposures have involved alpha emitters, whereas the ITRI work has been chiefly with beta emitters.

Typical dose-effect data from BPNWL and ITRI for a certain end point such as pulmonary neoplasia, indicate that the dose from beta emitters is up to a factor of 20 times that calculated for a comparable effect (carcinogenesis) from alpha emitters (43–45). These carcinogenic doses are in the hundreds and thousands of rads for alpha emitters and in the thousands and tens of thousands of rads for beta emitters. Assuming the concept of linear energy transfer (LET) (or RBE) to be based on firm premises, this difference between types of radiation is to be expected. Some of the types of lesions found do not appear to vary remarkably with the quality of the radiation. Typical of such lesions are:

- Fibrosarcoma.
- Squamous cell carcinoma.
- Bronchiolar carcinoma.
- Adenocarcinoma.
- Hemangiocarcinoma.
- Bronchioloalveolar carcinoma.

Some of these findings described by various authors may be morphologically similar and depend on the individual pathologist's interpretation.

In man, a limited number of lesions have been attributed to the deposition of radionuclides in the respiratory system. One of the most widely known occupations resulting in such findings is uranium mining. Internal biological lesions in this occupation are primarily the result of breathing radon (^{222}Rn), hence its ensuing decay products (46). The early miners in Schneeburg, Germany, and Joachimsthal, Czechoslovakia, suffered from a disease that was ultimately determined to be lung cancer (47, 48). In the United States the uranium miners in the early part of this century have been found to have a significant increase in lung cancer (49) This has been shown to correlate with smoking, and a greater incidence of lung cancer has been reported for smokers than for nonsmokers, among the miners (49–52). The lesions are bronchogenic and have been shown to correlate with estimated radiation doses (53). The incidences that have been reported for the increase over that expected from a "normal" nonsmoking population of individuals are variable, and no specific value is significant at this time. It has been estimated that a dose of 360 rads will double the incidence of normal lung cancer, as evidenced in these workers (53).

Dr. Saccomanno and associates (50) have derived a unique method for detection of cancerous lesions in the lung. His method will indicate, in cells taken from sputum samples, the predisposition to an invasive cancerous lesion in the bronchiolar tree. Squamous cell metaplasia of the bronchi is generally accepted as the forerunner to bronchogenic carcinoma. The average time for development of epidermoid carcinoma has been quoted to be approximately 15 years (54). The time of development from early metaplastic changes to marked atypia may be an average of 4 years. Thus cytologic investigation of sputum samples taken during this earlier phase may reflect this atypia and may lead to detection of a carcinomatous condition before it reaches the invasive state. This may allow performance of localized surgery, to eliminate the malignancy that would eventually lead to death.

2.2.2 Effects on Bone

"Bone seekers" are generally classified as two types, volume or surface, as described earlier. The primary cancerous lesion observed following radionuclide deposition in the skeleton is osteogenic sarcoma. One of the most thorough studies of this lesion versus estimated radiation dose has been carried out at the University of Utah over the past 25 years. The beagle dog has been the chief experimental subject in that laboratory, allowing for comparison with the inhalation studies described previously. In the Utah work the radionuclides (^{90}Sr, ^{241}Am, ^{228}Th, ^{226}Ra, or ^{239}Pu) have been injected intravenously (IV) in a soluble form, primarily the citrate. This route has desirable properties because it gives a greater probability of deposition in the organs of "choice" than perhaps would be predicted to occur with material entering the blood from the lung, and the doses thus delivered are administered with great accuracy. The choice of deposition site is determined by the physical-chemical properties of the radionuclide, as is the determination of whether a given material is to be a volume or surface seeker in bone.

Mays et al., have described the finding of bone sarcomas in mice and dogs (55) and have attempted to estimate risk to the skeleton using a linear model. Their paper compares the relative risks between ^{239}Pu and ^{226}Ra in experimental animals; Figures 9.3 through 9.5 plot the results. Also shown are data from the radium dial painters, with the predicted linear extrapolation to what the risk of bone sarcoma may be in humans from ^{239}Pu (Figure 9.6). An estimate of the cumulative risk from ^{239}Pu using a linear model is about 200 bone sarcomas per 10^6 person-rads. This bone sarcoma risk estimate is based on the results of the German ankylosing spondylitic patients that received ^{224}Ra as therapy (Figure 9.7). The paper by Spiess and Mays (29) is a thorough summary of bone sarcoma incidence versus radiation dose from many sources and is a recommended reference, not only for the value of the authors' interpretation but for the complete bibliography associated with this field.

2.2.3 Effects on Soft Tissues

As stated, radionulcides will localize in an organ or tissue to an extent dictated by their physical-chemical properties. Thus by choosing an element, one could feasibly obtain a specific concentration in an organ of choice and confine the resulting biological effect to that area. In fact, if this relationship were as simple as stated the medical field would be able to localize the proper radionuclide at the site of a tumor, and with the ensuing radiation, destroy the abnormal cells. However localization is not usually relegated to the tissue of concern alone but also to other, nonaffected tissues. Localization in the tumor-bearing organ subjects normal cells to the radiation as well, and the radiation

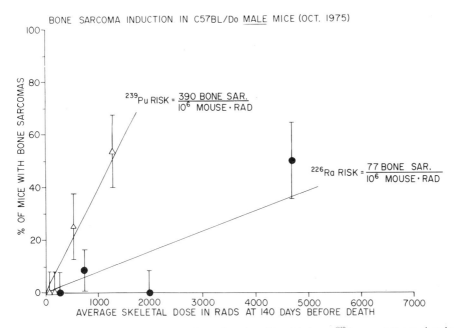

Figure 9.3 Incidence of bone sarcomas in male mice. The risk from ^{239}Pu was 5 times that from ^{226}Ra. From Mays et al. (55).

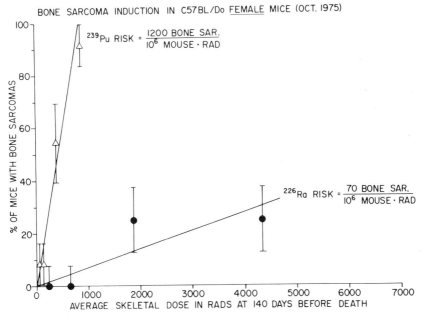

Figure 9.4 Incidence of bone sarcomas in female mice. The risk from ^{239}Pu was 17 times that from ^{226}Ra, assuming negligible control incidence. From Mays et al. (55).

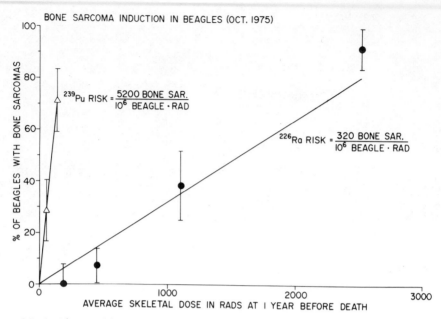

Figure 9.5 Incidence of bone sarcomas in beagles. Linear fits constrained to pass through zero incidence at zero dose were made from 0 to 135 rads for ^{239}Pu and from 0 to 2500 rads for ^{226}Ra. The derived risk from ^{239}Pu was 16 times that from ^{226}Ra. From Mays et al. (55).

dose that may cause subsidence of the malignancy will also be destructive to the normally functioning tissue. With this diversion on the behavior of internal emitters as background, it is now feasible to discuss a few examples of specific tissues that are primarily damaged by the entrance of specific radionuclides to the body.

The Reticuloendothelial System (RES). The RES broadly consists of the liver, spleen, bone marrow, lymph nodes, and lung. One normally thinks of the RES components of these tissues as containing cells that are available to the circulation (lymph or blood) and have the ability to react to any material that is recognized as foreign to the body. Colloidal or particulate substances are particularly subject to being acted on by cells of the RES. Phagocytosis is the primary means of detoxifying such foreign bodies, with the destructive action occurring within the phagocytic cell. For instance, when colloidal polonium is injected IV into experimental animals, it follows a body distribution pattern that is essentially analogous to the organs of the RES (56). The larger the particulate entity, the more rapidly it is removed from the circulation, and this can take place during the first pass through the circulatory system. The spleen and the bone marrow, for example, are two organs that are highly affected following ^{210}Po administration in this manner. The effects on these organs are somewhat comparable, and because they are considered to be at least partly a segment of the hematopoietic system, the resulting damage is included under that discussion.

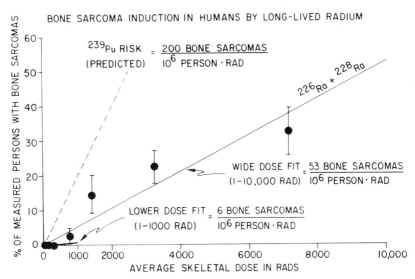

Figure 9.6 Incidence of bone sarcomas in persons (mostly dial painters) exposed to ^{226}Ra and ^{228}Ra. Below 10,000 rads the fitted slope is 53 bone sarcomas/10^6 person-rad, but this slope is based on only one reported bone sarcoma. If the true risk from low doses of long-lived radium is between these values, the predicted risk from ^{239}Pu in man is between 4 and 33 times that from long-lived radium. From Mays et al. (55).

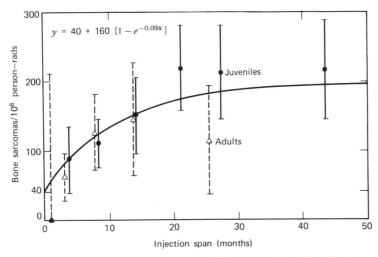

Figure 9.7 Risk of bone sarcomas in humans versus ^{224}Ra protraction. The risk rises to about 200 bone sarcomas/10^6 person-rad at long protraction. The standard deviation of each point is shown. From Spiess and Mays (29).

The lymph nodes are a major sink for the deposition of foreign materials that find their way into the lymphatic channels. Drainage from the lung is one of the more common examples of the defense mechanism exercised by the lymphatics, especially for particles deposited in the alveolar region of the lung. These particles are filtered out in the regional lymph nodes, where they reside for a very long period (16). Although large deposits in lymph nodes have been observed after inhalation of a radioactive aerosol (57), little biological damage of significance to life-span, due to the radioactive content of the regional nodes, has been observed. In uranium studies (58) the lymph nodes of monkeys were seen to concentrate far greater than 50 times the amount in lung after repeated exposure to UO_2 aerosol. A similar pattern has been described for many other radionuclide particles (16). The effect is generally one of fibrotic change, with some alteration of the cellularization of the germinal centers. However the cells involved have shown little tendency for tumor formation, and the lymph nodes under this circumstance are not considered by most to be a critical organ for radiation damage.

The Hematopoietic System. It is known that the blood cell forming tissues are subject to radiation when the nuclide is administered in such a form as to deposit in the reticuloendothelial cells associated with this function. There is also a tendency for some elements to localize in the functional cells directly related to blood cell production. Whatever the mode of localization, the general effect is essentially the same. The lymphocytes are the most sensitive of the circulating blood cells, but the precursor cells in the bone marrow and spleen (particularly under conditions of extramedullary hematopoiesis) are also extremely sensitive to radiation. The erythrocyte precursors in the marrow cavity are the key to any subsequent reduction in red cells. Thus what one may readily observe in circulating blood as a result of radionuclide deposition in the hematopoietic system is a rapid decline in the white cell population followed by a much slower decline in circulating erythrocytes. Reduction in the former, naturally, allows the onset of infection, as the ability to cope with outside disturbances is dramatically curtailed. If the animal (or person) survives this ordeal, the lack of red cell production will result in an anemia and, if the dose is sufficiently high, in the death of the subject.

There are many detailed biological effects inflicted on the hematopoietic system by deposited radionuclides, but the general pattern is as described.

Specific Tissue Localization. The thyroid is an excellent example of a tissue that will localize a particular internal emitter, in this case, iodine. A fine review of this subject with particular regard to the practical aspects of deposition of radioiodine in the thyroid gland has been published by the National Council on Radiation Protection and Measurements (NCRP) (59). The effects of ^{131}I irradiation of the thyroid are aptly described in this report and compared to those observed with X-rays. Because radiation effect on this particular organ has been so thoroughly studied, it is valuable to present two tables from the NCRP reference that generally summarize the findings to date (Tables 9.2 and 9.3). These tables list the risk estimates for various thyroid insults and compare the effects between children and adults.

Table 9.2 Relationship Between Low Dose Exposure to ^{131}I in Children and Subsequent Hypothyroidism[a]

Number of Subjects	Thyroid Absorbed Dose Range (rads)	Estimated Mean Thyroid Absorbed Dose (rads)	Number of Hypothyroid Subjects	Incidence of Hypothyroidism (%/year)
146	10–30	18	0	0
146	31–80	52	3	0.15
151	81–1900	233	5	0.23

[a] Preliminary results: Hamilton and Tompkins (60).

Table 9.3 Absolute Individual Risk of Thyroid Abnormalities After Exposure to Ionizing Radiation

Type of Abnormality and Population Surveyed	Mean Absorbed Dose or Dose Range for Which Data Were Available (rads)	Absolute Individual Risk (10^{-6} rad^{-1} y^{-1})	Statistical Risk Range[a] (10^{-6} rad^{-1} y^{-1})
	Internal Irradiation (^{131}I)		
Thyroid nodularity			
Children	9000	0.23	0 to 0.52
Adults	8755	0.18	0.13 to 0.23
Thyroid cancer			
Children	9000	0.06	0 to 0.158
Adults	8755	0.06	0.044 to 0.075
Hypothyroidism			
"Low dose" children	10–1900	4.9[b]	3.9 to 22.9
"High dose" adults[c]	2500–20,000	4.6[b]	2.8 to 7.8[d]
	External Irradiation		
Thyroid nodularity in children	0–1500	12.4	4 to 47.4[d]
Thyroid cancer in children	0–1500	4.3	1.6 to 17.3[d]
Hypothyroidism in adults	1640	10.2	0 to 24.8

[a] Unless otherwise indicated, the range of risk was determined by assuming that the number of cases, n, out of the population at risk represents the true mean of a Poisson distribution. The range is then estimated by using $\pm 2n$ as the 95 percent confidence level.

[b] Threshold of 20 rads.

[c] See Figure 2 of Reference 61.

[d] In these cases the risk was determined from the slope of the linear regression line. The range was estimated from the extreme data points, which provide the lowest and highest slopes.

3 STANDARDS AND GUIDELINES

The establishment of guidelines for the protection of workers and the public is a complicated endeavor involving many philosophies and some facts. This section includes the most relevant subdivisions in a relatively recent report by the NCRP (62). A brief statement is made regarding each subdivision, as put forth by the committee that generated the general reference (62). Most of the material derives from step-by-step analysis of the information presented in Part II of that handbook. The comments to follow are the authors' interpretation and may not represent that of the full committee.

3.1 Sources of Information

Low dose effects are based on some selected type of extrapolation from biological effects observed at high doses, to dose levels approaching background, where no significant data exist or will be obtained in experimental mammals. For conservative planning, for engineering purposes, it is acceptable to think in terms of an essentially zero dosage threshold, but this approach is not acceptable for predicting radiobiological effects. Genetic effects caused by radiation are known to depend on many factors such as dose rate and fractionation, and tying these into the human population presents an astronomical task. Genetic effects in the population entail geographical factors, types of employment, the quality factor of the radiation, the age of the subject, and a host of other complicated variables. Incorporated into these multifaceted factors (and not specifically stated in the reference) is the factor of acceptance: if the public and workers are allowed to be exposed to materials not as easily detected as radiation, how is this factor (acceptance) incorporated into the rest of life itself? These factors fall under the concept of what comprises an acceptable risk to the individual and the population.

3.2 Dose Level and Priorities

Many terms have been used to define the so-called acceptable levels or the recommendations of such bodies as the NCRP and the ICRP. These terms include tolerance, tolerance dose, permissible dose, maximum permissible dose, dose limits, and lowest practical level. The current tendency in radiation is toward the last term, which means the lowest level attainable under a specific set of conditions. The potential source of radiation should become, along with other less easily definable factors, the guiding influence in the use of this criterion.

3.3 Administrative Philosophies

A number of factors have to be used when the radiation worker is compared to the population at large. The worker is not accepted in employment unless in suitable physical condition; on the other hand, the population represents a spectrum of health conditions. Sex and age of the worker are included in employee selection for obvious

EVALUATION OF EXPOSURE TO IONIZING RADIATION

reasons, and these restrictions automatically involve possible genetic effects on the worker as well as with generations to follow, although these may not be applicable to the population at large.

3.4 Interpretation by Professionals

There are many reasons for basing guidelines on some sort of time span, such as the protraction of dosage over fractions of a one year potential exposure period. Limits such as 5 rems/year or 3 rems/quarter are arbitrary values, just as 18 years is arbitrary as the starting age for a worker in atomic energy related industries. One of the problems with such values is that the practicing administration (primarily health physics oriented) often places credibility in these numbers as if they were determined from some accurate base. This tends to happen in all fields where such limitations and guidelines are established, and some individuals may at times assume that these numbers, generally derived after excellent and very well-informed thought, are to be used as if they were unequivocal, instead of realizing that they represent only the best numbers that can be derived at the time.

3.5 Critical Organ Concept

The critical organ concept has been under discussion for many years. As the reference suggests, it might be more intelligent to consider the "critical" organ as the one most highly radiosensitive, and the "dose-limiting" organ as the one that may receive the greatest radiation dose. Thyroid is used as an example of the latter. The NCRP states,

For the development of protection guides, the concept of identifying the minimum number of organs and tissues that are limiting for dose consideration is the essential simplifying step. For general irradiation of the whole body, such critical organs and tissues are:

1. Gonads (fertility, genetic effects).
2. Blood-forming organs, or specifically red bone marrow (leukemia).
3. Lens of the eye (cataracts).

For external irradiation of restricted parts of the body, an additional critical organ may be:

4. Skin (skin cancer).

For irradiation from internally deposited sources alone or combined with irradiations from external sources, additional limiting organs are determined more by the metabolic pathways of invading nuclides, their concentration in organs, and their effective residence times, than by some inherent sensitivity factors. They include:

1. Gastrointestinal tract.
2. Lung.

3. Bone.
4. Thyroid.
5. Kidney, spleen, pancreas, or prostate.
6. Muscle tissue or fatty tissues (62).

This is a well constructed and brief description of the "critical organ concept" and it indicates the problems involved in proper organ selection criteria under the many conditions of possible radiation exposure.

3.6 Modifying Dosage Factors

The terms quality factor (QF), relative biological effectiveness (RBE), linear energy transfer (LET) and dose equivalent (DE) represent quantities that tend to compound the potential individual radiation situation and should be studied thoroughly before use.

Table 9.4 Summary of Dose-Limiting Guidelines (62)

Maximum permissible dose equivalent for occupational exposure	
Combined whole-body occupational exposure	
Prospective annual limit	5 rems in any one year
Retrospective annual limit	10–15 rems in any one year
Long-term accumulation to age N years	$(N - 18) \times 5$ rems
Skin	15 rems in any one year
Hands	75 rems in any one year (25/quarter)
Forearms	30 rems in any one year (10/quarter)
Other organs, tissues, and organ systems	15 rems in any one year (5/quarter)
Fertile women (with respect to fetus)	0.5 rem in gestation period
Dose limits for the public, or occasionally exposed individuals	
Individual or occasional	0.5 rem in any one year
Students	0.1 rem in any one year
Population dose limits	
Genetic	0.17 rem average per year
Somatic	0.17 rem average per year
Emergency dose limits: life saving	
Individual (older than 45 years if possible)	100 rems
Hands and forearms	200 rems, additional (300 rems, total)
Emergency dose limits: less urgent	
Individual	25 rems
Hands and forearms	100 rems, total
Family of radioactive patients	
Individual (under age 45)	0.5 rem in any one year
Individual (over age 45)	5 rems in any one year

3.7 Dose-Limiting Recommendations and Guidance for Special Cases

The topic of dose-limiting recommendations in routine and extreme situations is covered very well in the reference handbook (62) and can be well understood by referring to the main points indicated in the summary table (Table 9.4).

4 THERAPEUTIC MEASURES

It is possible to modify the effects of external irradiation by means of certain sulfur-containing compounds, including cysteine and cysteamine, provided the compounds are administered prior to irradiation (63). It is also considered possible to ameliorate radiation effects by bone marrow transfusion, preferably using bone marrow provided by the exposed individual prior to exposure. The practicality of measures such as these is so severely limited, because human exposure or contamination is accidental, that they cannot be considered as viable therapeutic measures.

The decision to apply therapeutic measures to decrease the effects of exposure to radiation usually must be made promptly and under the supervision of a physician and must always be made carefully. This decision must incorporate information on the potential risk due to the radiation and, often, on the potential risk due to therapy. The very best measurements of exposure conditions can serve only as good estimates of the radiation dose received, in the case of external radiation, or likely to be received, in the case of internal contamination. Therapeutic risks that must be considered include physical injury due to invasive techniques such as injection or surgery, the toxic effects of certain chemicals, the risk associated with general anesthesia or blood transfusion, and the psychological impact of therapy (particularly heroic measures), on the exposed individual and on the family.

The basis of therapy is quite different for radiation from external sources as contrasted with internal sources. In the former case therapy is initiated after the total radiation dose has been received. The dose cannot be lessened; therefore the objective of therapy is modification of effect by treatment of radiation sickness, prevention of secondary infection, or supplementation of dwindling hematopoietic elements. When the radiation dose will be protracted because of the internal presence of the radiation emitter, the goal of therapy is reduction of the quantity of the emitter. This may be accomplished by enhancing excretion of the emitter or by other physical means of removal.

4.1 External Radiation Exposure

The use of therapeutic means to combat exposure to external radiation is an area in which there has been little opportunity to gain human experience. Personnel involved in worker protection in this segment of toxicology should be proud of their record for maintaining such a low number of exposure incidents. A summary of the Lockport,

New York, accidental exposures (1) will give sufficient insight into the type of care and therapy that appear to be workable for external whole-body or partial-body irradiation.

The Lockport radiation incident provides knowledge of medical treatment following external radiation exposure that is classical insofar as almost everything, therapeutically, appears to have been done properly. Nine persons were exposed to radiation from an unshielded klystron tube at an air force radar site (1). Three of these personnel received doses of X-rays in the range of 1200 to 1500 R over certain areas of the body. Because exposure occurred during a period of 60 to 120 min in the working area, it was extremely difficult to estimate the dose to a given region of any individual.

Table 9.5 lists the basic symptoms shown by the exposed individuals, in descending order of estimated dose received. At the highest exposure level most of the classical signs of acute radiation damage are indicated. Nausea and vomiting were prevalent as well as fatigue and drowsiness. Erythema was a positive indicator of radiation exposure in every case. The first step in therapy was to admit the patients to hospital ward rooms that had been thoroughly cleaned, including the culturing of samples from throughout the room (air, furniture, etc.) for pathogens. This procedure was necessary because of the suspected lowered bacterial resistance of the patients.

All personnel entering the ward were required to wear face masks and to observe other contagion precautions. Visitors were limited to the immediate family. Only laboratory studies deemed necessary for making crucial decisions were allowed. Early after admission the decision was made to postpone any bone marrow transplants because the exposures were of the partial body only. Also, no antibiotics or transfusions were administered; it was thought best to hold off these procedures until the advent of a complication of infection or bleeding.

This represents a proper and sensible approach to the medical treatment of individuals following accidental exposure to relatively high levels of external radiation. The methods available to protect the exposed individual if administered prior to exposure, previously mentioned, are hardly applicable to the accidental situation.

4.2 Internal Radiation Exposure

Therapy for exposure from internally deposited radiation emitters consists of reducing the quantity, or body burden, of the deposited radioactive material. Until recently, the use of chelating agents, introduced primarily by IV injection, has been the only productive method of such reduction. Because the effectiveness of chelation therapy is limited to use with internal emitters in a form that is soluble in body fluids, this approach has not been successful for the removal of insoluble materials such as inhaled insoluble particles that are deposited in the lung. In 1972 an accidentally exposed individual underwent bronchopulmonary lavage for removal of inhaled, deposited ^{239}Pu (64). The positive results of that treatment, together with results of experiments with animals, indicate that bronchopulmonary lavage is a promising procedure for removing inhaled insoluble radioactive substances. With a choice of effective therapeutic measures, a sometimes dif-

Table 9.5 Signs and Symptoms in Victims of the Lockport Incident[a]

						Symptoms							
Patient	Tinnitus, Parotid Swelling	Temperomandibular Tenderness	Forehead Swelling	Headache	Abdominal Pain	Nausea	Vomiting	Anorexia	Chills and Fever	Erythema	Conjunctival Reddening	Dry Mucous Membrane	Lassitude and Somnolence
1	+	+	0	+	0	+	+	0	+	+	+	+	+
2	0	+	0	+	0	+	+	0	+	+	+	+	+
3	0	0	0	0	+	+	+	0	+	+	+	+	+
4	0	0	0	+	0	0	+	0	0	+	+	+	+
5	0	0	0	0	0	0	0	0	+	+	+	+	+
6	0	0	+	0	0	0	0	0	0	+	0	+	+
7	0	0	0	0	0	+	0	0	+	+	+	+	+

Source: Howland et al. (1).

ficult problem is that of determining the relative solubility of the internal contaminant to ensure that the most beneficial treatment is chosen.

4.2.1 Chelation Therapy

Chelation is natural to biological systems. Many metabolically formed compounds, including citric acid and gluconic acid, form chelates, and several amino acids are active chelators. Citric acid chelates calcium and is used by blood banks; the citrate-calcium chelate helps to prevent blood coagulation. Chelation therapy for heavy metal contamination has been in use for more than 30 years, and a variety of chelating agents have been tried (63, 65). Whereas citric acid is a very efficient binder of calcium, the well-known chelator ethylenediamine tetraacetic acid (EDTA) has a greater affinity for polyvalent metals such as lead, zinc, tin, yttrium, and plutonium. The effectiveness, thus the proper choice, of a chelator depends on its biological stability and the stability of the materials to be removed. (Diethylenetriamine pentacetate (DTPA) forms a more stable chelate with the rare earths than does EDTA, and DTPA has a residual effect that is desirable in terms of lowering the frequency of treatments.

Chelators are generally poorly absorbed from the gastrointestinal tract; thus they are not of much use when administered orally except in cases of ingestion of the contaminant. Although chelator absorption may be quite good following intramuscular (IM) injection, use of the IM route of administration is rare, partly because of the greater likelihood of local irritation associated with this route. Chelator administration by IV injection is most often reported because of its greater efficiency; however much greater potential for toxicity exists when the IV route is used (63, 66). The greater stability of DTPA and its greater efficiency for removal of heavy metals from the body than EDTA, render DTPA more useful at lower dosages, therefore less potentially toxic than EDTA. DTPA has become the chelator of choice for treatment of internal contamination of humans with polyvalent radionuclides. Information on the behavior and effects of chelating agents in man is available, because numerous experiments have been carried out using tracer quantities of radioactive rare earths in humans (66, 67).

Chelation therapy has been useful in the removal of plutonium from contaminated individuals, as described in the following brief case studies. In one reported case (68) a worker was sprayed with an acid solution of plutonium chloride and plutonium nitrate. Inhalation and ingestion, as well as skin contamination, occurred. The skin, except for burned areas, was decontaminated with dilute sodium hypochlorite solution. Eleven 1 g, IV, DTPA treatments were administered, beginning 1 hr after the accident and at intervals through 17 days. Burn scabs were removed 2 weeks following the accident and they were found to contain most of the plutonium. The combination of treatment methods used in this case was considered highly effective. Similar effectiveness of prompt DTPA treatment was reported for another plutonium-contamination incident involving an acid burn (69). Twenty-seven daily 1 g, IV, DTPA treatments, beginning 1 hr after the accident, resulted in elimination of more than 96 percent of the estimated systemic burden. As in the previous case, much of the contamination was removed with the burn scabs.

The effectiveness of DTPA treatment using a regime similar to those just reported, was considered inconclusive in the case of a wound to the thumb from a plutonium-contaminated metal sliver (68). Initially the sliver was removed without tissue excision. Subsequently tissue excisions were performed at the points of entrance and exit of the sliver; excised tissue from the point of sliver entrance contained about 98 percent of the plutonium removed by excision. Wound counting performed over several months indicated movement of the remaining embedded plutonium toward the skin surface. Nodules that formed concurrently with increased wound counts were excised, and these were found to contain essentially all the plutonium estimated to remain in the thumb. Although DTPA was effective in removing plutonium from the blood, the possible influence of DTPA in mobilizing the plutonium from the wound into the blood was noted.

In another puncture wound episide (70), DTPA was administered IV 4 days per week for about 11 weeks followed by a 30 week period of no treatment, then by a 90 week period in which DTPA was administered either IV or by aerosol 2 or 4 days per month. During the early treatment period, oral (tablet) EDTA was substituted for DTPA as a matter of convenience to the employee; the EDTA was found to be very ineffective in enhancing urinary plutonium excretion. Additional cases of puncture wound injury have been reported (69, 71). In each instance of puncture wound, the greatest efficiency of treatment has resulted from prompt DTPA therapy and one or more tissue excisions. In burn cases surface decontamination except in immediate burn areas, and prompt DTPA therapy, are recommended, followed by careful removal of the burn scabs. In all cases wound counting and/or whole-body counting has been used to determine the plutonium burden.

Several instances of human plutonium contamination by inhalation have been reported. In an early study (72) exposed individuals were treated with calcium EDTA, administered IV in 1 g doses twice a day, for the initial treatment, then administered orally. The IV treatment resulted in a tenfold increase in urinary plutonium excretion; however the oral doses caused only minor increases and were not considered to be of value. Chelation therapy was deemed successful, although the total urinary excretion amounted to only 10 percent of the body plutonium content. The key to effective chelation therapy is promptness.

4.2.2 Pulmonary Lavage

Pulmonary lavage is a relatively new technique that is applied primarily to patients with alveolar proteinosis. Dr. Kylstra at Duke University Medical Center has been working in this field for years and is doubtless one of the world's experts (73, 74). In recent years the application of pulmonary lavage to animals has been successful in removing radionuclides after inhalation exposure (6, 75). In addition, one human case of radionuclide exposure has been somewhat successfully treated (64).

The manner of treatment is to place a Carlens catheter (76) into the patient's respiratory tract such that one lung is supplied with oxygen while the other has access to washing fluid. In general, physiological saline is sufficient for the washing process. Approximately a tidal volume of saline is introduced into the open lung, then removed. This procedure may be repeated up to 5 to 10 times on the same lung during one session. The catheter is withdrawn, and after 24 to 48 hr the procedure is repeated on the other lung. With this procedure the surfactant that lines the alveolar regions is partially removed, and with it, any cells or other materials that reside there. Thus an insoluble material (radionuclide) in the deep lung, whether free or within macrophages,

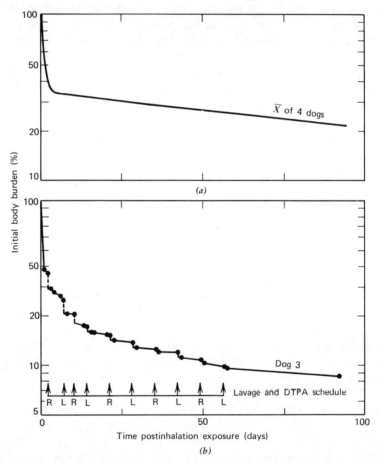

Figure 9.8 Whole-body retention of ^{141}Ce in dogs exposed by inhalation to ^{144}Ce in fused clay (a) Dogs received no postexposure treatment; dog 3 was lavaged 10 times during the first 56 days after exposure. (b) Treated dogs: DTPA was given intravenously after each lavage: solid points represent whole-body count. The vertical steps represent the ^{144}Ce removed in the lavage fluid with each treatment. Data are uncorrected for physical decay of ^{144}Ce. From Boecker et al. (77).

EVALUATION OF EXPOSURE TO IONIZING RADIATION

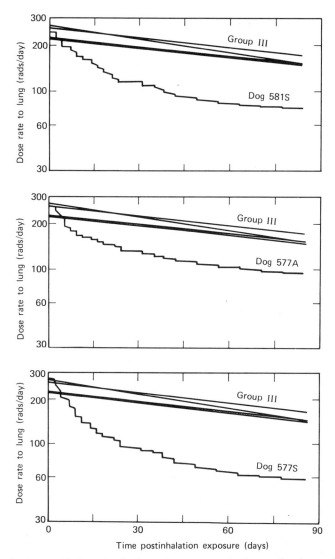

Figure 9.9 Reduction, with time, in the beta dose rate to the lungs of unlavaged group III dogs and lavaged group II dogs 581S, 577A, and 577S. From Silbaugh et al. (78).

is removed to some extent. The following experimental data illustrate the effectiveness of the procedure.

At the Inhalation Toxicology Research Institute, Albuquerque, New Mexico, many studies have been performed on the effectiveness of lavage in removing inhaled particles from experimental animals. One aerosol used for this purpose in beagle dogs has been

^{144}Ce in fused aluminosilicate aerosol particles. A study by Boecker (77) showed how effective multiple lavage can be in removing the particles deposited in the respiratory tract. Figure 9.8 gives an example. The reduction in radiation effect on the lung by systematic removal of the radiation source (particles of ^{144}Ce in this case) is obvious. The lavage fluid was physiological saline. DTPA treatment was also used in conjunction with the lavage procedure with some of the dogs, but the overall effectiveness of the added chelator was minimal. A similar inhalation experiment was performed with ^{144}Ce in fused clay aerosol particles in beagle dogs, and those data are expressed in terms of reduction in radiation dose to the lung in Figure 9.9. According to Silbaugh et al. (78) no impairment of health was seen at long times postexposure (14 months) in the lavaged dogs, whereas untreated dogs receiving those initial doses would have been sick or died.

5 BIOASSAY TECHNIQUES

It is difficult to treat irradiated individuals for potential damage if the extent of the projected biological insult cannot be roughly estimated. Bioassay has been one of the descriptive terms applied to this field of making such dosimetric approximations. As will be seen, this is a very difficult task and needs much more experimental advancement before bioassay becomes the exact science that is desirable. It seems to be more effective with internal emitters, as discussed in Section 5.2.

5.1 External Radiation Exposures

There are few techniques for detection of human external radiation exposures, although many have been explored and attempted during the past few decades. Biological indicators of exposure must be sensitive to low level radiation if the practical "in-plant" situation is to be monitored. This criterion eliminates chemical reactions that are quickly restored or require large doses to bring about measurable reaction products. It also essentially eliminates the old practice in therapy of using an erythema reaction to estimate the amount of radiation delivered.

An extremely troublesome factor in biological systems is the tremendous number of different chemical species present in any one finite sphere of interaction (ionization). If one had an ideal chemical oxidation-reduction reaction that would respond to low radiation doses, and if the primary constituents were present in necessary abundance the volume of tissue of interest (e.g., a cell), the chemical entities would still be overwhelmed by other chemical species. If these "foreign" substances were reactive either to the radiation or to the products being formed and measured in the primary event, the accuracy would immediately be challenged. Such is the case with chemically induced bioassay schemes for detecting external exposure to ionizing radiation *in vivo*.

Perhaps of some potential are schemes that measure actual effects on the biological macrosystems. The genetic materials, chromosomes, offer one possibility as a dosimeter for practical radiation exposure. Although measurement of chromosomal damage through breakage is tedious, this damage nevertheless may prevail for periods of time, lending it practical significance. One drawback is the recombination of chromosomes that have been broken by the radiation, to a normal condition. This can readily occur under the influence of ionizing events or of other toxic reactions that affect chromosomes. However background quantitation for the probability of recombination is possible (and feasible) and can become part of the overall scheme for dosimetry purposes. Samples of peripheral blood are readily obtainable from workers for use in this technique of quantitating chromosome damage due to ionizing radiation.

Lymphocytes are very sensitive to radiation and prove to be good indicators of damage. In Figure 9.1 it is obvious that many hematological parameters (cell numbers) show a dramatic change as a result of ionizing radiation. Therefore, one may be able to estimate the radiation dose in a rough manner, as was attempted with the Lockport cases (1). One of the more sensitive indicators using blood samples is the increased abundance of bilobed lymphocytes, a technique first explored and implemented by Ingram (79, 80). This rather specific indicator of radiation damage was first applied to workers associated with a cyclotron at the University of Rochester. Bilobed lymphocytes tend to be slightly larger than normal, and they contain two distinct nuclear masses. Each is a typical nucleus of normal lymphocytes, but they occur two to a cell. The main drawback to this type of personnel monitoring is the tedium of microscopic examination of so many cells in a preparation. The low incidence of these altered cells requires that statistical analyses be performed for the detection of low doses. However the accuracy is such that this indicator provides a suitable bioassay procedure for low doses of either electromagnetic radiation or neutrons (79–81).

5.2 Internal Radiation Exposure

Radionuclides deposited in the body also point up the gap between ease of detection and interpretation of results. The radioactivity of a material present in a biological system provides one of the most sensitive means that will ever be known for localizing an element involved in the body's metabolic scheme. Measurement of a radionuclide's effect and its metabolic parameters, as related to the overall *in vivo* system, is a difficult problem, however, and almost insurmountable under some circumstances. As an example, detection (quantitation) of an alpha particle emitter in a section of tissue can be carried out with great precision. Equally credible measurements can be made on the contaminated excreta resulting from deposition of this material. However the intervening biological scheme at play between the deposit of the radionuclide and its elimination is generally an unknown entity. It is this factor that leads to frustration where practical bioassay programs for determination of industrial exposures are concerned.

5.2.1 Inhalation Exposures

Many problems are inherent in bioassay of materials deposited in the lung following inhalation. The solubility of the inhaled radionuclide has an immediate, important role in the effectiveness of use of bioassay techniques. A "completely soluble" (in body fluids) radionuclide may enter the blood and either deposit in the tissue of choice or be excreted, primarily in the urine. If the release from the designated site of localization is with constant and known kinetics, appearance of the material in the urine may be an excellent source of information for extrapolation to an estimated body burden. An example of such a material is tritium. When inhaled, this isotope of hydrogen enters the normal body pool and in a short period becomes part of the body's metabolic scheme. Measurement of $_1H^3$ in the urine, and application of isotope dilution principles as related to body characteristics (e.g., weight, height), serves as a suitable bioassay technique (82, 84). Tritium assay presents some difficulties, which are not unusual with the bioassay of internally deposited radionuclides, in that tritium leaves the body with multiple exponents of excretion that may depend on environmental factors such as season of the year (84). However tritium represents the ideal case (soluble), and this bioassay method is generally not applicable in inhalation exposure cases.

When the inhaled radionuclide is not in a soluble form, the situation with regard to bioassay becomes nightmarish. If the radionuclide is "completely insoluble" (a misnomer, since it holds true for no material in body fluids), there is no bioassay technique that is quantitative through normal metabolic processes. A radionuclide in this form, which has some sort of detectable radiation (X-ray or gamma ray) outside the body, may be quantitated by chest counting. This is a reasonably accurate method except as the energy of emission approaches 100 keV and lower. At these low energies the rib cage becomes a troublesome factor because it absorbs the photon. The degree of absorption depends on the size of the rib cage and associated tissue, both in thickness and variable density, as well as the photon energy. Accurate calibration for a specific individual is essentially impossible, since mockups rarely match the subject. Thus estimates of the biological parameters are made through height, weight, lean mass–body weight ratios, and so on. Ultrasonic measurements are made at many locations about the chest cavity, but still, determination of the degree of self-absorption for an individual is close at best. Whole-body counting does remain the only practical scheme, however, for estimating lung burden of an insoluble radionuclide. One problem encountered with this and all chest counting is the inability to distinguish lymph node accumulation of radionuclide from that in lung tissue. Some experimentation has been carried out to evaluate the use of minute detectors that can be lowered into the respiratory tract using a device similar to a bronchioscope, but to date this technique has not found widespread use (85). Pinpoint detection will remain a problem until schemes for precision of measurements have been more thoroughly tested.

The case of exposure to somewhat soluble or slowly solubilized radionuclides is perhaps the most difficult from the viewpoint of bioassay (see discussion of the kinetics of loss from the lung to internal organs in Section 1.3.3). If one knows all the parameters of loss from the lung to blood to urine, a back-calculation readily produces the lung burden. Unfortunately, however, this is rarely the case, particularly in that an aerosol breathed by the same or different persons, at the same location, under seemingly identical conditions, will lead to variable results. Thus any bioassay procedure attempted under conditions of partial solubility will meet with frustration and, probably erroneous results and erroneous interpretation.

EVALUATION OF EXPOSURE TO IONIZING RADIATION

Many references exist on whole-body counting and excretory techniques for determining the radioactivity in man, and these represent a starting place for those interested in pursuing this field (86–88). Also, directories have been published that contain information on the laboratories that operate whole-body monitors, with information on the technical design and performance of the individual apparatus (89).

5.2.2 Puncture Wounds

One of the most common means of receiving accidental body burdens of radionuclides is through skin absorption or puncture. The situation is similar to inhalation exposure and can be considered in the same light. The wound burden can be assumed to be comparable to the lung burden, and rate of loss from the site is given the same consideration. If there is a detectable radioactive emission, a wound counter can be utlized in the same manner as a chest counter (68). However the advantage with wounds is that excision or amputation of the contaminated body part can be put to use in relieving the problem; this is less feasible in the case of inhalation exposure. Estimation of the wound radioactivity even can be made through biopsy, a technique that is not as satisfactory where the lung is concerned. Urine analysis may be satisfactory under certain conditions of wound contamination, but external counting, where feasible, still remains the most practical and successful method of assaying for body burdens.

6 RADIATION AND ENERGY SOURCES

Present predictions indicate that fossil forms of energy will allow the United States to satisfy its energy needs for several decades. Also, there appears to be considerable optimism that solar, geothermal, or other sources may increase this period indefinitely, and many research teams are working in these areas. In addition to these energy sources, however, are those from which radiation is an obvious by-product. The sections to follow briefly indicate the possible problems associated with them.

6.1 Radiation Sources

One of the more promising future sources of energy is the practical application of nuclear fusion. If this source is exploited to an economically feasible state, it can be the answer to the future's energy problems. In the fusion process atomic nuclei (light atomic weight) are brought together in such a manner that energy above the amount of the particles' binding energy is released and can be captured and utilized as a practical energy source. Since nuclear fusion would not produce as much atmospheric contamination with radioactive materials as would nuclear fission, the process itself might be acceptable to the factions in the public that are alienated by the term "radiation," regardless of the context in which it is used. However the fusion process, if and when it is feasibly developed as a practical energy source, may not be without radiation problems. There will be waste disposal problems as with nuclear reactors, and the neutrons that are emitted as part of the energy release could present an industrial radiation problem. The practicability of fusion as a source of energy is becoming a political and socioeconomic topic for discussion, and some recent articles face many of these issues (90–93).

Nuclear fission is the most realistic source of energy at present, and it is currently being used in many countries of the world. An article that appeared in 1976 indicates that France planned an increase from 8 to 70 percent electricity production from nuclear power by 1985, as a result of the embargo by major petroleum producers and increased oil prices (94). Various types of power reactor may be used as sources of energy, some more appealing than others from the standpoint of potential release of airborne radioactivity. Modern design of most proposed nuclear power reactors is such that even an internal malfunction (i.e., within the reactor) would not result in atmospheric contamination. Reutilization of the repaired initial plant, or substitution of a new one, could take place to restore the same site as a source of power. Engineering accomplishments to maintain any radiation problems within the confinement of the reactor complex itself indicate that any exposure to the surrounding area would be of little significance, probably zero. Because this energy source represents a horrifying (Hiroshima-like) scene to some of the American public, it will take considerable time and public education before the nuclear reactor is accepted as a safe major source of energy in this country. An aspect of concern to proponents of nuclear sources of energy is that 10 to 20 years may be required from the time that a new power station is begun until it can assume routine operation.

6.2 Mixed Potential Hazards

Potential hazards may be involved in the event that a radiation source combines with an environmental chemical source in the public atmosphere. One must consider the possibility, however slight, that a nuclear power source could give rise to some atmospheric contamination as the result of an accidental release. This release could be in the form of particulates and could be acted on by such factors as local meteorology, temperature, terrain, and altitude.

Synergism in inhalation toxicity, particularly concerning a source of radiation and a potential chemical cocarcinogen, has not been studied in depth.

A study with experimental animals at the Los Alamos Scientific Laboratory recently indicated a sizable increase in lung cancer incidence when Fe_2O_3 was added to PuO_2 particles deposited in the respiratory tract (95). The latter insult alone (PuO_2) was not tumorigenic under similar conditions of administration to the animal species used (Syrian golden hamsters) (96, 97). The public is informed almost daily in newspapers and journals of the possible atmospheric contamination arising from common sources. Materials such as asbestos from automobile brakes, oxides of sulfur, oxides of nitrogen, carbon monoxide, dust, and high humidity are popular subjects for news media and other concerned factions. These materials are generally measured and indexed by local (state and city) environmental regulatory organizations, and when certain limits are surpassed, an attempt is made to reduce these to some established "allowable" level. Radiation is treated in the same manner, but the measured values are required to be essentially zero (natural background for the area), or the lowest practicable.

6.3 Mining as a Factor

Hazards associated with the mining of uranium have been recognized for centuries. Within the past 50 years the origin of lung cancer among these miners has been quite firmly attributed to the radiation associated with the radioactive daughters of radon, the noble radioactive gas from uranium. There is also an external source of radiation associated with the uranium decay products, and this can range from 0.02 to 4 mrem/hr. As Holady (98) points out, these rates of gamma radiation have been reported for mines in France, Japan, Mexico, Spain, Australia, and the United States. It was determined a few years ago that the respiratory tumorigenicity associated with the miners was closely related to cigarette smoking in conjunction with the deposited radioactive particles of the radon daughters (see Section 2.2.1).

Mining problems have assumed new importance since the energy crisis has directed industry into evaluating the extensive procurement, refinement, and use of potential sources of underground fossil fuels. Oil shale mining is an excellent example, and certainly there are potential chemical carcinogens associated with the material mined, processed, refined, and utilized for energy. Radon and radon daughters are prevalent in coal mines, and there is every reason to believe that any underground mining, in the Rocky Mountain region particularly, will be a source of radon and its daughters. With newer, stricter regulations and guidelines, the cost of eliminating any such measurable materials can become enormous. In addition to the potentially small radiation problem, the dusty atmospheres associated with mining may contain potential chemical carcinogens that will be inhaled along with the radioactive sources. In fact, it is likely that if workers or the population receive any exposures to potentially toxic levels, the toxicants will be of chemical origin, not the low levels of radiation that are present in the mines.

REFERENCES

1. J. W. Howland, M.D., M. Ingram, M.D., H. Mermagen, and C. L. Hansen, Jr., M.D., Lt. Col. (USAF), "The Lockport Incident: Accidental Partial Body Exposure of Humans to Large Doses of X-Irradiation," in: *Diagnosis and Treatment of Acute Radiation Injury, Proceedings of a Scientific Meeting Jointly Sponsored by the International Atomic Energy Agency and the World Health Organization,* Geneva, October 17–21, 1960, WHO, Geneva, 1961, p. 11.
2. D. M. Ross, "A Statistical Summary of United States Atomic Energy Commission Contractors' Internal Exposure Experience 1957–1966," in: *Diagnosis and Treatment of Deposited Radionuclides, Proceedings of a Symposium,* Richland, Wash., May 15–17, 1967, p. 427.
3. H. C. Hodge, J. N. Stannard, and J. B. Hursh, Eds., *Uranium · Plutonium · Transplutonic Elements,* Springer-Verlag, New York, 1973, p. 643.

4. E. C. Anderson, *J. Bacteriol.*, **76,** 471 (1958).
5. T. T. Mercer, M. I. Tillery, and H. Y. Chow, *Am. Ind. Hyg. Assoc. J.*, **29,** 66 (1968).
6. B. A. Muggenburg, S. A. Felicetti, and S. A. Silbaugh, *Health Phys.*, **33,** 213 (1977).
7. K. J. Caplan, L. J. Doemeny, and S. D. Sorenson, *Am. Ind. Hyg. Assoc. J.*, **38,** 162 (1977).
8. K. J. Caplan, L. J. Doemeny, and S. D. Sorenson, *Am. Ind. Hyg. Assoc. J.*, **38,** 83 (1977).
9. M. Lippmnn, "Aerosol Sampling For Inhalation Hazard Evaluation," in: *Assessment of Airborne Particles*, T. T. Mercer, P. E. Morrow and W. Stöber, Eds., Thomas, Springfield, Ill., 1972, p. 449.
10. *American Industrial Hygiene Association Journal*, **31,** 133 (1970).
11. M. Lippmann, *Am. Ind. Hyg. Assoc. J.*, **31,** 138 (1970).
12. D. A. Holaday, "Uranium Mining Hazards," in: *Uranium · Plutonium · Transplutonic Elements*, H. C. Hodge, J. N. Stannard, and J. B. Hursh, Eds., Springer-Verlag, New York, 1973, p. 301.
13. International Commission on Radiation Units and Measurements, *Neutron Dosimetry for Biology and Medicine*, ICRU Report 26, 1977.
14. O. G. Raabe, "Generation and Characterization of Aerosols," in: *Inhalation Carcinogenesis*, AEC Symposium Series 18, 1970, p. 123.
15. Task Group on Lung Dynamics, *Health Phys.*, **12,** 173 (1966).
16. R. G. Thomas, "Uptake Kinetics of Relatively Insoluble Particles by Tracheobronchial Lymph Nodes," in *Radiation and the Lymphatic System, Proceedings of Symposium*, Richland, Wash., ERDA Conference CONF-740930, 1974, p. 67.
17. P. Nettesheim, M. G. Hanna, Jr., and J. W. Deatherage, Jr., Eds., *Morphology of Experimental Respiratory Carcinogenesis*, AEC Symposium Series 21, 1970, p. 417.
18. R. G. Thomas, "Tracheobronchial Lymph Node Involvement Following Inhalation of Alpha Emitters," in: *Radiobiology of Plutonium*, B. J. Stover and W. S. S. Jee, Eds., J. W. Press, University of Utah, Salt Lake City, 1972, p. 231.
19. W. S. S. Jee, Ed. *The Health Effects of Plutonium and Radium*, J. W. Press, University of Utah, Salt Lake City, 1976, p. 169.
20. U.S. Department of Commerce, National Bureau of Standards, *Maximum Permissible Body Burdens and Maximum Permissible Concentrations of Radionuclides in Air and in Water for Occupational Exposure*, NBS Handbook 69, Government Printing Office, Washington, D.C., 1959.
21. International Commission on Radiation Protection, Committee II, *Health Phys.*, **3,** June 1960.
22. P. W. Durbin, *Health Phys.*, **8,** 665 (1962).
23. International Commission on Radiation Protection Task Group on Biological Effects of Inhaled Radionuclides (to be published).
24. S. Okada, Ed., *J. Radiat. Res., Japan Radiat. Res. Soc.*, **16,** Supplement (1975).
25. G. M. Matanoski, R. Seltser, P. E. Sartwell, E. L. Diamond, and E. A. Elliott, *Am. J. Epidemiol.* **101,** 188 (1975).
26. A. C. Upton, "Effects of Radiation on Man," in: *Annual Review of Nuclear Science*, Vol. 18, Annual Reviews, Palo Alto, Calif., 1968, p. 495.
26a. G. A. Andrews, B. W. Sitterson, A. L. Kretchmar, and M. Brucer, "Criticality Accident at the Y-12 Plant," in: *Diagnosis and Treatment of Acute Radiation Injury, Proceedings of a Scientific Meeting Jointly Sponsored by the International Atomic Energy Agency and the World Health Organization* Geneva, October 17–21, 1960, WHO, Geneva 1961, p. 27.
27. E. Lorenz, *Am. J. Roentgenol.*, **63,** 176 (1950).
28. R. E. Albert, F. J. Burns, and P. Bennett, *J. Nat. Cancer Inst.*, **49,** 1131 (1972).
29. H. Spiess and C. W. Mays, "Protraction Effect on Bone-Sarcoma Induction of ^{224}Ra in Children and

Adults," in: *Radionuclide Carcinogenesis, Proceedings of the 12th Annual Hanford Biology Symposium,* 1973, p. 437.
30. W. M. Court Brown and R. Doll, "Leukemia and Aplastic Anemia in Patients Irradiated for Ankylosing Spondylitis," *Medical Research Council Special Report* Series 295, Her Majesty's Stationery Office, London, 1957, p. 21.
31. W. M. Court Brown and R. Doll, *Br. Med. J.,* **II,** 1327 (1965).
32. R. H. Mole, *Br. J. Radiol.,* **48,** 157 (1975).
33. S. Jablon and H. Kato, *Radiat. Res.,* **50,** 649 (1972).
34. R. E. Rowland, A. T. Keane, and H. F. Lucas, Jr., "A Preliminary Comparison of the Carcinogenicity of ^{226}Ra and ^{228}Ra in Man," in: *Radionuclide Carcinogenesis,* AEC Symposium Series 29, CONF-720505, 1973, p. 406.
35. J. D. Abbatt, "Human Leukemic Risk Data Derived from Portuguese Thorotrast Experience" in: *Radionuclide Carcinogenesis,* AEC Symposium Series 29, CONF-720505, 1973, p. 451.
36. R. E. Rowland and P. W. Durbin, "Survival, Causes of Death, and Estimated Tissue Doses in a Group of Human Beings Injected with Plutonium," in: *The Health Effects of Plutonium and Radium,* W. S. S. Jee, Ed., J. W. Press, University of Utah, Salt Lake City, 1976, p. 329.
37. G. L. Voelz, *Health Phys.,* **29,** 551 (1975).
38. J. H. Marshall, J. Rundo, and G. E. Harrison, *Radiat. Res.,* **39,** 445 (1969).
39. W. P. Norris, T. W. Speckman, and P. F. Gustafson, *Am. J. Roentgenol.,* **73,** 785 (1955).
40. C. E. Miller and A. J. Finkel, *Am. J. Roentgenol.,* **103,** 871 (1968).
41. G. E. Dagle, J. E. Lund and J. F. Park, "Pulmonary Lesions Induced by Inhaled Plutonium in Beagles," in: *The Health Effects of Plutonium and Radium,* W. S. S. Jee, Ed., J. W. Press, University of Utah, Salt Lake City, 1976, p. 161.
42. R. G. Cuddihy, R. O. McClellan, J. A. Mewhinney, and B. A. Muggenburg, "Correlations Between the Metabolic Behavior of Inhaled and Intravenously Injected Plutonium in Beagle Dogs," in: *The Health Effects of Plutonium and Radium,* W. S. S. Jee, Ed., J. W. Press, University of Utah, Salt Lake City, 1976, p. 169.
43. R. O. McClellan, S. A. Benjamin, B. B. Boecker, F. F. Hahn, C. H. Hobbs, R. K. Jones, and D. L. Lundgren, "Influence of Variations in Dose and Dose Rates on Biological Effects of Inhaled Beta-Emitting Radionuclides," in: *Biological and Environmental Effects of Low-Level Radiation, Proceedings of the International Atomic Energy Agency Symposium,* Chicago, November 3–7, 1975, Vol. 2, 1976, p. 3.
44. C. L. Sanders, G. E. Dagle, W. C. Cannon, D. K. Craig, G. J. Powers, and D. M. Meier, *Radiat. Res.,* **68,** 349 (1976).
45. F. F. Hahn, S. A. Benjamin, B. B. Boecker, T. L. Chiffelle, C. H. Hobbs, R. K. Jones, R. O. McClellan, and H. C. Redman, "Induction of Pulmonary Neoplasia in Beagle Dogs by Inhaled ^{144}Ce Fused-Clay Particles," in: *Biological and Environmental Effects of Low-Level Radiation, Proceedings of the International Atomic Energy Agency Symposium,* Chicago, November 3–7, 1975, Vol. 2, 1976, p. 201.
46. D. A. Holaday, "Uranium Mining Hazards," in: *Uranium · Plutonium · Transplutonic Elements,* H. C. Hodge, J. N. Stannard, and J. B. Hursh, Eds., Springer-Verlag, New York, 1973, p. 296.
47. C. D. Stewart and S. D. Simpson, "The Hazards of Inhaling Radon-222 and Its Short-lived Daughters: Consideration of Proposed Maximum Permissible Concentrations in Air," in: *Radiological Health and Safety in Mining and Milling of Nuclear Materials,* IAEA Symposium Series, Vol. 1, 1964, p. 333.
48. D. A. Holaday, *Health Phys.,* **16,** 547 (1969).
49. V. E. Archer, J. K. Wagoner, and F. E. Lundin, *Health Phys.,* **25,** 351 (1973).
50. G. Saccomanno, V. E. Archer, R. P. Saunders, O. Auerbach, and M. G. Klein, "Early Indices of

Cancer Risk Among Uranium Miners with Reference to Modifying Factors," in: *Occupational Carcinogenesis*, U. Saffiotti and J. K. Wagoner, Eds., New York Academy of Sciences, New York, 1976, Vol. 271, p. 377.
51. V. E. Archer, J. K. Wagoner, and F. E. Lundin, *J. Occup. Med.*, **15**, 204 (1973).
52. F. E. Lundin, J. W. Lloyd, E. M. Smith, V. E. Archer, and D. A. Holaday, *Health Phys.*, **16**, 571 (1969).
53. V. E. Archer and F. E. Lundin, *Environ. Res.*, **1**, 370 (1967).
54. R. A. Lemen, W. M. Johnson, J. K. Wagoner, V. E. Archer, and G. Saccomanno, "Cytologic Observations and Cancer Incidence Following Exposure to BCME," in: *Occupational Carcinogenesis*, U. Saffiotti and J. K. Wagoner, Eds., New York Academy of Sciences, New York, 1976, Vol. 271, p. 71.
55. C. W. Mays, H. Spiess, G. N. Taylor, R. D. Lloyd, W. S. S. Jee, S. S. McFarland, D. H. Taysum, T. W. Brammer, D. Brammer, and T. A. Pollard, "Estimated Risk to Human Bone from ^{239}Pu," in: *The Health Effects of Plutonium and Radium*, W. S. S. Jee, Ed., J. W. Press, University of Utah, Salt Lake City, 1976, p. 343.
56. R. G. Thomas and J. N. Stannard, *Radiat. Res. Suppl.*, **5**, 16 (1964).
57. W. J. Bair, J. E. Ballou, J. F. Park, and C. L. Sanders, "Plutonium in Soft Tissues with Emphasis on the Respiratory Tract," in: *Uranium·Plutonium·Transplutonic Elements*, H. C. Hodge, J. N. Stannard, and J. B. Hursh, Eds., Springer-Verlag, New York, 1973, p. 503.
58. L. J. Leach, E. A. Maynard, H. C. Hodge, J. K. Scott, C. L. Yuile, G. E. Sylvester, and H. B. Wilson, *Health Phys.*, **18**, 599 (1970).
59. National Council on Radiation Protection and Measurements, "Protection of the Thyroid Gland in the Event of Releases of Radioiodine," NCRP Report 55, 1977.
60. P. Hamilton and E. A. Tompkins, 1975, personal communication (Bureau of Radiological Health, Department of Health, Education and Welfare, Food and Drug Administration, Washington, and Oak Ridge Associated Universities, Oak Ridge, Tenn.).
61. U.S. Nuclear Regulatory Commission, "Reactor Safety Study: An Assessment of Accident Risks in U.S. Commercial Nuclear Power Plants, Appendix VI. Calculations of Reactor Accident Consequences," Report WASH-1400, NUREG-75/014, (U.S. NRC, Washington, D.C.), 1975. [Available from National Technical Information Service, Department of Commerce, Springfield, Va.]
62. National Council on Radiation Protection and Measurements, "Basic Radiation Protection Criteria," NCRP Report 39, 1971.
63. A. Catsch, *Radioactive Metal Mobilization in Medicine*, Thomas, Springfield, Ill., 1964.
64. R. O. McClellan, H. A. Boyd, S. A. Benjamin, R. G. Cuddihy, F. F. Hahn, R. K. Jones, J. L. Mauderly, J. A. Mewhinney, B. A. Muggenburg, and R. C. Pfleger, *Health Phys.*, **23**, 426 (1972).
65. A. Catsch and A. E. Harmuth-Hoene, *Biochem. Pharmacol.*, **24**, 1557 (1975).
66. A. Soffer, *Chelation Therapy*, Thomas, Springfield, Ill., 1964.
67. H. Spencer and B. Rosoff, *Health Phys.*, **11**, 1181 (1965).
68. C. R. Lagerquist, S. E. Hammond, E. A. Putzier, and C. W. Piltingsrud, *Health Phys.*, **11**, 117 (1965).
69. C. R. Lagerquist, E. A. Putzier, and C. W. Piltingsrud, *Health Phys.*, **13**, 965 (1967).
70. L. Jolly, H. A. McClearen, G. A. Poda, and W. P. Walke, *Health Phys.*, **23**, 333 (1972).
71. F. Swanberg and R. C. Henle, *J. Occup. Med.*, **6**, 174 (1964).
72. W. D. Norwood, P. A. Fuqua, R. H. Wilson, and J. W. Healy, "Treatment of Plutonium Inhalation Case Studies," in: *Experience in Radiological Protection, Proceedings of the Second United Nation International Conference on the Peaceful Uses of Atomic Energy*, Geneva, September 1–13, 1958, Vol. 23, 1958, p. 434.

73. J. A. Kylstra, D. C. Rausch, K. D. Hall, and A. Spock, *Am. Rev. Respir. Dis.*, **103**, 651 (1971).
74. J. A. Kylstra, W. H. Schoenfisch, J. M. Herron, and G. D. Blenkarn, *J. Appl. Physiol.*, **35**, 136 (1973).
75. K. E. McDonald, J. F. Park, G. E. Dagle, C. L. Sanders, and R. J. Olson, *Health Phys.*, **29**, 804 (1975).
76. E. Carlens, *J. Thorac. Surg.*, **18**, 742 (1949).
77. B. B. Boecker, B. A. Muggenburg, R. O. McClellan, S. P. Clarkson, F. J. Mares, and S. A. Benjamin, *Health Phys.*, **26**, 505 (1974).
78. S. A. Silbaugh, S. A. Felicetti, B. A. Muggenburg, and B. B. Boecker, *Health Phys.*, **29**, 81 (1975).
79. M. Ingram, "Lymphocytes with Bilobed Nuclei as Indicators of Radiation Exposures in the Tolerance Range," in: *Legal, Administrative, Health and Safety Aspects of Large-Scale Use of Nuclear Energy, Proceedings of the International Conference on the Peaceful Uses of Atomic Energy*, Geneva, August 8–20, 1955, Vol. 13, 1956, p. 210.
80. M. Ingram, "The Occurrence and Significance of Binucleate Lymphocytes in Peripheral Blood After Small Radiation Exposures," in: *Immediate and Low Level Effects of Ionizing Radiations*, A. A. Buzzati-Traverso, Ed. (Proceedings of Symposium June 22–26, 1956, UNESCO, IAEA, and CNRN), Venice, 1960, p. 233.
81. R. Lowry Dobson, "Binucleated Lymphocytes and Low-level Radiation Exposure," in: *Immediate and Low Level Effects of Ionizing Radiations*, A. A. Buzzati-Traverso, Ed. (Proceedings of Symposium June 22–26, 1956, UNESCO, IAEA, and CNRN), Venice, 1960, p. 247.
82. H. G. Jones and B. E. Lambert, "The Radiation Hazard to Workers Using Tritiated Luminous Compounds," in: *Assessment of Radioactivity in Man, Proceedings of the Symposium by International Atomic Energy Agency, the International Labour Organisation, and the World Health Organization*, Heidelberg, May 11–16, 1964, Vol. 2, 1964, p. 419.
83. F. E. Butler, "Assessment of Tritium in Production Workers," in: *Assessment of Radioactivity in Man, Proceedings of the Symposium by the International Atomic Energy Agency, the International Labour Organisation, and the World Health Organization*, Heidelberg, May 11–16, 1964, Vol. 2, 1964, p. 431.
84. A. A. Moghissi, M. W. Carter, and E. W. Bretthauer, *Health Phys.*, **23**, 805 (1972).
85. K. L. Swinth, J. F. Park, and P. J. Moldofsky, *Health Phys.*, **22**, 899 (1972).
86. International Atomic Energy Agency, *Assessment of Radioactive Contamination in Man, Proceedings of a Symposium by the IAEA and the World Health Organization*, Stockholm, November 22–26, 1971, 1972.
87. International Atomic Energy Agency, *Assessment of Radioactive Contamination in Man, Proceedings of a Symposium by the IAEA and the World Health Organization*, Stockholm, November 22–26, 1971, 1972.
88. G. R. Meneely and S. M. Linde, Eds., *Radioactivity in Man, Proceedings of the Symposium on Whole-Body Counting*, International Atomic Energy Agency, Vienna, June 12–16, 1961, 1962.
89. International Atomic Energy Agency, *Directory of Whole-Body Radioactivity Monitors*, IAEA, Vienna, 1970.
90. P. H. Abelson, *Science*, **193: 4250**, 279 (1976).
91. W. D. Metz, *Science*, **192: 4246**, 1320 (1976).
92. W. D. Metz, *Science*, **193: 4247**, 38 (1976).
93. W. D. Metz, *Science*, **193: 4250**, 307 (1976).
94. J. Walsh, *Science*, **193: 4250**, 305 (1976).
95. D. M. Smith, "Ferric Oxide and Plutonium Zirconium Oxide Induced Pulmonary Carcinogenesis," (in preparation).

96. E. C. Anderson, L. M. Holland, J. R. Prine, and C. R. Richmond, "Lung Irradiation with Static Plutonium Microspheres," in: *Experimental Lung Cancer: Carcinogenesis and Bioassays,* E. Karbe and J. F. Park, Eds., Springer-Verlag, New York, 1974, p. 432.
97. L. M. Holland, J. R. Prine, D. M. Smith, and E. C. Anderson, "Irradiation of the Lung with Static Plutonium Microemboli," in: *The Health Effects of Plutonium and Radium,* W. S. S. Jee, Ed., J. W. Press, University of Utah, Salt Lake City, 1976, p. 127.
98. D. A. Holaday, "Uranium Mining Hazards," in: *Uranium · Plutonium · Transplutonic Elements,* H. C. Hodge, J. N. Stannard, and J. B. Hursh, Eds., Springer-Verlag, New York, 1973, p. 300.

CHAPTER TEN

Evaluation of Exposure to Nonionizing Radiation

CHARLES R. McHENRY

1 INTRODUCTION

Man has been exposed to nonionizing radiation since the inception of time. Sources of this radiation are the earth, the atmosphere surrounding the earth, and extraterrestrial (primarily the sun). With the development of the electrical and atomic industries, we have greatly increased our exposure to nonionizing radiation in the home, in the workplace, and out of doors. The variety and intensity of radiation in the environment is increasing at a vastly accelerating rate. When electromagnetic radiation impinges on the human body or any biological system, it produces electrical and magnetic forces and generates heat. If intense enough, the heat and the forces may produce profound effects on biological systems. Occasionally these effects are beneficial; however they are potentially dangerous, and their interaction must be examined with care.

The theory and rationale behind the interaction of nonionizing radiation with biological systems is not fully understood, but there is a rapidly developing body of literature scattered throughout the journals of many disciplines, including biophysics, physics, physiology, neurobiology, and behavior studies. This chapter can serve as an overview to help industrial hygienists or other persons responsible for health protection to arrive at a basic understanding about how nonionizing radiation is developed, how it interacts with biological systems, and what steps can be taken to minimize the potential risks associated with exposures.

2 WHAT IS NONIONIZING RADIATION?

Nonionizing radiation is energy emitted by a body in the form of quanta or photons, each quantum having associated with it an electromagnetic wave possessing a frequency and a wavelength. For convenience this area covers the shortest wavelengths of ultraviolet radiation: visible light, infrared, radio or hertzian, and electric waves. Although these waves are seemingly very different in properties, they are actually identical in every respect except wavelength. They are not electrically charged and they do not possess mass. All have the same velocity of propagation that is, 30×10^9 cm/sec. They all may be refracted, reflected, diffracted, and polarized. Figure 10.1 presents a graphic display of nonionizing radiation in the more commonly used expressions of photon energy: frequency in megahertz (MHz) and in angstrom units (Å).

Radiation is sometimes viewed as a continuous electromagnetic field propagating through a medium and sometimes as particle or photon radiation. This dual description is necessary to explain the various ways that radiation interacts with matter. The wave description is useful in the explanation of the principles associated with diffraction, propagation, interference, refraction, polarization, and other radiation phenomena. The principle of particulate radiation is useful in explaining the concepts associated with photoelectric effect, photon emission, and absorption.

One of the reasons for our attribution of wave motion to light is that it is possible to measure the wavelengths of light. The wavelength of any wave motion is the distance along the direction of travel between corresponding points of successive waves. The usual symbol for wavelength is the lowercase Greek letter lambda (λ).

The number of waves that pass a fixed point during an interval of time is known as wave frequency. Frequency is normally measured as the number of wave crests that pass the fixed point in one second. This is designated in terms of hertz (Hz) (formerly

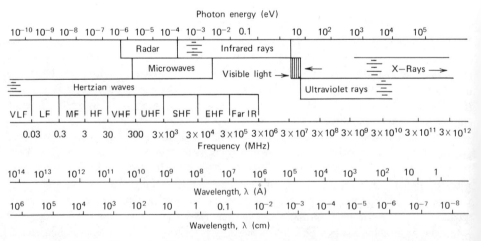

Figure 10.1 The nonionizing radiation portion of the electromagnetic spectrum.

"cycles per second"). One wave or cycle passing a fixed point in one second would have a frequency of 1 Hz.

Other characteristics of electromagnetic radiation include electric (E) and magnetic (B) fields, which travel together. These fields are propagated by the changing magnetic field inducing an electric field and the changing electric field inducing a magnetic field. The two fields propagate in a direction perpendicular to both and at a velocity depending on the electrical and magnetic properties of the medium of travel. In electromagnetic propagation the planes of vibration of the E-field and B-field are at right angles and both perpendicular (or transverse) to the direction of travel. The direction of the E-field is called the wave polarization.

3 ULTRAVIOLET RADIATION

3.1 Physical Characteristics

Ultraviolet radiation is that portion of the electromagnetic spectrum that extends from the longest portion of the X-ray spectrum to the shortest wavelength of visible light. There is no sharp delineation on either end of the band. For sake of convenience this portion of the spectrum ranges from 100 to 4000 Å. Any body with a color temperature of 2500°K or otherwise excited to the corresponding energy level may emit ultraviolet radiation. This emission is produced when excited atomic orbital electrons lose their energy and return to the ground state. The energy ranges of ultraviolet radiation is 3.26 to 123 eV (3000 to 1000 Å). As a practical matter, radiation below 1800 Å is of very little practical significance to the industrial hygienist, since it is readily absorbed by air.

3.2 Sources

Ultraviolet radiation is emitted from highly incandescent bodies and from ionized gases. Examples of hot bodies that emit radiation include the sun, arc welding plasmas, and compressed gas torches operated at high temperatures. The spectrum and intensity of radiation is a direct function of the absolute temperature of the hot body. Examples of ultraviolet produced by sources of ionized gases include pulsed and continuous wave (CW) lasers, gas discharge lamps such as xenon and mercury arcs, low pressure discharge lamps that emit radiation in a line spectrum, medium pressure lamps (1 to 3 atm) that produce line and band spectra, and very high pressure discharge devices ($>$ 3 atm) that produce broad band and continuum spectra.

3.3 Biological Effects

The first law of photochemistry requires that a quantum of energy be absorbed to produce a chemical reaction or effect. The results of numerous studies demonstrate that biological and nonbiological matter is highly specific in its ability to absorb ultraviolet and in the effects that are produced by such absorption. "Absorption" and "action" spectrums have been developed for a very large number of biological and nonbiological materials. An absorption spectrum relates the relative effectiveness of absorption by the medium to wavelength. The action spectrum relates relative effectiveness in producing an effect in the medium to wavelength. Figure 10.2 presents the action spectra of biological significance to the industrial hygienist.

The contemporary concept of the interaction of ultraviolet with biological tissue postulates a chemical or configurational change at the cellular level. This is accomplished either by affecting the electron energy level of the atoms or by changing the rotational, vibrational, and transitional energies of molecules. The ultraviolet spectrum is commonly separated into three separate regions: vacuum (1000 to 1900 Å), which is absorbed by water and air; far (1900 to 3000 Å), which is strongly absorbed by biological molecules and has been demonstrated to produce mutagenetic effects; and near (3000 to 3800 Å), which can be absorbed by some biological molecules.

The harmful effects of ultraviolet energy on living systems appear to be related to its absorption by either the nucleic acid or unconjugated protein of the cell. Ultraviolet radiation below 2900 Å is absorbed entirely in the epidermis. Less than 10 percent of the energy in the wavelength between 2900 and 3200 Å penetrate to the dermis. This shallow penetration of ultraviolet radiation into the body limits the areas of concern primarily to the skin and eyes.

The normal reaction of the skin to ultraviolet is first indicated by a dilation of the minute blood vessels of the corium that manifests itself as a reddening or erythema. Not all regions of the ultraviolet spectrum are capable of producing this effect. Peak effectiveness has been demonstrated at 2967 Å (Figure 10.2). There is a rapid fall-off in

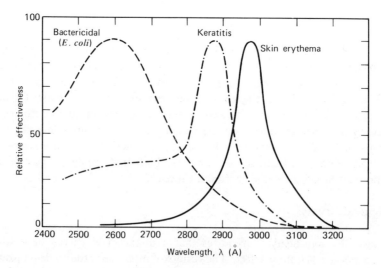

Figure 10.2 Action spectra.

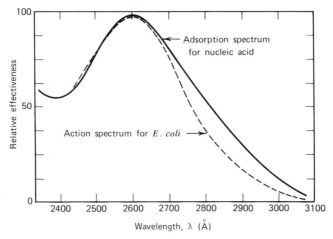

Figure 10.3 Absorption and action spectra.

effectiveness for the production of erythema below 2800 Å and above 3200 Å. The degree of erythema is a direct function of exposure dose. The erythema-producing wavelengths are almost completely absorbed in the avascular epidermis, which averages approximately 0.1 mm in thickness. Therefore it is speculated that erythema results from some chemical agent that is produced photochemically in the epidermal cells that migrate down to the capillary layer and there act on the blood vessels. The decreasing effectiveness for wavelengths above 3000 Å in causing erythema is undoubtedly related to the inability of these energies to produce photochemical effect. It has been shown that photon energies above 4 eV are required to break the molecular bond in the protein molecule.

The effect of ultraviolet on the eye is similar to that for skin. The absorption spectra for the conjunctiva and the cornea have roughly the same general action spectra as does the skin in the production of erythema. The term "keratitis" is used to describe ultraviolet damage produced to the cornea. It is usually manifest as a superficial ulceration that is regressive with time. It appears after a latent exposure period of approximately 24 hr. Figure 10.2 shows the action spectrum for keratitis: the peak activity is at 2880 Å, with rapid fall-off above and below 3005 and 2060 Å. Maximum absorption by the cornea occurs at 2650 Å. Since this is below the peak of the action spectrum, perhaps the basic mechanism involved is not a simple photochemical reaction. Figure 10.3 indicates that the maximum absorption spectrum of 2650 Å corresponds to the absorption spectrum for nucleic acid.

To date, no clear-cut action spectrum has been developed for carcinogenic effects from exposure to ultraviolet irradiation. It has been shown that a wavelength of 2537 Å is relatively effective in producing skin tumors. However this wavelength does not correspond with the absorption spectrum for nucleic acid. The basic mechanism for the production of skin cancer by ultraviolet is not fully understood. It can be speculated that it is related to erythema effect or to secondary chemical changes of hormone, and/or possibly linked to genetic considerations.

In 1928 Findlay reported findings from his study indicating that when coal tar was applied to the skin, it greatly enhanced the ability of ultraviolet to produce damage. Since that time the influence of chemical agents on animal skin photoreactivity has been tested extensively; however the mechanism for basic interactions that result in carcinogenic effects still is not fully understood.

3.4 Measurements

Laboratory instrumentation is available for measurement of ultraviolet irradiation, however there are a limited number of instruments for field measurement. The basic principles of instrumentation include photography, phosphorescence, photoemissivity, chemical action, ionization, and photoelectric action. Field instruments utilize the photoelectric principles for detection. All photoelectric detectors rely on their property of releasing electrons by absorption of radiation, which is considered a quantum of energy. The three types of photoelectric detector commonly in use employ the principles of photoemissivity and photoconductivity, and photovoltaic cells.

The photoemissive principle involves the use of a photosensitive cathode from which electrons are ejected as the result of photon absorption. The ejected photons are collected on a cathode and measured by a meter. Photoconductive cell measurement involves the principle of absorption of photons, causing the displacement of electrons from the semiconductor crystal lattice, leaving "holes" into which other electrons may migrate. The electrical resistivity of such a cell is the function of irradiance. The photovoltaic cell contains a semiconductor film sandwiched between two electrodes and generates an electromagnetic force when irradiated. It requires no external power supply and can operate as a microammeter directly.

There is a strong dependence on wavelengths in total photoelectric yield. Photoelectric receivers are therefore very selective. This makes these types of detector very desirable for studying narrow bands of ultraviolet irradiation. Examples of field instruments currently in use include the International Light Company's Instrument Model IL 600 and the Ultraviolet Products Company's "Black Ray" Meter. The latter is suitable for evaluating biological exposures, since its response is similar to the biologically significant action spectrum.

The problems associated with field instruments include lack of ability for resolving narrow wavelengths and inability to separate visible from ultraviolet radiation. Careful attention must be given to calibration of any field measuring device to assure accurate results. Standardization and calibration techniques are specified in the National Bureau of Standards in Special Publication 300, Vol. 7, "Precision Measurement and Calibration, Radiometry and Photometry."

3.5 Standards for Exposure

Hazard criteria for defining safe exposure levels to ultraviolet irradiation have been established to provide protection against the development of acute effects of erythema and photokeratitis primarily. There are no specified standards for prevention of chronic injury including the prevention of carcinogenic effects. The Committee on Physical Agents of the American Conference of Governmental Industrial Hygienists (ACGIH) have developed threshold limit values for ultraviolet radiation.

3.6 Control Measures for Exposure

The recommended control measures include complete enclosure of the source, selective shielding, personal protective equipment, and protective eye wear. The following factors must be considered in determining the appropriate protective eye wear.

1. Wavelength of the source.
2. Radiant exposure or irradiance.
3. Optical density of the eye wear at the specific wavelengths of interest.
4. Maximum permissible exposure as specified by the safety criteria.
5. Visible light transmission requirement for the "seeing" task.
6. Radiant exposure or irradiance at which the safety eye wear can be damaged.

4 VISIBLE LIGHT- COHERENT AND NONCOHERENT

4.1 Physical Characteristics

Visible light is the portion of the electromagnetic spectrum between 3800 and 7500 Å. By definition it is capable of stimulating the photoreceptors of the eye and producing the physiological response known as vision. The photon energies range from 3.1 to 1.6 eV. These are relatively low energy photons; however they initiate photochemical reactions in the retina that produce the visual sensation.

4.2 Sources of Light

There are many types of visible radiation that occur naturally and as a result of man-made activities. The following is a brief enumeration of the basic principles involved in the production of light.

1. *Thermoradiation temperature.* Most solid and liquid materials emit visible radiation at a wavelength of 7800 Å. At temperatures above 773°K the quantity of radiation emitted is a sole function of temperature. The kinetic energy of heat motion of the molecules of the substance translates into electronic vibrational and rotational motion through collision transfer, which results in emission of a continuum of radiation.
2. *Electrical discharges.* This radiation is produced by arc and spark discharges in gases under reduced pressures, radiation from gases, liquids, and solids under ionizing radiation bombardment such as cathode ray, X-rays and gamma rays, and corpuscular radiation.

3. *Chemiluminescence.* This type of radiation is produced by exergonic chemical reaction. Examples include oxidation of higher alcohols, aldehydes, fatty acids, amides, polyphenols, sugars, urea, and a large variety of other organic compounds by action of ozone or hydrogen peroxide in alkaline or alcoholic potassium hydroxide solutions.

4. *Photoluminescence.* This type of radiation is produced in accordance with the principle of absorption of photon energy and reradiation as resonance energy, fluorescence, or phosphorescence. Examples of radiation produced in this manner include light produced when dye solutions of anthracene, chlorophyll a, and chlorophyll b are irradiated with ultraviolet light.

5. *Radiation produced by high local electric fields.* This principle of light production is similar to that described under electric discharges. The general types fall into the following classes. (*a*) *Triboluminescence* is produced when cane sugar or uranyl nitrate is fractured by mechanical means and a faint spark occurs at the fracture line. It is presumed that the rupture of electrical bonds gives rise to this phenomenon. (*b*) *Sonoluminescence* involves light produced as a result of cavitations caused by sound waves. It is believed that frictional forces give rise to electrostatic fields with the resulting production of light. (*c*) *Crystalloluminescence* is produced during crystal formation of compounds such as strontium bromate. (*d*) *Galvanoluminescence* is produced at the cathode of an electric cell as reduction takes place, possibly different from chemoluminescence due to the radical formation. (*e*) *Electroluminsecence* is produced by specially doped crystals when placed in alternating electric fields. (*f*) *Recombination radiation* is produced by molecules upon neutralization of their respective ions.

4.3 Biological Effects

When a system absorbs light radiation it utilizes the energy in several ways. It may be converted into heat energy or reradiated in the form of resonance radiation, fluorescence, or phosphorescence. In addition, the energy can be used to accelerate or inhibit reactions. Biological response to light has been observed in an extensive number of biological systems. Some of these are photosynthesis, chlorophyll synthesis, chloroplast constitution, seed germination, vegetable growth, flowering, phototropism, chromosome damage, photoreaction, photoprotection, modification to biological "clocks," visual photoreception, and bactericidal action.

There are several laws in photochemistry that govern the interaction of light with biological systems. The two most significant are covered in the following generalizations.

1. The Grotthus-Draper law, which states that only radiations that are absorbed by a reacting system are effective in producing chemical changes.

2. The Stark-Einstein law, which states the photochemical equivalant as follows: each molecule taking part in a chemical reaction induced by exposure to light absorbs one quantum of the radiation, causing the reaction.

Each biological system varies in its ability to absorb light and therefore has a specific absorption spectra. Whenever light interacts with biological systems, the primary process is initiated when light is absorbed by the ground state molecule, providing sufficient free energy either to initiate or to alter the rate of a chemical reaction.

Since light does not penetrate very deeply into the intact human body, the eye and the skin become the prime end organs of concern to the industrial hygienist. The basic discussion of the interaction of light on these organs, with a major emphasis on pathophysiology, is presented below.

Light rays pass through the cornea at the surface of the eye through the liquid medium (aqueous humor) and impinge on the interior chamber of the eye where the retina forms a collecting mechanism that receives the visual impulse. The retina is composed of rods and cones that transduce the light into neuroelectric phenomena. All the tissue in the front of the eye has a very low absorption spectra for visible light, hence no significant photic interaction. The back of the eye has a high absorption spectrum, thus is significantly affected by visible light. The reader is referred to texts on vision physiology for a discussion of the beneficial effects of light on the eye.

The pathological effects of light on the eye have been well demonstrated. The primary action mechanisms involve thermal and photochemical processes. Injuries to the eye from observing the eclipse of the sun have been known since earliest history. Reports of such injuries date back to Hippocrates (460–300 B.C.). There appear to be three predominant factors controlling the potential from such injuries to the eye, namely, wavelength and intensity of radiation, pupillary diameter, and length of exposure. Thermal injury to the eye has been demonstrated by use of a large variety of pulsed and continuous wave light sources, including laser. The method of injury is primarily photocoagulation, such as would result from contact with a hot object.

Photochemical damage to the eye is much more complex than damage produced by thermal injury. Again, wavelength and intensity of the radiation, pupillary diameter, and length of exposure are significant factors that contribute to the effect. The absorption and interaction phenomena of the end cell or perhaps a specific chemical molecule become very important in the understanding of the interaction of light at the micromolecular layer. For example, it has recently been reported that the eye is more susceptible to injury from light of blue wavelengths than other portions of the spectrum. Further investigations are required before a clear understanding is developed on the intraocular action of light at the molecular level.

Visible light of an appropriate wavelength intensity and duration of exposure can produce heat lesions in the skin. The basic mechanism of interaction is believed to be thermocoagulation of tissue similar to that experienced when a hot object is placed in direct contact with the skin. In recent years studies have been demonstrated that DNA repair at the cellular level is affected by visible radiation; however it has not been proved that such changes occur in the intact skin of humans. "Secondary" light effects produce epileptiform responses in seizure-prone individuals who are exposed to pulsing light. It has been demonstrated that the pulse repetition rate should approximate the alpha rhythm frequency of the electroencephalogram to produce this effect.

4.4 Measurement

There are two types of radiation measurement, namely, relative and absolute. Measurement is accomplished by the use of a detector that indicates in any fashion the action of electromagnetic radiation of a specified wavelength range. When no adequate data are available for establishing the ratio of the value of the indicated energy to the energy of the radiation, the indications are essentially "relative" measurement. The absolute measurement of the incident light as a finite number of photons per centimeter squared per second, is quite difficult. The two commonly employed methods involve the principles of thermoelectric and photoelectric radiation detectors. In both cases the detectors must be calibrated for "absolute" measurement against a standard radiating source such as certified standard lamps, which are available from the National Bureau of Standards. The precise measurements are quite time-consuming and require good physical instrumentation. An excellent review article dealing with such a problem is found in *Measurement of Optical Radiations,* a book by George Bauer (1965).

The thermoradiation detector uses the heat engendered by the irradiation to create a change in temperature in the detector. A change in some other value (the pressure, thermoelectromotive force, capacity, length, volume, etc.) that is caused by the change in temperature is then measured. The bolometer is a thermoelectric detector. The principle involved in detection is based on a change in electrical resistance as a function of heat input from a light source.

All photoelectric radiation detectors operate by using their property of releasing electrons by absorption of radiation. Examples of detectors employing the photoelectric principle are photocells and photomultipliers, photovoltaic cells, and photoconductive cells. These are discussed in turn.

1. *Photocell-photomultiplier.* The principle of operation involves the photoelectric effect of an alkaline metal such as sodium, potassium, or cesium. The cell is made by placing a wire gauze or ring electrode on a light-sensitive layer forming the outer electrodes. The ring of the wire or gauze has a positive potential of about 100 V above the support layer, therefore serves as the anodes. When light strikes the cathode layer, electrons are released and migrate to the anodes. The number of electrons released per unit time is proportional to the incident radiation flux over a wide range.

2. *Barrier layer photovoltaic cells.* It is well known that an electric field is produced at the junction between two semiconductors or between a metal and a semiconductor. This field results in the production of a junction with an electrical resistance that is dependent on the direction of current flow. When electrons and the barrier layer are released by radiation absorption, they move under the influence of the electrical field. This produces an increased charge on one side of the barrier and results in the so called photoelectromotive force.

3. *Photoconductive cells.* These detectors take advantage of the action of photoelectrically released electrons in a photosensitive substrate. The released electrons change conduction or resistance of substrate, and the change in resistance is measured for the incident flux of radiation.

4.5 Standard for Control of Exposure

Since the eye is the most sensitive end organ, it has been used to establish so-called safe exposure levels as well as serving as the basis for developing legal requirements under Occupational Safety and Health Act of 1970. The intent is to provide protection against retinal thermal injury. There is a growing body of information to indicate potential retinal end injury by photochemical effect, but the data are inconclusive, and no specific standards or legal requirements have been developed to date. The "Threshold Limit Values for Chemical Substances and Physical Agents in the Workroom Environment" published by the ACGIH summarizes the currently accepted safe levels for exposure to visible light and near-infrared radiation.

4.6 Control of Personnel Exposure

The primary method for protecting the skin and the eyes is to keep exposure below the physiologically significant levels. This is accomplished by enclosures and the use of safety interlocks and attenuation of the radiation by means of selective filters.

5 INFRARED RADIATION

5.1 Physical Characteristics

Infrared radiation is the portion of the electromagnetic spectrum between 750 and 10^6 nm. The short wavelength borders on the limit of visible perception in the deep red, and the long wavelength overlaps the microwave spectrum in the millimeter wave range. The infrared portion of the electromagnetic spectrum occupies a rather wide area. It has been subdivided arbitrarily into three regions. The exact cutoffs of the regions vary, depending on the author. The criteria for such subdivisions vary too and are classified by such factors as type of radiation detectors employed and the regions that affect the rotational-vibrational and higher harmonic oscillations of molecules. The more commonly presented divisions of infrared radiation are as follows:

Near-infrared region	750–2500 nm
Intermediate infrared radiation	2500–25,000 nm
Far infrared radiation	Over 25,000 nm

5.2 Sources of Infrared Radiation

The primary sources of infrared radiation are the atoms and molecules in which the radiation is produced by "oscillations" of electrically charged orbital electrons. In this process the peripheral electrons ("unexcited") receive energy from external sources and pass to an outer or higher energy orbit. In returning to the normal or unexcited state, the electron gives up the absorbed energy in the form of radiation. Infrared sources can be divided into three major types, namely, heat or thermal sources, luminescent sources, and electromagnetic sources.

1. *Thermal sources.* Theoretically every body with a temperature exceeding 0°K is a potential source of radiation. When matter in any physical state is heated, radiation is emitted in the form of heat radiation. The quantity of radiation emitted is a direct function of the heat input.

2. *Luminescent sources.* These sources of infrared radiation are produced by absorption of another radiation or from the potential difference across a discharge path. Examples of luminescent sources are the mercury quartz electrical discharge lamp and the infrared laser, which was produced by Schawlow and Townes in 1958.

3. *Electromagnetic sources.* Radiation is produced by electronic generators of the oscillator type. Examples include hertzian oscillators and magnetrons and klystron generators.

5.3 Biological Effects

Most biological materials are considered to be opaque to energies above 1500 nm because of the almost complete absorption of these radiations by water. Radiant energies in the shorter wavelength regions can be transmitted into deeper tissues of the dermis and the eye; therefore the interest of the industrial hygienist in the biological effects focuses on the effects produced in these organs. Because of the long photon energy of infrared radiation below 1.5 eV, it is believed that such radiation does not enter into photochemical reactions and biological systems.

Following the principles of photon interaction with matter, there is absorption with an increase in the kinetic energy of the system, which results in the production of radiant energy; therefore the primary response from exposure to infrared radiation is a thermal one. The absorption of infrared radiation in the skin is a function of wavelength. Frequencies above 1500 nm are completely absorbed in the surface layer, where the heat is rapidly dissipated by conduction, convection, and the circulatory system. In the spectral region of 700 to 1300 nm there is good transmission into the epidermis as well as the surface of the cornea. The depth of penetration is 5 m into the dermis. This value may be significantly increased where there is high pigmentation of the skin.

The predominant effects of infrared radiation on the skin include the production of acute skin burn, increased vasodilation of the capillary beds, and an increased pigmentation of the epidermis. Where there is continuous or repeated exposure to high intensities of radiation, the erythematous appearance due to vasodilation may become permanent. The exact interaction of infrared light with the epidermis is not fully understood.

The cornea of the eye is highly transparent to energies between 700 and 1300 nm. It becomes opaque to radiant energies above 2000 nm. The occurrence of cataracts due to heat was first recorded among workmen early in the eighteenth century. It was not until early in the twentieth century that the disease was studied. The pathological changes taking place in the production of cataracts appear to occur in the posterior cortex of the lens. The long wave infrared rays are almost completely absorbed by the cornea. The short infrared rays are transmitted through the cornea and are absorbed in the lens. Only a very small portion of the short infrared rays reach the retina. The basic mechanism for injury is believed to be the absorption of energy by atoms and molecules causing an increased vibrational state with the subsequent production of heat energy, which causes coagulation of tissue.

5.4 Measurements

Instruments utilized in measuring infrared radiation employ the following categories of detector: photochemical, luminescence (especially quenching, or its converse, stimulation of phosphorescence), photoelectric, and thermal (temperature increases of the substance being measured or indicated by various methods).

Detectors based on the first three phenomena are sensitive in a comparatively narrow range of the infrared spectrum. The output signal determined by the given phenomenon is not only dependent on the intensity of the incident radiation but also on its wavelength. However these detectors exhibit greater sensitivity for specific wavelength than do the noncollective detectors (broad band nonspecific). In the case of thermal detectors the signal generated by the detector is independent of the wavelength of incident radiation, hence nonselective. It can be used over the entire range of the infrared spectrum as well as in the visible and ultraviolet regions. Section 4.4 discusses the interaction of electromagnetic radiation in the visible region for the four phenomena cited.

The thermal detector system of measurement is the method of choice in quantifying infrared radiation. Instruments of this type are based on the principle of nonselective or selective detectors, as applicable. Nonselective detectors are thermometer "bulbs" coated with a layer of "black" substance that acts to absorb thoroughly the incident radiation. An equilibrium increase in temperature in the "bulb" is due to incident radiation, therefore represents a measure of the incident radiant flux. The several detectors of the "thermometer-bulb type" include "black" body absorbers, bolometers, semiconducting bolometers (thermistors), golay pneumatic cells, thermocouples, and thermopiles.

Selective infrared radiation instrumentation is based exclusively on the principle of photoeffect occurring in a semiconductor. The principle of operation involves the absorption of energy by electrons in the outer shell of specific atoms uniformly distributed throughout a crystalline solid substance. When energy is absorbed, an internal photoeffect is produced. Depending on the type of detector involved, the response will be photoconductive, photovoltaic, or photoelectromagnetic. This changes the conductivity of the crystal, which therefore can be calibrated as a quantitative detector of infrared radiation. Examples of selective detectors are silver-cesium photocathodes, lead-telluride photoconductive cells, lead-selenide and lead-telluride photoconductive cells, indium-antimide detector cells, and doped germanium photoconductive cells.

5.5 Standards for Exposure

Since the skin and the eyes are the prime organs of interest, it is not surprising to see legal and recommended threshold limits specified for safety requirements. The philosophy underlying the values selected is to prevent thermal injury either to the skin or to the eyes. The two most widely quoted standards of recommended good practice are the American National Standards Institute Standard for the Safe Use of Lasers (10) and the ACGIH Threshold Limit Value for Chemical Substances and Physical Agents in the Work Environment (6). Several states have enacted laws based on the recommendations of ANSI Z136.1.

5.6 Control Measures for Exposure

The recommended control measures include complete enclosure of the source, and use of selective shielding, personal protective equipment, and protective eye wear. The factors that must be considered in determining the appropriate protective eye wear are the same as those given in Section 3.6.

6 HERTZIAN (RADIO) WAVES

6.1 Physical Characteristics

Hertzian waves or radiowaves are the portion of the electromagnetic spectrum between frequencies of 0.01 and 3×10^6 MHz. This arbitrary division overlaps a portion of the infrared band in the 0.01 MHz (very low frequency) range. This portion of the spectrum is further divided as shown in Table 10.1. Clearly the radiofrequency spectrum includes many frequencies. In addition to the band designations given in Table 10.1, there is a further commonly used code letter delineation of the portion of the spectrum employed for radar operation. Table 10.2 lists the band designations for

Table 10.1 Radiofrequency Bands

Frequency (MHz)	Band	Description
Below 0.03	VLF	Very low frequency
0.03–0.3	LF	Low frequency
0.3–3	MF	Medium frequency
3–30	HF	High frequency
30–300	VHF	Very high frequency
300–3,000	UHF	Ultra high frequency
3,000–30,000	SHF	Super high frequency
30,000–300,000	EHF	Extremely high frequency

EVALUATION OF EXPOSURE TO NONIONIZING RADIATION

Table 10.2 Radar Band Designations

Band Designation	Frequency (MHz)	Wavelength (cm)
P	220–390	133.3–76.9
L	390–1,550	76.9–19.3
S	1,550–5,200	19.3–5.77
C	3,900–6,200	7.69–4.84
X	5,200–10,900	5.77–2.75
K	10,900–36,000	2.75–0.834
Q	36,000–46,000	0.834–0.652
V	46,000–56,000	0.652–0.536

radar. The common term "microwave" is used to define another somewhat arbitrary band of frequencies between 300 and 300,000 MHz, which includes most of the frequencies normally used by radar.

All hertzian bands can be generated and propagated as a continuous wave or as a pulsed wave. Continuous wave is more commonly associated with communications, transmitting devices, and consumer products, whereas pulsed waves are usually employed for environmental scanning and industrial and medical uses.

6.2 Sources

Man is exposed to hertzian radiation as a result of both extraterrestrial radiation and the activities of people. Extraterrestrial radiation is extremely weak, therefore insignificant from an industrial hygiene viewpoint. The principles involved in generating hertzian waves can briefly be stated as follows. Most man-made hertzian radiation systems involve a transmitter and a receiver. Frequently the system incorporates a device to measure the time interval between pulse and echo, the latter being associated with ranging systems. In the most common hertzian generating system, the pulse finds its beginning in the flow of electrons in an electron tube or a semiconductor. Some of the more prominent generators are magnetrons, klystrons, and backward wave oscillators. Some of the semiconductor devices used in hertzian generators are diodes, Gunn effect diodes, and tunnel diodes. The generated radiation may be emitted either as a continuous wave or a pulsed wave. Among the devices that emit hertzian radiation are communication transmission systems, medical diathermy equipment, microthermy machines, radar units, and microwave heating systems.

6.3 Biological Effects

Absorbed hertzian energy is transformed into increased kinetic energy in the absorbing molecules. This results in the production of heat in the tissue. Such heating results primarily from ionic conduction and vibration of the dipole molecules of water and protein. Absorption of such radiation is dependent on specified electrical properties of the absorbing media, mainly its dielectric constant and electroconductivity. These properties change as a function of frequency of applied electric field.

Wavelengths of less than 3 cm (SHF and EHF bands) are absorbed primarily in the skin. Those from 3 to 10 cm penetrate more deeply. Wavelengths from 25 to 200 cm (VHF and UHF bands) penetrate deepest and involve internal body organs. It is generally agreed that the body is essentially transparent to wavelengths greater than 200 cm; thus there is no significant absorption from such radiation.

The thermal effect of hertzian radiation on biological systems has been well demonstrated. Documentations from some investigators suggest nonthermal effects at the microscopic, biochemical, and neurological levels. Reports of these investigations indicate that biological systems are affected through field force effects, excitation of biological membranes, and macromolecular resonance. There is considerable questioning about the existence of nonthermal effects by many researchers in the field.

A unit of measurement has not been developed that satisfactorily relates the quantity of hertzian energy absorbed to the expected biological effects, as has been done in the case of ionizing radiation. Some quantitative relationships have been developed to correlate the observed biological effect with that of the wavelength and power density of the radiation. These studies have demonstrated that biological effects are more closely related to the average power dose than to the peak power dose. The "dosage unit" is expressed in terms of watts per square centimeter (W/cm^2).

In studying the biological effects produced by hertzian radiation, one must consider the frequency of the energy and its relationship to the physical dimensions of the object being exposed. Generally it is accepted that to receive significant absorption, the physical size of the object being irradiated must be equivalent to at least $1/10$ of a wavelength of the instant radiation. The human body is a three-dimensional mass having width, depth, and height and can be thought of as a receiving antenna when placed in a radiation field. As the frequency of the radiation increases, the wavelength decreases and man's height represents an increasing number of "electrical wavelengths." Conversely, as the frequency is decreased, the wavelength increases, and man becomes a less significant object in the radiation field. For example, in a 30 MHz field, man's height of 1.7 m (5 ft 7 in.) represents 0.17 wavelength. At frequencies below 3.0 MHz, this represents less than 0.017 wavelengths (see Table 10.3).

As previously noted, absorbed hertzian radiation produces a heat rise in tissue. It is therefore not surprising to see that the organs in the body that have poor heat flow control—namely, the eyes, lungs, testes, gallbladder, urinary tract, and a portion of the gastrointestinal tract—are the most critical ones from the damage viewpoint.

The lens appears to be the most susceptible part of the eye to hertzian radiation damage. It has a very inefficient circulatory system, which provides for heat exchange to the surrounding tissues. It is believed that temperature elevation within the eye alters the cellular and molecular structures of the lens. This damage is manifest in the form of

Table 10.3 Comparison of Frequency, Wavelengths, and Equivalent Number of Wavelengths of Man 1.7 m Tall

Frequency (MHz)	Wavelengths (m)	Wavelengths (cm)	Equivalent Number of Wavelengths
3	100	100,000	0.017
30	10	10,000	0.17
300	1	100	1.7
3,000	0.1	10	17.0
10,000	0.03	3	56.6

increasing lens opacity, which results in the formation of cataracts. A temperature rise of 4°C is believed to be required to produce such damage.

The effects of hertzian radiation on testes has been studied by several investigators. Low level exposure of 5 mW/cm² for an indefinite exposure has been reported to produce damage to the morphology to the testes. Exposures of 250 mW/cm² have produced gross damage such as edema, enlargement of the testes, atrophy, cibrosis, and coagulation and necrosis of the seminiferous tubules. It is generally acknowledged that damage from such intense exposure results primarily from thermal injury.

Several investigators have reported nonthermal injury from hertzian radiation with a primary emphasis on the production of chromosome abnormality in cell reproduction. Other nonthermal effects of hertzian radiation have been demonstrated, such as increased clotting time of blood when animals are exposed to radiation levels that result in near-normal body temperature.

A phenomena known as "pearl chain formation" results when milk and human blood are exposed to hertzian radiation. Pearl chain formation occurs when oil or fat droplets in these liquids are exposed to hertzian radiation. Before the exposure the oil droplets are arranged in a random pattern. After exposure they align themselves in a "chain-like" formation resembling a string of pearls. When the radiation is turned off, the oil droplets return to the random pattern.

As noted previously, dimensional resonance of the human body can result in heating as a thermal effect from hertzian radiation. When other resonance that is not dimensional depends primarily on the material irradiated in its molecular structure, the electrons orbiting about the molecule can resonate; also the molecule itself can resonate and orient itself in respect to the energy field. Such resonance could result in the movement of constituents of the molecule, which would tend to stretch or strain the bond between them. It is conceivable that breakage of the bonds could occur; therefore it is postulated that such molecular breakup could result in the formation of different molcules and result in denaturation of living tissue.

6.4 Measurement

The hertzian radiation field exhibits both spatial and temporal variations. These are a function of the physical environment as well as the nature of the radiating source. The objects in the field of radiation, both natural and man-made, cause nonuniform power densities at any given point downrange. This spatial variation is caused by the interaction of the flow of energy reflected from the objects in the field of radiation. This tends to produce peaks in power level where common wave interaction takes place. The introduction of test equipment for personnel into a field gives further rise to perturbations to the field strength.

To assure personnel safety it is recommended that theoretical calculations be made to determine power level before any field measurement is attempted. Methods for computing power density with emphasis in the microwave region is given in the American National Standards Institute Standard C95.3-1973, "Techniques and Instrumentation for Measurement of Potentially Hazardous Electromagnetic Radiation at Microwave Frequencies" (14).

Radiofrequency power measurements are accomplished by use of a suitable antenna and associated circuitry to measure and report the incident energy. Three common types of power meter are presently employed. They involve the use of calorimetoric, bolometric, and electronic effects. The bolometer operates on the principle of absorption of energy and conversion of a fraction of the energy into heat. This can be done either dynamically or in a static manner. The basic static calorimeter consists of a termally isolated load that converts the absorbed electromagnetic energy into heat and a device for measuring the load-temperature rise. As with most meters, prior calibration is required for accurate measurement. The dynamic or flow calorimeter operates by converting the electromagnetic energy into heat in a liquid; essential components are a system for circulating the liquid, and a means for measuring temperature difference between circulating liquid and the load.

Bolometric power meters employ a temperature-sensitive resistant type element. The temperature-resistant element is heated from absorbed radiofrequency power. The change in temperature affects the electrical resistance and the current flow, which can be calibrated as a function of radiofrequency energy. Examples of the bolometer are as follows.

1. *Barretter.* This consists of a Wollaston resistance wire mounted in a dielectric capsule with metal incaps. It is used to measure radiofrequency power directly and has a fast response time (50 to 400 μsec).

2. *Thermistor.* This divert measurement instrument consists of a small bead of semiconducting material deposited on thin lead wires. It is similar in operation to barretter except that it is operated by a negative temperature coefficient. The thin film bolometer principle of operation is based on the heating of a thin metallic film that is vacuum deposited onto a glass or microsubstrate. Increased temperature increases resistance to current flow and can be calibrated as a function of power density.

3. *Load lamp.* This is similar to barretter except that a much larger wire is used for a greater power handling capacity.

4. *Wave guide wall bolometer (enthrakometer).* The basic principal of operation involves the use of a temperature-sensitive resistive film deposited on an insulated substrate that forms part of the narrow wall of a rectangular wave guide. The film absorbs a small quantity of radiofrequency power. The resultant changes in film resistance to current flow are measured using "bridge" techniques.

Three meters employing the electronic effect are briefly noted as follows:

1. *Thermocouple and thermoelement power meters.* The basic principle involves the use of a thermocouple that when heated directly by radiofrequency radiation develops a resistance to current flow, which can be calibrated to give precise measurements.
2. *Crystal diode power meters.* The diodes detect the radiofrequency voltage across the transmission lines. Power level is read directly by a meter.
3. *Mechanical and Hall effect power meters.* The basic principle calls for devices for converting radiating pressure into torque and vibration.

6.5 Standards for Exposure

Standards for personal protection against hertzian radiation have been developed for the microwave portion of the electromagnetic spectrum. Most of the standards commonly in use in the Western world are based on the original United States of America Standards Institute* standard, which was tentatively adopted about 18 years ago. The standard was based on the amount of exogenous heat the body could tolerate and dissipate without any rise in body temperature. This tolerance level was calculated to be 10 mW/cm^2 for continuous exposure. The current standard for acceptable exposure can be found in the ACGIH TLVs for Chemical Substances and Physical Agents in the Workroom Environment (6).

6.6 Control Measures for Exposure

The recommended control measures include complete enclosure of the source, selective shielding, personal protective equipment, and protective eye wear. The following factors must be considered in selecting the appropriate protective wear:

1. Wavelength of the source.
2. Field strength.
3. Relative effectiveness of the shielding material for the length of interest.
4. Maximum permissible exposure as specified by the safety criteria.

* Now the American National Standards Institute.

REFERENCES

1. A. Hollaender, *Radiation Biology,* Vols. 2, 3, McGraw-Hill, New York, 1955.
2. M. N. Varma and E. A. Traboulay, Jr., *Biological Effects of Non-Ionizing Radiation.* Howard University, Washington, D.C., 1975.
3. S. M. Michaelson, "Human Exposure to Non-Ionizing Radiant Energy—Potential Hazards and Safety Standards," *Proc. IEEE,* **60:** 4, 389–421 (1972).
4. A. C. Giese, *Photophysiology,* Vols. 1, 2. Academic Press, New York, 1964.
5. National Institute for Occupational Safety and Health, "A Recommended Standard for Occupational Exposure to Ultraviolet Radiation," Government Printing Office, Washington, D.C., 1977.
6. American Conference of Governmental Industrial Hygienists, "Threshold Limit Values for Chemical Substances and Physical Agents in the Workroom Environments with Intended Changes for 1977," ACGIH, P.O. Box 1937, Cincinnati, Ohio 45201, 1977.
7. I. Matelsky, *Non-Ionizing Radiations: Industrial Hygiene Highlights,* Vol. 1. Industrial Hygiene Foundation of America Inc., Pittsburgh, Pa., 1968.
8. A. J. Estine, *Precision Measurement and Calibration Electricity—Radiofrequency,* Vol. 4. Government Printing Office, Washington, D.C., 1970.
9. R. Mavrodineanu, J. I. Schultz, and O. Menis, "Accuracy in Spectro-Photometry and Luminescence Measurements." U.S. Department of Commerce, National Bureau of Standards, Gaithersburg, Md., 1973.
10. American National Standards Institute, ANSI Z136.1, American National Standard for the Safe Use of Lasers, ANSI, New York, 1976.
11. H. H. Seliger and W. D. McElory, *Light: Physical and Biological Action,* Academic Press, New York, 1965.
12. A. Vasko, *Infra-Red Radiation,* Chemical Rubber Co., Cleveland, 1968.
13. I. Simon, *Infrared Radiation.* Van Nostrand, Princeton, N.J., 1966.
14. American National Standards Institute, "American National Standard Technique and Instrumentation for Measurement of Potentially Hazardous Electromagnetic Radiation and Microwave Frequencies," No. C95.93-1973. Institute of Electrical and Electronics Engineers, New York, 1973.
15. S. M. Michaelson, "Effects of Exposure to Microwaves: Problems and Perspectives," *Environ. Health Perspect.,* **8,** 133–156 (1974).
16. C. C. Johnson and M. L. Shore, Eds., *Biological Effects of Electromagnetic Waves,* Vols. 1, 2, Government Printing Office, Washington, D.C., 1976.

CHAPTER ELEVEN

Evaluation of Exposure to Noise

VAUGHN H. HILL

1 PHYSIOLOGICAL EFFECTS OF NOISE ON WORKERS

1.1 Individual Reaction

Like most physiological stresses, noise causes reactions that can differ markedly between individuals. For example, one individual might consider a level of 80 to 85 dBA intolerable, whereas another might not object to a level exceeding 100 dBA. In addition, the hearing impairment that one individual might suffer because of long exposure at, say, 100 dBA, might differ markedly from that of another. This difference in individual reaction must be recognized when setting standards because setting a standard at a certain level does not guarantee that all people will be protected. Individual audiometric testing is the only way to identify individual reaction. If standards are not to be set at unrealistically low levels, protection of 100 percent of the population will not be accomplished unless periodic audiometric testing is a part of an effective hearing conservation program.

To illustrate how unrealistic it is to expect to set a standard to protect everyone against impairment of hearing due to exposure to excessive noise, consider the experience of the U.S. Environmental Protection Agency (EPA). Congress charged this agency with the responsibility of recommending maximum levels that would protect 100 percent of the population with a margin of safety. EPA's major medical consultant testified that someone might suffer hearing damage at a noise level as low as 55 dBA (1). If EPA were to take this charge seriously, it would obviously have to set the maximum allowable level at less than 55 dBA. This is in the sleeping level range and could not be met in most commercial areas and certainly not in industrial areas. This means that some risk is involved in setting a limit on noise, and probably the best way to minimize this risk is by means of the audiometric testing program as specified by the proposed rules of the Occupational Safety and Health Administration (OSHA).

1.2 Nonauditory Effects of Noise Exposure

Exposure to excessive noise can cause physiological stresses in areas other than hearing acuity. It can interfere with one's sense of balance (vertigo) and can cause nausea. A sudden noise will produce a startle reaction in anyone, even if it is expected. This "startle" effect cannot be voluntarily controlled. The blood vessels contract, the blood pressure increases, the pupils of the eyes dilate, and the voluntary and involuntary muscles become tense. After a few minutes these reactions return to normal and apparently cause no permanent damage. However since these changes occur with every repetition of the noise, even during sleep, and since these stresses are similar to other physiological stress reactions that in general are considered harmful if continued for long periods, it may be that the stresses due to excessive noise are harmful, especially to supersensitive individuals.

It has been pointed out that one of the effects of noise is to stimulate the adrenal glands to produce catecholamines, which are also seen in abnormally large amounts in certain heart and other circulatory diseases. Therefore noise exposure might contribute to the incidence of heart diseases. The small amount of research that has been done in an attempt to verify such possibilities is inconclusive. In these tests noise has been only one of a number of physiological stresses, and the effects of noise alone were not established. In addition, the noise levels at which these nonauditory physiological effects occur are usually considerably above those necessary to avoid hearing impairment. Therefore if noise is controlled at a level to prevent hearing impairment, the chances that it will cause these other serious physiological stresses are remote and in general are not important as far as occupational noise is concerned (2, 3). Likewise, if noise is controlled to prevent hearing impairment, it is not likely to be significant in combination with other physiological stresses.

1.3 Occupational Noise Exposure Criteria

It has been recognized for more than 100 years that worker exposure to excessive noise will result in hearing impairment. However it has been only recently that the problem has received serious study. Yet much remains to be learned about the relation of hearing loss to noise exposure. The reason for this is the difficulty in making a scientifically controlled study for which, at least, the following factors or conditions would be required:

1. Test subjects with normal hearing.
2. Test subjects who would not be subjected to excessive off-the-job noise during the test period.

3. Test duration of 5 years minimum.
4. Definable occupational noise exposures that would remain constant for the test period.
5. For a steady state noise study, the noise level should not vary by more than ±2 dBA.
6. For an intermittent noise study, the on-off noise bursts should be uniform or repetitious so that the noise dose could be accurately determined.
7. There should be at least 1000 test subjects available (both men and women).

Such test conditions cannot, or at least have not, been met for the following reasons:

1. Hearing impairment from exposure to excessive noise is permanent and irreversible, therefore subjects are not readily available.
2. Government regulations prohibit exposure to excessive noise.
3. Off-the-job noise is becoming a more and more serious problem because of the high level experienced almost everywhere.
4. Occupational noise exposures do not remain constant long enough.
5. Workers do not stay on the same job long enough.
6. Occupational noise levels are not constant enough.
7. Long-time exposure (years) is required for change in hearing acuity to become measurable.

In spite of all these difficulties, three major studies have resulted in very similar conclusions.

1. "Noise Control—Percent of Population Protected" (4) and "The Risk of Hearing Impairment as a Function of Noise Exposure," (5) by W. L. Baughn.
2. "Hearing and Noise in Industry" (6) by W. Burns and D. W. Robinson.
3. Inter-Industry Noise Study (7).

All these studies concluded that most workers would not suffer significant hearing impairment if occupational noise exposure did not exceed 90 dBA during the daily work shift. A significant hearing impairment is defined as an average hearing loss greater than 15 dB in the frequencies 500, 1000, and 2000 Hz using the American Standards Association* 1951 audiometric zero, or 25 dB average at the same frequencies using the ISO 1964 audiometric zero.

Baughn, Burns, and Robinson agree that the noise exposure levels in their studies were more variable than desired. The object of the more recent Inter-Industry Noise Study was to test more definable and controlled noise exposures. However as this study progressed it became apparent that the noise exposures could not be held as constant as

---

* Superseded by USASI, and ANSI.

desired. It is doubtful that more accurate tests ever can be obtained, for the reasons mentioned above.

Long-term experience in industry, such as reported by the Dow Chemical Company (8) and the Du Pont Company (9, 10) support the adequacy of the 90 dBA criterion, especially when coupled with the hearing testing program for those exposed to noise levels between 85 and 90 dBA.

To clarify widespread confusion and misunderstanding about governmental standards, the existing and proposed standards are included here.

1.4 Existing Government Regulation (OSHA)

The existing OSHA regulation is as follows:

29 CFR 1910.95 Occupational Noise Exposure

a. Protection against the effects of noise exposure shall be provided when the sound levels exceed those shown in Table 11.1 [of this section] when measured on the A Scale of a standard sound level meter at slow response.

b. When employees are subjected to sound exceeding those listed in Table 11, feasible administrative or engineering controls shall be utilized. If such controls fail to reduce sound levels

Table 11.1 Permissible Noise Exposures[a]

Duration per day (hr)	Sound Level (dBA, slow response)
8	90
6	92
4	95
3	97
2	100
1½	102
1	105
½	110
¼ or less	115

[a] When the daily noise exposure is composed of two or more periods of noise exposure of different levels, their combined effect should be considered, rather than the individual effect of each. If the sum of the following fractions: $C_1/T_1 + C_2/T_2 + \cdots + C_n/T_n$ exceeds unity, then the mixed exposure should be considered to exceed the limit value; C_n indicates the total time of exposure at a specified noise level, and T_n indicates the total time of exposure permitted at that level.

EVALUATION OF EXPOSURE TO NOISE 429

within the levels of the table, personal protective equipment shall be provided and used to reduce sound levels to within the levels of the table.

c. If the variations in noise level involve maxima at intervals of 1 second or less, it is to be considered continuous.

d. In all cases where the sound levels exceed the values shown herein, a continuing, effective hearing conservation program shall be administered.

Exposure to impulsive or impact noise should not exceed 140 dB peak sound pressure level.

This is a good regulation, but it does have weaknesses which makes enforcement difficult and therefore worker protection is less than desirable. For instance the 140 dB peak impact limit does not provide adequate protection. In addition, an effective hearing conservation program is not defined. At least such additional subjects as those listed below need clarification.

1. Hearing protector ratings and use.
2. Hearing testing equipment and procedures.
3. Evaluating hearing tests.
4. Frequency of hearing tests.
5. Plant level monitoring of worker environment and use of hearing protectors.
6. Information and warnings signs.

To strengthen and make the existing regulation more definitive, the DOL in 1974 issued proposed rules as follows:

PROPOSED RULES

Occupational Noise Exposure

a. *Application and purpose.* This section applies to occupational noise exposures in employments covered in this part. The purpose of this standard is to establish requirements and procedures that will minimize the risk of permanent hearing impairment from exposure to hazardous levels of noise in workplaces.

b. *Definitions.* "Administrative controls" means any procedure which limits daily noise exposure by control of the work schedule. Hearing protectors do not constitute administrative controls.

"Assistant Secretary," the Assistant Secretary of Labor for Occupational Safety and Health, U.S. Department of Labor, or his designee.

"Audiogram," a graph or table of hearing level as a function of frequency that is obtained from an audiometric examination.

"Baseline audiogram," the first audiogram taken during employment with the current employer.

"Certified audiometric technician," an individual who meets the training requirements specified by the Intersociety Committee on Audiometric Technician Training [*Am. Ind. Hyg. Assoc. J.*, **27**, 303–304 (May–June, 1966)] or who is certified by the Council of Accreditation in Occupational Hearing Conservation.

"Daily noise dose" (D), the cumulative noise exposure of an employee during a working day.

"dBA" (decibels—A-weighted), a unit of measurement of sound level corrected to the A-weighted scale, as defined in ANSI S1.4-1971, using a reference level of 20 micropascals (2 × 10^{-5} newtons per square meter).

"Director," the Director, National Institute for Occupational Safety and Health, U.S. Department of Health, Education and Welfare, or his designee.

"Engineering control," any design procedure that reduces the sound level.

"Hearing level" the amount, in decibels, by which the threshold of audibility for an ear differs from the standard audiometric reference level.

"Peak sound pressure level," the peak instantaneous pressure expressed in decibels, using a reference level of 20 micropascals.

"Workplace sound level," the sound level measured at the employee's point of exposure.

"Impulse or impact noise"—a sound with a rise time of not more than 35 milliseconds to peak intensity and a duration of not more than 500 milliseconds to the time when the level is 20 dB below the peak. If the impulses recur at intervals of less than one-half second, they shall be considered as continuous sound.

"Significant threshold shift," an average shift of more than 10 dB at frequencies of 2000, 3000, and 4000 Hz relative to the baseline audiogram in either ear.

c. *Permissible exposure limits.*

1. Steady state noise, single level.

    i. The permissible exposure to continuous noise shall not exceed an 8 hr time-weighed average of 90 dBA with a doubling rate of 5 dBA. For discrete permissible time and exposure limits, refer to Table 11.2, which is computed from the formula in paragraph (c)(1)(ii) of this section.
    ii. Where Table 11.2 does not reflect actual exposure times and levels, the permissible exposure to continuous noise at a single level shall not exceed a time amount "T" (in hours) computed by the formula:

$$T = \frac{16}{2[0.2(L - 85)]}$$

where L is the workplace sound level measured in dBA on the slow scale of a standard sound level meter. The relationship between time and sound level is depicted in Figure 11.1.

2. *Steady state noise—two or more levels.* Exposures to continuous noise at two or more levels may not exceed a daily noise dose D of unity (i) where D is computed by the formula:

$$D = \frac{C_1}{T_1} + \frac{C_2}{T_2} + \cdots + \frac{C_n}{T_n}$$

where C is the actual duration of exposure (in hours) at a given steady state noise level; and T is

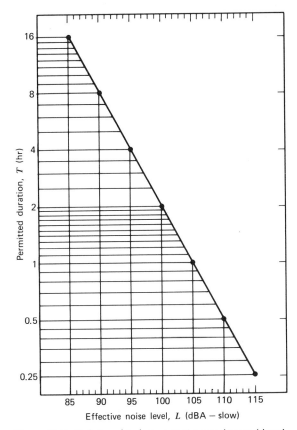

Figure 11.1 Relationship between time and sound level.

the noise exposure limit (in hours) for the level present during the time C, computed by the formula in paragraph (c)(1)(ii) of this section.

3. *Maximum steady state noise.* Exposures to continuous noise shall not exceed 115 dBA, regardless of any value computed in paragraphs (c)(1) or (c)(2) of this section.

4. *Impulse or impact noise.*

   i. Exposures to impulse or impact noise shall not exceed a peak sound pressure level of 140 dB.
   ii. Exposures to impulses of 140 dB shall not exceed 100 such impulses per day. For each decrease of 10 dB in the peak sound pressure level of the impulse, the number of impulses to which employees are exposed may be increased by a factor of 10.

d. *Monitoring.*

1. *Duty.* Each employer shall determine if any employee is exposed to a daily noise level dose of 0.5 or above, and shall determine if any employee is exposed to impulse or impact noise in

Table 11.2

Sound Level (dBA)	Time Permitted (hr, min)
85	16 0
86	13 56
87	12 8
88	10 34
89	9 11
90	8 0
91	6 58
92	6 4
93	5 17
94	4 36
95	4 0
96	3 29
97	3 2
98	2 50
99	2 15
100	2 0
101	1 44
102	1 31
103	1 19
104	1 9
105	1 0
106	0 52
107	0 46
108	0 40
109	0 34
110	0 30
111	0 26
112	0 23
113	2 20
114	0 17
115	0 15

excess of the exposure permitted by paragraph (c)(4) of this section. Such determinations shall be made:

    i. At least annually.
    ii. Within 30 days of any change or modification of equipment or process, or other workplace or work practice modifications affecting the noise level.

2. *Procedure.* If determinations made pursuant to paragraph (d)(1) of this section reveal

EVALUATION OF EXPOSURE TO NOISE

any employee exposure to a daily noise dose of 0.5 or above, or exposure to impulse or impact noise in excess of that permitted by paragraph (c)(4) of this section, the employer shall:

   i. Identify all employees who may be so exposed.
   ii. Measure the exposure of the employees so identified.
   iii. Make all noise level measurements with the microphone of the sound measuring instrument at a position which most closely approximates the noise levels at the head position of the employee during normal operations.

3. *Equipment.*

   i. Measurements of steady state noise exposures shall be made with a sound level meter conforming as a minimun to the requirements of ANSI S1.4-1971, Type 2, and set to an A-weighted slow response or with an audiodosimeter of equivalent accuracy and precision. The unit of measurement shall be decibels re 20 micropascals A-weighted.
   ii. Measurements of impulse or impact noise exposures shall be made with a sound level meter conforming as a minimum to the requirements of the ANSI S1.4-1971, Type 1 or Type 2, with a peak hold capability or accessory. For peak hold measurements, the rise time of the instrumentation shall be not more than 50 microseconds. The decay rate for the peak hold feature shall be less than 0.05 decibel per second. The unit of measurement shall be decibels peak sound pressure level re 20 micropascals.

4. *Calibration of equipment.* An acoustical calibrator accurate to within plus or minus one decibel shall be used to verify the before and after calibration of the sound measuring instrument on each day noise measurements are taken.

5. *Observation of monitoring*

   i. *Duty.* The employer shall give employees or their representatives an opportunity to observe any monitoring of the noise levels in the workplace which is conducted pursuant to this section.
   ii. *Notification of employee right.* Written notice of the opportunity to observe the monitoring required by this section shall be prominently posted in a place regularly visited by affected employees and where notices to employees are usually posted. The employer shall take steps to insure that this notice is not altered, defaced, or covered by other material.

      A. The notice shall be posted at least three working days before monitoring is scheduled to occur.
      B. The notice shall list the time and place where monitoring will take place.
      C. The employer may require the employee or the employee representative to give advance written notification of intent to observe such monitoring.

   iii. *Exercise of opportunity to observe monitoring.*

      A. When observation of the monitoring of the workplace for noise levels requires entry into an area where the use of personal protective devices is required, the

employer shall use such equipment and comply with all other applicable safety procedures.
B. Observers shall be given an explanation of the procedure to be followed in measuring the workplace noise level.
C. Observers shall be permitted, without interference with persons performing the monitoring, to:

1. Visually observe all steps related to the collecting and evaluation of the noise level data that are being performed at the time.
2. Record the results obtained.
3. Have a demonstration of the calibration function tests of the monitoring equipment when the calibrations are performed at the worksite before monitoring; where the calibrations are not performed at the worksite, the techniques shall be explained.

e. *Methods of compliance.*

1. Whenever employees are exposed to workplace sound levels exceeding those permitted by paragraph (c) of this section, engineering and administrative controls shall be utilized to reduce employee noise exposure to within permissible limits, except to the extent that such controls are not feasible. If such controls fail to reduce sound levels to within the permissible limits of paragraph (c) of this section, they shall be used to reduce the sound levels to the lowest level feasible and shall be supplemented by personal protective equipment in accordance with paragraph (f) of this section to further reduce the noise exposure to within permissible limits. Where the engineering and administrative controls which have been implemented do not reduce the sound levels to within the permissible limits of paragraph (c) of this section, the employer shall continue to develop and implement engineering and administrative controls as they become feasible.
2. A program shall be established and implemented to reduce exposures to within the permissible exposure limit, or to the greatest extent feasible, solely by means of engineering controls. Written plans for such a program shall be developed and furnished upon request to authorized representatives of the Assistant Secretary and the Director.
3. *Exception.* Hearing protectors may be provided to, and used by an employee, to limit noise exposures in lieu of feasible engineering and administrative controls if the employee's exposure occurs on no more than one day per week.

f. *Hearing protectors.*

1. Hearing protectors shall be provided to, and used by the following:

   i. Employees receiving a daily noise dose between 0.5 and 1.0 (a daily noise dose of 0.5 is equivalent to an 8 hr time-weighted exposure of 85 dBA) if their audiograms show any significant threshold shift:
   ii. Employees who receive noise exposures in excess of the limits prescribed in paragraph (c) of this section:

   A. During the period required for the implementation of feasible engineering and administrative controls.

EVALUATION OF EXPOSURE TO NOISE

    B. In instances where engineering and administrative controls are feasible only to a limited extent.
    C. In instances where engineering and administrative controls have been shown to be infeasible.

2. Hearing protectors shall reduce employee noise exposure to within the limits prescribed in paragraph (c) of this section.
3. Procedures shall be established and implemented to assure proper issuance, maintenance, and training in the use of hearing protectors.

g. *Hearing conservation.*

1. *General.*

    i. A hearing conservation program shall be established and maintained for employees who:

        A. Receive a daily noise dose equal to or exceeding 0.5; or
        B. Are required to wear hearing protectors pursuant to paragraph (f) of this section.

    ii. The hearing conservation program shall include at least an annual audiometric test for affected employees at no cost to such employees.
    iii. If no previous baseline audiogram exists, a baseline audiogram shall be taken within 90 days for each employee (A) who receives a daily noise dose of 0.5 or above; or (B) who is required to wear hearing protectors pursuant to paragraph (f) of this section.
    iv. Each employee's annual audiogram shall be examined to determine if any significant threshold shift in either ear has occurred relative to the baseline audiogram.

        A. If a significant threshold shift is present, the employee shall be retested within one month.
        B. If the shift persists:

            1. Employees not having hearing protectors shall be provided with them in accordance with paragraph (f) of this section.
            2. Employees already having hearing protectors shall be retrained and reinstructed in the use of hearing protectors.
            3. The employee shall be notified of the shift in hearing level.

2. *Audiometric testing.*

    i. Audiometric tests shall be administered by a certified audiometric technician or an individual with equivalent training and experience.
    ii. Audiometric tests shall be preceded by a period of at least 14 hr during which there is no exposure to workplace sound levels in excess of 80 dBA. This requirement may be

        met by wearing hearing protectors which reduce the employee noise exposure level to below 80 dBA.
- iii. Audiometric tests shall be pure tone, air conduction, hearing threshold examinations, with test frequencies including as a minimum, 500, 1000, 2000, 3000, 4000, and 6000 Hz and shall be taken separately for each ear.
- iv. The functional operation of the audiometer shall be checked prior to each period of use to ensure that it is in proper operating order.
- v. Equipment, calibration, and facilities shall meet the specifications set forth in the Appendix.

h. *Information and warning.*

1. *Signs.* Clearly worded signs shall be posted at entrances to, or on the periphery of, areas where employees may be exposed to noise levels in excess of the limits prescribed in paragraph (c) of this section. These signs shall describe the hazards involved and required protective actions.

2. *Notification.* Each employee exposed to noise levels which exceed the limits prescribed in paragraph (c) of this section shall be notified in writing of such excessive exposure within 5 days of the time the employer discovers such exposure. Such notification shall inform the affected employee of the corrective action being taken.

- i. *Records.*
  Noise exposure measurements. The employer shall keep an accurate record of all noise exposure measurements made pursuant to paragraph (d) of this section.
- ii. The record shall include the following information.

  - A. Name of employee, social security number and daily noise dose.
  - B. Location, date, and time of measurement and levels obtained.
  - C. Name of person making measurement.
  - D. Type, model and date of calibration of measuring equipment.

- iii. These records shall be maintained for a period of at least 5 years.

2. *Audiometric tests.*

- i. The employer shall keep an accurate record of all employee audiograms taken pursuant to paragraph (g) of this section.
- ii. The record shall include the following information:

  - A. Name of employee and social security number.
  - B. Job location of employee.
  - C. Date of the audiogram.
  - D. The examiner's name and certification.
  - E. Model, make and serial number of the audiometer.
  - F. Date of the last calibration of the audiometric test equipment.

- iii. These records shall be maintained for the duration of the affected employee's employment plus 5 years.

EVALUATION OF EXPOSURE TO NOISE 437

3. *Calibration of Audiometers*

   i. A biological calibration shall be made at least once each month and shall consist of testing a person having a known stable audiometric curve that does not exceed 25 dB hearing level at any frequency between 500 and 6000 Hz and comparing the test results with the subject's known baseline audiogram.
   ii. If the results of a biological calibration indicate hearing-level differences greater than 5 dB at any frequency, if the signal is distorted, or there are attenuator or tone switch transients, then the audiometer shall be subjected to a periodic calibration.
   iii. A periodic calibration shall be performed at least annually. The accuracy of the calibrating equipment shall be sufficient to assure that the audiometer is within the tolerances permitted by ANSI S3.6-1969. The following measurements shall be performed.:

      A. With the audiometer set at 70 dB hearing threshold level, measure the sound pressure levels of test tones using a National Bureau of Standards Type 9A coupler, for both earphones and at all test frequencies.
      B. At 1000 Hz, for both earphones, measure the earphone decibel levels of the audiometer for 10 dB graduations in the range 70 to 10 dB hearing threshold level. This measurement may be made acoustically with a National Bureau of Standards Type 9A coupler or electrically at the earphone terminals.
      C. Measure the test tone frequencies between 500 and 6000 Hz with the audiometer set at 80 dB hearing threshold level, for one earphone.
      D. A careful listening test, more extensive than that required for biological calibration, shall be made in order to ensure that the audiometer displays no evidence of distortion, unwanted sound, or other technical problems.
      E. The functional operation of the audiometer shall be checked to ensure that it is in proper operating order.

   iv. An exhaustive calibration shall be performed at least every 5 years. This shall include testing at all settings for both earphones. The test results shall demonstrate that the audiometer meets specific requirements stated in the applicable sections of ANSI S3.6-1969 as listed below:

      A. [Sections 4.1.2 and 4.1.4.3] Accuracy of decibel level settings of all test tones.
      B. [Section 4.1.2] Accuracy of test tone frequencies.
      C. [Section 4.1.3] Harmonic distortion of test tones.
      D. [Section 4.5] Tone-envelope characteristics (i.e., rise and decay times, overshoot, "Off" level).
      E. [Section 4.4.2] Sound from second earphone.
      F. [Section 4.4.1] Sound from test earphone.
      G. [Section 4.4.3] Other unwanted sound.

4. *Access to records.*

   i. All records required to be maintained by this section shall be made available upon request to authorized representatives of the Assistant Secretary and the Director.

ii. Records of noise exposure measurements required to be maintained by this section shall be made available to employees and former employees and their designated representatives.

iii. Employee audiometric data required to be maintained by this section shall be made available upon written request to the employee or former employee.

j. *Reference.*

1. ANSI S1.4-1971, American National Standard Specifications for Sound Level Meters, S1.4-1971, American National Standards Institute, 1430 Broadway, New York, New York 10018.
2. ANSI S3.6-1969, American National Standard Specifications for Audiometers, S3.6-1969, American National Standards Institute, 1430 Broadway, New York, New York 10018.
3. ANSI S1.11-1971, American National Standard Specification for Octave, Half-Octave, and Third-Octave Band Filter Sets S1.11-1966 (Reaffirmed 1971), American National Standards Institute, 1430 Broadway, New York, New York 10018.
4. ANSI Z24.22-1957 (R 1971), American National Standard Method for the Measurement of the Real-Ear Attenuation of Ear Protectors at Threshold, Z24.22-1957, American National Standards Institute, 1430 Broadway, New York, New York 10018.
5. *American Industrial Hygiene Association Journal,* **27,** 303–304 (May–June, 1966). American Industrial Hygiene Association, 475 Wolf Ledges Parkway, Akron, Ohio 44311
6. Council for Accreditation in Occupational Hearing Conservation, 1619 Chestnut Avenue, Haddon Heights, New Jersey, 08035.
7. ISO R389-1964, International Organization for Standardization Recommendation R389-1964, Standard Reference Zero for the Calibration of Pure Tone Audiometers, including Addendum 1-1970. Available from the American National Standards Institute, 1430 Broadway, New York, New York 10018.

APPENDIX

Audiometric Equipment and Facilities

1. *Audiometric test rooms.* Rooms used for audiometric testing shall not have sound pressure levels exceeding those in Table 11.3 when measured by equipment conforming to the requirements of ANSI S1.4-1971, Type 1 or Type 2, and ANSI S1.11-1971.

Table 11.3 Maximum Allowable Sound Pressure Levels for Audiometer Rooms

Octave band center frequency (Hz):	500	1000	2000	4000	8000
Sound pressure level (dB):	40	40	47	52	62

EVALUATION OF EXPOSURE TO NOISE

2. *Audiometric measuring instruments.*

   i. Instruments used for measurements required in paragraph (g) of this section shall be of the discrete frequency type which meet the requirements for limited range pure tone audiometers prescribed in ANSI S3.6-1969.
   ii. In the event that pulsed tone audiometers are used, they shall have a tone on time of at least 200 milliseconds.
   iii. Self-recording audiometers shall comply with the following requirements:

      A. The chart upon which the audiogram is traced shall have lines at positions corresponding to all multiples of 10 dB hearing level within the intensity range spanned by the audiometer. The lines shall be equally spaced and shall be separated by at least ¼ in. Additional gradations are optional. The audiogram per tracings shall not exceed 2 dB in width.
      B. It shall be possible to set the stylus manually at the 10 dB gradation lines for calibration purposes.
      C. The slewing rate for the audiometer attenuator shall not be more than 6 dB/sec except that an initial slewing rate greater than 6 dB/sec is permitted at the beginning of each new test frequency, but only until the second subject response.
      D. The audiometer shall remain at each required test frequency for 30 sec (± 3 sec). The audiogram shall be clearly marked at each change of frequency and the actual frequency change of the audiometer shall not deviate from the frequency boundaries marked on the audiogram by more than ± 3 sec.
      E. For audiograms taken with a self-recording audiometer, it must be possible at each test frequency to place a horizontal line segment parallel to the time axis on the audiogram, such that the audiometric tracing crosses the line segment at least six times at that test frequency. At each test frequency the threshold shall be the average of the midpoints of the tracing excursions.

2 COMMENTARY ON PROPOSED RULES

Rules such as these probably do not completely satisfy anyone. However they are basically the recommendations of the U.S. Department of Labor's Advisory Committee on Noise. This committee was composed of representatives of the industrial, labor, medical, academic, and governmental communities. Given the various backgrounds and interests of members of this committee, unanimous agreement was not possible. Compromise was required by all. The results (the Proposed Rules) are probably as close to agreement as can be reached at this time.

2.1 Nuisance Noise

Noise levels between the impairment range (i.e., 90 dBA) and the speech communication range can be classed as nuisance noise. Nobody likes nuisance noise, but it is difficult cost-wise to justify doing anything about it because it apparently does not harm physiologically and has no effect on worker efficiency.

2.2 Speech Communication

For the great majority of speech communication problems, the limits can be defined by NC45 and NC55 of Figure 11.2. For offices and conference rooms, NC45 should be used as the design goal. The levels indicated by NC55 should not be exceeded because speech intelligibility would become unreliable. Levels as low as NC35 are not necessary as far as speech communications are concerned.

The conditions specified above might not satisfy those with hearing impairment or poor speech projection, but they are the practical design goals in common use.

2.3 Neighborhood Noise

Neighbors are not concerned with what creates the noise, but merely the sound level, the frequency spectrum, and time of day that the noise occurs. Since neighbor reaction

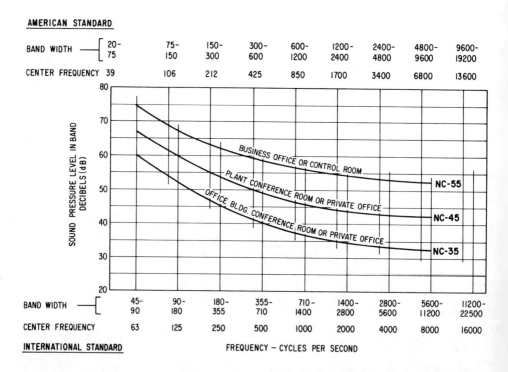

Figure 11.2 Criteria for speech communication. From L. I. Beranek, *Noise and Vibration Control*, McGraw-Hill Book Co., New York, 1971.

EVALUATION OF EXPOSURE TO NOISE

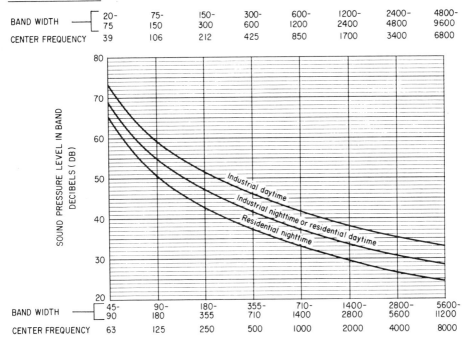

Figure 11.3 Recommended maximum neighborhood noise level. From W. A. Rosenblith and K. N. Stevens, *Handbook of Acoustical Noise Control,* Vol. 2, *Noise and Man,* WADC Technical Report 52-204, June 1953.

(resistance) varies markedly with the frequency of the sound, dBA measurements appear to be inadequate. However in 1956 MIT developed the octave band criterion for neighborhood noise as shown by Figure 11.3. This criterion appears to be adequate. Special interest groups have proposed many criteria since then, but it is doubtful whether they provide a better estimate of neighborhood reaction. The emphasis has been placed on more accurate measurement when the actual variables of concern are people and political reaction. Therefore the criterion of Figure11.3 is recommended.

3 LEGAL ASPECTS

3.1 Worker Compensation

Originally compensation was limited to cases where injury resulted in a loss of income to the employee. However today if an employee suffers some loss of bodily function, regardless of whether it results in any loss of wages, the individual is entitled to some remuneration. In the case of noise, the medical profession defines impairment as a function of hearing loss. Such an impairment is a hearing loss that begins to interfere with the understanding of speech in sentence form. This definition does not imply the ability to recognize every syllable or even every word. Impairment is the beginning of the experience of difficulty in getting the meaning of the sentence as a whole. The most generally accepted method of determining impairment was developed by the American Academy of Ophthalmology and Otolaryngology (AAOO) in 1959 and approved by the American Medical Association (AMA) in 1961 (updated in 1971). Impairment is described as an average hearing level for the three speech frequencies in excess of 25 dB (re ANSI 1969). The details of determining impairment are beyond the scope of this discussion but can be found in Reference 11.

3.2 Presbycusis

Some deterioration in the ability to hear is the result of aging alone. Where this is acknowledged, usually ½ dB is deducted from the average hearing level at the three speech frequencies for every year of the persons, age beyond 40.

3.3 Waiting Period

Following exposure to noise, a person experiences a certain amount of temporary hearing threshold shift. So that the employer is not held responsible for a hearing loss that will eventually be recovered, the law usually requires that an employee be removed from exposure to noise for a certain time before filing a claim. This waiting period is usually 6 months.

3.4 Date of Injury

Since occupational hearing loss occurs over an extended period, it is impossible to establish an exact date on which the impairment occurred. To avoid complications with the statutes of limitation and to protect the employee, some arbitrary decision is required to set the date of injury. Usually this date is recorded as 6 months following the employee's removal from the noise exposure.

EVALUATION OF EXPOSURE TO NOISE

3.5 Apportionment

Apportionment is the term used to identify that portion of a worker's hearing loss for which the employer is responsible. The employer is not penalized for any hearing loss acquired by an employee in the course of prior employment. Usually compensation law makes provision for apportionment, but the employer must show evidence to limit the liability. The last employer, without convincing evidence such as a preemployment audiogram, will probably be held responsible for a claimant's total hearing impairment.

3.6 Hearing Protection

OSHA does not like to consider hearing protection (such as earplugs and earmuffs) a legitimate means of noise control because the agency does not have confidence in the implementation and enforcement of safety measures in general in much of industry. It has been proved that hearing protection can be effective if rules and procedures are rigidly enforced. If hearing protectors are properly used and supported with a hearing testing program, and if the cost of noise reduction is high, chances are good that the use of hearing protectors will be acceptable. However it will probably be necessary to resort to legal action to obtain approval of this approach.

3.7 Citations

If one is cited for not meeting the noise standard, it is not necessarily cause for panic. Almost all industrial noise problems can be solved at reasonable cost if trained and experienced people are assigned to the job. It should be remembered that noise problems have been accumulating for 20 to 30 years, and noise control is not going to be a great yearly cost once control has been established. OSHA recognizes that noise control costs might have to be spread over several years to avoid the large cost of immediate total implementation. It should be obvious, however, that a large company should be able to afford a higher cost noise reduction program than a small company on the basis of more equipment (and production) involved alone. Actual citation fines are usually insignificant.

3.8 Administrative Control

Administrative control, that is, control of exposure time as well as "average" exposure levels, should be avoided because these factors are so difficult to monitor. Experience has shown that in most cases, and in the long run, it is less expensive and more positive to reduce the noise levels in all operating areas to 90 dBA or less. This also has the advantage of avoiding a continuous confrontation with regulations, as well as much expensive recordkeeping.

3.9 Compliance Evaluation

Now that we have considered most of the basic information necessary to administer noise control in general, let us discuss specific exposures. If a worker's exposure is constant throughout the workday, the noise dose can be determined with a type II sound level meter at slow response. The sound level meter reading is an instantaneous indication of the worker's average noise exposure and can be evaluated against the table of the existing OSHA regulations. An exposure totaling 6 hr or more per day is considered continuous. If the exposure is at a constant level but is intermittent, the dose can be determined with a type II sound level meter at slow response together with a simple time study as described in the existing OSHA regulation. However in most cases either the sound source is continually varying or the worker is moving in and out of various sound fields. In either case the worker's noise exposure is always varying. To determine a worker's noise dose under these continually varying noise exposures is tedious, time-consuming, expensive, and likely to be unreliable if done with a sound level meter and stopwatch.

To overcome these difficulties, the audiodosimeter has been developed. The instrument automatically integrates the relation between sound pressure level and time at that level so that at the end of the test period, the noise dose (total sound energy exposure) can be determined. The instrument also indicates whether the 115 dBA limit has been exceeded. There has been some criticism regarding the accuracy of audiodosimeters. However they are manufactured according to type II sound level meter specifications; therefore they are as accurate as the law requires. It certainly would be desirable to upgrade audiodosimeters to meet type I specifications.

In comparing audiodosimeter measurements with sound level meter–stopwatch time study measurements, it seems reasonable to expect audiodosimeter measurements to be more accurate. The exposure level versus time relationship is continuously and automatically being integrated by the audiodosimeter. This eliminates the very real possibility of carelessness or inaccuracy of a technician when making the necessary detailed time study with sound level meter and stopwatch. This is not to imply that accurate noise dose measurements cannot be made with a sound level meter and stopwatch. However experience in trying to find definable noise exposures for use in the Inter-Industry Noise Study revealed that most industrial exposures are so continually varying that the sound level meter approach is impractical and the audiodosimeter measurement is preferred.

3.10 Greatest Present Need

Contrary to some analyses, the great need today is not knowing how to control noise. This information is available for the great majority of noise problems. Neither is it the lack of a governmental regulation. The existing OSHA regulation has some loopholes that need attention, but in general it could be effective. The delay in adopting the new OSHA proposed regulation permits the opportunity to postpone implementing corrective measures, but it should not afford OSHA the means for not enforcing the existing regulation. It has been over four years since the U.S. Department of Labor (DOL) Advisory Committee on Noise presented its recommendations to the department. The DOL has presented its "Proposed Rules" and has held the necessary public hearing. However no new regulation has emerged. Even if opponents of a new regulation attempt to block it by resorting to court action, the sooner this action starts, the sooner it can be resolved. Putting a stop to the "stall" seems long overdue. As far back as 1952 James Sterner (12) tried to develop a proposal for a limit on industrial noise exposure. Unfortunately, this effort was not successful. If it had been, noise probably would have been largely controlled by now. It is doubtful whether "experts" are any closer to agreement on criteria today than they were in 1952. In other words, this is a stalemate and will probably remain so. In the meantime, workers are being exposed to excessive noise. The real needs today then are:

1. Compliance with the existing regulation.
2. Expedition of the OSHA "Proposed Rules," which clarify the generalities of the existing regulation.

REFERENCES

1. K. D. Kryter, Written presentation to U.S. Department of Labor Advisory Committee on Noise, 1973.
2. A. Glorig, "Non-Auditory Effects of Noise Exposure," *Sound Vibr.,* **5,** 28 (1971).
3. American Industrial Hygiene Association, *Industrial Noise Manual,* 3rd ed., "Effects of Noise on Man," A.H.A., Akron, Ohio, 1975, p. 83.
4. W. L. Baughn, "Noise Control—Percent of Population Protected," *Int. Audiol.,* **3,** 331 (1966).
5. W. L. Baughn, "The Risk of Hearing Impairment as a Function of Noise Exposure," *Proceedings of the 1971 Purdue Noise Control Conference,* Purdue University, Lafayette, Ind., 1972.
6. W. Burns and D. W. Robinson, *Hearing and Noise in Industry,* Her Majesty's Stationery Office, London, 1970.
7. Occupational Safety and Health Administration, "Inter-Industry Noise Study—Preliminary Findings, Occupational Hazards," July 1977.
8. E. J. Schneider, J. E. Peterson, H. R. Hoyle, E. H. Ode, and B. B. Holder, "Correlation of Industrial Noise Exposures with Audiometric Findings," *Am. Ind. Hyg. Assoc. J.,* **22,** 245 (1961).
9. S. Pell, "An Evaluation of a Hearing Conservation Program," *Am. Ind. Hyg. Assoc. J.,* **33,** 60 (1972).
10. S. Pell, "An Evaluation of a Hearing Conservation Program—A Five Year Longitudinal Study," *Am. Ind. Hyg. Assoc. J.,* **34,** 82 (1973).
11. American Industrial Hygiene Association, *Industrial Noise Manual,* 3rd ed., A.H.A., Akron, Ohio, 1975, pp. 154-155.
12. J. H. Sterner, *Standards of Noise Tolerance,* "University of Michigan Institute of Industrial Hygiene, 1952, pp. 161-166.

CHAPTER TWELVE

Evaluation of Exposures to Hot and Cold Environments

STEVEN M. HORVATH, Ph.D.

1 INTRODUCTION

The responses of man to cold and hot environments have been studied extensively. Man is considered to be a subtropical animal who has developed physiological adaptations and mechanisms that result in maximal increases in his body heat content. This ability to acclimatize to cold environments is limited. Even at an ambient temperature of 26° C an unclothed man cannot maintain his body heat stores. However he does have a much greater capacity to acclimatize to hot environments. Man's response to thermal changes is dependent on a number of factors.

In any consideration of these responses to either cold or hot environments, the possibility that man has either adapted, acclimated, or acclimatized must be taken into account. "Adaptation" implies changes of a genetic nature that have occurred because of natural selection processes. "Acclimation" refers to physiological changes in response to ambient temperature changes. "Acclimatization" refers to any change in physiological mechanisms that better enables an individual to perform and/or survive in a particular environment. In general, most studies on man are concerned with acclimatization processes, and the primary emphasis of this chapter is on these processes and the manner in which they enable man to improve his performance.

2 PHYSIOLOGICAL RESPONSES TO COLD

The successful response of man to a cold ambient environment is dependent on behavioral and physiological mechanisms. Behavioral factors involve a complex interplay utilizing other elements of the environment to protect against the impacting one. The employment of external energy sources, properly designed insulative clothing, reduction of effective area for radiation, convective and conductive losses, limited exposure times, and so on, illustrate the many behavioral responses utilized by man. Physiological responses, that is, modification of internal adaptive mechanisms, are also involved. These physiological alterations have been studied during short (acute) and prolonged exposures to lowered ambient temperatures. The experimental approaches utilized to study this problem have been made on three types of exposure: subjects living in polar regions, those exposed to cold (minutes to weeks) in temperature-regulated chambers, and those exposed to normally occurring seasonal environmental cold.

Several groups such as the Australian aborigines and the Kalahari Bushmen appear to demonstrate some genetic selection. Their responses to a cold stress differ from those of Caucasians in that the former apparently are capable of allowing their core temperature to fall while keeping their surface (skin) temperatures relatively high, with no significant increase of metabolic rate (energy output). This is in contrast to Caucasians, who have a high metabolic rate, a maintained or higher core temperature, and a decreased surface temperature under similar cold stress.

The responses of man to cold environments are integrated in the thermoregulatory centers in the brain (hypothalamus). In general, the brain operates through closely coordinated actions mediated through three effector outputs: the endocrine system, the autonomic nervous system, and the skeletal-muscular system.

1. Behavioral responses by way of the skeletal-muscular system provide numerous options: moving out of the cold, building a fire, putting on heavier and warmer clothing, reducing surface area.
2. The responses of the autonomic nervous system include diminishing peripheral blood flow, thus minimizing heat loss.
3. The endocrine system can trigger increased liberation of catecholamines, thyroxin, and other substances.
4. Further examples of skeletal-muscular activity are shivering, pilomotor activity, and active muscular work.

Involuntary muscular activity during cold exposure appears to be of a consistent pattern for a particular subject. The activity observed varies from fibrillation in single muscles or small muscle groups to clonic and tonic spasms that induce marked skeletal movement. Most of this muscular activity is limited to the thorax and upper portion of the lower extremities. The cooler distal portions of the body show only pilomotor activity. It is rather suggestive that the areas suffering most markedly from the local chilling are not involved in the protective reaction of shivering.

Shivering originates either in the upper thorax (pectoral muscles) or in the thigh (gluteal and vastus group). The spread of shivering activity may be localized to the area of origin, or it may extend throughout the area of shivering muscles. Such extension, a so-called march of shivering, is observed only rarely. If massive shivering occurs, it originates almost simultaneously in both the thorax and the thighs. Periodic fluctuations in shivering activity are always observed. The intervals separating shivering bouts are far from consistent. They can vary from seconds to minutes.

Respiratory movements (end of inspiration) or yawning appeared to be related to the occurrence of shivering bouts. In some individuals this relationship is limited to the earlier, more rapid cooling phase, although some subjects retain this correlation even when further heat loss from the body and changes in magnitude of peripheral vasoconstriction were minimal or nonexistent. Body size or obesity (fat) shows no relation to shivering activity.

The increased oxygen uptake (metabolism), which may be as high as 3 to 5 times normal standard resting rates during cold exposure, is related to either (*a*) shivering, (*b*) increased voluntary movement caused by cold discomfort, and (*c*) nonshivering thermogenesis. Shivering probably represents the first line of defense against cold, and nonshivering thermogenesis a defense that goes into action as a result of prolonged, continuous exposure to cold. Shivering is a relatively inefficient process in comparison to active muscular work (1) and results in an impression of fatigue beyond its metabolic cost. A severe cold stress, probably more than man would be willing to tolerate, is required to induce nonshivering thermogenesis (2).

Davis and Johnson (3) have suggested that in cold adaptation man seems to rely less on shivering and more on nonshivering and a decrease in the metabolic rate. When young men, dressed in shorts and sneakers only, were continuously exposed to an 8°C environment for 3 to 10 days (4), they initially responded by extensive and intensive shivering, which persisted more or less continuously night and day despite sleeping under a single blanket at night. The increased heat production (twofold to fourfold) remained high even during sleep. Body temperature remained normal. In some instances the subjects ceased vigorous shivering during the night and their body temperatures fell to low levels. These subjects doubled their metabolic rate, had an increased heart rate, and increased their urinary nitrogen losses. It was suggested that these changes were the resultant of hormonal changes brought about by the cold stress. Some of these subjects suffered ischemic cold injury of their feet, although the ambient temperature did not approach the freezing temperature of the tissue. This increased metabolism may be related to the calorigenic effects of the catecholamines, which are known to be released during cold exposure; but only one study has been made on men who have been subjected to acclimatization procedures.

Whereas increased heat production is dependent on metabolic processes, the heat loss from the body is largely dependent on the physical factors of the environment. The laws of heat flux from the skin to the environment have been well described by Hardy (5). Burton and Edholm (6) have emphasized the importance of the applicability of Newton's law of cooling, which states that the rate of dry heat loss from the exposed skin is directly proportional to the temperature gradient between the skin and its environment and is independent of the humidity. Engineers refer to the proportionality constant as the heat transfer coefficient C and have described coefficients for any number of substances and forms over a wide range of air velocities. The reciprocal of C is the thermal insulation of the system. The physiological analogue to the cooling constant is the combined radiative and convective heat transfer coefficient h, which describes the rate of dry heat exchange between the skin and the environment that can occur at any constant air velocity. The effect of such factors as ambient water vapor pressure, air velocity, and effective surface area, which also modify the rate of total heat loss, have been more accurately defined. Relative humidity had no significant effect on heat loss and the subjective sensation of cold in naked or lightly clothed men. Damp cold is not colder than dry cold (7, 8), although subjectively damp cold appears to be colder.

Precise measurement of the rate of evaporative heat loss is not of great importance during nude or nearly nude cold exposure, since in nonsweating, resting conditions, loss from the skin has a relatively constant value between 6 and 10 W/m^2 and loss from the respiratory tract can vary between 3 and 6 W/m^2. However during conditions of exercise, even in very cold environments, total evaporative heat loss can be increased as a result of the active secretion of sweat onto the skin surface (9, 10) and the increased ventilation subsequent to the elevated oxygen uptake.

Tolerance to cold exposure can be considered either as an aspect of cold acclimatization or merely as the consequence of the physiological response to a cold stress. Cold tolerance can be assessed by one of several means, including the time at which shivering begins, a change in the magnitude of shivering, the time at which metabolic heat production becomes significantly elevated, the state of discomfort of the subject, and/or some evaluation of peripheral circulatory activity. These assessments take advantage of the two primary physiological responses to cold. The first line of defense is an increase in peripheral resistance, causing a reduced peripheral blood flow and a reduction in convective heat transfer from the body core to the skin. There is consequently a reduction in heat loss from skin to environment by minimizing the skin-to-ambient thermal gradient. The second line of physiological defense is the increase in heat production. Another most important factor in cold tolerance is related to the quality and consistency of behavioral adaptations, which relate to knowhow, experience, and probably the state of physical fitness.

The importance of the level of physical fitness as a modifier of man's response to cold was shown by Adams and Heberling (11). Comparisons were made of the responses of men to a standardized cold stress before and after a 3 week period of intensive physical training. In the posttraining test average levels of heat production were 15 kcal/(hr)/(m²) higher, mean rectal temperatures were 0.5°C lower, average skin temperatures were 1.0°C higher, and foot and toe temperatures were 3.0 and 4.0°C higher, respectively, with no significant differences in average body temperatures. Although these observations are suggestive in that they imply some improvement in tolerance to cold, they need further confirmation.

Saltin (12) has found that submaximal and maximal work during short-term cold exposure ($-5°C$) can be performed at the same levels of circulatory capacity observed in 20°C ambients. Pulmonary ventilation, however, was somewhat higher when working in the cold.

Because the heat production response is an important one, measurement of oxygen uptake for determination of the metabolic heat production is a frequently utilized technique in the assessment of cold tolerance. One rationale for measuring oxygen uptake is that the heat production above basal level will be related to the intolerance of the cold stress. The test protocols using oxygen uptake have tended to involve several variations. One variation has been the determination of the basal metabolic rate (BMR) in neutral conditions to evaluate whether previously cold-exposed subjects or populations differ from an appropriate control group. Any adaptation affecting the BMR (hence stimulating a form of nonshivering thermogenesis) has yet to be adequately shown.

Kang et al. (13) have examined Korean Ama (women divers), previously noted as having an elevated shivering threshold, greater tissue insulation index, and elevated BMR (14). They found no differences in BMR between Ama and their matched controls during summer and winter. Furthermore, there was almost no calorigenic response to norepinephrine in either group, indicating either that nonshivering thermogenesis is difficult to demonstrate in humans or that the Ama were not cold acclimatized. Although Hammel (15) has described the Alacaluf Indians, a primitive group living on Tierra del Fuego, as demonstrating a metabolic adaptation to cold (i.e., elevated metabolic rate during warm exposure), he did not stipulate whether this could be considered to be nonshivering thermogenesis.

Finally, several investigators have shown a significant seasonal variation in BMR. Osiba (16) has concluded that the elevated BMR in the winter months seen in Japanese subjects was due to an increase in thyroid activity permitting the Japanese to endure cold more comfortably. However the role of the endocrine system during cold stress remains to be clarified. Davis and Johnson (3) and Girling (17) have reported that hourly cold exposures induced a greater heat production response to the cold stress in the summer than during the winter, suggesting that some degree of cold acclimation had occurred. Both these studies were performed under conditions of marked seasonal variation in ambient conditions. In Santa Barbara, California, however, where the average ambient temperature variability is small over the year, monthly exposures of 2 hr to 4.5°C revealed no seasonal variation in metabolic response to the cold in seven subjects (18). Previously Hammel et al. (19) had demonstrated that no metabolic adaptation occurred in the presumably cold-acclimatized Australian aborigine. Thus the question of man's ability to become cold adapted by way of some metabolic process or processes remains to be answered.

Another index of cold intolerance has involved the determination of the oxygen uptake response during prolonged cold exposure. This method has clear advantages, since comparisons in metabolic response between an experimental and a control group can readily be made at any period of cold exposure, at any level of internal body temperatures, or at any combination of body temperatures. Raven and Horvath (20) have recently emphasized that an attempt to assess the metabolic response to cold requires continuous measurements over greater than a 1 hr exposure. They found that the metabolic heat production response became relatively constant only during the final 30 min of a 2 hr exposure to 5°C, and it was only during this period that the rate of change in mean body temperature became constant and minimal. Wyndham et al. (21) have also proposed that a 2 hr cold exposure should serve as a standard procedure. Yoshimura and Yoshimura (22) found that 2.5 hr was required to attain consistent values of metabolic rates and internal temperature in a 10°C ambient. O'Hanlon and Horvath (23) also reported that a 2 hr cold exposure was necessary to reach a steady state.

The scarcity of information on other physiological changes during cold exposure induced O'Hanlon and Horvath (23) to evaluate the heart rate response. They found that the heart rate remained constant or was slightly increased during cold exposures. This constancy of heart rate and the marked increase in oxygen uptake during these cold stresses suggested that this high oxygen uptake could be accomplished only by an elevation in cardiac output, greater oxygen extraction, or a combination of both. Raven et al. (24), whose subjects were exposed to 5°C for 2 hr, reported that the cardiac output was elevated 95 percent above levels found in a neutral environment. This increased cardiac output was due primarily to an increased stroke volume.

There have been only a few investigations regarding physiological adaptations to long-term cold exposure, mainly because of the inherent difficulties in conducting such experiments. Convincing evidence for acclimatization to cold has been obtained from studies on small animals (rats, etc.). Unfortunately, such evidence has not been adequately documented for the human. Horvath et al. (25) studied five subjects who had lived continuously for 8 days in an ambient temperature of −29°C. They could find no evidence for acclimatization after that period. Carlson et al. (26) exposed seven subjects to −6°C for 16 to 18 hr daily for 14 days and reported that their subjects had reduced their heat elimination rate after this period. Davis (27) reported that men artificially acclimatized to cold in the summer had decreased heat production and lower skin and rectal temperatures as a consequence of a series of 8 hr daily exposures to an ambient temperature of 13.5°C. Other studies by Davis and Johnson (3) exposed nude subjects to a constant cold stress (14°C) for a period of one hour each month over a period of one year. Metabolic heat production and shivering decreased as a consequence of this habituation. Girling (17) similarly demonstrated a decreased metabolism and shivering with no change in mean skin or rectal temperature. There was a time lag in the metabolic response when compared with the minimal environmental temperature.

Yoshimura (16) has also reported seasonal changes in acclimatization to cold in Japanese subjects. Cold acclimatization was accomplished by lowering mean skin temperature and consequently decreasing the heat loss, which in turn minimized the metabolic cost of the thermal stress in winter. A second type of adaptation (i.e. an acceleration of metabolic heat production in cold environment), was evident in the work of Scholander et al. (28). Their subjects were able to maintain a warm skin by raising their heat production.

Local, primarily peripheral acclimatization to cold is much better documented than the previously discussed total body acclimatization. Deep sea fishermen can work with their hands immersed in cold water. Similar exposure of the hands of normal (sic) individuals would result in incapacitation because of numbness and pain. Fishermen apparently have a greater blood flow than other men in this situation (29, 30). Functional capacity was attained at the expense of additional heat loss. Similar adaptations have been reported for racial groups accustomed to local cold stress [Eskimos (31), Arctic Indians (32)], British fish filleters (33), and polar explorers (34). This adaptation is probably an acquired characteristic.

Additional studies on local acclimatization have utilized the "hunting" reaction* (35). Alterations in the intensity of the hunting response have been utilized also to investigate the differences between ethnic groups (36) and groups exposed to cold (37). Meehan (38) compared the hunting reaction of Caucasians, Negroids, and Alaskan natives. Krog et al. (39) studied this reaction in Lapps, Norwegian fishermen, and Norwegian controls. Adams and Smith (40) demonstrated an acceleration of the hunting reaction of the index finger that had been exposed repeatedly to cold for long periods and emphasized the existence of "local cold conditioning."

Most methodology employed for the study of cold tolerance has emphasized the relationship between a particular response variable and the time of cold exposure. This approach has tended to create ambiguous conclusions regarding cold tolerance. The optimal evaluation of cold tolerance should be attainable from a complete understanding of the means for physiological regulation of body temperature. The body has elements that are capable of sensing cold (i.e., thermosensors near or on the skin surface and in the hypothalamus), and an integrating center that receives these neural messages and directs an appropriate themoregulatory output. More recently, however, experimentation involving both water bath immersion (41) and exposure to a wide range of ambient air environments (42, 43) has shown that the regulation of the metabolic heat production response is a function of both skin and internal temperatures. Brown and Brengelmann (41) used different driving functions for skin temperature to identify both transient and steady state relationships between skin and internal temperatures in the determination of metabolic heat production. They concluded that the regulation was primarily resultant from the multiplication of thermal signals from the body core and skin.

Nadel et al. (42, 43) required their subjects to ingest different volumes of ice cream and/or hot pudding to change internal temperature by different amounts. They performed these studies in a wide range of ambient environments, thereby investigating the response to a change in internal temperature at many different levels of skin temperature. They extended the Stolwijk and Hardy (44) mathematical model, which described the primary thermal signal as multiplicative in the determination of the response, as follows:

$$\Delta M = 42(36.5 - T_{in})(32.2 - \bar{T}_s) + 8(32.2 - \bar{T}_s) \quad \text{in W/m}^2$$

where ΔM = metabolic increase from resting level (W/m^2)

---

* The "hunting" response represents the periodic fluctuation in blood flow to an extremity immersed in cold water.

Sympathicoadrenal hormones have been implicated as playing an integral part in homoiotherms' response to cold exposure (45). Release of catecholamines from the sympathicoadrenal medullary system has been associated with both metabolic and cardiovascular responses to cold. Apparently adrenal corticosteroids in association with other hormones act directly on the thermoregulatory centers of the brain. Wilkerson et al. (46) have recently shown that acute cold stresses caused significant alteration in the plasma concentration of the sympathicoadrenal hormones. An ambient temperature of 15°C or less appeared to be necessary to induce increased plasma levels and changes in urinary excretion rates of the catecholamines and cortisol. These data imply that adrenal medullary and adrenal cortical functions are interrelated with peripheral sympathetic nervous system functioning to provide the necessary substrates to allow augmented heat production in response to cold exposure and to reduce peripheral heat loss in these conditions. Earlier Joy (47) studied the responses of cold-acclimatized men to infusion of norepinephrine. His results indicated that after long-term cold exposure there was a change in sensitivity to norepinephrine, with a decrease in vasopressor response and the development of a calorigenic response. His data suggest that norepinephrine may be a mediator of a nonshivering thermogenesis occurring with cold acclimatization.

The sensitivity of skin receptors is diminished when tissue temperatures drop. Irving (48) has shown that sensitivity was reduced sixfold when skin temperature dropped from 35 to 20°C. Stuart et al. (49) found that muscle spindles show an increased sensitivity at moderately lower muscle temperatures, but when muscle temperatures were 27°C the activity in response to a standardized stimulus was 50 percent of normal; it was completely abolished when temperatures reached 15 to 20°C. In general numbing and loss of tactile sensitivity occurs at finger temperatures of approximately 8°C, and manual performance may be decreased because of loss of finger and hand dexterity. This probably explains the loss of fine coordinated movements noted to occur in individuals attempting to work in cold environments, as well as their lack of appreciation of the degree of cold experienced.

Mills (50) reported that the loss of tactile discrimination was roughly inversely proportional to the change in mean finger skin temperature. Rewarming of the hands resulted in a return of tactile sensitivity, but the degree of recovery lagged behind the return of skin temperature. Clark and Cohen (51) demonstrated a marked reduction in finger dexterity (to about 75 percent) when hand skin temperature was quickly cooled from 22.2 to 7.3°C, but dexterity was further reduced (to about 50 percent) if cooling time was extended. This finding reflects the more extensive cooling of deeper tissues consequent to the prolonged period of vasoconstriction.

Dexterity of the fingers and grip strength (hand) was markedly diminished by exposure to low ambient temperatures (52). Grip strength decreased some 28 percent, even though cold exposure was of relatively short duration. Subjective impressions of the severity of cold hands were not related to the degree of impairment that occurred, even when hands felt relatively comfortable. Some variability in loss was observed, with some subjects demonstrating a 70 percent decrement in strength. Davies et al. (53) have shown that hypothermia produced by immersion in cold water results in a reduced capacity to perform work. When body temperatures were approximately 35°C, maximal aerobic capacity was reduced and the oxygen cost of submaximal work increased, suggesting a considerable loss in efficiency.

Older individuals appear to exhibit a slightly different response to a cold environment; they allow their deep body temperature to fall and do not increase their metabolic rate to the same degree, in contrast to younger subjects who raise their oxygen consumption and maintain normal or elevated rectal temperatures (54, 55). Older subjects also appear to be less aware that they are cold. This may indicate they have carried the normal physiological process of habituation to cold to the stage where it becomes detrimental to their welfare.

It has been occasionally reported that some individuals have an extraordinary sensitivity to cold exposure manifested by an essentially allergic response. This is exhibited by the symptoms of urticaria and syncope. An early review of selected cases was given by Horton et al. (56) and a more recent one by Kelly and Wise (57). The types of reaction vary from localized urticaria at the point of contact to systemic reactions including headache, flushing, fall in arterial pressure, and even syncope. The latter symptom has been related to the death of some individuals who have been swimming in cold water. Another important facet of cold exposure is related to the aggravation of anginal symptoms in patients with angina pectoris (58). These investigators suggest that the increase in peripheral resistance consequent to cold exposure would result in augmenting myocardial oxygen requirements, thus more readily would provoke an attack of angina.

Much of the concern regarding the combined effects of cold and alcohol on the human organism are based on theoretical considerations of the effects of alcohol on subjects in thermally neutral environments. Anderson et al. (59) reported that moderate doses of alcohol had no deteriorative effect on heat balance during prolonged mild cold exposures. The ambient temperatures were 20 or 15°C, and the naked subjects were studied while sleeping. The typical vasodilating effects of alcohol were not observed in these subjects, probably because the vasoconstrictor drive consequent to cold counteracted the vasodilating drive of the alcohol. It should be noted that hypothermia, followed by death, has been frequently reported in highly intoxicated individuals exposed concurrently to a cold environment.

The potential development of hypothermia in man exposed to cold is not discussed, although it can be a problem under certain conditions. An extensive literature (60, 61) on this topic is available in the clinical literature, both in the surgical area (where hypothermia has been extensively employed) and in the internal medicine field (where it has been utilized in treatment of venal disorders, burns, etc., and more particularly in the occasional instances where a combination of alcohol ingestion and a cold ambient has resulted in hypothermic states). This problem is one of degree of cold exposure and usually reflects a marked drop in deep body temperature. Physiological actions to combat cold are not effective when deep body temperatures fall below 31 to 29°C. Hypothermia is often observed in older individuals (62), reflecting their inability to maintain normal thermal homeostasis.

3 PHYSIOLOGICAL RESPONSES TO HEAT

Man lives and functions in a physical environment wherein thermal and other stressors influence the function of his body. This physical environment may produce significantly intense strains resulting in either adaptation or altered health status. Perception of the environment may affect one's performance, efficiency, and adaptative potential as much as the actual environment itself. Nonetheless, the most important factor(s) in man's response to environmental conditions are related to the basic physiological mechanisms necessary for man's adjustments and adaptations to the ambient stressor. The manner and degree to which these physiological mechanisms are involved in response to heat stress are dependent on the degree, duration, and intensity of the exposure. Individuals who work continuously in the heat each day benefit from the acclimatization process. Other persons who are only sporadically exposed will develop less effective acclimatization, but they may tolerate short heat exposures. Individuals who are only occasionally exposed to heat stress may be able to withstand the influence of increased body heat stress, but a limitation to their performance capacity will be determined by the degree to which their body temperature is elevated. In all cases the final determination is made by the effectiveness of thermoregulatory processes and the interplay with the circulatory, sweating, hormonal, and nervous systems.

According to the principle of thermoregulation, body temperature is maintained constant within narrow limits. However it is probable that the body is never in complete thermal equilibrium. Man's core temperature exhibits a diurnal rhythm wherein the rectal temperature usually reaches a peak in the late afternoon or early evening and falls to its lowest level during the early morning hours. Other physiological parameters that may be involved in temperature regulation also exhibit 24 hr cycles (i.e., skin temperature, peripheral blood flow, metabolic rate, and heart rate). The exact and precise correlations of these various rhythmic alterations have been only roughly related to each other. Phase shifts can also be modified by altering the daily routine, by long distance flights, or by entraining the core temperature to other environmental conditions.

Environmental heat leads to definite reactions in man. The increased body heat content leads to alterations in the circulatory, respiratory, sweating, endocrine, and temperature regulating systems. Man's thermoregulatory system serves to maintain equilibrium of the body core within a narrow limit. If this core temperature is to be so precisely maintained, the amount of heat gained by the body must be equal to the amount of heat lost. Ultimately, the amount of heat exchanged between the body and its environment depends on the differences in temperature and of vapor pressure present between the skin and its environment. Calculation of this exchange can be obtained from the heat balance equation: $(M - W) \pm C \pm C_0 \pm R - E = \pm S$, where M = metabolic heat production; W = work performed; C, C_0, R, and E = convective, conductive, radiant, and evaporative heat exchanges, respectively; and S = amount of heat stored or lost in the tissues. If the body maintains thermal equilibrium, S is zero.

The foregoing physical relationships are affected by physiological mechanisms regulating primarily the cardiovascular, fluid balance, and sweating systems. The physiological strain will be related to the total heat stress (environmental and work loads). Heart rate has been frequently utilized to determine the magnitude of the cardiovascular stress. However this simple measure provides an inadequate indicator of the stress. The cardiac output for a given work load is similar in hot or thermally neutral environments. However stroke volume is markedly reduced in hot situations. Furthermore exercise reduces perfusion of internal organs such as the splanchnic and renal beds. This usually reflects the need to provide more blood to active muscles. In hot environments additional blood flow to the skin is needed to assist in thermoregulation.

The other major factor involved in man's ability to work in hot environments is his sweating mechanism. The ability to wet the skin and provide for evaporative cooling determines much of man's capability to perform effectively in hot conditions. Sweating also involves consideration of the water and salt balance of the body, as well as the ultimate capacity of the sweat glands to continue their function. It is common for males to sweat at a rate of 1.5 to 2.0 liters/hr. Observations on industrial workers show that they can produce at least a liter of sweat per hour over an 8 hr work day. The highest sweat production, 2 liters per half-hour, was recorded during a laboratory experiment. In environments with high vapor pressure, the effectiveness of evaporative cooling is impaired. It should be emphasized that clothing worn by individuals must be wetted to enable evaporative cooling to occur. The degree to which clothing of different types can be wetted will influence the capability of man to attain and maintain thermal equilibrium.

The signs and symptoms that can be expected from water and salt depletion have been documented by Adolph (63) and Marriott (64). Water deficits of approximately 2 percent of body weight lead primarily to symptoms of thirst. However a water deficit of 6 percent of body weight leads to thirst, oliguria, irritability, and aggressiveness. Water deficits in excess of 6 percent result in marked impairment of physical and mental performance. Salt depletion produces serious symptoms. A deficit of 0.5 g per kilogram of body weight results in lassitude, giddiness, fainting, and mild muscle cramps. A continuous slow decrease in body weight should arouse concern that a chronic salt deficit is developing. The safest procedure to prevent water and salt deficits is to provide adequate replacement of water and additional salt at meal times. Most diets contain sufficient excess of salt to meet most needs if water is adequately replaced. In situations where salt may be in deficit, replacement by drinking 0.3 percent saline is preferable to the indiscriminate use of salt tablets.

Repeated exposure of man to hot environments induces changes in a number of physiological systems, resulting in the development of improved tolerance to the heat load. Individuals with no recent history of exposure to hot environments show marked differences in their tolerance to a standardized heat exposure. Some individuals cannot develop an adequate level of acclimatization. The process by which this acclimatization to heat occurs has been considered by many investigators, but the mechanisms involved are still not clear. The symptoms most commonly evaluated are the circulatory, sweating, and body fluid control systems. Classical acclimatization responses observed are decreases in rectal temperature, skin temperature, and working heart rates, while sweat rates and work time increase and orthostatic tolerance improves (65, 66). Improved heat tolerance is obtained by increased cardiovascular efficiency early in heat exposures (67, 68). Changes in heart rate, stroke output, and rectal temperature occur before sweat rates significantly increase. Maximal venoconstriction and minimized arteriolar dilatation were observed by the third or fourth day of acclimatization (69). The chloride concentration of sweat decreases with continued heat exposure, and the total salt loss by sweating may be decreased with time. Plasma volume increases concomitantly with these early cardiovascular changes, and Senay et al. (70) believe that the preexposure increases in plasma volume observed are responsible for this increased cardiovascular stability. The 23 percent increase in plasma volume could be sufficient to raise right atrial pressure to account for an increased stroke output. However Rowell et al. (71) postulate that the drop in body temperature leads directly to the decreased heart rate, consequently to an increased stroke volume. These differences of opinion may be attributable to the type of heat stressor employed by these investigators (i.e., wet humid heat vs. dry heat).

It appears that acclimatization to heat occurs in stages. Initially the high body temperatures result in maximum dilatation of the cutaneous blood vessels, which facilitates an expansion of the plasma volume. This, in turn, results in a restabilization of the cardiovascular system with decreased heart rates, larger stroke volumes, and lowering of body temperature. At the same time sweat production increases, but there may not be a more efficient evaporation of available sweat. With continued heat exposure, evaporation rate increases despite no further increase in sweat production. Sweating occurs earlier in the heat exposure as acclimatization proceeds. These processes occur respectively in a time frame of 1 to 4, 6 to 10, and 10+ days of daily consecutive exposure to the hot environment.

Mere exposure to heat without work confers little acclimatization. Acclimatization to heat is the dramatic improvement in ability to work in a hot environment. It is possible to accelerate this acclimatization process. Acclimatization to 1 hr of work at 48.9/33.9°C (dry-bulb/wet-bulb ambient temperatures) enabled men to do 4 hr of work at 48.9/32.2 or 48.9/31.1°C (72). Thus acclimatization to work at a higher environmental load ensures full acclimatization to a similar work load for 4 hr at any lower heat stress. Robinson et al. (73) restudied men (44 to 60 years of age) who had been evaluated 21 years earlier for their responses to heat exposure. These older individuals exhibited the same degree of strain to their first exposure to heat and acclimatized as well as they had done when they were younger.

Acclimatization to work in a hot environment is not retained indefinitely. There can be a rapid loss over so short a period as a weekend, with recovery to the prior level of acclimatization by the second day of heat exposure. One week without heat exposure may require 4 days of exposure to reattain acclimatization. After 1 week of no heat stress there was a 50 percent loss in terms of the heart rate and sweat production responses. It appears that acclimatization is almost completely lost after 3 to 4 weeks in a cool environment and that a full reacclimatization program is required to return the individual to full acclimatization.

Various factors may modify the processes of acclimatization. Individuals who are engaged in hard physical work that induces a rise in deep body temperature are apparently partly heat acclimatized. They require a smaller number of repeated heat exposures to develop full acclimatization. Any situation that leads to dehydration of an individual will result in significant losses of acclimatization. Convalescing from debilitating illness, alcoholic hangovers, loss of sleep, and so on, have been shown to result in such losses. Strydom et al. (74, 75) have suggested that ascorbic acid supplementation has a significant and beneficial influence on body temperature response during heat stress, and additionally it also enhanced the rate of heat acclimatization.

Sex differences in response to thermal stress are thought to exist, but there are many questions about the precise nature of these differences. The number of studies that have investigated heat strain on females (i.e., their physiological reactions and adaptive capabilities) are few in number, and their results are inconclusive. Kawahata (76) has reported that females have a greater number of sweat glands than males but that the single male gland has greater activity. The onset of sweating to reflex heat stimulation or general heat stress is slower in young females than in young males. Haslag and Hertzman (77) reported that when resting males and females were exposed to progressively increasing ambient temperatures, 25°C for 1 hr, then an increase of 6.6°C per hour the next 3 hr, there was no significant difference in regional sweat rates or the level of body temperature at the time of onset of sweating.

Hardy and DuBois (78) found that men sweat more than women and begin sweating at a lower environmental temperature, men at 29°C and women at 32°C. These investigators also reported that in a warm environment there was a fall in heat production in women but not in men. This observation was not confirmed by Cleland et al. (79), who also discussed the various reasons for a similar confusion in the results obtained by various investigators in studies on young men. The working metabolism of women has been said to increase relatively more than men in the heat (80), but others (81) have not found this to be the case. Various investigators (79-83) have examined the response of men and women to a single bout of exercise in the heat (variable degrees of heat strain) and have found lower sweat rates, higher rectal temperatures, and higher heart rates in females. Some of these observations suggest that women have a reduced tolerance to heat stress.

Young adult women apparently follow the same pattern of acclimatization to heat that has been observed for young males. Lowered body core temperatures and lowered heart rates were observed during work in women during acclimatization. However the magnitude of these decreases varied considerably in the various studies (81, 84-87). A similar discrepancy was noted for total body sweating. Different investigators have reported levels of sweating to be similar, less than, or equivalent to those observed in male subjects. The small total number of subjects studied and the widely different ambient hot environments (from hot-dry to hot-wet) might explain these discrepancies, but it appears obvious that certain striking physiological differences need to be explained. Wyndham et al. (87) have reported some rather bizarre behavior of women during heat exposures. Whether these responses were related to the stage of the menstrual cycle, where somewhat similar psychological behavior patterns have been reported, remains to be determined.

Age has a very definite effect on the responses of women to heat, as Cleland et al. (79) demonstrated in a study of young (mean age, 21.3 years) and elderly (mean age, 67.6 years) women doing light work at 35°C T_a and 28°C T_{wb}. Cardiorespiratory differences were similar to those that had been noted previously at normal ambient temperatures: a higher systolic, diastolic, and pulse pressure in the elderly, as well as higher values for ventilatory efficiency and the respiratory exchange ratio. The most striking differences were in the temperature responses. The elderly women had a higher core temperature and a lower mean skin temperature than the young girls. At what point in life these differences become apparent and why they occur is not known.

Kawahata (76) studied the relation of menstruation and the ability to perspire. Since he had earlier found that testosterone was sudorific and oestradiol inhibitory, he thought that it was highly probable that the menstrual cycle affected the ability to perspire in females. He found rather large changes in the latent period for thermal sweating measured at different times during the menstrual cycle. The latent period was much shorter near the time of ovulation. These observations were not confirmed by Sargent and Weinman (88), since they failed to find any significant differences in the activity of the eccrine sweat gland during the menstrual cycle. The rate of sweating and the concentrations of sodium, potassium, and chloride in sweat were studied in natural desert environments (89) utilizing girls 9 to 18 years of age and two young women (in addition to boys and men), and it was shown that the rate of sweating depended on body surface, metabolic rate, and ambient temperature, not on sex or age. The sweating response apparently is directly related to the absolute metabolic rate (90).

The role of aerobic power has been generally ignored when studies on heat tolerance of females were conducted. Most investigators assigned the same work task to both men and women. Since core temperature and heart rate, two commonly used indicators of heat strain, vary in proportion to relative work load, it was not surprising that women generally appeared to be under greater stress. Even when both sexes work at the same relative work load, individuals with a high degree of cardiovascular fitness perform more adequately. The fit individual is better able to cope with acute exposure to work in the heat, since the cardiovascular system can meet the competing demands for muscular and peripheral blood flow while maintaining an adequate venous return to the heart.

Highly conditioned female athletes (91) have an earlier onset of sweating during heat exposures. This response results in lower body temperatures, thus delaying the increase in cutaneous venous capacitance and diverting less blood from the central to the peripheral circulation. Senay (92) indicated that females do not hemodilute (i.e. expand their plasma volume) to the same extent as males, ascribing this to inherent differences in skin surface to blood volume ratios and to the inability of females to maintain their vascular volume during heat exposure. Nonetheless, the capacity of females to respond to a heat stress at percentages of their aerobic capacity equivalent to that of males appears to be very similar. However since females generally have a maximum aerobic capacity 20 to 25 percent less than males, when a work load requiring a fixed caloric expenditure must be performed, females may be at a disadvantage. That is, females may have to utilize a greater percentage of their capacity to accomplish this fixed task.

There is a need in industry for some general indication of the comparative stresses of different warm environments. Heat stress indices have been developed by a number of investigators in an attempt to reduce the thermal components of an environment to a single figure reflecting the heat stress on the individual. Combining the physiological characteristics of man with the physical factors of the environment is not a simple task. None of the proposed heat stress indices meets all the requirements. A comprehensive analysis of the most popular indices has been made by Kerslake (93), and those interested should refer to this source. Any hope of an index simple enough for everyday use awaits future research.

Various types of inadequate response to heat stress have been described in the medical literature. Thse have been categorized by the World Health Organization (Table 12.1). No description of the various symptoms associated with these disorders is given here: a complete and precise description has been presented by Minard (94). It should be noted that heat deaths are a common occurrence when a sudden natural heat wave develops. The individuals most susceptible to heat morbidity and mortality are the young and the aged. These are the individuals who have the most difficulty in attaining an adequate and rapid acclimatization to heat. Ellis et al. (95) suggested that the inability to

Table 12.1 Classification of Heat Disorders

I.C.D.* Code	Clinical Designation	Etiologic Category
992.0	Heat stroke	Thermoregulatory failure
—	Heat hyperpyrexia	
992.1	Heat syncope	Orthostatic hypotension
992.2	Heat cramps	Salt and water imbalance
992.3	Heat exhaustion, water depletion	
992.4	Heat exhaustion, salt depletion	
992.5	Heat exhaustion, unspecified	
992.6	Heat fatigue, transient	Behavioral disorder
—	Heat fatigue, chronic	
705.1	Heat rash	Skin disorders and sweat gland injury
—	Anhydrotic heat exhaustion	

* International Classification of Diseases.

sweat sufficiently and to benefit from evaporative cooling may be a factor in the high mortality of the very young and the elderly.

Heat stress remains as a potent stressor to man, not only in his natural environment but in the artificial environment presented by certain industrial situations. Successful resistance to this stressor depends on the development and utilization of adequate physiological mechanisms.

REFERENCES

1. S. M. Horvath, G. B. Spurr, B. K. Hutt, and L. H. Hamilton, *J. Appl. Physiol.*, **8**, 595 (1956).
2. A. Hemingway, *Physiol. Rev.*, **43**, 397 (1963).
3. T. R. A. Davis and D. R. Johnson, *J. Appl. Physiol.*, **16**, 231 (1961).
4. K. Rodahl, S. M. Horvath, N. C. Birkhead, and B. Issekutz, Jr., *J. Appl. Physiol.*, **17**, 763 (1962).
5. J. D. Hardy, in: *Physiology of Heat Regulation and the Science of Clothing*, L. H. Newburgh, Ed., Saunders, Philadelphia, 1948.
6. A. C. Burton and O. G. Edholm., *Man in a Cold Environment*, Edward Arnold Ltd., London, 1955.
7. A. C. Burton, P. A. Snyder, and W. G. Leach, *J. Appl. Physiol.*, **8**, 269 (1955).
8. P. F. Iampietro and E. R. Buskirk, *J. Appl. Physiol.*, **15**, 212 (1960).
9. E. R. Nadel, J. W. Mitchell, and J. A. J. Stolwijk, *Int. J. Biometeorol.*, **15**, 201 (1971).
10. B. Saltin, A. P. Gagge, and J. A. J. Stolwijk, *J. Appl. Physiol.*, **28**, 318 (1970).
11. T. Adams and E. J. Heberling, *J. Appl. Physiol.*, **13**, 226 (1958).
12. B. Saltin, in: *The Physiology of Work in Cold and Altitude*, C. Helfferich, Ed., Arctic Aeromedical Laboratory, Fort Wainwright, Alaska, 1966.
13. B. S. Kang, D. S. Han, K. S. Paik, Y. S. Park, J. K. Kim, D. W. Rennie, and S. K. Hong., *J. Appl. Physiol.*, **29**, 6 (1970).
14. S. K. Hong, in: *Physiology of Breath-hold Diving and the Ama of Japan*, H. Rahn, Ed., National Academy Sciences–National Research Council, Washington, D.C., 1965.
15. H. T. Hammel, in: *Adaptation to the Environment, Handbook of Physiology*, Section 4, D. B. Dill, Ed., American Physiological Society, Washington, D.C., 1964.
16. H. Yoshimura, in: *Essential Problems in Climatic Physiology*, H. Yoshimura, K. Ogata, and S. Itoh, Eds., Nankoda Publishing, Kyoto, 1960.
17. F. Girling, *Can. J. Physiol. Pharmacol.*, **45**, 13 (1967).
18. S. M. Horvath, in: *Advances in Climatic Physiology*, S. Ito, K. Ogata, and H. Yoshimura, Eds., Igaku Shoin Ltd., Tokyo, 1972.
19. H. T. Hammel, R. W. Elsner, D. H. LeMessurier, H. T. Anderson, and F. A. Milan, *J. Appl. Physiol.*, **14**, 605 (1959).
20. P. B. Raven and S. M. Horvath, *Int. J. Biometerorol.*, **14**, 309 (1970).
21. C. H. Wyndham et al., *J. Appl. Physiol.*, **19**, 583 (1964).
22. M. Yoshimura and H. Yoshimura, *Int. J. Biometeorol.*, **13**, 163 (1969).
23. J. F. O'Hanlon, Jr., and S. M. Horvath, *Can J. Physiol. Pharmacol.*, **48**, 1 (1970).
24. P. B. Raven, I. Niki, T. E. Dahms, and S. M. Horvath, *J. Appl. Physiol.*, **29**, 417 (1970).
25. S. M. Horvath, A. Freedman, and H. Golden, *Am. J. Physiol.*, **150**, 99 (1947).
26. L. D. Carlson, H. L. Burns, T. H. Holmes, and P. P. Webb, *J. Appl. Physiol.*, **5**, 672 (1953).

27. T. R. A. Davis, *J. Appl. Physiol.*, **17**, 751 (1962).
28. P. F. Scholander, H. T. Hammel, J. S. Hart, D. H. LeMessurier, and J. Steen, *J. Appl. Physiol.*, **13**, 211 (1958).
29. J. Leblanc, J. A. Hildes, and O. Heroux, *J. Appl. Physiol.*, **15**, 1031 (1960).
30. J. Leblanc, *J. Appl. Physiol.*, **17**, 950 (1962).
31. G. M. Brown and J. Page, *J. Appl. Physiol.*, **5**, 221 (1952).
32. R. W. Elsner, J. D. Nelms, and L. Irving, *J. Appl. Physiol.*, **15**, 662 (1960).
33. J. D. Nelms and J. G. Saper, *J. Appl. Physiol.*, **17**, 444 (1962).
34. I. F. G. Hampton, *Fed. Proc.*, **28**, 1129 (1969).
35. T. Lewis, *Heart*, **15**, 177 (1930).
36. K. Hirai, S. M. Horvath, and V. Weinstein, *Angiology*, **21**, 502 (1970).
37. H. Yoshimura and T. Iida, *Jap. J. Physiol.*, **1**, 147 (1950).
38. J. P. Meehan, *Mil. Med.*, **116**, 330 (1955).
39. J. Krog, B. Fokon, R. H. Fox, and K. L. Anderson, *J. Appl. Physiol.*, **15**, 654 (1960).
40. T. Adams and R. E. Smith, *J. Appl. Physiol.*, **17**, 317 (1962).
41. A. C. Brown and G. C. Brengelmann, in: *Physiological and Behavioral Temperature Regulation*, J. D. Hardy, A. P. Gagge, and J. A. J. Stolwijk, Eds., Thomas, Springfield, Ill., 1970.
42. E. R. Nadel and S. M. Horvath, *J. Appl. Physiol.*, **27**, 484 (1969).
43. E. R. Nadel, S. M. Horvath, C. A. Dawson, and A. Tucker, *J. Appl. Physiol.*, **29**, 603 (1970).
44. J. A. J. Stolwijk and J. D. Hardy, *Pflügers Arch. Ges. Physiol.*, **291**, 129 (1966).
45. E. L. Arnett and D. T. Watts, *J. Appl. Physiol.*, **15**, 499 (1960).
46. J. E. Wilkerson, P. B. Raven, N. W. Bolduan, and S. M. Horvath, *J. Appl. Physiol.*, **36**, 183 (1974).
47. R. J. T. Joy, *J. Appl. Physiol.*, **18**, 1209 (1963).
48. L. Irving, *Sci. Am.*, **214**, 94 (1966).
49. D. G. Stuart, E. Eldred, A. Hemingway, and Y. Kawamura, in: *Temperature: Its Measurement and Control in Science and Industry*, Vol. 3, J. D. Hardy, Ed., Reinhold, New York, 1963, p. 545.
50. A. W. Mills, *J. Appl. Physiol.*, **9**, 447 (1956).
51. R. E. Clark and A. Cohen, *J. Appl. Physiol.*, **15**, 496 (1960).
52. S. M. Horvath and A. Freedman, *J. Aviat. Med.*, **18**, 158 (1947).
53. M. Davies, B. Ekblom, U. Bergh, and I. L. Kanstrup-Jensen, *Acta Physiol. Scand.*, **95**, 201 (1975).
54. S. M. Horvath, C. E. Radcliffe, B. K. Hutt, and G. B. Spurr, *J. Appl. Physiol.*, **8**, 145 (1955).
55. A. J. Watts, *Environ. Res.*, **5**, 119 (1972).
56. B. T. Horton, G. E. Brown, and G. M. Roth, *JAMA*, **107**, 1265 (1936).
57. F. J. Kelly and R. A. Wise, *Am. J. Med.*, **15**, 431 (1953).
58. S. E. Epstein, M. Stampfer, G. D. Beiser, R. E. Goldstein, and E. Braunwald, *New Engl. J. Med.*, **280**, 7 (1969).
59. K. L. Anderson, B. Hillstrøm, and F. V. Lorentzen, *J. Appl. Physiol.*, **18**, 975 (1963).
60. J. H. Talbott, *New Engl. J. Med.*, **224**, 281 (1941).
61. *Cold Injury*. Transactions of the First to Sixth Conferences, Josiah Macy Foundation, New York, 1952–1958.
62. S. M. Horvath and R. D. Rochelle, *Environ. Health Perspect.*, **20**, 127 (1977).
63. E. F. Adolph, *Physiology of Man in the Desert*, Wiley-Interscience, New York, 1947.
64. H. L. Marriott, *Water and Salt Depletion*, Thomas, Springfield, Ill., 1950.

65. S. Robinson, E. S. Turrell, H. S. Belding, and S. M. Horvath, *Am. J. Physiol.*, **140,** 168 (1943).
66. N. Nelson, L. W. Eichna, S. M. Horvath, W. Shelly, and T. F. Hatch, *Am. J. Physiol.*, **151,** 626 (1947).
67. H. L. Taylor, A. F. Henschel, and A. Keys, *Am. J. Physiol.*, **139,** 583 (1943).
68. C. H. Wyndham, G. G. Rogers, L. C. Senay, and D. Mitchell, *J. Appl. Physiol.*, **40,** 779 (1976).
69. J. E. Wood and D. E. Bass, *J. Clin. Invest.*, **39,** 825 (1960).
70. L. C. Senay, D. Mitchell, and C. H. Wyndham, *J. Appl. Physiol.*, **40,** 786 (1976).
71. L. B. Rowell, K. K. Kraning, J. W. Kennedy, and T. D. Evans, *J. Appl. Physiol.*, **22,** 509 (1967).
72. S. M. Horvath and W. B. Shelley, *Am. J. Physiol.*, **146,** 336 (1946).
73. S. Robinson, H. S. Belding, F. C. Consolazio, S. M. Horvath, and E. S. Turrell, *J. Appl. Physiol.*, **20,** 583 (1965).
74. N. B. Strydom, H. F. Kotze, W. H. van der Walt, and G. G. Rogers, *J. Appl. Physiol.*, **41,** 202 (1976).
75. H. F. Kotze, W. H. van der Walt, G. G. Rogers, and N. B. Strydom, *J. Appl. Physiol.*, **42,** 711 (1977).
76. A. Kawahata, in: *Essential Problems in Climatic Physiology,* H. Yoshimura, K. Ogata, and S. Itoh, Eds., Nankodo Publishing, Kyoto, 1960.
77. W. M. Haslag and A. R. Hertzman, *J. Appl. Physiol.*, **20,** 1283 (1965).
78. J. D. Hardy and E. F. DuBois, *Proc. Nat. Acad. Sci. (U.S.),* **26,** 389 (1940).
79. T. S. Cleland, J. C. Bachman, and S. M. Horvath, *J. Gerontol.,* (Submitted).
80. T. Morimoto, Z. Slabochova, R. K. Naman, and F. Sargent, II, *J. Appl. Physiol.*, **22,** 526 (1967).
81. T. S. Cleland, S. M. Horvath, and M. Phillips, *Int. Z. Angew. Physiol.*, **27,** 15 (1969).
82. L. Brouha, P. E. Smith, Jr., R. DeLanne, and M. E. Maxfield, *J. Appl. Physiol.*, **16,** 133 (1960).
83. R. H. Fox, B. E. Lofstedt, P. M. Woodward, E. Eriksson, and B. Werkstrom, *J. Appl. Physiol.*, **26,** 444 (1969).
84. O. Bar-Or, H. M. Lundegren, and E. R. Buskirk, *J. Appl. Physiol.*, **26,** 403 (1969).
85. B. A. Hertig, H. S. Belding, K. K. Kraning, D. L. Batterton, C. R. Smith, and F. Sargent, II, *J. Appl. Physiol.*, **18,** 383 (1963).
86. K. P. Weinman, Z. Slabochova, E. M. Bernauer, T. Morimoto, and F. Sargent, II, *J. Appl. Physiol.*, **22,** 533 (1967).
87. C. H. Wyndham, J. F. Morrison, and C. G. Williams, *J. Appl. Physiol.*, **20,** 357 (1965).
88. F. Sargent, II and K. P. Weinman, *J. Appl. Physiol.*, **21,** 1685 (1966).
89. D. B. Dill, S. M. Horvath, W. Van Beaumont, G. Gehlsen, and K. Burrus, *J. Appl. Physiol.*, **23,** 746 (1967).
90. B. L. Drinkwater, J. E. Denton, I. C. Kupprat, T. S. Talag, and S. M. Horvath, *J. Appl. Physiol.*, **41,** 815 (1976).
91. B. L. Drinkwater, I. C. Kupprat, J. E. Denton, and S. M. Horvath, *Ann. N.Y. Acad. Sci.*, **301,** 777 (1977).
92. J. C. Senay, Jr., *J. Physiol., Lond.,* **2325,** 209 (1972).
93. D. McK. Kerslake, *The Stress of Hot Environments.* University Press, Cambridge, England, 1972, p. 317.
94. D. Minard, "Heat Disorders: A Tabular Presentation", in: *Standards for Occupational Exposures to Hot Environments—Proceedings of Symposium,* S. M. Horvath, Ed.-in-Chief, and R. C. Jensen, Ed., National Institute for Occupational Safety and Health, Cincinnati, Ohio, 1976, pp. 21-25.
95. F. P. Ellis, A. N. Exton-Smith, K. G. Foster, J. F. Weiner, *Isr. J. Med. Sci.,* **12,** 815 (1976).

CHAPTER THIRTEEN

Evaluation of Exposures to Vibrations

JOHN C. GUIGNARD, M.B., Ch.B.

1 INTRODUCTION

Mechanical vibration is ever present in factories, in moving vehicles on land, at sea, and in the air, and in the vicinity of working machinery of all kinds. Like acoustical noise, with which it is commonly associated, vibration is encountered, at intensities ranging from the barely perceptible to the intolerably severe, on construction sites, in mines and quarries, around manufacturing equipment, in shipyards, on the farm and in the forest, and wherever man harnesses power to get work done.

In the context of occupational medicine, mechanical vibration may be defined as any continuous or intermittent oscillating mechanical force or motion affecting man at work through the mediation of structures and receptors other than the organ of hearing. (Such a definition excludes present consideration of the effects of noise, i.e., airborne vibration at audible frequencies, which are dealt with elsewhere in Chapter 11.) In engineering physics, vibration, a ubiquitous phenomenon, has received a succinct definition by Crede (1) as a series of reversals of velocity. This definition reminds us that in any vibratory process, both displacement of matter and acceleration (change of velocity, requiring the generation of mechanical force) necessarily take place. The reversal of velocity may occur at regular intervals, that is, with a specific frequency, in which case the vibration is called periodic; or it may occur irregularly and unpredictably, in which case the vibration is called nonperiodic, or random.

Occupational exposure to vibration arises in a variety of ways and it can reach the worker at intensities disturbing to comfort, efficiency, safety, or health through several routes of mechanical transmission (2, Ch. 29; 3). It may affect workers principally by transmission through a supporting or contacting surface that is vibrating, such as the deck of a ship, the seat or floor of a vehicle shaken by roadway and engine vibrations, the floor of a workplace shaken by nearby industrial machinery, or a workplace on a machine mounted by a worker, in which vibration is intentionally generated to grade, move, or compact such materials as coal, ore, or concrete. These are all examples of whole-body vibration.

In certain occupations in many industries, the chief route of entry of vibration into the human body is through the hands and arms of the worker. This occurs when he is using a handheld powered tool such as a road breaker, chipping hammer, grinder, power drill, or chain saw. As the major portions of this chapter demonstrate, it is practical and convenient to treat whole-body and hand-transmitted vibration separately as distinct provinces of study. This is recognized both in the domain of research and in the fields of engineering and regulatory approaches to the protection of the worker from adverse effects of occupational vibration exposure. In accordance with international expert consensus, separate international advisory standards have been issued or drafted governing human exposure to whole-body and hand-transmitted vibration (4, 5).

In some working situations, for instance, when operating heavy equipment or driving tractors and the like, vibration enters the body through several routes, including the operator's seat, the floor or controls on which the feet are placed, and the steering wheel or other hand controls. Mechanical vibration can also be disturbing, fatiguing, and prejudicial to occupational safety indirectly, without actually entering the body, for example, when vibration of the dials and pointers of gauges or other instruments in a vehicle or plant makes them difficult and at times impossible to read.

1.1 The Description and Measurement of Vibration

The physical description and measurement of vibration affecting man have much in common with the measurement and evaluation of noise and, in many applications, make use of similar instrumentation and analytical techniques. Several descriptive parameters are necessary to characterize completely a mechanical vibration affecting man. The most important ones are frequency, complexity, amplitude (or intensity), direction of application (with respect to anatomical axes), and duration and time course (including intermittency) of the vibration exposure.

1.1.1 Frequency

The frequency of a periodic (i.e., wavelike) vibration is the number of complete cycles of oscillation occurring in unit time. The international standard unit of vibration frequency is, as in acoustics, the hertz (Hz), which is one cycle of oscillation per second. An older unit, the cycle per second (c/s) is still in relatively common use and will be found in many of the publications cited in this chapter. However the hertz is now preferred (1 Hz = 1 c/s).

Traditionally the frequency of a vibration (or the component frequencies of a complex vibration) is determined by methods of graphical analysis, that is, inspection of a plot or recording of the vibration wave form against a time base. In most modern applications, however, vibration is measured by means of electronic transducers and analytical instrumentation yielding a power spectrum or some cognate mathematical representation of the power, energy, or intensity of a vibration as a function of frequency or inverse time.

Although accelerometers or other special transducers that convert mechanical force or motion into proportional electrical signals are needed to detect vibration in the low frequency range (below 20 Hz) that is mainly of interest in the case of whole-body exposure, the techniques of magnetic tape recording and electronic signal processing permit the kind of equipment now generally available commercially for noise analysis to be used for vibration analysis in a similar manner.

When the vibration frequencies of interest lie in the subaudible range, below the range normally handled by analytical equipment developed for acoustical work, the vibration frequencies to be analyzed can be shifted into a higher part of the spectrum for analytical purposes by replaying the tape-recorded data at an appropriate speed. Nowadays rapidly increasing use is being made of digital computers to analyze vibration data and to generate such informative functions as power spectra and vibration transmission factors. Such computational procedures rely on the analogue-to-digital conversion of the "raw," continuous vibration signals generated by the measuring transducer to a form that can be handled by the computer.

1.1.2 Complexity

Vibration affecting workers is almost always complex; that is, it is composed of more than one frequency and is multidirectional. Just as the pure tone of a tuning fork or an electronic oscillator is rarely heard among the sounds and noises of everyday life, so single-frequency (sinusoidal) vibration is rarely, if ever, encountered outside the laboratory. (Sinusoidal vibration, however, is a common and useful laboratory tool for gaining insight into the human biodynamic and physiologic response to vibration, just as the pure tone of the audiometer, being precisely definable, measurable, and repeatable, is an ideal stimulus for determining the response of the organ of hearing).

Approximately (if impure) sinusoidal vibration can be felt on board ship at a frequency determined by the product of the revolutions per minute of the propeller shaft and the number of blades on the screw. Such vibration is particularly strong in single-screw vessels and when the engine revolutions pass through a critical speed, that is, a speed at which the driving frequency coincides with one of the major natural frequencies of structural vibration of the ship's hull or superstructure, which is then set into sympathetic vibration (resonance). Heavy vibration at the rotating wing blade passage frequency (again, a product of the engine rpm and the number of blades in the lifting rotor) can also be a considerable problem in helicopters; and in a fixed plant, troublesome single-frequency vibration can be transmitted to workplaces when faulty design, installation, or maintenance of a machine permits resonant vibration of the machine on its mountings, or of the building frame or structures in the vicinity.

Vibration in vehicles and from industrial machines is often complex, irregular, or essentially random (e.g., the lurching of a tractor or heavy equipment operated over rough ground) and accordingly is lacking in obvious periodicity. Nevertheless, using the techniques of frequency (spectral) analysis, it is still possible, and in many applications appropriate, to describe and evaluate the motion in terms of frequency or spectrum. A vibration spectrum, by analogy with the spectra of electromagnetic radiations, is a graph or table of vibrational energy, power, or intensity against frequency.

1.1.3 Intensity

Another important descriptive parameter of vibration is its amplitude or intensity, in other words, the extent or severity of the motion at any given frequency. When the vibration is of the simplest kind, namely, sinusoidal or simple harmonic oscillation, resembling the swing of a pendulum, the amplitude is defined as the maximum (sometimes called the "peak") displacement of the vibrating system from its midposition of rest or equilibrium. A velocity of vibrational motion, and an acceleration due to the restoring or impressed force of the vibration, are always proportionally associated with the instantaneous displacement in the vibratory process. For sinusoidal vibration, or any single-frequency component of a complex vibration, when the magnitude of the displacement is fixed, the vibrational velocity associated with it rises proportionally with frequency; and the vibrational acceleration increases with the square of frequency. For this reason very large displacement amplitudes of motion (perhaps several meters, in the case of the heave motion of a large ship or an oil-drilling platform induced by ocean waves at frequencies well below 1 Hz) are required to generate appreciable levels of acceleration at low frequencies; yet higher up the frequency spectrum—at vibration frequencies, for example, associated with the peak output of vibrational energy by handheld power tools (typically ranging from around 30 Hz to several hundred hertz)—enormous accelerations can be generated by vibrations of very small displacement-amplitude (fractions of a millimeter).

Using the international system of units, vibrational displacement is properly measured in meters, although fractional metric units may be used for convenience to express the tiny displacements characteristic of high frequency vibrations. It is usually convenient, particularly when modern electronic vibration-measuring instrumentation is used, to measure the intensity of vibration in terms of acceleration, which is the second derivative of displacement with respect to time and is directly related to the force of vibration.

Acceleration measurements are becoming standard practice where vibrations affecting man are concerned (3–5). A variety of small transducers (accelerometers) now marketed generate an electrical output directly proportional to the mechanical vibrational acceleration of a vibrating machine or, in field and laboratory research applications, the human body. These instruments permit the direct measurement of vibrational acceleration.

For some applications it is convenient to use other kinds of transducer or vibration "pickup" that respond to displacement, pressure, or velocity rather than to vibrational acceleration. Signal processing (e.g., differentiation of the velocity signal with respect to time) is then required to obtain a measurement of vibration acceleration. It is important to make sure that the measuring system can handle such operations on the signal adequately over the bandwidth of concern in any given application.

Vibrational acceleration is measured in meters per second per second (m/sec^2.) For many practical purposes, however, and commonly in biomedical work, it is convenient and acceptable (4) to express the acceleration associated with vibrating forces nondimensionally in terms of g, that is, as multiples or submultiples of the standard acceleration of gravity at the earth's surface, where $1g$ is approximately equivalent to 9.81 m/sec^2.

An instantaneous or peak value of acceleration has little meaning when describing a continuous vibration that is complex or random. It is accordingly common practice to measure or compute an average or root mean square (rms) value of the accelerations recorded instantaneously over a finite data sample. Many electronic instruments used to measure vibration or noise (e.g., the sound level meters widely used in noise surveying and monitoring) are designed to yield an output that is approximately proportional to the rms value of the measured quantity. The rms value of a varying quantity such as vibrational acceleration or sound pressure is mathematically equivalent to the standard deviation of the instantaneous values of the quantity. In the case of a sinusoidally varying quantity such as a pure harmonic vibration, the rms value is 0.707 ($\sqrt{2}/2$) times the maximum (peak) value.

Although rms measurements are an adequate approximation for many applications such as the comparative evaluation of high frequency vibrations or noises, caution should be exercised when applying such an intensity-averaging technique to the low frequency domain. That is because the biological effectiveness of mechanical vibrations or whole-body motion in the subaudible range depends on additional factors, such as the relative intensities and the phase relationships of the component frequencies of a complex vibration, which are ignored by rms measurements. Human experimentation on vibration tolerance in the range 1 to 30 Hz (6) and on motion sickness incidence induced by complex harmonic motion in the range below 1 Hz (7) has shown that it can be fallacious to assume that the severity of combinations of sinusoidal motions can be equated solely in terms of their rms acceleration.

1.1.4 Direction

The mechanical response of man to whole-body vibration, and the corresponding sensory and psychophysiological reactions to the motion, are heavily dependent on the direction in which vibrating forces are applied to the body or its parts. For instance, the subjective perception and tolerance of vibration in the range 1 to 10 Hz, and the degree of interference with working efficiency that such vibration can provoke, are quite different when the subject's seat vibrates from side to side (y-axis vibration) and when it vibrates with the same amplitude vertically (z-axis vibration). That is mainly because the human body as a complex vibrating system exhibits resonance at different frequencies, corresponding with different modes of vibration, when vibrated in the transverse as opposed to the longitudinal axis in that range of frequency. The International Organization for Standardization (ISO) has accordingly promulgated separate standards (4) for transverse (x- and y-axis) and longitudinal (z-axis or cephalocaudal) human whole-body vibration exposure in the range 1 to 80 Hz. The same authority (8) is also drafting a standard on coordinate systems for use in biodynamics: properly defined anatomical coordinate systems are essential frames of reference for the unequivocal specification, measurement, and comparative valuation of force and motion inputs to the human body.

1.1.5 Duration and Time Course

The human response to vibration depends in a complex manner on the duration of a continuous or steady state vibration or on the time course of a fluctuating (nonstationary), transient, or intermittent vibration. The time factor in morbidity associated with prolonged exposure to whole-body or hand-transmitted vibration repeated occupationally on a daily basis is mentioned in more detail later in the chapter. In setting standards governing human exposure to whole-body vibration in the short term (i.e., discrete exposures, or exposures repeated daily but lasting less than a day), the ISO (4) has presumed, for want of substantive data, that a general biological principle applies, namely, that human tolerance of steady state vibration declines with increasing duration of exposure. The degree to which such a tendency is mitigated by specific adaptation or general habituation to vibration stress remains an open question, for very little definitive research has yet been devoted to it.

2 WHOLE-BODY VIBRATION

Whole-body vibration is commonly experienced in land, sea, and air vehicles, and in buildings and workplaces where the floor, ground, or structures in the vicinity are shaken by the action of machinery. The entire body may be vibrated by powerful handheld machines (e.g., road breakers, tamping machines, and rock drills), which generate substantial amounts of vibrational energy at low frequencies (below 20 Hz). The vibration of such machines, however, is mainly of concern as a threat to the hands and upper limb joints of the operator, which are considered in the next section. Whole-body vibration is usually considered to enter the body through a supporting surface, but it can be received by other routes, for instance, from certain kinds of motorized appliance (e.g., crop sprayers and dusters, hedge trimmers) supported or carried on the person as a backpack.

Vibration at work is generally thought of as, at best, an inescapable irritant or nuisance (akin to noise, with which it is almost universally associated), which causes concern when it reaches such an intensity, constancy, or regularity of action as to prejudice the individual's working efficiency, safety, rest, or health. It may be noted in passing that not all mechanical vibration felt by people is unwanted or harmful. Naturally occurring or self-induced oscillations of the body or its parts are normally present, albeit generally at very low intensities, throughout life, generated by the heartbeat, locomotion, and other internal sources. Ground vibration, occasionally at alarming or threatening levels (9), can be encountered from natural sources, such as the action of wind, water, or seismic forces. Artificially generated rhythmic motion or vibration, sometimes at quite intense levels, is used by many people for amusement, relaxation, pleasure, or quasi-therapeutic purposes, as is shown by the widespread popularity of a variety of motion or vibration generating devices, ranging from fairground machines to personal vibrators.

Be that as it may, whole-body vibration experienced occupationally is a matter of concern in many industries where it occurs at intensities sufficient to undermine working efficiency and safety; and where whole-body vibration is part of the composite stress of the working environment, distinct patterns of morbidity may be causally associated with chronic, daily exposure to the stress. Whole-body vibration in the range 1 to 20 Hz is the most disturbing and hazardous to workers in most such industries (10, 11). Many kinds of commercial and working vehicles, as well as mobile heavy equipment driven or ridden by the operator, produce whole-body vibration (predominantly vertical) mainly in the range 1 to 20 Hz (10, 12). In that range current standards or guidelines for limiting human whole-body vibration exposure may frequently be exceeded substantially (10). It has been estimated that in the United States alone, nearly 7 million workers are exposed occupationally to whole-body vibration (12).

When applied to the whole body, mechanical vibrations at frequencies above 20 or 30 Hz probably do not make a substantial contribution to morbidity in most situations where vibration is of concern. That is because vehicles and other generators of whole-body vibration do not impress substantial amounts of vibrational energy on man at high frequencies (unlike handheld power tools, discussed later in this chapter); and since high frequency vibrations are substantially attenuated by the human body surface and by seating and flooring materials, much of the energy of such vibrations does not enter the body. In many respects whole-body vibration at frequencies above 30 Hz, though it may contribute to discomfort and fatigue, as well as distraction from work, is more akin to irradiation by noise, and, at high frequencies, it may be regarded as part of the acoustic environment. Because of such considerations, the current international standard on human exposure to whole-body vibration (4) is limited to the frequency band 1 to 80 Hz.

Whole-body mechanical oscillation of man at frequencies below 1 Hz does not cause internal vibratory movements of the body, which moves in response to such very low frequency motion as a single mass. Nevertheless, such motion does exert important psychophysiological effects, of which the most widely experienced is motion sickness (kinetosis). Large amplitude, low frequency oscillations are a perennial hazard to the crews of ships, aircraft, and floating structures such as offshore drilling platforms, interfering with both work and rest. In heavy seas the heaving, pitching, and rolling of ships or floating rigs subject the crew to severe complex motion in the band 0.1 to 1 Hz that renders hazardous and not infrequently impossible the performance of seaborne tasks, and even simple locomotion aboard the craft, when a storm is at its height and the risk of falls or other physical injury increases. Similar hazards can beset the aviator (particularly cabin staff who are not strapped into their seats in passenger aircraft) when an aircraft encounters heavy turbulence (3).

If sufficiently intense and prolonged, whole-body motion in the band 0.1 to 1 Hz can induce *motion sickness* in susceptible people. The individual response is strongly conditioned but requires oscillatory stimulation of the vestibular organs (13). Vertical ("heave") motion in the band 0.1 to 0.6 Hz appears to be the most provocative in man, with the maximum susceptibility occurring during oscillation at around 0.17 Hz (14). (Such a frequency corresponds to a heave period of some 6 sec, not uncommonly encountered in medium-sized ships at sea.) Certain complex motions in the band 0.17 to 0.50 Hz can induce higher incidence of motion sickness than is predicted from a knowledge of the nauseogenicity of the component single-frequency oscillations (7, 15). Although motion sickness (also known popularly as air-, sea-, or travel sickness) is often the subject of unthinking amusement or ridicule, it can prove not only prejudicial to efficiency and safety at sea and in the air but dangerous to life in certain circumstances—for instance, when an entire ship's complement or a critical proportion of the crew succumb, when a survivor awaiting rescue becomes seasick in a liferaft, or when a bad ride in an ambulance induces vomiting in a critically ill or injured patient.

Motion sickness, however, is best viewed as a psychophysiological reflex manifestation in response to passive whole-body motion, rather than as a pathological condition associated with any known lasting morbidity or direct fatality. Detailed reference to the condition is therefore omitted from this chapter. It is a self-limiting condition, controllable in most cases by habituation to the motion and, when necessary and appropriate, by medication (16); and it is rapidly alleviated, without lasting residual effects, by removal from the provocative motion. The interested reader may refer to a number of recent definitive treatises on the topic (16–19).

Wasserman et al. (12) have recently reported that vibration acceleration spectra recorded from mobile heavy equipment operated on rough, offroad sites contain substantial amounts of energy at frequencies below 1 Hz, presumably because of the slow motion of the machines traversing uneven terrain. No operators of such equipment are known to have been troubled by motion sickness at work, however, and the authors have pointed out that unlike the nauseogenic motion of a ship at sea, which can be sustained for hours or days, the low frequency oscillations transmitted to heavy equipment operators are transient and their vibration exposure frequently interrupted during relatively brief spells of work.

2.1 Effects of Whole-Body Vibration on Man

Several detailed accounts of the human response to whole-body vibration have been published in recent years (3; 20–22), and there are also reviews with particular reference to occupational exposure (23, 24).

2.1.1 Biodynamics

The physiological and psychological effects of mechanical vibration in man are caused by oscillatory displacement or deformation of the structures, organs, and tissues of the body so as to disturb their normal functioning and stimulate the distributed mechanoreceptors mediating the vibration sense. Vibration, particularly in the range 1 to 30 Hz, can also degrade or disrupt the performance of tasks by forcing differential motion to take place between man and his visual or physical points of contact with the task (3, Ch. 6). The living body dynamically is a complex vibrating system capable of resonance at certain low frequencies between 1 and 30 Hz; thus many biological effects of whole-body vibration are strongly frequency dependent.

Human body resonance is the condition in which a forcing vibration is applied to the body at such a frequency that some anatomical structure, part, or organ is set into measurable or subjectively noticeable oscillation greater than that of related structures (3, Ch. 4). In terms of physical biodynamics, the body can be visualized, and modeled analogically, as a complex system comprising several masses linked by elastic and damping elements. To that extent, the human body resembles engineering structures such as an airframe, a ship's hull, or a steel-framed building. All such structures exhibit resonant modes of vibration, of which the characteristic frequencies and magnification factors (i.e., the extent to which impressed vibration is mechanically amplified at the resonant frequencies) are determined by the ratios between mass, elasticity, and damping (3, Ch. 1). A large part of the practice of vibration engineering is devoted to preventing or suppressing conditions of resonance, and similar principles can be applied to the protection of man from the adverse effects of low frequency, whole-body vibration.

The human body exhibits several modes of resonant vibration in response to whole-body motion in the range 1 to 50 Hz (3, Ch. 4; 25). The absorption of vibrational energy at low frequencies by the human body, and the excitation of particular modes of vibration in man by continuous vibration or by impact forces, depend on several intrinsic and external biodynamic factors; Table 13.1 summarizes the most important ones.

The principal resonance of the seated, standing, or recumbent human body vibrated in the z-axis (cephalocaudally) occurs in the region of 5 Hz, with substantial amplification of impressed vibrations being observable in the band 4 to 8 Hz. This response is reflected in the frequency dependence of many physiological and psychological human reactions to vibration (3), and it corresponds to a minimum in human subjective tolerance of whole-body vibration in the z-axis, particularly at severe levels (26, 27). Accordingly, current recommended limits of human exposure to whole-body vibration (4), expressed in terms of acceleration as a function of frequency between 1 and 80 Hz, are set at their lowest in the band 4 to 8 Hz.

When a person is vibrated in the x- (anteroposterior) or y- (lateral) axis—for example, in a swaying vehicle—the principal mode of resonant vibration, with oscillatory flexion of the trunk at the hips or lower back, occurs in a different frequency band, namely, 1 to 2 Hz. Such motion is not uncommonly experienced by passengers in

Table 13.1 Biodynamic Factors Influencing the Absorption and Transmission of Mechanical Vibration Entering the Human Body[a]

External Factors

Intensity (force) of vibration
Direction, site, and area of application of vibration
Nature (including contour) of support and restraint of the person
Resilience and damping of seat, cushion, or other structure through which vibration is transmitted
Nature, distribution, and weight (mass) of any external load on body (e.g., work clothes; man-mounted equipment)

Internal Factors

Individual build (height, weight, fatness, etc.)
Nonlinearity of tissue stiffness and damping
Posture (relative position of body segments and limbs)
Activity of the subject
Degree of muscular tone or tension

[a] Adapted from Guignard (2, Ch. 29).

trains, which tend to sway when passing over uneven track. In contrast to z-axis vibration, which as mentioned is most disturbing at frequencies in the 4 to 8 Hz band, transverse whole-body vibration is least well tolerated in the lower band below 2 Hz. This difference is again reflected in the setting of current standards on human exposure to whole-body vibration (3, 4).

During whole-body vibration at frequencies much above 50 Hz, the mechanical response of the human body can be visualized as that of a continuous viscoelastic medium of energy propagation rather than as a system containing discrete masses (although resonance of certain structures can still be elicited at high frequencies). As the frequency rises into the kilohertz range, the propagation of mechanical vibration within the body and its tissues progressively becomes essentially acoustical; that is, at high frequencies most of the vibrational energy entering the body, through whatever surface, is propagated through the tissues as compressional waves (28).

2.1.2 Physiology of Whole-Body Vibration

The physiological effects of moderately intense whole-body vibration fall into two broad categories (2, Ch. 29; 3, Ch. 5). First, vibration elicits frequency-dependent responses directly related to the differential oscillatory motion (especially at resonance) of body organs and tissues. Second, vibration (especially when the exposure is prolonged or repeated) elicits nonspecific generalized reactions of the kind seen as a response to stress in general, that is, not specific to the physical nature of the vibration stimulus. Similar reactions appear in response to loud noise and other environmental stressors. Such reactions to vibration are not markedly frequency dependent but seem to be related to the overall cumulative severity of the vibration exposure. Industrial vibration, very often associated with noise, is regarded by Soviet authorities as an important centrally acting stressor in the working environment, which may adversely affect the health of the worker through the mediation of the autonomic nervous system and neuroendocrinological mechanisms (29–31).

Low frequency, whole-body vibration of moderate intensity (in the region of 2 to 20 Hz at intensities of 0.1 to 1 g_{rms}) elicits a general cardiopulmonary response resembling the vegetative manifestations of moderate exercise or alarm, with variable increases in heart and respiration rates, cardiac output, pulmonary ventilation, and oxygen uptake (2, Ch. 29; 3, Ch. 5; 32). Blood pressure may also show slight to moderate increases, but the response is likely to be very variable. Such changes are attributable to raised metabolic activity associated with increased activity in the skeletal musculature provoked by vibration. According to Liedtke and Schmid (33), low frequency, whole-body vibration of the dog can induce a vasodilator response in the forelimb, apparently mediated by the vibratory excitation of muscle and tendon stretch receptors (but note the action of chronic exposure to intense hand-transmitted vibration in man, discussed later in this chapter).

In certain conditions strong z-axis (vertical) whole-body vibration of a seated person can induce hyperventilation, sometimes accompanied by symptoms and signs of hypocapnia. The hyperventilation does not appear to be explainable by oscillatory forced ventilation of the lungs as might be supposed but is probably due to the widespread vibratory stimulation of somatic mechanoreceptors, including those in the lungs and in the respiratory passages (34–36).

Various changes in the cellular and biochemical constituents of urine and blood have been observed in animals and in man in response to moderate or severe whole-body vibration. Generally speaking these changes appear to reflect a nonspecific response to mechanical vibration as a stressor. Certain endocrinological changes, involving morphological changes in endocrine glands (37) observed experimentally in animals subjected to prolonged vibration, also appear to reflect a generalized stress response.

Acute vibratory trauma, including contusion and abrasion of internal organs and tissues, with physiological or pathological consequences, can also occur. Severe vibration of experimental animals has been shown to cause hemorrhagic damage, sometimes leading to degenerative changes in chronic preparations, in many glandular and other organs and systems (38–40).

Brief, intense (up to $3g$) whole-body z-axis (cephalocaudal) vibration of animals, including the dog and the pig, in the frequency band associated with major body resonance effects (3 to 10 Hz) can be shown to produce gross mechanical interference with the hemodynamics of central (aortic) and regional arterial blood flow (41–43). Edwards et al. (41) postulated that by such mechanisms severe whole-body vibration in rough-riding vehicles could disturb cerebral blood flow in drivers, leading to decrements in performance: such a consequence has yet to be demonstrated in man, however.

2.1.3 Sensory and Neuromuscular Effects of Vibration

Mechanical vibration is perceived over a much broader range of frequency than that (approximately 16 to 20,000 Hz) spanned by the sensation of hearing in man (3, 9). The sensation of oscillatory motion and vibration, unlike the sense of hearing, is not mediated solely by one kind of receptor organ but is mediated by several sensory modalities subserved by a variety of receptors distributed throughout the body. The chief vibration-detecting organs in man are the cutaneous and related receptors subserving the vibrotactile sense (44, 45); the mechanoreceptors arrayed in deeper tissues, particularly the muscles, tendons, and periarticular structures; and the viscera and their attachments.

Sensations of oscillatory motion at very low frequencies below about 10 Hz are enhanced by stimulation of the vestibular (specifically the otolithic) receptors. Vestibular stimulation, augmented by visual cues, becomes paramount at frequencies below 2 Hz. The several kinds of diffusely distributed somatic mechanoreceptor, by contrast, respond mainly to vibration at higher frequencies (above 40 Hz, in the case of Pacinian corpuscles), in various overlapping frequency bands. These receptors and their afferents differ both in the effective bandwidth of their unit responses and in the degree of temporal integration of the information that they transmit to the central nervous system.

The threshold of sensation of whole-body vibration varies systematically with frequency and, at vibration intensities above the threshold, the strength of the vibratory sensation as a function of physical intensity of vibration at any given frequency appears to vary in a manner somewhat akin to the power law governing the strength of the sensation (loudness) of sound (22, 46–49).

Transient disequilibrium and enhanced postural sway and manual tremor have been reported following exposure to whole-body vibration of moderate duration and intensity, both in everyday experience (e.g., lengthy drives over bumpy roads; flights through rough air) and in laboratory studies (2, Ch. 5; 50). Workers exposed daily to vibration in factories or in operating mobile equipment such as large cranes have also been reported to exhibit such symptoms or signs (51, 52). Johnston (53) has observed that whole-body vibration above 2 Hz can adversely affect the speed of orientation. It can be reasoned that such effects in man could be prejudicial to safety at work.

The physiological basis of disturbances of equilibrium and postural regulation caused by whole-body vibration is as yet ill-defined. Possibly the effects are due to vibratory overstimulation of the receptors, particularly in muscle, and to competition in the neural pathways and their central connections that subserve both the regulation of posture and the low frequency somatic and vestibular vibration senses. However similar responses (particularly enhanced manual tremor) are observed during or following conditions other than vibration exposure and accordingly cannot be regarded as specific to motion or vibration exposure. Enhanced postural sway and tremor are seen in states of high arousal and in fatigue associated with sustained demanding work load and environmental stress not accompanied by strong motion stimuli (54).

Mechanical vibration of the whole body or of individual postural muscles or their tendons increases tonicity, while phasic spinal reflexes (e.g., tendon jerks) sometimes appear to be depressed or inhibited. These phenomena, observable over a wide frequency range from below 10 Hz to over 200 Hz, have been observed in man as well as in animals (including decerebrate preparations). Tendon vibration in man can interfere with the position sense in the limb, resulting in feelings of weakness, displacement, or other distortions of that sense in the arm or leg (55, 56).

Vibration of limb muscle tendons at relatively high frequencies (100 to 200 Hz) elicits or enhances a tonic stretch reflex (a spinal reflex, as can be demonstrated in the decerebrate cat) (57, 58). The tonic reflex contraction or tonic vibration reflex is mediated by receptors in muscle itself, chiefly (but not necessarily solely) the primary spindle endings (57, 59). In man, the tonic vibration reflex can be shown to gain in strength as the initial length of the muscle is increased (56, 60). (Vibration can be used therapeutically to facilitate volitional contraction in cases of spasticity.) The tonic vibration reflex apparently can be modified by a polysynaptic pathway involving higher centers including the cerebellum; accordingly it can be influenced by various factors operating supraspinally. Moreover, a degree of voluntary inhibition can be achieved (61). Low frequency vibration of postural muscles in man does not appear to alter the reflex excitability of the muscle, nor the character or strength of the maximal volitional response (62).

Varied findings have been reported concerning the effects of vibration on the central nervous system. Investigators in the Soviet Union in particular maintain that chronic exposure to whole-body vibration causes generalized debilitating effects mediated by the central nervous system. Melkumova and Russkikh (31), in experimental studies of the dog, cat, and rabbit, have reported a syndrome including debilitation, lack of coordination, and weight loss. Post mortem examination of the animals following several months of intermittent exposure to 50 Hz vibration was said to have shown cerebral edema and dystrophic changes in nerve fibers. Luk'yanova and Kazanskaya (63) have reported that vibration stress in experimental animals may be associated with fluctuations in the oxygen uptake of cerebral tissue. The lethality of intense ($10g$ at 25 Hz) whole-body vibration in mice is enhanced by centrally acting stimulants (dextroamphetamine) and reduced by central depressants (chlordiazepoxide, reserpine, barbiturates) (64).

Qualitative observations suggest that vibration can alter the level of arousal in more than one way (as can noise), depending on the physical characteristics of the stimulus and the nature of the subject's activity at the time of exposure. Low frequency (1 to 2 Hz), whole-body oscillations at moderate intensities (as in swings and rocking chairs) can be relaxing and soporific in man; however higher frequencies, higher intensities, and inconstancy of the stimulus are arousing. A considerable degree of adaptation or habituation to steady state vibration (e.g., the drone of aircraft or shipboard noise and vibration) can be achieved, provided the stimulus is regular (6) and is not changed or interrupted. Habituation can be so complete that one is alarmed if the vibration unexpectedly ceases, as when a ship stops at sea. Habituation to vibration is probably a central phenomenon, although some adaptation may occur at the receptor level. The habituated person can rest and apparently sleep well during quite strong vibration and noise in aircraft, ships, and vehicles, once accustomed to the particular motion.

The electroencephalogram (EEG) can be recorded during whole-body vibration at moderate acceleration levels, provided care is taken in the placement of electrodes and in other measures to minimize recording artifacts and electrical noise. However no specific EEG changes are known to occur in man during vibration exposure. It is difficult to distinguish synchronous activity of neural origin from mechanoelectrical recording artifacts when the EEG is recorded from vibrated subjects (65).

2.1.4 Effects of Vibration on Task Performance

Intense vibration and oscillatory motion of man can degrade the performance of tasks, or render tasks more difficult to perform satisfactorily, by both central and peripheral mechanisms (2, Ch. 29; 3, Ch. 6; 21; 22). Disruption of performance by motion and vibration is stressful, fatiguing, and sometimes hazardous. Peripherally, vibration mechanically disrupts or interferes with one's application to a task by degrading both the sharpness of vision and the precision of manipulation of tools or controls: such effects are immediate and are likely to be strongly frequency dependent, being related to resonance phenomena in the vibrated body (3). In tasks of some kinds, especially those requiring precise eye-hand coordination, skill (accuracy of performance) can be maintained during moderate vibration at the expense of speed.

It is known from flight experience (3), as well as various laboratory experiments, that heavy vibration can also degrade skilled performance centrally, acting as a distracting, thought-disrupting, and fatiguing agent (as does noise); but this mechanism is not as easy to demonstrate experimentally. The effect is more related to the intensity and duration, and to the demands of the task, than to the frequency of vibration. Military experience (66) has been extended by Caiger (67) and others to civil air transport flying, to show that heavy air turbulence encounters and associated aircraft vibration cause distraction, disruption of instrument reading and control movements, impairment of quick thinking and decision taking, and sometimes spatial disorientation.

It is of course very difficult to gather other than anecdotal data about such effects in flight, but the problem can be studied using flight simulators (68) and laboratory vibration machines (3, 22). Pilots and navigators find that in the main, turbulence-induced oscillation below 2 Hz produces discomfort and progressive muscular fatigue, increased effort by the pilot to avoid or correct inadvertent control movements, difficulty in swiftly using navigational instruments (so that fewer "fixes" can be obtained in a given time period), and difficulty in the prompt interpretation of information presented by the flight instruments.

Higher frequency (2 to 10 Hz) structural vibrations, which can be particularly strong in helicopters (as in surface vehicles, ships, and around fixed plant and factories), are associated with difficulty in reading instruments or print; interference with fine manipulative tasks, such as setting cursors on navigational aids, making adjustments to electronic equipment, and writing; and general discomfort, irritability, and fatigue that worsen progressively during lengthy missions. Vibration in that frequency range can also affect communication (a potential threat to safety in some circumstances) by altering the pattern and degrading the clarity of human speech when the speaker is being vibrated (69, 70).

Air pressure oscillations (acoustic waves) at frequencies too low to stimulate the human ear in the normal manner to produce the sensation of sound, that is, below about 16 Hz, are called *infrasound*. Such waves, impinging on the human body surface with sufficient intensity, can set it into vibration and provoke responses akin to those caused by whole-body mechanical vibration of equivalent intensity and frequency (3, Chs. 11, 14; 20; 72, 73, 126). High levels (more than 110 dB SPL) of infrasound in the range 5 to 20 Hz can be generated in diesel-engined ships and vehicles and in fixed industrial plants (74). Infrasound can be a nuisance in any large space in which the air can be set into low frequency vibration by panel vibrations (e.g., in the structures of ships or large factory buildings) excited by the action of machinery, wind flow, or the movement of the air in large duct air-conditioning systems.

In most circumstances it remains a moot point whether infrasound is a serious threat to health or safety, although it can produce unpleasant subjective effects (vibration, "fluttering," gagging or choking sensations in the abdomen, chest, and throat; occasionally disequilibrium and feelings of displacement), irritation, and fatigue. Recent animal studies, using genetically deaf subjects (75) indicate that high levels of infrasound can produce effects in the central nervous system mediated by nonauditory receptors and pathways.

Tentative limits for short-term (up to 8 min) human exposure to intense infrasound (1 to 20 Hz) have been published by von Gierke et al. (20), based on admittedly meager experimental data. The proposed limiting sound pressure levels (SPL) are: 150 dB from 1 to 7 Hz; 145 dB from 8 to 11 Hz, and 140 dB from 12 to 20 Hz.

2.2 The Pathology of Whole-Body Vibration

Intense whole-body vibration can cause pain and injury. Acute traumatic effects, depending mainly on the frequency, intensity, and direction of application of the vibration, are most likely to occur when severe vibration is applied to the unprotected person at frequencies related to the principal organ and system resonances. Prolonged and repeated exposure to moderately severe but not immediately damaging levels of vibration can also prejudice the health of the worker in certain pursuits and occupations.

2.2.1 Severe Acute Whole-Body Vibration Exposure

Animal studies on the lethality of intense whole-body vibration (acceleration amplitudes up to $20g$ at frequencies up to 50 Hz) have shown that severe shaking can cause hemorrhagic injury to the viscera and other parts of the living body. Roman (76) demonstrated in the mouse that the lethality of severe whole-body vibration is strongly frequency dependent, being greatest at frequencies of major internal resonance. Pathological changes commonly seen in animals killed by vibration are hemorrhagic damage to the lung and to the myocardium, and bleeding into the gastrointestinal tract; less commonly, there is superficial hemorrhage of kidney and brain.

Whole-body vibration of man at levels sufficient to cause acute injury is probably very rare in industrial and even military experience. However I have heard anecdotal accounts of spinal injuries resulting from severe vibrations or repetitive impacts sustained recreationally in riders of power boats speeding through choppy seas and snowmobiles raced over rough, snow-covered terrain. It is of interest that case histories and general observations have been presented by Pichard (77) and by Snook (78) to show that bad riding characteristics in ambulances driven at speed over bumpy roads can seriously aggravate the injuries or medical conditions of patients on their way to the emergency room, as well as making it difficult and hazardous for ambulance attendants to minister to the patient during the ride.

2.2.2 Chronic Occupational Exposure to Whole-Body Vibration

Certain disorders of the spine and internal systems have long been attributed to chronic occupational exposure to the rough motion of working vehicles such as farm tractors (23, 79), haulage trucks (80), and mobile heavy equipment (81, 82). In recent years the constant heavy vibration experienced in helicopter flying has been recognized as potentially harmful to the spine in professional military and civilian helicopter pilots (83, 84). The etiology of the disorders reported, and the causal link between the mechanical vibration exposure and the supposedly vibration-related disorder, remain obscure and have proved difficult to establish epidemiologically and experimentally.

Certain factors apart from the vibration exposure (e.g., climatically adverse working conditions; bad ergonomic factors in the design, construction, and method of operation of the vehicle; fatigue and nonspecific work load stress) may figure prominently, if not predominantly, in the etiology. The extent to which chronic exposure to moderate levels of whole-body vibration, repeated daily over many years, can injure the otherwise healthy body (as opposed to merely aggravating preexisting weaknesses) in the absence of compounding factors, remains an open question.

The constant vibration and jolting experienced on agricultural tractors has for many years been blamed by farm workers and physicians for an undue prevalence of gastrointestinal, anorectal, and spinal ailments, especially exacerbations of peptic ulceration, and low back pain. From a systematic study of 371 farm tractor drivers, Rosegger and Rosegger (79) concluded that in many cases the cumulative microtraumatic effects of the constant vibration and jolting were aggravated by coincidental factors, including faulty posture—especially twisting round in the seat to watch the tow (Figure 13.1)—bad seating, in terms of both seat design and seat maintenance; working long and irregular hours during busy seasons and bad weather, with inadequate rest breaks and mealtimes; and certain secondary influences such as alcohol and nicotine consumption.

EVALUATION OF EXPOSURES TO VIBRATIONS

Figure 13.1 When driving a tractor whole-body vibration (mainly vertical) is transmitted to the driver principally through the seat but also through the floor, wheel, and controls. Drivers often twist the trunk to watch the rig being drawn: this may aggravate the action of the seat vibration on the spine. Note that this driver is also smoking at work. Author's photo.

Similar uncertainty regarding the complex etiology of vibration-related occupational disorders exists in kindred industries, such as long-distance bus and truck driving, and the operation of mobile heavy equipment. In a study of the periodic physical examination records of 1448 interstate bus drivers, Gruber and Ziperman (85) found that a number of chronic ailments occurred more frequently among drivers with more than 15 years' experience than in control populations. Some of the disorders (particularly those affecting the gastrointestinal, respiratory, venous, and muscular systems, and the lumbar spine) might have been caused by the drivers' chronic exposure to whole-body vibration and jolting on the job; however it was not possible epidemiologically to distinguish vibration as a principal etiological factor from several others also present in the working situation. Such additional factors included poor seating and posture in a cramped cab, questionable dietary habits, long and sometimes irregular hours, the general job stress of dealing with the public and managing a busload of passengers in traffic, and possibly combinations of other stressors such as noise and excessive carbon monoxide levels.

As in the case of the bus drivers just cited, certain ailments appear to be more prevalent in long-distance truck drivers than in control populations subjected to comparable job stress (80). Nevertheless, although some of the chronic afflictions reported (e.g., low back pain, spinal deformities, gastrointestinal troubles, hemorrhoids) could be ascribed to the influence of prolonged and repeated exposure to whole-body vibration, other etiological factors cannot be ruled out as possible contributory or even primary causes. Such factors again include bad driving posture and deficient seat design or maintenance; driving excessive hours in cramped and uncomfortable conditions; heavy cargo handling (bad lifting and handling techniques can cause or aggravate back ailments in particular); rushed meals at odd hours; and the job stress of high anxiety and heavy physical demands in managing a large and valuable vehicle and its load in heavy traffic.

Seris, Auffret, and their colleagues in France (83), looking at the vibration factor in aviation, consider low back pain to be a relatively common and serious malady of professional helicopter pilots flying prolonged and repeated sorties (83, 84). The risk is attributable to a combination of factors capable of amelioration: these include not only the frequent gust loads (due to low altitude air turbulence) and heavy low frequency vibration sustained in helicopter flying (3, Ch. 2) but also such ergonomic factors as bad seating, and the general stress of the occupation, which is in many respects akin to that of the bus and truck drivers described by Gruber and his colleagues (80).

Epidemiological studies undertaken to identify unusual patterns or incidences of morbidity in the operators of rough-riding vehicles are likely to prove inconclusive for a number of reasons (80–82, 85). Spear et al. (82), who have studied insurance claims for medical services in various categories of disability affecting heavy equipment drivers, have drawn attention to a tendency for the statistics to show an initial increase in the risk of certain afflictions attributed with varying degrees of confidence to rough riding, followed by a decrease with experience (equated with years in the occupation). Spear and his co-workers have pointed out that this tendency suggests a selection process by which workers in the study groups exposed to rough riding tend to quit the occupation as they become afflicted with the ailments in question.

Spear et al. (81) made a follow-up study to earlier work on morbidity patterns among heavy equipment operators, to test whether selection out of jobs involving whole-body vibration exposure influenced the incidence of medical service insurance claims for certain occupationally related diseases (particularly ischemic heart disease and musculoskeletal afflictions). They observed that there may be some ailments or conditions whose onset is hastened by vibration but whose overall incidence does not differ significantly between vibration-exposed and nonexposed workers in the construction industry.

In the Soviet Union and other countries in Eastern Europe, many investigators have taken the view that continuous whole-body vibration in factories, in various kinds of mobile working equipment such as cranes, concrete compactors, and other equipment used in the construction industry, and in ships' engine rooms (86), can produce debilitating stress and malaise mediated centrally by neuroendocrinological mechanisms. Particular attention has been focused on more or less steady state vibration, generally occurring at somewhat higher frequencies than the low frequency jolting and shaking experienced in rough-riding vehicles, acting in concert with industrial noise and general job stress (30). Such occupational exposure is apparently recognized in the Soviet Union as the cause of various "vegetative-vascular," polyneuritic, asthenic, and vestibular syndromes to the detriment of working capacity and, in some cases, occasioning a change of job or compensation for reduced earning capacity (87).

Fatigue, irritability, and the impairment of dexterity and postural function have been observed not only in vehicle drivers but in workers in other occupations in which, in addition to whole-body vibration, there is a high job stress due to the degree of manual control skill and responsibility that must be regularly exercised for prolonged periods, for example, in the operation of large tower cranes used in construction work (51). Stabilometric measurements at the end of the work shift in dump truck drivers exposed to severe whole-body vibration and jolting have been reported to show increased postural instability (ascribed to "otolithic excitability") in comparison with control groups of miners not exposed to the heavy jolting (52). This tendency has been found by Nurbaev (52) to worsen with increasing length of service (up to 8 years) in dump truck driving. Chernyuk (88), however, has stated that 15 min rest breaks interposed between half-hour stints of rough-riding vehicle driving reduces substantially the vestibular and central nervous system excitability engendered by the vibration. Eklund (89) has described mechanisms that may be involved in the causation of neuromuscularly mediated vibration effects on the sense of balance and postural stability.

Although vibration has been reported by Soviet investigators (90) to affect renal function in human sufferers from vibration disease, and "kidney trouble" (frequently a layman's synonym for low back pain) apocryphally attributed by drivers to long hours of truck driving in the United States, the question whether chronic whole-body vibration exposure in vehicle driving or industrial activity poses a significant threat to the kidney remains open.

Experimental studies have yielded negative results. In protracted experimental exposures of rhesus monkeys to an intensity of low frequency (12 Hz) vibration sufficient to cause hemorrhagic damage to the gastrointestinal mucosa, Sturges et al. (40) failed to find evidence of renal damage. In human volunteers exposed experimentally to sinusoidal whole-body (z-axis) vibration at moderate levels in the range 2 to 32 Hz for up to 8 hr, Guignard et al. (91) found no significant changes in urinary constituents. The intensity of vibration in those exposures was, however, limited by the ISO

"Fatigue-Decreased Proficiency Boundary" (4), and a variety of other examinations of the subjects' physiological state and task performance also proved negative.

Specific attention has been drawn by investigators in the Soviet Union to the question of particular susceptibility of women to whole-body vibration in industry. Parlyuk (92) has claimed that the motions and vibrations to which female workers were exposed in their daily work operating bridge cranes led to neurocirculatory and vestibular disorders. The symptomatology included vertigo, unsteadiness of gait, and a heightened susceptibility to motion sickness. Lysina and Parlyuk (93) have reported hemodynamic and cardiographic changes, which they attributed to the irregular vibration and jolting experienced on the job in a series of 33 female bridge crane operators, all of whom had spent 10 years or more in the occupation.

However a causal link between the physical agents in the industrial environment, as opposed to job stress, and the symptoms reported in such cases has yet to be established. When the symptomatology is of a kind not peculiar to one sex, it cannot be presumed that disorders affecting a group of workers who happen to be women are related to sex differences in susceptibility to vibration stress unless it can be shown that comparable populations of male workers exposed to the same conditions are significantly less affected. Comparative data relating to this question are lacking and, accordingly, current standards on human exposure to vibration are deemed to apply equally to both sexes.

Gratianskaya and her colleagues (94) have drawn attention to a possible risk of increased incidence of disorders of menstruation and parturition in female workers employed in the vibrocompaction of concrete. In that job the women were exposed to whole-body vibration in the range 40 to 55 Hz while placing concrete mix for settlement by vibrocompacting machines in construction work. Apart from this report, claims that vibration affects menstrual function or pregnancy remain mostly apocryphal; although many years ago in Germany mechanical vibration was tried, apparently with some success, as a means of inducing labor (95, cited by Guignard in Reference 2, Ch. 29). It is probably prudent for women in advanced stages of pregnancy, particularly when the pregnancy has been threatened by obstetric complications, to avoid exposure to whole-body vibration at work and in travel.

2.2.3 Vibration Combined with Other Stressors, Especially Noise

It was mentioned earlier that in the Soviet Union, whole-body vibration, particularly machinery vibration at moderately high frequencies of the kind experienced continuously in factories, is generally regarded as part of a composite industrial environment that can adversely affect the worker's physiological state, metabolism, and health through the mediation of central mechanisms involving the central and autonomic nervous systems and the endocrines (30, 96). Noise, especially, is an important stressor that is almost always present in factories and other work situations where man is simultaneously exposed to continuous mechanical vibration. In addition to its possible enhancement of the auditory hazard associated with industrial noise (97,

98), whole-body vibration may also act additively with noise to cause stress and industrial fatigue and to degrade vigilance and the performance of skilled tasks (3, 99).

Workers exposed daily to industrial noise and vibration can be shown to have high stress indices, such as enhanced urinary excretion of vanillyl mandelic acid and 17-ketosteroids, when compared with control groups working in quieter environments (100). Work situations remote from the factory may also be contaminated by undesirable levels of combined vibration and noise. Noise and vibration are constantly present in ships, for example, and can be a serious problem, with high intensities occurring at low frequencies, on board very large vessels with poor structural damping qualities, such as large oil tankers (101). A peculiar environmental problem of life at sea is that ship noise and vibration are continuously present throughout the voyage, so that the ship's crew not only are exposed to the stress while on watch but when off duty also, finding little respite from the pervasive rumble of the ship's machinery when they retire to their quarters for recreation or sleep. Gibbons et al. (102) have shown enhanced 17-ketosteroid levels are measurable in ships' officers at sea, compared with the levels found following shore leave. Of course, such a sign is not specific to vibration and noise stress but may be indicative of the combined stress of the shipboard environment and the demands of responsible duty at sea.

Soviet work has indicated that vibration stress may in some circumstances enhance specific damage to the organism from other physical or chemical agents, including damage from ionizing radiation (103, 104) and heavy metal toxicity (particularly from mercury) (105). Such observations, however, have been based mainly on animal experiments, and firm data establishing a causal association for man between vibration and other noxious agents hazardous to occupational health are not yet available.

2.3 Protection of Man from Effects of Whole-Body Vibration

Four main steps must be taken to protect man from adverse effects of vibration, namely:

1. Predict or measure the vibration exposure. Ground rules for the proper measurement and evaluation of human exposure to whole-body vibration in the range 1 to 80 Hz have been published by the ISO.
2. Select and apply an appropriate criterion of vibration control and corresponding limits of exposure. Here again, the current international consensus regarding human whole-body vibration exposure criteria and limits is embodied in the ISO standard (4) (Table 13.2).
3. Determine the type and amount of vibration control (usually reduction) required. This is a computational step, based on the data of steps 1 and 2.
4. Select and apply the most effective and economical means of control available for the protection of the worker.

In step 4, adopting the classical approach of vibration or acoustical engineering, three main points can be distinguished at which to attack vibration disturbing to man (3, Ch. 9). These are:

1. At the source of vibration.
2. In the route of transmission of vibration from the source to man.
3. At the receiving point (i.e., in man himself).

Reduction of vibration at source, particularly when the source is machinery running in factories or propelling ships, vehicles, or aircraft, is a matter of engineering design and practice. Vibration very commonly arises from unbalance forces generated in the cycle of operation of revolving or reciprocating machinery. It can be prevented or reduced by good design and balancing. Quite often, the vibration originating in motors and machinery is worsened by bad installation or poor maintenance and can be reduced by proper attention to such matters.

Table 13.2 International Standard (ISO) Values of "Fatigue-Decreased Proficiency Boundary" for Human Whole-Body Exposure to Vibrational Acceleration in the Range 1 to 80 Hz, as Functions of Frequency and Exposure Time[a]

Frequency, or Center Frequency of Third-Octave Band (Hz)	Boundary Values for z-Axis (seat- or foot-to-head) Vibration							
	Root-Mean-Square Acceleration [m/sec^2]							
	Notional Exposure Times on an Occasional or Daily Basis							
	24 hr	16 hr	8 hr	4 hr	2.5 hr	1 hr	25 min	16 min	1 min
1.0	0.224	0.315	0.63	1.06	1.40	2.36	3.55	4.25	5.60
1.25	0.200	0.280	0.56	0.95	1.26	2.12	3.15	3.75	5.00
1.6	0.180	0.250	0.50	0.85	1.12	1.90	2.80	3.35	4.50
2.0	0.160	0.224	0.45	0.75	1.00	1.70	2.50	3.00	4.00
2.5	0.140	0.200	0.40	0.67	0.90	1.50	2.24	2.65	3.55
3.15	0.125	0.180	0.355	0.60	0.80	1.32	2.00	2.35	3.15
4.0	0.112	0.160	0.315	0.53	0.71	1.18	1.80	2.12	2.80
5.0	0.112	0.160	0.315	0.53	0.71	1.18	1.80	2.12	2.80
6.3	0.112	0.100	0.315	0.53	0.71	1.18	1.80	2.12	2.80
8.0	0.112	0.160	0.315	0.53	0.71	1.18	1.80	2.12	2.80
10.0	0.140	0.200	0.40	0.67	0.90	1.50	2.24	2.65	3.55
12.5	0.180	0.250	0.50	0.85	1.12	1.90	2.80	3.35	4.50
16.0	0.224	0.315	0.63	1.06	1.40	2.36	3.55	4.25	5.60
20.0	0.280	0.400	0.80	1.32	1.80	3.00	4.50	5.30	7.10
25.0	0.355	0.500	1.0	1.70	2.24	3.75	5.60	6.70	9.00
31.5	0.450	0.630	1.25	2.12	2.80	4.75	7.10	8.50	11.2
40.0	0.560	0.800	1.60	2.65	3.55	6.00	9.00	10.6	14.0
50.0	0.710	1.000	2.0	3.35	4.50	7.50	11.2	13.2	18.0
63.0	0.900	1.250	2.5	4.25	5.60	9.50	14.0	17.0	22.4
80.0	1.120	1.600	3.15	5.30	7.10	11.8	18.0	21.2	28.0

EVALUATION OF EXPOSURES TO VIBRATIONS

Table 13.2 (Continued)

Boundary Values for x- or y-Axis (transverse) Vibration

Frequency, or Center Frequency of Third-Octave Band (Hz)	Root-Mean-Square Acceleration [m/sec²]								
	Notional Exposure Times on an Occasional or Daily Basis								
	24 hr	16 hr	8 hr	4 hr	2.5 hr	1 hr	25 min	16 min	1 min
1.0	0.100	0.150	0.224	0.355	0.50	0.85	1.25	1.50	2.0
1.25	0.100	0.150	0.224	0.355	0.50	0.85	1.25	1.50	2.0
1.6	0.100	0.150	0.224	0.355	0.50	0.85	1.25	1.50	2.0
2.0	0.100	0.150	0.224	0.355	0.50	0.85	1.25	1.50	2.0
2.5	0.125	0.190	0.280	0.450	0.63	1.06	1.6	1.9	2.5
3.15	0.160	0.236	0.355	0.560	0.8	1.32	2.0	2.36	3.15
4.0	0.200	0.300	0.450	0.710	1.0	1.70	2.5	3.0	4.0
5.0	0.250	0.375	0.560	0.900	1.25	2.12	3.15	3.75	5.0
6.3	0.315	0.475	0.710	1.12	1.6	2.65	4.0	4.75	6.3
8.0	0.40	0.60	0.900	1.40	2.0	3.35	5.0	6.0	8.0
10.0	0.50	0.75	1.12	1.80	2.5	4.25	6.3	7.5	10
12.5	0.63	0.95	1.40	2.24	3.15	5.30	8.0	9.5	12.5
16.0	0.80	1.18	1.80	2.80	4.0	6.70	10	11.8	16
20.0	1.00	1.50	2.24	3.55	5.0	8.5	12.5	15	20
25.0	1.25	1.90	2.80	4.50	6.3	10.6	16	19	25
31.5	1.60	2.36	3.55	5.60	8.0	13.2	20	23.6	31.5
40.0	2.00	3.00	4.50	7.10	10.0	17.0	25	30	40
50.0	2.50	3.75	5.60	9.00	12.5	21.2	31.5	37.5	50
63.0	3.15	4.75	7.10	11.2	16.0	26.5	40	45.7	63
80.0	4.00	6.00	9.00	14.0	20	33.5	50	60	80

[a] To obtain the ISO "exposure limit," these values of acceleration are multiplied by a factor of 2. To obtain the "reduced-comfort boundary" they are correspondingly divided by a factor of 3.15 (4).

2.3.1 Standards

International standard ISO 2631-1974 (4) was issued by the ISO in 1974 as an attempt to achieve international consensus regarding the evaluation and limitation of human exposure to whole-body vibration. It superseded a confusing multiplicity of guidelines, many based on few and unreliable data, that had been written in many countries for separate industries and applications (2, Ch. 29; 3, Ch. 6). This document prescribes standard methods of evaluating whole-body vibration disturbing to man at work or in transportation and provides guidelines for regulating human exposure to such vibration in the range 1 to 80 Hz.

Three criteria (reasons for human protection) of vibration exposure are recognized, namely, (1) "reduced comfort" (limits set according to this criterion are intended mainly for application in the design of good riding quality in passenger-carrying vehicles); (2) "fatigue-decreased proficiency," with corresponding limits intended to preserve human working efficiency (skilled performance of tasks) and safety during vibration; and (3) the "exposure limit," intended to protect man from physical injury or disease due to whole-body vibration to which he may be exposed on a daily basis. Values (rms acceleration) of the reduced comfort "boundary" (limit) and the exposure limit as functions of vibration frequency and exposure duration are derived from the corresponding values of the "fatigue-decreased proficiency boundary" (Table 13.2) by dividing them by 3 or multiplying them by 2, respectively. Separate standards have been included for vertical (z-axis) as opposed to horizontal or transverse (x- and y-axis) whole-body vibration.

It has been demonstrated that ISO 2631-1974 is protective of human performance during z-axis vibration in the range 2 to 32 Hz for exposures up to 8 hr (91, 106). Current work of the relevant subcommittee (ISO/TC 108/SC4) is directed toward revising and amplifying these standards and extending them to frequencies below 1 Hz.

2.3.2 Controlling Vibration Between the Source and Man

The application of guidelines such as ISO 2631-1974 enables the type and amount of vibration reduction needed for the protection of man to be worked out. Several engineering solutions can then be sought to minimize the recipient's vibration exposure, including (3, Ch. 9):

1. Minimization of the vibration of structures.
2. Isolation of sources of vibration.
3. Location of recipient with respect to pathways of transmission.
4. Isolation of the recipient.
5. Prevention of flanking transmission.

A detailed discussion of these approaches is outside the scope of this chapter. It should be noted, however, that knowledge of the frequency dependence of human reactions to vibration, especially the biodynamic response in the range 1 to 20 Hz, is most important in the design of isolation systems to protect man from floor-, ground-, or seat-transmitted vibration (item 4).

Many kinds of vehicle and some fixed-base machines produce disturbing whole-body vibration in the range above 1 Hz. Current international and military standards for limiting human whole-body vibration exposure in that range are frequently exceeded substantially (10, 12). Well-designed suspension or isolation systems, based on the

EVALUATION OF EXPOSURES TO VIBRATIONS

interposition of springing or some resilient and damping element between the source of vibration and the human recipient, can materially reduce the exposure. Such solutions include spring-mounted seats for rough-riding vehicles and aircraft, as well as the familiar suspension systems fitted to road, track, and land vehicles. The main principle of design of such devices is that a resilient suspension attenuates vibration of the suspended mass of the rider or the vehicle body at frequencies above the resonant frequency of the loaded spring (3).

High frequency vibrations of the kind that can be troublesome around machinery running in factories and other fixed workplaces, and in ships, are sometimes amenable to treatment by means of mats or resilient floors to protect the worker or crewman. Jaworski (107) has shown that the selection of footwear incorporating suitable damping properties can protect the worker against the absorption of relatively high frequency floor vibrations in the vicinity of industrial plant. However such a solution is of little help in attenuating relatively large amplitude vibrations at frequencies below about 20 Hz.

Most vehicle and seat suspensions are of the passive kind; that is, they make use of the inherent mechanical properties (elasticity and damping) of springs and resilient materials. For specialized applications (e.g., in high performance aircraft and in rough-riding off road vehicles and mobile heavy equipment), where riders are exposed to severe large amplitude motion and vibration in the range 1 to 10 Hz, it is feasible to use "active" systems in which an accelerometer or other sensor monitors the impressed forces reaching the vehicle or seat and generates a signal that is used to drive an actuator (usually electrohydraulic) to oppose the motion (108–110). Such systems are loosely analogous to the automatic stabilizers sometimes used to reduce the roll of ships. All vibration-isolating systems and devices have their engineering limitations and frequently represent a best engineering compromise between the ideal solutions for conflicting problems—for instance, high frequency shock isolation versus low frequency continuous vibration isolation, or vertical as opposed to transverse vibrations (111). Rectilinearity of seat motion can be advantageous (112).

2.3.3 Minimizing Adverse Effects of Vibration Reaching Man

When an irreducible amount of vibration still gets through to man in his vehicle seat or workplace, a number of possibilities remain for the mitigation of vibration effects in the recipient. These include (3, Ch. 9):

1. Minimizing the duration of exposure to vibration on an absolute or daily basis.
2. Allowing adequate rest periods between spells of work.
3. Ergonomic design of displays, controls, and workplaces for best use during unavoidable vibration.
4. Training and experience of workers necessarily exposed to severe motion or vibration (this includes the possibility of habituation to the stress).

5. Physical fitness of the worker. Discouragement of smoking.

6. Avoidance or removal of physical, chemical, or other agents that might potentiate susceptibility to vibration.

7. Selection procedures, when indicated, and exclusion of persons medically unfit for occupational exposure to severe vibration or motion.

8. Engineering design or treatment to minimize excessive vibration of clothing, headgear, and man-mounted equipment (e.g., as worn by aircrew or workers wearing protective gear or other specialized clothing or equipment).

9. Biomechanical devices to restrain or suppress body oscillation (usually the last resort in cases of exposure to severe and hazardous levels of whole-body vibration).

Several of these items again entail essentially engineering solutions, elaboration of which is outside the scope of this chapter, but some of them fall within the purview of the industrial or ship's medical officer or flight surgeon, who may be in a position to recommend steps to reduce human vibration exposure. For example, it is known that provision of adequate rest breaks at work, as well as periods of recuperation, can reduce the stress of work or duty in vibrating environments in work situations varying from factories (100) to sea duty (102). In aviation, Delahaye et al (84) have recommended such ameliorative methods as exercise regimes for the spine and skeletal musculature of helicopter pilots, and the elimination from selection programs of pilot candidates having certain congenital abnormalities of the spine or preexisting lower back injury. A number of other medical and surgical conditions may generally be deemed to render it inadvisable for those afflicted to undergo severe, prolonged, repeated, or unnecessary whole-body vibration exposure (including experimental exposure in the capacity of a voluntary subject for human vibration experimentation) (3, Ch. 9).

Except for remedies for motion sickness caused by motion at frequencies below 1 Hz, alluded to earlier, no drugs or medications are known to increase human tolerance of whole-body vibration and accordingly none can be recommended.

3 HARMFUL EFFECTS OF HAND-TRANSMITTED VIBRATION: VIBRATION SYNDROME

It is becoming widely recognized that the intense vibratory energy generated by handheld electric, pneumatic, and motor-driven tools of the kind used habitually by shipbuilders, construction workers, miners and quarrymen, foundry workers, vehicle assembly workers, lumbermen, and people in many other industies can, when transmitted to the fingers, hands, and arms of the operator by the tool or workpiece, produce vasospastic, neuromuscular, and arthritic disorders of the hand and upper limb. The effects of such disorders on the worker can range from mild and occasional discomfort or inconvenience to severe social and occupational disability.

Vibrating tools are used so universally in industry, and in such a variety of occupations (Figures 13.2 to 13.5) that it is very difficult to estimate the numbers of workers at risk—they may number several million in North America alone—and the problem is

EVALUATION OF EXPOSURES TO VIBRATIONS

regarded as serious by clinical and other authorities familiar with it in most industrialized countries around the world. It has received particular research and administrative attention in countries where substantial numbers of workers have to use powered handheld tools out of doors (e.g., in shipbuilding, construction, and forestry work) in inclement weather, or in other workplaces, such as some mines, where cold, damp conditions prevail. The urgency of the problem has been highlighted by the recent

Figure 13.2 Rock-drilling postures. (a) Downward drilling, in which heavy vibration is transmitted through the hands to the upper limb girdle and upper torso. This is also a typical posture in road breaking and ground breaking for construction work. (The cables leading from the handle of this machine carry signals from instrumentation used to measure vibration levels on the job.) (b) Rock drilling underground in subway construction. Note left arm supporting the weight of the tool while it is guided by the right hand. Photos courtesy of Dr. I.-M. Lidström, Stockholm.

Figure 13.2 (*Continued*)

publication of several extensive reviews including the proceedings of major conferences on the topic (29, 113–117).

Nevertheless, despite the conclusions of numerous scientific authors, medical authorities, and administrators in industry and government in many countries regarding the seriousness of the problem, considerable reluctance to acknowledge vibration syndrome as a compensatable industrial disease persists within industry and among occupational safety and health legislators. That is largely because there is still a paucity of firm epidemiological and experimental evidence to prove a direct and unequivocal causal relationship between cumulative exposure to hand-transmitted vibration at work and the associated chronic disorders, which are generally difficult to distinguish diagnostically from similar disorders of idiopathic or other etiological origin not associated with occupational exposure to vibration.

In the United Kingdom, for instance, since the early 1950s, efforts have been made repeatedly by medical authorities to persuade the Department of Health and Social Security to recognize Raynaud's phenomenon (also known as "vibration white finger" or traumatic vasospastic disease) in workers with a history of using vibrating tools as a compensatable occupational disease. However prescription as such has not yet been granted, mainly on the ground that in most cases, the symptoms are relatively trivial and not commonly an impediment to the worker's continued employment (118).

In certain industries, however, the condition is far from being trivial, with substantial majorities of workers in some jobs succumbing to serious symptoms within a few years

EVALUATION OF EXPOSURES TO VIBRATIONS

of entering the occupation. The occupations of chain sawing and pedestal grinding provide particularly notorious examples (114, 119) and this observation has been repeated in several countries. In some job situations prevalences of symptoms exceeding 90 percent have been recorded before remedial measures were taken.

3.1 Characteristics of Hand-Transmitted Vibration

Most handheld and hand-guided power tools generate strong mechanical vibration distributed over a broad spectrum: in many types of tool the vibration transmitted to the

Figure 13.3 Pneumatically driven tools such as this casting scaler (also pneumatic chippers, riveters, etc.) are powerful sources of hand-transmitted vibration of relatively low frequency. Several factors (e.g., method of handling, proximity of the hand or fingers to the working point, inclusion of padding) determine the severity of vibration exposure, which frequently differs substantially between the two hands. Photo courtesy of Dr. W. Taylor, Dundee.

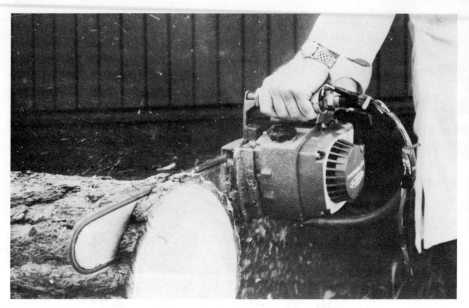

Figure 13.4 Occupational use of the motor-driven chain saw is strongly associated with the risk of vibration syndrome (commonly, Raynaud's phenomenon). Unbalance forces from the power unit are the chief source of vibration in these machines, affecting both hands through the carrying and guiding handles (the machine illustrated is fitted with a specially instrumented handle for measuring vibration exposure at work). The risk of vibration disease in chain sawyers can be reduced by the introduction of modern saws incorporating antivibration (A/V) devices. Photo courtesy of Dr. I.-M. Lidström, Stockholm.

hands of the operator through the handles of the applicance or through the workpiece is complexly periodic, the spectrum showing numerous peaks at frequencies related to the cycle of rotation or reciprocation of the tool. In some tools (e.g., concrete breakers, chipping tools, riveting hammers) the vibration is at source a rapid series of impacts delivered to the workpiece, but the impacts set up a complex series of harmonic vibrations because of ringing at the multiple resonance frequencies of the machine body and its moving parts. In certain kinds of continuously operating tool, notably the chain saw (120), the vibration arises mainly from unbalance forces generated by the engine and transmission and can accordingly be ameliorated by engineering treatments. [Contrary to popular belief, most chain saw vibration, at least in properly maintained equipment, arises from the revolutions of the motor and transmission, not primarily from the cutting action of the sawteeth (121).]

In most kinds of powered handheld tool, the most important components of the vibration spectra lie in the range 10 to 1000 Hz; and within that range frequencies between about 30 and 300 Hz have been considered by several investigators to be those most strongly associated with the development of Raynaud's phenomen and allied vasospastic disorders in susceptible workers (113, 121, 122–125). The reason for that observation appears to be complex: vibrations in the band 30 to 300 Hz not only contain the maximum vibrational energy output of many vibrating tools but appear also to be preferentially absorbed by the tissues of the hand (71, 127). It has also been postulated (125) that mechanoreceptors (especially Pacinian corpuscles) in the hand and upper limb, which are particularly responsive to vibration in that frequency band, are important mediators of the repeated vibratory activation of the sympathetic nervous system, leading eventually to alterations in vasomotor tone which are the precursors of vasospastic disease. This is considered further below.

Figure 13.5 Working posture, particularly if it is cramped or awkward, may play a secondary etiological role in the incidence of vibration-related disorders. The postures adopted by workers on the job are often necessarily far from ideal. (a) Using a grinding wheel to finish the interior of a welded pipe. (b) Working on an automobile body. Photos courtesy of Dr. I.-M. Lidström, Stockholm.

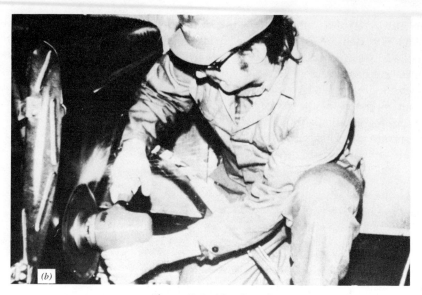

Figure 13.5 (Continued)

The ISO (5) is accordingly considering a draft proposal for a standard that prescribes methods of evaluating hand-transmitted vibration and guidelines for limiting human exposure to hand-transmitted vibration in the band 6.4 to 1000 Hz. Outside that band, very low frequency vibrations below about 8 Hz (of the kind generated by heavy road breakers, compacting machines, etc.) are transmitted up the arm to the upper limb girdle and the whole body, with very little of the energy being absorbed in the hands. Such low frequency components are probably not significant in the causation of vibration syndrome. At the other extreme, most vibrating tools probably do not generate much energy at frequencies above 1000 Hz; and such energy in the form of mechanical vibration of minute amplitude is heavily attenuated at the interface between the handle of the vibrating tool and the skin of the hand.

It should be noted, however, that the vibration spectra of powered hand tools extend well into the audiofrequency range. The moving parts and housings of such machines, therefore, as well as workpieces (particularly metal castings, riveted plates, etc.) act as radiators of acoustic noise. Intense noise can also emanate from the exhausts of pneumatic and motor-driven appliances, so that the user is subjected to the double hazard of vibration syndrome and noise-induced hearing loss (128).

3.2 Varieties of Vibration Syndrome

Raynaud's phenomenon secondary to an altered vasomotor tone in the hands is the disorder most frequently associated with the occupational use of vibrating tools and is accordingly considered in some detail below. However other troublesome and sometimes incapacitating afflictions may be the presenting complaint in power tool workers. These

EVALUATION OF EXPOSURES TO VIBRATIONS

neurovascular, sensory, and musculoskeletal manifestations (including Raynaud's phenomenon of occupational origin), collectively known as the vibration syndrome, may share a common underlying physical etiology but depend for their variety on differing predisposing factors in the nature of the occupation and the individual (Table 13.3).

The vibration syndrome may include vasomotor disorders not associated with

Table 13.3 Factors Influencing Risk of Vibration Syndrome

External and Occupational Factors

Physical factors
    Intensity of vibration in the range 8 to 1000 Hz
    Directionality (one or more axes) of vibration
    Spectrum/principal frequencies of vibration
    Incorporation or absence of antivibration devices
Operational and ergonomic factors
    Work cycle/temporal pattern of vibration exposure (i.e., frequency and relative duration of periods of work at full power, idling, rest breaks, etc.)
    Method of handling and using tool
    Weight (mass) of tool supported by the user
    Configuration of handle(s)
    Pressure required to guide or advance tool
    Necessity to support, hold, or guide workpiece
    Cramped or awkward working posture
    Use of gloves or padding; clothing worn at work
    State of maintenance or repair of tool (e.g., state of cutting or drilling edges, bits, etc.; tuning of internal combustion engines; condition of exhaust systems of motor-driven and pneumatic tools; balancing of rotating or reciprocating parts)
Environmental factors
    Prevailing climate at worker's home and workplaces
    Temperature ranges encountered commuting and at work
    Humidity/dampness
    Noise

Internal and Personal Factors

Experience, training, and exposure history of individual
    Age of entry into work with vibrating tools
    Time since entry into occupation (months or years on the job)
    Level of training, skill, and familiarity with the tool (affecting individual's propensity to tightly grip or easily guide the tool)
    Vibration exposure outside work (e.g., from domestic and recreational tools and appliances)
Vasoconstrictive agents affecting the individual
    Smoking
    Predisposing disease or injury (including previous trauma, vascular, cardiopulmonary, neuropathic, endocrinological, rheumatic or other disorder affecting peripheral nerve or blood flow: cf. Table 13.4)
    Medication
    Dietary factors and meal habits

manifest Raynaud's phenomenon, for instance, erythrocyanotic changes accompanied by pain in the hands (129); neurosensory and neuromuscular disorders, including the loss of tactile sensitivity and manual dexterity (130); pain and stiffness in the joints of the fingers and hand; and bone and joint changes, visible radiographically, affecting the hand and upper limb (131). Some who work with vibrating tools experience lacerating pains in the hands, awakening the victims at night and robbing them of sleep (113).

The pattern of disease in the individual can be strongly dependent on the spectrum, intensity, and distribution of vibrational energy entering the hand-arm system; and the vibratory input is in turn strongly influenced by physical, mechanical, anatomical, and physiological factors of the kind listed in Table 13.3 and considered in more detail below.

3.3 Raynaud's Phenomenon of Occupational Origin

3.3.1 Symptoms and Differential Diagnosis

It is necessary to distinguish Raynaud's phenomenon of occupational origin,* that is, the symptom associated with habitual exposure to intense mechanical vibration or repetitive impact applied to the hands, from other causes of secondary Raynaud's phenomenon; and secondary Raynaud's must itself be distinguished from primary or idiopathic Raynaud's disease. When in 1862 Maurice Raynaud presented his M.D. thesis on "Local Asphyxia and Symmetrical Gangrene of the Extremities," he described the clinical manifestation of paroxysmal blanching and numbness of one or more fingers of either hand (Figures 13.6 to 13.8).

As Ashe and his co-workers (132, 133) reported from experimental observations a century later, this is typically provoked by whole-body rather than local chilling. Up to 8 percent of a community could be afflicted, according to Raynaud, more than 90 percent of sufferers from the idiopathic disease—that is, patients without a history of mechanical injury to the hands or other known etiological factor—being female. (It should be remarked in this connection that no definitive study has been published to show whether female workers using vibrating tools are more likely than their male counterparts to contract Raynaud's phenomenon of occupational origin.)

Raynaud noted incidentally that an emotional or psychosomatic factor may not uncommonly underlie the syndrome: some patients with idiopathic Raynaud's disease report increased frequency or severity of white-finger attacks during periods of emotional or general stress. Although the symptoms remain trivial in many cases, the condition can become severe in a small proportion (up to 3 percent) of sufferers, progressing over several years to skin atrophy, sometimes with ulceration, and, rarely, to digital gangrene requiring surgery. In the severe case concomitant changes become established in the intimal lining and muscular layers of the digital arterial wall. It is unlikely that such changes in the arterial wall and in the skin of the affected digits, once established, can be reversed by medical treatment (Figures 13.9 to 13.11).

* Also called vibration-induced white finger (VWF) and, popularly, "dead" or "wax" finger.

EVALUATION OF EXPOSURES TO VIBRATIONS

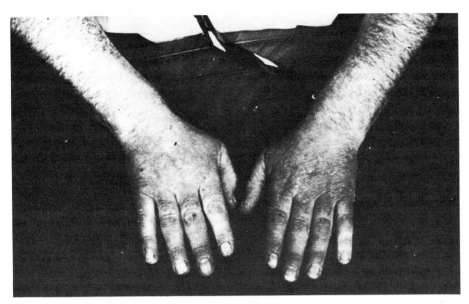

Figure 13.6 A mild case of Raynaud's phenomenon (vasospastic response to chilling). Note particularly the fifth digit of this patient's left hand. Photo courtesy of Dr. W. Taylor, Dundee.

3.3.2 Symptomatology and Progression of VWF

The vascular symptoms and signs of secondary Raynaud's of occupational origin resemble those of Raynaud's disease. The classical attacks of sudden blanching and numbness of digits distal to a sharp line of demarcation (Figures 13.6 to 13.8) are often provoked by general chilling. Except in severe cases, these digital vasoconstrictive paroxysms usually last some 5 to 15 min and may be relieved by warmth and massage of the affected extremity (113).

The affliction is progressive as long as the patient continues exposure to hand-transmitted vibration on a regular occupational basis. An initially symptom-free period of vibration exposure follows first entry into the occupation, before the first experience of a white-finger attack. This period is called the latent period or interval: according to Taylor and his coauthors (114–116), the length of the latent period is inversely related to the severity of the vibration exposure. Moreover, it is said that the shorter the latent period (which may vary from several years to a matter of weeks in extremely severe instances of occupational exposure), the more severe and rapidly progressive is the resulting VWF in the individual, and the higher is the incidence of VWF on a group basis. During the latent period some patients may experience early warning nonvascular symptoms, such as unusual tingling in the fingers following work with a vibrating tool, but others experience no warning before the first attack of VWF.

Although a variety of diagnostic and screening tests (reviewed below) have been proposed for the vibration syndrome, the diagnosis of VWF rests essentially on the

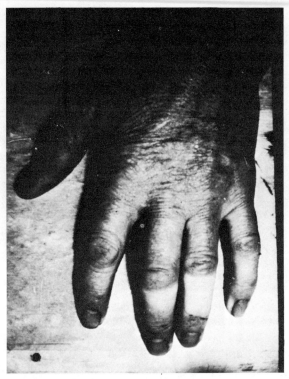

Figure 13.7 Raynaud's phenomenon affecting three digits in the right hand of a patient with a history of occupational exposure to hand-transmitted vibration. Note the sharp demarcation of the areas of blanching. Photo courtesy of Dr. W. Taylor, Dundee.

patient's history: as Okada et al. (134) have remarked, Raynaud's phenomenon is difficult to diagnose between attacks. A miscellany of clinical conditions enter into the differential diagnosis of secondary Raynaud's (Table 13.4), most of them being disorders that on a permanent or intermittent basis, restrict or modulate the regional or local arterial blood supply to the hand or digits.

Taylor and other authors have described the progression of classical VWF in the individual case (115, 116). After the latent period the frequency of VWF attacks increases over the months or years of cumulative exposure to hand-transmitted vibration. At first attacks occur mainly in the winter in cold countries, and especially during the early morning, either at home or on the way to work when the man is exposed to cold conditions (e.g., shaving in a chilly bathroom, bicycling to work on a frosty or cold, damp morning, getting the car out of an unheated garage). Employees working outside in all weathers (e.g., lumbermen) are most prone to early morning attacks. It is of etiological interest that circumstantial changes in the daily habits of workers with

EVALUATION OF EXPOSURES TO VIBRATIONS

established VWF can reduce the incidence of attacks. Futatsuka (135, 136) has reported a diminished incidence in chain sawyers who changed from bicycling to work to using a newly introduced bus service: presumably the warmer environment in the bus lessened the probability of provocation of attacks in the individual and also meant that the workers, as a group, were warmer at the start of their daily work.

Workers suffering from VWF quite commonly report progressive interference with activities outside work (e.g., gardening, home or car maintenance, car washing, fishing, meetings in ill-heated buildings such as church halls) before succumbing to attacks while at work. All these activities tend to have as a factor in common a reduced environmental temperature. Taylor and his coauthors (115, 116) have introduced a scale of stages for gauging the progression or severity of cases of VWF (Table 13.5).

In stage 2 there is interference with hobbies, social and recreational activities, while in stage 3 there is cessation of certain such activities, which may be interpreted as a curtailment of the patient's enjoyment of leisure time. In stage 3 there may also be interference with work or rest breaks at work, particularly in outside jobs such as forestry. In stage 4 the severity of the affliction has become such that serious interference with work, social activities, and hobbies induces the patient to change occupation. By the time this stage has been reached, it is likely that established changes have taken place in arteries of the affected digits, of the kind illustrated in Figure 13.9.

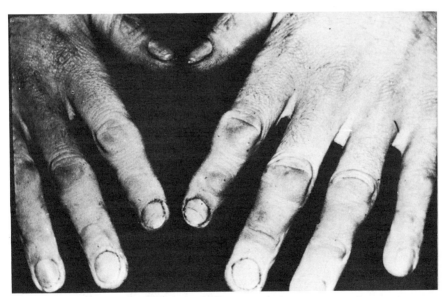

Figure 13.8 Hands of a patient affected by Raynaud's phenomenon involving distal phalanges of fingers, associated with a history of working with percussive and vibrating tools over several years. Note also callus formation over several of the joints of the fingers of the same hands. Photo courtesy of Dr. W. Taylor, Dundee.

Table 13.4 Differential Diagnosis of Secondary (Nonidiopathic) Raynaud's Phenomenon: Conditions Predisposing or Leading to Paroxysmal Digital Vasospastic Attacks or Blanching of Fingers

Disorders resulting from physical agents
    Traumatic vasospastic disease
        Vibration syndrome
        Vasospasm following mechanical injury to hand or digits
    Cold injury/frostbite
Mechanical compression of upper limb vessels
    Thoracic outlet syndromes
    Costoclavicular syndromes
Occlusive vascular disorders
    Arteriosclerosis
    Thromboangiitis obliterans
    Arterial embolism
Primary pulmonary hypertension
Neurological lesions
    Neurological diseases of the central nervous system
        Poliomyelitis
        Syringomyelia
        Hemiplegia (cerebrovascular)
    Peripheral neuropathies: diabetes
Connective tissue disorders
    Scleroderma/progressive systemic sclerosis
    Polyarteritis nodosa
    Rheumatoid arthritis
    Systemic lupus erythematosus
    Dermatomyositis
Endocrinological and metabolic disorders
    Dysproteinemias
    Myxedema
Vasoconstrictive medications, drugs, and chemical intoxications
    Ergot/methysergide
    Polyvinyl chloride
    Nicotine

Jepson (137), in a selective survey of cases of Raynaud's phenomenon associated with various occupations, concluded that the severity of the condition was likely to reach a maximum within a few years of its first appearance, after which it would not worsen with further exposure (at least in workers who remained in the occupation: an apparent limitation of the incidence of VWF with time may be seen on a group basis because severely afflicted workers tend to leave the job).

EVALUATION OF EXPOSURES TO VIBRATIONS

The traditional view, expressed by Jepson (137), has been that once Raynaud's phenomenon has developed, no abatement of symptoms occurs, even with removal of the individual from work with vibrating tools. More recent opinion, however (115, 116), has tended to moderate this pessimistic view: some patients in the earlier stages of the disease apparently can respond favorably to removal from vibration exposure and possibly to medical treatment. Preventive and ameliorative treatment for vibration syndrome is considered below.

3.4 Other Varieties of Vibration Syndrome

3.4.1 Vasomotor Symptoms Other than Raynaud's Phenomenon

Although VWF is the most frequently presenting consequence of occupational exposure to hand-transmitted vibration, other vasomotor symptoms can be the presenting and most important ones in particular industries. For instance, Agate and Druett (122, 123) suggested that appliances producing their peak vibrational energy output in the range 40 to 125 Hz were associated particularly with the causation of classical VWF (as is the chainsaw—not in widespread occupational use at the time of Agate and Druett's studies—which produces its most intense vibration in the same frequency range).

Table 13.5 Taylor's Grading System for the Severity of Vibration-Related Secondary Raynaud's Phenomenon ("VWF")

Stage of Disorder	Symptoms and Signs	
	Response of Digits	Work and Social Interference
0	No blanching	No complaints
0_T	Intermittent tingling without blanching	No interference
0_N	Intermittent numbness without blanching	
1	Blanching of a fingertip with or without tingling and/or numbness	No interference
2	Blanching of one or more complete fingers, usually during winter	Interference at home or in social activities but not at work
3	Extensive blanching of all fingers of one or both hands; attacks during all seasons	Definite interference at work, at home and in social activities; restriction of hobbies
4	Extensive blanching of all fingers, both winter and summer	Occupation changed because of severity of signs and symptoms of VWF

[a] Adapted from Wasserman, et al. (116).

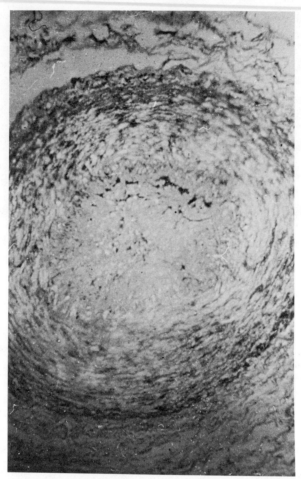

Figure 13.9 Photomicrograph of a section of a digital artery from a case of occupationally related Raynaud's phenomenon, showing degeneration of the intima and hypertrophy with infiltration of the muscular layer of the arterial wall. Photo courtesy of Dr. W. Taylor, Dundee.

Lighter, higher speed electric drills of a type widely used in the aircraft industry during World War II, and whose principal output of vibrational energy lay in the range 166 to 833 Hz, however, were reported by Dart (129) to be associated with a different kind of vasomotor disorder. The commonest presenting symptom in Dart's cases was pain, associated with swelling and erythrocyanotic discolorations in the hands. The patients did not report blanching of the hands in response to chilling.

3.4.2 Neurological Effects

Apart from vasomotor reactions, other types of disorder or injury are associated with the handling of vibrating or impacting power tools, including both peripheral and

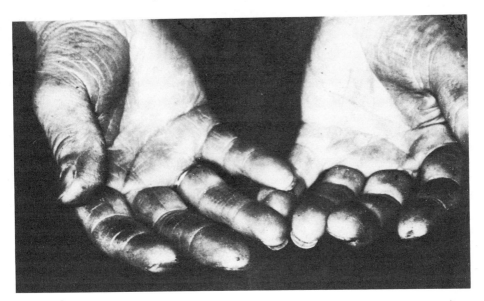

Figure 13.10 Hands of a patient suffering from an advanced case of traumatic vasospastic disease, showing areas of ischemic necrosis of the skin of the fingertips. From a color photograph by Dr. W. Taylor, Dundee.

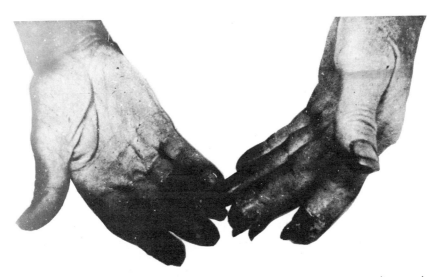

Figure 13.11 Hands of a patient exhibiting severe case of traumatic vasospastic disease, showing ischemic gangrenous necrosis of distal parts of affected fingers. From a color photograph by Dr. W. Taylor, Dundee.

central nervous symptoms, and disorders of the musculoskeletal system. In addition to attacks of VWF, patients in Taylor's stage 3 of the vibration syndrome may experience difficulty in undertaking fine work such as electronics and difficulty in recognizing small objects by touching and picking them up. They may also complain of difficulty in doing up and undoing buttons, or carrying out other domestic or hobby activities calling for fine tactile and manipulative skill, because of loss of tactile discrimination, clumsiness of the fingers, and increasing stiffness of the finger joints. Banister and Smith (130) have reported the association of VWF with loss of manipulative dexterity (measured using the Purdue pegboard test) as a consequence of vibrating tool usage. They reported that in the chain sawyers examined the dominant hand showed the most marked loss of skill.

Agate and Druett (122, 123) had shown incidentally that in hammerers and grinders, the left hand, which usually was used to guide the workpiece and therefore received the worst vibration, was often the hand most afflicted by VWF. It has proved difficult to establish, and it is not yet known, whether the nerve, skeletal, and muscular disorders seen in the vibration syndrome develop independently of or are secondary to the vasospastic response (138).

Sevčik et al. (139), reporting an incidence of 65.5 percent (120 cases) of vibration disease in 183 chain sawyers in southern Bohemia, found 83 cases (45.5 percent) of traumatic vasoneurosis in those workers, who had an average age of 36 years and had been using chain saws in forestry work for an average of 6.6 years. Some two-thirds of the cases examined showed neuropathological symptoms, and 24 percent showed radiographical changes at the wrist. Drobny et al. (140) has also reported neurological changes in foresters using chain saws in Czechoslovakia. Lukas and Kuzel (141) recorded pathological changes in nerve conduction velocity in either or both of the ulnar and median nerves in 47.4 percent of 137 miners who had been using pneumatic tools for an average working life of 14 years. Bjerker et al. (142) have demonstrated changes in the vibrotactile threshold of the hand following exposures to powerful vibration. These authors have proposed the adoption of tests of vibrotactile temporary threshold shift in the hands of workers with vibrating tools as a diagnostic and screening test for the vibration syndrome.

3.4.3 Musculoskeletal Changes

Somewhat less commonly than vasomotor or neurological disorders, skeletal injuries are reported in workers who have habitually used vibrating tools. Cystic or vacuolar areas of radiographically manifest decalcification of bone have been described (122, 123, 143), particularly in the region of the wrist joint. Horváth and Kákosy (143) found 45 cases of atrophic decalcification (particularly involving the lunate bone and the distal epiphysis of the radius) and 75 cases showing "pseudocysts" a few millimeters in diameter in a series of 274 foresters who had used chain saws for up to 15 years. The etiology of such effects is unknown (see below).

Heavy, low frequency vibration is transmitted through the hands to more proximal structures, including the elbow and shoulder joints, where the repeated mechanical assault can induce occupationally related arthritic changes and periarticular disorders. Iwata (131) reported substantial incidences of upper limb joint (46.9 percent) and muscle (36.7 percent) pain in a survey of 529 metal ore miners who used rock drills. The elbow joint was that most commonly afflicted, followed by the wrist joint. In many cases articular abnormalities were seen radiographically, although a clear relationship between symptomatology and radiological findings was not established. Pyykkö (113) mentions that working with vibrating tools is likely to provoke pain in joints that have previously sustained an injury. Previous injury to a digit may also predispose that digit to attacks of VWF.

Some workers with vibrating tools complain mainly of muscular weakness or stiffness in the hands, particularly on waking in the morning or at the start of a work shift. I know of a foundry worker aged 32 who had used swaging tools on castings for several years. He frequently found on awakening in the morning that his hands were stiffened in flexion so that the fingertips approximated the palm: it was necessary for the patient's wife to help him open the hands by gentle extension of the fingers and massage of the hands before he could get ready to go to work. Dupuytren's contracture involving the palmar fascia is sometimes seen in workers with vibrating tools but is not necessarily causally related to the vibration exposure (113).

3.5 Etiology of Vibration Syndrome

Several hypotheses have been advanced to account for the manifestation of Raynaud's phenomenon and the associated disorders constituting the vibration syndrome, but none has been universally accepted. Many etiological factors apart from the vibration exposure may be associated with the condition (Table 13.3). However some of these factors may be only coincidentally related to the occupation, and others may not be related to it at all. That is why it has proved difficult to establish a firm causal link between hand-transmitted vibration at work and compensatable disability, particularly on an individual basis.

Pyykkö (113) has stated succinctly that in traumatic vasospastic disease (Raynaud's phenomenon of occupational origin) vibration is the etiological factor and cold the provocative one. After the latent period the cumulative trauma causes an enhanced reactivity of the finer arterial subdivisions in the vascular system of the hand distal to the brachial artery (133), accompanied in the established case by anatomical changes that may be irreversible. Ashe and his co-workers (133) found that this begins with muscular hypertrophy in the medial layer of the arterial wall. Later the intima becomes altered also.

According to Koradecka (144), vibration syndrome is the result of a disorder of the thermoregulatory function of the circulation in the skin, although this hypothesis is open to question. Thermoregulatory disorders may appear early or late in the vibration syndrome (145).

Climatic conditions undoubtedly influence the rapidity of onset and the pattern and severity of vasospastic vibration syndrome. In a comparative study of groups of riveters doing similar work in climatically distinct regions of the Soviet Union, Mirzaeva (146) found that in the hot climate of Uzbekistan, the latent period of the syndrome was about twice as long as in a comparable group of riveters living and working in Moscow. Moreover the warm climate group experienced a milder syndrome, with neurological rather than vasospastic symptoms predominating, and without classical Raynaud's attacks (finger pallor) occurring during cold weather. Iwata (131) has also noted the absence of classical Raynaud's phenomenon in workers using rock drills in a warm climate. He has drawn attention to the significance of whole-body chilling in a comparative study of the effects of rock drill vibrations on miners in pits having varying degrees of coldness combined with high humidity.

The vasospastic or vascular form of the vibration syndrome (Raynaud's phenomenon) has been characterized by Langauer-Lewowicka (147) as a "cryopathy of occupational origin"; its sufferers can be shown to exhibit certain immunological manifestations. On a group basis patients with Raynaud's showed increases in β-globulins, γ-globulins, and antiglobulin antibodies in comparison with vibration-exposed subjects not suffering from Raynaud's phenomenon. Serum concentrations of Immunoglobulin M (IgM) were also significantly raised in the group with early Raynaud's, as were IgG immunoglobulins in cases with established disease. These findings appear to support those of Okada et al. (134). Both Okada and his colleagues and Langauer-Lewowicka have accordingly postulated that vibration-induced Raynaud's may be the result of serum protein disturbances caused by a change in "reactivity" in the body, presumably mediated centrally by the sympathetic nervous system (125).

Nerem (148, 149) has drawn attention to a possible mechanism underlying arterial disease associated with exposure to mechanical vibration. Using ^{131}I-albumen for *in vivo* and *in vitro* studies of blood–arterial wall macromolecular transport in the dog, he has demonstrated that heavy vibration at 10 Hz induces an arterial wall shear stress that enhances a shear-dependent transport process that in turn increases albumen uptake by the arterial wall. Such a process may underlie endothelial damage and changes in the medial layer of arteries.

Neuromuscular reflex mechanisms may play a contributory role. Vibration of the upper limb in man excites the tonic vibration reflex (150), with an increase in tonicity of vibrated muscle. Jansen (151) has claimed that noise may be a predisposing factor in vibration syndrome, increasing the incidence of Raynaud's phenomenon of occupational origin. According to Pyykkö (113), vibration, noise, and cold may act synergistically to maintain peripheral vasoconstriction.

Pyykkö (113) has postulated the following local mechanism for Raynaud's phenomenon: cold causes vasoconstriction, and because of the hypertrophic changes in the vessel wall, arterial blood flow in the affected digit falls very low. When the intraluminal pressure falls below the critical closing pressure for that vessel, the vessel collapses and a white-finger attack is precipitated. Stewart (152, 153) has advanced an alternative mechanical hypothesis, not generally supported: namely, that Raynaud's phenomenon of occupational origin results from constriction of the vessels due to increased digital skin thickness and callus formation in workers exposed to vibration.

3.5.1 Prevalence of Vibration Syndrome

In the long term, changing trends in both directions may be seen in the incidence and pattern of vibration syndrome in industry. For instance, Yershova (154) has reported that in the decade 1958–1968 the incidence of vibration syndrome in the Ukraine increased for complex epidemiological reasons, including the increasing use and power of vibrating tools—and also the improved awareness and diagnosis of vibration-related disorders. Since 1968, however, the trend has reversed, the incidence of vibration syndrome having been reduced by the introduction of preventive programs, improved tool design and technology, and the introduction of remote control of some industrial processes, effectively removing the worker from harmful vibration exposure. Some examples of new technology leading to a reduction in the prevalence or severity of vibration syndrome are the introduction of rock drilling rigs and trestles, the displacement of riveted by welded structures in shipbuilding, and the introduction of antivibration chain saws in forestry.

3.6 Biodynamic Aspects of Vibration Syndrome

Depending on the nature and magnitude of the vibrating forces transmitted by a handheld tool or workpiece, and the frequency content of the impressed motion, vibration and shock loads are transmitted through the hands of the worker and up the arms to the upper limb joints. The transmission and distribution of vibrational energy in the hand-arm system is highly frequency dependent: as a general rule, high frequency vibrational energy is strongly absorbed and dissipated in the skin and deeper tissues of the fingers and hand grasping the source of vibration, whereas low frequency vibrations are propagated up the arm (6, 127, 131, 155–158). Using data obtained from mechanical impedance and cognate measurements of hand-transmitted vibration, it is possible to construct analogical models of the human upper limb as a vibration receiver for research and testing purposes (127, 159, 160).

Vibrational energy absorption or dissipation can be measured in the human hand, operating various types of powered tool, in a standard manner applicable in both laboratory and field studies, using small vibration transducers fixed in specially prepared and fitted handles (126, 161, 162). The principal location of vibrational energy absorption depends on many factors in addition to frequency (71), for example, the strength of the grip applied to the tool (163, 164). Several such factors were listed in Table 13.3. Such factors probably account in large measure for the substantial differences in the latent period, incidence, severity, and pattern of vibration disorder seen in different industries.

In a carefully conducted comparative study in three occupations (rock drilling, chiseling, and grinding) characterized by different patterns of hand-transmitted vibration exposure, Lidström (161, 162) examined the correlation between vibration disorder (determined, according to specific criteria, from the history and a laboratory test of individual susceptibility to vibration-induced vibrotactile sensory threshold shift) and the

Table 13.6 Apparent Relationship Between Incidence of Vibration Disorder and Rate of Energy Absorption from Hand-Transmitted Vibration in Three Occupations (162)

Occupation	Frequency of Vibration Disorder (%)	Rate of Energy Absorption (nm/s)
Rock drilling	72	21.0
Chiseling	53	2.7
Grinding	21	0.07

vibrational energy absorption per unit time (determined from force and velocity measurements on the tools in question, using an instrumented standard handle). She found that there was a tendency for the incidence of vibration disorder to rise with the rate of vibrational energy absorption by the hand, depending on the type of tool and method of use (Table 13.6).

3.7 Diagnostic and Screening Tests for Vibration Syndrome

A variety of measurements and clinical tests have been advocated for use in research or medical practice concerning the diagnosis and assessment of vibration syndrome (113, 115, 144, 165, 166). None has yet been generally accepted for use in clinical industrial practice or in epidemiological studies, because all are open to criticism on the grounds that they lack reliability (i.e., the test yields too many false positive or false negative responses), specificity, or repeatability, especially on an individual basis; or because they are taxing or difficult to use in general practice or in field conditions (113). The best known tests currently in clinical or research use are summarized next.

3.7.1 Tests of Peripheral Vascular Function

Koradecka (144, 166) and other investigators have advocated a number of tests of peripheral circulation, and particularly digital blood flow measurements, for the diagnostic screening and periodical monitoring of workers complaining of vibration syndrome or at risk from exposure to hand-transmitted vibration. Several techniques have been tried, with varying degrees of confidence, including digital (occlusive) plethysmography, skin thermometry, capillary microscopy, skin and muscular rheography, and arteriography.

Mentioning the problem of the lack of specificity of tests for peripheral arterial disease, Zweifler (167) has presented case histories showing that finger ischemia (demonstrable by plethysmographic and arteriographic techniques) may result from occupational traumata of varied origin. Various forms of cold provocation test, in which the autonomically mediated peripheral vascular response to sudden cold is examined

EVALUATION OF EXPOSURES TO VIBRATIONS 511

and compared with supposedly normative data, have been tried. But the results of several investigators (115, 138, 168, 169), were disappointingly unreliable.

The technique of infrared thermography, already in diagnostic use in cases of primary Raynaud's disease, has been adapted by Tichauer (164) for the examination of patients suffering from various forms of "work stress" affecting the back or upper or lower limbs in jobs involving repeated volitional or imposed motion or vibration (e.g., operating treadles, handling vibrating appliances). According to Tichauer, the technique shows promise for diagnostic and prognostic purposes in relation to vibration syndrome, although Tiilila (169) had previously expressed disappointment with it.

Buzalo (170), in advocating the inclusion of a thorough ophthalmological examination in the clinical evaluation of patients with vibration syndrome, has stated that disturbances of eye vessels, retinal and iridal structures, and the visual fields (perimetry) may be observed in such cases.

3.7.2 Neurological Tests

Recognition that neurological symptoms and signs occur in a substantial proportion of cases of vibration syndrome has led to the introduction of a number of tests of peripheral nerve function for their diagnosis and evaluation, including measurements of conduction velocity by the method of Hopf (171) and of peripheral sensory function. Some success has been claimed for the clinical and research use of tests of tactile discrimination and vibrotactile sensitivity (pallesthesiometry) and vibrotactile sensory threshold shift following exposure of the hands to vibration (115, 161, 162, 172).

Radziukevich and Mikulinskiy (173) have pointed out, however, that considerable variance is to be expected in pallesthesiograms even in normal subjects not exposed to occupational vibration, and that even in nonvibrated control subjects, there is a systematic rise in vibration threshold with age between 18 and 60 years, for which a correction factor would have to be included if the test were used for the assessment of vibration-related disability. (The problem is somewhat akin to that of prebyacusis in the audiometric evaluation of noise-induced hearing loss.) Seppäläinen (171) has reported disappointing results with neurological tests that failed to distinguish between shipyard workers and a comparison group not occupationally exposed to hand-transmitted vibration.

3.7.3 Radiological Examination of Bones and Joints

As mentioned earlier in this chapter, a fraction of patients with vibration syndrome reveal radiographic changes in the bones of the upper limb, particularly in the neighborhood of the wrist joint (122, 123, 128). Bone densitometry has been used in attempts to assess such cases, sometimes with disappointingly equivocal results (174). Again, a problem arising with regard to the diagnosis and assessment of vibration syndrome on the basis of radiological examinations is the lack of specificity and quantifiability of the bone changes observed.

3.7.4 Physical Measurements and Electromyography

Electromyography has been advocated, principally as a research tool, for the study of human reactions to hand-transmitted vibration by several investigators (155, 157, 175, 176). The level of electromyographic activity probably can be correlated in a quantifiable manner with segmental vibration inputs and vibration propagation in the human forelimb (157, 165) and it is possibly more strongly indicative of the skill and effort required to handle and operate different kinds of vibrating tool in various work situations and postures (156).

3.8 Prevention and Treatment of Vibration Syndrome

3.8.1 Standards

Following several attempts to set regulatory standards or guidelines for the protection of workers exposed to hand-transmitted vibration in a number of European countries (177–180), ISO Subcommittee ISO/TC 108/SC4 (Human Exposure to Mechanical Vibration and Shock) prepared a guide (5) for the measurement and evaluation of human exposure to hand-transmitted vibration. In addition to presenting an international consensus regarding the proper manner in which to measure hand-transmitted vibration for occupational safety and health purposes, this document (a proposed international standard) includes recommendations (Table 13.7) for limiting the intensity (rms acceleration and velocity limits are specified) of vibration reaching the hands in the frequency range 6.4 to 1000 Hz.

The limits proposed are admittedly tentative, because of the paucity of firm quantitative data relating risk of vibration syndrome to measured human exposure to hand-transmitted vibration. The draft standard is heavily if not entirely based on subjective data [particularly the work of Miwa and his co-workers (47, 181) on the magnitude estimation of hand-transmitted vibration], and its introduction as a preventive guideline without supporting biomedical data is accordingly open to criticism (127).

3.8.2 Engineering Solutions: Vibration Reduction in Power Tools

Assuming that an accepted standard can be established to limit human exposure to hand-transmitted vibration, vibrating tools and appliances can be designed or modified to meet it in several ways. Recent years have seen remarkable improvements in the design of the chain saw to minimize the vibrational energy transmitted to the hands of foresters by these motor-driven machines (182–187). The most important innovation has been the introduction of the antivibration (A/V) saw, in which part of the mass of the machine, including the fuel tank, is resiliently suspended to act as a dynamic vibra-

EVALUATION OF EXPOSURES TO VIBRATIONS

tion absorber. The introduction of A/V saws has been reflected in a falling incidence of VWF among forest workers in countries where state forestry commissions have made the use of such saws the rule in their large work forces, notably Great Britain (172, 184) and Sweden (182).

Regrettably, the same administrative and engineering effort that has so markedly improved the lot of the chain sawyer has not yet been devoted to other kinds of vibrating tool such as pneumatic hammers and grinding and scaling machines: many such

Table 13.7 Proposed International Standard Acceleration and Velocity Exposure Limits for Hand-Transmitted Vibration as Functions of Frequency and Notional Daily Exposure Time (5)

Proposed Exposure Limits for Single-Frequency or Narrow Band[a] Vibration (Third-Octave Band Analysis), Where Daily Exposure is Essentially Uninterrupted for 4 to 8 hr

Frequency, or Center Frequency of Third-Octave Band (Hz)	Maximum rms Intensity of Hand-Transmitted Vibration in Any Direction (Axis)	
	Acceleration [m/sec^2]	Velocity (m/s)
6.4	0.8	0.016
8	0.8	0.016
10	0.8	0.013
12.5	0.8	0.010
16	0.8	0.008
20	1	0.008
25	1.3	0.008
31.5	1.6	0.008
40	2	0.008
50	2.5	0.008
63	3.2	0.008
80	4	0.008
100	5	0.008
125	6.3	0.008
160	8	0.008
200	10	0.008
250	12.5	0.008
315	16	0.008
400	20	0.008
500	25	0.008
630	31.5	0.008
800	40	0.008
1000	50	0.008

Table 13.7 Continued

Proposed Exposure Limits for Broad Band Vibration (Octave-Band Analysis), Where Daily Exposure is Essentially Uninterrupted for 4 to 8 hr

Center Frequency of Octave Band (Hz)	Maximum rms Intensity of Hand-Transmitted Vibration in Any Direction (Axis)	
	Acceleration [m/sec^2]	Velocity (m/s)
8	1.4	0.027
16	1.4	0.014
31.5	2.7	0.014
63	5.4	0.014
125	10.7	0.014
250	21.3	0.014
500	42.5	0.014
1000	85	0.014

Tentative Correction Factors for Hand-Transmitted Vibration Exposure Limits Allowing for Interruption of Exposure and for Daily Exposure Durations Less than 4 hr

Exposure Duration in Working Shift (more than 8 hr is not recommended)	Type of Vibration Exposure and Interruption Time (min/working hr)				
	Uninterrupted or Occasionally Interrupted	Regularly Interrupted			
	10	10–20	20–30	30–40	40
Up to 30 min	5	—	—	—	—
30 min–1 hr	4	—	—	—	—
1–2 hr	3	3	4	5	5
2–4 hr	2	2	3	4	5
4–8 hr	1	1	2	3	4

[a] Third-octave band analysis is generally recommended, in which case the more stringent limits for single-frequency or narrow band vibration apply.

machines remain in use in industry after many years, indeed, decades, their design having changed little if at all since the early years of this century (Figures 13.2, 13.3, 13.5).

3.8.3 Attention to Operational and Ergonomic Factors

Perusal of Table 13.3 will have suggested to the reader several ways in which the problem of vibration syndrome can be attacked. The beneficial influence of frequent and adequate breaks from regular work with vibrating tools such as the chain saw has long been recognized and advocated (124, 188). It is desirable that the unsupported weight of vibrating tools not exceed 10 kg, nor the effort required to guide or advance the tool a force of 20 kg (124, 189, 190).

In some occupations, notably chain sawing, in some countries the use of gloves is recommended (121), both to keep the hands warm and to provide some damping or attenuation of the vibration reaching the hands. In other occupations, such as that of chipping or swaging castings, padding (even old rags or cloths) is sometimes interposed between the worker's hand and the tool or workpiece. The value of such devices is debatable, for as a rule padding of the handles of vibrating tools or the wearing of gloves provides relatively little attenuation of vibration below about 200 Hz (191). Proper care and maintenance of powered tools is to be recommended in all industries, both for reasons of safety and good practice and also because the vibration levels in such tools are likely to creep up if the equipment is not regularly cleaned, examined, and overhauled (192).

Keeping warm at home and on the job (and also on the way to work) is generally agreed to be beneficial: this may include prewarming the hands and person before beginning a stint of work in cold, damp conditions, such as in forestry (121, 124, 188). Guidance given to chain saw operators by the United Kingdom Forestry Commission embraces many points of good practice, including those just mentioned, and deserves requoting from Taylor et al. (188)

There are a number of ways in which you can reduce the amount of vibration passing into your hands, when you are using any saw, and these points apply whether or not the saw has antivibration handles:

1. Good techniques for felling, cross-cutting, and snedding with lightweight saws include resting the saw as much as possible on the tree (or occasionally on your thigh); this means that some of the vibration is absorbed by the tree or the large muscles of your thigh. Holding the saw as lightly as possible when it is at full throttle, without, of course, reducing effective control of the saw, will also reduce vibration absorbed into your hands.

2. Wearing chain saw gloves spreads the grip over a larger area of your hands. (Often the first sign of white fingers is on one or two finger joints which have taken most of the vibration due to too tight a grip.)

3. Good blood circulation to the arms and hands gives maximum protection to the flesh, nerves, and bones of the hands, and this is achieved by warming up *before* starting the saw and wearing suitable clothing and gloves. Thus, it is better to be too warm than cold.

4. Sprockets, guide bars, and chains should be well maintained, and chains should be correctly sharpened with the recommended clearance for the depth gauge. Poor maintenance increases vibration by as much as one-third of the normal level for the saw.

5. The "safe" limits are based on continuous use of the saw, and every time the saw is idling or stopped gives your hands and arms a chance to recover from the effects of vibration. The more evenly breaks in saw usage can be spread throughout the day, the less the risk of any discomfort in your hands; try to organize the stops for fuel, sharpening, meals, piling of timber, or other work so that the saw is switched off for at least 10 minutes as often as possible during the day rather than a few longer stoppages.

3.8.4 Experience, Training, and Skill

Clutching a power tool or workpiece in an overly tight grip is characteristic of the novice or the anxious person. A tight grip restricts the circulation in the hand and may also enhance the transmission of vibration into the hand (E. Tichauer, comment in Reference 116, p. 208; 183). It is desirable that workers with vibrating tools learn to handle the equipment in a reasonably relaxed manner, avoiding a tight grip or an awkward, rigid work posture. Workers at risk or suffering from vibration syndrome should also exercise caution regarding exposure to intense hand-transmitted vibration away from work (e.g., in long motorcycle riding; riding or guiding motor lawn mowers; using power saws, drills, sanders, etc. in hobbies or in car or home maintenance), particularly if the activity is carried on in cold or damp conditions. The industrial medical officer or practitioner responsible for patients exhibiting prodromal or established symptoms and signs of the vibration syndrome may also wisely advise the patient to consider the beneficial effects of adequate exercise, regular meals and suitable diet, eliminating or cutting down on smoking, and care of the general health.

3.8.5 Medical Treatment for Vibration Syndrome

If not in an advanced stage, VWF usually is abated and does not progress after cessation or reduction of the patient's vibration exposure. Provided the disorder has not progressed to the stage of arterial damage and trophic changes in the digits, there is nothing to treat when an attack is not in progress. Severe or progressive Raynaud's phenomenon of occupational origin may be ameliorated by drugs such as are used in the treatment of primary Raynaud's disease (e.g., reserpine, methyldopa, mild sedatives), but the results are likely to be disappointing and success short-lived in cases of secondary Raynaud's if the underlying cause is not dealt with.

It has been claimed that medication to reduce the coagulability of blood can reduce the frequency and severity of symptoms in sufferers from severe vibration-related Raynaud's phenomenon. This is based on the observation that such patients not infrequently show an increased tendency to hemocoagulation. Demin et al. (193), noting that at least a third of patients suffering from vibration disease showed such a tendency, prescribed corrective heparin therapy. They reported that 43 out of 50 cases showed substantial improvement and the remainder some improvement (in relation to the coagulability of blood, if not symptomatically).

In Germany Ehrly (194) has reported some success using the drug Ancrod (a purified fraction obtained from the venom of the Malayan pit viper) in a series of six patients with severe secondary Raynaud's. The basis of the treatment was to reduce blood viscosity, thus enhancing peripheral blood flow. The drug was administered subcutaneously (1 unit per kilogram of body weight daily). Symptoms reportedly abated over several days, and a clinical improvement in skin temperature and color with healing of ischemic ulceration ensued. Apparently the beneficial effects of treatment, including the cessation of ischemic pain and skin ulceration, persisted for several months after short courses of this treatment.

ACKNOWLEDGMENTS

I am indebted to Dr. W. Taylor, Dundee, and Dr. I.-M. Lidström, Stockholm, for helpful comments and illustrative material. Views expressed in this chapter are my own and are not necessarily endorsed by my current employer, the United States Navy.

REFERENCES

1. C. E. Crede, "Principles of Vibration Control," in: *Handbook of Noise Control,* C. M. Harris, Ed., McGraw-Hill, New York, 1957, Chapter 12.
2. J. C. Guignard, "Vibration," in: *A Textbook of Aviation Physiology,* J. A. Gillies, Ed., Pergamon, Oxford, 1965, Chapter 29.
3. J. C. Guignard, "Vibration," in *Aeromedical Aspects of Vibration and Noise,* J. C. Guignard and P. F. King, AGARDograph AG-151, NATO/AGARD, Neuilly-sur-Seine, France, 1972, Part 1, pp. 2–113.
4. International Organization for Standardization, *Guide for the Evaluation of Human Exposure to Whole-Body Vibration,* ISO 2631–1974, ISO, Geneva, 1974.
5. International Organization for Standardization, *Guide for the Measurement and the Evaluation of Human Exposure to Vibration Transmitted to the Hand,* ISO Draft Proposal 5349, (Secr. Doc. ISO/108/4/WG3 20), ISO/TC 108/SC4, Berlin, May 1978.
6. H. Dupuis, E. Hartung, and L. Louda, *Ergonomics,* **15,** 237–265 (1972).
7. J. C. Guignard and M. E. McCauley, "Motion Sickness Incidence Induced by Complex Periodic Waveforms," Paper to the 21st Annual Meeting of the Human Factors Society, San Francisco, October 1977.
8. International Organization for Standardization, *Draft Proposal on Standard Biodynamic Co-Ordinate Systems,* Secr. Doc. ISO/108/4, 61, Secretariat of ISO/TC 108/SC4, Berlin, July 1978.
9. J. C. Guignard, *J. Sound Vib.,* **15,** 11–16 (1971).
10. L. F. Stikeleather, G. O. Hall, and A. O. Radke, "A Study of Vehicle Vibration Spectra as Related to Seating Dynamics, SAE Paper 720001, Society of Automotive Engineers, New York, January 1972.
11. B. Hellstrøm, "Measurement of Occupational Vibration and Its Biological Effects," in Reference 117, pp. 89–105.
12. D. E. Wasserman, T. E. Doyle, and W. C. Asburry, *Whole-Body Vibration Exposure of Workers during Heavy Equipment Operation,* U.S. Department of Health, Education and Welfare (NIOSH) publication 78-153 National Institute for Occupational Safety and Health, Cincinnati, Ohio, April 1978.
13. R. S. Kennedy, A. Graybiel, R. C. McDonough, and Fr. D. Beckwith, *Acta Otolaryngol.* **66,** 533–540 (1968).
14. J. F. O'Hanlon and M. E. McCauley, *Aerosp. Med.,* **34,** 366–369 (1974).
15. D. Goto and H. Kanda, *Motion Sickness Incidence in the Actual Environment,* Document ISO/TC 108/SC4/WG2 63, Secretariat of ISO/TC 108/SC4, International Organization for Standardization, Berlin, 1977.
16. K. E. Money, *Physiol. Revs.,* **50,** 1–39 (1970).
17. T. G. Dobie, *Airsickness in Aircrew,* AGARDograph AGARD-AG-177, NATO/AGARD, Neuilly-sur-Seine, France, February 1974.
18. A. N. Razumeyev, "Kinetoses," in: *Pathological Physiology of Extreme Conditions,* P. D. Gorizontov and N. N. Sirotinin, Eds., Meditsina Press, Moscow, 1973, pp. 332–348 (National Aeronautics and Space Administration Tech. Trans. NASA TT F-15, 324, Washington, D.C., March 1974).
19. J. T. Reason and J. J. Brand, *Motion Sickness,* Academic Press, London, New York, 1975.
20. H. E. von Gierke, C. W. Nixon, and J. C. Guignard, "Noise and Vibration," in: *Foundations of Space*

Biology and Medicine, Vol. II, Book 1, M. Calvin and O. G. Gazenko, Eds., National Aeronautics and Space Administration, Joint U.S.A/U.S.S.R Publication, Washington, D.C., 1975, Chapter 9, pp. 355–405.
21. R. J. Hornick, "Vibration," in: *Bioastronoautics Data Book,* 2nd ed., J. F. Parker and V. R. West, Eds., NASA SP-3006, National Aeronautics and Space Administration, Washington, D.C., 1973, pp. 297–348.
22. R. W. Shoenberger, *Percept. Mot. Skills, Monogr. Suppl.,* **1-V34** (1972).
23. J. Matthews, *J. Agric. Eng. Res.,* **9,** 3–31 (1964).
24. J. Hasan, *Work-Environ.-Health,* **6,** 19–45 (1970).
25. H. E. von Gierke, *Appl. Mech. Rev.,* **17,** 951–958 (1964).
26. E. B. Magid, R. R. Coermann, and G. H. Ziegenruecker, *Aerosp. Med.,* **31,** 915–924 (1960).
27. M. J. Mandel and R. D. Lowry, *One-Minute Tolerance in Man to Vertical Sinusoidal Vibration in the Sitting Position,* 6570th Aerospace Medical Research Laboratory Technical Document Report AMRL-TDR-62-121, Wright-Patterson Air Force Base, Ohio, October 1962.
28. H. E. von Gierke, H. L. Oestreicher, E. K. Franke, H. O. Parrack, and W. W. von Wittern, *J. Appl. Physiol.,* **4,** 886–900 (1952).
29. E. Ts. Andreeva-Galanina, E. A. Drogichina, and V. G. Artamonova, *Vibration Disease (Vibratsionnaya Bolezn'),* U.S.S.R. State Publishing House, Leningrad, 1961.
30. E. Ts. Andreeva-Galanina, *Gig. Tr. Prof. Zabol.,* **8:**8, 3–7 (1964).
31. A. S. Melkumova abd V. V. Russkikh, *Byul. Eksp. Biol. Med.,* **1973:**9, 28–31 (1973).
32. W. B. Hood, R. H. Murray, C. W. Urschel, J. A. Bowers, and J. G. Clark, *J. Appl. Physiol.,* **21,** 1725–1731 (1966).
33. A. J. Liedtke and P. G. Schmid, *J. Appl. Physiol.* **26,** 95–100, 1969.
34. J. Ernsting and J. C. Guignard, *Respiratory Effects of Whole Body Vibration,* Royal Air Force Institute of Aviation Medicine, Report RAF-IAM-179, Farnborough, England, 1961. (Also published by J. Ernsting as Air Ministry (London) Flying Personnel Committee Report FPRC-1164.)
35. L. R. Duffner, L. H. Hamilton, and M. A. Schmitz, *J. Appl. Physiol.,* **17,** 913–916 (1962).
36. T. W. Lamb, K. H. Falchuk, J. C. Mithoefer, and S. M. Tenney, *J. Appl. Physiol.,* **21,** 399–403 (1966); T. W. Lamb and S. M. Tenney, *ibid.,* **21,** 404–410 (1966).
37. A. Ya. Rakhimov and V. Sh. Belkin, *Arkh. Anat. Gistol. Entomol. (Leningrad),* **1970:**11, 43–49 (1970).
38. E. Ts. Andreeva-Galanina, *Gig. Tr. Prof. Zabol.,* **15:**12, 22–25 (1971).
39. T. G. Yakubovich and N. M. Zhukova *Gig. Sanit.,* **1970:**12, 98–100 (1970).
40. D. V. Sturges, D. W. Badger, R. N. Slarve, and D. E. Wasserman, "Laboratory Studies on Chronic Effects of Vibration Exposure," in: *Vibration and Combined Stresses in Advanced Systems,* H. E. von Gierke, Ed., AGARD Conference Proceedings AGARD-CP-145, Paper B10, NATO/AGARD, Neuilly-sur-Seine, France, March 1975. (See also D. W. Badger *et al., loc. cit.,* Paper B11.)
41. R. G. Edwards, E. P. McCutcheon, and C. F. Knapp, *J. Appl. Physiol.,* **32,** 384–390 (1972).
42. J. Demange, R. Auffret, and B. Vettes, "Effects of Low-Frequency Vibrations on the Human Cardiovascular System," in: *AGARD Conference Proceedings* AGARD-CP-145 (1975).
43. E. P. McCutcheon, "Effects of Vibration Stress on the Cardiovascular System of Animals," in: *Vibration and Combined Stresses in Advanced Systems,* H. E. von Gierke, Ed., Conference Proceedings AGARD-CP-145, Paper B9, NATO/AGARD, Neuilly-sur-Seine, France, 1975.
44. A. K. McIntyre, *Proc. Aust. Assoc. Neurol.,* **3,** 71–75, 1965.
45. T. J. Moore, *IEEE Trans. Man-Mach. Syst.* **MMS-11,** 79–84 (1970).
46. R. W. Shoenberger and C. S. Harris, *Hum. Factors,* **13,** 41–50 (1971).

47. T. Miwa, *Ind. Health (Jap.)*, **6**, 1–10, 11–17, and 18–27 (1968).
48. T. Miwa, *Ind. Health (Jap.)*, **7**, 89–115 and 116–126 (1969).
49. T. Miwa, *Ind. Health (Jap.)*, **8**, 116–126 (1970).
50. J. R. McKay *A Study of the Effects of Whole-Body $\pm a_z$ Vibration on Postural Sway*, 6750th Aerospace Medical Research Laboratory Technical Report AMRL-TR-71-121, Wright-Patterson Air Force Base, Ohio, 1972.
51. A. A. Menshov, D. V. Vinogradov, and A. M. Baron, *Gig. Sanit.*, **35**, 32–36 (1970).
52. S. K. Nurbaev, *Gig. Tr. Prof. Zabol.*, **19**:5, 32–35 (1975).
53. W. L. Johnston, "An Investigation of the Effects of Low-Frequency Vibration on Whole Body Orientation," Doctoral Dissertation (Order No. 70-1474) Texas Technical College, 1969.
54. A. N. Nicholson, L. E. Hill, R. G. Borland, and H. M. Ferres, *Aerosp. Med.*, **41**, 436–446 (1970).
55. B. Craske, *Science*, **196**, 71–73 (1977).
56. K.-E. Hagbarth and G. Eklund, *Scand. J. Rehabil. Med.*, **1**, 26–34 (1969).
57. P. B. C. Matthews, "Vibration and the Stretch Reflex," in: *Myotatic, Kinesthetic and Vestibular Mechanisms*, A. V. S. de Rueck, Ed., Section I: Myostatic and Kinesthetic Mechanisms, Little, Brown, Boston, 1967.
58. D. Burke, C. J. Andrews, and J. W. Lance, *J. Neurol. Neurosurg. Psychiatr.*, **35**, 477–486 (1972).
59. K.-E. Hagbarth and G. Eklund, "Motor Effects of Vibratory Muscle Stimuli in Man," in: *Muscular Afferents and Motor Control*, R. Granit, Ed., Almqvist and Wiksell, Stockholm; Wiley, New York, 1966, pp. 177–186.
60. G. Eklund, *Acta Soc. Med. Upsalien*, **74**, 113–117 (1969).
61. C. D. Marsden, J. C. Meadows, and H. J. F. Hodgson, *Brain*, **92**, 829–846 (1969).
62. J. C. Guignard and P. R. Travers, *Effect of Vibration of the Head and of the Whole Body on the Electromyographic Activity of Postural Muscles in Man*, Air Ministry Flying Personnel Research Committee Memo FPRC/Memo 120, London 1959.
63. L. D. Luk'yanova and Ye. P. Kazanskaya, *Fiziol. Zh.*, **53**, 563–570 (1967).
64. R. Aston and V. L. Roberts, *Arch. Int. Pharmacodyn.*, **155**:2, 289–299 (1965).
65. A. N. Nicholson and J. C. Guignard, *EEG Clin. Neurophysiol.*, **20**, 494–505 (1966).
66. G. J. Hurt, *Rough-Air Effect on Crew Performance during a Simulated Low-Altitude High-Speed Surveillance Mission*, National Aeronautics and Space Administration Technical Note D-1924, Washington, D.C. April 1963.
67. B. Caiger, *Some Problems in Control Arising from Operational Experiences with Jet Transports*, National Aeronautics Establishment Report NAE MISC 41, Ottawa, August 1966.
68. P. Lecomte, Chairman, and Contributors, *Simulation*, AGARD Conference Proceedings AGARD-CP-79-70, NATO/AGARD, Neuilly-sur-Seine, France, 1970.
69. C. W. Nixon and H. C. Sommer, *Influence of Selected Vibrations upon Speech (Range of 2 cps–20 cps and Random)*, 6570th Aerospace Medical Research Laboratory Technical Documentary Report AMRL-TDR-63-49, Wright-Patterson Air Force Base, Ohio, 1963.
70. C. W. Nixon and H. C. Sommer, *Aerosp. Med.*, **34**, 1012–1017, 1963.
71. I. Johnson, *A Method for the Measurement of Energy Absorption in the Human Hand Using Various Types of Hand Tools*, FFA Memo 97, Aeronautics Research Institute of Sweden, Stockholm, 1975.
72. R. N. Slarve and D. L. Johnson, *Aviat. Space Environ. Med.*, **46**, 428–431, (1975).
73. R. W. B. Stephens, *Rev. Acust.*, **2**, 48–55 (1971).
74. R. A. Hood and J. Leventhall, *Acustica*, **21**, 10–13 (1971).
75. R.-G. Busnel and A.-G. Lehmann, *J. Acoust. Soc. Am.* **63**, 974–977 (1978).

76. J. A. Roman, *Effects of Severe Whole Body Vibration on Mice and Methods of Protection from Vibration Injury,* Technical Report WADC-TR 58-107, U.S. Air Force Wright Air Development Center, Ohio, 1958.
77. E. Pichard, *Rev. Corps Sant Armees,* **1970**:10, 611–635 (1970).
78. R. Snook, *Br. Med. J.,* **1972**:iii, 574–578 (1972).
79. R. Rosegger and S. Rosegger, *J. Agric. Eng. Res.,* **5**, 241–275 (1960).
80. G. J. Gruber, *Relationships between Whole Body Vibration and Morbidity Patterns among Interstate Truck Drivers,* U.S. Department of Health Education and Welfare (NIOSH) Publication 77-167, National Institute of Occupational Safety and Health, Cincinnati, Ohio, 1976.
81. R. C. Spear, C. A. Keller, V. Behrens, M. Hudes, and D. Tarter, *Morbidity Patterns among Heavy Equipment Operators Exposed to Whole-Body Vibration-1975: Followup to a 1974 Study,* U.S. Department of Health Education and Welfare (NIOSH) Publication 177-20, National Institute of Occupational Safety and Health, Cincinnati, Ohio, November 1976.
82. R. C. Spear, C. A. Keller, and T. H. Milby, *Arch. Environ. Health,* **22**, 141–145 (1976).
83. H. Seris and R. Auffret, Measurement of Low Frequency Vibrations in Big Helicopters and Their Transmission to the Pilot," Paper to the 22nd Annual Meeting of the AGARD Aerospace Medical Panel, Munich, September 1965. (Translated from the French by John F. Holman & Co., NASA Tech. Trans., NASA TT F-471, 1967.)
84. R. P. Delahaye, H. Seris, R. Auffret, R. Jolly, G. Gueffier, and P. J. Metges, *Rev. Med. Aeron. Spat.* **38**, 99–102 (1971).
85. G. J. Gruber and H. H. Ziperman, *Relationship between Whole-Body Vibration and Morbidity Patterns among Motor Coach Operators,* U.S. Department of Health, Education and Welfare (NIOSH) Publication 75-104, National Institute of Occupational Safety and Health, Cincinnati, Ohio, 1974.
86. G. I. Rumyantsev and D. A. Mikhelson, *Gig. Sanit.,* **1971**:9, 25–27 (1971).
87. V. M. Gornik, *Gig. Tr. Prof. Zabol.,* **20**:4, 24–28 (1976).
88. V. I. Chernyuk, *Gig. Tr. Prof. Zabol.,* **19**:6, 19–23 (1975).
89. G. Eklund, *Upsala J. Med. Sci.,* **77**, 112–124 (1972).
90. V. I. Dynik, *Gig. Tr. Prof. Zabol.,* **19**:9, 32–36 (1975).
91. J. C. Guignard, G. J. Landrum, and R. E. Reardon, *Experimental Evaluation of International Standard (ISO 2631-1974) for Whole-Body Vibration Exposures. Final Report to the National Institute of Occupational Safety and Health,* University of Dayton Research Institute Technical Report UDRI-TR-76-79, Dayton, Ohio, 1976.
92. A. F. Parlyuk, *Vrach. Delo,* **1970**:7, 122–126 (1970).
93. G. G. Lysina and A. F. Parlyuk, *Vrach. Delo,* **1**, 124–128 (1973).
94. L. N. Gratianskaya, *Gig. Tr. Prof. Zabol.,* **18**:8, 7–10 (1974).
95. C. J. Gauss, *Zbl. Gynäkol.,* **50**, 13ff, 1926 (cited in Reference 2).
96. E. Ts. Andreeva-Galanina, S. V. Alekseyev, A. V. Kadyskin, and G. A. Suvorov, *Noise and Noise Sickness,* Meditsina Press, Leningrad, 1972. (National Aeronautics and Space Administration Technical Translation NASA TT F-748, Washington, D.C., July 1973.)
97. A. Begor, *Transp. Med. Vest.,* **14**:2, 22–29, 1969.
98. H. C. Sommer, *The Combined Effects of Vibration, Noise, and Exposure Duration on Auditory Temporary Threshold Shift,* 6570th Aerospace Medical Research Laboratory Technical Report AMRL-TR-73-34, Wright-Patterson Air Force Base, Ohio, 1973.
99. J. C. Guignard, *Combined Effects of Noise and Vibration on Man,* University of Dayton Research Institute Technical Report UDRI-TR-73-51, 1973.
100. C. Anitesco and C. Contulesco, *Arch. Mal. Prof.,* **33**, 365–371 (1972).

101. A. B. Lewis, *Noise Control Eng.,* **7,** 132-139 (1976).
102. S. L. Gibbons, A. B. Lewis, and P. Lord, *J. Sound Vib.,* **43,** 253-261 (1975).
103. N. N. Livshits, Ed, *Effects of Ionizing Radiation and of Dynamic Factors on the Functions of the Central Nervous System—Problems of Space Physiology,* National Aeronautics and Space Administration Technical Translation NASA TT-F-354 from the Russian, Washington, D.C., 1965 Nauka, Moscow, 1964.
104. T. S. L'vova, "The Influence of Vibration on the Course and Outcome of Radiation Injury in Animals, in: *Problems in Aerospace Medicine,* V. V. Parin, Ed., Moscow, pp. 347-348, 1966. (National Aeronautics and Space Administration translation from the Russian, JPRS-38272, Washington, D.C.)
105. L. Ya. Tartakovskaya, *Gig. Tr. Prof. Zabol.,* **20:**4, 32-36 (1976).
106. N. N. Malinskaya, *Gig. Tr. Prof. Zabol,* **19:**7, 16-20, (1975).
107. S. Jaworski, *Pr. Cent. Inst. Ochr. Pr.* **25,** 325-341. 1975.
108. F. P. Dimasi, R. E. Allen, and P. C. Calcaterra, *Effect of Vertical Active Vibration Isolation on Tracking Performance and Ride Qualities,* NASA Contractor Report CR 2146, National Aeronautics and Space Administration, Washington, D.C. November 1972.
109. L. F. Stikeleather and C. W. Suggs, *Trans. Am. Soc. Agric. Eng.,* **13,** 99-106. 1970.
110. R. E. Young and C. W. Suggs, *An Active Seat Suspension System for Isolation of Roll and Pitch in Off-Road Vehicles,* American Society of Agricultural Engineers Paper 73-156, American Society of Agricultural Engineers, St. Joseph, Michigan, 1973.
111. L. Sjøflot and C. W. Suggs, *Ergonomics,* **16,** 455-468 (1973).
112. D. Dieckmann, *Ergonomics,* **11,** 347-355 (1958).
113. I. Pyykkö, "Vibration Syndrome: A Review," in Reference 117, pp. 1-24.
114. W. Taylor Ed., *The Vibration Syndrome,* British Acoustical Society Special Volume 2, Academic Press, London, New York, 1974.
115. W. Taylor and P. L. Pelmear, *Vibration White Finger in Industry,* Academic Press, London, New York, 1975.
116. D. E. Wasserman, W. Taylor, and M. G. Curry, Eds., *Proceedings of the International Hand-Arm Vibration Conference,* Cincinnati, Ohio, October 1975, U.S. Department of Health, Education, and Welfare, (NIOSH) Publication 77-170, National Institute of Occupational Safety and Health, Cincinnati, Ohio. April 1977.
117. O. Korhonen, Ed., *Vibration and Work, Proceedings of the Finnish-Soviet-Scandinavian Vibration Symposium,* Helsinki, March 10-13, 1975, Helsinki, Institute for Occupational Health, 1976.
118. Department of Health and Social Security (United Kingdom), *Vibration Syndrome,* Report to Parliament by the Industrial Injuries Advisory Council, Cmnd 4430, Her Majesty's Stationery Office, London, July 1970.
119. W. Taylor, "Opening Remarks," in Reference 116, pp. 3-4.
120. D. D. Reynolds and W. Soedel, *J. Sound Vib.,* **44,** 513-523 (1976).
121. P. M. Allingham and R. D. Firth, *N. Z. Med J.,* **76,** 317-321 (1972).
122. J. N. Agate and H. A. Druett, *Br. J. Ind. Med.,* **3,** 159-166 (1946).
123. J. N. Agate and H. A. Druett, *Br. J. Ind. Med.,* **4,** 141-163 (1947).
124. S.-Å. Axelsson, "Analysis of Vibration in Power Saws," *Stud. For. Suec.,* **59,** 1-47 (1968).
125. J. Hyvärinen, I. Pyykkö, and S. Sundberg, *Lancet,* **i,** 791-794, 1973.
126. D. L. Johnson, "Auditory and Physiological Effects of Infrasound," *Proceedings Inter-Noise 75,* International Conference on Noise Control Engineering, Sendai, Japan, August 27-29 1975, Institute of Noise Control Engineering, September 1975.

127. D. D. Reynolds, "Hand-Arm Vibration: A Review of 3 Years' Research," in Reference 116, pp. 99-128.
128. H. Rafalski, K. Bernacki, and T. Switoniak, "The Diagnostics and Epidemiology of Vibration Disease and Hearing Impairment in Motor Sawyers, in Reference 116 pp. 84-88.
129. E. E. Dart, *Occup. Med.*, **1**, 515-550 (1946).
130. P. A. Banister and F. V. Smith, *Br. J. Ind. Med.*, **29**, 264-267 (1972).
131. H. Iwata, *Ind. Health (Jap.)*, **6**, 28-36 and 47-58 (1968).
132. W. F. Ashe, *Physiological and Pathological Effects of Mechanical Vibration on Animals and Man.* Ohio State University Research Foundation Report 862-4, Columbus, Ohio, 1961.
133. W. F., Ashe, W. T. Cook, and J. W. Old, *Arch. Environ. Health*, **5**, 333-343 (1962).
134. A. Okada, T. Yamashita, C. Nagano, T. Ikeda, A. Yachi, and S. Shibata, *Br. J. Ind. Med.*, **28**, 353-357 (1971).
135. Futatsuka, M. *Jap. J. Ind. Health*, **15**, 371-377 (1973).
136. Futatsuka, M. *Jap. J. Hyg.*, **30**, 266 (1975).
137. R. P. Jepson, *Br. J. Ind. Med.*, **11**, 180-185 (1954).
138. A. Okada, T. Yamashita, and T. Ikeda, *Am. Ind. Hyg. Assoc J.*, **33**, 476-482 (1972).
139. M. Sevčik, *Pracov Lek.*, **25**, 46-50 (1973).
140. Drobny, M. *Pracov Lek.*, **25**, 185-189 (1973).
141. E. Lukas and V. Kuzel, *Int. Arch. Arbeitsmed.*, **28**, 239-249 (1971).
142. N. Bjerker, B. Kylin, and I.-M. Lidström, *Ergonomics*, **15**, 399-406 (1972).
143. F. Horváth and T. Kákosy, *Zeitschr. Orthop. Grenzgeb.*, **107**, 482-494 (1970).
144. D. Koradecka, "Peripheral Blood Circulation under the Influence of Occupational Exposure to Hand-Transmitted Vibration," in Reference 116 pp. 21-36.
145. T. Banaszkiewicz, J. Gwozdziewicz, and J. Waskiewicz, *Biul. Inst. Med. Morsk. Gdansk.*, **21**, 147-162 (1970).
146. A. G. Mirzaeva, *Med. Zh. Uzb.*, **1971**:5, 26-29 (1971).
147. H. Langauer-Lewowicka, *Int. Arch. Occup. Environ. Health*, **36**, 206-216 (1976).
148. R. M. Nerem, *Arch. Environ. Health*, **26**, 105-110 (1973).
149. R. M. Nerem, "Vibration Enhancement of Blood-Arterial Wall Macromolecular Transport, in Reference 116 pp. 37-43.
150. G. Eklund and K.-E. Hagbarth, *Exp. Neurol.*, **16**, 80-92 (1966).
151. G. Jansen, *Arch. Gewerbepath. Gewerbehyg.*, **17**, 238-261 (1959).
152. A. M. Stewart, *Br. J. Occup. Safety*, **7**, 454-455 (1968).
153. A. M. Stewart and D. F. Goda, *Br. J. Ind. Med.*, **27**, 19-27 (1970).
154. M. A. Yershova, *Vrach. Delo*, **1970**:7, 116-119 (1970).
155. S. Carlsöö and J. Mayr, *Work-Environ.-Health*, **11**, 32-38 (1974).
156. H. Dupuis, E. Hartung, and W. Hammer, *Int. Archiv. Arbeits. Umweltmed.*, **37**, 9-34 (1976).
157. H. Iwata, H. Dupuis, and E. Hartung, *Int. Arch. Arbeitsmed.*, **30**, 313-328 (1972).
158. D. D. Reynolds and W. Soedel, *J. Sound Vib.*, **21**, 339-353 (1972).
159. C. W. Suggs and J. W. Mishoe, "Hand-Arm Vibration: Implications Drawn From Lumped Parameter Models," in Reference 116, pp. 136-141.
160. L. A. Wood and C. W. Suggs, "A Distributed Parameter Dynamic Model of the Human Forearm," in Reference 116, pp. 142-145.
161. I.-M. Lidström, "Vibration Injury among Rock Drillers, Chiselers, and Grinders," in Reference 117, pp. 81-88.

162. I.-M. Lidström, "Vibration Injury in Rock Drillers, Chiselers, and Grinders," in Reference 116, pp. 77–83.
163. E. Lukáš and L. Louda, *Cesk. Neurol. Neurochir.*, **37**, 258–262 (1974).
164. E. R. Tichauer, "Thermography in the Diagnosis of Work Stress due to Vibrating Implements," in Reference 116, pp. 160–168; also a comment on p. 208.
165. H. Gage, "Correlation of Segmental Vibration with Occupational Disease," Reference 116, pp. 239–243.
116. D. Koradecka, *Pr. Cent. Inst. Ochr. Pr.*, **20**, 147–160 (1970).
167. A. J. Zweifler, "Detection of Occlusive Arterial Disease in the Hand and its relevance to Occupational Hand Disease," in Reference 116, pp. 12–20.
168. B. Hellstrøm, *Int. Z. Angew. Physiol. Einschl. Arbeitsphysiol.*, **29**, 18–28 (1970).
169. M. Tiilila, *Work-Environ.-Health*, **7**, 85–87 (1970).
170. A. F. Buzalo, *Oftal'mol. Zh.*, **25**, 434–437 (1970).
171. A. M. Seppäläinen, "Neurophysiological Detection of Vibration Syndrome in the Shipbuilding Industry," in Reference 117, pp. 67–71.
172. W. Taylor, J. C. G. Pearson, and G. D. Keighley, "A Longitudinal Study of Raynaud's Phenomenon in Chain Saw Operators," in Reference 116, pp. 69–76.
173. T. M. Radziukevich and A. M. Mikulinskii, *Gig. Tr. Prof. Zabol*, **16**:7, 16–20 (1972).
174. T. Kumlin, *Work-Environ.-Health*, **7**, 57–58 (1970).
175. D. B. Chaffin, *J. Occup. Med.*, **11**, 109–115 (1969).
176. E. R. Tichauer, *Occupational Biomechanics*, New York University Rehabilitation Monograph 51, New York, 1975.
177. "A Czechoslovakian Proposal for a Guide to the Evaluation of Human Exposure of the Hand-Arm System to Vibration," *Work-Environ.-Health*, **7**, 88–90 (1970).
178. *Portable Powered Tools—Permitted Levels of Vibration*, USSR Standard (GOST) 17770-72, July 1, 1972.
179. British Standards Institution, *Guide to the Evaluation of Exposure of the Human Hand-Arm System to Vibration*, Draft for Development DD 43, BSI, London, 1975.
180. J. Starck, "A Finnish Recommendation for Maximum Vibration Levels," in Reference 117, (1976), pp. 117–120.
181. T. Miwa, *Ind. Health. (Jap)*, **5**, 182–205, 206–212 and (Hand-Transmitted Vibration) 213–220 (1967).
182. S. A. Axelsson, "Progress in Solving the Problem of Hand-Arm Vibration for Chain Saw Operators in Sweden, 1967 to Date," in Reference 116, (1977), pp. 218–224.
183. J. R. Bailey, "Chain Saws—Problems Associated with Vibration Measurements," in Reference 116, pp. 187–208.
184. G. D. Keighley, FAO/ECE/ILO Resolution on hand-arm vibration for modern antivibration chain saw: recommendation for medical monitoring. in Reference 116, pp. 233–235.
185. U. Naeslund, "The Theory Behind a new measurement for hand-arm vibration," in Reference 116, pp. 225–229.
186. A. P. Politschuk and V. N. Oblivin, "Methods of Reducing the Effects of Noise and Vibration on Power Saw Operators," in Reference 116, pp. 230–232.
187. S. Yamawaki, "Reduction of Vibration in Power Saws in Japan," in Reference 116, pp. 209–217.
188. W. Taylor, J. Pearson, R. L. Kell, and G. D. Keighley, *Br. J. Ind. Med.*, **28**, 83–89 (1971).
189. Z. M. Butkovskaya, *Gig. Tr. Prof. Zabol.*, **16**: 3, 19–24 (1972).
190. P. Lyarskiy, *Sanitary Standards and Regulations to Restrict the Vibration of Work-Places*, Standard

627-66 drawn up by F F Erisman Research Institute of Hygiene, Moscow, U.S.S.R. 1966. (Translated from the Russian by W. Linnard, Forestry Commission, U.K.)
191. L. Louda and E. Lukas, "Hygienic Aspects of Occupational Hand-Arm Vibration," in Reference 116, pp. 60–66.
192. A. P. Guk, *Bezop. Tr. Prom-St.*, **1970**:2, 16–17 (1970).
193. A. A. Demin, A. Ia. Khrupina, and G. P. Vasilenko, *Gig. Tr. Prof. Zabol.*, **15**:10, 16–20 (1971).
194. A. M. Ehrly, "Treatment of Severe Secondary Raynaud's Disease," (sic.) in Reference 116, pp. 44–46.

CHAPTER FOURTEEN

Evaluation of Exposure to Abnormal Pressure

JEFFERSON C. DAVIS, M.D.

1 INTRODUCTION

The abnormal pressures to be considered are decreased barometric pressure at altitude in the atmosphere and in space flight and barometric pressure greater than sea level in undersea work, exploration, and recreation. An emerging use of pressure greater than sea level is the use of high dose, short-term oxygen inhalation therapy (hyperbaric oxygen) in treatment of a limited group of medical disorders. Once the realm of a limited group of flight surgeons, diving medical officers, divers, and tunnel and bridge building engineers, altered barometric pressure today affects the lives of workers, passengers in aircraft, and recreational scuba divers and aviators. Excursions of sea level dwellers to mountainous regions for business or recreation is another important exposure.

Medical uses of hyperbaric oxygen have introduced a new dimension to the need for full engineering safety aspects related to exposures of patients and medical staff to altered barometric pressure as well as associated gas purity standards and fire and structural safety concerns. Because Chapter 9, Volume I, presented the biomedical factors of undersea work so completely, this chapter concentrates on the altitude environment and medical hyperbaric chambers in more detail. The central theme is represented by the interrelated physiological effects of the continuum of altered barometric pressure—higher or lower than sea level.

A recent unusual industrial exposure to low barometric pressure serves to demonstrate the variety of considerations involved. A metal processing plant used 6 × 6 × 6 ft unmanned vacuum chambers for a part of its procedures. For reasons not clearly defined, a man was inside one of the chambers when accidental activation of the vacuum source decompressed him to 40 mm Hg (equivalent to 20,117 m or 66,000 ft). About 3 to 5 min was required to return him to ground level, where he was unconscious. At the emergency room he showed X-ray evidence of lung damage and clinical signs of severe brain injury. Subsequent hyperbaric oxygen therapy and intensive pulmonary care resulted in full recovery and return to normal status.

The importance for this chapter is that this case occurred in a most unexpected place—a factory—and the spectrum of pathophysiology covered the entire range of severe effects of altered barometric pressure: Boyle's law mechanical effects on the lungs, severe hypoxia, acute decompression sickness, and "ebullism" (1) or boiling of body fluids at body temperature as the critical Armstrong line, 19,200 m (63,000 ft), is exceeded.

1.1 Physics of Abnormal Environmental Pressure

The weight of atmospheric air and water at any point on or above the surface of the earth or beneath the surface of the water may be expressed in various units of pressure. The major difference between pressure changes at altitude and at depth is the curvilinear change of altitude compared to the linear pressure changes at depth. On ascent to altitude, we pass through the blanket of air that extends from sea level to an altitude of about 430 miles, where the number of molecules is so small that no collisions occur. At sea level the atmosphere weighs 14.7 pounds per square inch (psi) or 760 mm Hg (760 torr). Because pressure at any altitude is simply the weight of air above, pressure changes are greater at low altitudes. For example, with ascent from sea level to 3048 m (10,000 ft), a pressure reduction of 237 mm Hg from 760 to 523 mm Hg is seen, whereas the same 3048 m altitude change from 9146 m (30,000 ft) to 12,192 m (40,000 ft) gives a pressure reduction of only 85 mm Hg from 226 to 141 mm Hg. At depth, pressure change is linear because each foot of seawater weighs 0.445 psi; thus 14.7 ÷ 0.445 = 33 ft of seawater or about 10 m.

The barometric pressure (P_B) at depth can be referred to in terms of gauge pressure (that in excess of sea level pressure) or more commonly, as absolute pressure (the total of sea level atmospheric pressure plus the water pressure). The most commonly used pressure units are millimeters of mercury (mm Hg), feet of seawater (FSW), meters of seawater, or atmospheres absolute (ATA). An example of the linear pressure changes at depth is that at 20 m (66 FSW), the total pressure is 3 ATA or 2280 mm Hg and at 30 m (99 FSW), the total pressure is 4 ATA or 3040 mm Hg. Barometric pressure at any altitude is available from standard altitude charts and pressure at any depth can be easily derived.

2 ALTITUDE PHYSIOLOGY

Air at altitude maintains a relatively constant percentage of gases in the mixture with oxygen about 21 percent and nitrogen about 79 percent. Traces of carbon dioxide, water vapor, and other inert gases are so small that they are included with the nitrogen percentage for calculation. Dalton's law states that in a mixture of gases, the total pressure is the sum of partial pressures of gases in the mixture. For example, in air at sea level: partial pressure of oxygen (P_{O_2}) equals 0.21 × 760 mm Hg or 160 mm Hg and partial pressure of nitrogen (P_{N_2}) equals 0.79 × 760 mm Hg or 600 mm Hg. The key point is that at altitude, gas percentages remain constant but partial pressures of gases change. For example, at 18,000 ft where barometric pressure is 380 mm Hg or one-half an atmosphere, P_{O_2} = 0.21 × 380 mm Hg or 80 mm Hg and P_{N_2} = 300 mm Hg.

2.1 Changes in Inspired Gas Composition

At normal body temperature the vapor pressure of water is 47 mm Hg; thus by the time inspired dry air reaches the trachea, it has equilibrated at a P_{H_2O} of 47 mm Hg, which then remains constant in the alveolar air. Venous blood returns carbon dioxide from the body's normal metabolism to the pulmonary circulation at P_{CO_2} of 47 mm Hg. Admixture of carbon dioxide diffused across the alveolar-capillary membrane into the alveolus results in an alveolar P_{CO_2} of about 40 mm Hg in the normal person. Thus to calculate the approximate alveolar partial pressure of oxygen (P_{AO_2}) at sea level or at any altitude, the simplified alveolar gas equation may be used if we assume the respiratory quotient (RQ) to be 1.

$$RQ = \frac{CO_2 \text{ output (ml/min)}}{O_2 \text{ consumption (ml/min)}}$$

The simplified alveolar gas equation is

$$P_{AO_2} = F_{IO_2}(P_B - P_{AH_2O}) - P_{ACO_2}$$

where P_{AO_2} = alveolar P_{O_2}
F_{IO_2} = fraction of inspired O_2
P_B = total barometric pressure
P_{AH_2O} = alveolar P_{H_2O}
P_{ACO_2} = alveolar P_{CO_2}

For example:

- At sea level breathing air

$$P_{AO_2} = 0.21(760 - 47) - 40$$
$$P_{AO_2} = 109 \text{ mm Hg}$$

- Breathing air at 18,000 ft

$$P_{AO_2} = 0.21(380 - 47) - 40$$
$$P_{AO_2} = 30 \text{ mm Hg}$$

The same equation can be used to determine the F_{IO_2} needed to remain above hypoxic levels at altitude.

For example:

- Breathing 100 percent O_2 at 18,000 ft
$$P_{AO_2} = 1.0\,(380 - 47) - 40$$
$$P_{AO_2} = 293 \text{ mm Hg}$$

- Nitrogen is inert and maintains equilibrium between alveolar, blood, and tissue P_{N_2}. At sea level, for example:
$$\text{Alveolar } P_{N_2} = 760 - (P_{AO_2} + P_{ACO_2} + P_{AH_2O})$$
$$P_{N_2} = 760 - (109 + 40 + 47)$$
$$P_{N_2} = 564 \text{ mm Hg}$$

2.2 Diffusion

The transfer of gases at the alveolar-capillary membrane is by the process of diffusion and can be represented by Figure 14.1, a schematic model alveolus and capillary.

2.3 Oxygen Transport from Lungs to Tissue

The unique properties of hemoglobin and the way it combines with and releases oxygen constitute a key element in altitude physiology. Hemoglobin concentration in the blood

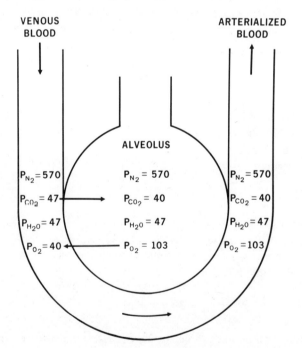

Figure 14.1 Model alveolus and pulmonary capillary.

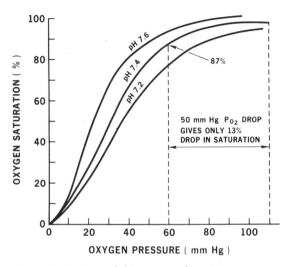

Figure 14.2 Hemoglobin-oxygen dissociation curve.

is expressed as grams-percent or grams of hemoglobin (Hb) per 100 ml of blood (2). An average, healthy adult male has about 15 g of hemoglobin per 100 ml of blood or 15 g-%. Each gram of hemoglobin can combine with 1.34 ml of oxygen. Thus a person with 15 g-% of hemoglobin can carry $1.34 \times 15 = 20.1$ ml of oxygen per 100 ml of blood or 20.1 volumes-percent.

Hemoglobin uptake and release of oxygen is described by the hemoglobin-oxygen dissociation curve (Figure 14.2) whose features are of great significance in altitude physiology (3).

The flat portion of the curve between 60 and 110 mm Hg allows a significant drop in alveolar P_{O_2} before there is a physiologically important drop in blood oxygen saturation. It is this characteristic of the curve that allows ascent to 3048 m (10,000 ft) altitude breathing air, where alveolar P_{O_2} is 60 mm Hg, without symptoms of hypoxia. Further ascent above 3048 m (10,000 ft) results in a rapid drop in arterial oxygen saturation as small drops in oxygen pressure result in large decreases in arterial saturation along the steep portion of the curve.

The curve also shows the profound effect of acidosis or alkalosis as the shape of the curve is altered by changes in pH. Thus, for example, hyperventilation, a normal compensatory mechanism to hypoxia, results in a lowering of P_{CO_2} and a resultant rise in pH. The shift of the curve to the left illustrates the body's attempt to increase efficiency of oxygen uptake at the lung level and release of oxygen to the tissues at higher oxygen pressures. On the other hand, the elevated P_{CO_2} and acidosis seen in exercise at altitude aggravates hypoxia by shifting the curve to the right.

Figure 14.3 is a simplification of the Krough-Erlang tissue oxygen model (4). The cylinder of tissue supplied by a given capillary is seen as a spectrum of oxygen tensions regulated by the arterial oxygen partial pressure supplied, distance from the capillary,

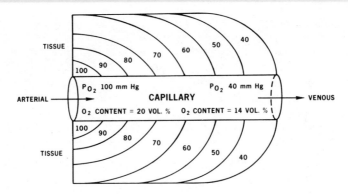

Figure 14.3 Capillary-tissue oxygen model.

and location along the capillary as oxygen is extracted during flow of blood toward the venous end.

3 ACUTE ADAPTATION TO ALTITUDE EXPOSURE

Upon rapid exposure to altitude, whether in an unpressurized aircraft or balloon or because of accidental decompression of a pressurized aircraft, decompression in an altitude chamber, or rapid ascent of a mountain, a sequence of physiological mechanisms attempts to compensate.

With decrease in arterial P_{O_2} there is hyperventilation caused by hypoxic stimulus to the chemoreceptors, the carotid and aortic bodies (5). Hyperventilation gives a decrease in alveolar and arterial P_{CO_2}, tissue alkalosis, and the favorable shift to the left of the oxygen-hemoglobin curve. There is also unfavorable cerebral vascular vasoconstriction as arterial P_{CO_2} drops, since carbon dioxide is a potent cerebral vasodilator (6). This effect is overridden by the hypoxia-induced vasodilation that occurs when venous P_{O_2} drops to about 30 mm Hg (6).

Also, with decreased P_{AO_2}, the carotid and aortic bodies mediate bradycardia, vasoconstriction in the extremities, systemic hypertension, and increased pulmonary vascular resistance. The net effect is hyperventilation, pulmonary hypertension, increased heart rate, and beneficial redistribution of blood from extremities to the brain and heart.

4 HYPOXIA

Without supplemental oxygen, the physiological effects of the reduced atmospheric partial pressure of oxygen (P_{O_2}) at altitude are described as hypoxic hypoxia. Other factors that lead to tissue hypoxia can coexist and be additive. Thus a patient whose

EVALUATION OF EXPOSURE TO ABNORMAL PRESSURE

oxygen transport capacity is impaired by anemia or carbon monoxide bound hemoglobin ("hypemic hypoxia") or whose tissue oxygen utilization is impaired by cyanide or ethanol ("histotoxic hypoxia") will be more susceptible to hypoxia at altitude. A special potentiating factor in aviation is "stagnant hypoxia" in an aviator exposed to high head-to-foot acceleration forces as in recovering from an aircraft dive (7). The increased weight of the column of blood reduces cerebral perfusion with resultant cerebral hypoxia. Although these factors can be important contributors to hypoxia in aviation, this discussion centers on hypoxia caused by reduction in P_{O_2} at the decreased barometric pressures at altitude.

As previously discussed, the shape of the oxygen-hemoglobin dissociation curve is beneficial up to 3048 m (10,000 ft). Except for a decrement in night vision at about 1220 to 1829 m (4000 to 6000 ft) and above, there are no significant effects of hypoxia in healthy people below 3048 m (10,000 ft). In general, pressurized aircraft cabins are maintained well below that equivalent air pressure, and no supplemental oxygen is required. However patients with reduced cardiac reserve may have difficulty tolerating even this small drop in arterial oxygen saturation and will require supplemental oxygen during flight.

4.1 Manifestations

Table 14.1 summarizes the major manifestations of altitude hypoxia with corresponding P_{AO_2} breathing ambient air.

Table 14.1 Responses to Hypoxic Hypoxia

Altitude		Alveolar P_{O_2} (mm Hg)
3,048 m (10,000 ft)	Impaired judgment and ability to perform calculations; increased heart rate and respiratory rate	60
3,658 m (12,000 ft)	Shortness of breath, impaired ability to perform complex tasks, headache, nausea, decreased visual acuity	52
4,573 m (15,000 ft)	Decrease in auditory acuity, constriction of visual fields, impaired judgment, irritability; exercise can lead to unconsciousness	46
5,486 m (18,000 ft)	Threshold for loss of consciousness in resting unacclimatized individuals after several hours exposure	40
6,706 m (22,000 ft)	Almost all individuals unconscious after sufficient exposure time	30
12,802 m (42,000 ft)	Inability to perform useful function in 15 sec or less	—

The time required for the onset of symptoms in Table 14.1 varies depending on age, fitness, acclimatization, and individual susceptibility. At higher altitudes the progress of symptoms to the point of inability to perform useful functions (time of useful consciousness, or TUC) is quite rapid and within a narrower range of individual variation. For example, the TUC at 9146 m (30,000 ft) is about 90 sec and at 13,106 m (43,000 ft) is about 15 sec or less.

4.2 Prevention

The use of pressurized cabins to stay within the physiologically comfortable zone below 3048 m (10,000 ft) is the main preventive factor in modern aviation. When aviators must be exposed to higher cabin altitude, the percentage of supplemental oxygen required to maintain an alveolar P_{O_2} no less than that at 3048 m (10,000 ft) breathing air can be derived by varying the F_{IO_2} in the simplified alveolar gas equation. For example, 100 percent oxygen is required at 12,192 m (40,000 ft) where the barometric pressure is 141 mm Hg and P_{ACO_2} is somewhat lower because of hyperventilation:

$$P_{AO_2} = 1.0\,(141 - 47) - 35$$
$$P_{AO_2} = 59 \text{ mm Hg}$$

This is equivalent to breathing air at 3048 m (10,000 ft) and is the maximum altitude at which 100 percent oxygen breathing is adequate to prevent hypoxia. At higher altitudes positive pressure breathing and, finally, pressure suits are required (8).

5 AIRCRAFT CABIN PRESSURIZATION

The most commonly employed method to avoid the risks of exposure of aircrews and passengers to the low barometric pressure and partial pressures of oxygen at altitude is to pressurize the aircraft cabin. There are two basic factors in cabin pressurization: (a) air compressors powered by the aircraft engines force outside air continuously into the cabin, whose structural integrity (b) is able to safely withstand differential pressures of 8 psi or more, with compensatory relief valves to continuously dump air overboard (9). A commonly used example in aviation is a differential pressure (cabin above ambient pressure) at flight level (FL) of 8.19 psi, so that at FL 12,000 m (40,000 ft; 2.72 psi) the addition of the 8.19 psi pressurization gives a total cabin pressure of 10.91 psia (pounds per square inch absolute) equivalent to 2400 m (8000 ft). From previous discussions, it is clear that 2400 m (8000 ft) is within the physiologically safe zone where no supplemental oxygen is required to maintain sufficient arterial oxygen saturation.

5.1 Limitations of the Pressurized Cabin

As the atmosphere density decreases at great altitude, a limitation is finally reached where it is no longer feasible to compress the ambient gas molecules efficiently to pressurize the aircraft cabin. This practical limitation is reached at altitudes of about 21,000 to 24,000 m (70,000 to 80,000 ft) where only very high aircraft velocity with sufficient ram pressure could achieve adequate aircraft pressurization (9). Even so, adiabatic compression of gas so rare results in very high cabin temperatures so that at FL above 24,000 m (80,000 ft), the sealed cabin with self-contained life support systems becomes necessary. Thus spacecraft carry onboard gas supplies to pressurize the craft to habitable cabin altitudes and gas composition.

5.2 Cabin Decompression

The remote but ever-present risk in aircraft flying at high altitude with cabins pressurized as just described is that of accidental decompression. This could occur in case of loss of aircraft structural integrity as in failure of a door or window. The severity of cabin decompression is dependent on several factors including volume of the cabin, area of the opening into the pressurized compartment, and cabin pressure–ambient pressure differential pressure. Hazards include physical harm to occupants by being sucked out of the opening or being struck by loose objects flying about the cabin because of the rush of pressurized air out of the cabin, hypoxia upon exposure to ambient pressure, and low temperature.

Excellent quality control in construction of modern aircraft has made this event quite rare, but provision of emergency oxygen supplies for crew and passengers, wearing of restraint belts or harnesses during flight, routine securing of loose objects, and training of flight crews in emergency procedures are necessary backup provisions.

6 ACCLIMATIZATION TO ALTITUDE AND MOUNTAIN SICKNESS

The acute physiological adaptations to altitude exposure presented earlier are ineffective in preventing serious symptoms of hypoxia upon acute exposure to altitude. Therefore the main protection lies in ensuring adequate oxygen equipment to provide supplemental oxygen.

Yet people born in mountainous regions of the world live and work for a lifetime at altitudes that would produce serious symptoms in the sea level dweller exposed to such altitudes. An understanding of practical aspects of altitude acclimatization as well as risks to workers and travelers who must do business in mountainous cities of the world is pertinent to this text.

6.1 Altitude Acclimatization

The goals of physiological adaptive processes can be summarized (10):

1. To facilitate acquisition of oxygen by venous blood entering the lungs.
2. To increase the capacity of blood to transport oxygen to tissues.
3. To promote oxygen utilization at the cellular level.

Table 14.2 Comparison of Mean Alveolar and Blood Oxygen Partial Pressures (mm Hg)

Parameter	Lima	Morococha
Alveolar P_{O_2}	96	46
Arterial P_{O_2}	87	45
Mixed venous P_{O_2}	42	35

Hurtado (11) did extensive studies on natural acclimatization among natives living at 4540 m (14,000 ft) in Peru. He compared 50 subjects living in Lima (150 m or 500 ft) to 40 subjects living in Morococha at 4540 m (14,000 ft).

He found that the Morococha residents had a mean resting ventilation (liters per minute) of 18 percent above the Lima subjects. Table 14.2 summarizes his findings of alveolar and blood oxygen partial pressures.

The increased efficiency of gas exchange in the lung among acclimatized mountain residents is indicated by the resting differential of alveolar to arterial P_{O_2}. In the Lima group this difference was 8 to 12 mm Hg, whereas in the Morococha acclimatized subjects, it was only about 1 mm Hg. The mechanism that accounts for this efficient diffusion of alveolar oxygen may be dilatation of pulmonary capillaries.

Table 14.2 indicates that Morocochan natives live on the steep slope of the oxygen-hemoglobin dissociation curve. Thus more oxygen is delivered to tissue with only a small change in oxygen tension. Table 14.3 summarizes hematologic factors in altitude acclimatized natives (11).

Acclimatization of near sea level dwellers to altitude never is as complete or efficient as it is for those who reside at altitude. The level of ability to perform work at altitude seen in natives does not occur in newcomers even after prolonged residence. One reason may be the higher number of capillaries in muscle observed in animals and natives of high altitude (11).

For newcomers to a high altitude region, the time required and best schedule for acclimatization is of importance. A summary of the literature on this subject can be stated:

1. There is individual variability in the ability for acclimatization.
2. Acclimatization starts at low altitudes and continues to about 18,000 ft.
3. The limiting altitude for sea level dwellers to achieve significant acclimatization is about 18,000 ft. Again, there is individual variability.
4. It is possible for mountaineers to achieve partial acclimatization up to 23,000 ft, but deterioration begins at about 20,000 ft.
5. One recommended schedule (10) for acclimatization is:

$$6,000\text{–}7,000 \text{ ft}$$
$$9,000\text{–}10,000 \text{ ft}$$
$$12,000\text{–}13,000 \text{ ft}$$

Table 14.3 Summary of Hematological Findings by
Hurtado Among Morococha Natives and Lima Natives
(mean of 41 to 250 subjects each observation)

Hematocrit (%)	46 (40–52)	59 (49–79.2)
Hemoglobin (g-%)	15.6	20
Reticulocytes (1000/mm^3)	17.9	45.5
Total blood volume (liters)	4.77	5.70
Plasma volume (liters)	2.52	2.23

Spend 10 days at each level with planned exercise and daily excursions to 2000 ft higher, but reside at the prescribed altitude for time indicated.

6. Drug therapy to improve altitude tolerance has been the subject of many studies. Ammonium chloride to increase blood acidity and offset respiratory alkalosis has been used (12), but there is no convincing evidence of its efficacy. Acetazolamide (Diamox®) is a carbonic anhydrase inhibitor that causes acidosis. Cain and Dunn (13) showed that the arterial P_{O_2} of dogs pretreated with Diamox before 21,000 ft exposure was 9 mm Hg higher than untreated controls. The mechanism for these changes is accelerated renal excretion of base and correction of respiratory alkalosis. In a human study (14) they showed that low doses of Diamox did increase tolerance to exposure to 16,000 ft.

7 MOUNTAIN SICKNESS

Acute exposure of near sea level, unacclimatized people to high altitude may produce a spectrum of serious clinical problems.

7.1 Acute Mountain Sickness

Soon after arrival at high altitude, the sea level dweller may note subjective difficult respiration, and increased rate and depth of breathing can be measured because of hypoxic stimulus to carotid and aortic body chemoreceptors (10). There may be Cheyne-Stokes respiration and persistent cough. The pulse rate is elevated. The subject may have decreased visual discrimination, impaired judgment, emotional lability, and even hallucinations. Headache, insomnia, giddiness, eyeball pain, visual blurring, and decreased night vision are noted. There may be loss of appetite, intolerance to fatty foods, nausea, vomiting, and dehydration. Patients with underlying disorders like hypertension or angina pectoris may have aggravation of their condition.

7.2 High Altitude Pulmonary Edema (HAPE)

HAPE is the most serious of the mountain sickness syndromes. It can occur in the first few days of residence at altitude if acclimatization has not been followed (10). It can occur in acclimatized persons who undertake very strenuous exercises at altitude, or even in natives who return to their mountain home after a sojourn at lower altitude. During their low altitude residence, about 4 to 8 weeks is required for full deacclimatization, and individuals experience a drop in hemoglobin and red blood cell count. Upon return to the mountain home, the hypertrophied heart muscle cannot receive adequate oxygenation because of the drop in hemoglobin.

The mechanism of HAPE is "leakiness" of pulmonary capillaries with resultant pulmonary edema. Several factors may contribute. The increased pulmonary artery pressure seen in acute hypoxia is coupled with hypoxic capillary damage and loss of integrity and depressed amount of surfactant, lowering osmotic pressure in alveoli with fluid drawn into alveoli.

HAPE may be recognized by onset within a few hours after arrival at altitude of insomnia, chest discomfort, cough productive of frothy, blood-stained sputum, shortness of breath, becoming worse on recumbency, nausea and vomiting, irritability, delirium, and finally coma. The patient will by cyanotic, with rapid pulse rate and systemic blood pressure elevated and fine rales heard on auscultation of the chest.

The major preventive measures are gradual altitude acclimatization and regulating exercise at altitude to avoid breathlessness for the first few days.

7.3 Chronic Mountain Sickness ("Monge's Disease")

After apparently complete acclimatization and perhaps after months of residence at high altitude, low grade symptoms of dimming of vision, loosening of teeth, weight loss, decreased ability to work, loose stools, indigestion, and difficulty with mental concentration may occur (10). These symptoms of chronic mountain sickness clear within 3 to 4 weeks of returning to lower altitude.

8 DECOMPRESSION SICKNESS

Decompression sickness refers to the plethora of clinical manifestations first described in the middle 1800s among caisson workers who were decompressed to sea level after working in compressed air to keep water and mud out of tunnel construction. (Hence the previous term, "caisson disease.") Military, commercial and, more recently, sport diving interests brought emphasis to studies of this complex disorder to determine safe decompression schedules and treatment of casualties. Since the 1930s it has been recognized that decompression from sea level to altitude can produce the same disorder. Since 1960 treatment of altitude decompression sickness persisting at ground level has been compression to greater than sea level pressure in chambers, as has been the treatment of divers and caisson workers for decades.

8.1 Pathophysiology

Chapter 9, Volume I, contains a very complete discussion of all aspects of decompression sickness, hence only a summary is given here.

The compressed gas respired at increased atmospheric pressure (diving) must be a mixture of oxygen and an inert gas diluent to avoid oxygen toxicity. This inert gas is usually nitrogen for shallow depths and helium at greater depths to prevent the narcotic effects of nitrogen at high partial pressures. When compressed gas is breathed underwater to balance the hydrostatic pressure of water on the thoracic wall or from the environment of a compressed air pressurized chamber (hyperbaric chamber or compression chamber), the alveolar partial pressure of inert gas rises according to barometric pressure (e.g., at 6 ATA or 165 FSW, absolute barometric pressure is 4560 mm Hg and breathing air the inspired P_{N_2} is $0.79 \times 4560 = 3602$ mm Hg). Allowing for water vapor, carbon dioxide, and oxygen ($P_{O_2} = 0.21 \times 4560 = 948$ mm Hg), the alveolar P_{N_2} approximates 3525 mm Hg.

According to Henry's law (the amount of gas in solution varies directly with the partial pressure of gas in contact with the solution), there is rapid diffusion of nitrogen across the alveolar-capillary membrane to physical solution in plasma. Upon reaching the tissues, the high P_{N_2} dissolved in arterial blood is diffused into tissues at a rate dependent on the perfusion rate of various tissues. The total nitrogen uptake by a given tissue is determined by its composition, with lipid-rich tissues taking on larger amounts of nitrogen. The body's uptake of inert gas can thus be seen as a complex of uptake curves that results in variable loading of different tissues, increasing with time of exposure to the elevated pressures until full equilibration is achieved at approximately 24 hr (saturation).

Upon decompression from the high pressure environment to lower pressures and finally to sea level, the reverse set of gradients is established for offloading of inert gas from tissues to venous blood, to the lung. Inert gas will stay in solution in tissues and blood within strictly defined pressure reductions, but if critical limits are exceeded, it comes out of solution into the gas phase as bubbles in blood and tissues. The limits of safe decompression are discussed in Chapter 9, Volume I. Once formed, bubbles in tissues and blood cause a series of pathophysiological events that can culminate in permanent paralysis or death. Local tissue distortion by bubbles may cause pain, and venous gas emboli may produce congestive infarction of the spinal cord or gas pulmonary embolism. Platelet aggregation has been seen at the blood-bubble interface, and with platelet damage there is release of vasoactive substances such as serotonin and epinephrine, complicating the picture with vasoconstriction. Furthermore, alteration of the platelet membrane makes phospholipid (platelet factor 3) available and accelerates clotting. Untreated, bubble-induced tissue ischemia and hypoxia result in loss of capillary integrity with resultant edema and hemoconcentration. Therapy is directed toward this entire spectrum of events.

The sea level dweller is in equilibrium (saturated) with the nitrogen partial pressure in the atmosphere. Upon rapid exposure to decreased barometric pressure at altitude, a series of events comparable to that just described can occur, with evolution of nitrogen bubbles. Because of the relatively higher proportion of carbon dioxide and water vapor at altitude, these gases diffuse into bubbles and play a larger role than in diving. The precise critical altitude for bubble formation has not been defined, but it is of clinical importance that the lowest documented case of altitude decompression sickness occurred at 5640 m (18,500 ft). The incidence rises sharply with increasing altitude and with time at altitude. Most cases recover upon recompression to sea level, but those that do not require emergency treatment in a compression chamber to resolve bubbles that persist at sea level because of growth while at altitude. Combining the two environments (i.e., ascent to altitude after diving and while tissues still contain excess inert gas) can produce bubbles at much lower altitudes. Cases of decompression sickness have been seen at as little as 1220 m (4000 ft) after safe decompression from diving.

8.2 Clinical Manifestations

The onset of symptoms and signs of decompression sickness tend to be gradual and progressive, beginning minutes to hours after an inciting dive or exposure to altitude. With certain exceptions, to be noted, the manifestations resulting from both environments are the same and can be divided into type I (minor manifestations) and type II (serious manifestations).

Type I (minor manifestations) include bends, involving mild to severe deep, boring pain in single or multiple joints. The pain is usually aggravated by exercise and is relieved by application of local pressure over the joint. There may be local swelling due to bubble blockade of lymphatic drainage. The other type I mainfestations include unusual profound fatigue following diving or exposure to altitude and itching or mottling of the skin.

Type II (serious manifestations) include the very dangerous "chokes" with shortness of breath, substernal pain, cough, and cyanosis. The basic mechanism of massive bubbling in the venous blood, with obstruction of the pulmonary circulation, hemoconcentration, and platelet damage, may lead to profound circulatory collapse or shock.

Neurological manifestations may involve the brain or spinal cord, with the former more common in altitude decompression sickness and spinal cord involvement more common in divers. Brain manifestations include visual disturbances, spotty numbness and/or weakness, paralysis, speech impairment, decreased mental alertness, severe headaches, or seizures. Spinal cord manifestations are usually heralded by severe low back or abdominal pain followed by varying degrees of numbness, weakness, or paralysis below the level of the cord involvement. There may be loss of anal sphincter tone and inability to urinate. Untreated, patients in these cases can remain permanently paralyzed.

A late manifestation is the so-called dysbaric osteonecrosis of divers and caisson workers. The exact etiology of this aseptic necrosis of bone in divers is still unclear, as is its relationship to untreated previous episodes of decompression sickness. This subject is covered in Chapter 9, Volume I.

8.3 Prevention and Treatment

Decompression sickness usually responds rapidly to early and adequate treatment by compression. Descent from altitude produces sufficient recompression to abolish the early manifestations, but slowly developing serious forms of decompression sickness can develop later. In such cases widespread bubble formation presumably already has occurred, and descent is entirely analogous to inadequate compression therapy during which symptoms subside only temporarily.

Divers can prevent decompression sickness by knowing and following established limits for depth and time at depth. Adequate denitrogenation by breathing 100 percent oxygen before and during ascent is an effective preventive measure in fliers.

9 HYPERBARIC CHAMBERS USED FOR OTHER MEDICAL INDICATIONS

The expanding use of hyperbaric chambers to provide short-term, high dose oxygen inhalation therapy calls for a description of chamber types and safety provisions for their use. Besides their role in treatment of altitude and diving decompression accidents, hyperbaric chambers are finding use in treatment of gas gangrene, carbon monoxide poisoning, and certain wound healing problems (15). Two general classes of hyperbaric chamber are in use:

1. The multiplace hyperbaric chamber in which the patient and medical attendants are pressurized together with compressed air to greater than sea level pressure and the patient is given oxygen by mask.
2. The monoplace hyperbaric chamber, where the patient is compressed in oxygen in a small chamber and there is no attendant. The patient breathes therapeutic oxygen from the chamber atmosphere.

9.1 Multiplace Hyperbaric Chambers

All multiplace hyperbaric chambers are, in principle, only a simple hermetically sealed volume (Figure 14.4). However they are complicated, and their safe operation requires that a number of rules be meticulously observed. Hazards arise from (a) the enormous pressures exerted on the chamber walls (at 3 ATA this is 20 tons/m^2); (b) the increased concentration of oxygen and the risk of fire; (c) the danger of decompression sickness among air breathing attendants: patients breathing oxygen by mask are not at risk; (d) air embolism; (e) oxygen toxicity; (f) contamination of the breathing mixtures; and (g) barotrauma of middle ears and paranasal sinuses.

The design of a chamber depends on its purpose, and there are numerous designs and sizes. Usually the chambers are steel cylinders with hemispheres on the ends, equipped with numerous penetrations to provide entrance and exit of breathing gas, water, air, electrical cables, and viewing ports. Alteration of the chamber walls requires complete retesting and safety certification of the chamber.

Most frequently used are chambers with a cylinder whose axis is on a horizontal plane. A "long" chamber is convenient, since strong hermetic partitions can be used to divide it into several intercommunicating lock compartments.

Figure 14.4 Two hyperbaric chambers controlled from a central console. Compressed air from an accumulator is used to pressurize these large chambers, each with internal volume adequate for several patients and attendants.

These locks serve primarily for the entry and exit of personnel from the chamber without reducing pressure in the main chamber. Hatches connect the lock with the main pressure chamber and with the outside world. To enter the main pressure chamber when it is at high pressure, one enters the lock, the outer hatch closes, and the pressure is raised in the lock compartment. When pressure in the lock becomes equal to that in the pressure chamber, the connecting hatch opens to allow passage from the lock to the main chamber.

Each compartment in the pressure chambers must be equipped with permanent safety valves set to prevent the possibility of accidental overpressurization of the chamber.

Each compartment in a chamber has air ducts to allow air to enter and to exit. Since the noise level of the air entering or leaving these ducts is high, and since the chamber itself is a good resonator, the ducts should have mufflers. If the noise level is still excessive, chamber occupants must be provided with noise protection devices for the ears. If muff-type ear defenders are used, a ventilation hole must be drilled in them to prevent a relative vacuum between muff and ear when the pressure is increasing.

Air compressors for hyperbaric chambers must provide exceptionally pure air and must be free from any possibilities of oil contaminating the air. As an emergency backup

EVALUATION OF EXPOSURE TO ABNORMAL PRESSURE

to the compressor, there should be cylinders containing enough compressed air to complete any normal treatment or decompression schedule. The breathing oxygen enters the chamber independent of the chamber pressurization source and is supplied to the individuals by a well-sealed face mask or hood. Exhaled oxygen must be dumped outside the chamber to avoid buildup in the chamber with increased fire risk.

Strict fire prevention procedures must be developed because of the increased oxygen and the concomitant increased danger of fires. All possible ignition sources must be removed. All but absolutely essential electrical connections must be outside the chamber. Lighting must be either isolated or, preferably, located outside the chamber. Sources of static electricity need to be eliminated, and metal objects that could strike the chamber wall and produce sparks are not used. Combustibles inside the chamber should be kept to a minimum, and fire retardant material is used for clothing and bedding. There are several fire-resistant materials that can be used, such as PBI, Durette Gold, and fiber glass fabrics.

The oxygen percentage in the chamber should be monitored continuously and kept as close to normal as possible; it must never exceed 25 percent. This can be controlled by adequate and frequent air ventilation of the chamber. Ventilation is accomplished by letting air into and out of the chamber simultaneously and in equal amounts. This method is also used to control the carbon dioxide level and to maintain optimum temperature and humidity. In case all precautions fail and a fire does ensue, a rapid, adequate fire suppression system is essential. A pressurized water deluge system forces water into the chamber through numerous nozzles to cover all areas.

Suction for removing secretions is provided by the difference of pressure inside the chamber and the outside world. A simple penetration through the chamber wall provided with tubing, a control valve, and a gauge are used to avoid excess suction.

Close observation of those inside the chamber must be maintained by outside operators. This can be done through viewing ports, over closed circuit television, and by instruments measuring physiological parameters. Besides visual contact with those inside the chamber, voice communication must be maintained.

All large pressure chambers need to have a medical lock for the passing in and out of medications, dressings, instruments, food, drinks, and other small objects.

All greases, fats, oils should be kept out of the chamber, since these may become explosive in oxygen enriched environments.

To ensure the safety of all personnel in the chamber, each person should have a breathing mask, and there must be a breathing mixture supplied from outside the chamber for use in case of smoke or other chamber atmosphere contamination.

The relative humidity in the chamber should be left high, at least 60 percent for comfort and to reduce the possibility of static electricity. Since the humidity of the breathing mixture is low, on-line humidification for patients should be provided.

9.2 Monoplace Hyperbaric Oxygen Chambers

Most of the safety principles described in Section 9.1 apply to monoplace chambers. The major difference is that these chambers are totally pressurized and purged with pure oxygen. Thus it is absolutely essential to prevent at all times the admittance of volatiles, greases, ointments, or oil (15).

REFERENCES

1. J. E. Ward, *J. Aviat. Med.*, **27**, 429 (1956).
2. F. M. G. Holmstrom, "Hypoxia," in: *Aerospace Medicine*, 2nd ed., H. W. Randel, Ed., Williams & Wilkins, Baltimore, 1971, p. 60.
3. F. J. W. Roughton, "Respiratory Functions of Blood," in: *Handbook of Respiratory Physiology*, W. M. Boothby, Ed., USAF School of Aviation Medicine, Randolph Air Force Base, Texas, 1954, pp. 51–102.
4. A. Krough, *J. Physiol.*, **52**, 409, 1919.
5. F. M. G. Holmstrom, "Hypoxia," in: *Aerospace Medicine*, 2nd ed., H. W. Randel, Ed., Williams & Wilkins, Baltimore, 1971, p. 63.
6. A. C. Guyton, *Textbook of Medical Physiology*, 5th ed., Saunders, Philadelphia, 1976, pp. 373–374.
7. F. M. G. Holmstrom, "Hypoxia," in: *Aerospace Medicine*, 2nd ed., H. W. Randel, Ed., Williams & Wilkins, Baltimore, 1971, p. 57.
8. J. Ernsting, "The Principles of Pressure Suit Design," in: *A Textbook of Aviation Physiology*, J. A. Gillies, Ed., Pergamon Press, London, 1965, pp. 374–405.
9. R. W. Bancroft, "Pressure Cabins and Rapid Decompression," in: *Aerospace Medicine*, 2nd ed., H. W. Randel, Ed., Williams & Wilkins, Baltimore, 1971, pp. 337–363.
10. B. Bhattacharjya, Ed., "Problems for Consideration, Acclimatization" in: *Mountain Sickness*, John Wright and Sons, Ltd., Bristol, 1964.
11. A. Hurtado, "Mechanisms of Natural Acclimatization," School of Aviation Medicine Reports 56-1, Randolph Air Force Base, Texas, 1956.
12. A. L. Barach, M. Eckman, E. Ginsbury, A. E. Johnson, and R. D. Brookes, *J. Aviat. Med.*, **17**, 123 (1945).
13. S. M. Cain and J. E. Dunn, II, *J. Appl. Physiol.*, **20**, 882 (1965).
14. S. M. Cain and J. E. Dunn, II, *J. Appl. Physiol.*, **21**, 1195 (1966).
15. J. C. Davis and T. K. Hunt, Eds., *Hyperbaric Oxygen Therapy*, Undersea Medical Society, Bethesda, Md., 1977.

CHAPTER FIFTEEN

Evaluation of Exposure to Biological Agents

JOHN S. CHAPMAN, M.D.

1 INTRODUCTION

1.1 Types of Agent

Agents under consideration in this chapter are of two types: in the first group are nonliving substances, products of fungal or bacterial metabolism plus perhaps degradation products of natural fibers altered by bacterial or fungal activity; the second type consists of living organisms encountered in the processes and activities of an industry. Most such infections fall in the category of zoönoses—diseases of animals transmissible to man.

1.2 Modes of Exposure

1.2.1 Cutaneous

Continuous exposure of skin to all the materials just cited is likely to occur under usual working conditions. Dusts of various kinds settle out and tend to stick to sweaty skin where they may set up various inflammatory and allergic responses. The immediate agent may not be readily identifiable. Viable pathogenic organisms may be inoculated through minor wounds or may enter through preexisting abrasions or other breaks in the integument.

1.2.2 Gastrointestinal

The gastrointestinal portal of entry is probably the least likely under industrial conditions. Improper cleaning of the hands or exposure of food or water to some of the agents may permit entry by way of the gastrointestinal tract. In addition, larger particulates caught on the mucous lining of the respiratory tract may be swallowed as a result of ciliary activity. Nonliving products of microbial metabolism, however, would probably be destroyed by digestive enzymes and rendered harmless.

1.2.3 Respiratory

The respiratory tract constitutes the main portal of entry for most of the biological agents, either as viable organisms or as products of their multiplication on various fibers. Larger particles are filtered out in the nose or deposited on the mucous membranes of the bronchi. Particles of 5 μm or less mean diameter will reach alveoli, whose surface, approximately 70 m², provides a huge area of exposure to allergenic material. Dependent on the size of particulate matter, therefore, the site of reaction in the respiratory tract may be primarily bronchial, with symptoms suggestive of asthma, or pulmonary, with various types of response within the alveoli.

1.3 Contributing Factors

As this chapter reveals, symptoms and injury appear to be dose dependent. Thus the amount of material per unit volume, the amount of increased pulmonary ventilation required by the type of physical activity, and the number of years of exposure enter into any expression of risk. In addition, there is variability in individual susceptibility, for which we possess no useful criteria of predictability.

In all studies the smoker appears to be more prone to develop symptoms than the nonsmoker. Since smoking itself results in bronchitis, the importance of this factor is obvious. But with respect to particulates the smoker is at further disadvantage, since the efficiency of the mucociliary escalator for removal of larger particles is less than normal (1); hence retention of injurious material is both greater in amount and longer in duration than in the nonsmoker. At the alveolar level, moreover, the impairment of phagocytic pneumocytes that results from smoking leads to less efficient removal of particulates.

The presence of inert particulates may also affect the response of the host, whether the particles are viable microorganisms or allergenic materials (2). Such material as carbon or other nonfibrogenic dusts may impair defense mechanisms at either the bronchial or the alveolar site, and irritant gases such as hydrocarbons may affect the protective mechanisms of the bronchi or alveoli.

Because these secondary factors are so numerous and variable—and often unquantifiable or not measured—the following discussion deals with the principal agent of disease as if it were present alone. It should always be remembered, however, that in a specific individual secondary factors may be of considerable importance.

2 NONLIVING MATERIALS

2.1 Nature of the Materials

The products of bacterial and fungal metabolism are complex and numerous. Chemically they consist of a wide variety of split proteins, lipoproteins, and mucopolysaccharides, the relative amounts of which are affected by temperature, humidity, and possibly by the kind of substrate (in this case, fibers) on which they grow. Though proteins probably produce the greater number of immunological effects, terminal sugars may also be antigenic (3), and bacterial and fungal material and vegetable fibers contain a wide array of exotic mono- and polysaccharides (4).

2.2 Modes of Action of These Substances

2.2.1 Direct Effects

Some of the substances produced by bacteria or molds may have a pharmacological effect. Pernis and associates demonstrated substances in samples of cotton that seemed to resemble bacterial endotoxin and postulated that these compounds probably produced fever and tightness of the chest (5). Cavagna et al., after sensitizing rabbits to cotton dusts, concluded that an endotoxinlike substance in the challenge material released serotonin and resulted in a type of chronic bronchitis with acute bronchospasm (6). This view of the mechanism of production of byssinosis was further amplified by Rylander and associates (7). Nicholls, however, disputed this mechanism on the basis of the very small amount of endotoxin that could be demonstrated, maintaining that a histaminelike material in the stems and pericarps of cotton more adequately accounted for symptoms (8).

Bouhuys and co-workers also recovered substances from hemp and cotton dust that seemed to release histamine in experimental preparations (9). Subsequently they isolated from cotton brachts a compound they identified as methyl piperonylate. This compound released histamine and produced symptoms that could be inhibited by isoproterenol (10). Whatever the precise compound or mechanism of its action, it is evident that a direct pharmacological effect may be produced by chemical substances contained in cotton, and—as the various investigators found—other vegetable fibers as well.

2.2.2 Immunologically Mediated Effects

Pepys in a review of "farmer's lung" emphasized that in individuals who manifest this reaction, it is possible to detect at least two types of immune reaction to an extract of moldy hay or products of *Thermopolyspora faeni*. The first was atopic, as manifested by immediate wheal. The second was an Arthus phenomenon with induration 4 to 6 hr after injection. The Arthus response was associated with demonstrable precipitins in serum against his antigenic materials. In some asthmatics with pulmonary infiltrations, in cows with "fog fever," and in humans with bird fancier's disease, Pepys was also able both to elicit an Arthus reaction and to demonstrate precipitins against antigens of *Aspergillus fumigatus*. In many patients biopsies revealed also the presence of epithelioid tubercles (11). It is evident that cell-mediated immunity is involved in a number of instances. Other features of the infiltrations include the presence of fair numbers of eosinophiles in some cases, of large numbers of lymphocytes and fibrocytes in others. Regardless of the substrate, when pulmonary infiltrations develop as a result of these various immune mechanisms, the reaction should be regarded as a hypersensitivity pneumonitis or, in Pepys's term, "extrinsic allergic alveolitis" (12).

It appears, moreover, that neither textile nor cordage fiber is essential to growth of molds. Fink et al. report the isolation of *Micropolyspora faeni* from ducts of a circulating warm air system to which a humidifier had been added. It is thus evident that the mold rather than compounds derived from the fiber is of paramount significance, though fiber-derived substances may alter or enhance the reaction of the alveoli or may participate in transition of skin reactions from wheal to tuberculin-type induration (13).

Since the range of reaction of tissue is so varied in all these responses to allergenic material, some kind of relationship may exist between these hypersensitivity reactions and the much less common diffuse fibrosing alveolitis (14). Alternatively a similarity between these responses and so-called thesaurosis is manifest, though in thesaurosis (15, 16) it should be possible to observe particulate foreign material.

In any event it is clear that the range of reaction to nonliving biological substances extends from acute reactions, similar to that to methyl piperonylate, to the other extreme of tubercle formation and fibrosis.

2.3 Textile and Cordage Fibers

2.3.1 Cotton (Byssinosis)

Byssinosis is marked by a sensation of tightness of the chest and cough, usually worse upon return to work after the weekend, but also worsening in the course of the workday. Workers in cotton mills are particularly at risk, though other workers involved in the processing of cotton, from gin to finished goods, may occasionally manifest symptoms. In cotton mills the highest incidence has been found among those in areas of preparation (opening, blending, picking, and carding) and in workers with more than 18 years of exposure. Males are affected about twice as often as females, and smokers much more than nonsmokers (17).

Byssinosis has been classified by Schilling, and a national conference has accepted the following grades of severity, which should be employed in every survey of a population at risk.

    0. No symptoms of respiratory difficulty.
    ½. Occasional tightness of the chest on the first day of the working week.
    1. Chest tightness on the first working day of each week.

2. Tightness of the chest on every working day.
3. Grade 2, plus evidence of permanent (irreversible) loss of exercise tolerance, with or without reduced ventilatory functions (18).

It is obvious that except in grade 3, evaluation of impairment from exposure to cotton dust is on the basis of complaints or symptoms. The degree to which workers in a particular mill may be affected will be reflected in absenteeism, complaints, or respiratory insufficiency in those with greatest seniority. Preliminary evaluation may take the form of a questionnaire.

Quantitative evaluation depends on dust concentrations in various parts of the building and on the results of simple pulmonary ventilatory tests, which should be performed under specified conditions. Imbus and Suhs used FEV_1 (forced expiratory volume, 1 sec) as the most suitable test under operating conditions (17). Hamilton et al, also employed this measurement and found that byssinotic workers exhibited a decrease of 10 percent from morning values in the course of the working day, whereas only 3 percent nonbyssinotic workers experienced a decrease of similar proportion (19). Merchant and associates, after comparing several ventilatory tests, also concluded that FEV_1 furnished the most consistent and accurate discrimination between byssinotic and unaffected workers (20). McKerrow and others, however, preferred the indirect MBC (maximum breathing capacity), which they found to decrease linearly through the working day (21). Airways resistance also varied in accord with the complaint of tightness of the chest. In studies under other conditions of exposure Bouhuys et al decided that the FEF_{25-75} (forced expiratory flow) provided the most sensitive evidence of change of function (22).

Probably for most industrial surveys the VC (vital capacity) and the FEV_1 will be found to be effective and useful. Regardless of the test that may be selected, it should be carried out under identical conditions, with the same instrument (which must be maintained in good order and calibrated from time to time), and by the same technicians, who should be well trained to urge maximum effort and to recognize poor performance. Tests ideally ought to be carried out at the beginning of each first working day of the week and repeated at the conclusion of the shift.

All the recommended tests for pulmonary function are based on the presumption that nonliving biological agents produce, as their first effect, varying degrees of bronchospasm. They further imply that continued exposure results in chronic changes in the bronchi, with progressively irreversible obstruction and ultimate fibrosis. The real test of obstruction is the rate of flow of gases during expiration, and FEV_1, (forced expiratory volume, one second), MVV(MBC) (maximum voluntary ventilation, maximum breathing capacity), and FEF_{25-75} (forced expiratory flow at 25-75% of vital capacity) all deal with this aspect of ventilation; FEF_{25-75} is regarded as representing more specifically obstruction of smaller bronchial branches.

Note that these remarks with respect to tests of pulmonary function apply to all exposures to nonliving biological agents discussed in this section.

In surveys of various stations for amount of cotton dust in the atmosphere, particles smaller than 15 µm are of primary concern. McKerrow and associates found that improvement in symptoms and in measured function was apparent when concentration of these particles was reduced to 2.4 mg/m^3. However a national conference has recommended that a suitable threshold limit value (TLV) for such particles should be 1 mg/m^3.

Various measures of control have been advised and attempted. Filtration of atmosphere is both difficult and expensive, and McNall advocates electrostatic precipitation as an economical and feasible measure (23). Prewashing of cotton before other processing was effective but interfered considerably with spinning (24). Steaming, which may be carried out as early as ginning, materially reduces byssinotic effects and does not alter characteristics of the fiber for spinning and weaving (24). McKerrow and colleagues (25) recommend vacuum extraction of dust, which reduces fiber content of the atmosphere by nearly 50 percent. In the recommendations of a national committee, the most easily achieved and satisfactory means of maintaining an acceptable TLV consisted of partial isolation of the procedures established as the most dusty. A necessary feature of partial isolation is effective local exhaust ventilation (17).

In individual instances it is recommended that workers who present grade 2 or 3 symptoms, or who show a decrease of 10 percent in FEV_1 during a working day, should be removed from further exposure to cotton dust. It is also recommended that applicants who present initial abnormal functions, who have abnormal roentgenograms of the chest, or who show a decrease in FEV_1 after a short period of employment, should be assigned to nondusty procedures.

There is disagreement over the possibility of progression after exposure ceases. It seems reasonable to conclude that in a few individuals changes may be irreversible, and even that further progression of fibrosis may occur (17, 22). If this view is correct, either the pharmacological effect results in permanent alteration of the bronchi, or additional immunological responses are involved.

2.3.2 Hemp, Jute, and Sisal

In their study of workers in hemp, Bouhuys and co-workers (9) concluded that hemp fiber produced a disease similar to byssinosis. It occurred particularly among workers engaged in bolting and hackling by hand. They found only a fair correlation between FEV_1 and symptoms, and workers exhibited a wide variability in sensitivity to the fiber. Heavily exposed workers produced urines with a high content of histamine metabolites. As they studied hemp workers in 1966 the authors concluded that many changes in the airways might prove to be irreversible, and the follow-up work in 1974 confirmed this earlier impression (22).

Masks afforded no protection to the exposed individuals, who had more cough and greater dyspnea than controls. However removal of gums from the raw product materially reduced respiratory complaints. The authors, however, do not make it clear that degumming represents a technologically feasible or acceptable procedure in the industry (9).

Jute seems to be less noxious than hemp, at least in biological tests (7), and sisal, another fiber employed in production of cordage, seems not to produce symptoms of byssinosis or to result in significant alterations of ventilatory indices (25).

Flax chiefly affects workers engaged in preliminary preparation, such as hackling. Reported symptoms resemble those of byssinosis (26). Correction of the situation depends on enclosing the dustier procedures and providing effective local exhaust ventilation.

2.4 Farmer's Lung

The term "farmer's lung" is applied to a syndrome marked by chills and fever following disturbance of moldy hay (or other moldy material.) Associated with the clinical syndrome are evanescent infiltrations of the lung. Usually the first episode subsides without appreciable effect on pulmonary function, but as attacks recur on subsequent exposures, portions of the acute infiltrate fail to resolve and gradually, organize. Eventually extensive fibrosis of the lung, marked impairment of function, and respiratory insufficiency develop.

The pathogenesis of the condition consists of the inhalation of airborne substances produced by the growth of molds in contaminated hay. Classically these molds are thermophilic actinomycetes. The process does not involve invasion of tissue but results from immunological reactions against the fungal antigens. In Great Britain the organism responsible seems to be *Micropolyspora faeni*. Pepys has demonstrated that the reaction involves an Arthus phenomenon, with circulating antibodies against extracts of moldy hay (11). Tissues reveal the presence of noncaseating tuberculoid granulomas; the process clearly is not a foreign body response. Although Pepys's concept of pathogenesis has been rather widely accepted, Harris and co-workers have presented findings that suggest that cell-associated rather than circulating antibodies may better explain the response in tissue (27).

Apparently the molds responsible for farmer's lung grow in hay that is improperly cured or subject to repeated moistening; hence the disease is much more common in cold and humid areas such as Great Britain and the region around the Great Lakes. However conditions adequate for the growth of thermophilic organisms may result from storage of improperly cured or wet hay in almost any area. Since the intensity of response to the antigen is proportionate to dose, storage in a tight barn offers the best conditions for development of the syndrome. It should be noted, however, that thermophilic molds do not require the presence of vegetable fiber, since Fink et al. have reported finding these organisms in the duct of a warm air heating system provided with a humidifier (13).

Diagnosis of farmer's lung depends on adequate occupational and environmental history supplemented by appropriate skin tests and the demonstration of specific precipitins, or other antibodies (28). Since suitable antigens are not commercially available, recognition rests largely on history, characteristic roentgenograms of the chest, sometimes biopsy of the lung, and the exclusion of other causes of a fibrosing and granulomatous disease.

Treatment depends on the severity of manifestation. In acute exacerbations, full doses of corticosteroids are necessary, and measures applied in any form of acute respiratory distress may be required. In late fibrotic disease it is probable that no measures will be effective other than such as may be required for intercurrent bronchitis.

Prevention of course calls for interdiction of further contact with hay. Complete curing of hay before storage might be effective, but in many parts of the world this measure would not be possible. In nonagricultural exposures such as that described by Fink and colleagues, the removal of the humidifier, together with disinfection by formaldehyde, might be adequate; otherwide an entirely new system of ductwork would be required.

2.5 Bagassosis

The symptoms of bagassosis are entirely similar to those of farmer's lung; the exposure, however, is to certain batches of bagasse, the dried stalks of sugarcane. Since only certain batches appear to be associated with symptoms, it seems probable that the mode of storage, perhaps the degree of extraction of juice, and the opportunity for wetting after storage, may determine the growth or nongrowth of molds (29).

Bagasse usually is shipped in bales, quite like baled hay. The material serves as a binder for many products, such as wallboard and insulating boards. Processing begins with the breaking open of bales and the grinding of the fiber. In both processes a very large amount of dust is generated, and it is at these sites that workers receive the heaviest exposure.

The many similarities to farmer's lung suggest a similar pathogenesis, but Pepys reported that precipitins against his antigens were far less consistently present. It is possible that other types of antibody or different varieties of molds or both may account for his lack of success. The pulmonary pathology is identical.

Progression with recurrent bouts of fever and pulmonary infiltrations resembles the pattern observed in farmer's lung. Restrictive defects in ventilation appear to be most characteristic, for Weill and associates reported a decrease in both vital capacity and residual lung volumes in 17 of 20 repeatedly exposed workers. DL_{co} (diffusing capacity of lung by carbon monoxide method) was correspondingly reduced in these workers. FEV_1 was reduced in most patients, and it is interesting that of those who had persistent respiratory difficulty, the predominant manifestation was obstruction (29). In a limited outbreak workmen were seen within 2 to 4 weeks after the first occurrence of the condition. There was a wide range of symptoms, from the slightest respiratory complaints to severe cyanosis with acute right ventricular failure. Those most severely ill had residual respiratory functional abnormality (30).

Management of the acute and chronic forms of the disease does not differ from corresponding stages of farmer's lung. Affected workmen must be removed from further exposure. Prevention would apparently call for isolation of the original processes of opening, chopping, or grinding from the rest of the plant and the provision of adequate exhaust ventilation.

2.6 "Grain Fever"

Apparently similar in pathogenesis and pathology to byssinosis, grain fever has been observed in handlers of grain in all stages of production from harvest and combining to drying and storage in elevators. doPico and co-workers consider both the fever and the pulmonary infiltrations to be mediated by antibodies (31). Wheal response was observed by this group and by Darke and associates (32), but the latter found no consistent correlation between the presence of precipitins in serum and the presence or absence of symptoms or of pulmonary infiltrations. In all exposed workmen they demonstrated the presence of circulating antibodies, regardless of whether symptoms were present.

doPico's group found that symptoms occured more frequently among smokers than nonsmokers, with an overall incidence of 37 percent. In the Darke study in England, examination was limited to men engaged in harvesting, in whom the incidence was 25 percent and development of the syndrome was related to duration of exposure. The English study further demonstrated an exposure during harvesting to a very high concentration of molds, as many as 200 spores and fragments of hyphae per cubic meter. Cultures were not reported, and no specific agent can be incriminated (32).

Physiological effects on the lung seem to consist especially of bronchospasm, or possibly of infiltrations of the submucosa of finer bronchioles, since the study of doPico et al. found that the midflow rate best demonstrated early effects of exposure.

Since most of the work takes place in the open, unless there is entry into elevators or storage bins, the most feasible control consists of the removal of sensitive workers from further exposure.

2.7 "Bird Fancier's Disease"

2.7.1 Pathology and Pathogenesis

The name "bird fancier's disease" has been applied to extrinsic allergic alveolitis as observed in some individuals who have been exposed to birds. The condition must be distinguished from invasive aspergillosis of the lung, particularly that reported from France, and it probably differs from the type of granuloma Pepys has encountered in some cases of asthma. Bird fancier's disease is encountered both in an acute form indistinguishable from the acute form of farmer's lung, and also in a chronic phase manifested by extensive fibrosis. Such reactions have occured not only in those individuals directly involved in the care and feeding of birds (in England, chiefly pigeons and budgerigars) but even in members of the household who have had no direct association with birds at all (33).

In considering pathogenesis one has to take account not only of the feed (usually grain), but also antigens of avian origin. Hensley et al. were able to show in sensitized monkeys a specific pneumonitis induced through aerosol challenge with pigeon serum

(34). In addition in experiments involving passive transfer of lymphocytes they were able to prove the presence of cell-associated antibodies. Circulating antibodies in high titer also are known to be involved in the reaction (33).

Pathological changes in the lung of bird fancier's disease seem to differ from those of other forms of allergic alveolitis. The monkeys in the experiments of Hensley et al. revealed hemorrhagic alveolitis 6 hr after challenge. Undoubtedly the degree of original hypersensitivity of the individual, the duration of exposure, and the concentration (and perhaps the character) of the antigenic material affect the kind of response observed in tissue.

In two patients with milder exposure and insidious onset, Riley and Soldana found the alveolar spaces filled with desquamated septal cells and foamy macrophages, while lymphocytes and plasma cells infiltrated the interstitial tissue. One of the tissues revealed a much more striking interstitial infiltration, with little alveolar material.

In a study of pulmonary tissue from 33 patients Hargreave and associates reported that the acute disease appears as a histiocytic granuloma of the interstitium, with occasional nonnecrotic tubercles containing multinucleated giant cells. Within these giant cells they observed lanceolate spaces not observed in other forms of extrinsic allergic alveolitis (35). In the more chronic forms of the disease there were dense fibrosis and honeycombing.

To Hensley and his associates, however, it seemed that the most striking feature was the presence of remarkable numbers of foamy histiocytes. In a study of the histology of the disease these investigators observed in one instance many foamy histiocytes in the alveoli and interstitial tissue, while in another workman histiocytes were associated with considerable infiltration of lymphocytes. In sections from both sources they observed occasional sarcoidlike granulomas. In their second patient, in whom exposure had been much heavier and more protracted, they found severe obliterative bronchitis and dense interstitial fibrosis (36).

2.7.2 Symptoms

The onset of symptoms may be either dramatic or insidious. Allen and others (37) believe that the age of the patient and the duration of exposure constitute the principal factors affecting the degree of distress and the reversibility of changes. In all forms of the disease the principal complaint is breathlessness or tightness of the chest, with more or less cough. Wheezing, an expression of the degree of bronchitic or bronchiolitic hypertrophy observed histologically, may be a prominent feature.

2.7.3 Roentgenographic Findings

In some roentgenographic studies nodulation is regarded as the earliest evidence of disease. Later in the course the finer nodulation becomes confluent, contraction and fibrosis become very prominent, and honeycombing is evident (35).

EVALUATION OF EXPOSURE TO BIOLOGICAL AGENTS

2.7.4 Physiological Effects

Physiological effects differ with the type of pathology. Allen et al. (37) observed that four of nine patients, who had exhibited mild obstructive and restrictive ventilatory defects early in the course of exposure, regained normal function after exposure ceased. Three of the remaining five showed progressive obstruction of the smaller airways, and in one greatly reduced elastic recoil of the lung was observed (37). Hargreave and associates (35) encountered reduction of all pulmonary volumes in their patients, and associated with this a reduction of DL_{co}. Riley and Soldana (33) observed both restrictive and obstructive defects in one of their patients, a pure and severe restrictive defect in the other.

In one study it was observed that removal from further exposure resulted in improvement in 37 of 41 patients (35). In severely symptomatic patients, such as those of Riley and Soldana, the administration of adrenal corticosteroids resulted in considerable improvement. The principles of treatment for acute symptoms and withdrawal from further exposure are therefore indicated in bird fancier's disease as in most other causes of extrinsic allergic alveolitis.

2.7.5 Specific Exposures

Turkeys. Reports establish that indiviudals employed in the raising and processing of turkeys face the same problems as those exposed to pigeons. In 142 of 205 subjects Boyer et al. observed symptoms within one hour after commencing a shift. The principal complaints were tightness of the chest and cough. In 13 workers, 4 to 8 hr after the reexposure began, symptoms progressed to include myalgia, fever, and dyspnea. These individuals, as well as some with less severe symptoms, demonstrated both positive skin tests and circulating antibodies against antigens derived from feathers or serum (38).

Chickens. Very similar symptoms and objective findings occur in workers engaged in processing chickens. They exhibit both immediate and Arthus responses to skin tests; precipitins are directed against antigens derived from feather, serum, and droppings, Warren and Tse found precipitins in bronchial mucus to be especially impressive; these were most manifest in response to the antigens contained in feathers, which may constitute the most important inhalant (39).

2.8 Wood Dust Disease

2.8.1 General Features

Pulmonary disease, in some respects resembling farmer's lung, in others predominantly of an asthmatic type, has been observed in individuals engaged in woodworking industry, but not in logging. Direct chemical effect, particulate structure, and immunological responses all play greater or less roles in the production of symptoms and structural changes in the lung. Sometimes the effect is related to the type of wood or the presence of various fungal contaminants.

A direct effect of wood dusts (type not specified but presumably from several species) appears in a report by Michaels (40) of disease in two workers, one a carpenter, the other employed in wood pulping industry. The tissues of each revealed peribronchiolar fibrosis and histiocytes and foreign body giant cells in alveolar spaces. Basophilic inclusions that had properties resembling those of wood were found in the giant cells in each instance.

2.8.2 Sequoiosis

Cohen et al. (41) encountered similar histological changes in a patient exposed to sawdust from redwood, a pulmonary reaction to which they gave the name sequoiosis. Although they observed in giant cells particles apparently derived from wood, they also demonstrated precipitating antibodies in the patient's serum. Chemically the antigens of redwood were identified as polysaccharides (42). Roentgenograms revealed diffuse pulmonary infiltrations, and the prinipal physiological impairment consisted of a restrictive defect associated with decreased DL_{co}.

2.8.3 Cork

Similar roentgenographic changes, with nodulation or occasional reticular patterns, appeared in the examination of workmen exposed to cork dust. Most of these abnormalities occured only after 15 to 20 years of exposure. Physiological findings included both restrictive and obstructive defects, with an especially striking increase in residual volumes.

Pathological findings consisted of histiocytes, interalveolar fibrosis, and chronic inflammation. There is no mention of inclusions or giant cells.

2.8.4 Maple Bark Disease

According to Wenzel and Emanuel (44), workmen engaged in the production of chips from maple logs develop a pattern of illness in most respects identical to those of farmer's lung. In so-called maple bark disease the source of sensitizing antigen was shown not to be the wood itself but a heavy infection of the bark by *Cryptosoma corticale*, which was dispersed in the atmosphere in clouds. The process is marked by the usual symptoms of cough, tightness of the chest, chills, and fever, by patchy infiltration of the lung, and by precipitins against the fungus. As in most such exposures, a number of the individuals at risk developed precipitins without clinical or roentgenographic abnormalities.

2.8.5 Red Cedar

In workmen exposed to the dust produced in processing of red cedar, Gandevia (45) observed asthma in three persons who previously had been nonasthmatic, and rhinitis in four men. In the remaining workmen FEV_1 decreased by at least as much as 100 cc in the course of the working day.

2.9 Carcinoma in Furniture Manufacturing

Finally it should be mentioned that adenocarcinoma of the nasal passages and paranasal sinuses has been observed in a disproportionately high percentage of workmen engaged in the manufacture of furniture, but not in carpenters or cabinetmakers. A similar high incidence was observed among leather workers (46).

2.10 Miscellaneous Problems

2.10.1 Tobacco

Among nonsmoking women who worked in a tobacco-processing plant there were complaints of chest tightness and wheezing. Respirable dust ranged from 0.3 to 3.6 mg/m^3. These workers exhibited no chronic changes in pulmonary function, but acute decreases in FEV_1 took place during the shift (47).

2.10.2 Coffee

Van Toorn encountered among coffee workers in a plant a case of extrinsic allergic alveolitis that closely resembled farmer's lung. A 46-year-old male who complained of breathlessness was found to respond to skin tests prepared from coffee beans, and precipitins were also demonstrable. Roentgenograms revealed nodulations that became confluent in the lower lung fields. Pulmonary biopsy presented vasculitis, and alveoli were lined with granular histiocytes and the alveolar walls were infiltrated with round cells (48).

2.10.3 Mushrooms

Stewart reported the occurrence of a syndrome of tightness of the chest among six workmen who were exposed to pasteurized compost in the spawn sheds for the culture of mushrooms (49). In some workers onset of symptoms occurred within a few hours of exposure, but in others exposure was much longer. Neither skin tests nor precipitins provided useful information. Roentgenograms revealed pulmonary infiltrations, but no tissue was available for histological examination.

2.10.4 Subtilin

Certain manufacturers added enzymes of *Bacillus subtilis* to various types of washing powders. Tightness of the chest and breathlessness were common complaints among workers exposed to this substance. Roentgenograms of the chest were normal, and neither skin tests nor search for antibodies proved useful (50, 51). In nearly half the employees at one such plant, pulmonary ventilatory functions were decreased, as was DL_{co} (50). In the study of Dijkman et al. bronchial challenge resulted in prompt reduction of VC and FEV_1, and dyspnea persisted for 6 to 8 hr after challenge (51). This condition seems to be more nearly related to papain-induced bronchitis than to either pharmacological or immunological effects.

2.11 Recognition of a Problem

Undoubtedly many other substances of biological origin may give rise to symptoms within the reported range of reactions. Mechanisms may vary from direct injury, through atopy, to pulmonary Arthus reaction, to the production of granulomas. Some agents may have direct pharmacological effects, whereas others may exert their effects through release of histamine. It is not improbable that some substances may stimulate all these reactions in varying degrees of intensity.

It is obvious that if complaints of tightness of the chest, wheezing, or breathlessness occur with any frequency in any workplace in which materials of biological origin are used, an investigation is required. The part-time industrial physician faces a difficult problem in recognizing that materials in the plant may be responsible for a hazard to health. Since in early phases the pulmonary changes are reversible, it is obviously the function of the physician to prevent the occurrence of late and irreversible damage. Furthermore, the development of symptoms is frequently insidious and complaints are not alarming. It is the duty of the physician to learn whether nonliving biological products or possibly contaminated materials may be in use. Upon discovery that potentially injurious substances appear in the plant, the physician should investigate to see whether these materials are producing a problem.

In pursuing an investigation of this type, the physician should bear in mind that susceptibility varies from individual to individual, that those exposed for the longest time are most likely to be symptomatic, and that after these, those in the dustiest portion of the work are also most at risk. Wheezing, cough, tightness in the chest, and possibly fever, most often striking on Mondays and diminishing later in the week, are evidence that biological products of any type may be producing illness. At this point conformatory testing as described in Section 2.12 is indicated.

2.12 Approaches to the Problem

The initial step should consist of a questionnaire, perhaps modified somewhat from the chronic bronchitis questionnaire of the British Medical Research Council. This should be supplemented by physical examination for the presence of wheezing or dry, crackling

EVALUATION OF EXPOSURE TO BIOLOGICAL AGENTS

rales. Samples of air, with analysis of contents, at various sites in the workplace should follow.

If questionnaire, examination, and samples of the atmosphere indicate that a problem exists, the severity of damage that may result must be measured. Ventilatory tests, properly conducted under standard conditions by well-trained technicians, constitute the appropriate next step. These tests should include as a minimum VC, FEV_1 and probably FEF_{25-75} at the beginning and at the conclusion of the shift. Results obtained should be tabulated separately for smokers and nonsmokers, for males and females, and—if the group is at all large—arranged by decade of age. (Due allowance for smaller ventilatory figures for blacks should be kept in mind.) Other more sophisticated measurements of pulmonary function may be desirable, depending on the types of ventilatory defects discovered.

Skin testing materials, methods for bronchial challenge, and studies of serum for the presence of antibodies are usually not available under conditions of industrial surveys. In some circumstances however, it may be possible and important to arrange a limited application of some of these techniques. If the problem appears to be quite troublesome, the advice and assistance of pulmonary physiologists or immunologists may become necessary.

Standard 6 ft roentgenograms of the chest will be a necessary part of the survey. Preferably the roentgenologist should be certified as an A reader by the National Institute for Occupational Safety and Health (NIOSH).

2.13 Control

Methods of control, essentially preventive medicine, vary somewhat with the materials known or suspected to be provocative. In almost any situation the most important step comprises control of dust. Provision of masks may have limited application, since masks that fit properly and exclude particles of respirable diameter may be uncomfortable when worn over long periods of time. Masks should be used only for protection against high dust levels during short exposure periods.

Exclusion of the most dusty procedure from the general atmosphere of the plant constitutes the best remedy. In certain problems, as they have been mentioned under specific substances, pretreatment of material may reduce dust or eliminate the pathogenic materials. In others it is necessary to isolate the process and provide effective exhaust ventilation. Symptomatic individuals or those who show loss of FEV_1 greater than 100 cc in the workday should be transferred to areas in which dust exposure is minimal.

3 LIVING AGENTS OF DISEASE

3.1 Vegetative Organisms

Industrial disease as a result of specific infectious agents is relatively uncommon. Two types of agent constitute particular risks. The first is the organism able to assume a vegetative form that permits long survival—and possibly transportation over long distances—notably spore formers and certain fungi. The second class comprises the zoönoses, organisms derived from living animals or the carcasses of freshly killed animals. Vector-borne diseases have no specific association with industry, unless farming and the raising of livestock is regarded as industrial. Laboratory workers in microbiology, of course, undergo extensive exposure to agents of many types, but although this represents occupational exposure, it can hardly be regarded as industrial. The same statement applies to the risks of veterinarians, physicians, and employees of hospitals.

3.1.1 *Coccidioides immitis*

Exposure to *C. immitis* takes place most frequently in the southwestern United States and the adjacent states of Mexico. Distribution of the organism in this area is not uniform, but is largely restricted to the Lower Sonoran Life Zone where vegetative forms survive in the soil. As a result of high wind or mechanical disturbance of the soil, the organism becomes airborne and may be inhaled. Operators of graders, bulldozers, and other heavy equipment are particularly at risk.

Inhalation of a viable unit results in a pulmonary infiltration, followed very soon by enlargement of bronchial and tracheobronchial lymph nodes. After 2 or 3 weeks cavitation of the central part of the pulmonic infiltration may occur. Symptoms are similar to those of influenza and include fever, malaise, and weakness, which persist usually for 3 or 4 months. In dark-skinned ethnic groups the primary infection may rapidly lead on to widespread dissemination with lesions of bone, skin, and meninges (52).

In the endemic area infections have occurred in oil field workers, in highway maintenance and construction crews, among drivers of heavy trucks, and occasionally among train crews on frequent runs through this area. Passive transport of the organism to distant areas may occur on trucks or automobiles; thus repairmen and mechanics may encounter the organism.

The prevalence of coccidioidal infection in a population may be measured by the coccidioidin skin test. Complement fixation studies are generally available through state health departments. Titers ranging in one dilution around 1:16 suggest recent infection, but those of 1:64 or higher almost always denote hematogenous dissemination. It is important that serum for complement fixation be drawn in advance of the skin test, which tends to produce an elevation of the complement fixation. If the rate of positive reactions to coccidioidin 1:100 exceeds 1 to 2 percent among nonresidents in a group of workers in the endemic area, a problem of occupational infection exists. (In the Air Force during World War II, in some sites, infections as measured by the skin test ran as high as 20 to 25 percent.)

The only measure found to be of any value in reducing the rate of infection appears to be oiling of the soil (53). This form of control may be undertaken when an oil field crew is likely to be at work in the same site for several weeks or months, but obviously no method can affect transient or irregular exposure. Since it was shown that the highest rate of incidence takes place in summer and autumn (53), the safest measure for maintenance and construction crews of highways and railroads would be to schedule work in such a way that most of it could be done during the winter and spring.

3.1.2 Histoplasma capsulatum

H. capsulatum survives in soil in much the manner of *C. immitis*, but its geographic distribution is quite dissimilar. Infection occurs endemically along the valleys of the Mississippi River and its main tributaries, but histoplasmosis is also associated with exposure to chicken and bird droppings (54, 55). The guano of bats, in such places as caves and rock shelters, also provides suitable conditions for the survival of the organism (56).

H. capsulatum produces clinical symptoms and signs like those of *C. immitis*. Roentgenographic changes are similar, but often are very extensive and may resemble scattered bronchopneumonic infiltrations or even assume a miliary pattern. Histologically the lesions resemble those of tuberculosis, though there is usually less necrosis; with suitable stains the organism can be identified in the lesions. Skin tests and serological reactions provide additional information—subject to very much the same restrictions and features as those described for similar tests for *C. immitis*.

Conditions of exposure are similar to those for *C. immitis*, except for geographic differences, although raising of large flocks of poultry poses a different and equal risk. Any type of industrial activity that results in disturbance of soil may be followed by an outbreak of histoplasmal infection among workers. Residents of adjacent areas will also be at risk.

The skin test with 1:100 histoplasmin affords a rapid and fairly effective means of screening a population, but has the disadvantages of coccidioidin with respect to effect on serological results. If the skin test is employed, the frequency of response in the test population should be compared with figures for the region as a whole, usually obtainable from health departments. Cultures of soil can reveal sites of very heavy contamination. For such sites, if they are not very large, Powell and associates have recommended three sprayings of 3 percent formalin (55).

3.1.3 Bacillus anthracis

Inhalation anthrax, known for many years as "wool sorter's disease," has become exceedingly rare, but it still occurs occasionally in textile mills where imported goat hair is prepared for incorporation into fabrics of mixed fibers (57). On the basis of airborne infection of monkeys, Brachman and associates conclude that exposure to 1000 spores over a period of 3 to 5 days should correspond with an infection rate of 10 percent in a plant (58). Fortunately conditions for so serious an outbreak occur very rarely.

The result of inhalation of an adequate number of organisms is a highly lethal pneumonia. The occurrence of such a disease in an employee of a plant in which imported goat hair is used calls for immediate shutdown and disinfection of the entire plant.

Cleanup of an infected mill is difficult and expensive. Young et al. removed and burned all wooden material, dismantled all machinery and cleaned it with hydrocarbon solvents, and sanded and repainted all painted metal surfaces, first having exposed the entire building to vaporized formalin for 2 days. In Great Britain all goat hair is treated with formalin before it leaves the pier (57).

3.2 Zoönoses

3.2.1 Brucella

Although traditionally brucellosis has been associated with farm workers and particularly with consumption of unpasteurized milk, this mode of infection has diminished very strikingly in the United States. In Texas a review of cases reported to the Health Department establishes that after 1957 this mode of infection applied to only one-fifth of the cases (59). In Scotland, however, 62 percent of infections occurred among farm workers (60), and in Great Britain as a whole a review in 1975 still showed 70 percent of brucellosis to occur in farm workers (61).

In contrast in 1969 Busch and Parker reported that 68 percent of all cases in the United States had involved packing house employees; infections were derived from both swine and cattle (62). In a study of an outbreak of the disease in an abattoir in Illinois, Schnurrenberger and associates found 13 clinical cases of brucellosis during a 6 month period. In addition there were 71 serological reactions of 1:25 or more among 551 personnel. In this outbreak infected swine were the source of the human infection.

The review of Buchanan et al. (64) underscores the magnitude of the problem: among packing house employees between 1960 and 1971 there were 1644 symptomatic cases. In Iowa and Virginia the attack rate was highest, more than 20 human cases per million slaughtered animals. Study of the activities and the histories of the patients suggested that some cases developed as a result of infection through breaks in the skin, that probably the majority were airborne, and that perhaps a number may have occurred through conjunctival inoculation.

The development of acute illness among employees at a slaughter site calls attention to the possibility of brucellosis. Unfortunately the symptoms of this disease are quite nonspecific and the diagnosis may be overlooked unless a high degree of suspicion exists. Since brucellosis appears somewhat sporadically, and since its symptoms are so variable, the figures above may underrepresent actual rates of infection.

Though serological studies are of limited clinical value, they serve a useful purpose in surveys. The discovery of titers in an intermediate range, 1:32 to 1:128 among a group of employees specifically exposed would indicate that a problem exists. Office and sales personnel might serve as controls for the particular region of the country, since their jobs lead to little or no exposure.

Control is most difficult, since *Brucella suis* infection among pigs is often undetectable. Eradication of brucellosis from cattle herds is possible but is not yet complete. If the principal mode of infection is airborne or conjunctival, there is no conceivable protection

EVALUATION OF EXPOSURE TO BIOLOGICAL AGENTS

under operating conditions. The use of protective goggles might be considered, and exhaust or dilution ventilation of the atmosphere of the slaughter area might afford some protection. Complete eradication of infections among animals represents the only final protection for exposed humans.

3.2.2 Ornithosis

Employees engaged in poultry packing plants are at risk of ornithosis, which may be transmitted by birds of many kinds. At higest risk are workers engaged in plucking and evisceration. Durfee found that in the United States in 1971 71.7 percent of all reported cases occurred among the employees of two turkey processing plants, most of them during a period of only 2 months (65). In their review in 1976 Dickerson and associates found that 13 outbreaks of ornithosis occurred in the United States between 1948 and 1972, all of them in association with the slaughter and processing of turkeys. Of the 494 verified cases that developed, seven were fatal. Infection was traced back to specific flocks, from which birds were transported widely, and small epidemics occurred in Nebraska and Missouri (66). In Sweden Jernehus et al. traced two outbreaks in poultry plants to infected flocks of geese and ducks (67).

These reports suggest a higher attack rate and a much closer grouping of human cases than is characteristic of brucellosis. Evidence of a small epidemic should put one on guard, but the symptoms are such that the real nature of the problem might be masked by such diagnoses as "flu" or "atypical pneumonia."

Presumably most infections with ornithosis take place through inhalation, though inoculation cannot be entirely excluded. Control of the human infection depends on eradication of the disease among birds destined for human consumption. When it is appreciated that a flock poses a problem, quarantine and antibiotic treatment are necessary. Whether this management will later permit preparation of the birds for human consumption is uncertain. Complete eradication of the disease among domestic fowl is probably a practical impossibility: often infection of the birds is inapparent. Reinfection of flocks by wild birds makes control most difficult.

3.2.3 Q Fever

Q fever is a rickettsial infection caused by *Coxiella burnetti*, and it has appeared sporadically in the United States, chiefly as a disease of packing house employees. The organism has been shown to become airborne (69); as a particulate in the atmosphere it retains its viability and infectivity for long periods (70). Epidemiological studies in California also indicated that the organisms may survive on surfaces or in soil for long periods, only to become airborne again (71). Direct contact with infected animals (sheep, goats, and cattle) is not essential (69, 71); many cases evidently arise from organisms in the environment.

Infection among dairy cattle may be demonstrated by serological examinations for antibodies. In the Milwaukee area it was possible to isolate the organisms from individual animals in 42 of 50 serologically identified herds (72). Human infection, however, was not well correlated with the dairy or the cattle-raising industry. An outbreak in Texas, on the other hand, was marked by infection of 55 of 136 exposed workmen and was limited to handlers of cattle in the holding pens and to workers in the slaughter area of the abattoir (74).

Recognition of the disease may be delayed. It should be suspected if cases of severe pneumonia, sometimes associated with hepatosplenomegaly, occur among individuals engaged in the handling or slaughter of cattle, sheep, or goats.

Control measures ultimately depend on eradication of the infection among domestic animals. Since it appears possible that premises may remain infected for a long time after original contamination, immunization of employees at risk may become necessary. There is no experience with any method of environmental decontamination.

3.2.4 Mycobacterium marinum

M. marinum, an atypical mycobacterium, produces nodular, slowly necrosing lesions of the skin, usually that of the extremities, where inoculation takes place readily through current or previous abrasions and minor wounds. Most infections have occurred among children bathing in certain swimming pools or in adults who clean aquariums for tropical fish. However infections of commercial fishermen have occurred, both in the Gulf of Mexico and in Scandinavian waters (75).

Although the organism may produce fatal illness in tropical fish, it apparently results in no detectable disease in larger fish (or the cause may have been overlooked or not considered). Whether *M. marinum* may colonize the skin, fins, or mouths of fishes is uncertain, but a Swedish case suggests that the source of human infection was an inoculation from a fin (75) and another human infection followed the bite of a dolphin (76).

The disease should be suspected if commercial fishermen or workers at the dock develop slowly progressive nodular lesions of the extremities. These ultimately drain a small amount of material and heal spontaneously, but new crops of nodules may develop beyond the original sites. Tissue reveals necrotizing granulomas containing acid-fast bacilli.

Since lesions of the hands and forearms are so common among commercial fishermen, it is possible this infection has been overlooked. A survey of Gulf fishermen demonstrated that 10 percent reacted with positive skin tests to an antigen prepared from *M. marinum* (75). Even so, it appears that infections are probably quite rare, and since infection is limited to the skin and never spreads to deeper structures, it is of limited importance.

REFERENCES

1. A. Wanner, J. A. Hirsch, D. E. Greeneltch, E. W. Swenson, and T. Fore, *Arch. Environ. Health,* **27,** 370 (1973).

2. A. Tacquet, A. Collet, V. Macquet, J.-C. Martin, C. Gernez-Rieux, and A. Policard, *C. R. Acad. Sci. Paris,* **257,** 3103 (1963).
3. J. K. N. Lee, E. A. Pachtman, and A. M. Frumin, *Ann. N. Y. Acad. Sci.,* **234,** 161 (1974).
4. E. A. Kabat, in: *The Chemistry and Biology of the Mucopolysaccharides, A Ciba Symposium,* G. E. Wolstenholme and C. M. O'Connor, Eds., Little, Brown, Boston, 1958, p. 230; also G. F. Springer, in: *ibid.,* p. 230.
5. B. Pernis, E. C. Vigliani, C. Cavagna, and M. Finulli, *Br. J. Ind. Med.,* **18,** 120 (1961).
6. C. Cavagna, V. Foa, and E. C. Vigliani, *Br. J. Ind. Med.,* **26,** 314 (1969).
7. R. Rylander, A. Nordstran, and M. C. Snella, *Arch. Environ. Health,* **30,** 137 (1975).
8. P. J. Nicholls, *Br. J. Ind. Med.,* **19,** 33 (1962).
9. A. Bouhuys, A. Barbero, S.-E. Lindell, S. A. Roach, and R. S. F. Schilling, *Arch. Environ. Health,* **14,** 533 (1967).
10. M. Hitchcock, D. M. Piscitelli, and A. Bouhuys, *Arch. Environ., Health,* **26,** 177 (1973).
11. J. Pepys, *Ann. Intern. Med.,* **64,** 943 (1966).
12. J. Pepys, in: *New Concepts in Allergy and Immunology,* V. Serafini, A. W. Frankland, C. Musala, and J. M. Jamar, Eds., Amsterdam, Excerpta Medicine, Foundation, 1971, p. 136.
13. J. N. Fink, E. F. Banaszak, W. H. Thiede, and J. J. Barboriak, *Ann. Intern. Med.,* **74,** 80 (1971).
14. J. G. Scadding and K. F. W. Hinson, *Thorax,* **22,** 291 (1967).
15. G. W. H. Schepers, *JAMA,* **181,** 635 (1962).
16. J. M. Gowdy and M. J. Wagstaff, *Arch Environ. Health,* **25,** 101 (1972).
17. H. R. Imbus and W. M. Suh, *Arch. Environ. Health,* **26,** 183 (1973).
18. "The Status of Bissynosis in the United Sates, A Summary of the National Conference on Cotton Dust and Health," *Arch. Environ. Health,* **23,** 230 (1971).
19. J. D. Hamilton, G. M. Halprin, K. H. Kilburn, J. A. Merchant, and J. R. Ujda, *Arch. Environ. Health,* **26,** 120 (1973).
20. J. A. Merchant, G. M. Halprin, A. R. Hudson, K. H. Kilburn, W. N. McKenzie, Jr., D. J. Hurst, and P. Bermazohn, *Arch. Environ. Health,* **30,** 222 (1975).
21. C. B. McKerrow, M. McDermott, J. C. Gilson, and R. S. F. Schilling, *Br. J. Ind. Med.,* **15,** 75 (1958).
22. A. Bouhuys and E. Zuskin, *Ann. Intern. Med.,* **84,** 398 (1976).
23. P. E. McNall, Jr., *Arch. Environ. Health,* **30,** 552 (1975).
24. J. A. Merchant, J. C. Lumsden, K. H. Kilburn, V. H. Germino, J. D. Hamilton, W. S. Lynn, H. Byrd, and D. Baucom, *Br. J. Ind. Med.,* **30,** 237 (1973).
25. C. B. McKerrow, J. C. Gilson, R. S. F. Schilling, and J. W. Skidmore, *Br. J. Ind. Med.,* **22,** 204 (1965).
26. P. C. Elwood, J. D. Merrett, G. C. R. Carey, and I. R. McAulay, *Br. J. Ind. Med.,* **22,** 27 (1965).
27. J. O. Harris, D. Bice, and J. E. Salvaggio, *Am. Rev. Respir. Dis.,* **114,** 29 (1976).
28. G. A. doPico, W. G. Reddan, F. Chmelik, M. E. Peters, C. E. Reed, and J. Rankin, *Am. Rev. Respir. Dis.,* **113,** 451 (1976).
29. H. Weill, H. A. Buechner, E. Gonzales, S. J. Herbert, E. Aucoin, and M. M. Ziskind, *Ann. Intern. Med.,* **64,** 737 (1966).
30. D. P. Nicholson, *Am. Rev. Respir. Dis.,* **97,** 546 (1968).
31. G. A. doPico, W. Reddan, D. Flaherty, A. Tsiatis, M. E. Peters, P. Rao, and J. Rankin, *Am. Rev. Respir. Dis.,* **115,** 915, (1977).
32. C. S. Darke, J. Knowelden, J. Lacey, and A. M. Ward, *Thorax,* **31,** 294 (1976).
33. D. J. Riley and M. Saldana, *Am. Rev. Respir. Dis.,* **107,** 456 (1973).
34. G. T. Hensley, J. N. Fink, and J. J. Barboriak, *Arch. Pathol.,* **97,** 33 (1974).

35. F. Hargreave, K. F. Hinson, L. Reid, G. Simon, and D. S. McCarthy, *Clin. Radiol.*, **23**, 1 (1972).
36. G. T. Hensley, J. C. Garancis, G. D. Cherayil, and J. M. Fink, *Arch. Pathol.*, **87**, 572 (1969).
37. D. H. Allen, G. V. Williams, and A. J. Woolcock, *Am. Rev. Respir. Dis.*, **114**, 555 (1976).
38. R. S. Boyer, L. E. Klock, C. D. Schmidt, L. Hyland, K. Maxwell, R. M. Gardner, and A. D. Renzetti, Jr., *Am. Rev. Respir. Dis.*, **109**, 630 (1974).
39. C. P. W. Warren and K. S. Tse, *Am. Rev. Respir. Dis.*, **109**, 672 (1974).
40. L. Michaels, *Canad. Med. Assoc. J.*, **96**, 1150 (1967).
41. H. I. Cohen, T. C. Merigan, J. C. Kosek, and F. Eldridge, *Am. J. Med.*, **43**, 785 (1967).
42. H. I. Cohen, T. C. Merigan, and F. Eldridge, *Clin. Res.*, **13**, 346 (1965).
43. L. De Carvalho Cancella, *Ind. Med. Surg.*, **32**, 435 (1963).
44. F. J. Wenzel and D. A. Emanuel, *Arch. Environ. Health*, **14**, 385 (1967).
45. B. Gandevia, *Arch. Environ. Health*, **20**, 59 (1970).
46. E. D. Acheson, R. H. Cowdell, and E. Rang, *Br. J. Ind. Med.*, **29**, 21 (1972).
47. F. Valić, D. Ceritić, and D. Butković, *Am. Rev. Respir. Dis.*, **113**, 751 (1976).
48. D. W. Van Toorn, *Thorax*, **25**, 399 (1970).
49. C. J. Stewart, *Thorax*, **29**, 252 (1974).
50. T. Franz, K. D. McMurrain, S. Brooks, and I. L. Bernstein, *J. Allergy*, **47**, 170 (1971).
51. J. H. Dijkman, J. G. A. Borghans, P. J. Savelberg, and P. M. Arkenbout, *Am. Rev. Respir. Dis.*, **107**, 387 (1973).
52. M. L. Seviers, *Am. Rev. Respir. Dis.*, **109**, 602 (1974).
53. C. E. Smith, R. R. Beard, H. G. Rosenberger, and E. G. Whiting, *J. Am. Med. Assoc.*, **132**, 833 (1946).
54. D. J. D'Alessio, R. H. Heeren, S. L. Hendricks, P. Ogilvie, and M. L. Furcolow, *Am. Rev. Respir. Dis.*, **92**, 725 (1965).
55. K. E. Powell, K. J. Kammerman, B. A. Dahl, and F. E. Tosh, *Am. Rev. Respir. Dis.*, **107**, 374 (1973).
56. H. F. Hasenclever, M. H. Shacklette, R. V. Young, and G. A. Gelderman, *Am. J. Epidemiol.*, **86**, 238 (1967).
57. L. S. Young, J. C. Feeley, and P. S. Brachman, *Arch. Environ. Health*, **20**, 400 (1970).
58. P. S. Brachman, A. F. Kaufmann, and F. G. Dalldorf, *Bacteriol. Rev.*, **30**, 646 (1966).
59. S. J. Lerro, *Tex. Med.*, **67**, 60 (1971).
60. D. Reid, *Scot. Med., J.*, **21**, 125 (1976).
61. Anon., Editorial, *Lancet*, **1**, 436 (1975).
62. L. A. Busch and R. L. Parker, *J. Infect. Dis.*, **125**, 289 (1972).
63. P. R. Schnurrenberger, R. J. Martin, P. R. Wactor, and G. G. Jelly, *Arch. Environ. Health*, **24**, 337 (1972).
64. T. M. Buchanan, S. L. Hendricks, C. M. Patton, and R. A. Feldman, *Medicine*, **53**, 427 (1974).
65. P. T. Durfee, *J. Infect. Dis.*, **132**, 604 (1975).
66. M. S. Dickerson, W. R. Bilderback, and L. W. Pessarra, *Tex. Med.*, **72**, 57 (1976).
67. H. Jernelius, B. Pettersson, J. Schvarcz, and A. Vahlne, *Scand. J. Infect. Dis.*, **7**, 91 (1975).
68. C. Gale, *Proceedings of the 64th Annual Meeting, U.S. Livestock Sanitary Association*, 1960, p. 223. (No publisher or place given.)
69. E. H. Lennette and H. H. Welsh, *Am. J. Hyg.*, **54**, 44 (1951).
70. W. W. Spink, in: *Tropical Medicine,* 5th ed., Hunter, Schwarzwelder, and Clyde, Eds., Saunders, Philadelphia, 1976, pp. 188–192.

71. W. H. Clarke, E. H. Lennette, and M. S. Romer, *Am. J. Hyg.*, **54,** 319 (1951).
72. H. J. Wisniewski and E. R. Krumbiegel, *Arch. Environ. Health,* **21,** 58 (1970).
73. H. J. Wisniewski and E. R. Krumbiegel, *Arch. Environ. Health,* **21,** 66 (1971).
74. N. H. Topping, C. C. Shepard, and J. V. Irons, *J. Am. Med. Assoc.,* **133,** 813 (1947).
75. W. C. Miller and R. Toon, *Arch. Environ. Health,* **27,** 8 (1973).
76. D. J. Flowers, *J. Clin. Pathol.,* **23,** 475 (1970).

CHAPTER SIXTEEN

Toxicologic Data Extrapolation

PERRY J. GEHRING D.V.M., Ph.D., and
K. S. RAO, Ph.D.

1 INTRODUCTION

The ultimate objective of toxicological research on chemicals is to obtain information that will form a sound basis for recommending "safe" levels of exposure for humans contacting these substances during manufacture, use, and disposal. The information sought in such research is generally elucidation of the dose-response function that characterizes the untoward effects produced by selected doses of a chemical. Subsequently the "safe" level of exposure has been selected traditionally by judgment to be $1/10$ to $1/5000$ of the highest dose that produced no discernible effect (1, 2). Safety factors are selected in accordance with the seriousness and persistence of the adverse effects and the shape of the dose-response curve.

"Safe" as used here is a relative term. No matter what adverse effect is of concern, absolute safety to everyone regardless of conditions can never be assured for any chemical, or for that matter any activity in which man indulges.

Generally, data depicting the adverse response of humans to selected doses of a chemical are unavailable. In some incidences data from human experimentation are available; however the adverse effects selected for study are generally transient and mild, such as irritation or slight depression of central nervous system function. Rarely, epidemiological studies of people exposed during the manufacture and use of a chemical provide dose-response information for more severe adverse effects. When available, such information is most important.

Because human experimentation is not feasible to elucidate serious manifestations of toxicity, it is necessary to use experiments in animals to characterize such responses.

However using animals to predict the response of humans creates a difficulty. It is not possible to assure absolutely that people will not be more or less sensitive than the animal model studied. Also the human population is much more heterogeneous than are animal populations; therefore the range of doses producing a given effect in man may be much larger than in the animal species selected for experimentation. Indeed, experimental results from studies on small numbers of humans are not totally reliable for predicting the response of large population groups for the same reasons.

In spite of the foregoing deficiency, to satisfy the ultimate objective, data collected in studies using animals must be extrapolated to predict the response in man. This chapter provides insight to such extrapolation. The basic premise is that toxicity is manifest as a result of the presence of the toxic agent at a specified concentration at the receptor site in cells and tissues. Species differences in reactions to given exposures to a chemical occur generally because of differences in how the chemical is absorbed into the body, distributed and biotransformed once in the body, and excreted from the body. A difference in the reaction of the toxicant—the chemical per se or a product formed from it—with the receptor is perceived to be rare.

2 DOSE RESPONSE

Fundamental to the extrapolation process, whether intraspecies or interspecies, is the dose-response function. This concept was enunciated first in the sixteenth century by the physician-alchemist Theophrastus Bombastus von Hohenheim, better known as Paracelsus. In a Third Defense of the Principle written in 1538, he said: *"Was ist das nit gift ist? Alle ding sind gift und nichts ohn gift. Allein die dosis macht das ein ding kein gift ist."* ("What is it that is not poison? All things are poison and none without poison. Only the dose determines that a thing is poison.")

Figure 16.1 simulates for a normally distributed population, the percentage of individuals responding to a range of doses for two chemicals. For fictitious chemical I, the range of doses producing a discernible response is considerably smaller than for chemical II. Curves for both chemicals indicate that at some selected doses on the low side, only a few in a population will respond. At the other extreme are those few individuals in the population requiring large doses to elicit the response. Dose X for both chemicals will cause 50 percent of the population to respond, since the areas under the two curves to the left of this point constitute one-half of the total areas under the respective curves. The dose causing 50 percent of the population to respond is the effective dose, 50 percent or (ED_{50}); specifically for lethality the dose is termed LD_{50}.

The primary reason for showing simulated curves for two chemicals in Figure 16.1 is to emphasize the necessity of using different judgment to select a "safe" level of exposure. A sharply defined curve such as that for chemical I allows greater assurity that $\frac{1}{10}$ the dose not producing a response will be "safe." Because of the wide range of

TOXICOLOGIC DATA EXTRAPOLATION

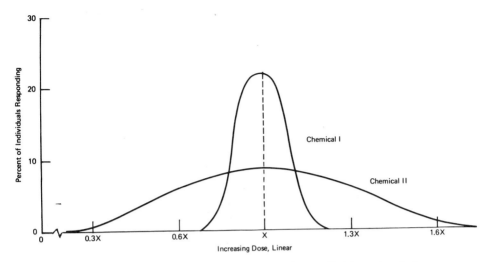

Figure 16.1 Illustrative dose-response curves for two fictitious chemicals.

doses producing a response to chemical II, a larger safety factor is needed. For this reason using an ED_{50} or LD_{50} to assess the relative hazard of a chemical is absolutely inadequate.

If the data in Figure 16.1 are plotted as the cumulative percentage of the population responding, a sigmoid curve will result, going from 0 to 100 percent. To provide greater sharpness of the sigmoid curve, the dose response is plotted frequently as the percentage responding versus the logarithm of the dose (see Figure 16.7, Section 6). Since straight rather than sigmoid line curves are preferred, the most desirable representation of the data is using a logarithmic probability display (3).

Utilizing a logarithmic probability display of data for the hepatotoxicity incurred from single doses of various chlorinated solvents to mice, Plaa et al. (4) published the results appearing in Figure 16.2. Obviously the ED_{50} values for the various compounds provide some indication of the relative hepatotoxicity. However with respect to judging what dose will not produce hepatotoxicity to any of the population, a much more conservative safety factor must be used for carbon tetrachloride and chloroform than for the other materials because of the shallow slope for these compounds.

Another important criterion for selection of a safety factor for a chemical lies in a comparison of the doses producing a subtle unnoticed effect with those producing an effect that can be readily discerned. To illustrate, refer to Figures 16.3 to 16.5, depicting the dose responses for hepatotoxicity, narcosis, and death for carbon tetrachloride, chloroform, and 1,1,1-trichloroethane, respectively (5).

For carbon tetrachloride and chloroform, unnoticeable hepatotoxicity can be occurring with exposure levels less than those that will produce narcosis. For 1,1,1-tri-

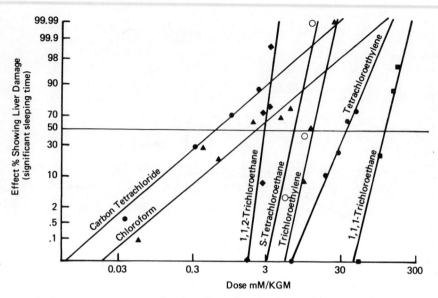

Figure 16.2 Dose-response curves for the effect of each of seven halogenated hydrocarbons on prolongation of pentobarbital sleeping time in mice.

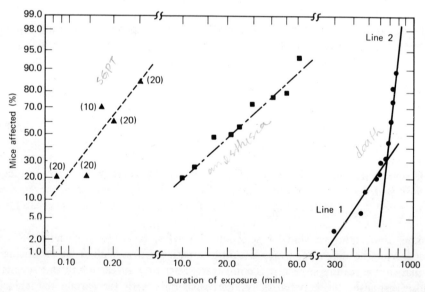

Figure 16.3 Carbon tetrachloride vapor, 8500 ppm. Percentage plotted on probability scale of mice anesthetized (squares) dead (circles), or having a significant serum glutamic acid–pyruvic acid–transaminase (SGPT) elevation (triangles), as a function of the $\log_{10}$ duration of exposure. Each point for anesthesia and lethality was obtained using a single group of 30 mice; the number in each group used to obtain the points for SGPT activity as given in parentheses.

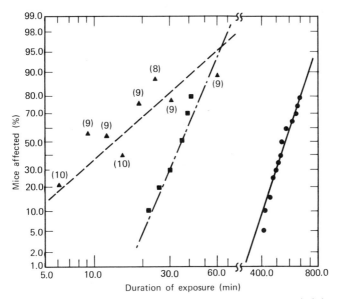

Figure 16.4 Chloroform vapor, 4500 ppm. Percentage plotted on probability scale, of mice anesthetized (squares) dead (circles), or having a significant SGPT elevation (triangles), as a function of the $\log_{10}$ duration of exposure. Each point for anesthesia was obtained using a single group of 10 mice, and each point for lethality was obtained using a single group of 20 mice; group size used for determining SGPT activity is indicated in parentheses.

chloroethane, however, injury to the liver occurs only at exposure levels in excess of those needed to produce narcosis. Indeed, liver injury occurs only when the exposure levels are sufficient to cause death to some individuals. For 1,1,1-trichloroethane, these results provide considerable confidence that prevention of narcosis, a noticeable effect, will preclude development of liver disease.

For carbon tetrachloride and chloroform, more restrictive judgments for tolerable exposures are needed to assure a nil hazard. The shallow curve depicting the hepatotoxicity incurred from various exposures to chloroform indicates more variability in the population; therefore a more conservative safety factor is justifiable for chloroform than for carbon tetrachloride. Saying it another way, flat dose-response curves indicates that susceptibility to the chemical is highly variable in the population and there is greater probability that a small number of individuals may be adversely affected even though most will be unaffected. Diethylene glycol is another substance that gives rise to a flat dose-response curve in most species. In humans this flatness has been reflected in human poisoning cases in which some individuals have died from ingestion of small doses, whereas others have survived relatively large doses (6). In contrast, Gonyoulex toxin, a substance occurring occasionally in clams and mussels, has a steep dose-response curve and one-fourth the LD_{50} can be eaten without measurable risk (7).

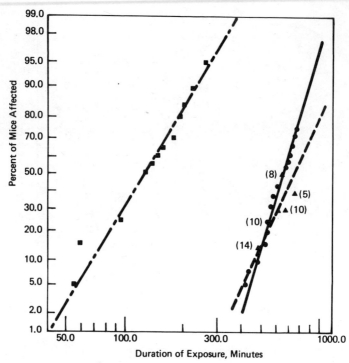

Figure 16.5 1,1,1-Trichloroethane vapor, 13,500 ppm. Percentage plotted on probability scale of mice anesthetized (squares), dead (circles), or having a significant SGPT elevation (triangles), as a function of the log₁₀ duration of exposure. Each experimental point for anesthesia and lethality was calculated using composite groups of 20 to 135 mice. Individual group sizes used to obtain SGPT activity are indicated in parentheses.

3 ACUTE TOXICITY

Acute toxicity of chemicals in animals is elucidated by administration of single doses by various routes of exposure. Although uncommon for exposure to most industrial chemicals, oral administration is used commonly as a first assessment of the potential of a chemical to cause toxicity. Too frequently, lethality is considered to be the most important adverse effect, and other signs of adverse effects such as the physical condition of the animals and damage to particular organs, are overlooked. Rigorous evaluation of these parameters frequently indicates the organ system affected by the chemical and on occasion the mechanism of toxicity.

In determining the acute oral toxicity, strict attention should be given to the persistence of the effects. For example, delayed deaths occurring 2 or more days following administration, or depression of body weight gain 1 to 2 weeks after administration, suggests either a persistence of the chemical in the body or slow repair of the damage incurred. In either case more conservative handling of the chemical to preclude adverse

TOXICOLOGIC DATA EXTRAPOLATION

effects on health is necessary. It may also be anticipated that repeated exposure to a chemical eliciting a delayed response will constitute a greater hazard than exposure to a chemical that does not.

Exposure to industrial chemicals occurs most frequently by way of contamination of the skin or inhalation of the vapor or dust. Contamination of skin may result in local damage and in some instances absorption into the body and systemic toxicity. Adverse local reactions of either skin or eyes are generally extrapolated directly to man, and appropriate measures are instituted to preclude contamination of the skin with injurious concentrations of chemicals.

Some chemicals penetrate the skin readily. And indication of ready penetration is an equivalency or near equivalency of acutely toxic doses, whether given orally or by skin application. Evidence of significant penetration of a chemical through the skin warrants institution of precautions to minimize skin contamination. The rigor of these precautions will be dictated by the degree of acute, subchronic, and chronic toxicity, usually revealed in studies relying on other routes for administration.

Exposure to chemical by way of inhalation is a major concern in the industrial environment. An initial assessment of the acute toxicity of a chemical by way of inhalation is made frequently by exposing animals to a concentrated atmosphere of the chemical for 6 to 8 hr. If injury or death occurs, the duration of exposure and the exposure concentrations are decreased progressively until the adverse effects disappear. The objective of such experimentation is to characterize the toxicity as a function of duration and concentration of exposure. Frequently this function can be visualized by plotting the log of the concentration versus the log of the exposure needed to produce a given effect or lack of effects.

To illustrate, consider the data of Adams et al. (8) depicting the acute toxicity of carbon tetrachloride (Figure 16.6). For rats, line *CD* represents the most severe exposures not resulting in death, and *EF* represents exposures causing no discernible effects. The large difference between exposures represented by these lines carries the same interpretive significance as discussed previously for the data in Figure 16.3.

In the absence of human experience, tolerable single exposures to carbon tetrachloride would be set tenfold or more less than those represented by line *EF*. Data for additional species—dog, monkey, rabbit—indicating lesser or equivalent susceptibility, increase confidence that a tenfold margin of safety is adequate.

For many chemicals such as carbon tetrachloride, human experience will augment the judgment. When Adams et al. (8) reported their results, sufficient human experience was available to support the conclusion that single exposures like those represented by *EF* did not produce discernible adverse health effects in man.

4 SUBCHRONIC AND CHRONIC TOXICITY

Only crude initial judgments are possible from acute toxicity tests. If carefully done, they provide insight into the potential toxicity on repeated exposure to a chemical, but

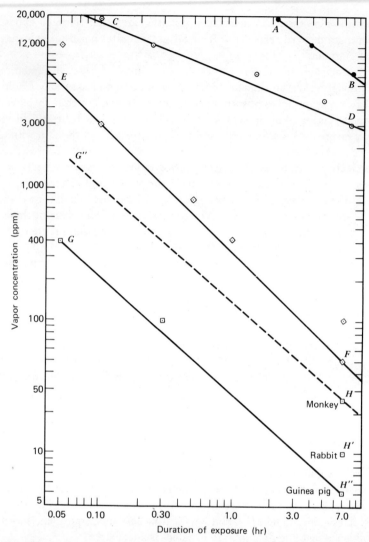

Figure 16.6 Vapor toxicity of carbon tetrachloride. Line *AB* represents the least severe single exposures causing death of all rats tested and line *CD* the most severe single exposures, permitting survival of all rats tested. Line *EF* represents the most severe single exposures without detectable adverse effect and line *GH* the most severe repeated exposures without detectable adverse effect in rats. Points, *H*, *H'*, and *H''* represent the most severe repeated exposures without detectable adverse effect in the guinea pig, the rabbit, and the monkey, respectively. Line *G''H''* represents the most severe repeated exposures with little probability of adverse effect in human subjects.

TOXICOLOGIC DATA EXTRAPOLATION

repeated exposure studies are needed to provide a more definitive assessment of subchronic and chronic toxicity.

In studies to characterize the potential subchronic and chronic toxicity, chemicals are administered to animals daily for 5 to 7 days per week for an entire lifetime or a fraction thereof. When administered orally, the chemicals are admixed with the diet when feasible. Inhalation exposure is conducted in chambers using typically a regimen of 6 to 8 hr per day, 5 days per week. Occasionally other routes of exposure are used (e.g., daily application to the skin or injection). Generally extensive ante mortem and post mortem evaluations are made using the most recent clinical, chemical, and pathological methods. Typically, more than 100 different parameters are assessed.

In the report of Adams et al. (8) the subchronic toxicity of carbon tetrachloride was also reported. Line GH in Figure 16.6 represents daily exposures that did not produce discernible adverse effects in rats over a duration of 6 months. Also shown as single points are no effect exposures for guinea pigs, rabbits, and monkeys exposed 7 hr per day, 5 days per week, for 6 months. In these studies rats and guinea pigs appeared to be most susceptible; rabbits and monkeys were intermediate and least susceptible, respectively. The liver was the target organ. Since human experience indicated that man was less susceptible than rats, a line, $G''H''$ parallel to GH was drawn to represent exposures judged to be safe in man. As a result, 25 ppm was recommended as the standard for the work environment for 7 hr daily exposures.

In the absence of human experience, exposures represented by line GH or possibly even lower exposures may have been selected. To preclude subchronic toxicity, a workroom standard of 5 ppm would have been justified in that this would have protected the most sensitive species, rats and guinea pigs. The volume of air breathed by these species per unit body weight is greater than for man. Thus the dose received by man is less, providing a built-in safety factor.

A key question in subchronic and chronic toxicology studies is, How long should the animals be exposed? Ideally, chronic toxicology studies should involve treatment of the animals for their lifetime, considered to be 2 years for rats and mice. When the chemical is administered by routes other than inclusion in diet, administration for a lifetime requires an extensive effort. For example, daily generation of atmospheres containing the vapor of a chemical and chemical analyses to assure the desired concentration is time-consuming and costly, as is handling of the animals. In addition, there is a dearth of physical facilities to perform inhalation studies, and tying up these facilities precludes studying other materials. Therefore chronic inhalation studies often are conducted by using daily exposures for a fraction of the lifetime (e.g., 6 to 18 months) and subsequently maintaining the animals for the duration of the lifetime. This protocol duplicates reasonably well exposures incurred by workers. Also the track record of this protocol for revealing carcinogenic effects of a chemical is good, although no definitive studies comparing the results of daily administration for a lifetime versus a fraction of a lifetime are available.

A number of authors have demonstrated that toxicology studies exceeding 3 months in duration are usually unnecessary to reveal the potential chronic toxicity of chemicals

(9–14). Weil and McCollister (15) compared the minimum dose causing a toxic effect and a maximum dose causing no effect for 33 chemicals in short-term (29 to 200 days) versus long-term studies. The ratios of the indicated dosage levels for short-term versus 2 year feeding studies were 2.0 or less for 50 percent of the chemicals. A ratio greater than 5 occurred for only three chemicals; the largest ratio for a maximum no-effect level was 12.0. These results suggest that in lieu of chronic toxicity data, selection of $\frac{1}{10}$ the no-effect level in a subchronic study as a no-effect for chronic toxicity may be appropriate until a definitive study is conducted. Again judgment should be tempered by human experience when available.

As indicated previously, subchronic toxicity studies frequently reveal adequately the potential chronic toxicity of a chemical. However as more chemicals are studied, exceptions are becoming too common to permit overreliance on this generality. Recent as yet unreported chronic toxicity studies on a few chemicals in the Toxicology Research Laboratory, Dow Chemical U.S.A., have revealed toxic manifestations not observed in prior studies of 6 months duration. Therefore subchronic toxicity studies should be considered as interim assessments of potential toxicity for chronic, repeated, low level exposures.

As indicated previously, chronic studies by way of inhalation exposure constitute a major effort. For many chemicals, only studies utilizing exposure by way of ingestion may be available. To estimate the maximum dose that will not cause an effect by way of inhalation, it may be assumed that a worker will inhale 10 m³ of air per 8 hr workday (16). Since not all the chemical inhaled is retained, a retention factor is suggested as follows (16, 17):

	Water Solubility (%)	Retention Factor (%)
Essentially insoluble	<0.1	10
Poorly soluble	0.1–5	30
Moderately soluble	5–50	50
Highly soluble	50–100	80

The estimated dose is calculated in accordance with the equation

$$\text{estimated dose} = \frac{(10 \text{ m}^3)(\text{mg/m}^3)(\text{retention factor})}{\text{kg body weight}}$$

Estimation in accordance with the foregoing is frequently useful. However it is not recommended that such an estimation be used to satisfy the need for more definitive studies (i.e., inhalation exposure or pharmacokinetic evaluation) to characterize the fate of inhaled versus ingested chemical.

Assessment of the chronic toxicity potential of chemicals must be considered the ultimate objective to provide a basis for setting tolerable exposure levels for man. Typically, in chronic toxicity studies, 50 to 100 animals per sex per species are exposed to each of three to four levels of the agent, together with an equal or greater number of controls. Two species are studied frequently. Unfortunately tunnel vision has resulted too frequently in using such studies to study only carcinogenicity. Chronic toxicity studies should be aimed at assessing as best possible any manifestation of an untoward effect. Furthermore, interpretation of a carcinogenic response without assessing other manifestations of toxicity is an unscientific exercise in futility, since secondarily induced carcinogenesis occurs (18) and since other manifestations of toxicity may be equally serious with respect to worker's health.

5 RELIABILITY OF ANIMAL STUDIES FOR PREDICTING TOXICITY IN MAN

There is no comprehensive, definitive assessment of the reliability of predicting toxicity or lack thereof in man using toxicity data collected in animals. Although surprising initially, this apparent deficiency is understandable because such an assessment requires dose-response data for humans as well as animals. Doses of chemicals causing frank signs of toxicity cannot be administered intentionally to people. Even when human studies are conducted, reliances on symptoms such as nausea, headache, anxiety, depression, disorientation, weakness, or irritation as perceived by the subjects precludes, for the most part, a correlation with animal studies in which the parameters are limited to those that can be observed and measured by the investigator: signs versus symptoms.

Correlations of the adverse effects in man with those in animals exist predominantly for drugs. Review of this extensive literature will cause considerable anxiety because it contains many instances of studies in animals that have revealed both false positive and false negative results with respect to man (19, 20).

In the study of six unnamed drugs, Litchfield (21) found a significant relationship between the signs of toxicity in man, rats, and dogs. However 23 of 234 signs of toxicity were seen only in man. Rats and dogs did not predict symptoms unique to humans—headache, loss of libido, and so on. Dogs were found to be somewhat more useful in predicting the drugs effect in man than rats.

Classical examples of drugs that induce false positive results in animals are fluroxene, an anesthetic that has been used uneventfully in man but kills dogs, cats, and rabbits; and penicillin, which in doses that would be therapeutic for humans, produces lethality to guinea pigs (22).

Rall (23) demonstrated an excellent one-to-one correlation between toxicity data for 18 anticancer agents in mouse and man when the dose was based on milligrams per square meter of body surface area (mg/m^2). When examined on a milligram per kilogram basis, an excellent correlation also existed; however the mouse was consistently twelve fold less susceptible than man. In the original publication of the correlation of toxic responses in man and animals for the antineoplastic agents, monkeys, dogs, and rats were also considered (24). Assessments in these and other animal species added little except to substantiate better the existence of good quantitative correlation when the dose is expressed as milligrams per square meter rather than milligrams per kilogram.

Administration of biologically active chemicals, such as antineoplastic agents, frequently induces equivalent responses when the dose is administered in proportion to body surface area (25). This relationship is gaining recognition in estimating the risk incurred by man from exposure to chemical in the environment (26). In utilizing this relationship, however, it must be recognized that the original relationship was developed for biologically active agents. Since metabolism and other physiological processes involved in detoxification are generally more active in smaller animals, the dose of a biologically active chemical per unit mass required to produce a given effect increases as the body mass decreases, while the dose per unit surface area remains relatively constant. For a chemical requiring activation to the biologically active toxic form, however, the opposite is likely to be generally true because the amount transformed will be roughly proportional to the body surface area. Thus for a chemical requiring conversion to the active toxicant, a greater dose of the active form may be received on a milligrams per kilogram or concentration basis because the smaller animal has a greater surface area to body weight ratio.

In reviewing animal experimental data for assessing the carcinogenicity of chemicals in man, Wands and Broome (27) reported that studies in animals have revealed the carcinogenicity of all known human carcinogens except arsenic. Even for arsenic an increased incidence of leukemia and malignant lymphomas in pregnant Swiss mice has been reported (28). Also pointed out by Wands and Broome (27) was the existence of positive carcinogenic responses in animals for roughly 1000 compounds having no evidence for a carcinogenic response in man.

With respect to the foregoing discussion of the apparent lack of reliability of data from animal experiments for predicting toxicity in man, how can the usefulness of animal experimentation be advocated for this purpose? It must be recognized that exceptions to a correlation rather than correlations tend to be reported. Existence of toxicity data in animals and application of precautions to avoid toxicity in man for literally thousands of chemicals preclude visibility of a correlation.

Undoubtedly the "false positive" finding of carcinogenic activity in animals for roughly 1000 chemicals is attributable to the subjects' limited exposure to these agents. Exposure of people to amounts of these agents found to be carcinogenic in animals would very likely change the status of most of these apparent "false positives." For many of these "false positive" chemicals, the doses used to elicit a response were sufficient to overwhelm detoxification mechanisms and to cause prominent manifestation of toxicity other than carcinogenicity.

For drugs, it must be recognized that the apparent weakness, not absence, of a correlation for signs of toxicity is frequently qualitative. This is not surprising because administration of therapeutic or near therapeutic doses of biologically active agents to highly integrated biological systems may be expected to produce variable qualitative intra- and interspecies responses. Such differences in the qualitative manifestations of toxicity in response to toxic doses of a chemical should engender less concern than the absence of any manifestations of toxicity in animals given a dose of a chemical later found to be toxic to man. Although of less concern, qualitative manifestations of toxicity

TOXICOLOGIC DATA EXTRAPOLATION

in animal experimentation are important because they provide a basis for selecting parameters to be monitored in an exposed population of people.

To conceptualize further the problems inherent to extrapolating toxicity from animal experimentation to predict the hazard of man, the dynamics of toxicity must be realized. The toxicity of a chemical is a function of absorption of the chemical into the body, distribution, and biotransformation of the chemical once in the body, reaction of the chemical per se or a biotransformation product with the biological receptor leading to the ultimate action, and excretion of the chemical per se or its biotransformation products. If each of these five processes is an independent variable and each has a correlation coefficient of .9 for man versus animals, the overall correlation coefficient will be .6. In spite of the unacceptable prediction of such a correlation, animal experiments have been much more effective in providing data that have been used as a basis of recommending tolerable exposure levels to chemicals. Application of adequate safety factors ranging from $\frac{1}{10}$ to $\frac{1}{5000}$ for most industrial chemicals may be largely responsible for this unpredictable success. For drugs, larger fractions of those producing toxic responses in animals are administered in carefully controlled clinical trials. Hence toxic manifestations have been observed more frequently in man for drugs than for other chemicals.

Regardless of the success experienced in using toxicity data from animal experimentation as a basis for recommending tolerable exposures for man, future development of the science of toxicology must be directed at improving the reliability of this extrapolation. Consideration of the five factors influencing the toxicodynamics of chemicals reveals immediately how this can be done.

First, it requires quantitative knowledge of the absorption, distribution, biotransformation, and excretion of a chemical or its biotransformation products in man versus animals. This may be accomplished in pharmacokinetic evaluations of the chemical, which are becoming more common. Second, the reaction of the active chemical with the biological receptor or the mechanism of action must be elucidated. The mechanism of the toxicity of a chemical is not, however, as easily revealed as is its pharmacokinetics. Modern technology and conceptualization are allowing mechanisms of toxicity to be elucidated for a few chemicals. In the meantime, elucidation of the pharmacokinetic parameters will lead to dramatic improvements of the already quite reliable extrapolation to man of toxicity data obtained in animals.

6 THRESHOLD

In typical dose-response curve (Figure 16.7), the solid line represents an observable increase in the percentage of individuals responding to increasing doses. This type of dose response occurs when the response of individuals within a population is distributed normally, the situation characteristic for most pharmacological and toxicological responses to chemicals. The concept of "threshold" has been accepted for most pharmacological and toxicological responses. Biologically "threshold" has been

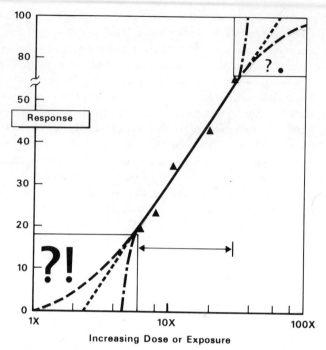

Figure 16.7 Simulated percentage of individuals responding adversely to the logarithm of selected doses. Measurable responses are represented by triangles. The sigmoid (dashed) curve represents a population described by normal distribution; in theory the percentage responding never reaches zero on the low end or 100 on the high end. The other curves represent a threshold for the response. The boxed-in portions represent regions in which prediction of incidence depends on stochastic, statistical projection.

interpreted to mean that there is a dose below which no response will occur in a population of animals. Hence when a dose is found that did not produce a toxic response in a reasonable number of subjects (laboratory animals or humans), it is assumed that the dose is subthreshold.

"Threshold" for an adverse response to a chemical differs with species and various physiological factors as well—age, sex, diet, stress. In a population where these variables have been eliminated, it is assumed that an infinite number of animals or people may be exposed to a subthreshold amount of the chemical without a response. Such an assumption renders much easier prediction of a "safe" exposure to a chemical for a population. After finding a dose eliciting no response in an experiment, a safety factor is merely applied to this dose to account for species and physiological differences. Subsequently, no experience of an adverse effect in a few individuals of a population is taken as confirmation of the appropriateness of the safety factor, and the existence of a subthreshold is assumed for the entire population.

A threshold concept is used to deem "safe" many events or materials in our lives other than chemicals. If a building is constructed to withstand a wind force of 100 mph, it is assumed that construction of millions of duplicates will not result in some that will be devastated by a 10 mph wind. In lieu of such a concept, a judgment of "safe" for any human endeavor is impossible.

For chemical carcinogenesis, teratogenesis, and mutagenesis, the threshold concept is not universally accepted. It is accepted generally that increasing doses of such agents will produce increased incidences of the response, represented by the solid line fitted to doses producing a discernible response in Figure 16.7. To establish existence of a threshold, however, statistically interpretable data must be acquired at doses below those that elicit experimentally discernible responses, represented by the boxed-in area containing the question mark and the exclamation point. The dilemma is whether the incidence of cancer, mutations, or terata, goes quickly to zero as the dose is decreased as in the case of the threshold, or whether the incidence will decrease as predicted by extrapolation of the experimentally discernible response function or, indeed, whether the incidence may be greater than that predicted by such extrapolation. Because of the incidence of spontaneous tumors in animals, experimental resolution of this dilemma is precluded by the need for thousands to millions of animals per experiment.

The argument that no threshold exists for carcinogens is essentially as follows: cancer is an expression of a permanent, replicable defect resulting from amplification of a defect initiated by reaction of a single molecule of a chemical carcinogen with a critical receptor. Primary support for this argument is that exhaustive experiments on radiation-induced cancer have failed to reveal a threshold within the realm of statistical reliability. Equating cancer induced by radiation and cancer induced by chemicals is tenuous, however, since the latter involves absorption, distribution, biotransformation, and excretion and the former does not. Even for radiation, Evan (29) has demonstrated clearly smaller incidences of cancer in individuals exposed to low doses of radium than were predicted by those exposed to large doses.

Gehring and Blau (18) have enumerated evidence for the existence of a threshold for chemical carcinogens. Summarizing:

1. Carcinogenesis is a multistage process. Interference with any of the processes precludes cancer development.
2. There is evidence of mechanisms that suppress development of cancer even though cells programmed to become cancer exist in the body.
3. As the dose of carcinogens is reduced, the latent period for development of cancer increases. A latent period longer than the life of individuals in a population is for all practical purposes a threshold.
4. Chronic inflammation causes an increased incidence of cancer, as does stress. Therefore it is not unreasonable to anticipate that administration of high, toxic doses of chemicals will also produce cancer secondarily.

5. There is a substantial and growing body of evidence that carcinogenesis is subject to immunosurveillance, particularly cell-mediated immunity.

6. Mechanisms exist to repair desoxyribonucleic acid that has been reacted with a chemical in a manner that has programmed it for cancer induction. For example, a genetic deficiency of repair mechanisms in people having xeroderma pigmentosa renders them very susceptible to radiation-induced cancer (30).

7. Man and animals are exposed to a sea of carcinogens, some man-made but most preexisting the contributions of our species. Overnutrition as well as overindulgence in exposure to chemicals increases the incidence of cancer. A much, much larger share of human cancer can be attributed to the former cause.

Arguments for the existence of a threshold or lack thereof for chemical carcinogens, mutagens, and teratogens will continue, since no totally definitive solution is possible in the foreseeable future. The same type of argument can be extended to almost all manifestations of chemical toxicity. Indeed, similar arguments can be used to deem unsafe almost all activities of man, since almost all constitute some degree of hazard. On the positive side, the crisis precipitated by the "threshold" concept for chemicals has provided a great motivation to assess more definitively the risks of chemicals. The resulting technological advancement will undoubtedly benefit not only the science of toxicology but other sciences devoted to the health of man and his environment as well.

7 PHARMACOKINETICS IN INTER- AND INTRASPECIES EXTRAPOLATIONS OF TOXICITY DATA

As indicated previously, the dynamics of toxicity includes absorption of the chemical into the body, distribution, biotransformation of the chemical in the body, and excretion of the chemical itself or of the biotransformation products formed from the chemical. Toxic manifestations produced by a chemical are functions of the concentrations of the toxic entity at the target sites and the duration of exposure of these sites to the toxic entity.

Hence to assess the hazard of a chemical to man as well as other species, it is essential to elucidate the kinetics for its absorption, distribution, biotransformation, and ultimate excretion, that is, its pharmacokinetics. Only with acquisition of such information can interspecies, intraspecies, and high dose to low dose extrapolations of potential toxicity be made definitively. Gehring et al. (31) should be referred to for a detailed discussion of the use of pharmacokinetics to assess the toxicological and environmental hazards of chemicals. Gehring et al. (32) provide insight of how knowledge of the pharmacokinetics of vinyl chloride assists in the risk assessment of this substance.

The utility of pharmacokinetics in the hazard evaluation of chemicals is illustrated here using two examples.

TOXICOLOGIC DATA EXTRAPOLATION

7.1 2,4-Dinitrophenol (DNP): Linear Pharmacokinetics to Elucidate the Difference in Species Susceptibility to the Cataractogenic Activity

DNP was known for some time to be cataractogenic in fowl and humans but not in laboratory animals used routinely to evaluate toxicity. Gehring and Buerge (33) showed that rabbits less than 62 days of age, but not older, developed cataracts when treated with DNP.

To resolve the species and age difference in susceptibility to the cataractogenic activity of DNP, a pharmacokinetic study was undertaken to determine the concentration-time relationship for DNP in the serum, aqueous humor, vitreous humor, and lens of mature and immature rabbits and ducklings. The results appear in Figures 16.8, 16.9, and 16.10, respectively. The rate constants for the apparent first-order elimination of DNP from the plasma of mature and immature rabbits were 0.82 and 0.15 hr^{-1}, respectively. For ducklings, elimination of DNP from plasma was biphasic and in accordance with a two-compartment open model system; elimination rates for the two phases were 0.25 and 0.11 hr^{-1}, respectively. Thus susceptibility correlated with a slower rate of clearance of DNP from plasma.

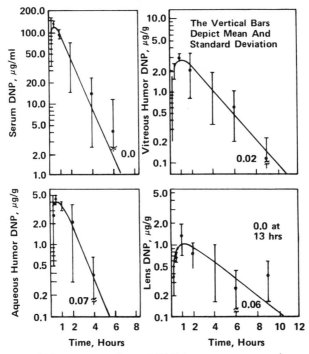

Figure 16.8 Mature rabbit: concentrations of DNP in serum, aqueous humor, vitreous humor, and lens as a function of time following intraperitoneal administration.

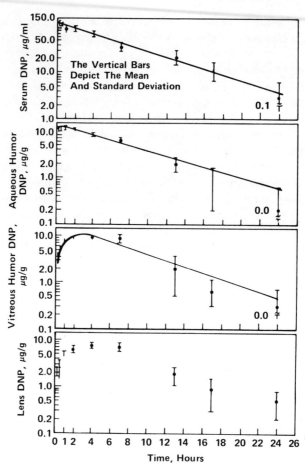

Figure 16.9 Immature rabbit: concentrations of DNP in serum, aqueous humor, vitreous humor, and lens as a function of time following intraperitoneal administration.

Associated with the slower rate of clearance were the higher concentration of DNP attained in the aqueous humor, vitreous humor, and lens. Furthermore, mathematical analyses of the data for the concentration-time relationships for DNP in plasma and aqueous humor allowed determination of the ratio of the rate of movement of DNP from aqueous humor to plasma k_{ef} to the rate of movement into aqueous humor from plasma k_{in}. This ratio is, in essence, an expression of the blood–aqueous humor barrier; the higher the value of k_{ef}/k_{in}, the more substantial the barrier. For mature rabbits, immature rabbits, and ducklings, the values were 15.8, 6.3, and 3.5, respectively.

The results of these pharmacokinetic studies elucidate the reason for species and age differences in the cataractogenic activity of DNP. Demonstration that the blood–aqueous humor barrier served to decrease the cataractogenic activity, suggesting

7.2 1,4-Dioxane: Saturation of Metabolism Correlated with Toxicity in Rats

Studies by Kociba et al. (34) demonstrated a small increased incidence of hepatomas and nasal carcinomas in rats maintained on drinking water containing sufficient dioxane to provide daily doses exceeding 1000 mg/kg. This dose supersedes that necessary to produce death in some rats and marked pathology of the liver and kidneys in all. Hepatic and renal damage but no tumors occurred in rats receiving 100 mg/kg daily. No untoward effects were discernible in rats receiving 10 mg/kg daily. Humans exposed to the current threshold limit value (TLV) established by the American

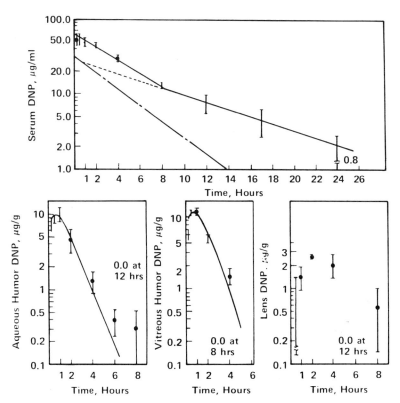

Figure 16.10 Duckling: concentrations of DNP in serum, aqueous humor, vitreous humor, and lens as a function of time following intraperitoneal administration. The lowest line depicting the disappearance of DNP from serum was determined by projecting the line representing the concentration of DNP in serum between 8 and 24 hr to the origin and subtracting the projected values from the experimentally determined values. Vertical bars depict means and standard deviations.

Conference of Governmental Industrial Hygienists in 1977—namely, 50 ppm (180 mg/m^3) for 6 hr—received a total dose of 5.4 ± 1.1 mg/kg (35). Thus the meaningfulness of toxicological data obtained at doses more than one hundredfold greater was called into question by the results of the pharmacokinetic data.

Figure 16.11 presents the plasma concentration versus time curves for ^{14}C-dioxane in rats given single intravenous doses of ^{14}C-dioxane ranging from 3 to 1000 mg/kg. The clearance of dioxane from plasma is markedly dose dependent and in accordance with Michaelis-Menten kinetics. The area under the curve increases disproportionately with dose, indicating that the elimination of dioxane is a saturable, dose-dependent, or nonlinear pharmacokinetic process. The pharmacokinetic model that described best the data was a parallel combination of Michaelis-Menten and first-order elimination. Parameters for this combination were V_d- 301 ± 41 ml; V_m = 13.3 ± 1.1 µg/(ml)(hr); K_m = 20.9 ± 2.0 µg/ml, and k_e = 0.0149 ± 0.0015 hr^{-1}. The maximum capacity for elimination of dioxane by the rat is 4003 µg/hr. Thus a dose of 1000 mg/kg to a 250 g rat exceeds 62 times the maximum capability of the rat to eliminate dioxane per hour.

The excretion of ^{14}C-activity by rats given various doses of ^{14}C-dioxane also demonstrated dose-dependent kinetics; as the dose was increased, more dioxane per se was eliminated by way of exhalation; at low doses essentially all was excreted rapidly as β-hydroxyethoxyacetic acid (HEAA) in the urine. Thus the biotransformation of 1,4-dioxane to the detoxification product HEAA is a saturable process that is overwhelmed by increasing the magnitude of the dose. The marked retention of dioxane with an increased dose led to the conclusion that the metabolism of dioxane must be induced markedly with repeated daily doses.

Figure 16.12 shows the body burden of radioactivity in rats given repeated daily oral doses of 10 or 1000 mg/kg for 17 days. The body burden was calculated by subtracting the total cumulative excretion by all routes from the total cumulative dose and expressing the results as a percentage of the total cumulative dose. The body burden in rats given 10 mg/kg of dioxane daily averaged about 5 percent and ranged between 2 and 9 percent with no apparent upward or downward trend. However, a striking decrease occurred in the body burden of rats given 1000 mg/kg of dioxane daily during the first four days of administration. This indicates that a daily dose of 1000 mg/kg but not 10 mg/kg caused a marked induction of the elimination of dioxane. In essence, the rats receiving 1000 mg/kg of dioxane daily had undergone marked biochemical alterations; their responses to dioxane, toxicological or carcinogenic, no longer can be extrapolated to rats receiving low doses.

Metabolic induction itself has been shown to increase tumorigenesis. Peraino et al. (36) demonstrated an enhancement of tumorigenesis in mice whose diet contained 0.05 percent phenobarbital, a well-known inducer of metabolism. Furthermore, induction of metabolism in rats by phenobarbital enhanced tumor production by 2-acetylaminofluorene (2AAF), a known hepatic carcinogen (37), suggesting that induction may enhance the expression of tumors by naturally occurring carcinogens.

To assess the potential hazard incurred by inhalation of dioxane, rats were exposed to 50 ppm of dioxane for 6 hr. After 6 hr of exposure, a steady state level of 7.3 µg/ml

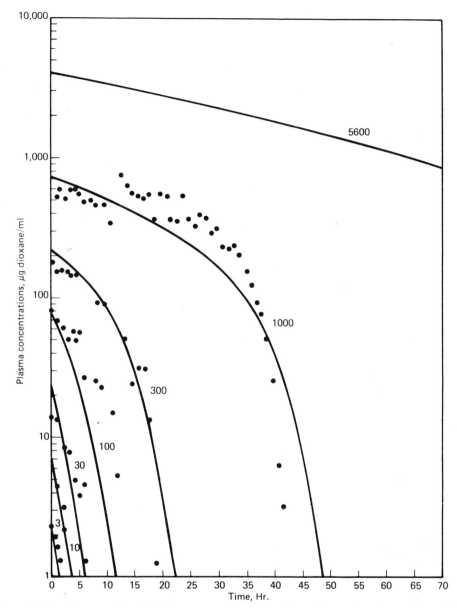

Figure 16.11 Concentration of dioxane per se in plasma of rats given various intravenous doses of dioxane.

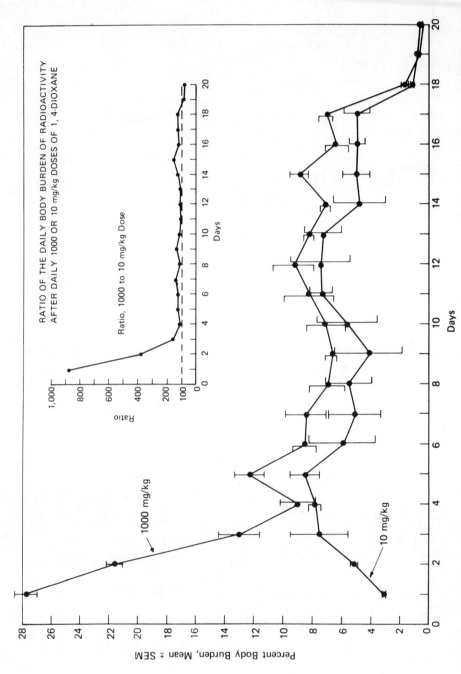

Figure 16.12 Daily body burden of ^{14}C-radioactivity of rats given oral doses of 10 or 1000 mg of dioxane per kilogram of body weight. The body burden was calculated by subtracting the total cumulative excretion by all routes from the total cumulative dose and expressing the results as a percentage of the total cumulative dose.

had been attained in plasma. Following exposure, the rate of elimination of dioxane from plasma was equivalent to that observed after low intravenous doses of dioxane, $t_{1/2}$ = 1.01 hr. Thus this level of exposure had not saturated the detoxification mechanism for dioxane.

To assist in extrapolation of toxicological data to man, four human volunteers were exposed to 50 ppm of dioxane for 6 hr (35). The concentrations of dioxane and of its metabolite HEAA in plasma during and following exposure are plotted in Figure 16.13. During exposure a steady state level was attained. After exposure the elimination of dioxane was apparent first-order kinetics having a $t_{1/2}$ of 1.0 hr^{-1}. As in the rat, essentially all the dioxane inhaled was eliminated as HEAA in the urine. This conclusion was reached because the amounts of HEAA excreted in the urine of the subjects was equivalent to the amount expected if all the dioxane in an assumed tidal volume of

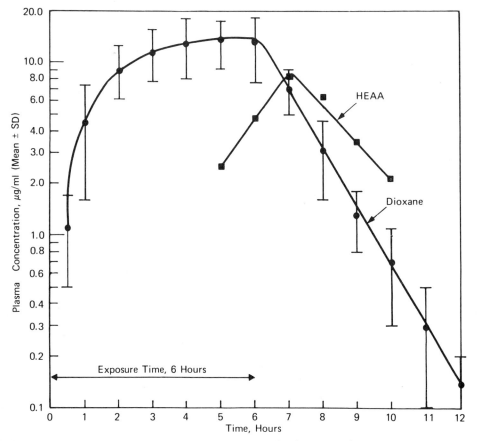

Figure 16.13 Plasma concentration versus time curves for dioxane and HEAA concentrations are averages for two to three individuals.

10 liters/min had been absorbed. Using the amount of HEAA excreted in the urine to estimate the total dose received, it was further determined that a human exposed to 50 ppm of dioxane receives, on a per kilogram basis, about one-thirteenth the dose received by a rat exposed to the same concentration. Thus for a material such as dioxane, which is absorbed readily upon inhalation and metabolized, experiments conducted in rats provide a thirteenfold safety factor.

Using the pharmacokinetic data together with the toxicological data, it is possible to conclude that in rats adverse effects are encountered only when doses of dioxane supersede those that can be detoxified readily without induction of morphological and biochemical changes. Since people exposed to 50 ppm dioxane detoxify dioxane readily, it is reasonable to conclude that it is highly unlikely that this level of exposure will be associated with untoward effects. It is highly significant that exposure of rats to 111 ppm dioxane, 7 hr/day, 5 days/week, for 2 years causes no untoward effects (38). In rats this level of exposure provides a daily dose on a per kilogram basis thirtyfold greater than that which would be received by a man exposed continuously to 50 ppm for 6 hr.

8 MECHANISM OF TOXICITY IN INTER- AND INTRASPECIES EXTRAPOLATION OF TOXICITY DATA

In addition to elucidation of pharmacokinetic parameters, future progress in increasing the reliability of extrapolation of toxicity resides in resolution of the mechanism of toxicity. To illustrate the usefulness of understanding the mechanism of toxicity, consider the dose-dependent clearance of O-6-methylguanine from the kidney of rats treated with the potent carcinogen dimethylnitrosamine (DMN). This work was conducted by Pegg et al. (39, 40) and is an extension of previous research by others (41–44). Administration of a large single dose of DMN induces kidney tumors in rats; single administration of DMN does not induce liver cancer. Following a single intraperitoneal dose of 2.5 or 20 mg/kg of DMN, alkylated bases of DNA were isolated from the liver and kidney. The clearance of N-7-methylguanine, expressed as its ratio to guanine, over 120 hr was similar in both liver and kidney (Figure 16.14). Furthermore, the N-7-methylguanine content in both liver and kidney appeared to increase in direct proportion with the increase in dose.

In contrast to this profile, the clearance of O-6-methylguanine was markedly dose-dependent in the kidney (Figure 16.15). Following of 20 mg/kg dose, the ability to eliminate the O-6-methylguanine, presumably by DNA repair mechanisms of the kidney, is markedly inhibited. This persistence of the O-6-methylguanine correlates with the induction of kidney tumors, suggesting that the inability to repair the O-6-alkylated guanine following a high dose may result in induction of cancer. Of the various sites of potential alkylation of DNA bases, the O-6 position of guanine is a particularly attractive candidate for causing mistakes in base pairing because alkylation of the oxygen forces guanine into its enol form (45–47). This may cause O-6-ethylgua-

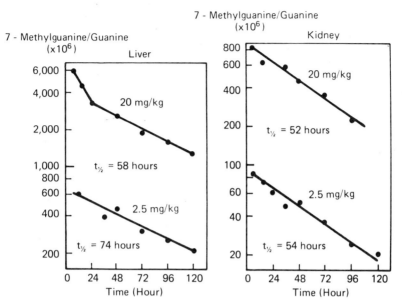

Figure 16.14 Mole fraction of N-7-methylguanine/guanine in DNA of liver and kidney as a function of time following 2.5 or 20 mg/kg intraperitoneal injection of dimethylnitrosamine in rats. From Pegg et al. (39).

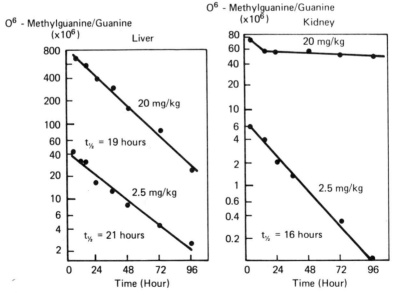

Figure 16.15 Mole fraction of O-6-methylguanine/guanine in DNA of liver and kidney as a function of time following 2.5 or 20 mg/kg intraperitoneal injection of dimethylnitrosamine. From Pegg et al. (39).

nine to pair with thymine rather than cytosine. It has been shown by Gerchman and Ludlum (48) that O-6 alkylation results in misincorporation of bases in a bacterial nucleic acid–polymerase system. In contrast N-7-methylguanine showed normal guanine-cytocine pairing in a similar system.

Assuming that O-6-methylguanine in DNA is the biochemical entity responsible for the carcinogenicity of DMN, the disproportionate decrease in this entity with decreasing dose is suggestive of a threshold or at least a disproportionate nonlinear decrease in the risk of incurring cancer.

9 CONCLUSIONS

Absolute safety is the goal of society in all endeavors. Absolute safety, however, is never achieved, whether in skyscraper construction or in environmental management. An imperfect system is not to be condoned, but constantly improved through experience and research, as rapidly as possible within limitations placed on the system by society itself.

In the evaluation of the safety of chemicals as in all other fields of human endeavor, some degree of risk must be considered acceptable to society. The alternative would be the needless prohibition of important benefits. The estimation of the safety (or hazard) of a chemical for a particular use can span an enormous range of complexity. Irrespective, the objective of all safety testing is to ensure attainment of the desired benefits of use without incurring needless risks. There must, of course, be some balance between the benefit and the cost of assessment, just as there must be a balance between benefit and acceptable risk.

Investigations of "potential hazard" as applied to chemicals include a demonstration of the toxicological properties of an agent by appropriate test procedures and a determination of an exposure level that does not produce detectable adverse effects by the same procedure. For man, the margin of safety of an agent is evaluated by relating to the predicted human exposure, the maximum exposure level in experimental animals that produces no detectable adverse effects. An agent with a small margin of safety has a high potential for producing adverse effects, and conversely, an agent with a large margin of safety has a low potential for producing adverse effects.

To provide optimum assurance of safety within the existing limitations of capabilities and skills for evaluation, safety must be concentrated on environmental situations in which there is a reasonable expectation that exposure to chemicals may cause real hazards. The evaluation of toxicity data for judging the safety or hazard associated with the material can be made only in the light of the anticipated amount and circumstances of human exposure. If the anticipated human exposure is very close to, or perhaps greater than, the maximum dose that produces no observable effect in an adequately sized group of experimental animals, consideration should be given to the type and extent of controls to be applied, to reduce human exposure to acceptable levels. The acceptable risk determination includes an evaluation of the benefits to the individual and to society of the proposed exposure and an evaluation of the nature and severity of the anticipated effects. To quote Golberg:

Neither neglect nor panic is the answer (for safety evaluation). Somewhere between these two extremes lies the course of reasonable action appropriate to each chemical-contaminant contingency. Given good science, good judgment, and above all, freedom from extraneous pressures, the right course can be found (49).

REFERENCES

1. J. A. Zapp, *J. Toxicol. Environ. Health,* **2,** 1425 (1977).
2. C. S. Weil, *Toxicol Appl. Pharmacol,* **21,** 454 (1972).
3. J. T. Litchfield and F. Wilcoxon, *J. Pharmacol. Exp. Ther.,* **96,** 99 (1949).
4. G. L. Plaa, E. A. Evans, and C. H. Hine, *J. Pharmacol. Expt. Ther.,* **123,** 224 (1958).
5. P. J. Gehring, *Toxicol. Appl. Pharmacol.* **13,** 287 (1968).
6. E. M. K. Geiling and P. R. Cannon, *JAMA.,* **111,** 919 (1938).
7. E. B. Dewberry, *Food Poisoning,* 3rd ed., Leonard Hill Limited, London, 1950, pp. 205-217.
8. E. M. Adams, H. C. Spencer, V. K. Rowe, D. D. McCollister, and D. D. Irish, *AMA Arch. Ind. Hyg. Occup. Med.,* **6,** 50 (1952).
9. J. M. Barnes and I. A. Denz, *Pharmacol. Rev.,* **6,** 191 (1954).
10. G. E. Paget, *Proc. Eur. Soc. Drug. Toxicol.,* **2,** 7 (1963).
11. H. J. Bein, *Proc. Eur. Soc. Drug. Toxicol.,* **2,** 15 (1963).
12. J. P. Frawley, *Food Cosmet. Toxicol.,* **5,** 293 (1967).
13. H. M. Peck, *Importance of Fundamental Principles in Drug Evaluation.* Raven Press, New York, 1968, pp. 449-471.
14. W. Hays, Jr., *Essays in Toxicology.* Vol. 3, Academic Press, New York, 1972, pp. 65-77.
15. C. S. Weil and D. D. McCollister, *J. Ag. Food Chem.,* **11,** 486 (1963).
16. H. E. Stokinger and R. L. Woodward, *J. Am. Water Works Assoc.,* **50,** 519 (1958).
17. D. D. McCollister, W. H. Beamer, G. J. Atchison, and H. C. Spencer, *J. Pharmacol. Exp. Ther.,* **102,** 112 (1951).
18. P. J. Gehring and G. E. Blau, *J. Environ. Pathol. Toxicol.,* **1,** 163 (1977).
19. S. Baker, J. Tripod, and J. Jacob, *Proc. Eur. Soc. Drug Toxicol.,* **11,** 9 (1970).
20. S. Baker, *Proc. Eur. Soc. Drug Toxicol.,* **12,** 81 (1971).
21. J. T. Litchfield, *JAMA,* **177,** 34 (1961).
22. W. M. Wardell, *J. Anesthesiol.,* **38,** 309 (1973).
23. D. P. Rall, *Environ. Res.,* **2,** 360 (1969).
24. E. J. Freireich, E. A. Gehan, D. P. Rall, L. H. Schmidt, and H. E. Spipper, *Cancer Chemother. Rep.,* **50,** 219 (1966).
25. D. Pinkel, *Cancer Res.,* **18,** 853 (1958).
26. *Committee on Safe Drinking Water, Report,* National Research Council-National Academy of Sciences, Washington, D.C., 1977.
27. R. C. Wands and J. H. Broome, *in-Proceeding of the Fifth Annual Conference on Environmental Toxicology,* 1974.

28. H. Osswald and K. Goertler, *Verh. Deut. Ges. Pathol.*, **55**, 289 (1971).
29. R. D. Evans, *Health Phys.*, **27**, 497 (1974).
30. J. E. Cleaver, *Genetics of Human Cancer*, Raven Press, New York, 1977, pp. 355–363.
31. P. J. Gehring, P. G. Watanabe, and G. E. Blau, *Advances in Modern Toxicology—New Concepts in Safety Evaluation*, Halstead Press, New York, pp. 195–270.
32. P. J. Gehring, P. G. Watanabe, and C. N. Park, *Toxicol. Appl. Pharmacol.* (in press).
33. P. J. Gehring and J. Buerge, *Toxicol Appl. Pharmacol.*, **14**, 475 (1969).
34. R. J. Kociba, S. B. McCollister, C. N. Park, T. R. Torkelson, and P. J. Gehring, *Toxicol. Appl. Pharmacol.* **30**, 275 (1974).
35. J. D. Young, M. B. Chenoweth, W. H. Braun, G. E. Blau, and L. W. Rampy, *J. Toxicol. Environ. Health*, **3**, 507 (1977).
36. C. Peraino, R. J. M. Fry, and E. Staffeldt, *J. Nat. Cancer Inst.*, **51**, 1349 (1973).
37. C. Peraino, R. J. M. Fry, E. Staffeldt, and W. E. Kisieleski, *Cancer Res.*, **33**, 2701 (1973).
38. T. R. Torkelson, B. K. J. Leong, R. J. Kociba, W. A. Richter, and P. J. Gehring, *Toxicol. Appl. Pharmacol.*, **30**, 275 (1974).
39. A. E. Pegg, J. W. Nicoll, P. N. Magee, and P. F. Swann, *Proc. Eur. Soc. Drug Toxicol.*, **17**, 39 (1976).
40. A. E. Pegg, *J. Nat. Cancer Inst.*, **58**, 681 (1977).
41. P. N. Magee, *Scientific Basis of Medicine*, Annual Review, Athlone Press, London, 1962, pp. 172–202.
42. P. N. Magee and E. Farber, *Biochem. J.*, **83**, 114 (1962).
43. P. F. Swann and P. N. Magee, *Biochem. J.*, **110**, 39 (1968).
44. P. D. Lawley, P. Brooks, P. N. Magee, V. M. Craddock, and P. F. Swann, *Biochim. Biophys. Acta.*, **157**, 646 (1968).
45. P. D. Lawley and C. J. Thatcher, *Biochem. J.*, **116**, 693 (1970).
46. P. D. Lawley and S. A. Shah, *Biochem. J.*, **128**, 117 (1972).
47. P. Kleihues and P. N. Magee, *J. Neurochem.*, **20**, 595 (1973).
48. L. L. Gerchman and D. B. Ludlum, *Biochim. Biophys. Acta*, **308**, 310 (1973).
49. L. Goldberg, *Food. Cosmet Toxicol.*, **9**, 65 (1971).

CHAPTER SEVENTEEN

Health Surveillance Programs in Industry

W. CLARK COOPER, M.D.

1 INTRODUCTION

Health surveillance programs in industry can be directed primarily at general health maintenance and the appropriate placement of workers, or the major emphasis can be on hazards in the workplace. In the latter case workers are periodically observed with the objective of prevention or early detection of harmful effects of such hazards (1).

This chapter emphasizes hazard-oriented medical surveillance. Such activity however, should be part of a broader program of health maintenance. Fragmentation of the medical management of employees should be avoided. The effective incorporation of specialized elements into a more general health examination program is the preferred approach. For this reason it is not feasible to discuss hazard-oriented examinations out of the context of examinations that detect preexisting conditions or intercurrent abnormalities that are unrelated to the work situation.

2 OBJECTIVES OF HEALTH SURVEILLANCE

2.1 General

An occupational health and safety program has many elements, and physical examinations of employees are only part of a comprehensive program. Examinations do not in themselves prevent illness or injury, though they may contribute directly to prevention. They should be designed and performed to secure maximum preventive benefits at minimal cost and inconvenience.

The following sections outline the types of examination that are commonly performed.

2.1.1 Preplacement Examinations

Preplacement examinations are those performed on all otherwise qualified applicants prior to initial employment, to aid in their proper job placement. However the exams often have been performed with the objective of obtaining the most physically fit work force that is available and to exclude individuals who have medical or psychological problems. This practice could be defended by pointing out the cost of training and indoctrinating individuals who could not perform available jobs, the risk of potential worker compensation costs, and the need to protect a medical care program from individuals with preexisting medical problems. Current policies aimed at keeping the handicapped gainfully employed are making such exclusions socially and legally unacceptable.

As subsequent sections point out, the preplacement or preemployment examination serves an essential function in health surveillance, providing a historical record of previous exposures, state of health prior to joining a work force, and a baseline for comparison with later health observations. To a core preemployment history, physical examination, and laboratory appraisal can be added elements tailored to specific hazards of the plant or job under consideration.

2.1.2 Preassignment Examinations

When an individual is being transferred from an operation to another with a known hazard, or when a new process is to be begun, it may be necessary to carry out special inquiries, examinations, or tests. These again have the objective of sizing up individuals with respect to susceptibility and of supplying baselines for later observations. They also offer an opportunity for worker education about potential health hazards associated with the new assignment or exposure.

2.1.3 Periodic Examinations

Periodic examinations are administered to detect incipient disease, physiologic changes, biochemical deviations, or evidences of absorption of toxic agents, and to establish interim reappraisals. They provide opportunities for reinforcing education of the worker. When the experiences of a group of workers having similar exposures are studied, effects may be identified that are not of sufficient magnitude to be significant in an individual worker but identify an increased risk when they occur in several members of a group. Thus periodic examinations can contribute to early diagnosis and treatment in the individual and may identify common factors in a group, thereby enhancing the value of preventive programs.

2.1.4 Termination Examinations

Examinations at the termination of employment are desirable in that they document the health status at the end of exposure and provide evidence of any changes that have occurred during the employment period. This, of course, does not rule out the possibility that effects may show up at a later date. For example, asbestosis can progress years after the last exposure to asbestos dust, and cancer from a chemical carcinogen may not appear for decades.

2.1.5 Special Purpose Examinations

Other examinations, which may or may not coincide with and be supplementary to preemployment or periodic examinations, fill the requirements of special purposes, such as the following:

1. Requirements of regulatory agencies, such as those of the Department of Transportation for vehicle operators in interstate commerce, or of the Federal Aviation Agency or other such body.
2. Evaluation of the effects, if any, of accidental overexposure.
3. Evaluation of recovery before return to work after an absence for illness or injury.
4. Evaluation of the health status of an employee who has difficulty in performing work satisfactorily, in the absence of specific exposures to any known hazard.
5. Determination of impairment of function or disability after complaint of such impairment or disability.
6. Certification of fitness to wear a respirator.

3 SURVEILLANCE FOR GENERAL HEALTH MAINTENANCE

3.1 Content and Scope

3.1.1 Multiphasic Screening

The cost-effectiveness and the long-term benefits of periodic examinations are still subjects for debate. Modern technology and the efficient use of paramedical personnel make it possible to obtain a great deal of information in a short time at relatively low cost. Space does not permit a critical review of the evidence, so the reader is advised to consider a range of opinions by others (2-4). Most thoughtful students of the subject agree that multiphasic screening, if used judiciously, can provide the occupational physician with a valuable tool. When used periodically it may occasionally provide life-promoting information, such as the early detection of a curable cancer or asymptomatic hypertension. It is important that such screening be regarded as a tool, with regard for quality control, an understanding of the so-called normal or average values, a relaxed attitude toward minor deviations from the average range, and vigorous follow-up of the findings that are important.

3.1.2 Educational Value

As in all medical examinations, the opportunity for education of the patient or worker is one of the major benefits of periodic examinations. These opportunities are inseparable from those for hazard-oriented examinations.

4 HAZARD-ORIENTED MEDICAL EXAMINATIONS

It is necessary to consider hazard-oriented medical surveillance in terms of (*a*) what is legally required, (*b*) what has been recommended by official agencies such as the National Institute for Occupational Safety and Health (NIOSH), and (*c*) what has persuasive medical justification. The resulting programs are not necessarily the same. The objectives of medical surveillance are seldom clearly defined in regulations. It would be useful to compare the relative advantages of providing the physician with guidelines on potential effects as compared with prescribing the details of the examination. One can question the wisdom of mandating detailed examinations of thousands of workers whose exposures are below a standard which is so low that hazard-derived abnormalities would be extremely rare. Nevertheless, if a mandatory medical examination is incorporated in a standard, the requirement must be met faithfully and with maximum attention to the quality of information and its relationship to the work exposures.

4.1 Legally Required Medical Surveillance

4.1.1 Current OSHA Requirements

All permanent standards of the Occupational Safety and Health Administration (OSHA) contain requirements for medical surveillance. As of mid-1978 these included standards for asbestos (5), vinyl chloride (6), coke oven emissions (7), 1,2,3-dibromo-3 chloropropane (8), benzene (9), acrylonitrile (10), arsenic (11), and 13 suspect or proved carcinogens (12). It is certain that other standards incorporating medical surveillance requirements will appear over the next few years.

4.1.2 Proposed OSHA Requirements

As of mid-1978 several proposed OSHA regulations included medical surveillance requirements—for example, those for asbestos (13), beryllium (14), and lead (15), as well as the proposed generic carcinogen standard (15). These documents tend to be very specific with respect to the content, frequency, and interpretation of medical examinations. The need for such specificity, when medical examinations are made mandatory in

HEALTH SURVEILLANCE PROGRAMS IN INDUSTRY

regulations, was brought out by court action on the carcinogen regulations of 1974 (12), which were struck down for their lack of specificty (17). A more logical response by OSHA probably would have been to draw back from mandatory examinations, rather than to move into detailed directions regarding the content of medical inquiry.

4.1.3 Requirements of Other Regulatory Agencies

The Federal Coal Mine Health and Safety Act of 1969 (18) was landmark legislation in its requirement for medical examination of coal miners. It set schedules for chest roentgenograms, required the Secretary of Health, Education and Welfare to prescribe classification schemes for radiographic interpretation, and tied film interpretations to compensability. Regulations developed under the act added provisions relating to the training and proficiency of physicians who interpreted films and specified methods of measuring pulmonary function, among other detailed requirements.

Occupational medical surveillance requirements will almost certainly be incorporated in the standards for other mining operations promulgated under the federal Mine Safety and Health Administration (19). Its legislative mandate (PL 95-164) includes the provision that its standards shall "where appropriate, prescribe the type and frequency of medical examinations or other tests which shall be made available by the operator."

4.2 Medical Surveillance Recommended by NIOSH

The NIOSH criteria documents, recommending standards for occupational exposures, always include recommendations to OSHA for medical surveillance. In nearly all cases these prescribe mandatory surveillance. The objectives (e.g., for epidemiologic studies, for early detection and treatment, for detection of group risk factors, or merely for good occupational medical practice) are rarely stated. These recommendations, which as of May 1978 covered more than 75 chemical or physical agents, do not have the force of regulations, but constitute a powerful coercive influence, suggesting as they do a level of good practice. They do not always appear to represent a careful consideration of costs and benefits.

In 1974 NIOSH commissioned the preparation of so-called mini-criteria documents, in which the salient toxicologic and occupational health control factors for nearly 400 chemicals in the occupational environment were summarized in Draft Technical Standards. Abbreviated recommendations for biologic monitoring and medical examinations were included. The latter (Appendix C) were released by NIOSH in 1978 in a volume entitled, *Medical Surveillance Guidelines for Chemical Hazards*.

4.3 Other Sources for Recommendations

The essential contents of the NIOSH mini-criteria documents have been published by Proctor and Hughes (20), who provide a valuable summary to aid the occupational physician in the preplacement, preemployment, or preassignment examination of workers,

...s for special tests and systems to be investigated in periodic surveillance. It is ... that for 180 of 398 substances, the authors do not see a need for periodic ...ical examinations solely on the basis of exposures to a specific agent. They stress ...at a brief interim history is usually sufficient for such agents.

4.4 Justification for Hazard-Oriented Medical Surveillance

There are several medical justifications for hazard-oriented surveillance, even though not all justify mandatory inclusion in standards. First, there are biologic indicators of absorption of toxic agents, based on analysis for the agent or a metabolic product in expired air, urine, or blood. Second, there can be the detection of minor physiologic changes, reductions of function, or early pathologic changes that may be early manifestations of toxicity in individual workers or groups of workers. Third, there may be indicators of hypersusceptibility, which can lead to special recommendations for certain individuals in the work force. Fourth, the history or findings may detect factors in the life-style of the worker, such as cigarette smoking, alcohol consumption, hobbies, or other activities, which should be discussed candidly with the individual as related to work exposures. Finally; every contact with the health establishment provides an opportunity for education of the workers.

4.4.1 Indicators of Absorption

Biologic monitoring is important in medical surveillance. Because the total exposure or absorption of a toxic chemical in an individual can result from a combination of inhalation, ingestion, and skin absorption on the job as well as absorption from nonoccupational sources, the measurements of blood or urinary concentrations can be very important for the individual. The most well-established case is lead, but other metals such as mercury and cadmium are also examples. The relationship between blood or urine levels and exposure needs to be understood by the physician or industrial hygienist. The experience of the group is often more significant than that of the individual but in either case abnormally high values show the need for more careful environmental evaluation.

Urine and blood are not the only biologic materials used for monitoring. Expired air can provide an index of absorption of a number of solvents. Hair can be an indicator of past exposures to heavy metals.

4.4.2 Indicators of Early Effects

Indicators of early effects include reduction of red blood cell cholinesterase in individuals exposed to organic phosphates, reduced erythrocyte δ-aminolevulinic acid dehydrase activity and increased nerve conduction time in lead workers, excessive amounts of low molecular weight proteins in the urine of cadmium workers, changes in the nasal mucosa of nickel workers, diminished sperm counts in those working with chemicals affecting fertility, changes in the gums of workers to exposed phosphorus, postexpo. reduction in expiratory flowrates or forced expiratory volumes in cotton workers a. those exposed to toluene-diisocyanate, and anemia in lead workers and benzene workers. The list can go on.

A major problem in interpreting early changes is that since normal variations may occur in most of the tests involved, serial tests are usually required. The detection of premonitory or early changes of cancer presents special problems, discussed in Section 4.4.4.

4.4.3 Indicators of Hypersusceptibility

All individuals exposed to a toxic agent do not respond alike. Nor are all these differences the result of different levels of exposure. Some individuals are hypersusceptible. In some cases preexisting disease may be the cause of hypersusceptibility. In others personal habits combined with occupational exposures may create synergistic effects, such as that between ethyl alcohol and chlorinated hydrocarbons, or between cigarette smoking and asbestos. There may also be inborn errors of metabolism that interfere with the detoxification of chemical toxins or augment their effects.

Although the reality of such inherited factors is indisputable, it has been difficult to establish firmly their importance in occupational medicine. Cooper in 1973 (21) reviewed the status of the most promising indicators of hypersusceptibility and could not recommend their application at present to routine screening. These included studies for sickle cell trait, glucose-6-phosphodehydrogenase (G6PD), α-1-antitrypsin, and cholinesterase deviants. All appeared to be appropriate subjects for controlled research, but not for mandatory inclusion in regulations or as positive indicators for exclusion from jobs. However they can offer important information to a physician in guiding job placement.

4.4.4 Early Detection of Cancer

A major problem for the concerned occupational health physician, and a recurrent consideration of those mandating medical surveillance of workers exposed to suspected or proved carcinogens, has been the types of medical examination that might be useful for early detection of cancer. Confusion is reflected in the provisions of the permanent standards so far promulgated, and in provisions of proposed standards.

The current asbestos standard (5) does not address itself to lung cancer or mesothelioma in its medical surveillance requirements, which are limited to annual chest films, measurements of pulmonary ventilatory function, and history. The proposed standard (13), however, includes a requirement for periodic sputum cytologic examination, further considered below.

The standard for vinyl chloride (6) contains provisions for a battery of liver function tests, presumably because vinyl chloride is a known hepatotoxin, and it has been suggested that such hepatotoxic effects would precede angiosarcomas. There is no reason to be sure that this occurs. In any event, if exposures are below 1 ppm it is probable that thousands of exposed workers would be examined before any vinyl chloride-related disorder were discovered. In the meantime many individuals with liver disorders resulting from other toxins, such as alcohol, would be detected and removed from exposure or advised to terminate employment.

The standards for coke oven emissions (7) and for arsenic (11) include provisions for sputum cytologic examination. The former also requires urine cytologic studies. Frequent sputum examinations in those who have worked in high risk areas for many years may detect an occasional operable lung cancer. This is probably an unproductive exercise in relation to individuals working under controlled conditions, even though it is required by law. Urine cytology is also unlikely to be a useful diagnostic tool in coke oven workers, in view of the relatively low incidence of urinary tract cancers in the group. It is a useful test in those heavily exposed in the past to proved bladder carcinogens, such as β-naphthylamine.

Medical surveillance requirements for several currently regulated carcinogens or suspect carcinogens (12), exclusive of asbestos, vinyl chloride, coke oven emissions, and arsenic, have been struck down by the courts for lack of specificity. The agents in question are 4-nitrobiphenyl, α-naphthylamine, β-naphthylamine, benzidine, and 4-aminodiphenyl (associated with bladder tumors), methyl chlormethyl ether and bischloromethyl ether (associated with human lung cancer), and 3-3'-dichlorobenzidine, ethyleneimine, β-propriolactone, 2-acetylaminofluorene, 2-methylaminoazobenzene, and N-nitrosodimethylamine, which have been associated with tumors only in lower animals.

Specifying medical surveillance, particularly for chemicals whose effects in humans are speculative and based solely on findings in experimental animals, clearly produced difficulties for regulators. The published standards include a requirement for preplacement and annual physical examinations with a "personal history of the employee, family, and occupational background, including genetic and environmental factors. In all physical examinations the examining physician should consider whether there exist conditions of increased risk reducing immunological competence, as undergoing treatment with steroids or cytotoxic agents, pregnancy, and cigarette smoking."

It is unclear how the examining physician would evaluate the information obtained or how it might affect employability. It is also questionable whether the interpretation would make any difference, if exposures were controlled to conform with standards.

Cytologic Examination. Frequent examination of sputum for abnormal cells, identifying both those with marked atypia regarded as a precursor of cancer and those showing malignant changes, is currently the best diagnostic tool available for the early detection of lung cancer. It should be reserved, however, for those at highest risk, particularly heavy cigarette smokers. Sputum examination is recommended as part of the periodic medical surveillance of asbestos workers, coke oven workers, chromate workers, heavily exposed arsenic workers, and cigarette smokers, if limited to those who

HEALTH SURVEILLANCE PROGRAMS IN INDUSTRY

have had a long and heavy exposure. This procedure is unlikely to be useful in relation to nonsmokers working under controlled conditions, and it can create many problems associated with the employability of individuals with atypical cells unrelated to their work exposures. Unfortunately, it has not been shown to detect many cases early enough to produce cures.

As stated earlier, sputum cytology has been included as a mandatory portion of the medical examination of workers exposed to coke oven emissions (10) and to arsenic (11) and has been proposed in a revised asbestos standard (13).

Search for abnormal cells in urine is indicated in individuals whose work exposures have in the past included α-naphthylamine, β-naphthylamine, benzidine, or other proved bladder carcinogens.

Biochemical Markers of Cancer. The role of such biochemical markers of cancer as carcinoembryonic antigen (CEA) α-fetoprotein (AFP), and acid phosphatase for prostatic cancer, deserve mention. None is at a stage of proved validity, specificity, or sensitivity to warrant inclusion in regulations or routine testing. It is desirable that groups at high risk because of past exposure be included in studies for the evaluation of these so-called early warning signs (22) as well as for cytogenetic surveillance (23).

5 RELATIONSHIP OF MEDICAL SURVEILLANCE AND EXPOSURE DATA

Meeting the requirements of regulations and providing adequate medical surveillance of workers necessitates interlocking of qualitative and quantitative information on exposures with medical programs. This can be a complex process in a large plant with numerous and multiple exposures. The reader is referred to several recent considerations of the problem (24–26).

6 RECORDKEEPING

6.1 Maintenance of Records

Because of the long latent periods between exposure and the appearance of chronic effects such as asbestosis or occupationally related cancers, there is a need for preservation of medical records longer than formerly was regarded as necessary. OSHA regulations and proposals, and NIOSH recommendations in criteria documents, contain a variety of record-retention periods, some as low as 5 years and many as long as 40 years from the termination of employment. The current trend is toward longer retention periods. It is recommended that systems be utilized that will permit all medical records to be kept for a minimum of 40 years. Duplicate storage in computer-accessible microfiche systems is currently the method that is most economical of space.

6.2 Confidentiality of Records

Records that contain personal information on workers must be protected from transmission to or perusal by those not responsible for medical services or care. Without assurance of confidentiality, it is impossible to maintain the professional relationship necessary to elicit needed information. This has not been easy, however, since management must know an individual's physical limitations as they relate to employment, and governmental agencies desire to know whether individual workers have been harmed in their employment. The reader is referred to some pertinent references (27, 28).

Full exploration of this issue is impossible in this section. In its simplest terms, the medical record should be retained by the physician, whether a member of a plant medical department, an external consultant, or part-time physician. He or she should provide the plant only with information that directly pertains to employment, and this should be done with the understanding of the employee. Any other information must be transmitted only with the express and written consent of the employee. Since in the occupational setting consent can be tantamount to a condition of employment, the physician must be zealous in protecting the worker's right to privacy.

6.3 Accessibility of Records

Provisions in regulations regarding accessibility and dissemination of medical records vary. Whereas the standard for asbestos requires that medical reports be sent to employers, later standards have been less explicit. It is clear that policy relating to this, and ultimately the law, is still evolving.

7 PROBLEM AREAS

7.1 Compulsory Versus Voluntary Examinations

All regulations so far promulgated provide that medical surveillance be made available by employers, but nowhere is it stated that employees shall be required to take the examinations. Individual employers may, however, make examinations a condition of employment. The issues here are obviously complex. If a regulatory agency believes that a given examination or test is so important to worker health that every employer must be prepared to provide it, how can it not require employees to take it if they are to work with a specified chemical agent? Present regulations leave the question of such medical surveillance at the level of negotiation between the workers or their representatives and the industry, and necessitate assurances relating to protection of job rights. For that part of the work force which is not organized, such protection would have to be guaranteed by law.

7.2 The Problem of Small Industries

The multiplying problems of medical surveillance strike with particular force on small industries. Major corporations with medical departments gradually adjust to regulations and recommendations requiring medical surveillance. Many of them have managed similar and effective programs for decades. The small employer must meet these needs from a different baseline of operation. The regulations now in force and being proposed and promulgated leave no choice. The employees must be informed of any hazardous materials with which they are working and must be provided with appropriate safeguards. Also arrangements must be made to carry out medical surveillance. At present the resources available to do physical examinations are unevenly distributed, and often necessary guidance is lacking.

7.3 Handling Abnormal Findings

A major problem for the occupational physician in the current climate of physical examinations is the proliferation of borderline abnormal findings. Regulations stipulate that a physician must certify that a worker has no condition that might be adversely affected by exposures on the job. It can be very difficult for a cautious physician to make such a certification if there are any abnormal findings. The presence of metaplastic cells in the sputum of an asbestos worker or a coke oven worker would make certification difficult, even though current exposures were low and controlled. Similarly, certification for wearing a respirator of an individual who had had a coronary occlusion would be difficult, even though the risk was very slight. Situations that can be handled easily in a doctor-patient relationship become matters with serious medicolegal implications because they have been the subject of certification under regulations.

7.4 Consequences of Overregulation

Inherent in all the foregoing discussion has been concern over the inclusion of detailed requirements for medical surveillance in regulations. Many practices that are highly desirable for physicians to follow in selected occupational groups, or for plants to carry out with full understanding of their implications, are not necessarily right for mass application mandated by law. When applied to individuals with very low exposures in adherence to environmental standards, they will result in an extremely low yield of abnormalities, a stultifying overuse of scarce physician time, and needless expense. Often it is former employees, no longer covered by regulations, who need the surveillance that is being carried out on new employees with relatively little exposure. Recourse to the judgment of industrial hygienists and occupational physicians would be desirable.

8 LEGAL CONSIDERATIONS

There are many legal burdens placed on the physician in carrying out medical surveillance of the worker. Some of these are discussed by Felton (29). No comprehensive or authoritative review of these is possible here, but some of the areas that can cause problems are discussed briefly.

8.1 Informing the Worker

There is general agreement at present that the worker must be given information about the hazards of the workplace. This is management's responsibility, but the physician must and will have a role. For many hazards the explanation is easy and readily understood by the worker. For carcinogens, particularly for suspect carcinogens, the message is difficult to impart. At present, the best that can be said is that the physician should be honest in giving an appraisal of the evidence, should create sufficient concern and anxiety to encourage observance of rules for containment and protection against known carcinogens, and should indicate the order of probability of low exposures and for weak or suspect carcinogens. The physician must also be candid with the worker about all results of the individual's physical and laboratory examinations and should place them in honest perspective.

8.2 Certification for Continued Employment

It is difficult to know the impact of regulations that require the examining physician to certify that workers will not be adversely affected by continued employment. There will be a tendency by some physicians to take no chances. There is relatively little information to say whether an individual with an elevated serum glutamic oxaloacetic transaminase (SGOT) or other liver enzyme would be harmed by exposure to 1 ppm of vinyl chloride. The probability is extremely high that it would make no difference. If, however, a worker should develop liver disease after certification, regardless of whether it was due to vinyl chloride, the certifying physician could conceivably be sued. A person certified to work with asbestos, even at low levels, after the finding of moderate metaplastic changes in his sputum could present a similar problem. While the questions involved are being worked out, examining physicians can only use their best clinical judgments, attempt to learn something of the actual exposures their patients are experiencing, and realize that depriving a person of a job unnecessarily is a very serious matter.

8.3 Wearing of Respirators

In all current and proposed regulations and in NIOSH criteria documents, physicians are given the responsibility for certifying whether workers are physically able to wear nonpowered respirators. Objective criteria of ability to wear a respirator are nonexistent, and the translation of pulmonary function tests results to respirator use is uncertain. Actual trial of individuals with respirators is the ultimate test. Differentiation must be made between situations when a respirator is worn briefly for emergency situations for a worker's own protection, when it is required for the worker to perform duties essential to the safety and health of others, and where the worker may be required to wear it over long periods of time.

9 SUMMARY

The health surveillance of workers requires preemployment evaluation and periodic examinations aimed both at general health maintenance and the prevention or early detection of effects from specific job hazards. Good practice calls for a comprehensive preemployment evaluation, with an occupational and medical history and review of all systems, baseline laboratory studies (including blood chemistry and urinalysis), study of visual and hearing acuity, simple tests of pulmonary ventilatory function, and a general physical examination. Periodic examinations should include a core examination directed toward general health maintenance and special hazard-oriented studies, such as biologic monitoring for determining levels of absorption of chemicals, early indicators of toxic or other biological effects, and audiometry when indicated by noise exposures. Proper scheduling and interpretation of hazard-oriented surveillance make it essential that environmental and medical data be closely interlocked.

There are no consistently effective methods for the early detection of most occupational cancers. Urinary cytologic screening can be useful in individuals who have been heavily exposed to bladder carcinogens. Sputum cytology, the only technique that is promising for the early detection of lung cancer, is recommended for those in high risk groups, but results to date have been discouraging.

An important part of a health surveillance program is the opportunity provided for periodic contact of the worker with a member of the health team. This should be utilized for education in the prevention of both occupational and nonoccupational disease.

REFERENCES

1. World Health Organization," Early Detection of Health Impairment in Occupational Exposure to Health Hazards," Technical Report Series 571, WHO, Geneva, 1975.
2. N. J. Robert, "The Values and Limitations of Periodic Health Examinations," *J. Chronic Dis.*, **9**, 95–116 (February 1959).
3. G. S. Siegel, "Periodic Health Examinations. Abstracts from the Literature," Public Health Service Publication 1010, March 1963, Government Printing Office, Washington, D.C., 1963.
4. W. K. C. Morgan, "The Annual Fiasco (American Style)," *Med. J. Aust.*, **2**, 923–925 (November 1 1969).

5. Department of Labor, Occupational Safety and Health Administration, *Standard for Exposure to Asbestos Dust, Fed. Reg.,* **37** (June 7, 1972).
6. Department of Labor, Occupational Safety and Health Administration, *Standard for Exposure to Vinyl Chloride, Fed. Reg.,* **39** (December 3, 1974) *Fed. Reg.* **40,** 23072 (May 28, 1975).
7. Department of Labor, Occupational Safety and Health Administration, Section 1910.1029. *Standard for Exposure to Coke Oven Emissions,* added at *Fed. Reg.,* **41,** 46784 (October 22, 1976), corrected at *Fed. Reg.,* **42,** 3304 (January 18, 1977).
8. Department of Labor, Occupational Safety and Health Administration, Part 1910, Occupational Safety and Health Standards, *Occupational Exposure to 1,2-3-Dibromo-3-Chloropropane (DBCP), Fed. Reg.,* **43,** 11527 (March 17, 1978).
9. Department of Labor, Occupational Safety and Health Administration, *Occupational Exposure to Benzene: Permanent Standard. Fed. Reg.,* **43,** 5963 (February 10, 1978).
10. Department of Labor, Occupational Safety and Health Administration, Section 1910.1045, *Standard for Exposure to Acrylonitrile,* added at *Fed. Reg.,* **43,** 2586 (Janary 17, 1978) as emergency temporary standard.
11. Department of Labor, Occupational Safety and Health Administration, Section 1910.1018, *Standard for Exposure to Inorganic Arsenic,* added at *Fed. Reg.,* **43,** 19624 (May 5, 1978).
12. Department of Labor, Occupational Safety and Health Administration, Occupational Safety and Health Standards, *Carcinogens, Fed. Reg.,* **39,** 3756 (January 29, 1974).
13. Department of Labor, Occupational Safety and Health Administration, *Occupational Exposure to Asbestos, Notice of Proposed Rulemaking, Fed. Reg.,* **40:** 47652–47665 (October 9, 1975).
14. Department of Labor, Occupational Safety and Health Administration, *Occupational Exposure to Beryllium, Notice of Proposed Rulemaking, Fed. Reg.,* **40,** 48814 (October 17, 1975).
15. Department of Labor, Occupational Safety and Health Administration, *Occupational Exposure to Lead, Notice of Proposed Rulemaking, Fed. Reg.,* **40,** 45934 (October 3, 1975).
16. Department of Labor, Occupational Safety and Health Administration, *Identification, Classification, and Regulation of Toxic Substances Posing a Potential Occupational Carcinogenic Risk, Fed. Reg.,* **42,** 54148 (October 4, 1977).
17. Department of Labor, Occupational Safety and Health Administration, *Carcinogens: 4,4'-Methylene bis(2-chloroaniline), Notice of Proposed Rulemaking, Fed. Reg.,* **40,** 4932 (February 3, 1975).
18. Federal Coal Mine Health and Safety Act of 1969, Public Law 91–173, December 30, 1969.
19. Federal Mine Safety and Health Amendments Act of 1977, Public Law 95–164, November 9, 1977.
20. N. H. Proctor and J. P. Hughes, *Chemical Hazards of the Workplace,* Lippincott, Philadelphia, 1978.
21. W. C. Cooper, "Indicators of Susceptibility to Industrial Chemicals," *J. Occup. Med.,* **15,** 355–359 (April 1973).
22. T. H. Maugh, II, "Biochemical Markers: Early Warning Signs of Cancer," *Sci.,* **197,** 543–545 (August 5, 1977).
23. D. J. Kilian and D. Picciano, "Cytogenetic Surveillance of Industrial Populations," *Chemical Mutagens,* Vol. 4, Plenum Press, New York, 1976, Ch. 43.
24. M. G. Ott, H. R. Hoyle, R. R. Langner, and H. C. Scharnweber, "Linking Industrial Hygiene and Health Records," *Am. Ind. Hyg. Assoc. J.,* **36,** 760–766 (October 1975).
25. M. G. Ott, "Linking Industrial Hygiene and Health Records," *J. Occup. Med.,* **19,** 388–390 (June 1977).
26. J. D. Forbes, J. P. Dunn, G. Hillman, L. L. Hipp, T. J. McDonagh, S. Pell, and G. F. Reichwein, "Utilization of Medical Information Systems in American Occupational Medicine, A Committee Report," *J. Occup. Med.,* **19,** 819–830 (December 1977).

27. G. J. Annas, "Legal Aspects of Medical Confidentiality in the Occupational Setting," *J. Occup. Med.,* **18,** 537–540 (August 1976).
28. A. McLean, "Management of Occupational Health Records," *J. Occup. Med.,* **18,** 530–533 (August 1976).
29. J. S. Felton, "Legal Implications of Physical Examinations," *West. J. Med.,* **128,** 266–273 (March 1978).

CHAPTER EIGHTEEN

Philosophy and Management of Engineering Controls

KNOWLTON J. CAPLAN

1 DEFINITION

Engineering controls for industrial hygiene purposes may be defined as an installation of equipment, or other physical facilities, including if necessary selection and arrangement of process equipment, that significantly reduces personal exposure to occupational hazards.

Common examples are local exhaust systems, general ventilation systems, enclosures around noisy machines, and substitution of a nondusting mixer for a dusty mixer. Substitution of a less hazardous material in the process involves many other disciplines besides industrial hygiene, but since the proposed substitute may also require process or handling modifications, it frequently becomes involved in the question of engineering controls.

Currently the Occupational Health and Safety Administration (OSHA) defines engineering controls "by difference." OSHA defines personal protective equipment as "anything worn on the person of the worker to reduce the exposure"; administrative controls as "any adjustment of the work schedule to reduce the exposure"; and engineering controls as "all other measures to reduce the exposure". This approach has obvious flaws. For example, the third definition would place biological monitoring in the category of an engineering control, clearly not a very good fit. Future developments in industrial hygiene also may evolve measures to reduce the hazard that are not compatible with the general concept of engineering.

2 DEGREE OF ENGINEERING CONTROL REQUIRED

The determination of what degree of engineering control is required to maintain a given concentration of air contaminant is a most difficult task when approached on a theoretical or general basis. In most industries the required degree of control has been arrived at through an evolutionary process wherein an increasing level of control, and control of increasingly minor sources, is pursued until the objective is reached. Thus for industries and operations that are fairly common, the required level of control is based on rather wide experience; and if there is adequate cooperation and care between the hygienist and the engineer, the results can be largely on target.

For industries and operations that are relatively unique (i.e., there are only a few plants or operations) or when a threshold limit value (TLV) is to be drastically reduced from prior practice, however, the determination of adequate controls is much less certain. There are several reasons for this.

1. The properties of the materials involved and the ways in which they are handled in the manufacturing process represent a bewildering variety of combinations. In addition to the rather well-defined physical and chemical properties of the materials (such as vapor pressure), there are many other, less readily quantified variables that can lead to the generation of more or less contaminant in the air. One obvious example is the difference between a light fluffy dust and a sticky dust. The sticky dust tends to adhere to surfaces and does not become airborne easily. Thus if reasonable housekeeping is maintained, contamination of surfaces is of little import to the airborne concentrations, whereas the opposite is true if the dust is not sticky and is light and easily airborne.

2. The amount of contaminant generation from the process that would be of hygienic significance is small enough to be completely insignificant from a material balance point of view. Therefore the only way in which the significance of a potential source can be quantified is by experimentally measuring the generation rate, which can be relatively simple in some few operations but is usually difficult and expensive.

3. When contaminant concentrations are to be drastically reduced, the major sources mask and obscure the effect of the minor sources. Since as noted previously the rate of generation from sources is usually not known, one is obliged to guess just what sources require control to reach a given target level of air contamination.

Obviously the best approach in terms of securing good results with engineering controls and not wasting resources on unneeded controls is the evolutionary, stepwise method. First the obvious sources are controlled and resulting concentrations determined. The masking effect of the major sources having been removed, the next tier of sources can be estimated and considered. This approach could well take several years of persistent and intelligent effort in the event that more than one or two rounds of application of engineering controls are required.

2.1 Hazard Classification System for Particulate

The control of airborne dust or fume is more difficult in general than is the control of gases or vapors. Dust that is not captured and removed from the environment settles on floors, work surfaces, and overhead ledges and becomes a secondary source of airborne dust, whereas this is not true of most gases and vapors. (A notable exception is mercury: this low vapor pressure material can slowly and continuously evaporate into the air from minor sources in amounts significant to the low TLV.)

Furthermore, the customary units used in reporting dust or particulate concentration versus gas and vapor concentrations (i.e., mg/m^3 or $\mu g/m^3$ vs. ppm) creates a subjective impression of concentrations which is entirely erroneous. A solvent vapor or gas that has a TLV of 1 ppm, which "sounds low" is, assuming a molecular weight of 100, actually 4080 $\mu g/m^3$ when expressed in those units.

After cautioning the reader not to treat the following as "design data" but rather as a concept level attempt at guidance, a method for estimating the level of control required for particulate matter in the form of dust and fumes can be described. The approach is to first select a "production factor" category, designated by a letter A, B, C; then to establish a hazard classification that is a combination of generation of contaminant (high, medium, or low) and the control level to be achieved.

The production factor is established as follows:

A. Plantwide processing of hazardous material; by-products as well as products may contain hazardous material.

B. Departmental processing of hazardous material.

C. Local processing of hazardous material; or intermittent, infrequent, or very small scale; or short-term.

The dust or fume generation category is selected according to the following description:

High. Vapor pressure of material as used would result in equilibrium vapor concentration over the TLV; a high degree of process heat is applied relative to melting point or boiling point; material is oxidized; violent physical dispersion; high volatility rate; dry, dusty powder; fluffy, easily airborne dust; soft, friable materials; dry bulk processing.

Medium. Low vapor pressure, moderate heat, less violent physical dispersion, low volatility rate; properties intermediate between "high" and "low."

Low. Negligible vapor pressure; gentle physical dispersion; hard, abrasion-resistant material; little or no oxidation; heavy or sticky dust not easily airborne.

Then the appropriate control level is selected. Since engineering controls, particularly ventilation, are physical in nature, the content of the target material in the process stream is important. For example, consider grinding on a 2 percent beryllium-copper

Table 18.1 Hazard Class for Dust and Fume

Control Level Target (mg/m³)	Generation category: High	Medium	Low
		Hazard Factor	
<0.05	1	1	2
0.05	2	2	2
0.10	2	2	3
0.25	3	3	3
0.5	3	3	3
1.0	3	3	4
3.0	4	4	4
5.0	4	4	4
>5.0	4	4	4

alloy. Leaving aside all arguments over the relative toxicity of beryllium in that form, assume that the element is to be controlled to 2 $\mu g/m^3$. This means that since the dust is only about 2 percent beryllium, the total dust concentration evolved from the process can be $100/2 = 50$ times the 2 $\mu g/m^3$, or 100 $\mu g/m^3$. If on the other hand the concern is lead oxide fume, the fume will be 92.8 percent lead and the TLV for lead itself would be governing.

Having selected the control level for the contaminant mix, enter Table 18.1 to select a hazard class. The production factor and the hazard factor are then combined to lead to the appropriate concept of engineering control required. The control levels applicable to the classification derived are as follows:

A1. Techniques similar to and approaching those used in nuclear fuel reprocessing facilities would be required. In general the source of contamination should be completely removed from contact with the worker. This requires complete enclosure or barrier walls, completely automated and mechanized processing, and perhaps mechanized maintenance. Even such facilities require human intervention at least for maintenance and troubleshooting, and an extraordinarily high degree of personal protection is required during such activities.

A2. Complete control of every observable or foreseeable source; glove-box, mechanized, or remote handling in ventilated enclosures where practical; general ventilation as adjunct to local control; higher ventilation rates than "minimum standards"; scrupulous housekeeping and general cleanliness a necessary adjunct to engineering controls.

A3. Good control of all except minor sources; generous general ventilation; good housekeeping highly advisable; ventilation rates standard or in excess of standards.

A4. Standard ventilation control of significant sources.

B1. As in A1, necessary only in affected department. Segregation or isolation of affected department allows standard industrial procedures in remainder of plant. Adjunct techniques of cleanliness necessary to prevent contamination of entire plant by physical means.

B2. As in B1, necessary only in affected department. Adjunct techniques of cleanliness necessary to prevent contamination of entire plant by physical means.

B3. As in A3, applied only to affected department.

B4. As in A4, applied only to affected department.

C1. Total enclosure, glove-box hoods, segregation and isolation of facility.

C2. Good local control; ventilation rates in excess of minimum standards; extent of control depends on whether there is a "life hazard" capable of causing death or severe injury on short exposure.

C3. Local control of significant sources using standard ventilation techniques; if sources are sufficiently small, general ventilation may suffice; if sources are sufficiently intermittent, respiratory protection of worker may suffice.

C4. Local control of major sources only, using standard ventilation rates. In some cases dilution ventilation alone may suffice.

3 INTERFACE WITH THE PROCESS

Even the most simple of engineering controls work better if there is interaction between the hygienist and the plant operators and plant engineering personnel. Even a simple exhaust hood should be designed to meet the following operating criteria.

1. It should not interfere unduly with normal operations; if some interference is required to achieve control, the acceptable design should be determined mutually with operating personnel.

2. Provisions should be made for easy opening and closing of access doors or easy removal and replacement of access panels for anticipated maintenance of enclosed equipment. "Easy removal and replacement" does *not* permit a multiplicity of nuts and bolts.

Such design cannot be successful if done "at a distance" but must be accomplished after adequate observation of the operations and face-to-face discussions with operators and engineers. Designs that show a simple symbol for a hood or a simple box, leaving the details to the sheet metal contractor, are almost never sucessful.

If a change in process or substitution of less toxic materials is a viable alternative, even greater interaction with production and engineering personnel is required. Depending on the nature of the project, research and marketing functions may also be involved.

Perhaps the most simple process change is that involving only material handling. For example, a double-cone blender or a Y-blender requires very difficult and expensive dust control if a low TLV is required. Much better dust control at lower total cost may be achieved by using a rotary blender. However the blending characteristics of these two machines are somewhat different; before such a substitution is made, it must be determined whether the proposed new blender will in fact perform the required blending properly. A somewhat similar situation would exist if it were proposed, for example, to subsitute a screw conveyor for a belt conveyor for a material that is difficult to handle.

If the proposed substitution or change in process goes deeper into the process chemistry and physics, the project is liable to assume some magnitude. In addition to the design of the process itself, the products may be changed in their characteristics and if so, there would be an impact on the market for the end products as well as on the manufacturing process.

It should be remembered that most industrial processes have a long evolutionary history. An existing process has evolved to its present state through gradual change and improvement during which each step has been adjusted and proved in production to ensure that the system is workable. Interaction with the marketplace and knowledge of actual production costs are also necessary in such evolution. It is true, however, that the process as evolved, and with certain quality control measures, yields a product of known and acceptable properties. The evolutionary process is a combination of the theoretical and the empirical. In lay language, "if we make a certain thing in a certain way, with certain controls, it produces a product with certain known attributes and limitations." There may well be parameters involved that although capable of measurement, have never been measured for one reason or another.

At one time in the manufacture of natural uranium metal fuel elements for graphite-moderated reactors, the metal was produced in rather small ingots that contained various impurities. These impurities were removed by recasting the ingots in high vacuum induction furnaces. In such vacuum melting the impurities boiled out of the molten uranium. The impurities included magnesium metal which, as it boiled out, swept with it the highly radioactive daughter products of natural uranium, the beta emitters U_{X1} and U_{X2}. These materials condensed on the water-cooled lid of the vacuum furnace. Since the magnesium was pyrophoric and the U_{X1} and U_{X2} no longer benefited from the self-shielding of the parent metal, the opening of the vacuum furnaces presented a severe fire and hygiene hazard. These problems were controlled by means of operating the top end of the furnaces in a total enclosure with robot handling of crucibles, ingots, and so on.

A new process was developed using the theoretical principles of metallurgy. It involved casting a very much larger ingot, and by controlled cooling, segregating the impurities in the top of the ingot and in the outside skin. These impurities could be removed by a simple machining operation, which was much less expensive than vacuum recasting and presented much less severe fire and hygiene problems. After machining, the interior of the ingot successfully passed the many quality control tests for the uranium metal. Since at the time a new plant was being built and an old plant abandoned, the new process was installed.

Within a few months numerous complaints came from the reactor operators. The new metal was causing an inappropriately high number of slug ruptures; that is, the aluminum jacket of the uranium metal was rupturing while in service in the reactor, releasing fission products into the cooling water, thus causing untold difficulties for the reactor operator. An immediate investigation was launched, but repeated and intensified quality control testing at the uranium plant failed to disclose the source of the problem. A statistical analysis of the batches of metal that caused slug ruptures revealed no correlation with operating parameters other than the weather. At about the same time the reactor operators reported that the cause of the slug ruptures was bubbles of hydrogen gas accumulating between the uranium and the jacket, impeding the necessary heat transfer. These two findings led to a new quality control test—the measurement of hydrogen in the uranium. Samples of metal made by the old process contained a maximum of 2 ppm hydrogen, whereas uranium made by the new process in humid weather contained 6 ppm.

This is a case in point. Uranium metal was manufactured under quality control procedures much more stringent than are employed for most commercial products; yet because hydrogen had never been a problem, the measurement of hydrogen had not been a subject of quality control. It was not even realized, before the bad experience with the new product, that the hydrogen content of the metal would be of any concern. Obviously such mistakes can be very costly. Thus major changes in industrial processes are usually slow, evolutionary, step-by-step procedures.

It not infrequently happens that the hygiene control of an existing process is very difficult and expensive, but simpler, more effective, and less expensive controls are anticipated with a new process. Assuming that hygiene control is to be accomplished for reasons of severe health hazard or perhaps because of legal enforcement actions, it seems appropriate to deduct the cost of hygiene controls for the old process from the projected cost of the new process. Frequently the new process will offer operating economies, but it is not at all infrequent that the capital cost of the new process cannot be amortized adequately by such advantages. In such a case the deduction of the hygiene control cost for the old process may be the deciding factor in whether the new process is implemented.

Once a new process has been developed and is being implemented, the hygienist runs the risk of getting out of touch. The new process seems to be so much better from a hygiene point of view that the problems seem minor, and perhaps complacency sets in. Furthermore, the new project is frequently so big that it is designed as a major engineering project over which the hygienist may not have adequate control. Under such circumstances it may be a bitter disappointment to find that the new process is not in good shape, that further retrofit controls are required anyway. Very seldom does a new process accomplish the complete transition from very difficult or unsolvable hygiene problems to no problems at all. The hygienist *must* keep in touch with the detailed design of the new process.

4 EXPERIMENTAL SAMPLING

Since the advent of personal sampling pumps and legal enforcement of standards based on 8-hr time-weighted average exposures, most industrial hygiene sampling is oriented toward determination of the 8 hr time-weighted exposure. Although these are the most appropriate data for evaluating exposures when short-term exposure is not important, they furnish little or no information leading to the intelligent application of engineering controls. The classical case of a production worker performing essentially the same task all day long is disappearing from the industrial scene. Therefore if hygiene controls are to be wisely applied, it is necessary to know what parts of the worker's job (i.e., which tasks) contribute what proportional amount of the day's exposure. It is also perfectly possible for significant sources of contaminant that are not directly connected with the worker's tasks to be responsible for the exposure.

The industrial hygiene of most fairly common industries has been quite well studied. In the foundry industry, for example, general process knowledge and experience may be adequate without experimental sampling. However if the emissions of concern are not visible, not odorous (or if odor fatigue sets in), and not well-known, experimental sampling may be necessary to determine the significance of various possible sources.

The first and perhaps easiest kind of experimental sampling is task oriented. Assuming that the worker does not perform the same duties continuously during the workday, task-oriented sampling consists of taking separate samples for the different tasks in which the individual is involved to determine the exposure separately for each task. This kind of sampling imposes added strains on the logistics of sample rate, sampling time, and analytical sensitivity, to ensure the validity of the samples, and attention to these details is essential. For example, it is possible to use a given sampling filter repeatedly if the sampling time for each round of a specific task is too short to produce definitive results. It is not uncommon for such a task-oriented sampling regime to reveal that it would be more beneficial to control some part of the job that actually is at lower concentration than some other part, as is evidenced by Table 18.2.

For the job analyzed in Table 18.2 presumably a single 8 hr sample would have shown a concentration of 0.21 mg/m^3. The task-oriented sampling, however, reveals several interesting things. Column 5 shows that tasks B and C are the major contributors to the day's exposure and that a significant reduction in the concentration at either of those tasks would be adequate to bring the 8 hr exposure well below the TLV. This is true even though task C in itself is below the TLV. In addition, it shows that task F, well above the TLV concentration, is of such short duration that significant improvement in that part of the exposure would not have a large effect on the 8 hr exposure.

Sampling of a more "experimental" nature is limited only by the ingenuity of the experimenter. By means of such experimental sampling it has been shown that transportation of dry pasted battery plates by fork truck on clean floors is not, in itself, of concern at a 100 µg/m^3 level of lead in air (1). By a more complex set of experimental samples it has been shown that in manual loading or unloading of an automatic stacking machine in a battery plant, the generation of lead dust at the rack of pasted plates is

PHILOSOPHY AND MANAGEMENT OF ENGINEERING CONTROLS

Table 18.2 Task-Oriented Sampling

Task	GA or BZ[a]	Minimum/day (mg/m³)	Concentration (mg/m³)	Minimum Concentration (mg/m³)
A. Charge pot	BZ	40	0.12	4.8
B. Unload pot	BZ	80	0.50	40.0
C. General survey	GA	250	0.16	40.0
D. Lab—sample trips	GA	20	0.05	1.0
E. Change room	GA	30	0.08	2.4
F. Pump room—repack	BZ	30	0.32	9.6
G. Lunch room	GA	30	0.07	2.1
		480		99.9

TLV = 0.2 mg/m³   Wt. avg. = $\frac{99.9}{480}$ = 0.21 mg/m³

[a] GA = general air; BZ = breathing zone

significant and requires control at the 150 µg/m³ level, even though historically control of the stacking machine alone has seemed to be adequate at the 200 µg/m³ level (2). For such experimental sampling it is necessary to follow the principles of any measurement process, and it is most difficult to measure something without disturbing that which is being measured. The general principles may be outlined as follows:

1. The operation should be as normal as can be contrived under the conditions of the experiment.
2. The sample taking should interfere as little as possible with the operator.
3. All other contaminant-generating operations should be in normal operation or perhaps alternatively, completely shut down.
4. If at all possible, it is advisable to estimate or measure the airflow patterns as well as the contaminant concentration, to permit the estimation of the total amount of the contaminant released.

5 EDUCATION AND PARTICIPATION OF WORKERS

Most hygiene controls, especially local exhaust ventilation hoods and enclosures, sound enclosures, and so on, require changes in the way the worker does the job. Usually the job is made somewhat more difficult. Ideally, however, good design can minimize the detractive aspects and make the hood easier rather than more difficult to live with; and sometimes the same project can include mechanical assists or other revisions that make the job easier. Nevertheless, changes to the job are introduced.

In the absence of any participation in the design by the worker, or of any information about the proposed changes, human nature automatically generates resistance to the change, and the new aspects of the job may well be resented. In addition to whatever design features prove to be less than optimum on an objective basis, the human aspects of the situation are even more important. Most of us can remember similar events from our own experience. The young fellow with the old wreck of a car, who really wants to keep it running, manages to do so with baling wire and chewing gum. The do-it-yourselfer who has created a gadget for home use will put up with major inconveniences to make it work. However the purchaser of a new car will complain bitterly of every rattle and minor inconvenience. The key is, Does the user (worker) *want* to make it work?

No engineering job can be perfect in its design or construction. Even though it may be almost perfect for normal operations, operations are not always normal. Accommodation must be made by the operators for unusual situations. If the design happens to be less than perfect for such an event, and if the workers are psychologically opposed to the installation in the first place, it will not be long before the cutting torches come out and the offending feature is removed, and not replaced and not reengineered.

This destined-for-failure sequence can be minimized, perhaps completely avoided, by engaging the participation of the workers in intermediate stages of design. Operating supervisors, busy as they are with all facets of production, may not even know of some of the small problems that occur on the job. Even supervisors are people, and people tend to tell the engineer or systems designer how things "should be" rather than how they actually are.

All these problems can be minimized by observations of the job over an adequate time period, talking with the operator, asking questions, and checking the observations and conclusions with the supervisor. Later in the design effort, when preliminary drawings are available, these should be reviewed with the operators as well as with the operating supervision. Many industrial organizations have frequent safety meetings or similar meetings, which usually provide a good forum for such discussions if the group is small enough. If such meetings are not routinely held, special meetings for the purpose of design review are well worthwhile.

Management sometimes fears that in such a situation workers will make requests or present ideas that are technically or economically unsound or cannot be accomplished within the budget for the job. It has been my experience that such questions or suggestions frequently arise, but if *all* suggestions are sincerely considered and at least some are adopted, no lingering resentment is detected over those that are declined or rejected for good reason. It has also been my experience in such sessions that invariably operating problems are brought to light or good ideas for improvement introduced that had escaped the design engineer's attention and had not ben mentioned by operating supervision. Some comments are as simple as noting that using an access door would be more convenient if the door opened to the right instead of to the left. Other suggestions, of course, are much more complex.

PHILOSOPHY AND MANAGEMENT OF ENGINEERING CONTROLS

In addition to whatever objective improvements are gained in the design of the project, the intangible gains are even more important. The workers have had a chance to get a better idea of what is forthcoming, their questions should have been answered, their ideas and suggestions have been seriously considered, and some of them have been adopted. The same meeting may also serve as a conduit for further explanations of the reason for doing the job and for answering questions concerning the hazard, therefore, by this means, also assisting the motivation of the worker to "make it work."

In today's industrial social climate it is fruitless for the engineer or the supervisor to complain that the workers will not use the hygiene control properly. Effective and practical mechanisms do not exist, in general, to enforce proper utilization on a broad base, especially if there are genuine defects in the design. Therefore the only successful alternative is to convince the workers to assist in proper utilization; and as a rule this can be achieved only by education and cooperation. It is also true that there is always a small scattering of recalcitrant individuals, for whom the standard labor relations type of disciplinary action may be required.

In short, the success of engineering controls for hygiene purposes depends on the same factors as does the success of engineering features related to production. The subject matter may not be as familiar to the operating supervisors and workers, but it certainly is not beyond comprehension. With top management backing, the same techniques of training and motivation that are applied to production will be successful if applied to hygiene.

6 DESIGN CONSIDERATIONS FOR EXHAUST VENTILATION SYSTEMS

Since ventilation is one of the major types of engineering control for hygiene problems, it serves as a good basis for discussion of some design philosophies that in priniciple also apply to any other kind of engineering control. This discussion assumes that the design is of good quality from a purely engineering standpoint; the remarks are addresssed to the specific problems of the interface with operations represented by most engineering controls.

6.1 Operating Convenience

It is essential that interference of exhaust system hoods with normal production operations be at the minimum extent practical while obtaining the desired control objective. Since aspects of this have been discussed in detail previously, the design requirements for operating convenience are merely summarized here. The elements of the design activity necessary to obtain operating convenience involve:

1. Adequate observations of operation.
2. Consultation with operators and operating supervision.

3. Interaction with workers to offer an opportunity for understanding the purpose of the system, obtaining worker input to aid in design convenience, and involvement of the worker in the design process. Again, all normal elements of design competence are assumed to be accomplished.

6.2 Maintenance Access

The ventilation equipment itself should be arranged and designed so that maintenance is reasonable and practical, as for any other mechanical equipment.

Special considerations concerning maintenance reflect awareness that exhaust system hoods, especially if they are partial or total enclosures, necessarily interfere with the maintenance of the contaminant-generating equipment being so enclosed. (The same considerations apply in many respects to noise enclosures.) It is obvious that there must be access to the enclosed equipment so that normal maintenance and lubrication can be provided easily; what is all too often forgotten is abnormal or major maintenance access. Merely specifying a removable panel that is held in place with many nuts and bolts or other fasteners is not sufficient.

When maintenance is required on production equipment, especially breakdown maintenance rather than preventive maintenance, time is usually at a premium. Time is also important even for preventive maintenance. Thus one can rest assured that access will be obtained to the equipment requiring repair. If removal of nuts and bolts is necessary, it will be performed; in the absence of bolted panels, access will be had with a cutting torch. The problem, from the hygiene point of view, is to make possible the quick and easy replacement of the panel after maintenance has been conducted. If a cutting torch has been employed to get to the equipment, rare indeed is the equipment maintenance crew who will apply a welding torch to repair the hood. Similarly, time is seldom available to get all the nuts and bolts back in place on a bolted panel. In addition, the multiplicity of nuts and bolts involved is in itself a problem: when the time comes to replace these fasteners, they will not all be found, much less reinstalled. It is a very unusual mechanic who will take time to go to the storeroom to get more nuts and bolts. Psychologically the offending panel is a hindrance to the accomplishment of the task at hand, rather than part of the task.

Improvement in this situation can be obtained by the application of a small set of principles.

1. A minimum number of tools should be required (preferably no tools) for any anticipated access, and at most a single tool such as a hammer or a single simple wrench.
2. A hinged door that swings completely out of the way when opened is better than a panel that must be completely removed and set aside, perhaps incurring damage or distortion in the process.
3. Any fasteners should preferably be both easy to use and rugged—a number of devices are available (various latches with handles, taper lock wedges, husky rubber

PHILOSOPHY AND MANAGEMENT OF ENGINEERING CONTROLS   623

tension fasteners, pins, etc.). Such fasteners should not be completely removable but should be retained on the piece, by means of its own design or by means of a short length of light chain.

4. The weight and convenience of the panel should be considered and handles provided if advisable. Some closures, especially if the design is such that gravity will keep the door closed, can well be provided with a rubber flap, such as a piece of conveyor belting, so that bending and distortion of light gauge sheet metal do not cause problems.

Special provisions often can be made for items that require frequent maintenance—for example, the lubrication fittings on the idler pulleys of conveyor belts, when the conveyor belt is to be enclosed. During installation of the enclosure, the lubrication fittings can be extended by means of a short piece of pipe and a lubrication fitting through a small hole in the enclosure wall so that nothing has to be removed for the oiler to gain access to the fitting.

An example of excellent design in this regard is the hammer mill enclosing hood illustrated in Figure 18.1. Hammer mills are a notorious source of dust. Although the dust is generated inside the machine, the joints in the machine and shaft seals do not always stay dust tight, and even if they do, the machine itself acts as an inefficient fan

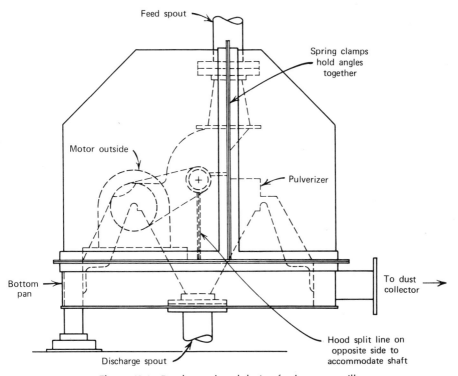

Figure 18.1 Breakaway hood design for hammer mill.

and creates a positive pressure on the downstream side, thus any receiving vessels must be airtight or provided with exhaust ventilation. If provided with exhaust, it frequently happens that the ventilation control removes an excessive amount of product and accordingly the airflow is throttled, resulting in insufficient hygiene control. The principles embodied in this hood design include appropriate interface with the process in that an exhaust hood is around the source but does not actually interfere with the process containment, containing and collecting as it does only material that leaks from the process. The maintenance and operational features of the design are illustrated by the following features:

1. The bottom pan is rigidly affixed to the discharge flange of the pulverizer and does not interfere with access to and maintenance of the machine itself; furthermore, a solid connection to the exhaust duct can be provided, since no movement is needed.
2. The upper part of the hood is self-supporting. When it is removed, no structural supports for hood panels are there to block access to the pulverizer.
3. The upper part of the hood is custom designed for the pulverizer in that it has a split line, offset as necessary. This permits accommodation of the feed spout and the drive shaft on the split line. Thus neither the feed spout nor the drive has to be disturbed to remove the hood.
4. The drive motor and the V-belt are outside the hood. If lubrication fittings are provided on the bearings of the pulverizer, they may be extended to the outside of the hood on the split line.
5. The fasteners that hold the hood in place are conveyor cover clips, chained to the hood itself. These can be removed by use of a screwdriver or other similar device as a pry bar, or by a hammer. They are replaced by a simple tap with a hammer.
6. The hood sections are shaped to have reasonable structural integrity; they will not be distorted by ordinary handling. Handles should be provided if the hood sections are so large that it would be awkward to remove the hood without them.

6.3 Blast Gates and Dampers

6.3.1 General

The *Manual on Industrial Ventilation* (3) published by the American Conference of Governmental Industrial Hygienists (ACGIH) offers a brief tabulation of advantages and disadvantages of using dampers or blast gates to balance exhaust systems. These points should be amplified and their import described. Many of the opinions expressed are based on my observations over many years but are not quantified on a broad basis and perhaps are not universally true. The need for this discourse is based on an observation of a persistent trend among plant engineers and vendors of equipment to prefer the blast gate (damper) method of balancing exhaust systems because it is "easier" and "more flexible". By contrast, it is hard to find an exhaust system balanced by blast gates that is controlling adequately the problem for which it had been installed.

Balancing an exhaust system implies that the correct amount of airflow is obtained in each of the several branches of the system. There are two general ways of achieving this goal, one is to use blast gates or other types of dampers to permit field adjustment of the air flows in each branch, and the other is to design the hoods, ductwork, and fan to achieve the appropriate balance without dampers. If the contaminant to be controlled is hazardous, the maintenance of proper exhaust volumes assumes some added importance. When the contaminant is particulate matter (dust or fume), a further constraint is placed on the system in that a minimum velocity must be maintained in each branch or the ductwork will plug. Many other possible design problems arise from time to time, such as condensation in the ductwork, erosion or corrosion, and undesirable chemical reactions. This discussion, however, is limited to the problem of balancing exhaust systems handling dust or fume.

Dampers may be used for several purposes:

1. As on-off valves where only some of the branches of the system need to be active at any given time.
2. To prevent motor overload on startup, for example, in a fabric filter system with new bags.
3. When required volume in a given branch is not known with certainty and a conservative estimate may lead to excessive removal of product.
4. For balancing the system to achieve proper airflow in each branch.

Of these four reasons; the first two are obviously legitimate and do not in face involve "balancing." Even in such circumstances, difficulty is frequently encountered when operators fail to open and shut dampers as required, and it is advisable to interconnect the damper operation with the process if a practical means of accomplishing this can be found. However, situations do arise in which the required exhaust volume is uncertain, and a conservative estimate may result in excessive volumes, which remove desired product from the process or unnecessarily overload the dust collector. A damper may be used for these situations, but there will be severe limitations in that closing the damper too far will reduce the duct velocity below the minimum; other means of compensating are available and are usually a better choice. In the third case the blast gate would be found only in one branch of the system, not in all branches. It is the fourth case, the general use of dampers or blast gates to balance the entire system, that requires elaboration with respect to its problems. It is obvious, however, that there are occasions when blast gates are advisable, and the use of engineering judgment is to be encouraged in these cases (as opposed to siding with the use or nonuse of blast gates as an act of faith).

The *Industrial Ventilation Manual* tabulates the relative advantages of using blast gates. That list is reproduced here for convenience.

Relative Advantages of Method A and Method B

Method A: Balance Without Blast Gates

1. Air volumes cannot be changed by workmen or at the whim of the operator.
2. Small degree of flexibility for future equipment changes or additions; the ductwork is "tailormade" for the job.

3. Choice of exhaust volumes for a new unknown operation may be incorrect; in such cases some ductwork revision is necessary.
4. No unusual erosion or accumulation problems.
5. Ductwork will not plug if velocities are chosen wisely.
6. Design calculation is more time-consuming than method B.
7. Total air volumes slightly greater than design air volume due to added air handled to achieve balance.
8. Poor choice of "branch of greatest resistance" will show up in design calculations.
9. Layout of system must be in complete detail, with all obstructions cleared and length of runs accurately determined. Installations must exactly follow layout.

Method B: Balance with Blast Gates

1. Air volumes may be changed relatively easily, though precautions are taken. Such changes are desirable where pickup of unnecessary quantities of material may affect process.
2. Greater degree of flexibility for future changes or additions.
3. Correction of improperly estimated exhaust volumes is easy, within certain ranges.
4. Partially closed blast gates may cause erosion to slides, thereby changing degree of restriction or may cause accumulations particularly of linty material.
5. Ductwork may plug if the blast gate adjustment has been tampered with by unauthorized persons.
6. Design calculations are relatively brief.
7. Balance may be achieved with design air volume.
8. Poor choice of "branch of greatest resistance" may remain undiscovered. In such case the branch or branches of greater resistance will be "starved".
9. Leeway is allowed for moderate variation in duct location to miss obstructions or interferences not known at time of layout.

All the points in the ACGIH list are valid, but in many cases elaboration and explanation is called for. The overall impression given by the list is improper, in that there are severe limitations to the advantages of blast gates, and a whole series of interrelated problems derives from their use.

It is perfectly true that a system can be designed and balanced with blast gates. However that is just the beginning of the story; the following sets of conditions prevail:

1. It is *not* less expensive to balance the system with dampers than it is to design it to be balanced.
2. The flexibility afforded by blast gates is severely limited in one direction by the necessity for maintaining carrying velocities, and in the other direction by the available suction in the system.
3. Changing the setting of any one blast gate affects the performance of all the other branches in the system.
4. The purported ease of adjustment with blast gates is in itself a severe disadvantage in that the limitations are not understood by operating and maintenance personnel. The availability of blast gate "adjustments" is what leads to patchwork and butchered systems, where changes have been made with no semblance of engineering technology.

PHILOSOPHY AND MANAGEMENT OF ENGINEERING CONTROLS

5. Last but not least, no method of adjusting blast gates and fixing them in place has yet been developed that can withstand the onslaughts of the operator, the maintenance department, or the night shift mechanic. These people naturally enough feel that the blast gates are there to be adjusted as a solution to whatever problems may be perceived to exist at any given time.

These five conditions are explained separately.

6.3.2 Ease of Balancing

For the *designer* of a system it is easier to call for balancing by damper adjustment than it is to devise a system that is balanced. This does not mean, however, that it is easier or cheaper for the owner to have the balancing achieved by way of dampers. The field balancing, even for a relatively simple system, requires a large number of iterations of trial-and-error balancing of each branch, since adjustment of any damper affects the rest of the entire system. It typically takes more man-hours to achieve balance by this method than by design. Depending on the local labor situation, the owner may have to pay for a craftsman, an apprentice, and a technician—three people—to achieve the same results that could have been produced on the drawing board.

Balancing of ductwork is so expensive that a number of commerical organizations base their business on that work, and there is even a trade association, the Air Balance Council, involved in this market. It is true that the bulk of such work concerns air-conditioning and air-supply ventilation systems that are controlled by dampers, and the technical considerations for such systems differ considerably from those for exhaust systems handling dust or fume. Air-conditioning systems are usually concealed, making access to the dampers difficult, and the premises are usually used by office personnel who have little inclination to tamper with such equipment. Moreover poor air balancing in the systems results in discomfort, and complaints are immediately forthcoming. In the case of ductwork involving excessive concentrations of harmful air contaminants, however, typically one must await industrial hygiene monitoring or OSHA inspection before the deficiency is discovered—and, more important, poor performance results in a health hazard, not merely in discomfort. All these factors taken together cause a shift in the engineering judgment involved, and the usual result is that *for the air-conditioning system,* balancing by use of dampers is appropriate.

6.3.3 Flexibility

The flexibility achievable by the use of blast gates is severely limited. For a system of given capacity, the power requirement increases with the square of the duct velocity; accordingly it is inappropriate to design the ductwork for velocities much higher than the minimum carrying velocity actually required. This means that as dampers are closed to reduce the volumes in a given duct, the amount of such throttling that can be accomplished without reducing the duct velocity below the minimum depends entirely on which branch is being throttled. If the branch is toward the far end of the system (i.e., the branch of greatest resistance or a branch of high resistance with respect to the available static pressure), the available latitude for such throttling will be minimal. If the branch enters the system where the static pressure in the main is greatly in excess of that required to operate that branch at design volume, there is much more latitude for throttling.

It is by no means obvious in most systems which branch is the branch of greatest resistance—it is not always the one farthest from the fan—and the subtleties of appropriate throttling of a given branch, combined with the affect on the rest of the system of changing one damper, make such adjustment a prime cause of system inadequacy.

If more air is desired in a given branch, opening the damper will work up to the available static pressure in the main where that branch joins, which again is highly variable and is just the reverse of the situation described for throttling a damper. Even worse, however, when a damper is opened, other branches in the system then draw less air than they were drawing before the adjustment.

The purported "ease of adjustment" and "flexibility" create an attitude in the minds of operating and maintenance personnel, indeed sometimes in engineering personnel, that the system is something like a utility water distribution system: "If you want more water, open the valve, and if you want less water, close the valve." In fact, only very minor changes can be made with dampers without engineering and redesign. The general attitude created, however, is that the flexibility exists to permit such adjustment, modifications, and additions, without recourse to engineering and design calculations. As explained earlier, such is not the case. If on the other hand the ductwork was not provided with dampers, and perforce changes would be required in duct size, orifices, and so on, recourse to engineering redesign would be unavoidable. It is true that the specific change or modification being made would cost more and take longer than would butchering the ductwork and moving the blast gates around; but in the long run the cost will be less because the performance of the system will not be ruined.

6.3.4 Fixing the Position

Many schemes have been proposed for fixing dampers and blast gates in position once a system has been balanced by their use. These include putting a bolt through the damper, clamping the damper, and tack welding the damper in position, all of them to no avail. Dust collection systems frequently receive inadequate maintenance; and any number of things can go wrong with the system. If dampers and blast gates are available, they will be the first choice for adjustment and will be the quickest available "remedy" for the problem as perceived by the operator or night shift mechanic. These people have the tools and the contacts necessary to get the inhibiting and offending lock removed, in the hope of finding a quick and easy cure for the problem. Furthermore, the locked-in-place damper is poor psychology. The "lock" itself is offensive to many of us. It further seems as though the engineer is imposing his decision regarding the proper damper position on the operator, and many fancy themselves to be amateur experts on the subject of ventilation.

PHILOSOPHY AND MANAGEMENT OF ENGINEERING CONTROLS 629

In another context, it is desirable to encourage an attitude among operators and mechanics that they can and should "fine tune" their production equipment to achieve better production, higher quality and so on. This management philosophy may be appropriate if the personnel sufficiently understand the production equipment to achieve those ends and are permitted the freedom, but it should not be applied to the dust collection system and its complexities. I have never seen a blast gate system that did not generate apologies on the part of the management or the engineering personnel because it had been tampered with and was no longer in balance.

6.3.5 Why So Popular?

If balancing by dampers and blasts gates is such a poor approach, why are there so many proponents of this system? The numerous proponents include plant engineers who are attracted to the "ease of adjustment" and "flexibility" of such systems, as indeed they should be if such advantages were significant. However most plant engineers are not sufficiently experienced with these systems to understand the pitfalls; moreover they tend to overrate the desirable features (flexibility, etc.).

The other two groups who usually strongly recommend the blast gate system are the equipment vendors and the contractors, both of whom would be expected to know their business. As a matter of fact they do know, and the blast gate system is the one that is the best for *their* business. The equipment vendors and contractors both furnish a degree of "free" engineering, which of course is included in the price of the equipment or contract and is necessarily limited for competitive reasons. To properly design a balanced system without blast gates, a degree of knowledge of the operations is required that may be beyond the expertise of the vendor or contractor. Even if it is not, time and effort are required that the supplier can ill afford under the umbrella of free engineering. Similarly, the design computations needed to balance the system would also be effort to be charged to the free engineering; but the equipment vendor can escape this cost almost completely by saying that a blast gate system is better.

The same motivation exists in a somewhat different way for the contractor: if he is required to do the balancing by blast gates, the work will add to the value of the contract in a visible way for which he can charge without complaint from the customer. Alternatively, if the customer is going to balance the system, it is a cost the the contractor escapes while at the same time adding slightly to the value of the contract by using the blast gates or dampers.

The foregoing comments may be excessively harsh in some cases but not in most. The worst situations occur when the architect-engineer, working at long distance from the project, merely shows a triangle for an exhaust hood; leaving the complete design of the hood up to the sheet metal subcontractor. At the other end of the spectrum are the contractors who have become so sophisticated that they not only install balanced systems without blast gates, but have developed computer programs for the layout of elbows and fittings for any pipe diameter without resort to shop patterns, so that they can easily and economically balance systems by proper choice of pipe sizes.

6.3.6 Summary

For most dust and fume control systems, balancing without dampers and blast gates is by far the preferred choice. A blast gate system that is working adequately is seldom encountered. The advantages of blast gates, although true, are highly exaggerated and the limitations of their use not generally recognized. They are advantageous to the vendor or contractor, not to the owner. A major exception to this position is the type of system in which dampers or blast gates are used as on-off valves, and even there, their operation should be interlocked with the process if practical.

6.3.7 Example

To provide an example of the limited flexibility gained by the use of a blast gate system, consider a modestly complex system, without an air cleaning device (merely to simplify the calculations required to determine what would happen if various changes were made in the system). The presence of an air cleaning device would not change the results significantly, except in the case of some kinds of equipment where changes in total system volume may change the performance of the air cleaning equipment itself.

The illustrative system (Figure 18.2) has five branches, designed as a blast gate system, with a blast gate in each branch. The depicted data are typical of such a system, showing the volume in each branch and each section of the main, and the pressure drop due to the duct and due to the blast gate adjustment "as balanced." Three cases of need for adjustment are summarized as follows. As in Figure 18.2, numbers in brackets are values after change and BG stands for blast gate.

Case I was chosen as a change that would be regarded as "likely to be successful" because the branch being adjusted is the one nearest the fan where the highest static pressure is available and there is a significant pressure drop across the blast gate, indicating that a great deal of upward adjustment should be available. It is desired to handle more air through branch 5D. How much more can be handled, and what will be the effect of that change on the rest of the system? By opening the blast gate wide, the entire 5.3 in. w. g. suction would be available at the branch entry, assuming that the system itself did not change, although of course it would. What the system "would be" if it did not otherwise change is called the "would be" operating situation.

The example calculation shows that if the "would be" system worked, a fan static pressure of 5.65 in. at 11,070 cfm would be required. Plotting this point on the fan system graph (Figure 18.3) and basing the plot of the system curve on the principle that pressure drop is proportional to the square of the volume, we have a new operating point for the system at 10,280 cfm, 5.4. in. w.g. fan static presssure. The system pressure would have about 0.1 in. less suction than formerly, with the result that branch 5D would handle 4090 cfm, a 17 percent increase over the initial design volume. This is truly a modest increase and in most cases would not be sufficient to improve materially the capture of the exhaust hood on that branch.

Case II is an example of an adjustment that is not very likely to succeed, although an unsophisticated user of the system might attempt it. Opening the blast gate wide on

PHILOSOPHY AND MANAGEMENT OF ENGINEERING CONTROLS

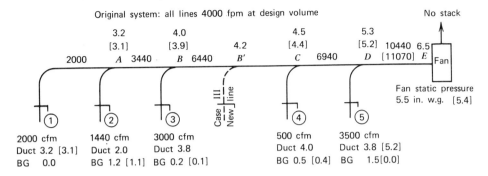

Case I. Open 5D wide (see calculation at end of this section).

[4900 cfm]   17 percent increase
[Duct 5.2]
[BG 0.0]

Open all other blast gates to design; 1A is below design (low static pressure).

Case II. Open 3B wide.

$$\text{cfm} = 3000 \sqrt{\frac{4.0}{3.8}}$$

$$= 3080 \text{ cfm} \quad 2.6 \text{ percent increase}$$

Negligible change in rest of system.

Case III. Add 1000 cfm new line at B'.

1000   cfm
Duct   4.2
BG   0.0

Cannot be done. See calculation at end of this section.

Figure 18.2 Example of blast gate system. Numbers after "BG" (blast gate) show ΔP across blast gate; numbers after "duct" show duct pressure drop; numbers at branch show static suction (in. w. g.); numbers in brackets are values after change.

branch 3B provides only a 2.6 percent increase in volume, too small to measure, and the rest of the branches are not significantly affected. Of course more air could be sucked through branch 3B if the other blast gates were further closed, but that would result in less than design volume and velocity in all other branches of the system.

Case III represents a not untypical modification to the system, requiring only about a 10 percent increase in volume by adding a new line. It is postulated that the new line would be desired to handle 1000 cfm. The new line would be connected to the system at B' where the existing static pressure is −4.2 in. The calculations for case III show that

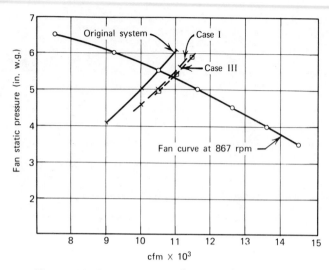

Figure 18.3 Fan system curve for use with Figure 18.2.

a significant additional pressure drop in the main "would be" required to handle this amount of air. Plotting a new system curve, we see that the system would operate at 10,900 cfm instead of the desired 11,440 cfm, and the fan static pressure would be 5.3 in. Furthermore, this would only provide a volume increase of 460 cfm, not 1000. The new branch would handle about half of what was desired even with the blast gate wide open. Branches 1, 3, and 4 will be below design because the blast gates cannot be opened enough to compensate for the lower suction at the branch entry, and branches 2 and 5 could be adjusted to original design. It is not worthwhile to work out the precise flows through each branch under these conditions, since the change would be so unsuccessful.

The three cases above, predictable examples of change in a relatively simple system, indicate a very limited degree of flexibility. What plant foreman or even plant superintendent is going to be sufficiently sophisticated in exhaust system design, or wants to be bothered in view of the "flexibility" he has with the blast gates, to make the calculations that have been made in this example? In the real world, the blast gates would simply be changed. Then the problems would start, and depending on the circumstances, a great deal of effort might be expended in trying to correct the problems.

Case I. Open blast gate $5D$. Changes system. Then system "would be"

$$\text{cfm } 5D = 3500 \sqrt{\frac{5.3}{3.8}} = 4130$$

$$\text{cfm } D\text{-}E = 6940 + 4130 = 11{,}070$$

$$\Delta P\ D\text{-}E = 1.2 \left(\frac{11{,}070}{10{,}440}\right)^2 = 1.35$$

PHILOSOPHY AND MANAGEMENT OF ENGINEERING CONTROLS 633

Fan static pressure = 5.5 + 0.15 = 5.65 in. w.g.

Plot new system curve at 11,070 cfm; 5.65 in. w.g. The operating point is 10,820 cfm; 5.4 in. w.g. (0.1 in. less). Available static pressure at D is about 5.2 in. w.g.

$$\text{cfm } 5D = 3500 \sqrt{\frac{5.2}{3.8}} = 4090 \text{ cfm (17 percent increase)}$$

Case III. Add new line at B', 1000 cfm required static pressure at this point, existing system, 4.2 in. Changes system. System "would be":

$$\text{cfm } B'\text{-}C = 6440 + 1000 = 7440 \text{ cfm}$$

$$\text{Friction } B'\text{-}C = 0.3 \left(\frac{7440}{6440}\right)^2 = 0.4 \text{ in. w.g.}$$

$$\text{New } VP = \left(\frac{7440}{6440}\right)^2 \times 1.0 = 1.33$$

$$\Delta VP = 1.33 - 1.00 \quad \underline{0.33}$$
$$\text{TOTAL} = 0.73$$

Static pressure at C becomes 4.5 + 0.7 = 5.2 in. w.g.

$$\text{Friction } C\text{-}D = 0.8 \left(\frac{7940}{6940}\right)^2 = 1.05$$

Static presssure at D becomes 5.2 + 1.0 = 6.2 in. w.g.

$$\text{Friction } D\text{-}E = 0.8 \left(\frac{11,440}{10,440}\right)^2 = 1.0$$

Static pressure at E becomes 6.2 + 1.0 = 7.2 in. w.g.

$$\text{Fan static pressure} = 7.2 - 1.33 = 5.9 \text{ in.}$$

System operates at 10.900 cfm; 5.3 in. w.g. (see fan curve).

Approximately: static pressure is 0.6 in. w.g. lower than required. Fan volume increases 460 cfm, not 1000. New branch runs between 460 and 1000 cfm (nearer 460). Branches 1, 3, and 4 are below design—blast gate cannot open enough to compensate for lower system pressure. Branches 2 and 5 can be adjusted.

6.4 Recirculation of Cleaned Air

Air cleaning equipment is available with adequate efficiency to clean recirculated air of many kinds of contaminant to a safe concentration. Typical is the fabric filter type of collection equipment when applied to a dry, easily collected dust or to a fume that coagulates and filters with high efficiency. Industrial hygienists in general and regulatory bodies in particular have been very reluctant to approve recirculation of air from such equipment. Although manufacturers claim and in many cases can prove that the cleaned air is cleaner than the workroom air, there always remains the vexatious problem of adequate maintenance and operation of the air cleaning equipment. Should it fail to receive the proper maintenance, as all too often happens, the recirculated air would contain dangerous or obnoxious quantities of contaminants.

The extensive history and background on this topic should be briefly reviewed. Until recently it has been a rather uniform policy of almost all official agencies and almost all professional industrial hygienists to recommend against the recirculation of air cleaned of toxic contaminants. It was recognized that the air cleaning devices, even if capable of cleaning the air to a safe degree, were not completely reliable and all too frequently received inadequate maintenance. With recirculation, therefore, there was a finite risk that the workroom air would be dosed with toxic materials by the ventilation system itself. In the days of cheap energy it was a correct decision and good policy to require that recirculation be prohibited and that the air exhausted from the building be replaced with outside air suitably heated or otherwise conditioned. It should also be remembered that historically, most proposals for such recirculation were based on the use of a single air cleaning device without any "backup" or redundancy to ensure that the recirculated air would be safe. Recirculation proposals were made on that basis because (in the days of cheap energy) the savings by recirculation would be negated if additional safeguards and monitoring devices were required for the recirculated air.

Now, however, the cost of energy is such that it becomes economically wise to spend the extra capital and maintenance cost to provide for recirculation of cleaned air *if* such a system can be designed and monitored so that it is reliable and safe.

The following discussion is based on the case of dry particulate (dust or fume) because that application is probably the easiest to design. Exhaust air containing gas-phase contaminants can also be cleaned and recirculated, using the same principles, but the implementation of the concept will probably be more difficult and costly.

The technology to permit safe recirculation of dry particulate dusts has been in existence for some time. The typical system would consist of a self-cleaning fabric filter as the first air cleaning device. For safety and reliability and more-than-adequate efficiency, the first filter would be followed by a second or backup filter, which would consist of a high efficiency filter of the high efficiency particulate air (HEPA) type. The second filter would not be self-cleaning but would require replacement if sufficient dust leaked through the first filter and built up resistance on the second filter. Filters of the class proposed for the second or backup filter can be tested in place to ensure that their efficiency is reliably high; there are no moving parts or other phenomena likely to destroy the efficiency of such a filter bank, once properly installed. Various other controls and devices can easily be installed to monitor the concentration of dust in the cleaned air, to automatically bypass the system in the event of failure, and so on.

Such a recirculation system, properly designed, can recover essentially 100 percent of the heat in the air exhausted from the workroom; in addition, it can recover essentially 100 percent of the energy used in moving the air and forcing it through filter resistances; and even can recapture essentially 100 percent of the wasted electrical energy represented by the inefficiency of the electric motors employed.

PHILOSOPHY AND MANAGEMENT OF ENGINEERING CONTROLS

In typical dusty operations only a slightly greater capital investment is required to permit such recirculation than is needed to provide makeup air. Fuel costs are such that the capital cost is typically recovered in less than one year. In view of the reasonable projection that energy costs will increase faster than other costs, such economic benefits can only become more attractive.

It can be reasonably argued that a well-designed recirculation system will provide a better working environment than would a conventional exhaust and makeup air system. The dust can be removed from the air so effectively that the concentration is as low or lower than the outside air concentration in the vicinity of the plant. Furthermore, because of economic pressures, any legal requirement that recirculation cannot be practiced would result in attempts to use minimal exhaust volume to reduce the volume of makeup air required and the cost of heating it. In that event better control at the capture points would be achieved if the more generous air volumes permitted by recirculation were used.

The price of fuel is not the sole consideration. The only fuels suitable for small installations for makeup air duty are natural gas, propane or LPG, or fuel oil. For these fuels price is not the only factor because they are at times literally unavailable, and unavailability may become increasingly common in the future. The next step up would be to purchase electricity for heating, which is typically 2 to 3 times as expensive as the fuels just named, or to provide a steam plant fired by coal. The capital cost of the coal-fired steam plant would have a significant economic impact.

In recognition of the oncoming "fuel crisis" the Committee on Industrial Ventilation of the ACGIH devoted several pages of its *Industrial Ventilation Manual* (13th and 14th editions, 1974 and 1976) to a description of ways in which recirculation of air, cleaned of toxic contaminants, could be achieved; including a list of pertinent considerations to ensure that the recirculation was safe. Furthermore, the National Institute for Occupational Safety and Health (NIOSH), Engineering Branch, recognizing the same factors, has conducted studies on the same subject resulting in a report, "Guidelines for the Recirculation of Industiral Exhaust" (Contract no. 210-76-0129).

One interesting aspect of current discussions and studies of recirculation is the emphasis on reliability. There can be no argument that a high degree of reliability, especially adequate provision for safety in event of failure of the recirculation feature, is required. The only difference between an ordinary system and a recirculation system in this regard concerns the elements of the system that are necessary to prevent recycling of contaminant to the workroom in amounts or for lengths of time that would be excessive. The concern for reliability of the recirculating system seems to extend itself to the total system, to a degree inappropriate compared to concern for the reliability of the nonrecirculating system, or for adequate design in the first place. General or total system reliability should be treated equally, whether the system is recirculating or nonrecirculating.

6.5 Heat Stress Relief Versus Exhaust Control

Unless exhaust system hoods are total enclosures, they function on the principle of establishing an air velocity into the hood that is sufficient to capture and contain the contaminants. Obviously if large, high velocity air currents are directed at the hood, as from a man cooling fan or a misplaced air supply fixture, the velocity of those currents will be high enough to completely overwhelm and spoil the capture velocities created by the hood. In some cases more air is blown into the face of the hood than is removed by the exhaust connection, thus causing the hood to "overflow" or spill contaminated air into the workroom. Almost all exhaust ventilation standards state that the absence of such disturbing high velocity room air currents is assumed. Achieving this happy state of affairs is easier said than done.

The situation is bad enough if the heat stress is due merely to warm ambient air as from summer conditions in the typical factory. In many cases, however, there is a concentrated heat load from the process or work being done at the hood, a common example being the shakeout of hot castings.

One of the best general methods of overcoming a heat stressful situation is the provision of "spot cooling." This involves the direction of a high velocity air current of cooler outside air (or possibly artificially cooled air) directly on the worker. The airflow over the worker's body increases the available maximum evaporation of sweat and increases the conduction of heat away from the body if the temperature of the airstream is lower than the temperature of the skin.

The behavior of air jets in open ambient air space, as in the workroom, is well known. Most texts or handbooks dealing with exhaust ventilation do not give data concerning such jets. Suitable information can be found in the *ASHRAE Handbook* (4). The properties of air jets must be understood by the designer and the hygienist when they confront a problem combining heat stress and contaminant control.

There is no solution to this type of problem known to be generally applicable. Each problem needs careful analysis, and frequently a compromise is the best that can be achieved. Potential solutions include possible heat stress relief by other means, such as radiant heat panels, and modification or mechanization of the process. Frequently the spot cooling air jet can be made more effective at lower velocity if the air is treated by evaporative cooling techniques. Evaporative cooling has a bad reputation in that it is not effective for comfort air conditioning in any but dry desert climates. Its effectiveness for the relief of heat stress is usually not recognized, but it is surprisingly effective in tyical humid midwestern climates (5).

Intelligent application of evaporative cooling principles can result in effective spot cooling using lower jet velocities, thus with less disturbance to the control of the exhaust hood. The exhaust volumes used on the hood and the corresponding capture and control velocities should be increased above the minimum standards because they are applicable only when there are no disturbing air currents. In some situations the "heat stressful" part of the work is not continuous, and a spot cooling or area cooling island of cool air can be provided at some little distance from the hot contaminant source. Thus the heat stress relief is obtained by the worker when standing back from the hot source, and the flow of cooling air does not directly interfere with the exhaust hood control.

PHILOSOPHY AND MANAGEMENT OF ENGINEERING CONTROLS

6.6 Maintenance and Inspection

An exhaust system that has been properly designed and properly installed and is exerting adequate control must be maintained properly, as well. Obviously and as previously described, reasonable access must be had to the equipment itself when maintenance is required. A number of other features can be considered, however, for routine inspection of the system to determine whether maintenance is needed.

One of the easiest things to check is whether the design airflow is flowing through each hood. That, combined with a visual check of the integrity of the hood itself, is not only important but also is easy. As described in Volume I, Chapter 18, there is a pressure drop when the air flows from the exhaust hood into the connecting duct, commonly called the "throat" of the hood. The suction at this point is calculated during the design of the system so that the "as designed" value should be available. When the system is installed and operating, a measurement of the suction at the throat of the hood provides an operational "quality control" datum point that is easily checked to determine whether the hood operation is within "quality control" limits.

This verification calls for a static pressure tap. The simplest static pressure tap is a small hole—for example, $1/16$ to $1/8$ in. diameter, drilled in the side wall of the duct in the region of the hood throat (i.e., about one duct diameter downstream of the connection to the hood itself)—and otherwise suitably located to prevent distorted readings (see Volume I, Chapter 18). A simple U-tube manometer or other static pressure measuring device can be connected to this static pressure tap by holding the open end of square-cut soft rubber tubing to the hole in the duct. This requires some care, however, and human error can easily introduce a large discrepancy in readings. Therefore a slightly more sophisticated fitting is recommended.

A simple pipe T can be welded or brazed to the outside of the duct so that the static pressure tap hole is encompassed, a fitting screwed into the leg of the T with a short stub of $1/4$ in. copper tubing, permitting the hose from the manometer to be slipped over it, assuring a reasonably airtight fit. A plug or appropriate valve can be inserted in the other arm of the T to provide an easy method for cleaning out the static pressure hole with a wire or pipe cleaner if it becomes plugged.

A manometer or other pressure gauge can be permanently installed at each static pressure tap. If the system is handling dust, corrosives, or condensables, however, this procedure is not recommended. The static pressure hole will become plugged and the gauge will give a false indication. The problem can be avoided by the use of a purge airflow as is commonly used in process instrumentation, but such an installation adds considerably to the cost and is usually not worthwhile except under unusual circumstances.

The simple pressure tap and rubber tube method assumes easy access to the desired physical location. Lacking easy access, however, extension of copper or other suitable tubing from the hood throat to some reasonably accessible point is advisable. If the pressure tap is inaccessible, so is the clean-out plug for cleaning the static pressure hole, and a different provision, such as a connection for blowing high pressure compressed air into the line to clear the static hole, may be used.

Provisions for routine inspection of the airflow through hoods is required by OSHA standards. With the attribute described above (i.e., the pressure drop through the hood throat being a function of the airflow), the hood throat can be considered to be an orifice-type flow measurement device for which correct reading can be established either from the design or from the experience gained in startup.

If the static suction at the hood throat is low, one of the following causes can be found.

1. If the system has blast gates, they are out of adjustment.
2. There is a constriction (partial plugging of the ductwork, damage to the ductwork, etc.) in the pipe between the hood throat and the exhaust fan.
3. There is major leakage in the piping between the hood throat and the fan.
4. Fan speed is low (viz., slipping belts on the fan drive).
5. There is a high pressure drop through the air cleaning device on the system.

If the static suction is high, the following factors may be the cause.

1. If the system has blast gates, they are misadjusted.
2. Other branches in the system are plugged or constricted.
3. There is a malfunction in the air cleaning device (e.g., broken bags in the fabric collector) causing its pressure drop to be less than normal.

More sophisticated devices may also be used if the proper operation and integrity of the system are important enough to warrant the cost. One such continuous monitoring feature could be a device that indicates that the system is indeed in operation—for example, by means of a pilot light indicating whether the electrical circuit to the fan motor is on. This example also illustrates the fallacy of many such control features in that what it measures is different from the desired parameter. In this case, what is desired is knowledge of whether there is significant airflow in the system, but all the pilot light reveals is whether there is voltage on the power circuit at the point where the pilot light tap-in is made. There could easily be voltage at this point and insufficient airflow through the system; prime causes of this kind of event are badly slipping drive belts or a burned out blower motor. If the variable to be checked is the airflow in the system, an airflow switch that is actuated directly by airflow should be installed.

Other and more sophisticated monitoring devices can be used, depending on the need. Such devices may be pressure recorders or differential pressure recorders, permanently installed centerline Pitot tubes with provisions for compressed air clean-out, and various electrical interlocks. Highly complex interlocks are not successful in the typical factory. The general environment is very different from that of flying an aircraft or spaceship. This matter is discussed in greater detail in Section 8 of this chapter.

6.7 Fan Noise

With surprising frequency, industrial exhaust systems that are installed to correct one hazard have the effect of generating a new hazard, excessive noise. Adequate information exists to predict the amount of noise created by a fan. This problem is unfortunately treated, to a degree, as a separate discipline and the design data are not published in the same sources as for exhaust hoods. I believe that any competent hygienist or engineer, with rudimentary training in noise technology, can also handle the average problem of fan noise. Data are available in References 6 to 9.

Fan noise may be a problem in the workroom or in the neighborhood of the plant. Some minimal attention should be paid to both possibilities when the exhaust system is being designed, and provisions should be made to prevent a hazard or a nuisance from this new potential source.

7 PURCHASING SPECIFICATIONS RELATED TO INDUSTRIAL HYGIENE

Purchasing specifications can be considered in two ways: in relation to the ever-present problem of specifying the appropriate quality for a machine regardless of whether its intended purpose is hygienic, and in relation to the special specifications associated with health and safety requirements that tend to creep in.

A simple example will suffice to illustrate the first category without getting into the depths of engineering expertise at the detail design level. In the purchase of an exhaust fan, the primary consideration is the air volume and static pressure required. A mere specification of that requirement on a "performance specification" basis will almost guarantee an unsatisfactory installation. There are many other aspects of performance or quality that must be considered. First the type of fan must be specified, since there are several basic types, some of which may not be appropriate for the intended use. Once that is done, however, there may be several duty classes of blower that would satisfy the particular specification. For example, a class 2 blower might be adequate for the specified duty. However there are finite limits to the speed and pressure at which a class 2 blower may be safely operated, and it is frequently advisable to buy a higher class blower to permit greater flexibility in this regard, or to permit more latitude in other subsequent potential uses of the blower.

Other questions would also arise- Is a drain plug needed? Is a clean-out door needed? Should the inlet and outlet be flanged? Should the blower be a type that permits rotation of the housing to change the direction of the discharge? Is a vibration limitation specification needed? Does the shaft require a heat slinger or a shaft seal? For any given requirement of cubic feet per minute and static pressure, the range of blowers that could be procured probably covers a cost span of 4 to 1 depending on whether a "pickup truck" or a "Euclid ore carrier" is the appropriate machine for the job. The same general considerations of detailed engineering specifications apply, of course, to all the major equipment involved in the hygiene control system.

The other category of specification tends to include such language as "to meet all applicable OSHA standards" or perhaps "to have a noise level no greater than 85 dBA at 5 ft from the machine." In this regard, the limitations of the vendor should be realized and the interaction between his equipment and the plant environment taken into account. The vendor has a moral and perhaps legal responsibility to predict the performance and characteristics of the machinery being sold. However interactions between that machine and the location and installation of that machine are beyond the vendor's control and perhaps beyond his knowledge. The single exception to this occurs when the vendor is privy to all the conditions surrounding the proposed installation of the machine; then in fact he is functioning as a design engineer rather than as an equipment vendor.

The noise specification affords a good case in point. An analogy would involve a common experience, the purchase of a light bulb. One can purchase a 100 W light bulb and be assured that its power consumption is reasonably close to 100 W. To be somewhat more technical, the wattage of the light bulb could be related to the lumen output of the bulb, and the item could be purchased on that basis. So far, we are within the duty of the vendor, that is, to predict and guarantee what the light bulb will do when appropriately furnished with 115 V, 60 Hz, AC power. The vendor of the light bulb, however, is unable to tell what level of illumination will appear on the buyer's desk because the installation environment has a high effect on the relationship between the wattage of the light bulb and the light on the desk. The light fixture, the color of the walls, the distances and geometrical relationships between the light bulb and the desk, all affect the resulting illumination level at the work surface.

A comparable situation exists with the noise from machinery. The manufacturer can and should be able to state the sound power level of the machine, that is, how much sound it generates. What the resulting sound pressure level field will be at any point in space depends on how the sound source is installed, whether it has vibration isolators, the nature and proximity of reflecting surfaces, and so on. There is no escape from the technology and engineering that needs to be applied to establish the relationship between the sound power level of the source and the sound pressure level at the receptor. If the vendor is to be burdened with this responsibility, he needs complete data concerning the installation environment, he becomes the design engineer for that portion of the project rather than solely an equipment vendor, and he should be and is (in the price of the equipment) compensated for that additional effort and responsibility.

Another problem arises in connection with the purchase or specification of innovative devices. One of the unfortunate aspects of our reasonably free market system is the appearance, when demand becomes heavy, of a number of innovative devices, many of which operate by some kind of magic that appeals to the unsophisticated purchaser. One example is the generation of "air ions" that charge the airborne dust particles and cause them to sit down on the floor and behave themselves. It is the same marketing freedom, however, that permits the easier and more rapid development of meritorious innovative devices. Fortunately most hygienists are fairly skeptical in their appreciation of new and magic gadgets. If the project is of any importance at all, it merits the expense of obtaining, from the vendor of the innovative device, a list of installations and contacts at those installations. Then the hygienist can visit these places to obtain firsthand information about the functioning of the device. Quantitative measurements of the performance are highly advisable if that can be arranged. The owners of such the innovative devices may be quite unsophisticated, and their opinion alone may or may not be valid. One instance involved the use of a new kind of dust collector, rated at 2000 cfm and 99.97 percent efficiency; when tested, the device proved to be handling 500 cfm and emitting a visible dust plume. However the owner expressed a high degree of satisfaction with the dust collector and was preparing to buy two more.

8 MONITORS, ALARMS, AND "FAIL-SAFE" DESIGN

When engineering controls are indicated but the design problem is difficult, there is a tendency to specify the use of a monitoring device that will sound an alarm, for example, if the contaminant concentration exceeds a specified level. All too frequently, this procedure can be labeled "the designer's cop-out." Having failed to conceive a reasonably effective and reliably engineered system for control, the designer discharges his immediate responsibility by specifying a monitor hanging on the wall to give the alarm in case the control system fails to work. At the current state of the art, these devices are not highly reliable and require a considerable amount of attention if dependability is to be ensured.

A philosophical question arises: If the maintenance capability of an establishment is not adequate to keep an exhaust fan running properly, who is going to keep the delicate electronics of a monitor functioning properly? Such instruments have a tendency to suffer from zero drift and other malfunctions, including false alarms; they need to be adjusted and calibrated frequently. Typically the occasion for the monitor sounding the alarm occurs rarely, and an unusual degree of discipline in preventive maintenance is required to assure that the monitor will be functioning correctly when it is needed. On the other hand, if it has a tendency to give false alarms, it will not be a surprise to find that it has been unplugged or turned off, although still hanging neatly on the wall, encouraging a feeling of false security. If the designer must solve problems by the use of such a monitor, the solution is valid only if arrangements are instituted for adequate maintenance and calibration on a preventive maintenance basis.

A somewhat similar situation exists in the typical factory if an intricate interlock system is designed and installed to prevent misoperation. Such interlocks work much better on paper than they do in real life. Consider a typical scenario. An intricate interlock system has been installed involving microswitches, limit switches, and pressure switches. Sooner or later one of the switches sticks or otherwise malfunctions at 3 a.m. Because of the malfunction, the process is not operable, and the foreman is pressuring the mechanic to get the machinery running again. The electrical schematics are available in two places: in the file cabinet in the locked-up maintenance office, and in the file cabinet in the locked-up engineering department. The night shift mechanic cannot figure out the interlock circuit, cannot locate the offending malfunction, and anyway replacement parts are not likely to be available. The solution: jumper the interlock circuit. If there was no malfunction other than in the interlock circuit, no harm is done. Less happily, the place blows up.

If interlocks are required, keep the design simple. If there is good training and attitude on the part of the workers, the opportunity for human intervention is more reliable than an intricate interlock system. The typical factory is not a space ship where the devotion of time and resources to proper automatic function is infintite, and two or three levels of redundancy are provided. Operating and maintenance personnel are accustomed to coping with malfunctions—motors that do not run, pumps that leak, valves that stick and so on. Efforts to educate, train, and motivate will be preferable to the intricate interlock approach.

If the interlock approach is chosen, it is workable only if the administrative procedures are instituted to ensure that the right information, the right parts, and the right talent are available, and the cost of such is justified. There is one alternative—also costly—which is in case of failure of the interlock system, the operation is shut down and stays down until the fault is found and corrected.

9 DEFINITION OF "FEASIBLE"

Existing OSHA standards require that to control hazardous air contaminants to or below the level of the standard, ". . . administrative or engineering controls must first be determined and implemented whenever feasible." Proposed new standards, although varying somewhat in wording, generally provide that, "engineering controls shall be instituted immediately . . . except to the extent not feasible" and even if "not sufficient to reduce exposures to or below the permissible . . . limit, they shall nonetheless be used . . . to the lowest practicable level."

Neither "feasible" nor "practicable" is defined.

Inexperienced OSHA compliance officers, motivated by natural zeal and encouraged by the language of both present and proposed standards, frequently issue weakly based citations for "failure to use feasible engineering controls." A contest results, and such contests are won by the employer more often than not. This is a waste of talent and resources and accomplishes nothing toward bettering the health of the workers.

Expert witnesses retained by OSHA sometimes recite the entire textbook list of engineering control principles as being feasible engineering controls: (a) substitution of a less hazardous substance, (b) change of process, (c) isolation, enclosure, or segregation, (d) local exhaust ventilation, (e) general or dilution ventilation—without making any realistic assessment of the practicality of applying such controls to the problem at hand, or the results that would be achieved, or the cost, or the benefits.

This "engineering controls" requirement as stated has led to many more contested citations than would have been required for enforcement of a better-defined standard. Unless a meaningful definition is provided, this situation will continue. Such excess litigation will continue to detract from the useful and constructive efforts of OSHA, thus actually diminishing the results obtained in terms of worker health.

In an effort to be constructive, a definition of feasibility is offered. It is not quantitative, but at least it includes all the basic parameters that should be considered in decid-

PHILOSOPHY AND MANAGEMENT OF ENGINEERING CONTROLS 643

ing whether an engineering control is feasible. For a specific case, the definition can be reasonably quantified, although a value judgment is still required. If such a definition were to be included in OSHA standards, it would encourage those thought processes in making an evaluation. The proposed definition is as follows:

"Feasible" means that the method or equipment is available on the market and has been used before with success in the same or closely similar applications, or that the technology exists to create the equipment and implement the method with reasonable assurance of success; that the method or equipment will result in reducing the exposure to or below the time-weighted average standard or is necessary to reduce the exposure to a level where the reasonable use of administrative controls or personal protective equipment which is not unduly onerous to the employee will adequately protect the health of the employee; that the number of employees exposed, and the frequency and severity of the exposure, are included in a cost-effectiveness consideration of the implementation of such controls; and that seriousness of the potential risk to the employee health is given due weight in all considerations.

Some application guidelines can also be furnished. To permit easier reference to the elements of the definition in analyzing a specific case, it is rewritten with parenthetical identifiers so that the text need not be frequently repeated, as follows:

"Feasible" means that the following conditions are met (in the case of items 1 and 2, only one statement in the set must be met):

1a. That the method or equipment is available on the market.

1b. That it has been used before with success in the same or closely similar applications.

1c. That the technology exists to create the equipment and implement the method with reasonable assurance of success.

2a. That the method or equipment will result in reducing the exposure to or below the time-weighted average standard.

2b. That the method or equipment is necessary to reduce the exposure to a level where the reasonable use of administrative controls or personal protective equipment which is not unduly onerous to the employee will adequately protect the health of the employee;

3. That the number of employees exposed and the frequency and severity of the exposure are included in a cost-effectiveness consideration of the implementation of such controls.

4. That the seriousness of the potential risk to the employee health is given due weight in all the considerations.

To be judged feasible:

1. The answer to either 1a, 1b, or 1c should be "yes."
2. The answer to either 2a or 2b should be "yes."

3. Analysis of 3 should show that the proposed control *is* cost-effective. (Obviously a value judgment is involved here.)
4. "Seriousness" of the hazard includes several elements. These are:
    a. The immediacy of the potential hazard to life. For example, the potential for high concentration (many multiples of permissible 8 hr TWA) of some agents such as carbon monoxide, hydrogen sulfide, or chlorine, can be fatal in a few minutes. (Obviously very serious; does not apply to TLV concentrations.)
    b. Whether a chronic effect of the hazard would be a debilitating, painful, irreversible disease, such as cancer (more serious); or would be instead reversible with little or no remaining disability, such as lead poisoning (less serious).
    c. Whether there exist biological monitoring techniques that if implemented, would give warning *before* any significant health effect had occurred. Examples are testing blood for lead, sputum cytology, and audiometric tests (less serious than if no such technique is available).
    d. Whether the effect of the hazard would be a debilitating disease, such as silicosis (more serious), or whether it would result in nondebilitating diminution of performance, such as hearing loss, peripheral neuropathy, or skin sensitization (less serious).

10 UNSOLVED PROBLEMS

The textbook list of engineering controls for hygiene problems is short and simple in its classifications. Segregation, isolation, substitution, change in process—are all valid concepts. Local exhaust ventilation with its subset of principles—total enclosure, partial enclosure, get close to the source of the contaminant—embody valid concepts.

However at the present state of the art the application of these concepts to many problems is unwise and wasteful, if included in the list of concepts is the idea that the installation should be reasonably workable and reasonably effective. The plant engineer may seem to be uncooperative and recalcitrant when approached by the hygienist to apply some reasonably clear-cut concept to a problem in the plant. The plant engineer may in some cases indeed be recalcitrant; but more likely he is merely being more skeptical and more practical. He has the problem of designing the application of the concept and living with its problems after it is installed. To the plant engineer, the project is a failure unless it is reasonably workable and reasonably effective. His career, and indeed the future of the company, cannot stand too many failures.

There are at least four categories of hygiene problems in which the application of the concept principles becomes exceedingly difficult and frequently impossible. These are moving sources of contaminant, handling of bulk materials that are heterogeneous and not susceptible to mechanized handling, manual contaminant-generating operations on very large objects, and the combination of heat stress and contaminant control hoods.

Moving contaminant sources can be controlled if they are very large—such as a ship-loading gantry crane—where an entire control system can be mounted on the moving object. They are also susceptible to control if they are quite small, as in the case of handling of beakers and flasks in a laboratory fume hood. In the intermediate size range effective control of moving sources varies from the difficult and expensive to the impossible. The Hawley Trav-L-Vent® concept sometimes can be applied; in other cases the kinds of motion and the available space do not permit its application. Flexible hoses and swivel joints have severe limitations.

If raw materials or by-poducts are to be used in amounts large enough to require bulk handling (i.e., more than can be handled in small, closed containers), the nature of the materials becomes important in choosing the type of material handling that is workable. If the material is difficult to handle because it is quite heterogeneous in size and shape of particles or objects, or very sticky, the available bulk material handling equipment that is amenable to contaminant control is usually not workable. An example of difficulty in handling is provided by the recycling of scrap metals, where the feed material may vary from a fine powder, to chunks and objects measuring several inches, to dross removed from pyrometallurgical processes and consisting of granular dusty material interspersed with beads or strings of frozen metals. For such material handling problems, only the most crude equipment such as front-end loaders and dump trucks is satisfactory. Even though the dumping points for a front-end loader can be provided with ventilation control, the other activities of the front-end loader cannot be.

When manual contaminant-generating operations are conducted on very large objects, engineering controls fail, at the present state of the art. For example, a very large object can be spray painted in a spray booth under good capture and control air velocities. The ventilation typically prevents the spread of the contaminants through the rest of the plant and prevents accumulation of explosive mixtures of solvent vapor, but is not sufficient to protect the person doing the spraying from toxic pigments. It is necessary to spray large surfaces where the back spray is significant, to spray sideways or upwind to get various parts of the equipment painted, and so on.

Similar problems exist with the manual tending of large furnaces or reactors which, if they cannot be made leakproof, generate contaminants in the area where the tender is working. Any enclosure or exhaust ventilation provisions would result in the person working inside the hood, which may be advantageous for the rest of the plant but does not protect the worker.

Where work stations involve a combination of heat stress and contaminants to be controlled by local exhaust hoods, the control situation is a difficult and many times is a somewhat unsatisfactory compromise, as discussed in detail in Section 6.5.

REFERENCES

1. K. J. Caplan and G. W. Knutson, "Generation of Dust During Plate Transportation," IBMA Fall Meeting Reports, October 1977, pp. 79–83.

2. K. J. Caplan and G. W. Knutson, "Dust Generation and Control at Rack Loading and Unloading," IBMA Fall Meeting Reports, October 1977, pp. 74–78.
3. American Conference of Governmental Industrial Hygienists, *Industrial Ventilation—A Manual of Recommended Practice,* ACGIH, Cincinnati, Ohio, current issue.
4. American Society of Heating, Refrigerating and Air Conditioning Engineers, *Handbook and Product Directory, Volume on Fundamental,* Ch. 30, "Space Air Diffusion," ASHRAE, New York, 1977.
5. American Industrial Hygiene Association, *Heating and Cooling for Man in Industry,* 2nd ed., AIHA, Akron, Ohio, 1975.
6. American Society of Heating, Refrigerating, and Air Conditioning Engineers, *ASHRAE Handbook and Product Directory,* ASHRAE, New York, current issues.
7. Air Moving and Conditioning Association, AMCA Standard 300-67, "Test Code for Sound Rating," AMCA, 1967.
8. Air Moving and Conditioning Association Publication 303, "Application of Sound Power Level Ratings for Ducted Air Moving Devices," AMCA, 1965.
9. American Industrial Hygiene Association, *Industrial Noise Manual,* 3rd ed., AIHA, Akron, Ohio, 1975.

CHAPTER NINETEEN

Personal Protection

BRUCE J. HELD

1 INTRODUCTION

The philosophical basis of the use of personal protective devices has been that first preventive measures should be taken to keep contamination from entering the workplace and from contacting workers. This is done if at all possible through the use of good engineering control measures. However, engineering measures cannot be used in all circumstances, and personal protective devices must be relied on in some cases.

The *Respiratory Protective Devices Manual* (1), the first book I know of to comprehensively cover the field of respiratory protection in industry in the United States, begins by stating: "In the control of those occupational diseases caused by breathing air contaminated with harmful dusts, fumes, mists, gases, or vapors, the primary objective should be to prevent the air from becoming contaminated. This is accomplished as far as possible by accepted engineering control measures; . . . ."

Similar wording is found in the American National Standard Z88.2-1969, "Practices for Respiratory Protection" (2). This standard, adopted verbatim by the Occupational Safety and Health Administration (OSHA), states that

[I]n the control of those occupational diseases caused by breathing air contaminated with harmful dusts, fogs, fumes, mists, gases, smokes, sprays, or vapors, the primary objective shall be to prevent atmospheric contamination. This shall be accomplished as far as feasible by accepted engineering control measures (for example, enclosure or confinement of the operation, general and local ventilation, and subtitution of less toxic materials). When effective engineering controls are not feasible or while they are being instituted, appropriate respirators shall be used pursuant to the following requirements.

In health standards and proposed health standards, OSHA further defines the use of respirators as a last line of defense for the worker in a contaminated atmosphere. The language varies from one standard to the next, but the regulatory agency basically limits respirator usage to the following conditions:

1. For routine operations while engineering controls are being instituted or evaluated.
2. When engineering controls are not technically feasible or cannot by themselves control a contaminant below an acceptable level.
3. For nonroutine operations that occur so infrequently that engineering controls would be completely impractical (e.g., certain maintenance operations).
4. For emergency use or other unplanned events where the possibility of an overexposure to a worker exists.

The attempt to avoid the use of personal protective equipment in place of engineering controls is readily understandable for several reasons. First, the effectiveness of personal protective devices depends on the actions of many people; each and every user is responsible for knowing the limitations of the devices, using them properly, and in fact, using them at all. Engineering controls, on the other hand, can be installed and operated by a few "experts"; they can be monitored and even can sound an alarm in the event of a failure so that workers can be removed from the area before an overexposure occurs. If a respirator fails and the air contaminant is not readily detectable by odor or one of the other senses, however, an overexposure will not be detected until after the fact, when it may be too late. In like manner, the wrong or improperly used laser glasses cannot prevent an eye injury from laser light, but proper shielding and interlocks on the laser source will.

Other undesirable features of respirators in particular are discussed in the referenced documents. Some of these are summarized in the proposed OSHA standards for benzene (4) and cotton dust (5) in the testimony given in defense of the limitations imposed on the use of respirators in lieu of engineering controls and include:

1. Fitting problems with persons with small faces such as many women, with persons with facial hair, such as beards or long sideburns, scars or growths which break the facepiece-to-face seal and with persons with pronounced wrinkling, a sunken nose bridge, deeply cleft chin, very narrow nose, a very wide or very narrow face, or very long or very short face.
2. Communications problems between workers wearing respirators. Any respirator will muffle voice communications to varying degrees. With many devices, the jaw movements used for speaking can break the facepiece-to-face seal and cause inleakage of contaminated air.
3. Vision problems can be numerous. Downward vision is obstructed with half- or full-face masks, thus increasing tripping or stumbling possibilities. The inability to look straight down could easily cause serious injury. Full-face respirators will restrict the wearer to tunnel vision. Persons who must wear prescription glasses must be fitted with special glasses kits for full-face masks, as the temples of regular glasses break the facepiece-to-face seal. Some persons also have problems wearing glasses with half masks, as the respirator either pushes the glasses above the

PERSONAL PROTECTION

eyes or the glasses come between the mask and the face around the nose, thus breaking the facepiece-to-face seal.

4. Fatigue and reduced efficiency occur more quickly among workers wearing any type of covering over the face or head. This is caused by many factors, among them being increased breathing resistance, heat, some possible feelings of anxiety from being closed in, and reduced vision. On many jobs we have permitted, or required, 15-minute breaks every one or two hours while wearing a respirator; rather than the usual one coffee break in the morning and one break in the afternoon.

5. Some persons, for reasons of health, would not be permitted to work if they had to wear a respirator. A person with a heart problem may not be able, for example, to wear a 30-pound self-contained breathing apparatus. Persons with some lung problems such as emphysema may not be able to stand the increased breathing resistance inherent in any respirator.

It is noteworthy that in 1963 the *Respiratory Protective Devices Manual* (1) states that respirators should be used only when engineering control measures are "uneconomical, inapplicable, impractical, or ineffective." A dozen years later, OSHA no longer considered the economic aspects of engineering controls in its standards. The belief now is that human health cannot be measured in dollars and cents, and the effectiveness of engineering controls is so superior to respirator usage that cost should not be a factor. This alone has probably caused the greatest amount of friction between industry and OSHA in the respiratory portions of the proposed standards. It is indeed difficult to define engineering feasibility solely in altruistic terms, since economics certainly plays a part at some point. Locating that point with precision, though, is the problem.

2 BASIS FOR PERFORMANCE TESTING OF PERSONAL PROTECTIVE EQUIPMENT

The National Institute for Occupational Safety and Health (NIOSH) under the Department of Health, Education and Welfare, has been established as the testing and approval agency for personal protective devices. At present NIOSH tests and approves only respiratory protective devices. Other protective devices for skin and eyes are not tested and approved by any agency, although it is hoped that eventually NIOSH will take on this responsibility.

2.1 Respiratory Protective Devices

2.1.1 Early Development of Devices

Prior to World War II the United States was notably lacking in good respiratory protective devices. Except for the mining industry, which imported European devices for mine rescue work, general industry and the fire service had very little in the way of protection for workers.

I have learned that general industry essentially used no respirators before the twentieth century. This was probably because occupational illness was generally unidentified as such, and when it was, the worker had no legal rights or compensation available. Thus employers did not feel obligated to offer protection.

The fire service had a more visible problem in that smoke inhalation caused an immediately identifiable effect: illness or death. Such effects led to a multitude of inventions of breathing apparatus of questionable value for the firefighter, but at least efforts were made. Figures 19.1 and 19.2 show two of the devices invented for fire service usage. Unfortunately the devices were often purchased and used for quite a while before the firefighters entered an immediately dangerous atmosphere, and the true worth of the device was discovered only when the "protected" firefighters died or sustained severe lung impairment. Some fire departments even budgeted for steam beer for their firefighters to be available after each firefighting episode to help "cleanse the lungs" (6). Other departments required that the firefighters have at least a 6 in. beard, which was to be dipped in water before entering a burning building. The beard was then folded up

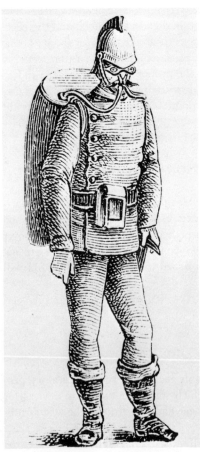

Figure 19.1 Galibert's apparatus of 1864. Exhaled air is returned to the bag, deteriorating the remaining air. From Magirus, *Das Feuerloschwesen,* 1877. The book was loaned to the author by the New York City Fire Museum.

PERSONAL PROTECTION 651

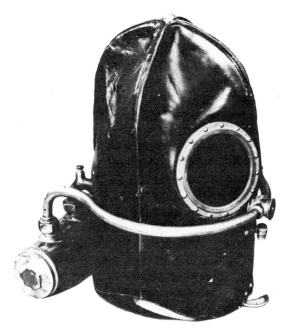

Figure 19.2 Servus "Eveready" smoke and ammonia helmet of 1919. Photo by the Sandia Livermore Laboratory Photography Department; device loaned to the author by Robert Foley, Napa, California, Fire Department.

and put in the mouth; the smoke was supposed to be filtered and cooled when breathed through the growth. Wet sponges were also sometimes used to filter and cool the smoky air (Figure 19.3).

The mining industry, however, had no problem in recognizing the inadequacies of contemporary equipment for breathing protection. Mine rescue teams had to enter oxygen-deficient atmospheres for long periods of time, and only good oxygen-supplying equipment would serve. Since none was available in the United States, the mining companies looked to Europe to fulfill their needs. Devices made by Siebe Gorman in England and Dräger and Westphalian in Germany were used by the rescue teams. Figure 19.4 depicts a 1916 Dräger device being worn by a miner.

The use of poison gas by Germany in World War I quickly illustrated America's woeful lag in producing adequate respiratory protection. An English device was brought to the United States, copied, and produced by the Bureau of Mines (B of M) for use by the Allied Expeditionary Forces. Gas mask development work was quickly started, resulting in devices such as the Connel, Kops, and Tissot masks.

2.1.2 Formation of Bureau of Mines Approval System

After World War I the B of M started studies of respirators for industrial uses, based on their earlier work on military devices (1). A method to test and approve self-

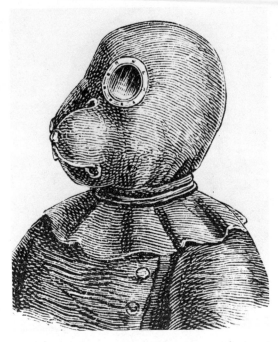

Figure 19.3 Kuhfuss smoke hood of 1862. A moist sponge is placed in the buttoned opening in front of the mouth. From Magirus, *Das Feuerloschwesen*, 1877.

contained breathing apparatus (SCBA) for mine rescue teams was developed, and the first device, the Gibbs Oxygen-Breathing Apparatus (Figure 19.5) was tested and approved in 1920. The method of operation of the B of M, and now NIOSH, was and is that the respirator manufacturer first perform all required tests and submit the results; the testing agency then attempts to confirm the results.

The first B of M approval schedule, number 13, for SCBA, went into effect on March 5, 1919. Five revisions of that schedule were made (Schedules 13A through 13E) (7) between 1913 and 1968 before NIOSH took over the testing functions from the B of M in 1972.

Most of the B of M tests for SCBA were designed as "man tests," that is tests that are performed while the unit is being worn by someone. The test subjects performed various exercises such as walking, running, crawling, weight lifting, and sawing wood while wearing the apparatus during a series of 15 types of test in an irrespirable atmosphere of formaldehyde. Pulse and breathing rates of the test subjects were measured throughout the testing periods, as well as inspired air temperature and carbon dioxide content of the atmosphere within the SCBA. Certain engineering qualifications were also placed on the unit—for example; fully charged devices could weigh no more than 40 lb, and bypass valves had to be provided.

PERSONAL PROTECTION 653

The tests were designed to provide answers to the following questions (1) and are still the objectives of the NIOSH test schedule:

1. Will the apparatus supply air or oxygen fast enough to meet the needs of the wearer?
2. Can the wearer breathe freely at all times?
3. Does the temperature of the inspired air remain within prescribed limits?
4. Are vital parts of the apparatus protected so as to prevent damage or excessive wear?
5. Is the concentration of carbon dioxide in the inspired air within prescribed limits?
6. Does the harness hold the apparatus on the wearer's body without undue discomfort to him?
7. For what period of time will the apparatus provide adequate respiratory protection for the wearer under the test conditions?

Gas Masks. Two types of approval test were necessary for the gas mask schedule. Like the SCBA schedule, man tests to determine the effectiveness of the complete system

Figure 19.4 Dräger apparatus of 1916. Negative 22795, album 0.0.3–0.021, U.S. Bureau of Mines Files, Pittsburgh. Photo loaned to the author by Edwin J. Kloos, Mining Enforcement and Safety Administration.

Figure 19.5 Fresno, California, Fire Department training class using Gibbs apparatus in 1920. Photo loand to the author by B. F. McDonald, San Clemente, California.

were required. In addition, tests to determine the reliability of the gas mask canisters were included.

The gas mask schedule, number 14, for approval went into effect in August 1919. Six revisions were made over the years from 1923–1955. Schedule 14F (8) was in effect when NIOSH took over the testing in 1972.

The canisters were machine tested against low concentrations of the contaminant of interest at low flowrates and at high concentrations at high flowrates. The low concentrations and flowrates gave information on the ability of the sorbent to hold the contaminant and also on how much of the contaminant the sorbent could hold without releasing any into the facepiece. The high concentration and flowrate tests determined whether the sorbent could react quickly enough to capture the contaminant before it could enter the facepiece. Both dry air and moist air tests were performed to determine effects on the sorbent.

The man tests were made in irrespirable atmospheres of ammonia to determine gross leakages on men with varying facial shapes and sizes, then in phosgene as an even more severe gas-tightness test. To pass these tests, the gas mask had to give complete protection to the men while doing various exercises; excessive eyepiece fogging could not occur, and undue discomfort could not be experienced by any of the subjects. Gas masks

PERSONAL PROTECTION 655

were approved for use in 2 percent maximum concentrations of acid gases, organic vapors, carbon monoxide, or 3 percent ammonia.

Supplied Air Respirators. Supplied air devices were first approved in 1932 under schedule 19. Only two revisions were made before NIOSH took over the testing, namely, 19A in 1937 and 19B in 1955 (9). Approvals were granted for units such as hose masks with or without blowers, continuous flow or demand air lines with air supplied by a compressor or tank, and continuous flow or demand abrasive blasting respirators.

As with the SCBAs, man tests were the primary testing form to determine facepiece, helmet, or hood fit, air distribution in the respiratory inlet covering, vision clarity, safety, and comfort factors. In addition, limitations on hose length were required (75 ft for a hose mask without blower, 150 ft for a hose mask with blower, and 250 ft for all other devices in this category), inhalation and exhalation resistance limitations were set, maximum air pressures at hose inlet was established, and strength, noncollapsibility, and resistance to liquid gasoline requirements on the air hoses were set. A sandblasting man test operation involving silica dust and airflow characteristics was also required for abrasive blasting respirators.

Dust, Fume, or Mist Respirators. Respirators for protection against particulate matter, dusts, fumes, or mists, were approved under schedule 21, which first became effective in 1934. Revisions were made in 1955 (Schedule 21A) and in 1965 (Schedule 21B) (10) before NIOSH took over the testing in 1972.

None of the dust, fume, or mist respirators approved by the B of M were for protection against dispersoids significantly more toxic than lead. The filters used in those respirators were tested against silica dust and litharge for dust respirators, and chromic acid mist and a mist formed by atomizing silica in a water suspension for mist respirators. Combinations of these tests could be used for approving respirators for protection against dusts and fumes, dusts and mists, fumes and mists, or all three, provided the respirator passed the requirements for each type. Men with varying facial sizes would also man test each respirator submitted for approval to check on its fitting qualities.

Chemical Cartridge Respirators. The last B of M test schedule was number 23, for chemical cartridge respirators. It became effective in 1944 and was set up to approve only organic vapor chemical cartridge devices, as were the two revisions 23A and 23B (11) promulgated in 1955 and 1959, respectively. The chemical cartridge respirators were tested in a manner very similar to that employed for their gas mask counterparts, but at a much lower test concentration. Carbon tetrachloride was used as the organic vapor test agent, but the approvals limited the respirator use to concentrations of organic vapor of 0.1 per cent (1000 ppm) or less. Isoamyl acetate was used as the test agent for the fitting quality tests.

Combination organic vapor and dust fume, or mist respirators were also approved, if the device passed the tests for each type of unit for which it was submitted. Paint spray respirators were also approved by testing the device against carbon tetrachloride, atomized lead paint, enamel, and lacquer.

2.1.3 NIOSH/MESA Approval System

In 1972 NIOSH took over the testing functions of respirators from the B of M. All testing was performed by NIOSH, the results reviewed by the B of M, and the approval was issued jointly by NIOSH/B of M. In 1973 the B of M was reorganized and the regulatory functions formed into a new agency, the Mining Enforcement and Safety Administration (MESA). Respirator approval certification went along into the new agency, and henceforth, the approvals were issued jointly between NIOSH/MESA.

NIOSH took the B of M test schedules, revised them as appropriate at the time, and reissued them (12). A new schedule, Subpart M, was added for the approval of pesticide respirators.

Subpart A: General Provisions. Subpart A lists the purpose of Part 11 and defines words and terms used throughout. In addition, requirements were added that limited respirators used for entry or escape from hazardous mine atmospheres to those meeting Part 11 requirements after March 30, 1974. Subpart A also stated that an "approved" respirator was no longer approved unless it was selected, fitted, used, and maintained in accordance with provisions given in ANSI Z88.2-1969 (2).

Subpart B: Application for Approval. The procedures for applying for approval, application contents, and the number of respirators the applicant must provide to NIOSH are outlined in Subpart B.

Subpart C: Fees. Subpart C gives the fee schedule for respirator approvals. Fees listed on publication ranged from $500 for a nuisance dust respirator to $4100 for a type N gas mask.

Subpart D: Approval and Disapproval. This subpart sets forth the requirements for issuing certificates of approval, for disapproving a respirator that has failed to meet the tests, for marking and labeling approved devices, for revoking an approval certificate, and for issuing a modification of a certificate of approval when the manufacturer wishes to make or change a feature of an approved device.

Subpart E: Quality Control. Subpart E was a major addition from the B of M schedules. It sets forth detailed requirements for a quality control plan that a respirator manufacturer must prepare, submit for approval, and follow when manufacturing an approved device. Also, detailed quality control records are required, which can be reviewed from time to time by NIOSH during the period an approved device is manufactured. Failure to follow the quality control plan submitted and approved, or failure to keep proper records, can result in revocation of the approval.

Subpart F: Classification of Approved Respirators; Scope of Approval; Atmospheric Hazards; Service Time. This subpart names the types of respirator and tells the scope of approval. It also classifies the respirator types according to hazards in which they may be used (e.g., entry and escape, escape only, oxygen deficiency), or the contaminants against which the device protects. Service times for each respirator type must be specified by the manufacturer according to a classification set forth in Part 11.

Subpart G: General Construction and Performance Requirements. Rather than repeat construction and performance requirements under each subpart for each type of respirator, subpart G lists all those that are common to all. These requirements include good engineering, no hazards to the user by the device, pretesting by the applicant, observers permitted, and other miscellaneous items.

Subpart H: Self-Contained Breathing Apparatus. The SCBA requirements remain essentially the same as in B of M Schedule 13E. Pressure-demand apparatus was added, and the total weight of the equipment was reduced to 35-lb. Under some conditions 40-lb is still permitted. Units rated for 15 min or longer are permitted for entry into a hazardous area, low temperature tests were modified, and air quality standards for cylinders were modified. The man test requirements remain essentially the same.

Subpart I: Gas Masks. The test schedule for gas masks also stayed essentially the same as B of M Schedule 14F, but additions were made. A chin-style gas mask was added that contains less sorbent, and the canister is attached directly to the facepiece instead of being mounted on the chest or back and attached with a breathing hose. These devices are permitted in air contaminant concentrations up to 0.5 percent (5000 ppm). Escape gas masks were also added, which consisted of a half mask or mouthpiece, a canister, and associated connections.

Masks to protect against acid gas and organic vapor gas were given an added restriction: they could not be used against contaminants with poor warning properties, generally regarded as those with an odor threshold above the threshold limit value (TLV). Nitrogen dioxide was also added as a test agent for acid gas approvals.

There is now some controversy concerning the use of gas masks. NIOSH has been somewhat reluctant to approve them and some manufactuers are purposefully not producing some types, such as type N or universal gas masks. Misuse of these devices can easily occur, since they are a respirator for use in highly dangerous concentrations of a contaminant and in emergency situations. Because air concentrations generally are not known in an emergency, the device could be incorrectly used, that is, employed as protection in concentrations beyond its capability and even in an oxygen-deficient atmosphere. Possible liability suits against manufacturers further discourage the production of such devices.

Users are generally cautioned to contact NIOSH for guidance in the use of any gas masks and are encouraged by most health and safety professionals to use an appropriate atmosphere-supplying device. The OSHA health standards likewise prohibit gas masks for use in emergency situations or in atmospheres immediately dangerous to life or health (IDLH). MESA has also withdrawn permission to use these devices, even for escape.

Subpart J: Supplied-Air Respirators. This section also is very similar to the B of M counterpart, except that the air quality portions have become more restrictive. Use of the type C supplied-air respirators in IDLH atmospheres is prohibited.

Although Subpart J still permits the use of a hose mask with blower in an IDLH atmosphere, the OSHA health standards do not, nor does the NIOSH/OSHA decision logic in use for the Standards Completion Program (13). Both types A and B supplied-air respirators generally are not permitted for use at all in the new OSHA health standards, since the disadvantages of using them coupled with the advantages of other devices now available make them obsolete.

Subpart K: Dust, Fume, and Mist Respirators. Provisions have been made in Subpart K to approve filter-type respirators for protection against particulates significantly more toxic than lead, for protection against radon daughters, for protection against asbestos, for single-use dust respirators as protection against pneumoconiosis fibrosis-producing dusts or dusts and mists, and for powered air-purifying respirators.

The high industrial usage of highly toxic particulates led to the need to approve respirators equipped with high efficiency particulate air (HEPA) filters, defined as being 99.97 percent efficient against airborne particles with a mass median aerodynamic diameter (MMAD) of 0.3 μm. The HEPA filter definition came about because filters were available with the 99.97 percent efficiency and a test aerosol was readily available, heated dioctyl phthalate (DOP), which produced a monodispersed aerosol of 0.3 μm MMAD. Respirators with such filters were in high demand for protection against radioactive particulates, beryllium dusts, and other highly toxic dispersoids.

Advances in aerosol technology will undoubtedly force changes in the test methods for other particulate filters as well. Problems in the reproducibility of test methods using silica dust and lead fumes will be alleviated when tests using sodium chloride or some other reliable aerosol come to replace the older dispersoid methods as originated by the B of M.

Subpart L: Chemical Cartridge Respirators. Subpart L expands the B of M chemical cartridge respirator approval system to include any gas or vapor, including acid gases, ammonia, or whatever the manufacturer stipulates. Again, as with gas masks, organic vapors with poor warning properties are not included in the approval for organic vapor respirators. Powered air-purifying respirators are also included in the test schedule.

Subpart M: Pesticide Respirators. This completely new test schedule replaces the listing of respirators that could be used for protection against pesticides published by

PERSONAL PROTECTION 659

the Department of Agriculture from 1950 through 1966. The subpart entails tests using carbon tetrachloride to check the sorbent capabilities and silica dust and lead fume for testing the filter portion of the cartridge or canister. Front- or back-mounted pesticide canister HEPA filters are tested with DOP.

One question that has been posed concerns the use of carbon tetrachloride as the only test for the sorbent. In studies by Nelson and Harder (18), a charcoal sorbent was found to be very effective for absorbing carbon tetrachloride but not for some other organic vapors. This could mean that if the vapor phase of a particular pesticide was not readily absorbed by the sorbent and if it also had poor warning properties, the user would not be aware of vapor breakthrough into the facepiece.

2.1.4 Research for New and Better Test Methods

Needs for new and better respirator testing methods are recognized by most people in the industrial hygiene field and in NIOSH. Toxicology studies resulting in the lowering of TLV or time-weighted averages (TWA) are appearing in the new OSHA health standards. Consequently, respirator reliability must be increased. What may have been an acceptable leak rate for a respirator that afforded protection against a substance with a given TWA may no longer be acceptable if the TWA has been lowered appreciably.

A complaint sometimes made by respirator manufacturers is that the test and certification schedules allow no flexibility for innovations by the manufacturer. For the manufacturer to obtain NIOSH/MESA approval on a respirator, the device must be designed to meet and pass the requirements in the approval schedule. Thus it is difficult to secure approval for a new respirator concept, either because there is no approval schedule or because the concept may not meet the engineering requirements in the existing schedules, hence cannot be approved. Since OSHA requires use of approved respirators, there is no sales market for a respirator designed around a new concept.

Numerous other problems exist in connection with the NIOSH test schedules. NIOSH is aware of these and is funding research projects, as the budget allows, to try to resolve many of them. However it may take several years for funds to be made available, research to be completed, proposed changes to be written and published in the *Federal Register,* public hearings held, differences resolved, and final changes published and made effective.

Current Research in Respirator Face Fitting. One major field that has been investigated over the past few years has been respirator face fitting. The Atomic Energy Commission originally funded the Los Alamos Scientific Laboratory (LASL) in 1969 to begin investigations in facepiece fitting (15). Additional funding came from NIOSH when it was organized in 1972, both to LASL for facepiece fitting studies (16, 17) and to Webb Associates to study the facial sizes of adult workers and determine which dimensions most affected facepiece fit (18). From these studies came information leading to the development of protection factors (19) and maximum use limits (20) (see Table 19.1, below).

The results of these studies will probably lead to a change in the NIOSH testing and certification schedules, whereby the randomly selected subjects who test facepiece fit subjectively, or qualitatively, will be replaced by a test panel of male and female subjects, selected anthropometrically to represent most of the facial sizes found in adult workers. The fitting tests themselves can also be made quantitatively to determine exact leak rates. The leak rate allowances then can be set within acceptable and measurable parameters.

Other research directed toward improving test methods for respiratory protective devices of all types include development of a sodium chloride aerosol system, a respirator exhalation valve test system, a head harness strap tension test system, a facepiece-to-face pressure measuring system (21), a gas facepiece-to-face fitting system, and a valve test unit (22).

Other NIOSH-sponsored projects at LASL to improve specific test schedules include development of fit-test procedures for powered air-purifying respirators and comparisons of sodium chloride filter test methods to silica dust and silica mist filter tests, to arrive at more reliable test methods for dust, fume, and mist respirators (17). Performance of single-use dust respirators are also under investigation along with other filter studies to more reliably test filter performance (17). At Lawrence Livermore Laboratory (LLL) an examination of chemical cartridge respirators (14) showed that many organic vapors broke through cartridges approved by NIOSH for organic vapors in minutes or even seconds. Use of the currently used carbon tetrachloride method in the approval schedules for organic vapors and pesticides obviously must be changed and new methods developed.

Respirator Face Fitting for Firefighters. One major concern has developed in the SCBA test schedule (Subpart H). The schedule now in effect originated in the B of M, since there was concern about SCBA usage by miners. The fire service is now one of the major users of SCBA in the United States, however, and the environments to which firefighters expose their equipment are not at all similar to a mine. Extremes of heat may be faced, along with radiant heat, high moisture, fire decomposition products, very cold temperatures in the winter in northern cities, and high vibration from riding on speeding fire engines. Preliminary studies at LLL (23) indicate the necessity of including ways to test these factors either in Subpart H or under a totally new test schedule for firefighters' breathing apparatus.

Along with a test schedule for the firefighters' breathing apparatus, changes in use methods and equipment used must be made (24). Firefighting is now considered one of the most hazardous occupations in the United States, and inhalation and heart injuries and deaths are major contributors to the high statistics (25). Most fire departments use demand apparatus, whereas in today's technological age of synthetic materials whose decomposition products may be very highly toxic, the use of pressure demand or safety pressure equipment should be mandatory (24). Although no fitting of facepieces for firefighters occurs and the SCBA of most fire departments are all from the same manufacturer, it is known that a large percentage of firefighters cannot be fitted if only one brand of facepiece is available (24). Since local fire departments do not fall automatically under the regulations of OSHA, which is federal, no regulatory action may be taken (unless the state has agreed to bind itself and its municipalities to OSHA standards). It is thus incumbent on the fire service itself to correct use problems. It is also necessary for NIOSH to provide a test and certification schedule that will assure the firefighter that the equipment will not fail, no matter what the exposure environment consists of.

2.1.5 Research for Better Respiratory Protective Devices

In addition to improving the methods used in the test schedules, NIOSH is also charged with trying to develop better protective devices. Of course, the respirator manufacturers themselves are also seeking improvements in present equipment.

Evaluation studies of existing equipment and use practices are usually conducted before any attempts are made to develop new equipment. One NIOSH-sponsored study that evaluated respirators used in paint spraying operations (26) led to several recommendations both for improvements of the respiratory equipment itself, as well as for better certification tests and methods. The published report was then made available to respirator manufacturers to encourage the development of better equipment. Similar studies were made on abrasive blasting equipment (27) and respirators used in coal mines (28, 29). Several other NIOSH-funded studies conducted at LASL on various types of particulate filter used in respirators resulted in recommendations for improved performance criteria (17, 21, 22).

Firefighters' breathing apparatus has also been studied in an effort to improve protection. Studies at LASL (30) show that the weight of personal protective equipment is the main contributor to fatigue of personnel using it. The National Aeronautics and Space Administration (NASA), in a cooperative program with the Fire Technology Division of the National Bureau of Standards and Public Technology, Inc., polled city fire departments on their needs for improved breathing apparatus and developed a prototype SCBA that was both smaller and 10 lb lighter than currently available units (31). One manufacturer is now producing a unit patterned after the NASA prototype, another is planning to manufacture a similar unit, and yet another has reduced its unit's weight by 10 lb by using a fiber-wrapped aluminum cylinder instead of the steel cylinder. The LLL studies (23) are now demonstrating problem areas where firefighters' SCBA need to be hardened for the fire environment, particularly against high air temperatures and radiant heat. Communications systems development is another area needing attention (24).

Studies at LLL by Nelson and Harder (14) indicated the need for improvement in the sorbent capabilities of chemical cartridge respirators used for protection against organic vapors. Before these studies were conducted, it was believed that any cartridge that could effectively absorb carbon tetrachloride would be adequate for any organic vapor. The LLL research showed that this was not true, that new sorbents, specific for certain organic vapors must be developed, and that the general all-purpose organic vapor cartridge now in use must be limited to certain substances. These limitations can also be found in the respirator selection tables used in the OSHA health standards and the NIOSH/OSHA Standards Completion Program.

Other areas that remain to be worked on include the need for better assurance that highly toxic and carcinogenic materials cannot enter the facepiece under any conditions, methods to limit chemical cartridges and canisters from being used beyond their useful capacity (e.g., through the use of color indicators, timers, or other means), continuation studies to determine psychological and physiological problems associated with respirator usage and development of devices to counter these problems, better assurance of respirator facepiece-to-face fit each time a device is used, reproducibility studies of facepiece fit each time a mask is donned, and development of low cost, more reliable fitting methods.

2.2 Skin Protective Devices

This section covers skin protective devices only to the extent that the device protects the person from a health standpoint; it does not discuss traumatic injuries caused by burns, abrasion, flying or falling objects, and so on.

Unfortunately there are no federal testing and approval schedules for skin protective equipment, nor are there any applicable accepted standards. Determining the effectiveness of such equipment is largely up to the individual health and safety professional.

2.2.1 Supplied-Air Suits

Occassionally complete body protection is required in addition to breathing protection to prevent skin absorption. In such conditions a supplied-air suit can provide the necessary protection. Since NIOSH does not have a test schedule for supplied-air suits, it is incumbent on the health and safety professional to determine that the suit does indeed protect the wearer adequately.

An adequate supply of air is mandatory. Too little air can cause the suit to go negative, drawing outside contaminated air inside the suit. On the other hand, too much air can cause the suit to inflate to the point where movement is difficult. Tests conducted at LASL showed that the noise levels in some suits could exceed those permitted under the OSHA standards and that the very method of donning the suit could determine whether the suit remains under positive or negative pressure while in use (30).

For highly toxic particulates, care must be taken in removing the suit to prevent the spread of contamination both to the wearer and to the surrounding environment and unprotected personnel in the area. The quality of air supplied to the suit must be monitored in the same manner as required in the OSHA standard for supplied-air respirators. Suits should never be worn in IDLH atmospheres unless adequate precautions are taken for escape in case of failure of the air supply or a tear occurring in the suit. The possibility of gas or vapor migration through the suit material, even when under positive pressure, must be recognized. It has been found that tritium gas and vapor can so migrate, and this phenomenon may occur with certain other gases and vapors.

2.2.2 Gloves, Suits, and Footwear

Any materials used in gloves, suits, or footwear to protect the wearer from contaminants should be tested for permeation. Solvents used as carriers for carcinogenic or mutagenic materials may pass through rubber or plastic gloves used in glove boxes. Various plastics and rubber formulations used by different manufacturers for gloves and suits may be impermeable to certain substances, but not to others. Unless the manufacturer can furnish test data, or data are available in the literature, permeation tests should be made by the health and safety professional before any material is permitted to be used for skin protection. Additional consideration must be made for the user's comfort, since some commercially available suits become very hot when worn even briefly.

2.3 Eye Protective Devices

As with skin protective equipment, there are as yet no federal approval schedules for eye protective devices used for nonionizing radiation sources.

NIOSH is having some research conducted in this area at the Southwest Research Institute to develop performance testing and user standards to recommend federal laws (32). ANSI Standard Z49.1 (33) recommends filter shades to be worn for different types of welding operation, ANSI Z87.1 (37) defines the allowable radiation transmission characteristics of these filter plates as a guide to manufacturers, and ANSI Z136.1 (35) recommends protection against laser beams.

The objective of eye protective devices is to attenuate to a safe level radiation that can cause ocular damage. Welding and cutting processes can emit ultraviolet (UV), visible, and infrared (IR) wavelengths, and various segments of these spectra can cause distinctly different modes of damage to the eyes. Ideally, the transmittance values for selection of appropriate filter shade numbers should be based on scientifically determined ocular TLVs. However the present transmittances were developed empirically and have remained essentially unchanged since they were generated by Coblenz and Stair at the Bureau of Standards in 1928.

The rationale for the original specification was quoted in a letter betweeen Astleford and Ralph Stair (32):

"At that time we had no medical information, except that a lot of welders had received great injury from ultraviolet and possibly infrared. We, therefore, set the ultraviolet and infrared values as low as practicable for the manufacture of the glass and for certification by our laboratory that met the standards we had set up. In our papers and in the Federal Specifications there are suggestions that certain shades are "intended" for certain operations, but the final answer rests with the operator to choose the proper shade for "best seeing."

These values then, are used in ANSI Z87.1. The questions that remain to be answered are whether the filter shade numbers recommended in Z49.1 attenuate the arc spectrum to sub-TLV levels in the UV, visible, and IR bands as given in Z87.1 and by the ACGIH (36). Moreover the visible radiation TLVs for the components of the ocular media where none have been established must be found before it can be determined whether existing eye protection is adequate.

Other questions are as follows: Is the TLV established by the ACGIH (36) for the near-UV (0.315 to 0.4 μm) of a maximum corneal dose of 1.0 mW/cm^2 for greater than 1000 sec necessary, since keratitis (snowblindness or arc-eye) occurs only in the 0.2 to 0.315 μm region, and if so, what type of eye protection might best reduce the exposure to sub-TLV levels? Since there are few data on an IR TLV for broad band or extended sources and a laser TLV is used for this area of the spectrum, is it too restrictive? Again, what is the best eye protection that should be used? Have sufficient animal studies been performed, then correlated to the human eye, to assure the user that laser glasses offer short- and long-term protection?

When these questions have been answered, it should be possible to promulgate a testing schedule to assure users that they are provided with adequate eye protection from nonionizing radiation, without sacrificing maximum visibility for the job.

3 BASIS FOR PROGRAM REQUIREMENTS

In 1959 ANSI issued the American National Standard Safety Code for Head, Eye, and Respiratory Protection, Z2.1. This was the first formal standard for personal protective equipment, although previously published books, circulars, and bulletins described program recommendations. By 1963 it was thought advisable to separate the Z2.1 standard into separate standards for each type of personal protective equipment. The Z88 Standards Committee was organized the same year, and subcommittees were established to rewrite the respirator standards (2). Committees were also established for head and eye protective devices, and Z87.1, Practice for Occupational and Educational Eye and Face Protection, was issued in 1968 (34).

3.1 Respiratory Protection

It was not until 1969 that the first separate respirator standards were issued by ANSI: Standard Z88.1-1969, Safety Guide for Respiratory Protection Against Radon Daughters, and Standard Z88.2-1969, Practices for Respiratory Protection. Selected portions of the latter standard were then made into law by being incorporated, verbatim, into the Code of Federal Regulations (3). Further details of respirator program requirements are spelled out in the OSHA standards for specific toxic substances.

3.1.1 Respirator Selection

Proper respirator selection can be one of the most difficult jobs for the health and safety professional in a respirator program. "The selection of a proper respirator for any given situation requires consideration of the following factors: (1) nature of the hazard; (2) extent of the hazard; (3) work requirements and conditions; and (4) characteristics and limitations of available respirators" (2).

Proper respirator selection is the choice of a device that fully protects the worker from the hazards to which he or she may be exposed and will permit the worker to perform the job with the least amount of discomfort and fatigue, and the greatest reliability possible. Cost of running the entire program, not just the initial investment, must also be considered, but only to the point to which the worker's health and job performance are not affected.

Consideration of the nature and extent of the hazard is easy in some instances and difficult in others. For example, determining whether an atmosphere is or may be oxygen deficient will immediately eliminate the use of devices that do not supply oxygen to the user if the atmosphere is deficient. However deciding whether an atmosphere is immediately dangerous to life or health may be far more difficult.

Oxygen Deficiency. Pritchard describes the dilemma of defining "oxygen deficiency" in the NIOSH publication "A Guide to Industrial Respiratory Protection" (37). Definitions of oxygen deficiency range from 16.0 to 19.5 percent. In the OSHA Standards (3) it is defined as 16 percent in Part 1910.134 and as 19.5 percent in Part 1910.94. Some standards take into consideration altitude or oxygen partial pressure, others do not. The rationale for the various definitions rests primarily with the persons or person doing the defining. The 16 percent figure is one below which definite physiological reactions can be expected to occur. The 19.5 percent figure is used by others because slight symptoms such as increased heart beat and respiration rate may be observable. The physiological response, plus the occurrence of something to cause the oxygen content to fall below the normal 20.95 percent, therefore possibly to continue to lower the percentage to unsafe levels, are believed by some to be sufficient justification to set a safety factor. Pritchard's recommendation (37) for this dilemma is perhaps the best.

The important thing is the respirator wearer's safety. If the legal definition of O_2 deficiency is above the O_2 level you can consider safe for humans, you are justified in following the legal definition. If the O_2 deficiency level as legally defined is less than the O_2 concentration you believe safe for human exposure, you must consider raising your minimum O_2 level above the legal limit. Although not infallible, the P_{O_2} limit of 60 mm Hg in the alveolar space should be the absolute minimum to which the O_2 level should be allowed to drop. This means that P_{O_2} in the ambient air should not drop below about 120 mm Hg. This problem is under study, and eventually oxygen deficient atmospheres will be redefined to eliminate the present discrepancies and account for the effect of altitude.

Immediately Dangerous to Life or Health Atmospheres. The decision of whether an atmosphere is IDLH can be difficult and the respirator selected will depend on the interpretation. ANSI Z88.2-1969 defines IDLH as including "conditions that pose an immediate threat to life or health and conditions that pose an immediate threat of severe exposure to contaminants such as radioactive materials which are likely to have adverse delayed effects on health" (2). The Joint NIOSH/OSHA Standards Completion Program Respirator Decision Logic (13) interpreted this definition to include

Escape without loss of life or irreversible health effects. Thirty minutes is considered the maximum permissible exposure time for escape. Severe eye or respiratory irritation or other reactions which would inhibit escape without injury. . . . Contaminant concentrations in excess of the lower flammable limit are considered to be IDLH. . . . Firefighting is defined . . . as being immediately dangerous to life. [Where only acute exposure animal data is available (30 min to 4 hr exposures], the lowest exposure concentration causing death or irreversible health effects in any species is determined to be the IDLH concentration.

Several people in the respiratory protection field, including myself, have questioned whether the lower flammable limit or eye irritation should be a consideration in the selection of respirators for an IDLH atmosphere, or whether respiratory effects alone should be the governing factor.

Using the decision logic interpretation, OSHA now requires that only full-face, pressure demand (safety pressure), SCBA or combination supplied-air respirator and auxiliary SCBA with full facepiece in positive pressure mode will be acceptable in IDLH atmospheres in all new standards promulgated or proposed.

ANSI Z88.2-1969 (and 29 CFR Part 1910.134) permitted the use of several types of respirator in IDLH atmospheres under certain conditions: gas masks, demand-flow SCBA, hose masks with blower, and air line respirators were included. Half mask facepieces were also permitted with air line respirators, provided the atmosphere in which they were being used did not cause eye irritation or injury (2, 3).

The decision by OSHA to not permit the other types of device in IDLH atmospheres was based on two factors. The first was the results of fitting studies done at LASL (15, 19) and elsewhere, which showed that respirators that had a positive pressure in the facepiece performed far better to prevent inward leakage of air contaminants for many workers than did respirators with negative pressure in the facepiece. Second, it was felt that a worker in an IDLH atmosphere should be capable of escaping by himself and should not have to depend on being rescued by a second person in the event of failure of the air supply. By the process of elimination then, only the SCBA, operated in a safety pressure mode, or a combination supplied-air SCBA (for escape if the air supply fails) would be acceptable. OSHA also reasoned that half mask facepieces on these devices would not be acceptable because they are somewhat less stable and stand a greater chance of being dislodged from the face than a full facepiece, do not protect the eyes from irritants, and provide a poorer facepeice-to-face seal on a larger percentage of the working population than does the full facepiece (13).

OSHA also believes that firefighting should fall into a separate category in the health standards and permits only the use of SCBA operated in a positive pressure mode. This decision was made because a firefighter has no way of determining whether a given

PERSONAL PROTECTION

atmosphere is IDLH, the worst case (i.e., that IDLH conditions exist) must be assumed; moreover, firefighters cannot be expected to drag air lines from combination supplied-air, SCBA operated in a positive pressure mode around with them.

It might be noted that the ANSI Z88.2 subcommittee, which was rewriting the standard at the time of this writing, did not agree with the NIOSH/OSHA respirator decision logic concerning the use of respirators in IDLH atmospheres. The draft revised standard, yet to be voted on by the full Z88 Committee, permitted the use of gas masks, combination positive or negative pressure air line devices with full facepieces and auxiliary self-contained air supply, and any self-contained breathing apparatus with any facepiece, operated in positive or negative pressure modes (see Table 19.1).

The subcommittee believed that if the respirator program recommended in the standard was followed, these other devices could be used safely in IDLH atmospheres, as they have been in the past. The subcommittee redefined the IDLH atmosphere to "include conditions that pose an immediate hazard to life or irreversible debilitating effects on health." If this standard is accepted by the full Z88 Committee as recommended by the subcommittee, OSHA may again review the restrictions and possibly alter its position on devices permitted in IDLH atmospheres.

MESA, which sets the safety regulations for the mining industry, does permit the use of SCBA with negative pressure in the facepiece in IDLH atmospheres (38, 39). Mine rescue teams use closed circuit devices that can be operated for an hour or longer, a necessity in mine rescue operations. Since the currently available SCBA with positive pressure in the facepiece is available only as 30 min maximum time duration, such devices are not practical for the mining industry. The same argument may also be justified for some industrial operations and for firefighting (e.g., below-deck fires on large ships or fires in highrise buildings).

Maximum Use Limits and Protection Factors. By the late 1960s it was becoming more apparent that respirators were not providing the best protection possible, even when companies that purchased them for their employees tried to establish good industrial safety programs. Overexposures to radioactive materials were detected, for example, even when workers were wearing respirators provided with good filtering materials. Researchers such as Ed Hyatt at LASL and Bill Burgess at Harvard University were beginning to identify some of the problem areas. The National Fire Protection Association recommended that firefighters no longer use the universal canister type of gas mask, but rather SCBA to prevent fatalities from oxygen-deficient atmospheres.

With funds provided by the Atomic Energy Commission (superseded by the Nuclear Regulatory Commission), Hyatt set up a respirator research section at Los Alamos. Some of the early work measured the fitting characteristics of respirators of various makes and types. From the fitting data obtained, fitting indices were established, and from these, more reliable protection factors (PFs) were established. The PFs were reported in several different LASL progress reports. Hyatt's final report, which appeared in 1976 (19), defined PFs as follows:

The overall protection afforded by a given respirator design may be defined in terms of its protection factor, which is defined as a ratio of the concentration of contaminant in the ambient atmosphere to that inside the facepiece under conditions of use. This definition is illustrated by the equation

$$PF = \frac{\text{ambient airborne concentration}}{\text{concentration inside facepiece or enclosure}}$$

... When both the ambient atmospheric concentration and the contaminant TLV are known, the protection factor may be used to select a respirator so that the concentration inhaled by the wearer will not exceed the appropriate limit. ... For example, a respirator with a PF of 10 may be selected for use where a maximum use concentration of $10\times$ the TLV exists. ...

The protection factors are a quantitative measurement of the leakage of an air contaminant into the facepiece of a man performing certain exercises while wearing a respirator; thus it was possible to determine what concentration of a contaminant can be present outside the respirator without the concentration inside the mask exceeding the permissible exposure level. Several hundred tests were run on different men wearing various types of respirator. The protection factors were then assigned to the various categories of devices based on the condition that at least 95 percent of the adult males tested could achieve that PF. The PFs are valid only when a good qualitative fitting program is conducted.

Protection factors used by the British are published in British Standard 4275 (40). In the United States the PF concept is used in all the later NIOSH criteria documents and OSHA standards and by the Nuclear Regulatory Commission in "Standards for Protection Against Radiation" (41). The user not only needs to have some knowledge of the PF of a given type of a respirator to make an intelligent selection of what category of device to use, but also assurance that the device fits well enough to provide that PF. This can only be done by means of a good fitting program. Volume I, Chapter 21, lists the "Respirator Protection Factor Table" as recommended by Hyatt.

The PF of 50 for full face masks under negative pressure recommended in the table of PFs has been questioned at OSHA standards hearings (41), and it has been recommended that a PF of 100 be sustituted. "The report of the tests run at Los Alamos states that 5 of 6 makes of masks had a PF of 100 or greater. Since the sixth did not even make a PF of 50, the report concludes that a PF of 50 should be assigned to the full facepieces." The next paragraph of the report states that six of eight half mask respirators had a PF of 10 or greater, but two did not. However since for practical purposes the remaining two models are not available, a PF of 10 was assigned. Using the same reasoning given for the half mask respirators, a PF of 100 should be assigned to full facepieces, since the one mask that failed to attain a PF of 100 is, for practical purposes, no longer available, as was also the case with the two half masks.

Further substantiation of a PF of at least 100 for full-face, negative pressure facepieces is available from several hundred tests run at the Energy Research and Development Administration's facility at Rocky Flats, Colorado, and has been published by Mr. Jack Leigh (42). Tests run at Lawrence Livermore Laboratory on 306 employees using even more rigorous test exercises than those used originally at Los Alamos also showed that 100 percent of these employees could also get a PF of at least 100 on at least one mask (43).

The new ANSI Z88.2 Subcommittee, which first met in 1975, was dissatisfied with the use of PFs as they were being applied by OSHA and NIOSH. It was felt that the PFs were a measurement of facepiece fitting only and should not be used to satisfy all respirator selection criteria. It was thought that because a facepiece did not perform well on one worker it should be limited on another worker where a higher PF could be obtained within certain limitations. Furthermore, conditions other than fitting existed but were not adequately covered by the PF table. A proposal submitted by a special committee to recommend a solution was accepted (44), with modifications that based respirator selection on maximum use limits (MUL's). The MULs incorporate PFs based on various fitting methods, cartridge or canister limitations, oxygen-deficient atmospheres, and IDLH atmospheres. Table 19.1 shows the MULs proposed by the Z88.2 Subcommittee but not yet voted on by the full Z88 Committee.

Approved or Accepted Respirators. OSHA requires that only approved or accepted respirators be selected for use by industry (3), by reference to ANSI Z88.2-1969 (2). The ANSI standard in turn defines approved respirators as those approved by the B of M or listed by the U.S. Department of Agriculture. Since the Department of Agriculture no longer lists respirators (for protection against pesticides) and the B of M testing and approval system was transferred to NIOSH/MESA,* the latter approvals now are required. Only supplied-air apparatus and SCBA approved by the B of M may be used (until March 31, 1980 and 1979, respectively); all other types of device have been grandfathered out (12).

Accepted devices are respiratory protective devices that cannot be approved by NIOSH/MESA "because it is outside their approval or testing authority" (2). I know of only one device that falls into this category, namely, supplied-air suits, for which there is no test schedule yet available. ANSI Z88.2-1969 (2) does state that for a device to be "accepted," the user should assure himself that the device is "adequate for the required service, and that quality control during manufacture can be expected. He should make or have made suitable tests of the respirators' effectiveness which, as far as feasible, simulate tests made by the official test agencies" (in this case, NIOSH).

The OSHA requirement that only NIOSH/MESA approved respirators be used is not restrictive to the user, in that all reliable respirator manufacturers have their devices so approved. The major advantage to the user is, of course, insurance that the device will operate properly within the limitations of the test procedures and as stated by the manufacturer.

---

* Now the "Mine Safety and Health Administration."

Table 19.1 Respirator Maximum Use Limits[a]

Respirator Type	Permitted for Use in Oxygen-Deficient Atmosphere	Permitted for Use in IDLH Atmosphere	Maximum Use Limit of Respirator According to Type of Fitting Test Method Used	
			Qualitative[b]	Quantitative[b]
1. Particulate filter, quarter mask, half mask facepiece[c,d]	No	No	10	Half the fitting index as measured on each person
2. Vapor- or gas-removing, quarter mask or half mask facepiece	No	No	10 or maximum use limit of cartridge, whichever is less	Half the fitting index as measured on each person or maximum use limit of cartridge, whichever is less
3. Particulate filter full facepiece	No	Yes	100	Half the fitting index as measured on each person.
4. Vapor- or gas-removing facepiece	No	Yes	100 or maximum use limit of cartridges or canister, whichever is less	Half the fitting index as measured on each person, or maximum use limit of cartridge or canister, whichever is less
5. Powered air-purifying respirator, any respiratory-inlet covering[c,d]	No	No	Maximum use limit of cartridge or canister	Maximum use limit of cartridge or canister
6. Air line, demand, quarter mask or half mask facepiece with or without escape provisions[e]	No	No	10	Half the fitting index as measured on each person
7. Air line, demand, full facepiece, without escape provisions	No	No	100	Half the fitting index as measured on each person

8.	Air line demand, full facepiece with escape provisions	Yes[a]	100	Half the fitting index as measured on each person
9.	Air line, pressure demand or continuous flow, any respiratory-inlet covering, without escape provisions	No	Not applicable[f]	Not applicable[f]
10.	Air line, pressure demand or continuous flow, any respiratory inlet covering with escape provisions[e]	Yes[a]	Yes	Not applicable[f]
11.	SCBA, open circuit demand or closed circuit, full facepiece or mouthpiece	Yes	100	Half the fitting index as measured on each person
12.	SCBA, open circuit pressure demand, quarter mask, half mask, or full facepiece[e]	Yes	Yes	Not applicable[f]
13.	Combination respirators not listed previously[e]	The type and mode of operation having the lowest maximum use limit shall be applied to the combination respirator		

[a] The value of the maximum use limit listed is the multiple of the permissible TWA concentration or the excursion limit concentration, whichever is applicable for the persons who meet the criterion of the particular respirator fitting test used.
[b] Respirators for which respirator fitting tests cannot be performed shall have a maximum use limit of 5.
[c] When the respirator is used for protection against airborne particulate matter having a permissible TWA concentration less than 0.05 mg particulate matter per cubic meter of air, or less than 2 million particles per cubic foot of air, the respirator shall be equipped with a high efficiency filter or filters.
[d] If eye irritation occurs, respirator wearers shall be permitted to use protective goggles or full facepieces.
[e] The escape provisions shall be an auxiliary self-contained air supply.
[f] No fitting index can be listed because the respiratory inlet covering is under positive pressure.

Other Selection Considerations. As stated earlier, many factors must be considered carefully in respirator selection. Determination of the nature and extent of the hazard may be accomplished by monitoring methods and the proper respirator used may be selected by applying the proper MUL. However the working conditions may prohibit the use of the selected respirator. High heat, humidity, cold, and/or other environmental conditions may limit the device's usefulness. A full facepiece with its inherent tunnel vision may be dangerous to the user in an area with tripping hazards.

Finally the respirator's limitations and characteristics must be weighed with care. For example, a NIOSH/MESA-approved organic vapor respirator can be used only for organic vapors that have good warning properties (e.g., the odor threshold is below the TLV). Nelson and Harder (14) found that many organic vapors "broke through" the organic vapor cartridge in a very short time. If the wearer could not detect the odor of the vapor at a level below the TLV, an overexposure could easily occur. A similar danger exists with respect to poorly fitting filter respirators used for protection against highly toxic or carcinogenic dusts.

All such factors and possibilities must be carefully examined by the health and safety professional when selecting any respiratory protective device.

3.1.2 Respirator Use

As with the matter of selection, many factors must also be considered in the use of a respirator. Too often, a respirator that will provide adequate protection is selected without consideration of whether it will be used properly.

Obviously the wearers must be well trained, not only in the use of the device, but in the reasons for which they are required to use a respirator. The health and safety professional must periodically survey the work area and discuss problems with respirator wearers to circumvent misuse. All too frequently, users do not wear the devices, either because of lack of understanding of their necessity or because of discomfort. Discomfort from wearing a respirator may be sufficient to drive the user to gamble with his or her health rather than suffer severly in performing a job. Work times in a respirator should be well planned so that the user has ample opportunity to retreat to a clean area and remove the device. As the work area temperature and/or relative humidity increases, the period during which the user can wear the respirator without undue discomfort decreases significantly.

For air-purifying devices, an adequate supply of filters or chemical cartridges or canisters must be available. A user of a filter device must be able to change the filter when the breathing resistance increases. The health and safety professional must also calculate the length of time that a chemical cartridge or canister can last without a breakthrough, provide an adequate margin of safety, and thoroughly train the user to replace the cartridge or canister at the proper times.

In summary, proper respirator use can be achieved only through teaching the wearer not only how to use the device but why it must be worn; continued follow-up by the health and safety professional is essential to be able to anticipate and alleviate misuse problems before they occur.

3.1.3 Respirator Program Administration

OSHA requires that a respirator program be administered by one qualified individual. When two or more people are responsible for different aspects of the program, they may not be consistent in their administration, thus confusing the user. Worse, each program administrator may tend to assume that the other is covering some aspect of the program, which could result in a major shortcoming. This does not mean to imply that the program administator may not delegate responsibilities to other qualified individuals, which, in all likelihood is necessary in a large program. However the ultimate responsibility for the program must be with one person only.

Standard Operating Procedures. Standard operating procedures (SOPs) are required by OSHA, covering the selection, use, and care of respirators. ANSI originally made this requirement to be sure that the administrator of the program would be forced to think out all aspects and to put them into writing. Written SOPs would also serve as a checklist for the administrator and others involved in the program, could be used easily by the employees during training, and would provide a source of continuity of the program if the administrator or others were replaced.

Though not required, it is often desirable to have the SOPs approved or endorsed "in writing by the company owner, president, or other person in high authority, to give them the emphasis they deserve" (45). Periodic reviews and updates as necessary should also be made of all SOPs.

Records. OSHA requires that records be maintained on the inspection dates, and findings of all respirators maintained for emergency use (emergency use respirators must be inspected monthly). This is to ensure that the equipment is operating properly and in a "go" state of readiness. OSHA also requires a record on the date of issuance of each respirator to an individual. This affords an easy way of checking whether the users are returning their equipment for maintenance, inspection, and cleaning as required.

"Other records that are advisable are training and fitting records, procurement records, inspection records, and maintenance records. The latter can provide valuable information on common failures of a particular brand of respirator which may prove uneconomical to keep in stock" (45).

Inspections. Finally, OSHA requires appropriate surveillance of work area conditions and employee exposure and stress, along with regular inspections and evaluations to determine that the respirator program remains effective.

Surveillance of the work area conditions calls for air sampling either as required by OSHA or in accordance with good industrial hygiene practice. The results can then be used to determine whether the respirator program is adequate, whether air concentrations in the work area have increased or decreased to the extent that the respirators being used have exceeded their MUL, or whether a less expensive or more comfortable device could be substituted.

When possible, bioassay results can be used to evaluate the program's effectiveness. A rise in blood or urine sample results from employees wearing respirators can indicate misuse or lack of use of respirators.

3.1.4 Cleaning and Sanitizing

Cleaning and sanitizing of respirators is necessary to prevent the spread of communicable diseases, to prevent dermatitis, and to encourage worker acceptance through aesthetics. OSHA requires at least one daily cleaning of respirators issued for the exclusive use of a worker, but more frequent cleaning may be necessary to prevent dermatitis from contaminants lodging at the facepiece-to-face seal and being held in contact with the skin, or from skin contact with a curing agent that may be coming off from the respirator itself (particularly with new respirators). Respirators used by more than one person and emergency use devices must be cleaned and disinfected after each use to prevent the spread of communicable diseases.

ANSI Z88.2-1969 recommends that wash water and drying temperatures not exceed 180°F. However the more delicate parts of many makes of respirators can be damaged at this temperature. I have found that in the absence of specific manufacturers' directions, washing and drying temperatures should never exceed 140°F. Sanitizing can be accomplished either as a separate rinsing in a disinfectant solution (again, care must be taken not to use a disinfectant that can warp or corrode parts), or by using a cleaner-sanitizer solution for initially cleaning the device, thus eliminating a step in the cleaning-drying process.

3.1.5 Inspection and Maintenance

Following cleaning, sanitizing, and drying of the respirator, the device must be thoroughly inspected for defects, missing parts, worn parts, and parts not operating properly. New equipment should be inspected to make sure that all parts are present and operating properly. The inspectors must make sure that the device will give the user the protection expected. For example, a missing or warped exhaust valve will let outside contaminants into the facepiece. A warped facepiece or headstrap will affect the user's fit, thus nullifying the fitting program. An improperly seated lens will also be a leak source for contaminants into the facepiece.

For more complicated devices such as supplied-air or self-contained breathing apparatus, the inspection procedure may have to include a testing facility to ensure proper regulator operation and airflows.

Maintenance operations would naturally follow where indicated by the inspection. If parts not manufactured specifically for the particular respirator being repaired are used,

PERSONAL PROTECTION

NIOSH approval is automatically voided. Furthermore some manufacturers of SCBA require that maintenance operations be performed only by persons certified by the respective manufacturers; failure to observe this requirement will void the NIOSH approval.

3.1.6 Storage

Respirators must be stored in such a way that they are protected against dust, sunlight, heat, extreme cold, excessive moisture, or damaging chemicals. Obviously, anything that can contaminate or cause damage to the respirator must be guarded against. Storage in plastic bags may be helpful.

The respirators should not be stored together or in a tight space. Each device must be positioned so that it retains its natural configuration. Synthetic materials and even rubber will take a "set" if stored in an unnatural shape, thus affecting the fitting characteristics of the facepiece.

Emergency use respirators must be stored at quickly accessible sites, in clearly marked compartments built for the purpose. In determining the location of the storage area, thought should be given to whether the respirators would be used for escape (in which case they must be near the workers) or for reentry (in which case they should be near the entry point, but not in an area that might iself be involved in the emergency, thus becoming inaccessible or contaminated).

Finally, care must be taken that the respirators will not become contaminated in the lunch time, break, or overnight storage areas that are chosen. Common mistakes are hanging respirators in an area with airborne contamination and placing them in lockers with contaminated work clothing.

3.1.7 Training and Fitting

ANSI Z88.2-1969 (2) requires that the training of respirator users include instruction in the nature of the hazard, what would happen if the respirator is not used, and why other control methods are not used. It must be explained also why the particular type of respirator provides protection, and the respirator's capabilities and limitations must be outlined. Instruction and training in actual use of the respirator, classroom and field training in recognizing and coping with emergencies, and other special training must be offered as needed.

Field studies (26–28) have shown that many users resist wearing respirators because they are uncomfortable and inconvenient. Proper training and a thorough understanding of the reasons for wearing the device can help overcome many of these objections. Motivation for proper use can be obtained only through training.

Respirator fitting studies have revealed the large facepiece-to-face seal leakages that can occur. Fitting methods that have been used, such as positive and negative pressure tests described in ANSI Z88.2-1969 (2), have been proved to be faulty and should be used only as a field check prior to each use. When performing those tests, the facepiece is pushed against the face with the hands, thus sealing off small leaks that will reccur at the conclusion of the test. Furthermore the skin is not sensitive to detect small leakages when performing a positive pressure test.

The proposed draft of ANSI Z88.2-1977 has eliminated the positive and negative tests and recommends only qualitative or quantitative fitting tests. The qualitative tests, using either a stannic chloride ventilation smoke tube or isopentyl acetate, provide a more accurate fitting method. Some persons cannot detect either substance by odor except in large quantities, and uncooperative subjects may simply lie: either not acknowledging detection of the odor to get the test over quickly, or claiming to smell something when they do not.

The preferred fitting method is quantitative. This permits the operator to detect the leakage accurately and quantitatively using either dioctyl phthalate or sodium chloride. The drawbacks are the expensive instrumentation and the need for a well-qualified operator. Some private consulting firms are now offering quantitative testing, which will put this superior form of fitting within reach of smaller companies and companies with small respirator programs.

Fitting should be performed at least annually to detect changes in fitting characteristics. Changes in an employee's weight, a new set of dentures, or new blemishes or scars can cause a respirator that once fit adequately to no longer be acceptable. Refresher training should be given when the refitting work is performed.

3.1.8 Medical Requirements

OSHA forbids the assignment of "persons . . . to tasks requiring use of respirators unless it has been determined that they are physically able to perform the work and use the equipment. The local physician shall determine what health and physical conditions are pertinent. The respirator user's medical status should be reviewed periodically (for instance, annually)" (3). This restriction has caused a considerable amount of confusion regarding what minimum tests should be performed, what constitutes inability to work while wearing a respirator, whether a worker's livelihood can or should be denied when a physician says the individual should not wear a respirator, and how often a physical examination must be given—the requirement states "periodically," but "annually" is suggested.

Some of the OSHA health standards and proposed health standards attempt to clarify some of these points, yet others add to the confusion.

I believe that the health and safety professional should educate the physician, (especially if the company physician is not an industrial physician by training) with respect to types of respirators, positive and negative pressures in the facepiece, breathing resistances, whether the employees will work in an IDLH atmosphere, the air contaminants that are present, and the alternatives that are available. By working together, the health and safety professional and the physician usually can work out satisfactory answers to problems involving physical conditions of employees who are required to wear respirators.

3.1.9 Program Evaluation

Regular respirator program evaluation and appropriate surveillance of respirator users' work area conditions and exposures or stress are OSHA requirements (3). Here again, sound professional judgment must be applied by the health and safety professional to adequately perform these tasks.

Changes in operations or any other occurrences that could affect air level concentrations must be monitored to determine whether the respirators being used are still permissible in accordance with the MUL or PF tables or in compliance to applicable OSHA health standards. Visual observation on respirator usage is necessary to determine whether the training program is effective. Discussions with users often can bring to light problem areas relating to discomfort from heat factors or other conditions, which may require reexamination for appropriateness of the respirators selected.

Respirator effectiveness can be measured for substances that can be detected by bioassay. The bioassays should be performed often enough to detect overexposures before they could become serious.

3.1.10 Other Standards and Regulations

The respirator program administrator may be required to follow regulations other than those of OSHA, or guidance on specialized respirator programs may be obtained from other specific standards. Examples of the latter include MESA regulations for metal, nonmetal, and coal mines (38, 39), Nuclear Regulatory Commission regulations (41); OSHA specific health standards, and other ANSI standards in the Z88 series. The peculiar nature of the hazards for which specific standards or regulations were written warranted the separate presentations. General standards such as ANSI Z88.2 could not cover all the applications for which respirators might be used.

3.2 Skin and Eye Protection

As mentioned previously, there are very few established program requirements for skin and eye protection as it relates to occupational health protection. OSHA (in 29 CFR 1910.132 and 1910.133) lists some general requirements pertaining to all personal protective equipment: personal protective equipment must be provided, used, and maintained in a sanitary and reliable condition whenever necessary; it must be of safe design and construction and reasonably comfortable; it must not interfere unduly with the user's movements; it must be durable and capable of disinfection and cleaning; it must be kept clean and in good repair; corrective lenses must be provided as applicable; each device must be marked to identify the manufacturer; manufacturer information must be transmitted to the user; and eye and face protection must conform to ANSI Z87.1-1968 requirement (34). Specific requirements may also appear in OSHA health standards as they are promulgated.

In addition to the requirements listed by OSHA, the health and safety professional should use common sense to provide adequate protection against skin and eye health hazards. Other applicable ANSI standards can be consulted, such as those for welding and cutting guidelines (33), and appropriate professional journals. The TLV list (36) for physical agents gives guidelines for reasonably safe exposures to many forms of nonionizing radiations, and protective devices should be selected to lower exposures within these limits. It is then incumbent on the health and safety professional to design a program so that the use of the personal protective device will protect the user within the prescribed limits.

As with respiratory protective devices, the use of positive engineering controls should be reflected in the choice and deployment of any personal protective equipment. Skin and eye protective devices should be used only if and when engineering controls are not feasible.

REFERENCES

1. American Industrial Hygiene Association, American Conference of Governmental Industrial Hygienists, *Respiratory Protective Devices Manual,* Braun and Brumfield, Ann Arbor, Mich., 1963.
2. American National Standards Institute, American National Standard Practices for Respiratory Protection, ANSI Z88.2-1969, ANSI, New York, 1969.
3. Occupational Safety and Health Administration, General Industry Safety and Health Regulations, Title 29, Code of Federal Regulations, Part 1910, *Fed. Reg.,* **39:** 125 (June 27, 1974).
4. B. J. Held, Respirator Testimony for Proposed Benzene Standard, Occupational Safety and Health Administration public hearing proceedings, July 20, 1977.
5. B. J. Held, Respirator Testimony for Proposed Cotton Dust Standard, Occupational Safety and Health Administration public hearing proceedings, April 6, 1977.
6. Robert Wells (a member of the 1910 Berkeley, California, Fire Department) and Robert Foley, Napa, California, private correspondence.
7. Bureau of Mines, "Respiratory Protective Apparatus, Self-contained Breathing Apparatus," Schedule 13E, Title 30, Code of Federal Regulations, Part 11, 1968.
8. U.S. Bureau of Mines, "Respiratory Protective Apparatus, Gas Masks," Schedule 14F, Title 30, Code of Federal Regulations, Part 13, 1955.
9. U.S. Bureau of Mines, "Respiratory Protective Apparatus, Supplied-air Respirators," Schedule 19B, Title 30, Code of Federal Regulations, Part 12, 1955.
10. U.S. Bureau of Mines, "Respiratory Protective Apparatus, Filter-type, Dust, Fume, and Mist Respirators," Schedule 21B, Title 30, Code of Federal Regulations, Part 14, 1965.
11. U.S. Bureau of Mines, "Respiratory Protective Apparatus, Nonemergency Gas Respirators (Chemical Cartridge Respirators, Including Paint Spray Respirators)," Schedule 23B, Title 30, Code of Federal Regulations, Part 14a, 1959.
12. U.S. Bureau of Mines–National Institute for Occupational Safety and Health, "Respiratory Protective Apparatus," Title 30, Code of Federal Regulations, Part 11, March 10, 1972.
13. National Institute for Occupational Health and Safety–Occupational Safety and Health Administration, *Respirator Decision Logic,* latest edition. Available from NIOSH or OSHA in computer printout form.
14. G. O. Nelson and C. A. Harder, "Respirator Cartridge Efficiency Studies: V. Effect of Solvent Vapor," *Am. Ind. Hyg. Assoc. J.,* **35,** 391 (1974).

15. E. C. Hyatt et al., "Respirator Efficiency Measurement Using Quantitative DOP Man Tests," *Am. Ind. Hyg. Assoc. J.,* **33**: 10 (1972).
16. A. L. Hack et al., "Selection of Respirator Test Panels Representative of U.S. Adult Facial Dimensions," LA-5488, Los Alamos Scientific Laboratory, Los Alamos, N.M., 1973.
17. P. L. Lowry et al., "Respirator Studies for the National Institute for Occupational Safety and Health," LA-6722-PR, Los Alamos Scientific Laboratory, Los Alamos, N.M., 1977.
18. J. T. McConville, "Ethnic Variability and Respirator Sizing," American Industrial Hygiene Association Conference, Paper 128, 1973.
19. E. C. Hyatt, "Respirator Protection Factors," LA-6084-MS Los Alamos Scientific Laboratory, Los Alamos, N.M., 1977.
20. Mine Safety Appliances Co., "Key Elements of a Sound Respiratory Protection Program," Bulletin 1000-16, MSA, Pittsburgh, 1977.
21. B. J. Held et al., "Respirator Studies for the National Institute for Occupational Safety and Health," LA-5805-PR, Los Alamos Scientific Laboratory, Los Alamos, N.M., 1974.
22. D. D. Douglas et al., "Respirator Studies for the National Institute for Occupational Safety and Health," LA-6386-PR, Los Alamos Scientific Laboratory, Los Alamos, N.M., 1976.
23. B. J. Held and C. P. Richards, "Hazards Control Progress Report 53, July–December 1976," UCRL-50007-76-2, Lawrence Livermore Laboratory, Livermore, Calif., 1977.
24. B. J. Held and C. P. Richards, "Research and Development Needs in Firefighter's Breathing Protection," *Fireline Mag.* (San Francisco), April–May 1977.
25. International Association of Fire Fighters, *1975 Annual Death and Injury Survey,* IAFF, Washington, D.C., 1976.
26. C. R. Toney and W. L. Barnhart, "Performance Evaluation of Respiratory Protective Equipment Used in Paint Spraying Operations," Department of Health, Education and Welfare Publication (NIOSH) 76-177, National Institute for Occupational Safety and Health, Cincinnati, Ohio, 1976.
27. A. Blair, "Abrasive Blasting Respiratory Protection Practices," Department of Health, Education and Welfare Publication (NIOSH) 74-104, National Institute for Occupational Safety and Health, Cincinnati, Ohio, 1976.
28. H. E. Harris and W. C. Di Sieghardt, "Factors Affecting Protection Obtained by Underground Coal Miners from Half-mask Dust Respirators," American Industrial Hygiene Association Conference, Paper 117, 1973.
29. H. E. Harris et al., "Respirator Usage and Effectiveness in Bituminous Coal Mining Operations," American Industrial Hygiene Association Conference, Paper 116, 1972.
30. T. O. Davis et al., "Respirator Studies for the ERDA Division of Safety, Standards, and Compliance," LA-6733-PR, Los Alamos Scientific Laboratory, Los Alamos, N.M., 1977.
31. National Aeronautics and Space Administration, Technology Utilization Program Report 1974, NASA, Washington, D.C., 1975.
32. T. Dunham, "Occupational Safety Research Specifically Related to Personal Protection, A Symposium," Department of Health, Education and Welfare Publication (NIOSH) 75-143, National Institute for Occupational Safety and Health, Cincinnati, Ohio, 1975, p. 65.
33. American National Standards Institute, American National Standard Practice for Safety in Welding and Cutting, ANSI Z49.1, ANSI, New York, 1973.
34. American National Standards Institute, American National Standard Practice for Occupational and Educational Eye and Face Protection, ANSI Z87.1, ANSI, New York, 1968.
35. American National Standards Institute, American National Standard for the Safe Use of Lasers, ANSI Z136.1, ANSI, New York, 1973.
36. American Conference of Governmental Industrial Hygienists, "Threshold Limit Values for Chemical

Substances and Physical Agents in the Workroom Environment with Intended Changes," ACGIH, Cincinnati, Ohio, 1976.
37. J. A. Pritchard, *A Guide to Industrial Respiratory Protection,* Department of Health, Education and Welfare Publication (NIOSH) 76-189, National Institute for Occupational Safety and Health, Cincinnati, Ohio, 1976.
38. Mining Enforcement and Safety Administration, "Metal and Nonmetal Mine Health and Safety Standards and Regulations," Title 30, Code of Federal Regulations, Part 57.5-5, latest edition.
39. Mining Enforcement and Safety Administration, "Coal Mine Health and Safety Standards and Regulations" Title 30, Code of Federal Regulations, Parts 70 and 75, latest edition.
40. British Standards Institution, "Recommendations for the Selection, Use and Maintenance of Respiratory Protective Equipment," BS 4275, BSI, London, 1974.
41. J. L. Caplin, B. J. Held, and R. J. Catlin, *Manual of Respiratory Protection Against Airborne Radioactive Materials,* NUREG-0041, U.S. Nuclear Regulatory Commission, Washington, D.C., 1976.
42. J. O. Leigh, "Quantitative Respirator Man Testing and Anthropometric Survey," RFP-2358, Dow Chemical Company, Rocky Flats Division, Golden, Colo., 1975.
43. B. J. Held et al., "Evaluation of One-Year Results of the Full-face Respirator Quantitative Man-test Fitting Program at the Lawrence Livermore Laboratory," UCRL-52187, Lawrence Livermore Laboratory, Livermore, Calif., 1976.
44. D. Bevis, B. J. Held, and C. P. Richards, unpublished report proposing respirator maximum use limits to the ANSI Z88.2 Subcommittee.
45. W. E. Ruch and B. J. Held, *Respiratory Protection OSHA and the Small Businessman,* Ann Arbor Science Publishers, Ann Arbor, Mich., 1975.

CHAPTER TWENTY

Job Safety and Health Law

MARTHA HARTLE MUNSCH, J.D., and
ROBERT L. POTTER, J.D.

The principal legislation relating to job safety and health is the federal Occupational Safety and Health Act of 1970 (OSHA). However, certain industries and/or portions of industries are subject to regulation by federal statutes other than OSHA. In addition, OSHA does not entirely preclude regulation of job safety and health by the states or their political subdivisions. Nevertheless, OSHA is clearly the most comprehensive legislative directive relating to workplace safety and health; accordingly this chapter and the next focus primarily on developments and requirements pursuant to this act.

1 LEGISLATIVE HISTORY AND BACKGROUND OF OSHA

The Occupational Safety and Health Act of 1970 was enacted by Congress on December 17, 1970, and became effective on April 28, 1971. It represents the first job safety and health law of nationwide scope (1). Passage of the act was preceded by a dramatic and bitter labor-management political fight. The legislative history of OSHA is summarized in *The Job Safety and Health Act of 1970* (1, pp. 13–21).

Congress enacted OSHA for the declared purpose of assuring "so far as possible every working man and woman in the Nation safe and healthful working conditions" [§ 2(b)]. The act is intended to *prevent* work-related injury, illness, and death.

1.1 Agencies Responsible for Implementing and Enforcing OSHA

The Department of Labor is responsible for implementing OSHA. On the date the act became effective, the Department of Labor created the Occupational Safety and Health

Administration (OSH Administration or OSHA) to carry out such responsibilities. The OSH Administration is headed by an assistant secretary of labor for occupational safety and health and is responsible, among other things, for promulgating rules and regulations, setting health and safety standards, evaluating and approving state plans, and overseeing enforcement of the act (2).

Section 12(a) of the act establishes the Occupational Safety and Health Review Commission (OSAHRC) as an independent agency to adjudicate enforcement actions brought by the Secretary of Labor. The commission is composed of three members appointed by the President for 6 year terms. The chairman of the commission is authorized to appoint such administrative law judges as he deems necessary to assist in the work of the commission [§ 12(d)].

Sections 20 and 21 of the act give the Secretary of Health, Education and Welfare (HEW) broad authority to conduct experimental research relating to occupational safety and health, to develop criteria for and recommend safety and health standards, and to conduct educational and training programs (3). Section 22 establishes the National Institute for Occupational Safety and Health (NIOSH) to perform the functions of the Secretary of HEW under Sections 20 and 21.

The act specifically directs NIOSH to develop criteria documents that describe safe levels of exposure to toxic materials and harmful physical agents and to forward recommended standards for such substances to the Secretary of Labor (4). The act also directs NIOSH to publish at least annually a list of all known toxic substances and the concentrations at which such toxicity is known to occur [§ 20(a)(6)].

Section 7(a) establishes a National Advisory Committee on Occupational Safety and Health (NACOSH), whose basic functions are to "advise, consult with and make recommendations to the Secretary [of Labor] and the Secretary of Health, Education, and Welfare on matters relating to the administration of [the act]." NACOSH consists of 12 members who represent management, labor, occupational safety and occupational health professions, and the public. The members are appointed by the Secretary of Labor, although four members are to be designated by the Secretary of HEW.

The Secretary of Labor is authorized by Section 7(b) to appoint other advisory committees to assist him in the formulation of standards under Section 6. For example, ad hoc advisory committees have been used to assist in developing standards for exposure to asbestos and coke oven emissions. The Secretary of Labor has also appointed various standing advisory committees (5).

1.2 Scope of OSHA's Coverage

The Occupational Safety and Health Act of 1970 applies to every private employer engaged in a business affecting commerce, regardless of the number of employees. It applies with respect to employment performed in a workplace in any of the 50 states, the District of Columbia, Puerto Rico, the Virgin Islands, American Samoa, Guam, the Trust Territory of the Pacific Islands, Wake Island, the Outer Continental Shelf Lands, Johnston Island, and the Canal Zone.

JOB SAFETY AND HEALTH LAW   683

The act's definition of "employer" does not include the states, political subdivisions of the states, or the United States. However Section 19 directs the head of each federal agency to establish and maintain an effective and comprehensive occupational safety and health program that is consistent with the standards required of private employers (6).

2 REGULATION OF JOB SAFETY AND HEALTH BY FEDERAL STATUTES OTHER THAN OSHA

Section 4(b)(1) of OSHA states that nothing in the act shall apply to working conditions of employees with respect to which other federal agencies exercise statutory authority to prescribe or enforce standards or regulations affecting occupational safety or health. Thus federal agencies other than the OSH Administration that are authorized by statute to regulate employee safety and health can continue to do so after the effective date of OSHA; in fact, the *exercise* of such authority preempts OSHA from regulating with respect to such working conditions.

Although Section 4(b)(1) seems to be self-defining, it has generated a tremendous volume of litigation. Three major interpretive questions have been raised:

1. What constitutes a sufficient exercise of regulatory authority to preempt OSHA regulation?
2. Does the exercise of authority by another federal agency in substantial areas of employee safety exempt the entire industry from OSHA standards?
3. Must the other federal agency's motivation in acting have been to protect workers?

It appears to be well settled that the mere existence of statutory authority to regulate safety or health is not sufficient to oust OSHA's regulatory scheme; some exercise of that authority is necessary. Furthermore, at least three federal courts of appeals have taken the position that speculative pronouncements of proposed regulations by a federal agency are not sufficient to warrant preemption of OSHA standards. Rather, it has been ruled that Section 4(b)(1) requires a concrete exercise of statutory authority (7).

The same courts of appeals have also rejected the notion that the exercise of statutory authority by another federal agency creates an industrywide exemption from OSHA regulations. Rather, the courts have agreed that the term "working conditions" in Section 4(b)(1) refers to something more limited than every aspect of an entire industry. Ambiguity remains, however, with respect to the scope of the displacing effect of another agency's regulation of a working condition.

For example, in *Southern Pacific Transportation Company* (7) the Fifth Circuit explained that the term "working conditions" has a technical meaning in the language of industrial relations; it encompasses both a worker's surroundings and the hazards incident to the work. The court stated that the displacing effect of Section 4(b)(1) would depend primarily upon the agency's articulation of its regulations:

Section 4(b)(1) means that any FRA [Federal Railway Administration] exercise directed at a working condition—defined either in terms of a "surrounding" or a "hazard"—displaces OSHA coverage of that working condition. Thus comprehensive FRA treatment of the general problem of railroad fire protection will displace all OSHA regulations on fire protection, even if the FRA activity does not encompass every detail of the OSHA fire protection standards, but FRA regulation of portable fire extinguishers will not displace OSHA standards on fire alarm signaling systems (8).

The Fourth Circuit defined "working conditions" as "the environmental area in which an employee customarily goes about his daily tasks." The court in *Southern Railway Company* (7, 3 OSHC at 1943) explained that OSHA would be displaced when another federal agency had exercised its statutory authority to prescribe standards affecting occupational safety or health for such an area.

The courts of appeals seem to indicate, at least implicitly, that regulation of a working condition by another federal agency need not be as effective or as stringent as an OSHA standard to preempt the OSHA standard [see, *Southern Pacific Transportation Company* (7) 4 OSHC at 1696]. But it remains unclear whether a decision by another federal agency that a particular aspect of an industry should not be regulated at all would preempt or preclude OSHA regulation of that same aspect. Resolution of these and other issues involving the scope of the displacement effect under Section 4(b)(1) awaits future litigation or legislation.

Finally the commission has held that to be cognizable under Section 4(b)(1), "a different statutory scheme and rules thereunder must have a policy or purpose that is consonant with that of the Occupational Safety and Health Act. That is, there must be a policy or purpose to include employees in the class of persons to be protected thereunder" (9). In *Organized Migrants in Community Action Inc.* v. *Brennan* (10) the court, although not deciding the issue, implicitly rejected the argument that preemption under Section 4(b)(1) exists only where the allegedly preempting statute was passed *primarily* for the protection of employees (11).

3 OVERVIEW OF FEDERAL REGULATORY SCHEMES OTHER THAN OSHA

The following material represents an overview of the major federal regulatory schemes other than OSHA that deal with or relate to job safety and health. The listing *is by no means exhaustive,* and employers are urged to consult specific statutory schemes in substantive areas relating to their respective industries.

3.1 Mine Health and Safety Legislation

As of November 1, 1977, occupational safety and health matters with respect to the nation's mining industry were regulated pursuant to the Metal and Non-Metallic Mine Safety Act of 1966 and the Coal Mine Health and Safety Act of 1969.

The Metal and Non-Metallic Mine Safety Act of 1966 (12) extended federal supervision of safety and health to mines of all types (other than coal mines), the products of which regularly enter commerce or the operations of which affect commerce. The Metal and Non-Metallic Act required the Secretary of Interior to promulgate mandatory health and safety standards to deal with conditions or practices of a kind that could reasonably be expected to cause death or serious physical harm. The act also authorized the Secretary of Interior to enforce such standards.

The Coal Mine Health and Safety Act of 1969 (13) and regulations promulgated pursuant to it prescribed health and safety standards for the protection of coal miners. The Coal Mine Act applied to all coal mines "the products of which enter commerce, or the operations or products of which affect commerce." The statute assigned to the Secretary of Interior responsibility for its administration and enforcement.

On November 9, 1977, President Carter signed into law the Federal Mine Safety and Health Amendments Act of 1977. Under the provisions of the 1977 act, the Metal and Non-Metallic Mine Safety Act of 1966 was repealed; a single mine safety and health law was established for all mining operations under a modified Coal Mine Health and Safety Act of 1969 (now called the "Federal Mine Safety and Health Act of 1977"); and the authority for the enforcement of mining health and safety was transferred from the Department of Interior to the Department of Labor. The 1977 Act became effective 120 days after signing (14).

3.2 Environmental Pesticide Control Act of 1972

The Federal Environmental Pesticide Control Act of 1972 (FEPCA) regulates the use of pesticides and makes misuse civilly and criminally punishable. The Court of Appeals for the District of Columbia has held that FEPCA authorizes the Environmental Protection Agency (EPA) to promulgate and enforce occupational health and safety standards with respect to farm workers' exposure to pesticides. The EPA has exercised that authority, and thus has preempted OSHA from regulating in that area [*Organized Migrants in Community Action* (Ref. 10); 15].

3.3 Railway Safety Act of 1970

The Railway Safety Act of 1970 authorizes the Federal Railway Administration (FRA) within the Department of Transportation (DOT) to promulgate regulations for all areas of railroad safety, including employee safety. To date, however, DOT has not adopted railroad occupational safety standards for all railroad working conditions or workplaces (16). The Department of Labor (OSHA) retains jurisdiction over safety and health of railroad employees with respect to those "working conditions" for which DOT has not adopted standards (17).

3.4 Federal Aviation Act of 1958

The Federal Aviation Act of 1958, as amended, empowers the Federal Aviation Administration (FAA) within the Department of Transportation "to promote safety of flight of civil aircraft in air commerce" (18). If the congressional mandate in that statute is deemed to include the safe working conditions of airline employees, OSHA would be precluded from exercising its jurisdiction with respect to the working conditions regulated by the FAA. At least one administrative law judge has determined that the FAA's mandate encompasses the safe working conditions of airline ground crews when performing aircraft maintenance work (19).

3.5 Hazardous Materials Transportation Act

The Hazardous Materials Transportation Act (HMTA) authorizes the Secretary of Transportation to issue regulations governing any safety aspect of the transportation of materials designated as hazardous by the secretary (20). The act encompasses shipments by rail, air, water, and highway. If worker safety is deemed to be a purpose of the HMTA, safety standards promulgated by DOT under the statute would trigger a preemption of OSHA jurisdiction with respect to the working conditions covered by such standards.

3.6. Natural Gas Pipeline Safety Act of 1968

The Natural Gas Pipeline Safety Act of 1968 (NGPSA) authorizes the Secretary of Transportation to establish minimum federal safety standards for pipeline facilities and the transportation of gas in commerce. In *Texas Eastern Transmission Corp.* (21) the Occupational Safety and Health Review Commission determined that the NGPSA was intended to affect occupational safety and health. Thus employers engaged in the transmission, sale, and storage of natural gas would be exempt from OSHA with respect to working conditions covered by DOT standards promulgated under NGPSA (22).

3.7 Federal Noise Control Act of 1972

Although a health and safety standard adopted pursuant to OSHA governs the level of noise to which a worker covered by the act may be exposed in the place of work (see Section 5.3), other federal statutes deal with noise abatement and control as well (23).

The first such enactment, a 1968 amendment to the Federal Aviation Act, required the Administrator of the FAA to include aircraft noise control as a factor in granting type certificates to aircraft under the act (24). To the extent that the Administrator of the FAA denies certification to an aircraft that produces noise in excess of the standards or prohibits the operation of a certified aircraft in a manner that violates his regulations, the general environmental noise level in workplaces covered by OSHA and located adjacent to airports and landing patterns is correspondingly reduced.

The first attempt to deal with noise on a nationwide basis, however, is the federal Noise Control Act of 1972. The control strategy of that act is generally as follows: the Administrator of the EPA is required to develop and publish criteria with regard to noise, reflecting present scientific knowledge of the effects on public health and welfare that are to be expected from different quantities or qualities of noise. The administrator then must identify products or kinds of products that in his opinion are major sources of environmental noise harmful to the public, and publish noise emission regulations where it is feasible to limit the amount of noise produced by such products (25).

Under the act, the administrator is further charged with publishing regulations identifying "low noise emission products." Such products, once so designated, must thereafter be purchased by federal agencies in preference to substitute products, provided the "low noise emission product" costs no more than 125 percent of the price of the substitute.

As the Administrator of the EPA identifies more and more products as major sources of noise, and subjects those products to noise emission standards adopted under the Noise Control Act of 1972, the noise levels found in workplaces covered by OSHA and in which such products are used should decrease.

3.8 Federal Toxic Substances Control Act of 1976

The Toxic Substances Control Act of 1976 establishes a broad, nationwide program for the federal regulation of the manufacture and distribution of toxic substances (26). The act divides all "chemical substances" and "mixtures" into two categories: the old and the new. The Administrator of the EPA is charged under Section 8(b) with the gargantuan task of compiling and publishing in the *Federal Register* an inventory or list of all chemical substances manufactured or processed in the United States.

Manufacturers of substances that appear on that inventory are at liberty to continue to manufacture and distribute such substances unless the administrator by rule promulgated under Section 4 of the act first requires that a designated substance be tested and data from the tests be submitted to the EPA. If EPA thereafter makes a determination under Section 6 that the continued manufacture, processing, or distribution in commerce of the substance presents an unreasonable risk of injury to health or the environment, the administrator may either prohibit altogether the manufacture of the chemical substance or may impose restrictions (limitations on the quantity manufactured, the use to which the chemical may be put, the concentrations in which it may be used, the labels and warnings that must accompany its sale, etc.).

Section 5 of the act provides a different treatment with respect to a "new chemical substance" or a "significant new use" of a substance that appears on the Section 8(b) inventory. The manufacturer is not at liberty to commence manufacture or distribution of such a "new" substance but must first submit a notice to the administrator of his intention to manufacture such a new substance or to engage in a significant new use. Then testing data that relate to the toxicity and the effect on health and on the environment of the substance must be submitted. If the administrator does not act within 90 days, the manufacturer is at liberty to proceed with manufacture or distribution. During the initial period, however, the administrator may extend his time for action an additional 90 days. If during the original period (or its extension) the administrator believes that the information available to him is inadequate to make a reasoned finding that the proposed new substance or use does not present an unreasonable risk of injury to health or the environment, he may prohibit or limit the manufacture of the substance and obtain an injunction in court for that purpose. It would appear that in the absence of testing data submitted in compliance with Section 4, this injunction against manufacture or distribution of the new substance or use would continue indefinitely. If, however, the administrator finds, based on information before him, that the proposed new substance or use does present an unreasonable risk to health and safety, the administrator must proceed by means of the provisions of Section 6 to prohibit manufacture or to impose restrictive conditions (27).

Although the Toxic Substances Control Act of 1976 promises to be a major weapon in the federal health, safety, and environmental arsenal, to date there are no significant developments under the act. The Administrator of the EPA has occupied himself with compilation of the chemical substances inventory required by Section 8(b) and has proposed regulations with respect to only one hazardous substance, polychlorinated biphenyls, the only group of chemicals with which the administrator is statutorily obligated to deal (28).

3.9 Federal Consumer Product Safety Act

The Consumer Product Safety Act of 1972 was drafted to apply only to "consumer products." That term is defined in the act in a manner that serves to exclude most products destined principally for use in workplaces covered by OSHA. Nevertheless, the act promises to provide increased protection of the American worker from hazardous products that by their nature are "consumer products" within the meaning of the act, yet are frequently found in the workplace (29).

The act created a Consumer Product Safety Commission and empowered that agency to promulgate "consumer product safety standards" applicable to consumer products found by the commission to present an unreasonable risk of injury. Such standards may be performance standards, or they may require that products not be sold without adequate warnings or instruction. Where no feasible safety standard that could be promulgated would eliminate an unreasonable risk of injury, the commission is empowered to ban the consumer product altogether from interstate sale or distribution.

In addition to publishing safety standards, the commission is empowered to file suit and seek the seizure of a consumer product believed to be "imminently hazardous," regardless of whether the product in question is covered by already promulgated consumer product safety standards. The commission is also authorized to find, after hearing, that a consumer product presents a "substantial product hazard." In the event of such a finding, the commission may order the manufacturer, distributor, or retailer of

JOB SAFETY AND HEALTH LAW

the product to give public notice of that finding, and to repair, replace, or refund the purchase price of the product affected (30).

3.10 Hazardous Substances Act

The federal Hazardous Substances Act provides a mechanism by means of which the Consumer Product Safety Commission may find that a substance distributed in interstate commerce is "hazardous." After such a finding the commission may either impose packaging and labeling requirements to protect public health and safety or, in the cases of hazardous substances intended for the use of children or likely to be subject to access by children, or substances intended for household use, prohibit distribution altogether ("banned hazardous substance") (31). Insofar as safety in the American workplace is concerned, the effect of the act is that hazardous substances distributed in interstate commerce and utilized by the American worker will arrive safely packaged and accompanied by appropriate warnings.

3.11 The Atomic Energy Act of 1954 and Other Statutory Sources of Radiation Control

As discussed later (Section 5.3), the Department of Labor has published occupational health and safety standards regulating exposure to ionizing (i.e., alpha, beta, gamma, X-ray, neutron, etc.) radiation and nonionizing (i.e., radiofrequency, electromagnetic) radiation. However the primary federal law regulating human exposure to radiation is not OSHA but rather the Atomic Energy Act of 1954, as amended. Exercising power under that statute, the Nuclear Regulatory Commission (NRC) has published "Standards for Protection Against Radiation" (32).

A detailed discussion of those regulations is beyond the scope of this chapter. The operation of the standards can be summarized briefly as follows, however. Any person holding a license issued under the Atomic Energy Act of 1954 and using "licensed material" (i.e., radioactive or radiation-emitting material) may not permit the exposure of individuals within a "restricted area" (i.e., an area in which radioactive materials are being used) to greater doses of radiation than are set forth in the regulations (33).

Although the primary thrust of the NRC regulations is to control ionizing radiation within the "restricted area," the NRC has also published regulations on permissible levels of radiation in unrestricted areas, in effluents discharged into unrestricted areas, and for the disposal of radioactive materials by release into sanitary sewerage systems (34).

The Administrator of the EPA, exercising authority under the Atomic Energy Act of 1954, which he acquired by means of the Reorganization Plan No. 3 of 1970, has also promulgated regulations limiting exposure of the general population to ionizing radiation produced during the operation of nuclear power plants licensed by the NRC. The EPA regulations, however, do not become effective until December 1, 1979 (35).

There are additional federal agencies empowered to set standards for ionizing radiation control within areas under their jurisdiction. The Department of Agriculture, for example, regulates the irradiation of certain food substances under the federal Meat Inspection Act (36). The Department of Interior has published regulations on exposure of miners to radiation in mines which had been covered by the Metal and Non-Metallic Mine Safety Act (37). And the Department of Labor has issued radiation standards for uranium mining conducted under the Walsh-Healy Public Contracts Act (38).

Radiation generated by devices and products that are not governed by the Atomic Energy Act and licensed by the NRC is regulated by the Federal Radiation Control for Health and Safety Act of 1968 (39).

3.12 Outer Continental Shelf Lands Act

The Outer Continental Shelf Lands Act authorizes the head of the department in which the Coast Guard is operating to promulgate and enforce "such reasonable regulations with respect to lights and other warning devices, safety equipment, and other matters relating to the promotion of safety of life and property" on the lands and structures referred to in the act or on the adjacent waters (40). If worker safety is deemed to be within the mandate of this statute, OSHA jurisdiction may be preempted with respect to working conditions that are governed by standards issued by the Coast Guard pursuant to this act.

4 REGULATION OF JOB SAFETY AND HEALTH BY THE STATES

One of the primary factors that induced Congress to enact the Occupational Safety and Health Act was the failure of many of the states adequately to regulate workplace safety and health (41). In passing OSHA, Congress hoped to ensure at least a minimum level of regulation of the conditions experienced by workers throughout the country.

The Occupational Safety and Health Act of 1970 preempts state regulation of job safety and health with respect to matters which OSHA regulates, even when a state has a more stringent regulation with respect to a particular hazard (42). However OSHA does not totally ban the states from developing and enforcing occupational safety and health standards. Pursuant to Section 18(b) of the act, a state may regain jurisdiction over development and enforcement of occupational safety and health standards by submitting to the federal government an effective state occupational safety and health plan. Final approval of a state plan can lead ultimately to exclusive authority by a state over the matters included in its plan.

The process of regaining jurisdiction over the regulation of occupational safety and health begins with the submission of a plan that sets forth specific procedures for ensuring workers' safety and health. According to the regulations of the Secretary of Labor, the states can submit either of two types of plan: a complete plan or a developmental plan.

A "complete" plan (43) is a plan that, upon submission, satisfies the criteria for plan approval set forth in Section 18(c) of the act, as well as certain additional criteria outlined by the Secretary of Labor in his administrative regulations (44). Complete plans are given "initial" approval by the Secretary of Labor upon submission. For at least 3 years following the "initial" approval, the Secretary of Labor will monitor the state plan to determine whether on the basis of the actual operations of the plan, the criteria set forth in Section 18(c) are being applied. If this determination (the "Section 18(e) determination") is favorable, the state plan will be granted "final approval" and the state will regain exclusive jurisdiction with respect to any occupational safety or health issue covered by the state plan. Federal (i.e., OSHA) standards continue to apply to hazards not covered by the state program; thus state plans need not address all hazards, yet gaps in protection are avoided.

A "developmental" plan (45) is a plan that, upon submission, does not fully meet the criteria set forth in the statute or in the regulations. A developmental plan may receive initial approval upon submission, however, if the plan contains "satisfactory assurances" that the state will take the necessary steps to bring its program into conformity within 3 years of the date of submission.

If the developmental plan satisfies all the statutory and administrative criteria within the 3-year "developmental period," the Secretary of Labor will so certify and will initiate an evaluation of the actual operations of the state plan for purposes of making a Section 18(e) determination. The evaluation must proceed for at least one year before such a determination can be made.

Plans that have received final approval will continue to be monitored and evaluated by the Secretary of Labor pursuant to Section 18(f) of the act, which authorizes the secretary to withdraw approval if a state fails to comply substantially with any provision of the state's plan.

Although a state does not regain exclusive jurisdiction over matters contained in its plan until the plan receives final approval, a state with initial plan approval may participate in the administration and enforcement of the act prior to final approval by satisfying the following four criteria (46):

1. The state must have enacted enabling legislation conforming to that specified in OSHA and the regulations.
2. The state plan must contain standards that are found to be at least as effective as the comparable federal standards.
3. The state plan must provide for a sufficient number of qualified personnel who will enforce the standards in accordance with the state's enabling legislation.
4. The plan's provisions for review of state citations and penalties (including the appointment of the reviewing authority and the promulgation of implementing regulations) must be in effect.

If the criteria above are met, the state plan is deemed to be "operational." Thereupon the federal government enters into an operational agreement with the state whereby the state is authorized to enforce safety and health standards under the state plan (46). During this period the act permits but does not require the federal government to retain enforcement activity in the state (47). Thus during this period an employer could be subject to enforcement activities by both the state and federal authorities. However the secretary's regulations provide that once a plan (either complete or developmental) becomes "operational," the state will conduct all enforcement activity, including inspections in response to employee complaints, and accordingly, the federal enforcement activity will be reduced and the emphasis will be placed on monitoring state activity (46).

As of November 1, 1977, no states had submitted "complete" plans. However the Secretary of Labor had certified the developmental plans of the following states: Alaska, California, Iowa, Minnesota, North Carolina, South Carolina, Utah, and Vermont. Seventeen additional states or territories had received initial approval of developmental plans as of that date: Arizona, Colorado, Connecticut, Hawaii, Indiana, Kentucky, Maryland, Michigan, Nevada, New Mexico, Oregon, Puerto Rico, Tennessee, Virgin Islands, Virginia, Washington, and Wyoming (48).

Thirteen states or territories plus the District of Columbia had submitted plans and were awaiting initial approval by the Secretary of Labor: Alabama, American Samoa, Arkansas, Delaware, Florida, Guam, Idaho, Massachusetts, Missouri, Oklahoma, Rhode Island, Texas, and West Virginia.

The following 11 states submitted plans at one time but have withdrawn them: Georgia, Illinois, Maine, Mississippi, Montana, New Hampshire, New Jersey, New York, North Dakota, Pennsylvania, and Wisconsin. Five states (Kansas, Louisiana, Nebraska, Ohio, and South Dakota) have never submitted plans to the Department of Labor (49).

Most states that are presently operating approved and/or certified plans have adopted standards that are substantially similar, if not identical, to the federal standards. At least five state plans, however, contain certain provisions that vary from the federal standards and have been approved as being "at least as effective" as the federal standards. The states are California, Hawaii, Michigan, Oregon, and Washington (49). In some instances these states have adopted standards that are more stringent than the analogous federal standards. Employers who are operating in more than one state should be aware that they may have to deal with different regulations, different enforcement procedures, and perhaps different interpretations of similar standards for purposes of complying with the applicable occupational safety and health laws (50).

4.1 State Jurisdiction in Areas Regulated by Federal Legislation Other than OSHA

If state legislation regulates a job safety or health issue that OSHA does not cover, the state may continue to enforce its relevant standards unless other applicable federal law has preempted state enforcement.

In some areas a federal regulatory scheme permits concurrent federal and state regulation of job safety and health. For example, the Federal Mine Safety and Health Act of 1977 states that no state law that was in effect on December 30, 1969, or may become effective thereafter shall be superseded by any provisions of the federal mine act unless the state law is in conflict with the mine act. State laws and rules that provide for standards more stringent than those of the mine act are deemed not to be in conflict (51).

On the other hand, the Railroad Safety Act of 1970 has essentially preempted the states' regulation of railroad safety and health, except in cases of a state having a more stringent law because of the need to eliminate or reduce an essentially local safety hazard (52).

5 EMPLOYERS' DUTIES UNDER THE OCCUPATIONAL SAFETY AND HEALTH ACT OF 1970

A private employer's primary duties under the Occupational Safety and Health Act of 1970 are found in Section 5(a), which provides that each employer

1. Shall furnish to each of his employees employment and a place of employment which are free from recognized hazards that are causing or are likely to cause death or serious physical harm to his employees.
2. Shall comply with occupational safety and health standards promulgated under the act.

5.1 The General Duty Clause [§ 5(a)(1)]

The essential elements of the so-called general duty clause of OSHA are the following: (1) the employer must render the workplace "free" of hazards that arise out of conditions of the employment, (2) the hazards must be "recognized," and (3) the hazards must be causing or likely to cause death or serious physical harm (53).

5.1.1 Failure to Render Workplace "Free" of Hazard

It is fairly well settled that Congress did not intend to make employers strictly liable for the presence of unsafe or unhealthful conditions on the job. The employer's general duty must be an achievable one (54). Thus the term "free" has been interpreted by the courts and the commission to mean something less than absolutely free of hazards. Instead the courts and the commission have held (*National Realty and Construction Company*, Reference 54) that the employer has a duty to render the workplace free only of hazards that are preventable.

The determination of whether a hazard is preventable generally is made in the context of an enforcement proceeding under the act when an employer asserts the inability of preventing the hazard as an affirmative defense to a proposed citation. The employer often contends that (1) the technology does not exist to prevent a hazard, or (2) even though the technology to prevent the hazard exists, the cost of the technology would be prohibitive, or (3) the hazard was created by an employee's misconduct that was so unusual that the employer could not reasonably prevent the existence of the hazard.

When technology does not exist to prevent a hazard, OSHA does not require prevention by shutting down the employer's operation. Rather, the Secretary of Labor must be able to show that "demonstrably feasible" measures would have materially reduced the hazard (*National Realty and Construction Company*, Reference 54). Similarly, it seems that measures that, even though technologically feasible, would have been so expensive as to bankrupt the employer, are not "demonstrably feasible" measures. (For further discussion relating to economic feasibility, see Section 5.2.3 of this chapter.)

Furthermore the courts and the commission have recognized that certain isolated or idiosyncratic acts by an employee that were not foreseeable by the employer could result in unpreventable hazards for which the employer should not be held liable. For example, in *National Realty and Construction Company*, the Court of Appeals stated:

Hazardous conduct is not preventable if it is so idiosyncratic and implausible in motive or means that conscientious experts, familiar with the industry, would not take it into account in prescribing a safety program. Nor is misconduct preventable if its elimination would require methods of hiring, training, monitoring, or sanctioning workers which are either so untested or so expensive that safety experts would substantially concur in thinking the methods infeasible.

The court in *National Realty* emphasized, however, that an employer does have a duty to attempt to prevent hazardous conduct by employees. Thus the employer must adopt demonstrably feasible measures concerning the hiring, training, supervising, and sanctioning of employees, to reduce materially the likelihood of employee misconduct (55).

5.1.2 The Hazard Must Be a "Recognized" Hazard

The general duty clause does not apply to all hazards but only to the hazards that are "recognized" as arising out of the employment. The test for determining a "recognized" hazard is whether the hazard is commonly known by the public in general or is recognized by the industry of which the employer is a part (56, 57). The employer's actual knowledge of the hazard is also sufficient proof that the condition is recognized as a hazard in the industry, but once the Secretary of Labor has proved that a condition is recognized as a hazard in the industry, he need not prove that the employer had actual knowledge (58).

In *American Smelting and Refining Company v. OSAHRC* (59) the Eighth Circuit held that the general duty clause is not limited to recognized hazards of types detectable only by the human senses but also encompasses hazards that can be detected only by instrumentation.

5.1.3 The Hazard Must Be Causing or Likely to Cause Death or Serious Physical Harm

It is not necessary that there be actual injury or death to trigger a violation of the general duty clause. The purpose of the act is to prevent accidents and injuries. Thus violation of the general duty clause arises from the existence of a statutory hazard, not from injury in fact (60).

Proof that a hazard is "causing or likely to cause death or serious physical harm" does not require a mathematical showing of probability. Rather, if evidence is presented that a practice could eventuate in serious physical harm upon other than a freakish or utterly implausible concurrence of circumstances, the commission's determination of likelihood will probably be accorded considerable deference by the courts (61).

The term "serious physical harm" is defined neither in the act nor in the secretary's regulations, but OSHA's Compliance Operations Manual defines it as follows:

Serious physical harm is that type of harm that would cause permanent or prolonged impairment of the body in that (1) a part of the body would be permanently removed or rendered functionally useless or substantially reduced in efficiency on or off the job (e.g., leg shattered so severely that mobility would be permanently reduced), or (2) a part of an internal bodily system would be inhibited in its normal performance to such a degree as to shorten life or cause reduction in physical or mental efficiency (e.g., lung impairment causing shortness of breath). On the other hand, breaks, cuts, bruises, concussions or similar injuries would not constitute serious physical harm.

5.2 The Specific Duty Clause [§ 5(a)(2)]

Section 5(a)(2) of OSHA imposes on employers a duty to comply with the occupational safety and health standards promulgated by the Secretary of Labor. These standards constitute the employers' so-called specific duties under the act. Specific promulgated standards preempt the general duty clause, but only with respect to hazards expressly covered by the specific standards (62).

5.2.1 Processes for Promulgating Standards

The act established processes for promulgating three types of occupational safety and health standards: interim, permanent, and emergency.

Interim standards consist of standards derived from (1) established federal standards, or (2) national consensus standards that were in existence on the effective date of OSHA (63). Section 6(a) of the act directed the Secretary of Labor to publish such standards in the *Federal Register* immediately after the act became effective or for a period of up to 2 years thereafter. These standards became effective as OSHA standards upon publication without regard to the notice, public comment, and hearing requirements of the Administrative Procedure Act (64).

The intent of the interim standards provisions was to give the secretary a mechanism by which to promulgate speedily standards with which industry was already familiar and to provide a nationwide floor of minimum health and safety standards (65). The secretary's 2-year authority to promulgate interim standards expired on April 29, 1973.

Pursuant to Section 6(b) of the act, the Secretary of Labor is authorized to adopt "permanent" occupational safety and health standards to serve the objectives of OSHA. With respect to standards relating to toxic materials or harmful physical agents, Section 6(b)(5) specifically directs the secretary to set the standard "which most adequately assures, to the extent feasible, on the basis of the best available evidence, that no employee will suffer material impairment of health or functional capacity even if such employee has a regular exposure to the hazard dealt with by such standard for the period of his working life."

The promulgation of these "permanent" occupational safety and health standards requires procedures similar to informal rule making under Section 4 of the Administrative Procedure Act. Upon determination that a rule should be issued promulgating such a standard, the secretary must first publish the proposed standard in the *Federal Register*. Publication is followed by a 30 day period during which interested persons may submit written data or comments or file written objections and requests for a public hearing on the proposed standard. If a hearing is requested, the secretary must publish in the *Federal Register* a notice specifying the standard objected to and setting a time and place for the hearing. Within 60 days after the period for filing comments, or, if a hearing has been timely requested, within 60 days of the hearing, the secretary may either issue a rule promulgating a standard or determine that no such rule should be issued (66).

Section 6(c)(1) of OSHA authorizes the secretary to issue emergency temporary standards if he determines (1) that employees are exposed to grave danger from exposure to substances or agents determined to be toxic or physically harmful or from new hazards, and (2) that such emergency standard is necessary to protect employees from such danger.

An emergency temporary standard may be issued without regard to the notice, public comment, and hearing provisions of the Administrative Procedure Act. It takes immediate effect upon publication in the *Federal Register*.

The key to the issuance of an emergency temporary standard is the necessity to protect employees from a grave danger, as defined in *Florida Peach Growers* (67). After issuing an emergency temporary standard, the secretary must commence the procedures for promulgation of a permanent standard, which must issue within 6 months of the emergency standard's publication in accordance with Section 6(c)(3).

5.2.2 Challenging the Validity of Standards

Any person who may be adversely affected by an OSHA standard may file a petition under Section 6(f) challenging its validity in the United States Court of Appeals in the circuit wherein such person resides or has his or her principal place of business. The petition may be filed at any time prior to the sixtieth day after the issuance of the standard. Unless otherwise ordered, the filing of a petition does not operate as a stay of the standard.

Section 6(f) of the act directs the courts to uphold the Secretary of Labor's determinations in promulgating standards if those determinations are "supported by substantial evidence in the record considered as a whole" (68). In practice, the courts have generally declined to apply a strict "substantial evidence" standard of review. Rather, they have essentially taken the position that the secretary's policy determinations must be substantiated by a detailed statement of reasons, which are subject to a test of reasonableness. The secretary's findings of fact, however, are generally reviewed pursuant to a substantial evidence test (69). The courts have adopted this approach with respect to emergency temporary standards as well as permanent standards [e.g., in *Florida Peach Growers* (67)].

In addition to a direct petition for review under Section 6(f), at least two courts of appeals have held that judicial review of OSHA standards is appropriate during enforcement proceedings under Section 11 (70). In fact the Third Circuit in *Atlantic & Gulf Stevedores* (70) stated that the validity of a standard may be challenged not only in a federal court of appeals as part of an appeal from an order of the commission, but in the commission proceedings themselves. The Third Circuit explained, however, that in an enforcement proceeding invalidity is an affirmative defense to a citation and the employer bears the burden of proof on the issue of the reasonableness of the adopted standard. To carry its burden the employer must produce evidence showing why the standard under review, *as applied to it*, is arbitrary, capricious, unreasonable, or contrary to law (71).

5.2.3 Economic and Technological Feasibility of Standards

In enacting OSHA, Congress did not intend to make employers strictly liable for unavoidable occupational hazards. Accordingly, the courts have generally held that feasibility of compliance is a factor the Secretary of Labor may properly consider in developing occupational safety and health standards.

In *American Federation of Labor, Etc. v. Brennan* (72) the Third Circuit Court of Appeals addressed the issue of feasibility in connection with a challenge to the secretary's promulgation of a revision of the safety standards applicable to mechanical power presses and held that the secretary must be permitted to consider technological feasibility of a proposed standard. The court pointed out, however, that "at least to a limited extent, OSHA is to be viewed as a technology-forcing piece of legislation. Thus the Secretary would not be justified in dismissing an alternative to a proposed health and safety standard as infeasible when the necessary technology looms on today's horizon" (73, 74).

The Third Circuit also held that OSHA permits the Secretary of Labor to take into account the likely economic impact of proposed occupational safety and health standards:

Congress did contemplate that the Secretary's rulemaking would put out of business some businesses so marginally efficient or productive as to be unable to follow standards otherwise universally feasible. But we will not impute to congressional silence a direction to the Secretary to disregard the possibility of massive economic dislocation caused by an unreasonable standard. An economically impossible standard would in all likelihood prove unenforceable, inducing employers faced with going out of business to evade rather than comply with the regulation (75).

In addition, Section 6(b)(5), dealing with standards for toxic materials and harmful physical agents, explicitly confines the secretary's rule making power within feasible boundaries. In *Industrial Union Department, AFL-CIO v. Hodgson* (4), the Court of Appeals applied Section 6(b)(5) in a case challenging OSHA's standard for exposure to asbestos dust and held that the secretary, in promulgating the standard, could properly consider problems of both economic and technological feasibility.

Furthermore, on November 27, 1974, President Gerald Ford issued Executive Order 11821, which directed that major proposals for regulations by executive branch agencies, including OSHA, be accompanied by a certification that the economic impact of the proposal has been evaluated. These evaluations must include (1) an analysis of the principal cost or the inflationary effects of the action on markets, consumers, businesses, or government; (2) a comparison of the benefits to be derived from the proposed action with the estimated costs and inflationary impacts; and (3) a review of alternatives to the proposed action that were considered, and their probable costs, benefits, risks, and inflationary impacts compared with those of the proposed action (76).

Some courts of appeals are willing to permit employers to raise the issue of infeasibility of compliance in proceedings brought to enforce promulgated health and safety standards (see Section 5.2.2, this chapter). In doing so, employers would be permitted to raise economic considerations with respect to their own culpability for a given violation but not with respect to the general economic consequences of enforcement of the standard (77).

In a series of enforcement proceedings regarding noise levels in the workplace, the Secretary of Labor has taken the position that economic considerations should enter into a determination of feasibility at the enforcement stage only if the cost of engineering controls to achieve compliance would "seriously jeopardize the financial condition of the company." The employers in such proceedings have argued that economic feasibility should include a determination of the costs of engineering controls as compared with the benefits which will be produced by such controls. The commission and at least one federal court of appeals to date have agreed with the employers (78).

Finally, in cases of violations of standards caused by employee disobedience or idiosyncratic behavior, the decisions of the courts and the commission have been similar to those rendered under the general duty clause (79; see also discussion in Section 5.1.1, this chapter).

For example, in *Brennan v. OSAHRC & Hendrix (d/b/a Alsea Lumber Co.)* (80) an employer was cited for violation of OSHA standards requiring workers to wear certain personal protective equipment. The record established that the violations resulted from individual employee choices, which were contrary to the employer's instructions. The Ninth Circuit affirmed the commission's decision vacating the citations, explaining as follows:

The legislative history of the Act indicates an intent not to relieve the employer of the general responsibility of assuring compliance by his employees. Nothing in the Act, however, makes an employer an insurer or guarantor of employee compliance therewith at all times. The employer's duty, even that under the general duty clause, must be one which is achievable. [See *National Realty, supra.*] We fail to see wherein charging an employer with a nonserious violation because of an individual, single act of an employee, of *which the employer had no knowledge* and which was contrary to the employer's instructions, contributes to achievement of the cooperation [between employer and employee] sought by the Congress. Fundamental fairness would require that one charged with and penalized for violation be shown to have caused, or at least to have knowingly acquiesced in, that violation [emphasis added] (81).

Nevertheless, even though Congress did not intend the employer to be held strictly liable for violations of OSHA standards, an employer is responsible if it knew or, with the exercise of reasonable diligence, should have known of the existence of a violation. Thus in *Brennan* v. *Butler Lime & Cement Co.* (82) the Seventh Circuit drew on general duty clause concepts from the *National Realty* case (54) and explained that a particular instance of hazardous employee conduct may be considered preventable even if no employer could have detected the conduct or its hazardous nature at the moment of its occurrence, where such conduct might have been precluded through feasible precautions concerning the hiring, training, or sanctioning of employees.

In *Atlantic & Gulf Stevedores, Inc.* the Third Circuit held that such feasible precautions include disciplining or dismissing workers who refuse to wear protective headgear, even where such employer action could subject the company to wildcat strikes by employees adamantly opposed to the regulation (83, 84).

Compare *Horne Plumbing and Heating Company* v. *OSAHRC*, where the Fifth Circuit found that an employer had taken virtually every conceivable precaution to ensure compliance with the law, short of remaining at the job site and directing the employees' operations himself. The court held that the final effort of personally directing the employees was not required by the act and that such an effort would be a "wholly unnecessary, unreasonable and infeasible requirement" (85, 86).

5.2.4 Environmental Impact of Standards

Section 102(2)(C) of the National Environmental Policy Act of 1969 (NEPA) (87) requires all federal agencies, including OSHA, to prepare a detailed environmental impact statement in connection with major federal actions significantly affecting the quality of the human environment. The Secretary of Labor has identified the promulgation, modification, or revocation of standards regulating environmental hazards to health in the workplace as possibly constituting such major action requiring the preparation of an environmental impact statement. The secretary's regulations regarding the procedure for preparation and circulation of environmental impact statements can be found in the Code of Federal Regulations (29 CFR § 1999).

5.2.5 Variances from Standards

Section 6(d) of OSHA provides that any affected employer may apply to the Secretary of Labor for a variance from an OSHA standard. To obtain a variance, the employer must show by a preponderance of the evidence submitted at a hearing that the conditions, practices, means, methods, operations, or processes used or proposed to be used by him will provide his employees with employment and places of employment that are as safe and healthful as those that would prevail if he complied with the standard.

If granted, the Section 6(d) variance may nevertheless be modified or revoked on application by an employer, employees, or by the Secretary of Labor on his own motion at any time after 6 months from its issuance. Affected employees are to be given notice of each application for a variance and an opportunity to participate in a hearing (88, 89).

The act also provides mechanisms to enable employers to obtain variances of a more temporary nature than those sought under Section 6(d). Section 6(b)(6)(A) provides for "temporary" variances upon application when an employer establishes, among other things, after notice and hearing that he is unable to comply with a standard by its effective date because of unavailability of professional or technical personnel or of materials and equipment needed to come into compliance with the standard or because necessary construction or alteration of facilities cannot be completed by the effective date.

Section 6(b)(6)(C) authorizes the secretary to grant a variance from any standard or portion thereof whenever he determines, or the Secretary of HEW certifies, that the variance is necessary to permit the employer to participate in an experiment approved by the Secretary of Labor or the Secretary of HEW designed to demonstrate or validate new and improved techniques to safeguard the health or safety of workers.

Finally, Section 16 permits the Secretary of Labor, after notice and an opportunity for hearing, to provide such reasonable limitations and rules and regulations allowing "reasonable variations, tolerances, and exemptions to and from any or all provisions of" the act as he may find necessary and proper to avoid serious impairment of the national defense.

5.3 Overview of Occupational Safety and Health Standards (90)

To date, the bulk of federal job safety and health standards deal with occupational safety rather than with occupational health (91).

The occupational safety standards promulgated by the Secretary of Labor pursuant to OSHA are voluminous, encompassing hundreds of pages in the Code of Federal Regulations. A comprehensive analysis of these safety standards, most of which were adopted in 1971 as interim standards, is accordingly beyond the scope of this chapter (92).

For purposes of simplification, however, OSHA's safety standards can be broken down into the following general categories: (1) requirements relating to hazardous materials (e.g., compressed gas, acetylene, hydrogen, oxygen) and related equipment, (2) requirements for personal protective equipment and first aid, (3) fire protection standards and the national electrical code, (4) design and maintenance requirements for industrial equipment and working surfaces, and (5) operational procedures and equipment utilization requirements for certain hazardous industrial operations such as welding, cutting and brazing, and materials handling. Certain industries are also subject to specialized safety standards (93).

As of December 15, 1977, OSHA was undertaking a major revision of its safety standards. In June 1977 the Carter administration established two OSHA task forces to compile a list of safety standards that could be considered to have little or no application to safety as well as a list of standards that could be treated as *de minimis* most of the time. On December 5, 1977, OSHA announced that more than 1100 safety regulations in 29 CFR 1910 had been judged to have no direct or immediate effect on worker safety and that such regulations would be proposed for revocation. The standards proposed for revocation were listed in the *Federal Register* (94).

The OSH Administration has also been reviewing its general industry standards relating to walking-working surfaces (Subpart D), fire protection (Subpart L), and machinery-machine guarding (Subpart O). A revised set of rules for walking-working surfaces had been scheduled tentatively for proposal by February 15, 1978; revised fire protection regulations were to be proposed by March 15, 1978; and revised standards relating to machinery and machine guarding were to be ready by June 1, 1978 (95).

Like the safety standards, the bulk of OSHA's health standards were promulgated in 1971 as interim standards under Section 6(a). Established federal standards set levels for workplace exposure to airborne concentrations of approximately 400 chemical substances based on an 8 hr time-weighted average of exposure. These levels are referred to as threshhold limit values (TLVs) and are expressed in terms of milligrams of substance per cubic meter of air and/or parts of vapor or gas per million parts of air. The established federal TLVs had been developed principally in 1968 by the American Conference of Governmental Industrial Hygienists (ACGIH) and subsequently incorporated into the Walsh-Healy Act (96).

Also under Section 6(a) the secretary promulgated certain ANSI health standards as national consensus standards (97). These ANSI standards established TLVs for 22 hazardous substances (98).

The existing TLVs have been sharply criticized. A congressional committee (99) has pointed out that the substances on the list of TLVs represent only a small portion of the estimated 19,000 unique toxic chemicals that are found in the workplace. In addition, constant changes in technology create new hazards and increase the difficulty of maintaining an up-to-date listing of toxic substance standards. Furthermore the TLVs were apparently developed to be used as guidelines rather than as standards. As they now stand, they do not include several key provisions that are required for standards promulgated under the permanent rule-making procedures of Section 6(b). For example, Section 6(b)(7) of the act states that standards promulgated under Section 6(b) should prescribe, among other things, suitable protective equipment, engineering controls, medical surveillance, monitoring of employee exposure to hazards (100), and the use of labels to apprise employees of hazards (101).

However the OSH Administration and NIOSH are engaged in a program to supplement the existing standards for up to 384 of the toxic substances listed as TLVs. The "Standards Completion Project" calls for the two agencies to complete standards for these substances by adding specific requirements, methods, and procedures for determining exposures, monitoring employee health, recordkeeping, and similar actions necessary to provide a complete standards package (102).

In addition to the TLVs, the Secretary of Labor also promulgated interim health standards relating to radiation, ventilation, and noise. The noise standard (103) requires protection against the effect of occupational noise exposure when the sound levels in the workplace exceed acceptable limits shown in the standard. With respect to radiation, employers are responsible for proper controls to prevent any employee from being exposed to either ionizing or electromagnetic radiation in excess of acceptable limits. The radiation standard also requires posting of radiation areas, monitoring of employee exposure to radiation, and keeping records of such monitoring (104). Ventilation standards are included for various industrial processes, such as abrasive blasting operations, grinding, polishing, buffing, and spray finishing (105).

Since the effective date of OSHA, the Secretary of Labor has promulgated four major health standards under the permanent rule-making procedures. These standards cover the following toxic substances: asbestos, vinyl chloride, 14 carcinogens, and coke oven emissions. Each of these permanent standards can be viewed as a "complete" standard, since each provides not only a specific value for the level of exposure to the toxic substance but also specifications as to monitoring, engineering controls, personal protective equipment, recordkeeping, medical surveillance, and other matters. A summary of these major health standard follows.

5.3.1 Asbestos

On December 7, 1971, the Secretary of Labor published an emergency temporary standard that limited the 8 hr time-weighted average of airborne concentration of asbestos to 5 fibers greater than 5 μm in length per milliliter of air (the "five-fiber standard"). On June 7, 1972, a permanent standard was promulgated that retained the five-fiber standard for 4 years, then required reduction to two fibers (the "two-fiber standard") (106).

In *Industrial Union Department, AFL-CIO* v. *Hodgson* (4) the unions whose members are affected by the health hazards of asbestos dust challenged the timetable established by the permanent standard for achievement of permissible levels of concentration of asbestos and also objected to other portions of the standard concerning methods of compliance, monitoring intervals and techniques, cautionary labels and notices, and medical examinations and records. The Court of Appeals upheld the standard with two exceptions. The court directed the secretary to reconsider the effective date for the two-fiber standard and to determine whether this date might be accelerated for all or some of the industries affected. The court also directed the secretary to review the standard's recordkeeping provision, which required only a 3-year retention period for exposure monitoring records.

In response to the court's order, the OSH Administration initiated a new rule-making proceeding and issued a *proposed* revised standard for asbestos on September 30, 1975. The proposal, which goes beyond the issues the secretary had been directed by the court to consider, would revise the existing standard by reducing the permissible exposure level to 0.5 fiber per cubic centimeter for all employments covered by the act except the construction industry. Other modifications were suggested, as well. A separate proposal will be issued for the construction industry (107).

5.3.2 Vinyl Chloride

On April 5, 1974, the secretary promulgated an emergency temporary standard for vinyl chloride. One of its requirements was that no worker be exposed to concentrations of vinyl chloride in excess of 50 ppm over any 8 hr period. The prior TLV had set an exposure limit of 500 parts vinyl chloride per million parts of air (108).

On October 1, 1974, a permanent standard was promulgated that calls for a permissible exposure limit no greater than one part per million averaged over an 8 hr period. The standard also requires feasible engineering and work practice controls to reduce exposure below the permissible level wherever possible; these practices are to be supplemented by respiratory protection (109).

In *Society of the Plastics Industry* the Court of Appeals for the Second Circuit upheld the standard and it became effective in April 1975 (110).

5.3.3 Carcinogens

The Department of Labor excluded the ACGIH's list of carcinogenic chemicals from its interim standards package, which it promulgated in 1971. After approximately one year of consulting with NIOSH and receiving data and commentary from interested groups, the secretary promulgated emergency temporary standards on May 3, 1973, for a list of 14 chemicals found to be carcinogenic (111). Permanent standards for the 14 carcinogens were issued on January 29, 1974.

In *Synthetic Organic Chemical Manufacturers Association* v. *Brennan* (112) the Third Circuit upheld the standards for 13 of the chemicals except for the provisions pertaining to medical examinations and to laboratory usage of said chemicals. The standard for 4,4 methylene bis(2-chloraniline) (MOCA) was remanded and as of November 1, 1977, had not been reissued.

5.3.4 Coke Oven Emissions

In October 1976, the Secretary of Labor issued a permanent standard regulating workers' exposure to coke oven emissions. The standard defines coke oven emissions as the benzene-soluble fraction of total particulate matter present during the destructive distillation of coal for the production of coke. It limits exposure to 150 μg of benzene-soluble fraction of total particulate matter per cubic meter of air averaged over an 8 hr period. The standard also mandates specific engineering controls and work practices, which are to be in use as soon as possible after the standard becomes effective and no later than January 20, 1980 (113).

The steel industry has challenged the validity of the coke oven standard, and all petitions for review have been consolidated in the Court of Appeals for the Third Circuit (114) and were pending as of December 1, 1977. The Third Circuit vacated an interim stay of certain provisions of the standard (115); thus it remained fully in effect pending the review.

5.3.5 Proposed Standards

The Secretary of Labor announced an emergency temporary standard for workplace exposure to benzene on April 29, 1977. The standard was to have gone into effect May 21, 1977, but it was temporarily stayed on May 20, 1977, by the Court of Appeals for the Fifth Circuit pending review.

A proposed permanent standard for benzene was issued by the secretary on May 27, 1977. Both the emergency standard and the proposed permanent standard would limit workplace exposure to one part of benzene per million parts of air and would require the use of engineering and work practice controls to reduce employee exposure to or below the permissible limit (116).

On September 9, 1977, the secretary issued an emergency temporary standard for exposure to 1,2 dibromo-3-chloropropane (DBCP). The emergency temporary standard set a permissible exposure level for DBCP at 10 parts of DBCP per billion parts of air. A proposed permanent standard for DBCP was issued on November 1, 1977. The proposed permanent standard reduces the permissible exposure to one part DBCP per billion parts of air. The proposal also requires employee exposure measurements, engineering controls, personal protective equipment and clothing, work practices, and so forth (117).

As of November 1, 1977, the OSH Administration was still completing work on permanent standards for several additional occupational health hazards. Final drafts of permanent standards for workplace exposure to lead, cotton dust, arsenic, benzene, beryllium, and nickel were to have been completed by the end of 1977. Public hearings began in May 1977 on a proposed standard for workplace exposure to sulfur dioxide. The OSH Administration has also issued proposed standards for occupational exposure to noise, ammonia, ammonia (anhydrous), toluene, and trichloroethylene. The Stan-

JOB SAFETY AND HEALTH LAW

dards Completion Project has issued proposed standards for six ketones [2-butanone, 2-pentanone, cyclohexanone, hexone, methyl (n-amyl) and ketone] and for 11 other toxic substances as well (118).

On October 3, 1977, the OSH Administration announced a proposal that if implemented, could greatly accelerate the rule-making process for workplace carcinogens. The proposal would treat cancer-causing substances on a generic rather than a substance-by-substance basis.

For purposes of OSHA rule-making four categories of chemical substances were proposed: (1) confirmed carcinogens, (2) suspected carcinogens, (3) substances for which more scientific data would be needed for final classification, and (4) carcinogens that are not currently found in American workplaces.

Classification as a confirmed carcinogen would result in immediate issuance of an emergency temporary standard. The emergency temporary standard would follow the format and content of the model emergency temporary standard set forth in the proposal. Within 60 days of classifying a substance as a confirmed carcinogen, the secretary would be required to issue a notice of proposed permanent rule making; the proposed permanent rule would follow the format and content of a model permanent rule.

A classification of "suspected" carcinogen would result in the initiation of permanent rule making for the substance within 60 days of the secretary's announcement of his intent to so classify the substance. The proposed rule would follow the format and content of still another model standard set forth in the proposal. In the case of substances classified in the third category, OSHA would notify other federal agencies and would request any information that could lead to a future reclassification. Substances classified in the fourth category would be subject to regulation if they were ever introduced into American workplaces.

The OSH Administration explained that once an overall generic policy is adopted, rule making for individual substances would be limited to questions such as the following - Is the substance's classification correct? What is the lowest feasible exposure level for the substance? What would be the environmental impact of the standard? Officials at OSHA projected that the proposal of October 3, 1977, could be adopted in less than one year (119).

5.3.6 Mine Industry Health Standards

The Secretary of Interior has promulgated health standards pursuant to the Coal Mine Health and Safety Act of 1969 and the Metal and Non-Metallic Mine Safety Act of 1966 focusing primarily on the following matters: (1) concentration of respirable dust in the coal mine atmosphere, (2) noise, (3) air contaminants; and (4) radiation (underground metallic and nonmetallic mines only). According to Section 301(b)(1) of the Federal Mine Safety and Health Act of 1977 (see Section 3.1, this chapter) such mandatory standards issued by the Secretary of Interior that are in effect on the date of enactment of the 1977 act are to remain in effect as mandatory health standards applicable to metal and nonmetallic mines and to coal or other mines, respectively, under the 1977 act until the Secretary of Labor has issued new health standards applicable to such mines or revised mandatory standards (120).

6 EMPLOYEE RIGHTS AND DUTIES UNDER JOB SAFETY AND HEALTH LAWS

6.1 Employee Duties

Section 5(b) of OSHA requires employees to "comply with occupational safety and health standards and all rules, regulations, and orders issued pursuant" to the act that are applicable to their own actions and conduct.

The act does not, however, expressly authorize the Secretary of Labor to sanction employees who disregard safety standards and other applicable orders. The United States Court of Appeals for the Third Circuit has held that although Section 5(b) would be devoid of content if not enforceable, Congress did not intend to confer on the secretary or the commission the power to sanction employees (121).

Section 110(g) of the Federal Mine Safety and Health Act of 1977 not only requires mine employees to comply with health and safety standards promulgated under that statute, it also authorizes the imposition of civil penalties on miners who "willfully violate the mandatory safety standards relating to smoking or the carrying of smoking materials, matches, or lighters".

6.2 Employee Rights

The Occupational Safety and Health Act grants numerous rights to employees and/or their authorized representatives. The most fundamental employee right is the right set forth in Section 5(a) to a safe and healthful employment and place of employment. Other significant rights of employees and/or their authorized representatives include the following:

1. The right to request a physical inspection of a workplace and to notify the Secretary of Labor of any violations that employees have reason to believe exist in the workplace [§ 8(f)].
2. The right to accompany the secretary during the physical inspection of a workplace [§ 8(a)].
3. The right to challenge the period of time fixed in a citation for abatement of a violation of the act and an opportunity to participate as a party in hearings relating to citations [§ 10(c)].
4. The right to be notified of possible imminent danger situations and the right to file an action to compel the secretary to seek relief in such situations if he has "arbitrarily and capriciously" failed to do so [§ 13(c)].

5. Various rights, including the right to notice, regarding an employer's application for either a temporary or permanent variance from an OSHA standard [§ 6(b)(6) and § 6(d)].

6. The right to observe monitoring of employee exposures to potentially toxic substances or harmful physical agents, the right of access to records thereof, and the right to be notified promptly of exposures to such substances in concentrations which exceed those prescribed in a standard [§ 8(c)].

7. The right to petition a court of appeals to review an OSHA standard within 60 days of its issuance [§ 6(f)].

Section 11(c) of OSHA makes it unlawful for any person to discharge or in any manner discriminate against an employee because the employee has exercised his rights under the act. This provision is designed to encourage employee participation in the enforcement of OSHA standards (122).

Employees who believe they have been discriminated against in violation of Section 11(c) must file a complaint with the Secretary of Labor within 30 days after such violation has occurred. The secretary will investigate the complaint, and if he determines that Section 11(c) has been violated, he is authorized to bring an action in federal district court for an order restraining the violation and for recovery of all appropriate relief, including rehiring or reinstatement of the employee to his former position with back pay. The statute authorizes only the Secretary of Labor to bring an action for violation of Section 11(c) (123).

The Federal Mine Safety and Health Act of 1977 also contains a broad anti-retaliation provision and grants other rights to mine employees as well (124).

An employee has no explicit right, under OSHA, to refuse a work assignment because of what he feels is a dangerous working condition. The Secretary of Labor, however, has issued an administrative regulation that interprets the act as implying such a right under certain limited circumstances (125). As of December 1, 1977, the Fifth Circuit Court of Appeals and at least three federal district courts had declared this regulation invalid (126). The regulation has been upheld by at least one federal district court (127).

It is fairly well settled that OSHA does not create a private right of action for damages suffered by an employee as the result of an employer's violation of the act [see, e.g., *Skidmore* v. *Travelers Insurance Co.* (128)]. Finally, Section 4 (b)(4) states that the act does not supersede or in any manner affect any worker's compensation law.

ACKNOWLEDGMENT

We acknowledge our gratitude for the research assistance provided by Murphy Goodman, a third year student at the University of Pittsburgh School of Law.

NOTES AND REFERENCES

1. See Bureau of National Affairs (BNA), *The Job Safety and Health Act of 1970*, Washington, D.C., 1971, p. 13. The Occupational Safety and Health Act is codified in Title 29 of the United States Code (USC), sections 651–678. References to the act in this chapter and the next are to the appropriate section of the statute itself and do not include a corresponding citation to the United States Code.
2. N. Ashford, *Crisis in the Workplace*, M.I.T. Press, Cambridge, Mass., 1976, pp. 141, 236–237.
3. See American Bar Association, *Report of the Committee on Occupational Safety and Health Law*, ABA Press, Chicago, 1975, p. 107.
4. For a discussion of the weight to be accorded by the Secretary of Labor to the NIOSH recommendations, see *Industrial Union Department, AFL-CIO* v. *Hodgson*, 499 F.2d 467, 476–477; 1 OSHC 1631 (D.C. Cir. 1974). [References to "F.2d" designate the volume (e.g., 499) and page (e.g. 467) of the *Federal Reporter*, Second Series, which contains the official reported decisions of the United States Courts of Appeals as published by the West Publishing Company. References to "OSHC" designate the same decision as reported in the "Occupational Safety and Health Cases" published by the Bureau of National Affairs (BNA).]
5. See Ashford, *Crisis in the Workplace* (2), pp. 249–251. Section 27(b) of the act established a National Commission on State Workmen's Compensation Laws, which was directed to study and evaluate such laws to determine whether they provide an adequate, prompt, and equitable system of compensation for injury or death arising out of or in the course of employment. The commission's tasks were completed in July 1972, and the commission was disbanded. For a description of the activities of the commission and an evaluation of its work, see Ashford, *Crisis in the Workplace*, pp.246, 289–292.
6. Numerous proposals have been introduced in Congress throughout the 1970s to amend the scope of OSHA's coverage. As of November 1, 1977, it seemed highly unlikely that any such amendments would be enacted in the near future. Several bills have been introduced to repeal OSHA altogether. A less drastic proposal is contained in a bill to exempt from coverage employers of 10 or fewer employees. This bill was introduced by Congressman Joe Skubitz and 100 cosponsors on March 22, 1977. The Skubitz bill (HR 5364) would also change the definition of employer to include only workers who had been employed for a period of at least 30 days. A few bills have also been introduced to extend OSHA's coverage. For example, a proposal (HR 1335) supported by Congressman Steiger, one of the principal sponsors of the 1970 act, would provide for federal authority over state and local government employees where an imminent danger or pattern or practice of unsafe or unhealthful working conditions exists. Congressman Pressler has introduced a bill (HR 4215) that would extend OSHA's coverage to Congress, the federal agencies, and the federal court system.
7. *Southern Pacific Transportation Co.* v. *Usery and OSAHRC*, 539 F.2d 386; 4 OSHC 1693 (5th Cir. 1976), *certiorari* denied, 5 OSHC 1888 (1977); *Southern Railway Company* v. *OSAHRC and Brennan*, 539 F.2d 335; 3 OSHC 1940 (4th Cir. 1976); *Baltimore & Ohio Railroad Co.* v. *OSAHRC*, 548 F.2d 1052; 4 OSHC 1917 (D.C. Cir. 1976).
8. 4 OSHC at 1696.
9. *Fineberg Packing Co.*, 1 OSHC 1598 (Rev. Comm. 1974).
10. 520 F.2d 1161; 3 OSHC 1566, 1572 (D.C. Cir. 1975).
11. American Bar Association, *Report of the Committee on Occupational Safety and Health Law*, ABA Press, Chicago, 1976, pp. 247–248.
12. 30 USC §§ 721–740.
13. 30 USC §§ 801–960.
14. See Bureau of National Affairs, *Occupational Safety and Health Reporter*, Current Report for November 10, 1977, at p. 827. [References to this source are hereafter cited as "BNA, *OSHR.*"] Pursuant to the 1977 amendments, the modified Coal Mine Health and Safety Act of 1969 is now to be cited

as the "Federal Mine Safety and Health Act of 1977." References to specific provisions of the mine safety and health legislation and standards promulgated pursuant thereto are made throughout this chapter and the next.

15. The FEPCA is codified at 7 USC §§ 136 et seq. It is a comprehensive revision of the Federal Insecticide, Fungicide and Rodenticide Act of 1970, 7 USC §§ 135 et seq. (1970). The regulations promulgated by EPA to protect farm workers from toxic exposure to pesticides are found in title 40 of the Code of Federal Regulations at §§ 170.1 et seq. [The Code of Federal Regulations is hereafter cited "CFR"].

16. The Secretary of Labor has acknowledged that under the Railway Safety Act the Department of Transportation (DOT) has *authority* to regulate all areas of employee safety for the railway industry. *Southern Railway Company* v. *OSAHRC* (Reference 7; 3 OSHC at 1941). However the scope of DOT's statutory authority to regulate matters relating to worker health is still unsettled. *Southern Pacific Transportation Company v. Usery* (Reference 7; 4 OSHC at 1694, n. 3).

    The Railway Safety Act is codified at 45 USC §§ 421 et seq. Other federal statutes dealing with railway safety include the Safety Appliance Acts, 45 USC §§ 1–14; the Signal Inspection Act, 45 USC § 26; the Train, Brakes Safety Appliance Act, 45 USC § 9; the Hours of Service Act, 45 USC §§ 61 et seq; and the Rail Passenger Safety Act, 45 USC §§ 502 et seq. Department of Transportation regulations relating to railway safety can be found in 49 CFR, Chapter II.

    The Department of Transportation also has statutory authority to regulate safety in modes of transportation other than rail. For example, Section 304(a)(1) of Part II of the Interstate Commerce Act states that the Interstate Commerce Commission has the duty to regulate common carriers by motor vehicle and to that end may establish reasonable requirements with respect to "continuous and adequate service . . . and safety and operation of equipment." The Department of Transportation Act transferred all powers and functions under Section 304(a)(1) to the Secretary of Transportation. See 49 USC § 1655(e)(6)(C). For additional areas of DOT jurisdiction, see Sections 3.4, 3.5 and 3.6 of this chapter.

17. See Section 2 of this chapter.

18. The provisions of the statute regarding safety regulation of civil aeronautics are codified at 49 USC §§ 1421 et seq.

19. See decision discussed in *Usery* v. *Northwest Orient Airlines, Inc.*, 5 OSHC 1617 (E.D.N.Y. 1977). ["E.D.N.Y." refers to the federal district court in the Eastern District of New York. Reported decisions of federal district courts dealing with job safety and health matters are reported in BNA's *Occupational Safety and Health Cases (OSHC)* and many are also reported in West Publishing Company's Federal Supplement (F. Supp.).] The administrative law judge determined that the FAA had exercised its authority by requiring each air carrier to maintain a maintenance manual that must include all instructions and information necessary for its ground maintenance crews to perform their duties and responsibilities with a high degree of safety.

20. The HMTA is codified at 49 USC §§ 1801 et seq. A table of materials that have been designated as hazardous by the secretary can be found at 49 CFR Part 172. The secretary's regulations prescribe the requirements for shipping papers, package marking, labeling, and transport vehicle placarding applicable to the shipment and transportation of those hazardous materials.

21. 3 OSHC 1601 (Rev. Comm. 1975).

22. The Secretary of Transportation has exercised statutory authority under NGPSA to promulgate safety standards for employees at natural gas facilities. These regulations can be found at 49 CFR Part 190. The NGPSA is codified at 49 USC §§ 1671 et seq.

23. There also exist thousands of state and local laws regulating noise, discussion of which is beyond the scope of this work. See *Compilation of State and Local Ordinances on Noise Control,* 115 Cong. Rec. 32178 (1969).

24. See 49 USC § 1431. In carrying out this task the Administrator of the FAA has adopted and published aircraft noise standards and regulations. See 14 CFR Part 36 (1977). Section 7 of the Noise Control Act of 1972, discussed below, amended the noise abatement and control provision of the Federal Aviation

Act to provide generally that standards adopted with respect to aircraft noise must have the prior approval of the Administrator of the EPA.

25. The Noise Control Act is codified at 42 USC §§ 4901 et seq. The products with which the administrator is statutorily authorized to deal must come from among three categories: construction equipment, transportation equipment (including recreational vehicles, motors, or engines, including any equipment of which an engine or motor is part), and electrical or electronic equipment.

    To date, the administrator has identified portable air compressors as major sources of construction equipment noise and has published noise emission regulations. See 40 CFR Part 204, Subpart B. He has similarly published noise regulations for railway locomotives and cars (40 CFR Part 201) and for heavy and medium trucks (40 CFR Part 205, Subpart B). On July 11, 1977 (42 *Federal Register* 35803), the Administrator proposed noise emission regulations for additional construction equipment (wheel loaders, crawler tractors, and wheel tractors). [The *Federal Register* is hereafter cited as *Fed. Reg.*]

    Pursuant to Section 8 of the Noise Control Act of 1972, the administrator has published proposed labeling requirements for labeling of products that produce noise (42 *Fed. Reg.* 31722) and labeling requirements to accompany products that are designed to reduce noise (42 *Fed. Reg.* 37130).

26. The Toxic Substances Control Act of 1976 is codified at 15 USC §§ 2601 et seq. Prior to 1976 there existed no enactment that authorized the federal government to regulate toxic substances generally. Although regulations published under OSHA did govern worker exposure to toxic substances in the workplace (see Section 5.3, this chapter), federal law lacked a general authority to prohibit or restrict the manufacture of such substances.

27. Section 7 of the act empowers the Administrator of EPA to commence a civil action in federal court for the purpose of seizing an "imminently hazardous substance," defined in the act as one that "presents an imminent and unreasonable risk of serious or widespread injury to health or the environment."

28. See Section 6 (e) of the act and 42 *Fed. Reg.* 26,564 (May 24, 1977).

29. The Consumer Product Safety Act is codified at 15 USC §§ 2051 et seq. The act was amended by the Consumer Product Safety Commission Improvements Act of 1976.

30. Provision is made in the act for suit by "any person who shall sustain injury by reason of any knowing (including willful) violation of a consumer product safety rule or any other rule or order issued by the Commission" against the responsible party in a federal district court. This right to sue is in addition to existing common law, federal, and state remedies.

    The regulations published by the Consumer Product Safety Commission can be found at 16 CFR Chapter II, Subchapters A and B.

31. The Hazardous Substances Act is codified at 15 USC §§ 1261–1274. Regulations published under this statute can be found at 16 CFR Subchapter C, Part 1500.

32. The Atomic Energy Act is codified at 42 USC §§ 2011 et seq. The Nuclear Regulatory Commission, an independent executive commission, was created by the Energy Reorganization Act of 1974, 42 USC § 5841(a), and all licensing and related regulatory functions of the Atomic Energy Commission were then transferred to the NRC. See 42 USC § 5841(f) and (g). The NRC's "Standards for Protection Against Radiation" can be found at 10 CFR Part 20.

33. The permissible dosage per calendar quarter within such a "restricted area" is 1.25 rems to the whole body, head and trunk, active bloodforming organs, lens of the eyes, or gonads; 18.75 rems to hands and forearms, feet and ankles; 7.5 rems to the skin of the whole body. Dosage standards are also set forth for the inhalation of radioactive substances. See 10 CFR § 20.103. Detailed personnel monitoring and reporting requirements are also included in the regulations.

34. See 10 CFR §§ 20.105, 20.106, and 20.303.

35. See 40 CFR Part 190.

36. The Meat Inspection Act is codified at 21 USC §§ 601 et seq. The relevant regulations can be found at 9 CFR § 318.19.

JOB SAFETY AND HEALTH LAW 711

37. See 30 CFR § 57.5 (1977). The Metal and Non-Metallic Mine Safety Act was repealed by the Federal Mine Safety and Health Amendments Act of 1977 (see Section 3.1, this chapter), but the mandatory standards relating to mines issued by the Secretary of Interior pursuant to that statute, which are in effect on the date of enactment of the 1977 act, are to remain in effect as mandatory health or safety standards applicable to metal and nonmetallic mines until such time as the Secretary of Labor shall issue new or revised mandatory health or safety standards applicable to metal and nonmetallic mines. Conference Report on Senate Bill 717, Federal Mine Safety and Health Amendments Act of 1977, 123 *Cong. Rec.* H 10481, H 10490 (daily ed., October 3, 1977).
38. The Walsh-Healy Act is codified at 41 USC §§ 35 *et seq.* The relevant regulations can be found in 41 CFR § 50-204. These standards were later promulgated by the Secretary of Labor as established federal standards under Section 6(a) of OSHA. See Section 5.2.1, this chapter.
39. That statute amended the Public Health Safety Act and is codified at 42 USC § 263. The regulations promulgated by the Food and Drug Administration under this statute can be found at 21 CFR Subchapter J (Radiological Health). See also the standard applicable to diagnostic X-ray systems at 42 USC § 201.
40. The Outer Continental Shelf Lands Act is codified at 43 USC §§ 1331 *et seq.*
41. See Ashford, *Crisis in the Workplace* (2), pp. 47–51.
42. See D. Currie, "OSHA," *Am. Bar Found. Res. J.*, 1976, 1107, 1111 (1976). Section 18(a) of OSHA explicitly directs, however, that the states may assert jurisdiction under state law with respect to occupational safety or health issues for which no federal standard is in effect. See Section 4.1, this chapter.
43. A so-called complete plan is described in an administrative regulation issued by the Secretary of Labor and codified at 29 CFR § 1902.2(a).
44. Pursuant to Section 18(c), the state plan must:

   a. Designate a State agency or agencies as the agency or agencies responsible for administering the plan throughout the state.
   b. Provide for the development and enforcement of safety and health standards relating to one or more safety or health issues, which standards (and the enforcement of which standards) are or will be at least as effective in providing safe and healthful employment and places of employment as the standards promulgated under Section 6 of OSHA which relate to the same issues.
   c. Provide for a right of entry and inspection of all workplaces subject to this chapter which is at least as effective as that provided in Section 8 of OSHA and include a prohibition on advance notice of inspections.
   d. Contain satisfactory assurances that such agency or agencies have or will have the legal authority and qualified personnel necessary for the enforcement of such standards.
   e. Give satisfactory assurances that such state will devote adequate funds to the administration and enforcement of such standards.
   f. Contain satisfactory assurances that such state will, to the extent permitted by its law, establish and maintain an effective and comprehensive occupational safety and health program applicable to all employees of public agencies of the state and its political subdivisions, which program is as effective as the standards contained in an approved plan.
   g. Require employers in the state to make reports to the Secretary in the same manner and to the same extent as if the plan were not in effect.
   h. Provide that the state agency will make such reports to the Secretary in such form and containing such information as the Secretary shall from time to time require.

   The additional criteria outlined in the secretary's regulations can be found at 29 CFR §§ 1902.3 and 1902.4.
45. CFR § 1902.2(b).
46. CFR § 1954.3(b).

47. The commission has held that OSHA is not precluded from exercising its own enforcement authority during this period. *Par Construction Co., Inc.,* 4 OSHC 1779 (Rev. Comm. 1976); *Winston-Salem Southbound Railway Co.,* 3 OSHC 1767 (Rev. Comm. 1975).
48. All the approved and/or certified state plans have become operational except for those of Arizona, Indiana, Michigan, Nevada, New Mexico, Virgin Islands, Virginia, and Wyoming. Although Michigan has not signed an operational agreement with OSHA, Michigan is apparently being permitted to enforce its state standards as specified in its developmental plan. (Interview with staff member in the OSH Administration's national office, Division of State Plans, August 17, 1977.)
49. A chart on the status of state plans can be found in BNA, *OSHR,* Reference File at 81:1003.
50. See P. Hamlar, "Operation and Effect of State Plans," in: *Proceedings of the American Bar Institute on Occupational Safety and Health Law,* ABA Press, Chicago, 1976, pp. 42–45.
51. See Sections 506(a) and 506(b) of the Federal Mine Safety and Health Act of 1977.
52. See Section 205 of the Railway Safety Act.
53. There is language in the legislative history of OSHA to indicate that the general duty clause merely restates the employer's common law duty to exercise reasonable care in providing a safe place for his employees to work. However the courts have generally characterized such statements as "misleading." For example, in *REA Express* v. *Brennan,* 495 F.2d 822, 825; 1 OSHC 1651 (2d Cir. 1974), the Second Circuit could not "accept the proposition that common law defenses such as assumption of the risk or contributory negligence will exculpate the employer who is charged with violating the Act."
54. *National Realty and Construction Company, Inc.* v. *OSAHRC,* 489 F.2d 1257; 1 OSHC 1422 (D.C. Cir. 1973); *Brennan* v. *OSAHRC and Canrad Precision Industries,* 502 F.2d 946; 2 OSHC 1137 (3rd Cir. 1974).
55. 489 F.2d 1266–1267; see also *Brennan* v. *OSAHRC and Canrad Precision Industries* (54), *Brennan* v. *OSAHRC and Republic Creosoting Company,* 501 F.2d 1196; 2 OSHC 1109, 1111–1112 (7th Cir. 1974). See also the discussion in Section 5.2.3, this chapter.
56. *Brennan* v. *OSAHRC and Vy Lactos Laboratories,* 494 F. 2d 460, 464; 1 OSHC 1623 (8th Cir. 1974).
57. American Bar Association, *1975 Report of the Committee on Occupational Safety and Health Law* (3), p. 11.
58. American Bar Association, *1975 Report of the Committee on Occupational Safety and Health Law* (3), p. 12.
59. 501 F. 2d 504; 2 OSHC 1041 (8th Cir. 1974).
60. *Brennan* v. *OSAHRC and Vy Lactos Laboratories* (56), 1 OSHC at 1624; R. Morey, "The General Duty Clause of the Occupational Safety and Health Act of 1970," *Harv. Law Rev.,* 86, 988, 991 (1973). The same is true for violations of the specific duty clause.
61. *National Realty, supra* [(54): 489 F.2d at 1265, n. 33]. "[T]he 'likely to cause' test should be whether reasonably foreseeable circumstances could lead to the perceived hazard's resulting in serious physical harm or death—or more simply, the proper test is plausibility, not probability." Morey, "The General Duty Clause of the Occupational Safety and Health Act of 1970" (60, pp. 997–998).
62. 29 CFR § 1910.5 (f).
63. An "established Federal standard" is defined in Section 3(10) of the act as "any operative occupational safety and health standard established by any agency of the United States and presently in effect, or contained in any Act of Congress in force on December 29, 1970." Section 4(b)(2) of the act listed several federal statutes from which established federal standards were to be derived, including the Walsh-Healy Act, 41 USC §§ 35–45, the Service Contract Act of 1965, 41 USC §§ 351–357, and the National Foundation on Arts and Humanities Act, 20 USC §§ 951–960.

A "national consensus" standard is defined in Section 3(9) of the act as any occupational safety and health standard which "(1) has been adopted and promulgated by a nationally recognized standards-producing organization under procedures whereby it can be determined by the Secretary that persons interested and affected by the scope or provisions of the standard have reached substantial agreement on its adoption, (2) was formulated in a manner which afforded an opportunity for diverse views to be considered, and (3) has been designated as such a standard by the Secretary, after consultation with other appropriate Federal agencies." The standards meeting these criteria were those of the American National Standards Institute (ANSI) and the National Fire Protection Association. *American Federation of Labor Etc.* v. *Brennan*, 530 F.2d 109, 111, 3 OSHC 1820, n. 2 (3rd Cir. 1975).

64. The Administrative Procedure Act was enacted by Congress in 1946 to impose some coherent system of procedural regularity on the growing regulatory bureaucracy of the federal government. It provides procedures for administrative "rule making" and administrative "adjudication," among other things. See generally H. Linde and G. Bunn, *Legislative and Administrative Processes*, Foundation Press, Mineola, N.Y., 1976, p. 814. The act is codified in 5 USC §§ 551 *et seq.*

65. *The Job Safety and Health Act of 1970*, BNA, Washington, D.C., 1971, p. 23.

66. See *Florida Peach Growers Association* v. *U.S. Department of Labor*, 489 F.2d 120, 124; 1 OSHC 1472 (5th Cir. 1974) and Sections 6(b)(1) to 6(b)(4) of OSHA. In *National Congress of Hispanic American Citizens* v. *Usery*, 554 F.2d 1196; 5 OSHC 1255 (D.C. Cir. 1977), the Court of Appeals for the District of Columbia Circuit held that the statutory deadlines in Sections 6(b)(1) to 6(b)(4) for the promulgation of permanent standards were discretionary rather than mandatory as long as the secretary's exercise of discretion was honest and fair.

67. *Florida Peach Growers Association, supra,* 489 F.2d at 124.

68. The "substantial evidence" standard of judicial review is traditionally conceived of as suited to adjudication or formal rulemaking. OSHA, however, calls for informal rule making which under the Administrative Procedure Act generally entails judicial review pursuant to the less stringent "arbitrary and capricious" test. This apparent anomaly can be explained historically as a legislative compromise. The Senate OSHA bill called for informal rule making, but the House version specified formal rule making and substantial evidence review. The House receded on the procedure for promulgating standards, but the substantial evidence standard of review was adopted. *Industrial Union Department, AFL-CIO* v. *Hodgson (4),* 499 F.2d at 473. For a more detailed discussion of these legislative events, see *Associated Industries of New York State, Inc.* v. *U.S. Department of Labor,* 487 F.2d 342; 1 OSHC 1340 (2d Cir. 1973).

69. See B. Fellner and D. Savelson, "Review by the Commission and the Courts," in: *Proceedings of the American Bar Association Institute on Occupational Safety and Health Law*, ABA Press, Chicago, 1976, pp. 113–114.

70. *Atlantic & Gulf Stevedores, Inc.* v. *OSAHRC,* 534 F.2d 541; 4 OSHC 1061 (3d Cir. 1976); *Arkansas–Best Freight System, Inc.* v. *OSAHRC,* 529 F.2d 649; 3 OSHC 1910 (8th Cir. 1976).

71. *Atlantic & Gulf Stevedores, supra,* at 551–552.

72. 530 F.2d 109; 3 OSHC 1820 (3rd Cir. 1975).

73. 530 F.2d at 126.

74. Accord: *Society of the Plastics Industry, Inc.* v. *OSHA,* 509 F.2d 1301; 2 OSHC 1496 (2d Cir. 1974), *certiorari* denied, 421 U.S. 992 (1975). ["U.S." refers to the official reported orders and decisions of the United States Supreme Court. Such orders and decisions are also reported by West Publishing Company's *Supreme Court Reporter,* hereafter cited as "S.Ct."]

The Third Circuit in *American Federation of Labor (72)* admitted that there are industrial activities involving hazards so great, yet offering so little social utility, that the secretary would be justified in concluding that their total prohibition is proper if there is no technologically feasible method of eliminating the operational hazard. "But although Congress gave the Secretary license to make such a determination in specific instances, it did not direct him to do so in every instance where total elimination of risk is beyond the reach of present technology." 530 F.2d at 121.

75. 530 F.2d at 123.

76. BNA, *OSHR*, Reference File at 21:5101. The Executive Order was due to expire in December 1976, but it was extended by President Ford until December 31, 1977, by Executive Order 11949, which also changed the title of the order from "inflation impact statement" to "economic impact statement." On November 18, 1977, President Carter published (42 *Fed. Reg.* 59740) a proposed Executive Order that would require federal regulatory agencies to prepare a regulatory analysis of proposed agency actions. The regulatory analysis called for in Carter's proposal would supersede the requirements of Executive Orders 11821 and 11949. According to the Carter proposal, each such regulatory analysis shall contain "a succinct statement dealing with the problem considered by the agency; an analysis of the economic consequences of the proposed regulation and major alternative approaches considered; and an assessment of the approach selected in relation to other alternatives." See also BNA, *OSHR*, Current Report for November 24, 1977, pp. 875, 901.

As of November 1, 1977, a lawsuit still was pending in federal district court in the District of Columbia challenging the legality of considering economic impact in the preparation of standards. *Oil, Chemical and Atomic Workers* v. *Marshall* (Civil Action No. 76-0365). For a description of the lawsuit, see BNA, *OSHR*, Current Report for March 11, 1976, p. 1331. Congressional proposals were pending as of that date to incorporate the requirement of economic impact statements into the statute.

77. *Atlantic & Gulf Stevedores, supra*, at 548.
78. In *Turner Company* v. *Secretary of Labor and OSAHRC*, 5 OSHC 1790 (7th Cir. 1977), the United States Court of Appeals for the Seventh Circuit explained that the commission must consider realistically the hazards presented by the excessive noise in the company's plant and determine whether the health benefits to employees who are already equipped with personal protection equipment (ear plugs) justify the cost to the company for engineering controls. For commission decisions, see, for example, *Secretary* v. *Continental Can Company*, 4 OSHC 1541 (Rev. Comm. 1976) (appeal pending in 9th Cir., No. 76-3229); *Secretary* v. *West Point Pepperell, Inc.*, 5 OSHC 1257 (Rev. Comm. 1977).
79. Ashford, *Crisis in the Workplace* (2), p. 169.
80. 511 F.2d 1139; 2 OSHC 1646 (9th Cir. 1975).
81. 511 F.2d at 1144–1145.
82. 520 F.2d 1011; 3 OSHC 1461 (7th Cir. 1975).
83. *Atlantic & Gulf Stevedores, supra*, at 555.
84. See Note, "Employee Noncompliance with OSHA Safety Standards," *Harv. Law Rev.*, **90,** 1041 (1977).
85. 528 F.2d 564, at 570; 3 OSHC 2060 (5th Cir. 1976).
86. For an administrative decision outlining the elements of an employer's defense based on employee misconduct under the California Occupational Safety and Health Act, see *Moore Production Service* (Docket No. 667-76) (summarized in BNA, *OSHR*, Current Report for May 5, 1977, at p. 1504).
87. 29 USC § 4332.
88. As of March 31, 1977, the secretary had received 1015 requests for Section 6(d) variances.
89. Variances had been granted to 74 applicants as of March 31, 1977. See BNA, *OSHR*, Current Report for May 19, 1977, at p. 1560.
90. Health and safety standards promulgated by the Secretary of Labor pursuant to OSHA can be found at 29 CFR Part 1910. A standards digest (OSHA Publication 2201) outlining the basic applicable standards is published in BNA, *OSHR*, Reference File at 31:4001.

Federal health and safety standards for the construction industry were initially promulgated under the Contract Work Hours and Safety Standards Act, 40 USC §§ 333 *et seq.* These standards were incorporated by reference under OSHA, are enforceable under both laws, and can be found at 29 CFR Part 1926. A standards digest (OSHA Publication 2202) outlining the basic applicable construction standards is published in BNA, *OSHR*, Reference File at 31:3001. Health and safety standards for ship repairing, shipbuilding, shipbreaking, and longshoring were initially promulgated pursuant to the Longshoremen's and Harbor Workers' Compensation Act, 33 USC §§ 901 *et seq.* These standards were incorporated by reference by OSHA, are enforceable under both laws, and can be found in 29 CFR Parts 1915-1918. Health and safety standards originally promulgated under the Walsh-Healy Public Contracts Act, the McNamara-O'Hara Service Contract Act of 1965, and the National Foundation on the Arts and Humanities Act of 1965, can be found in 41 CFR Part 50-204. The majority of these standards were also adopted as interim OSHA standards and are enforceable pursuant to that act. Standards promulgated under the aforementioned statutes will be superseded if corresponding standards that are promulgated under OSHA are determined by the Secretary of Labor to be more effective. See Section 4(b)(2) of OSHA.

Federal health and safety standards for coal mines were promulgated by the Department of Interior pursuant to the Federal Coal Mine Health and Safety Act of 1969. Health standards for underground coal mines can be found in 30 CFR Part 70; health standards for surface work areas of underground coal mines and surface coal mines are codified in 30 CFR Part 71. The safety standards for underground coal mines can be found in 30 CFR Part 75, and the safety standards for surface coal mines and surface work areas of underground coal mines are codified in 30 CFR Part 77. The Secretary of Interior promulgated health and safety standards for metal and nonmetallic mines pursuant to the federal Metal and Non-Metallic Mine Safety Act of 1966. The standards for open pit mines, for sand, gravel, and crushed stone operations, and for metal and nonmetallic underground mines can be found at 30 CFR Parts 55, 56, and 57 (1977), respectively. According to Section 301(b)(1) of the Federal Mine Safety and Health Act of 1977, (see Section 3.1, this chapter), such mandatory health and safety standards issued by the Secretary of Interior that are in effect on the date of enactment of the 1977 act are to remain in effect as mandatory standards applicable to metal and nonmetallic mines and to coal or other mines, respectively, under the 1977 act until such time as the Secretary of Labor shall issue new or revised mandatory health or safety standards applicable to such mines.

91. Safety standards generally focus on the time that an employee is actually working. The harm created by a safety hazard is generally immediate and violent. An occupational health hazard, on the other hand, is slow acting, cumulative, irreversible, and complicated by nonoccupational factors. Ashford, *Crisis in the Workplace* (2), pp. 68-83.

92. For a listing of the initial package of national consensus and established federal standards, see 36 *Fed. Reg.* 10466-10714. Since the effective date of OSHA, the secretary has promulgated safety standards under the permanent rule-making procedures of Section 6(b) with respect to the following matters: commercial diving operations, 29 CFR §§ 1910.401-441; agricultural operations, 29 CFR § 1928; helicopter operations, 29 CFR § 1910.183; telecommunications operations, 29 CFR § 1910.268; slings, 29 CFR § 1910.184; and ground-fault circuit interrupters, 29 CFR § 1910.309(c). The final standard for commercial diving operations was promulgated on July 22, 1977, and was scheduled to become effective October 20, 1977. The diving standard actually establishes mandatory safety *and health* requirements for all commercial diving operations within OSHA jurisdiction. A lawsuit challenging the standard was filed by the diving industry in the Fifth Circuit Court of Appeals on September 16, 1977. *Taylor Diving and Salvage Company et al.* v. *U.S. Department of Labor* (No. 77-2875). See BNA, *OSHR,* Current Report for September 29, 1977, p. 527.

93. Industries covered by specific OSHA regulations include pulp, paper, paperboard mills, textiles, bakery equipment, laundry machinery and operations, sawmills, pulpwood logging, telecommunications, agriculture, commercial diving, construction, ship repairing, shipbuilding, shipbreaking, and longshoring. The mining industry is subject to comprehensive safety regulations issued pursuant to the Federal Mine Safety and Health Act of 1977 (formerly the Coal Mine Health and Safety Act of 1969 and the Metal and Non-Metallic Mine Safety Act of 1966). See also Section 3, this chapter.

94. 42 *Fed. Reg.* 62734 (December 13, 1977).

95. See BNA, *OSHR,* Current Reports for June 9, 1977, June 16, 1977, June 23, 1977, November 17, 1977, and December 8, 1977 at pp. 54, 76, 107, 844, and 947, respectively. The safety standards revi-

sion projects will reportedly encompass a review and revision of the safety standards that account for almost 50 percent of the complaints and criticisms that OSHA has been receiving from industry on its safety standards.

96. See Ashford, *Crisis in the Workplace* (2), p. 154. The Secretary of Labor did not include the ACGIH's carcinogen standards in his Section 6(a) package but instead preferred to develop his own standards regarding carcinogens. Ashford, pp. 154, 247–248.
97. See 29 CFR § 1910.1499.
98. The TLVs that have been developed by the ACGIH and the ANSI and promulgated by OSHA can be found at 29 CFR § 1910.1000, Tables Z-1, Z-2, and Z-3.
99. U.S. Congress, Committee on Governmental Operations, "Chemical Dangers in the Workplace," House Report 1688, 94th Congress, 2d Session, 1976, pp. 15–16. In addition, see Ashford, *Crisis in the Workplace* (2), pp. 295–296.
100. For example, OSHA's vinyl chloride standard (see Section 5.3, this chapter) requires employers to undertake a program of initial monitoring and measurement to determine whether there is any employee exposure to vinyl chloride (without regard to the use of respirators) in excess of the action level specified in the standard. If there is such exposure, the employer must establish a program for determining exposures for each such employee. The standard specifies how frequently those employees must be monitored. The standard also (1) requires monitoring when the employer has made a change in production, process, or control that may result in an increase in release of vinyl chloride; (2) specifies the required accuracy of the method of monitoring and measurement; and (3) gives employees the right to observe the monitoring and measuring. 29 CFR § 1910.1017(d). Compare OSHA's coke oven emissions standard, which requires employers to notify each employee in writing of the exposure measurements that represent that employee's exposure and, if the exposure exceeds the permissible exposure limit, to so notify the employee and to inform the employee of the corrective action being taken to reduce exposure to or below the permissible exposure limit. 29 CFR § 1910.1029(e).
101. Labeling requirements in standards that have been promulgated under the permanent rule-making provisions of Section 6(b) generally tend to specify the language to be used in labels or other visual warning devices as well as the locations where such labels or other devices should be placed. For example, see the labeling requirements for OSHA's vinyl chloride standard, 29 CFR § 1910.1017(1).
102. See BNA, *OSHR,* Current Report for April 7, 1977, p. 1392.
103. The noise standard was promulgated as an established federal standard under Section 6(a) and can be found at 29 CFR § 1910.95. A more stringent noise standard has been proposed by the OSH Administration but has not yet been promulgated. See BNA, *OSHR,* Current Report for October 24, 1974, p. 587 and Current Report for June 3, 1977, p. 3. See also the discussion regarding noise in this chapter, Section 3.7.
104. The standard for ionizing radiation was substantially derived from an established federal standard and can be found at 29 CFR § 1910.96. The standard for electromagnetic (nonionizing) radiation was derived from an ANSI standard and can be found at 29 CFR § 1910.97. See also the discussion regarding radiation in this chapter, Section 3.11.
105. The ventilation standards were derived from ANSI standards and can be found at 29 CFR § 1910.94.
106. The OSHA asbestos standard can be found at 29 CFR § 1910.1001. Asbestos standards have also been promulgated by the Secretary of Interior pursuant to the Coal Mine Health and Safety Act of 1969 (30 CFR § 71.202) and the Metal and Non-Metallic Mine Safety Act of 1966 (30 CFR §§ 55.5, 56.5, and 57.5) (now the Federal Mine Safety and Health Act of 1977; see this chapter, Section 3.1).
107. For the text of the proposal, see BNA, *OSHR,* Current Report for October 16, 1975, p. 714. The existing asbestos standard remains in effect until the proposed revised version is promulgated by the secretary.
108. *Society of the Plastics Industry* v. *OSHA, supra* [(74) 2 OSHC at p. 1500].

109. *Society of the Plastics Industry, supra,* 2 OSHC at pp. 1500-1501.
110. The standard can be found at 29 CFR § 1910.1017.
111. *See Dry Color Manufacturers' Association, Inc. v. U.S. Department of Labor,* 486 F.2d 98; 1 OSHC 1331, 1332 (3d Cir. 1973). The 14 chemicals included in the carcinogen standard are 4-nitrobiphenyl, α-naphthylamine, 4,4-methylene(bis)(2-chloroaniline), methyl chloromethyl ether, 3,3-dichlorobenzidine, bis chloromethyl ether, β-naphthylamine, benzidine, 4-aminodiphenyl, ethyleneimine, β-propiolactone, 2-acetylaminofluorene, 4-dimethylaminoazobenzene, and N-nitrosodimethylamine. The permanent standards can be found at 29 CFR §§ 1910.1003-1910.1016.
112. 506 F.2d 385; 2 OSHC 1402 (3d Cir. 1974); *certiorari* denied, 95 S.Ct. 1396 (1975).
113. For a summary of the standard's other major requirements, see BNA, *OSHR,* Current Report for October 28, 1976, pp. 619-620. The entire text of the standard can be found at 29 CFR § 1910.1029.
114. *American Iron & Steel Institute et al. v. OSHA,* nos. 76-2358 et al.
115. BNA, *OSHR,* Current Report For May 12, 1977, p. 1544. *Note.* On March 28, 1978, the Third Circuit upheld the exposure limit and other key provisions of the coke oven standard. BNA, *OSHR,* Current Report for April 6, 1978, p. 1659.
116. BNA, *OSHR,* Current Report for June 2, 1977, p. 4. For the full text of the proposed permanent standard, see *idem,* p. 21. *Note.* A final permanent standard for benzene was issued by OSHA on February 2, 1978. BNA, *OSHR,* Current Report for February 9, 1978, p. 1363. The Fifth Circuit invalidated the benzene standard on October 5, 1978. BNA, OSHR, Current Report, for October 12, 1978, p. 635. The court's opinion is reported at 6 OSHC 1959.
117. The full text of the proposed permanent standard for DBCP can be found in BNA, *OSHR,* Current Report for November 3, 1977, p. 797. On March 15, 1978, OSHA promulgated a permanent standard regarding DBCP. BNA, *OSHR,* Current Report for March 16, 1978, p. 1564.
118. BNA, *OSHR,* Current Reports for June 2, 1977, p. 3, and May 5, 1977, p. 1496. A status chart of OSHA's proposed standards and NIOSH criteria documents can be found at BNA, *OSHR,* Reference File at 11:2155-2159. *Note.* On September 29, 1978, OSHA announced a permanent standard regarding workplace exposure to acrylonitrile. The standard sets a permissible exposure limit of 2 parts per million parts of air over an 8 hr period. BNA, *OSHR,* Current Report for October 5, 1978, p. 563-64. Further, on June 23, 1978, OSHA published its permanent standard regarding cotton dust. Enforcement of the cotton dust standard was stayed by the Fifth Circuit on October 20, 1978. BNA, OSHR, Current Reports for June 29, 1978, p. 101 and October 26, 1978, p. 700. On November 13, 1978, OSHA announced its permanent standard for lead. BNA, OSHR, Current Report for November 16, 1978, p. 851.
119. The OSHA proposal would be codified as 29 CFR 1990. The full text of the proposal appears in 42 *Fed. Reg.* 54148 (October 4, 1977) and also in BNA, *OSHR,* Current Report for October 6, 1977, p. 577. A draft of the OSHA proposal had been announced and was described in January 1977. See BNA, *OSHR,* Current Report for January 27, 1977, pp. 1107-1108. See also BNA, *OSHR,* Current Report for October 6, 1977, p. 555.
120. See note 90. The standards adopted by the Secretary of Interior regarding air contaminants are the TLVs developed by the ACGIH. With respect to coal mines, the TLVs apply only to surface coal mines and to surface work areas of underground coal mines. A separate standard has been promulgated by the Secretary of Interior for asbestos.
121. *Atlantic & Gulf Stevedores, supra* [(70), p. 553].
122. *Dunlop v. Trumbull Asphalt Company, Inc.,* 4 OSHC 1847 (E.D. Mo. 1976).
123. See *Powell v. Globe Industries, Inc.,* 5 OSHC 1250 (N.D. Ohio 1977). The National Labor Relations Board (NLRB) has concurrent jurisdiction over Section 11(c) cases. In 1975 the General Counsel of the NLRB and the Secretary of Labor entered into an understanding for the procedural coordination of litigation arising under Section 11(c) of OSHA and Section 8 of the National Labor Relations Act, to avoid duplicate litigation. See J. Irving, "Effect of OSHA on Industrial Relations and Collective Bargaining,"

in: *Proceedings of the ABA National Institute on Occupational Safety and Health Law,* ABA Press, Chicago, 1976, pp. 125–127.

124. The general antiretaliation provision in the 1977 Mine Safety and Health Act is set forth in Section 105(c) of that statute. The 1977 mine act also provides for immediate inspection of a coal mine at the request of a miner [§ 103(g)], the right of employees to accompany the inspector on his walk-around inspection of the coal mine [§ 103(f)], and limited payments to miners when a safety violation closes the mine [(§ 111)]. Black lung (coal worker's pneumoconiosis) benefits are provided to totally disabled coal miners and surviving dependents of coal miners whose deaths were due to black lung disease [(§§ 401 *et seq.*)].

125. 29 CFR § 1977.12(b)(2).

126. *Dunlop* v. *Daniel Construction Co., Inc.,* 4 OSHC 1125 (N.D. Ga. 1975), affirmed *sub nom., Marshall* v. *Daniel Construction Co., Inc.,* 6 OSHC 1031 (5th Cir. 1977); *Usery* v. *Whirlpool Corp.,* 4 OSHC 1391 (N.D. Ohio 1976); *Brennan* v. *Diamond International Corp.,* 5 OSHC 1049 (S.D. Ohio 1976).

127. *Usery* v. *The Babcock & Wilcox Company,* 4 OSHC 1857 (E.D. Mich. 1976). With respect to work stoppages over safety disputes in the context of collective bargaining agreements and the National Labor Relations Act, see *Gateway Coal Company* v. *United Mine Workers,* 414 U.S. 368; 1 OSHC 1461 (1974).

128. 483 F.2d 67; 1 OSHC 1294 (5th Cir. 1973).

CHAPTER TWENTY-ONE

Compliance and Projection

MARTHA HARTLE MUNSCH, J.D., and
ROBERT L. POTTER, J.D.

1 INVESTIGATIONS AND INSPECTIONS

With the enactment of the Occupational Safety and Health Act of 1970 (OSHA), Congress authorized the Secretary of Labor to enter, inspect, and investigate places of employment to discover possible violations of the employer's general and specific duties under the act.

Section 8(a) authorizes the secretary, upon presenting appropriate credentials to the owner, operator, or agent in charge,

1. To enter without delay and at reasonable times any factory, plant, establishment, construction site, or other area, workplace or environment where work is performed by an employee of an employer.
2. To inspect and investigate during regular working hours and at other reasonable times, and within reasonable limits and in a reasonable manner, any such place of employment and all pertinent conditions, structures, machines, apparatus, devices, equipment, and materials therein, and to question privately any such employer, owner, operator, agent or employee.

The Federal Mine Safety and Health Act of 1977 directs authorized representatives of the secretary to make frequent, unannounced inspections and investigations in coal or other mines each year. The purposes of these visits include determining whether an imminent danger exists and whether there is compliance with the mandatory health and safety standards issued under that statute (1).

1.1 The Constitutionality of Warrantless Inspections

On December 30, 1976, the entry and inspection provisions of Section 8(a) of OSHA were declared unconstitutional by a three-judge federal district court sitting in Idaho (2). The court held that the inspection provisions attempt to authorize warrantless inspections of those business establishments covered by the act and thus run afoul of the Fourth Amendment of the federal Constitution (3).

The United States Supreme Court affirmed the judgement of the three-judge District Court on May 23, 1978 (4). The Supreme Court explained, however, that the Secretary of Labor will not be required to demonstrate "probable cause to believe that conditions in violation of OSHA exist on the premises" in order to obtain a warrant to conduct an inspection. Rather, probable cause justifying the assurance of a warrant may be based on a showing that reasonable legislative or administrative standards for conducting an inspection are satisfied with respect to a particular establishment (5).

1.2 Inspection Procedures

It is well settled that OSHA inspections must be made at reasonable times in a reasonable manner and within reasonable limits pursuant to the act (6). The act also requires the inspector to present his credentials to the employer before beginning the inspection (7). The Seventh Circuit Court of Appeals has held, however, that even if the inspector failed to present his credentials, such failure cannot operate to exclude evidence obtained in the inspection when there is no showing that the employer was prejudiced thereby in any way (8).

Section 8(e) of OSHA requires that a representative of the employer and a representative authorized by his employees be given an opportunity to accompany an OSHA inspector during the physical inspection of any workplace. In *Chicago Bridge & Iron Company v. OSAHRC and Dunlop* (9) the Seventh Circuit held the dictates of Section 8(e) to be mandatory rather than merely directory. The court refused, however, to hold that the absence of a formalized offer of an opportunity to accompany the compliance officer on his inspection rendered the citations for violations observed during that inspection void *ab initio*. Rather, the court explained that when there has been substantial compliance with the mandate of the Act in regard to the granting of a walk-around right and the employer is unable to demonstrate that prejudice resulted from his nonparticipation in the inspection, citations issued as a result of the inspection are valid (10).

The court in *Accu-Namics* (8, at p. 833) did not reach the question of whether the language of Section 8(e) was mandatory or directory, but it did refuse to adopt a rule that would exclude all evidence obtained illegally, no matter how minor or technical the government's violation and no matter how egregious or harmful the employer's safety violation (11).

2 RECORDKEEPING AND REPORTING

Section 8(c)(1) of OSHA requires employers to make, keep, and preserve such records regarding their OSHA-related activities as the Secretary of Labor, in cooperation with the Secretary of Health, Education, and Welfare (HEW), may prescribe as being necessary or appropriate for the enforcement of the act or for developing information regarding the causes and prevention of occupational accidents and illnesses. Such records are also to be made available to the Secretaries of Labor and/or HEW.

Section 8(c)(2) more specifically directs the Secretary of Labor, in cooperation with the Secretary of HEW, to issue regulations requiring employers to maintain accurate records of, and to make periodic reports on, work-related deaths, injuries, and illnesses other than minor injuries requiring only first aid treatment and not involving medical treatment, loss of consciousness, restriction of work or motion, or transfer to another job.

Regulations promulgated by the Secretary of Labor implementing Sections 8(c)(1) and (2) and in effect as of July 1, 1977, require employers to keep the following records or the equivalent thereof:

OSHA Form 100 A log of all recordable occupational injuries and illnesses.
OSHA Form 101 A supplementary record for each occupational injury or illness.
OSHA Form 102 An annual summary of occupational injuries and illnesses, which is to be based on the information contained in the log and a copy of which is to be posted in each establishment in a conspicuous place or places where notices to employees are customarily posted (12).

The records required by forms 100, 101, and 102 are to be retained in each establishment for 5 years following the end of the year to which they relate. The regulations further require that within 48 hr after the occurrence of an employment accident that is fatal to one or more employees or results in hospitalization of five or more employees, the employer of such employees is to report the accident either orally or in writing to the nearest office of the Area Director of the OSH Administration.

In addition, Section 24 of the act directs the Secretary of Labor, in consultation with the Secretary of HEW, to develop and maintain a program of collection, compilation, and analysis of occupational safety and health statistics. This program requires employers to participate in periodic surveys of occupational injuries and illnesses. The survey form is OSHA form 103 and an employer who receives such a form has a duty to complete and return it promptly (13).

On July 19, 1977, however, Secretary of Labor Marshall announced a proposed redesign of the recordkeeping requirements to center around a new OSHA form 200 and eliminate forms 100, 102, and 103. The Labor Department estimated that the changes would result in a 50 percent reduction in the number of forms required of employers. The proposed changes would also (1) exempt from all recordkeeping requirements employers with 10 or fewer workers unless the employers were selected for the annual survey, (2) reduce the number of companies required to participate in the survey by approximately 85,000, and (3) give workers the right to see the health and safety records of plants in which they are employed (14).

Finally, with respect to toxic materials or harmful physical agents, the Secretary of Labor, in cooperation with the Secretary of HEW, is directed by Section 8(c)(3) of the act to issue regulations requiring employers to maintain accurate records of employee exposures to potentially toxic materials or harmful physical agents that are required to be monitored or measured under Section 6. Such regulations must also provide an employee or former employee with access to such records as will indicate his own exposure to such toxic materials or harmful physical agents. These statutory directives have been implemented by the Secretary of Labor in the specific permanent standards issued by the secretary for certain toxic materials and harmful physical agents (15).

3 SANCTIONS FOR VIOLATING SAFETY AND HEALTH LAWS

3.1 Citations

The Occupational Safety and Health Act authorizes the Secretary of Labor to issue citations and proposed penalties to employers who are believed to have violated the act or regulations promulgated pursuant thereto.

Section 9(a) directs the secretary to issue citations "with reasonable promptness" following an inspection or investigation. No citation may be issued after the expiration of 6 months from the occurrence of any violation [§ 9(c)].

Each citation is to be in writing and must describe "with particularity the nature of the violation, including a reference to the provision of the chapter, standard, rule, regulation, or order alleged to have been violated" (16). Section 9(a) states that each citation must also fix a reasonable time for the abatement of the violation.

Section 9(b) requires employers to post each citation prominently at or near each place where a violation referred to in the citation occurred. The mechanics of how, when, where, and how long to post the citations are set forth in regulations issued by the secretary (17).

3.2 Penalties

Within a reasonable time after a citation has been issued, the secretary is directed by Section 10(a) to notify the employer by certified mail of the penalty, if any, that will be assessed for the violation. The penalty will be based at least in part on the nature of the violation (18). Violations fall into the following general categories: serious, nonserious, *de minimis,* willful, repeated, and criminal.

3.2.1 Serious Violations

Section 17(k) of OSHA defines a serious violation as follows:

[A] serious violation shall be deemed to exist in a place of employment if there is a substantial probability that death or serious physical harm could result from a condition which exists, or from one or more practices, means, methods, operations, or processes which have been adopted or are in use, in such place of employment unless the employer did not, and could not with the exercise of reasonable diligence, know of the presence of the violation.

The *probability* of occurrence of an *accident* need not be shown to establish that a violation is serious (19). Rather, the court ruled in *California Stevedore and Ballast* (19) that a serious violation exists if any accident that should result from a violation of a regulation would have a substantial probability of resulting in death or serious physical harm. No actual death or physical injury is required to establish a serious violation (20).

Employer knowledge is clearly an element of a serious violation. The knowledge requirement in Section 17(k) deals with actual or constructive knowledge of practices or conditions that constitute violations of the act; it is not directed to knowledge of the law (21). The burden of proof is on the secretary to prove knowledge as well as the other elements of a serious violation of the act (22).

Section 17(b) provides that an employer who has received a citation for a serious violation of the act *must* be assessed a civil penalty of up to $1000 for each such violation.

3.2.2 Nonserious Violations

The original Senate version of the occupational safety and health bill treated all violations as "serious." As finally enacted, however, OSHA incorporated a House proposal for violations "determined not to be of a serious nature" (23).

The statute does not describe the elements of a nonserious violation and provides no guidelines for specifically determining when a violation is not serious. The Fifth Circuit has described nonserious violations as violations that do not create a substantial probability of serious physical harm (24). The commission has explained that serious and nonserious violations are distinguished on the basis of the seriousness of injuries that experience has shown are reasonably likely to result when an accident does arise from a particular set of circumstances (25). At least one federal court of appeals (22) has held that employer knowledge is an element of a nonserious violation.

When a violation is determined not to be serious, the assessment of a penalty is discretionary rather than mandatory. Section 17(c) states that for such nonserious violations, the employer *may* be assessed a civil penalty of up to $1000 for each such violation. In the Department of Labor's appropriations for fiscal years 1977 and 1978, however, Congress exempted employers from penalties for nonserious, first-instance violations unless 10 or more such violations were uncovered during the inspection (26).

Section 110(a) of the Federal Mine Safety and Health Act of 1977 does not allow discretionary penalties for so-called nonserious violations but instead requires the Secretary of Labor to assess a civil penalty of up to $10,000 for each violation of a mandatory health or safety standard under that act.

3.2.3 De minimis Violations

If noncompliance with an OSHA provision or standard presents no direct or immediate relationship to the safety or health of employees, the violation is *de minimis* and the Secretary of Labor may issue only a notice, not a citation, to the employer (27). The notice contains no proposed penalty.

3.2.4 Willful or Repeated Violations

The Occupational Safety and Health Act provides more stringent civil penalties for employers who "willfully or repeatedly" violate the act or any regulations promulgated pursuant thereto. Willful or repeated violations are subject under Section 17(a) to penalties of up to $10,000 for each violation (28).

The act contains no definition of either "willful" or "repeated" as applied to violations. Thus it is not surprising that the courts have been unable to agree on the elements of these types of violation.

In *Frank Irey Jr., Inc.* v. *OSAHRC* (29), the Third Circuit Court of Appeals held that "[w]illfulness connotes defiance or such reckless disregard of consequences as to be equivalent to a knowing, conscious, and deliberate flaunting of the Act. Willful means more than merely voluntary action or omission—it involves an element of obstinate refusal to comply."

The First and Fourth Circuits, on the other hand, have interpreted a willful action as a "conscious, intentional, voluntary decision," regardless of venial motive (30).

The Occupational Safety and Health Review Commission has declined to follow the *Frank Irey* definition of "willfulness" and instead has followed the contrary views of the Fourth and First Circuits (31).

The interpretation of "repeated" violation has also generated disagreement among employers, the courts and the Commission.

In *Bethlehem Steel Corp.* v. *OSAHRC* (32) the Third Circuit held that the commission can find a repeated violation only when it is established that the employer consciously ignored or flaunted the requirements of the act and was cited for a similar violation on at least *two prior* occasions. The word "flaunting" had been used by the Third Circuit in *Frank Irey Jr.* in determining whether a violation was properly classified as "willful." The court in *Bethlehem Steel* reasoned that a repeated violation, like a willful violation, must consist of particularly flagrant conduct. The court explained its "test" as follows:

The mere occurrence of a violation of a standard or regulation more than twice does not constitute that flaunting necessary to be found before a penalty can be assessed under Sec. [17(a)]. What acts constitute flaunting of the requirements of the Act must be determined, in the first instance, by the Secretary and the Commission, but they should be guided by our statements in *Frank Irey*. . . . It should be noted that Sec. [17(a)] can be applicable even if the same standard is never violated twice, if the general or specific duty clauses of Sec. [5(a)] are repeatedly violated in such a way as to demonstrate a flaunting disregard of the requirements of the Act. Among the factors the Commission should consider when determining whether a course of conduct is flaunting the requirements of the Act are the number, proximity in time, nature and extent of violations, their factual and legal relatedness, the degree of care of the employer in his efforts to prevent violations of the type involved, and the nature of the duties, standards or regulations violated (33).

The court further explained that in applying the "repeatedly" portion of the act, the commission must determine that the acts themselves flaunt the requirements of the statute and need not determine whether the acts were performed with an intent to flaunt the requirements of the statute. A "repeated" violation is established by proof of facts from which it can be inferred that an employer's conduct has constituted disregard of the act's requirements.

A majority of the members of the commission as well as several state tribunals have refused to adopt the Third Circuit's interpretation of "repeated" violation.

In *George Hyman Construction Co.* (34) the members of the commission expressed their individual views in separate opinions on this question (35).

Chairman Barnako agreed with the Third Circuit that a repeated violation is similar in degree to a willful violation. That is, it must involve a comparable type of aggravated conduct. Barnako rejected, however, the Third Circuit's conclusion that a repeated violation must be based on at least two prior violations (36).

Commissioner Cleary took the position that an employer need not have a particular state of mind or motive for flaunting the act, nor otherwise exhibit an aggravated form of misconduct. He stated that the Third Circuit's holding in *Bethlehem Steel* would for all practical purposes read the "repeated" violation out of the act (37).

Commissioner Moran (38) agreed with the Third Circuit's interpretation of "repeated" violation.

The Kentucky Occupational Safety and Health Review Commission refused to adopt the Third Circuit's test, stating that the requirement of establishing an element of culpability to prove a repeated violation of a standard tends to blur the distinctions between willful and repeated violations (39). Similarly, a referee of the Oregon Worker's Compensation Board declined to apply the Third Circuit's test to the Oregon job safety and health act (40). On the other hand, an administrative law judge in North Carolina applied the *Bethlehem Steel* test to that state's job safety and health law (41).

Disagreement and confusion surrounding "willful" and "repeated" violations will undoubtedly continue until definitive interpretations of those provisions are rendered by either the Supreme Court or Congress.

3.2.5 Criminal Sanctions

Job safety and health legislation also provides criminal sanctions for certain specified conduct. The most stringent criminal sanctions are set forth in the Federal Mine Safety and Health Act of 1977. Section 110(d) states that a mine operator can be subjected to a fine of up to $25,000 or a prison term of up to one year (or both) for willfully violating mandatory health or safety standards or for knowingly refusing to comply with certain orders issued under that statute.

Willful violations (42) of OSHA (or of any standard or rule promulgated pursuant thereto) that cause death to any employee can result in a fine of up to $10,000 or imprisonment for up to 6 months or both (43).

Criminal penalties may also be imposed for (1) knowingly making any false statement, representation, and so forth, in any document filed pursuant to or required to be maintained by OSHA or by the Mine Safety and Health Act of 1977; (2) giving advance notice of an OSHA inspection without the authority of the Secretary of Labor; (3) knowingly distributing, selling, and so on, in commerce any equipment for use in coal or other mines that is represented as complying with the Mine Safety and Health Act and does not do so; (4) killing an OSHA inspector or investigator on account of the performance of his duties.

3.2.6 Failure to Abate a Violation

The Occupational Safety and Health Act does not specify fixed periods within which violations must be remedied, but Section 9(a) does require that each citation "fix a reasonable time for the abatement of the violation" (44). An employer who fails to correct a violation within the period specified in the citation may receive an additional citation pursuant to Section 10(b) for "failure to abate." Failure to abate may result in the assessment of civil penalties of not more than $1000 per day for each day the violation continues. According to Section 17(d) the abatement period does not begin to run until the date of the final order of the commission as long as the review proceeding, if any, initiated by the employer was in good faith and not solely for delay or avoidance of penalties.

Notices of violations under Sections 104(a) and 104(b) of the Federal Mine Safety and Health Act of 1977 must similarly specify time periods for abatement of violations. Failure to abate under that statute can result in an order directing all persons to be withdrawn from the affected area of the mine until a representative of the secretary determines that the violation has been abated.

3.3 Contesting Citations and Penalties

Section 10(a) of OSHA and regulations promulgated thereunder provide a means for contesting citations and proposed penalties. After the employer has been notified of the penalty proposed to be assessed by the secretary, the employer has 15 working days within which to notify the secretary that he wishes to contest the citation or the proposed assessment of penalty. A failure to notify the secretary within 15 days of intent to contest the citation or proposed penalty will render the citation or penalty "a final order of the Commission and not subject to review by any court or agency."

The secretary's regulations (45) instruct the employer that "[e]very notice of intention to contest shall specify whether it is directed to the citation or to the proposed penalty, or both." Similarly, the courts have construed the OSHA enforcement scheme as mandating a distinction between a contest of a citation and a contest of a proposed penalty (46). Thus in *Dan J. Sheehan* the Fifth Circuit held (47) that an employer's letter that contested the proposed penalty but failed to contest the citation (in fact, the letter affirmatively admitted the violation) constituted waiver of the employer's right to challenge the citation on appeal. However the commission has stated that it will construe notices of contest that are limited to the penalty to include also a contest of the citation if a cited employer indicates later that it was his intent to also contest the citation (48).

If an employer files a timely notice of contest (or if within 15 working days of the issuance of a citation, a representative of his employees files a notice challenging the period of abatement specified in the citation), the secretary must immediately advise the Occupational Safety and Health Review Commission of such intent to contest. The commission then must afford an opportunity for an administrative hearing (49).

A commission hearing is conducted pursuant to the Administrative Procedure Act and is presided over by a single administrative law judge employed by the commission. After taking testimony, the judge writes an opinion, which is subject to review by the full three-member commission at its discretion (50). An aggrieved party may petition for discretionary review before the full commission, or any commission member may direct review of a case (51). If no commissioner directs review, or if a timely petition for review is not filed, the administrative law judge's decision becomes a final order of the commission (52).

Section 9(a) authorizes the commission to review either the citation or the proposed penalty or both. The commission's scope of review is set forth in Section 9(c), which provides:

The Commission shall thereafter [i.e., after hearing] issue an order, based on findings of fact, affirming, modifying or vacating the Secretary's citation or proposed penalty, or directing other appropriate relief, and such order shall become final thirty days after its issuance.

Furthermore Section 17(i) empowers the commission to assess civil penalties, giving due consideration to the appropriateness of the penalty with respect to the size of the business of the employer being charged, the gravity of the violation, the good faith of the employer, and the history of previous violations.

On several occasions the commission has taken the position that it may exercise its power under Section 17(i) to *increase* the secretary's proposed penalty after considering the factors outlined above. At least three courts of appeals have expressed the view that the commission may act in this manner (53).

The Ninth Circuit in *California Stevedore* (19) also sanctioned the commission's right to increase the degree of a violation from nonserious to serious. The commission has taken the view that it can reduce the degree of a violation as well (54).

The final stage of an OSHA enforcement proceeding is review in the Court of Appeals and thereafter discretionary review by the Supreme Court. Any person adversely affected or aggrieved by the commission's disposition may obtain review in the Court of Appeals pursuant to Section 11(a) of the act. The Secretary of Labor may also obtain review or enforcement of any final order of the commission by filing a petition for such relief in the appropriate Court of Appeals pursuant to Section 11(b). The reviewing court is bound by Section 11(a) to apply the "substantial evidence test" to the commission's findings of fact. The same section empowers the court to direct the commission to consider additional evidence if it is found that the evidence is material and that reasonable grounds existed for the failure to admit it in the hearing before the commission. The standard of judicial review regarding a proposed penalty is whether the commission abused its discretion, since the assessment of penalty is not a finding of fact but rather the exercise of a discretionary grant of power (55).

3.4 Imminent Danger Situations

Section 13(a) of OSHA confers jurisdiction on the United States District Courts, upon petition of the Secretary of Labor, to restrain any hazardous employment conditions or practices that create an imminent danger of death or serious physical harm that could not be eliminated through the act's other enforcement procedures.

As originally reported out of the House committee, the act contained a provision that would have permitted an OSHA inspector to close down an operation for up to 72 hr without a court order if he found that an imminent danger existed. The original Senate version of the bill also contained a provision allowing an inspector to close down an operation for 72 hr, but this provision was revised so that no shutdown could occur without the application and granting of a temporary restraining order by a federal district judge (56). Thus under OSHA an operation can be shut down only by court order.

The Federal Mine Safety and Health Act of 1977, however, permits coal or other mine operations to be shut down without a restraining order from a court. Section 107(a) of the act provides that where a federal mine inspector finds that an imminent danger exists in a coal or other mine, he shall order the withdrawal of all persons from a part or all of that mine until the imminent danger is no longer present (57).

3.5 Constitutional Challenges to the OSHA Enforcement Scheme

The citation and penalty scheme of OSHA has been subject to constitutional challenge on several fronts.

In *Atlas Roofing Company, Inc.* v. *OSAHRC* (58) a cited employer contended that the act was constitutionally defective because (1) civil penalties under OSHA are really penal and call for the constitutional protections of the Sixth Amendment and Article III; (2) even if the penalties are civil, OSHA violates the Seventh Amendment because of the absence of a jury trial for fact finding; (3) the act denies the employer his right to a Fifth Amendment "prejudgment" due process hearing, since commission orders are self-executing unless the employer affirmatively seeks review; and (4) the overall penalty structure of OSHA violates due process because it "chills" the employer's right to seek review of the citation and penalty.

The Fifth Circuit in *Atlas Roofing* rejected all the employer's constitutional contentions. The Supreme Court granted a petition for certiorari in that case, limited to the Seventh Amendment issue (59). On March 23, 1977, the Supreme Court upheld the act's provision for imposition of civil penalties without fact finding by a jury (60).

Several other Courts of Appeals have considered similar constitutional challenges to the act's enforcement scheme and most, if not all, have likewise rejected the constitutional contentions (61).

4 THE FUTURE OF JOB SAFETY AND HEALTH LAW

As this book was published the future of federal regulation of workplace safety and health was uncertain. The uncertainty was particularly acute with respect to occupational safety in light of an announcement by the Carter administration in the summer of 1977 recommending a system for regulating workplace safety hazards that would avoid "detailed safety precautions" (62).

Characterizing OSHA as the "leading national symbol of overregulation," a memorandum prepared by several high ranking officials in the Carter administration recommended "totally eliminating most safety regulations and replacing them with some form of economic incentives" (62). As of August 5, 1977, an interagency task force had been organized to study such possible changes in OSHA regulations. Alternatives that probably will be studied by the task force include the use of tax incentives, possible changes in worker's compensation laws, and a greater emphasis on consultation and training to improve safety conditions (62).

The task force's study probably will not encompass workplace health hazards, however, since health hazards are incurred over a long period of time and it would be difficult to attribute a worker's particular health problem to a certain employer (62).

It was estimated that the task force would complete its study in March 1978. Organized labor immediately announced its opposition to any effort to replace safety standards with economic penalties or improved worker's compensation (62).

In the meantime, OSHA has assigned the highest priority to an internal effort to identify and revise safety standards that could be deemed irrelevant to worker protection (see Section 5.3, Chapter 20) and also intends to emphasize the development of new standards that focus on important occupational health problems (63). The OSH Administration has also indicated an interest in promulgating generic standards for matters such as labeling, disclosure of employee medical conditions, and rate retention (64).

With respect to its compliance activities, OSHA has been sharply criticized ever since the act became effective. Much of the criticism has resulted from OSHA's apparent emphasis on enforcing "nitpicking" safety regulations, particularly in small business establishments. The Carter administration has indicated that it will move away from "nitpicking" enforcement of OSHA's standards and instead will increase the effort directed toward serious occupational health problems. The Carter administration's compliance priorities reportedly will couple "systematic inspection of high hazard industries with an 'aggressive' educational effort to 'foster self-compliance by employers and employees'" (63).

The administrative emphasis on education and voluntary self-compliance could be accompanied in the near future by a strengthened program of on-site consultation services. Such services consist generally of advice and technical assistance to employers during and after a visit to the workplace by a government consultant at the employer's request. There presently is no general federal on-site consultation program because under OSHA, on-site consultation could not be provided by the federal government without mandatory citations for violations observed by the consultants. Since 1972 Congress has been considering proposals to amend OSHA to require the Secretary of Labor to send a consultative officer to a worksite, at the request of an employer, to provide advice in complying with OSHA without threat of penalty. Such proposals were pending in both the Senate (Senate Bill 21) and the House of Representatives (HR 1327,2516 and 2517) as of July 1, 1977. The proposals would separate consultation services from compliance activities so that employers would not hesitate to request the consultation services (65).

The OSH Administration has also been sharply criticized for mismanagement, bureaucratic duplication, and waste. A report submitted by a private consulting firm to OSHA in July 1977 observed that "regardless of the degree of commitment to the health and safety of American workers, the good intentions of policies, or the competency of top staff, the OSHA organization will not reach its potential unless it is better managed" (66). The report identified several problem areas in OSHA's management, including problems with policy development and implementation, personnel and staffing, field coordination and communications, resource allocation, and management information systems (66). Organizational and operational changes at OSHA will undoubtedly follow in the wake of this report.

Similarly, OSHA has been criticized for the burdensome and often duplicative recordkeeping and reporting requirements imposed on employers, particularly small employers. On July 19, 1977, OSHA announced that employer recordkeeping requirements would be substantially reduced (see Section 2, this chapter). Furthermore on August 2, 1977, OSHA and three other federal agencies announced a program of cooperation to help eliminate waste and duplication in regulation by the federal government and emphasized the need to combine their efforts on regulations of chemicals that pose a threat to public health and the environment (67). Such programs should begin to improve OSHA's tarnished image among employers.

It is somewhat surprising, however, that neither the Carter administration nor Congress has seriously addressed the troublesome problem of allocation of jurisdiction over workplace safety and health among OSHA, other federal agencies, and the states. As this chapter has indicated, ambiguities and confusion remain with respect to the jurisdiction and preemption provisions of workplace safety and health laws. To date, the resolution of such problems has been left to the commission and the courts on a case-by-case basis. In the absence of definitive interpretations by either the Supreme Court or Congress, these jurisdiction and preemption issues will undoubtedly spawn much more litigation and confusion.

In sum, federal regulation of workplace safety and health has been with us for less than a decade and is still experiencing growing pains. As a result, this area of the law is unsettled and confusing and is likely to remain so for several years to come. Nevertheless, the early months of the Carter administration seem to indicate that good faith efforts will be made to resolve some of these problems as expeditiously as possible.

NOTES AND REFERENCES*

1. See Section 103(a). For discussion of "imminent danger," see Section 3.4 of this chapter. Underground coal or other mines must be inspected at least four times each year. Each surface coal or other mine must be inspected at least twice a year.
2. *Barlow's Inc.*, v. *Usery*, 424 F. Supp. 437; 4 OSHC 1887 (D. Idaho 1976).
3. *Labor Law J.*, **28,** 263 (1977). In addition to *Barlow's*, numerous other lower court decisions, state and federal, have dealt with the constitutionality of the entry and inspection provisions of OSHA. Rather than declaring the inspection provisions unconstitutional, however, several courts have interpreted Section 8(a) as authorizing nonconsensual inspections only pursuant to the authority of a warrant issued upon satisfaction of standards of probable cause in the area of administrative searches. For example: *Dunlop* v. *Sandia Die and Cartridge Company*, 4 OSHC 1569 (D. N.M. 1976); *Rupp Forge Company*, 4 OSHC 1487 (N.D. Ohio 1976); *Brennan* v. *Gibson's Products, Inc.*, 3 OSHC 1945 (E.D. Tex. 1976); *contra, Brennan* v. *Buckeye Industries, Inc.*, 374 F. Supp. 1350 (S.D. Ga. 1974). Several state courts have likewise held the inspection provisions of their respective state occupational safety and health acts unconstitutional. For example: *State* v. *Albuquerque Publishing Co.*, 5 OSHC 2034 (New Mexico Sup. Ct. 1977); *Woods & Rohde Inc., d/b/a Alaska Truss & Millwork* v. *State,* 5 OSHC 1530 (Alaska Sup. Ct. 1977); *Yocom* v. *Burnette Tractor Company, Inc.*, 5 OSHC 1465 (Kentucky Ct. of Appeals 1977).
4. *Marshall v. Barlow's, Inc.*, 6 OSHC 1571 (1978).
5. See BNA, *OSHR*, Current Report for June 1, 1978, pp. 3–5.
6. *Dunlop* v. *Able Contractors, Inc.*, 4 OSHC 1110 (D. Mont. 1975).
7. See Section 8(a). The secretary's regulations provide in relevant part as follows:

At the beginning of an inspection, Compliance Safety and Health Officers shall present their credentials to the owner, operator, or agent in charge at the establishment; explain the nature and purpose of the

* Abbreviations and forms of citation are explained in the Notes and References of Chapter 20.

inspection; and indicate generally the scope of the inspection and the records ... which they wish to review. 29 CFR § 1903.7(a).

The secretary's regulations further provide for a conference at the conclusion of the inspection wherein the Compliance Safety and Health Officer advises the employer of any apparent safety or health violations disclosed by the inspection and the employer is afforded an opportunity to bring to the attention of the Officer any pertinent information regarding conditions in the workplace [29 CFR § 1903.7(e)].

8. *Accu-Namics, Inc.* v. *OSAHRC and Dunlop,* 515 F.2d 828; 3 OSHC 1299 (5th Cir. 1975).
9. 535 F.2d 371; 4 OSHC 1181 (7th Cir. 1976).
10. 535 F.2d at 377. The Seventh Circuit found substantial compliance in the *Chicago Bridge & Iron Company* case because the on-site representative of Chicago Bridge & Iron had been informed of the pending inspection tour and was given literature setting forth the directives of the act.
11. In *Leone* v. *Mobil Oil Corporation,* 523 F.2d 1153 (5th Cir. 1975), the Fifth Circuit held that neither OSHA nor the Fair Labor Standards Act of 1938, 29 USC § 203(o), requires an employer to pay wages for time spent by employees in accompanying OSHA inspectors on walk-around inspections of the employer's plant. However on September 20, 1977, OSHA announced an amendment of its administrative regulations (29 CFR § 1977.21) to reflect a new policy that employees should be paid by their employers for time spent on such walk-around inspections. See BNA, *OSHR,* Current Report for September 22, 1977, p. 499. The full text of the amendment appears in 42 *Fed. Reg.* 47343 (September 20, 1977) and in BNA, *OSHR,* Current Report for September 22, 1977, p. 519. The Chamber of Commerce of the United States filed suit challenging the policy on October 25, 1977, in the United States District Court for the District of Columbia. BNA, *OSHR,* Current Report for October 27, 1977, p. 763.
12. These regulations are found in 29 CFR Part 1904. The secretary has defined "recordable occupational injuries or illnesses" as those that result in:

   a. Fatalities, regardless of the time between the injury and death, or the length of the illness.
   b. Lost workday cases, other than fatalities, that result in lost workdays.
   c. Nonfatal cases without lost workdays, which result in transfer to another job or termination of employment, or require medical treatment (other than first aid) or involve: loss of consciousness or restriction of work or motion. This category also includes any diagnosed occupational illnesses that are reported to the employer but are not classified as fatalities or lost workday cases. 29 CFR § 1904.12(c).

13. 29 CFR §§ 1904.20–1904.21.
14. BNA, *OSHR,* Current Report for July 21, 1977, p. 235. A final rule promulgating the exemption provisions was issued by OSHA on July 29, 1977. Employers with 10 or fewer workers still are required, however, to report accidents resulting in a fatality or multiple exposure. BNA, *OSHR,* Current Report for August 4, 1977, p. 314. The secretary's prior regulations had exempted employers with seven or fewer employees from the general record keeping requirements. 29 CFR § 1904.15 (1976).
    On October 18, 1977, OSHA issued a proposed amendment to its recordkeeping regulations that would provide access to the log of occupational injuries and illnesses to employees, including former employees, and their representatives. The full text of this proposal appears in BNA, *OSHR,* Current Report for October 20, 1977, p. 740.
    Recordkeeping and reporting requirements under the Federal Mine Safety and Health Act of 1977 are set forth in subsections 103(c)–103(e) and 103(h) of that statute.
15. See, for example, the OSHA standard regarding employee exposure to coke oven emissions, 29 CFR § 1910.1029(m).
16. See Section 9(a). Interpretations of the "particularity" requirement have generally dealt with (1) the precision of the reference to the standard allegedly violated, and (2) the adequacy of the description of the alleged violation. See American Bar Association, *Report of the Committee on Occupational Safety and Health Law,* ABA Press, Chicago, 1976, pp. 261–263.

17. American Bar Association, *Report of the Committee on Occupational Safety and Health Law,* ABA Press, Chicago, 1975, p. 57. The secretary's regulations can be found at 29 CFR § 1903.16.
18. In determining the amount of the penalty, the secretary is also directed to consider "the size of the business of the employer being charged, the gravity of the violation, the good faith of the employer, and the history of previous violations." 29 CFR § 1903.15(b).
19. *Dorey Electric Company* v. *OSAHRC,* 553 F.2d 357; 5 OSHC 1285 (4th Cir. 1977); *California Stevedore and Ballast Company* v. *OSAHRC,* 517 F.2d 986; 3 OSHC 1174 (9th Cir. 1975).
20. *Brennan* v. *OSAHRC and Vy Lactos Laboratories,* 494 F.2d 460; 1 OSHC 1623 (8th Cir. 1974).
21. *Mid-Plains Construction Company,* 3 OSHC 1484 (Rev. Comm. 1975); *Southwestern Acoustics & Specialty, Inc.,* 5 OSHC 1091 (Rev. Comm. 1977).
22. *Brennan* v. *OSAHRC and Hendrix (d/b/a Alsea Lumber Co.),* 511 F.2d 1139, 1142; 2 OSHC 1646 (9th Cir. 1975).
23. Conference Report 91-1765, 1970 USC, *Congressional and Administrative News,* p. 5237.
24. *Ryder Truck Lines, Inc.* v. *Brennan,* 497 F.2d 230, 233; 2 OSHC 1075 (5th Cir. 1974).
25. *Standard Glass and Supply Company,* 1 OSHC 1223-1224 (Rev. Comm. 1973).
26. BNA, *OSHR,* Current Report for August 11, 1977, p. 348.
27. *Lees Way Motor Freight, Inc.* v. *Secretary of Labor,* 511 F.2d 864, 869; 2 OSHC 1609 (10th Cir. 1975).
28. The Federal Mine Safety and Health Act of 1977 [Section 110(d)] does not provide more stringent civil penalties for willful violations of that act's safety and health standards but does impose criminal liability on coal mine operators who are found guilty of such willful conduct. The act does impose [Section 110(g)] a civil penalty on miners who willfully violate the safety standards relating to smoking or the carrying of smoking materials.
29. 519 F.2d 1200; 2 OSHC 1283, 1289 (3d Cir. 1974), affirmed on other points sub nom., *Atlas Roofing Co., Inc.* v. *OSAHRC,* 97 S.Ct. 1261 (1977).
30. *Messina Construction Co.* v. *OSAHRC,* 505 F.2d 701; 2 OSHC 1325 (1st Cir. 1974); *Intercounty Construction Co.* v. *OSAHRC,* 522 F.2d 777; 3 OSHC 1337 (4th Cir. 1975), *certiorari* denied 3 OSHC 1879 (1976).
31. *Kent Nowlin Construction Inc.,* 5 OSHC 1051, 1055 (Rev. Comm. 1977). For an interpretation of the California Occupational Safety and Health Act's "willful" violation provision, see *Rawly's Division of Merit Ends, Inc.* (Docket No. 823-75) (summarized in BNA, *OSHR,* Current Report for February 24, 1977, p. 1234).
32. 540 F.2d 157; 4 OSHC 1451 (3rd Cir. 1976).
33. 4 OSHC at 1454.
34. 5 OSHC 1318 (Rev. Comm. 1977) (appeal pending).
35. Prior to the *George Hyman* case, the commission had determined whether a violation was "repeated" on a case-by-case basis without developing and consistently applying a rule applicable to all cases (5 OSHC at 1319).
36. Commissioner Barnako summarized his test for a repeated violation as follows:

    [T]he Secretary establishes a prima facie case that a violation is repeated when he shows (1) a violation has been cited and has become a final order, and (2) a substantially similar violation occurs under the control of the supervisor who had responsibility for abating the first violation. I stress, however, that the ultimate test for whether an employer has committed a repeated violation is whether, under all the circumstances, an employer can be said to have disregarded the Act by failing to take the necessary steps after the occurrence of the initial violation to prevent its recurrence (5 OSHC at 1322).

37. Commissioner Cleary explained:

    If a "repeated" violation must be supported by a showing of two or more violations in a manner that flouts the Act, and proof of a willful violation requires only that an employer flout the Act, there would be no reason to have recourse to the provision for a "repeated" violation, because it would include the elements of a 'willful' violation (5 OSHC at 1326).

38. The Carter administration announced on August 31, 1977, that Commissioner Cleary would be appointed to replace Commissioner Barnako as chairman of the commission. Mr. Barnako has indicated that he will continue to serve on the commission until his term expires in 1981. BNA, *OSHR,* Current Report for September 8, 1977, p. 443.

    Commissioner Moran's term expired in April 1977. On September 7, 1977, President Carter nominated Bertram R. Cottine to replace Commissioner Moran. The Senate confirmed Mr. Cottine's nomination on April 27, 1978. BNA, *OSHR,* Current Report for May 4, 1978, p. 1803.

39. *Commissioner of Labor* v. *Bendix Corp., Heavy Systems Group* (Docket No. 289) (summarized in BNA, *OSHR,* Current Report for April 7, 1977, p. 1393).

40. *Globe Union, Inc.* (Docket No. SH-76-369) (summarized in BNA, *OSHR,* Current Report for April 21, 1977, p. 1445). In *Star-Kist Foods, Inc.* (Docket Nos. 795-76–801-75 and 829-75), the California Occupational Safety and Health Appeals Board held that to establish a repeated violation under the state act, the same standard must be involved and the circumstances of both violations must be "essentially the same." The decision is summarized in BNA, *OSHR,* Current Report for March 31, 1977, p. 1372.

41. *Baker-Cammack Hosiery Mills* (OSHANC Docket No. 77-178) (summarized in BNA, *OSHR,* Current Report for June 30, 1977, p. 143).

    Another interpretative problem that has arisen regarding "repeated" violations involves the treatment of transient as opposed to fixed work sites. The *OSHA Field Operations Manual,* Chapter VIII, sets forth the following guidelines regarding how these respective types of work sites should be treated:

    For purposes of considering whether a violation is repeated, citations issued to employers having fixed establishments (e.g., factories, terminals, stores) will be limited to the cited establishment. For employers engaged in businesses having no fixed establishments (construction, painting, excavation) repeated violations will be alleged based on prior violations occurring anywhere within the same State.

    In *Desarrollos Metropolitanos, Inc.* v. *OSAHRC,* 551 F.2d 874; 5 OSHC 1135 (1st Cir. 1977), the First Circuit held that these guidelines, insofar as they distinguish between employers with fixed and transient work sites, do not violate the guarantee of the equal protection clause of the Federal Constitution. The court also ruled that there is no constitutional infirmity concerning the guidelines insofar as they distinguish between construction work predicated on whether it is performed intra- or interstate. (See *George Hyman Construction Co., supra,* 5 OSHC at 1320, n. 6.).

42. For an interpretation of "willful" in the context of Section 17(e) of OSHA, see *United States* v. *Dye Construction Co.,* 510 F.2d 78 (10th Cir. 1975).

43. See Section 17(e) of OSHA. More than one conviction under Section 17(e) or under Section 110(d) of the Federal Mine Safety and Health Act of 1977 can result in much greater fines and/or prison terms.

44. American Bar Association, *1975 Report of the Committee on Occupational Safety and Health Law* (17), p. 68.

45. 29 CFR § 1903.17.

46. *Brennan* v. *OSAHRC and Bill Echols Trucking Co.,* 487 F.2d 230; 1 OSHC 1398 (5th Cir. 1973); *Dan J. Sheehan Company* v. *OSAHRC and Dunlop,* 520 F.2d 1036; 3 OSHC 1573 (5th Cir. 1975), *certiorari* denied 424 U.S. 965.

47. *Dan J. Sheehan, supra,* at 1039.

48. *Turnbull Millwork Co.,* 3 OSHC 1781 (Rev. Comm. 1975). This policy will apply whether the employer is represented by counsel or proceeds *pro se. Nilsen Smith Roofing & Sheet Metal Co.,* 4 OSHC 1765 (Rev. Comm. 1976); *Penn-Dixie Steel Corp.* v. *OSAHRC and Dunlop,* 553 F.2d 1078; 5 OSHC 1315 (7th Cir. 1977).

49. See Section 10(c) of OSHA. The rules governing practice before the commission can be found at 29 CFR §§ 2200.1–2200.110.
50. *Secretary of Labor* v. *OSAHRC and Interstate Glass Company*, 1 OSHC 1372, 1373 (8th Cir. 1973).
51. 29 CFR §§ 2200.91 and 2200.91a.
52. B. Fellner and D. Savelson, "Review by the Commission and the Courts," in: *Proceedings of the American Bar Association Institute on Occupational Safety and Health Law*, ABA Press, Chicago, 1976, p. 102.
53. *REA Express* v. *Brennan*, 495 F.2d 822; 1 OSHC 1651 (2nd Cir. 1974); *Brennan* v. *OSAHRC and Interstate Glass Co.*, 487 F.2d 438; 1 OSHC 1372 (8th Cir. 1973); *California Stevedore & Ballast Co.* v. *OSAHRC* (19). Compare *Dale M. Madden Construction, Inc.* v. *Hodgson*, 502 F.2d 278; 2 OSHC 1236 (9th Cir. 1974), where the court held that the commission has no authority to modify *settlements* made by the secretary and the cited employer.
54. See, for example, *Dixie Roofing and Metal Co.*, 2 OSHC 1566 (Rev. Comm. 1975).
55. *Secretary* v. *OSAHRC and Interstate Glass, supra* [(50), at 442]. The procedures for judicial review under the Federal Mine Safety and Health Act of 1977 are set forth in Section 106. For a summary of the review procedures available under the earlier law, the Coal Mine Health and Safety Act of 1969, see *National Independent Coal Operators Association* v. *Kleppe*, 96 S.Ct. 809 (1976).
56. See *Usery* v. *Whirlpool Corp.*, 4 OSHC 1391, 1392–1393 (N.D. Ohio, 1976).
57. Imminent danger is defined in the Federal Mine Safety and Health Act of 1977 as "the existence of any condition or practice in a coal or other mine which could reasonably be expected to cause death or serious physical harm before such condition or practice can be abated. . . ." In a case arising under the imminent danger provisions of the Coal Mine Health and Safety Act of 1969 (the definition of "imminent danger" in the 1969 coal mine act was almost identical to that which now appears in the 1977 act), the Seventh Circuit affirmed the Board of Mine Operations Appeals' interpretation of "imminent danger" as being a situation in which a reasonable man would estimate that if normal operations designed to extract coal in the disputed area should proceed, it is at least as probable as not that the feared accident or disaster would occur before the danger were eliminated. *Freeman Coal Mining Co.* v. *Interior Board of Mine Operations Appeals*, 504 F.2d 741; 2 OSHC 1310 (7th Cir. 1974). Accord: *Old Ben Coal Corp.* v. *Interior Board of Mine Operations Appeals*, 523 F.2d 25; 3 OSHC 1270 (7th Cir. 1975); *Eastern Assoc. Coal Corp.* v. *Interior Board of Mine Operations Appeals*, 491 F.2d 277 (4th Cir. 1974).
58. 518 F.2d 990; 3 OSHC 1490 (5th Cir. 1975).
59. The Supreme Court also granted certiorari in *Frank Irey, Jr., Inc.* v. *OSAHRC and Brennan*, 519 F.2d 1215; 3 OSHC 1329 (3d Cir. *en banc* 1975) to review the same issue.
60. 97 S.Ct. 1261.
61. See e.g. *Dan J. Sheehan Company* v. *OSAHRC* (46); *Clarkson Construction Company* v. *OSAHRC*, 531 F.2d 451; 3 OSHC 1880 (10th Cir. 1976).
62. BNA, *OSHR*, Current Reports for July 21, 1977, pp. 235–236, July 28, 1977, p. 267, and August 11, 1977, p. 339, respectively. The recommendation encompasses all federal worker safety regulations, including those enforced by the Department of Interior and the Department of Transportation.
63. BNA, *OSHR*, Current Report for June 9, 1977, p. 51.
64. BNA, *OSHR*, Current Report for February 3, 1977, pp. 1131–1132, and for June 2, 1977, p. 8. For example, OSHA's director of health standards announced on November 21, 1977, that OSHA intended to issue within the month a proposed standard for the labeling and identification of all chemical substances used in the workplace. BNA, *OSHR*, Current Report for November 24, 1977, p. 875. With respect to rate retention (or medical removal protection), see BNA, *OSHR*, Current Report for September 22, 1977, p. 499.

65. See N. Ashford, *Crisis in the Workplace,* MIT press, Cambridge, Mass., 1976, pp. 165-167; BNA, *OSHR,* Current Reports for January 20, 1977, p. 1081, and March 10, 1977, p. 1284. Many states with approved plans under Section 18(b) of OSHA do include on-site consultation (without threat of penalty) as a part of voluntary compliance programs under their plans. These state plan programs are eligible for up to 50 percent federal funding pursuant to Section 23(g) of the act. States without approved plans can, pursuant to an OSHA regulation promulgated May 20, 1975, enter into an agreement with the federal government whereby trained state personnel can provide on-site consultation services separate from federal investigators. The federal government will pay for 50 percent of the cost of such an on-site consultation service provided by the state (29 CFR § 1908). As of July 1, 1977, the OSH Administration had issued a proposed revision to Section 1908 whereby the level of federal funding would be increased to 90 percent. In addition, OSHA, to encourage states to maintain their approved plans and also to provide on-site consultation services, proposed to extend the eligibility of the Section 1908 program to all states, including states with approved plans. BNA, *OSHR,* Current Report for May 5, 1977, pp. 1492 and 1516.

The on-site consultation programs that presently exist under Section 1908 as well as those under approved state plans generally give priority for consultation to small businesses and hazardous workplaces. As of February 1977, 34 states had been providing on-site consultation to employers pursuant to one of these plans. BNA, *OSHR,* Current Report for February 24, 1977, p. 1227.

66. BNA, *OSHR,* Current Report for August 4, 1977, pp. 307-308.
67. BNA, *OSHR,* Current Report for August 4, 1977, p. 312. The other three federal agencies are the Environmental Protection Agency, the Food and Drug Administration, and the Consumer Product Safety Commission.

Index

Abnormal pressure, 525-542
Absorption, chemicals, *see* Routes of entry
Acclimatization, altitude, 533
   hot and cold environments, 447-464
Accuracy, analytical measurements, 191, 192
Action level, 220
Acute toxicity, 572
Adaptation, acute altitude, 530
   hot and cold environments, 447-464
Additive effects, 349
Aerodynamic equivalent diameter, 232
Air contaminants, masking, 612
Air pollution emission factors, 20-23
Air recirculation, 633
Aldehydes, biological monitoring, 345
n-Alkylbenzene compounds, metabolism, 276
Altitude, 343, 351, 530, 533, 535
Aluminum, hair, normal values, 308
   occurrence, 304
American Academy of Industrial Hygiene, 8
American Conference of Governmental Industrial Hygienists, education, 8
   Manual of Industrial Ventilation, 624, 635
   threshold limit values, 7, 220, 321, 354, 355
American Industrial Hygiene Association, education, 8
   Hygienic Guide Series, 354
   industrial hygiene, definition, 2
   Laboratory Accreditation Program, 322
   Respiratory Protective Devices Manual, 647
American National Standards Institute, eye protective devices, 663
   practices for respiratory protection, 647, 665, 666, 669, 674, 675
   safety code for head, eye and respiratory protection, 664
   standards, Z.37 committee, 354

Analysis, chemical, 191, 193, 203, 206
   arsenic, atomic absorption spectrophotometry, 207
   bias, 192
   calibration, 192
   classification of methods, 197
   coefficient of variation, 192, 199
   errors, 191, 192, 193
   lead, 207
   matrix effects, 193, 207
   methods, 197, 206
      definitive, 197
      format, 196
      industrial hygiene, 209
      NIOSH-OSHA, 208
      reference, 197
      validation criteria, 195
      working ranges, 209
   problems, 213
   proficiency testing, 197
   reference materials, 192, 195
   specificity required, 205
Analytical measurements, 191-214
Anodic stripping voltammetry, 206, 209
Antagonistic effect, 349
Anthrax, *see* Wool sorter's disease, 559
Antimony, hair, normal values, 308
Arsenic, analytical sensitivity, 207
   biological monitoring, 345
   exposure assessment, 227
   hair, normal value, 308
   occurrence in, 304
   urinary levels, 345
Asbestos, 330, 338, 598, 601, 702
   evaluation, old versus new methods, 234, 330
Ashing methods, 203, 204, 205
Atomic absorption spectrophotometry, 207, 209

Atomic Energy Act of 1954, 689
Atropine, therapy and risk, 261
Audiometric testing, 435, 436
Averaging time, 237

Bagassosis, 550
 agents, 550
 exposure sources, 550
 pathogenesis, 550
 physiological effects, 550
 prevention, 550
 symptoms, 550
 treatment, 550
 ventilatory decrease, 550
Barrier creams, 226
Bends, decompression sickness, 538
Benzene, biological monitoring, 345
 pulmonary excretion, 296
Benzoic acid, metabolism, 276
 occurrence in foodstuff, 276
Beryllium, air monitoring, 335
 hair, occurrence in, 304
 urine, analysis for, 293
Bias, analytical, 191, 192
Biological agents, exposure evaluation, approach to problem, 556
 air samples, 557
 antibody tests, 557
 physical examination, 557
 pulmonary ventilatory tests, 557
  forced expiratory flow (FEV_{25-75}), 547, 557
  forced expiratory volume, 1 sec. (FEV.), 547, 557
  maximum breathing capacity (MBC), 547, 557
  vital capacity (VC), 547, 557
 questionnaire, 556
 standard 6 ft. roentgenogram, 557
 contributing factors, 544
  dose dependency, 544
  individual susceptibility, 544
  pulmonary ventilation, 544
 control, 557
  exposures, 557
  medical, 557
 modes of exposure, 543
  cutaneous, 543
  gastrointestinal, 544
  respiratory, 544
 recognition of problem, 556

 complaints, 556
 confirmatory tests, 556
 symptoms, 556
Biological analysis, alveolar "end tidal" air, 297
 error sources, 299
 exhaled air, 296
Biological availability, chemical transport, factors affecting, 269, 270
 protein binding and, 268
Biological indicators, chemical dosage and burden, 257-318
Biological monitoring:
 advantages, 295
 exposure correlation with chemical index, urine, 286
 definition, 258
 direct, 5, 262
 disadvantages, 262, 303
 disease, effect on, 282
 factors to be considered, 294
 history, 260
 indirect, 262
 problems associated with, 263-286
 sex related effects, 282, 298
 unsuitability, measuring fetal exposure, 282
 usefulness, 5, 260, 261, 262, 345
Bird fancier's disease, 551
 pathogenesis, 551
 pathology, 551
 physiological effects, 553
 roentgenographic findings, 552
 specific exposures, 553
 symptoms, 552
Blast gates, 624
Blood flow, peripheral, vibration syndrome, 510
Blood volume, human, sex variation, 298
Body burden estimation, excreted chemicals, 260
Body fluids, free versus bound chemicals, transport, 268-270
Breakthrough capacity, air sampling, 197, 202
Breast milk, human, 310-312
 metabolite excretion, 311
 metals, 311
 pesticides reported in, 311
 Ph, milk and plasma, 311
 xentobiotics excreted in, 311
Breathing zone, measurement option, 238
Brucellosis, 560
 control, 560
 epidemiology, 560
 exposure sources, 560

INDEX

mode of infection, 560
symptoms, 560
Bureau of Mines, respirators, 651, 669
Byssinosis, 234, 546
   classification, 546
   controls, 548
   exposure sources, 546
   physiological effects, 547
   smoking, 546
   symptoms, 546

Cabin decompression, 533
Cabin pressurization, 532
   limitations, 532
Cadmium, hair, normal values, 308
   occurrence in, 304
   urine, analysis for, 293
Caisson disease, 536
Cancer, early detection of, 601
Carbon monoxide, biological monitoring, 345
Carbon tetrachloride, pulmonary excretion, 296
   following skin absorption, 298
Carcinogens, 342, 703
Chain saws, vibration effects, 494, 515
Chemical agents, see Xentobiotics
Chemicals, absorption, see Routes of entry
Chokes, decompression sickness, 538
Chromatography, gas, 207, 208, 209
   ion, 212
   liquid, 209
Chromic acid, 333
Chromium, urine analysis, 293
Chronic toxicity, 573
Coccidioidomycosis, exposure, 558
   pathogenesis, 558
   prevalence, 558
   prevention, 558
   symptoms, 558
Coefficient limits, normal distribution mean, 229
Coefficient of variation, biological measurements, 88, 89
   relative standard deviation, 229
Coffee dust, 555
Coke over emissions, 704
Cold environment, evaluation of exposure to, 447-464
   physiological response, 448-456
      acclimatization, seasonal changes, 452
      small animals, 452

adrenal corticosteroids, 454
behavioral response, 448
blood flow, 453
body temperature, 449
cardiac output, 452
catecholamine, 454
cold and alcohol, 455
finger dexterity, 454
genetic selection, 448
grip strength, 454
heart rate, 452
heat production, 451
"hunting" reaction, 453
hypothermia, 455
local acclimatization, 452
maximal aerobic capacity, 455
norepinephrine, 454
oxygen uptake, 449
physical factors, heat loss and, 449
physical fitness, 450
prolonged cold exposure, 451
seasonal variation, BMR, 451
sensitivity to, 455
skin, internal temperature, 453
   rectal temperature, 452
skin receptors, 454
sympathicoadrenal medullary system, 454
tactile discrimination, 454
Collaborative tests, analytical, charcoal tubes, 203
   classification of methods, 197
Compliance and projection, occupational safety and health laws, 719-736
   future of, 729
   investigations and inspections, 719
      constitutionality of warrentless inspection, 720
      inspection procedures, 720
   recordkeeping and reporting, 721
   sanctions for violating, 722-726
      citations, 722
      contesting citations and penalties, 726
      imminent danger situations, 728
Compliance decision theory, decision rules, full period sampling, 63
   short period random sampling, 65
   general, 47, 57
   sampling strategies, nomenclature, 57, 69
   size requirements, 57, 69
      ceiling, 69, 74
      time weighted average, 69

Computer equipment:
  central processing unit, 102
  desk top, type, 148
  hardware, 101
  input devices, 101, 107
  magnetic ink, 107
  magnetic tape, 107, 150
  output devices, 102, 107
  portable, 148
  punch cards, 107
  software, 105
  storage, 102
Computer program:
  access, 118
  acquisition methods, 118, 123, 139
  application, 114, 129, 130
  benefits, 152
  binary notation, 109
  central processing unit, 102
  codes, 110
  complex/shared base systems, 158
  concepts, 160
  control, data, 125
  data analysis, design, 111
  data base, 112
  data base design, 116, 117
  data processing selectronic, 111, 155
  data retrieval, 126
  desk top, type, 148
  glossary, 176-180
  hard copy, 150
  hardware, 101
  information, availability, 126
    search and recovery, 130
  input devices, 101, 107
  inquiries, 115
  instrumental analysis, 138
  key words, search, 135
  language, 109
  machine data, recording, 109
  management, data, 125
  manual data, acquisition, 119
  medical data, 144
  methods presentation, 127
  modify and extend system, 114
  pilot test, 114
  print out, 107, 147
  problem definition and sizing, 111
  programming, 113
  requirements, 160
  retrieval data, 126

  safeguards, data, 128
  screening, data, 127
  security, data, 128
  sensors, 120-125
  software, 105
  sources, data, 118
  task charting, 111
  task training, 151, 154
  testing and debugging system, 113
  testing program, ventilation, 129
  unauthorized access, 128
  visual displays, 150
  see also Data automation application; Data automation information sources
Confidence limits, 222, 249
Consumer Products Safety Act of 1972, 688
Consumer Products Safety Commission, 688
Control programs, administrative, recovery time, 336
  data acquisition, 333
  engineering, 334
  medical surveillance, 336
  monitoring, 337
Copper, hair, normal value, 308
    occurrence in, 304
Cotton dust, 233, 546
Covariance analysis, animal exposure studies, 81
Critical organ concept, ionizing radiation, 385

Dalton's law, altitude physiology, 527
Data analysis requirements, 43-97
Data automation, 99-189
  advantages, 152
  justification, 118
  need, 99
Data automation application, 114
  air pollution abatement, 140
  effluent monitoring, 140
  electronic library, 135
  energy management, 140
  industrial hygiene, 115, 128, 136, 160
    environmental sampling, 136, 140
    monitoring, chemical and physical agents, 136
    records keeping, 160
    ventilation systems, testing, 136
  information, search and recovery, 130
  laboratory automation, 138
    chemical monitoring, 136, 143
    computerization, 136
    instrumental analysis, 136

INDEX

medical surveillance, 144
  blood examination, 148
  cardiovascular system, 137, 138
  digestive system, 137, 138
  electroencephalograms, 138
  medical examinations, 144, 145, 146, 148
  physiological monitoring, 137, 138
  radionuclide scanning, 138
  records keeping, 160
  respiratory system, 137, 138
  urinalysis, 148
 users inquiry, 115
Data automation information sources:
  Environmental Protection Agency, 118
  facility engineering, 118
  industrial hygiene records, 118
  manufacturing engineering, 118
  medical records, 118
  Occupational Safety and Health Administration, 118
  personnel records, 118
  purchasing departments, 118
Data collection, evaluation, 7
 purpose, 6, 7
Decompression sickness, 536
 clinical manifestations, 538
 pathology, 536
 prevention and treatment, 538
Desorption, analytical, efficiency, 193, 195, 202
 methods, 204
Dichlorodiphenyltrichloroethane(DDT), biological monitoring, 345
Dicumerol, 266
Diet, effect on metabolism, 267
Diethyl ether, pulmonary excretion, 296
Dilution ventilation, 35-38
Direct reading colorimetric indicators, 239
Dispersion estimates, air pollution, 34
Distribution, industrial hygiene data, 48
 log normal distribution, 53
 normal distribution, 50
Dose-response, 45, 46, 568
 curve, 45, 46, 288
 low risk levels, models, 46
  Crump, 46
  Hartley-Sieken, 46
  Mantel-Bryan, 46
 threshold levels, 45
Dosimeters, 238
Dosimetry, ionizing radiation, 365-369

Drug abuse, 226, 332

Ear protection, *see* Hearing protection
Ebullism, 526
Education, management, 8, 9
 worker, 8, 9, 598, 606, 619
Electron spectroscopy, 209
Emission inventory, 11-41
 emission factors, 19-23
 estimates from records and reports, 26-34
 fugitive sources, 23-26, 223, 229
 identification, agents, 12, 13
  emission sites, 13
 quantitating emissions, 14-26
 records retention, 39, 40
 time factors, emission, 13, 14
 uses, inventory, 34-39
Employees' rights and duties, occupational safety and health laws, 706
Employer's duties, Occupational Health Safety and Health Act of 1970, 693
Enforcement, OSHA, 681, 728
Engineering controls, philosophy and management of, 611-646
 adequate controls, determination of, 612
 alarms, 641
 control level, 613
 evolution of, 612
 fail-safe design, 641
 feasibility, definition, 642
 hazard classification system, particulates, 613, 614
 heat stress, exhaust hoods, 644
 heterogeneous bulk materials, 644
 housekeeping, 325
 innovative devices, 640
 manual work, large projects, 644
 monitoring, 638, 641
 motivation, 621
 moving sources of contaminant, 644
 noise specifications, 640
 OSHA, definition, 611
 process change, 616
 process interface, 615
 process interlock, 638
 purchase specifications, 639
 spray painting, 645
 substitution, 616
 unsolved problems, 644
 worker participation, 619
Environmental exposure limits, 354

Environmental Pesticide Control Act of 1972, 685
Environmental Protection Agency, 14, 19, 425
Environmental variability, 227, 250
Enzyme, variation, 265
Epidemiological studies, 5, 44, 328, 345
  methodology, 44
  research needs, 45
Error, analytical measurement, 191, 193
  variance, 51, 251
Ethanol, metabolism, 271
    effects on trichloroethylene exposure, 231, 271, 333
Ethyl benzene, conversion to mandelic acid, 286
Evaluation, exposure to, abnormal pressures, 525-542
  biological agents, see Biological agents
  chemical agents, 319-357
  hot and cold environments, 447-464
  ionizing radiation, 359-404
  noise, 425-445
  nonionizing radiation, 405-424
  vibrations, 465-524
Evaporative cooling, 636
Exhaust ventilation systems, hoods, 637
  inspection of, 637
  maintenance, 637
  noise from, 638
  static pressure taps, 637
  trouble shooting, 638
Expected values (mean), data analysis, 50
Experimental design, 47
  animal exposure studies, 75
    experimental error control, 79
    experimental error estimates, 78
    philosophy, 76
    sources of variation, biological data, 77
    total error of treatment mean, 83
  role, statistician, 47
  subject matter specialist, 47
Exposures, general, confidence limits, 243
  episode, 225
  evaluation, 220, 340
  indices, 233
  ingestion, 226
  inhalation, 224
  inhalation and skin absorption, additive, 225, 226
  measurement, 219
  nonoccupational, 226
  phenol, 225
  poultice effect, 226
  regulation, 219
  skin absorption, 225, 226
  sources, 224-227
  specificity, 233
  time, 247
  work place, 35
Eye protective devices, 663

Factoral design, 230
Farmer's lung, 549
  agents, 549
  diagnosis, 549
  pathogenesis, 549
  pathology, 549
  prevalence, 549
  prevention, 550
  symptoms, 549
  treatment, 550
Fatigue, vibration exposure, 483, 488
Federal Aviation Act of 1958, 686
Federal Consumer Product Safety Act, 688
Federal Mine Safety and Health Act of 1977, 684
  Coal Mine Health and Safety Act of 1969, 684
  Metal and Non-Metal Mine Safety Act of 1966, 685
Federal Noise Control Act of 1972, 686
Federal Toxic Substances Control Act of 1976, 12, 687
Fetal exposure, xentobiotics, 280
Filters, 199
  cellulose ester, 200
  glass fiber, 199
  nuclepore, 212
  polycarbonate, 200
  polyvinyl chloride, 201
  polyvinyl-acrylonitrile, 201
  preparation for analysis, 203
  silver, membrane, 201
  teflon, 199
Fluoride, biological monitoring, 345
  urinary level versus exposure, 294
Footwear, 663. See also Personal protection
Fugitive emissions, 23, 223, 229
Future of job safety and health laws, 729

Gasoline, 236
General Air, measurement option, 238
General duty clause, elements of, OSHA of 1970, 693

INDEX

causing or likely to cause death or severe
physical harm, 695
"free" of hazard, National Realty and Construction Company, 693
"recognized" hazard, *American Smelting & Refining Company v. OSAHRC,* 694
General health, maintenance examinations, 597
multiphasic screening, 597
worker education, 598
Geometric mean, data analysis, 54, 229
Geometric standard deviation, data analysis, 54, 229
Gloves, 663. *See also* Personal protection
Glue sniffing, 227
Governmental regulations, chronological log, 324
compliance with, 321
personal protective devices, 324
recordkeeping and reporting, 322
Grab samples, 239, 247
Grain fever, 551
exposure sources, 551
physiological effects, 551
prevention, 551
symptoms, 551

Hair, factors affecting metal content, 305
growth, 304, 305
metals reported in, 304-310
Half-life, biological, 261
excretion variation, 284
sex, 283
sampling time, 247
Halogenated hydrocarbons, biological monitoring, 345
Hazard classification, particulates, 613, 614
Hazardous substances act, 689
Health hazards, detection, 4, 5
general, 1
measurement, 5, 6
Health surveillance programs, industry, 595-609
hazard-oriented, 595
health maintenance, general, 595
Hearing protection, 434, 443
Heat, 343, 456-462
Heavy equipment, vibration in, 471, 480
Helicopters, adverse effects, 478, 482
vibration in, 478, 482
Hemoglobin, altitude physiology, 528
Hemp, 548
Henry's law, abnormal pressure, 18, 537
Hertzian (radio) waves, 418-423
biological effects, 419

exposure control, 423
exposure standards, 423
measurements, 422
physical characteristics, 418
sources, 419
High altitude pulmonary edema, 535
Hippuric acid, delayed urinary appearance, 278, 284
formation from styrene, 278
normal concentration, urine, 275, 279
unsuitability, styrene exposure measurement, 278
urine levels, toluene exposure, 278
Histoplasmosis, 559
distribution, 559
exposure sources, 559
immunology, 559
roentgenography, 559
soil disinfection, 559
Hot environment, 456-462
physiological response, 456-462
acclimatization, 458
aerobic power, 460
age, 460
ascorbic acid, 459
cardiovascular system, 457
diurnal rhythm, 456
fluid balance system, 457
heat disorders, 461
heat stress indices, 461
menstruation, 460
physical relationships, 457
plasma volume, 458
salt depletion, 457
sex differences, 459
sweating system, 457
thermoregulation, 456
water deficits, 457
Human milk, *see* Breast milk
Hydrogen sulfide, monitoring, 335
Hyperbaric chambers, carbon monoxide poisoning, treatment, 539
gas gangrene, treatment, 539
monoplace, 539, 541
multiplace, 539
Hypoxia, altitude related, 530
manifestations, 531
prevention, 532

Imminent danger situations, 728

Immunological manifestations, vibration syndrome, 508
Index chemicals, 272
  concentration variations, 273
  definition, 258
  excretion levels, misleading, 272
  interferences, 284, 285
  normal concentration range, bioassay materials, 274, 275, 283
  normalizing urinary values, 274, 275
  progenitors, 276
  removal, interfering chemicals, 284
  urinary variation, sex related, 282
  urine, analysis for, 265, 286
  variations, analytical methods, 289
Inductively coupled plasma-optical emission spectroscopy, 206, 207
Industrial hygiene:
  analytical measurements, 191-215
  chemical dosage and burden, 257-318
  compliance and projection, 719-736
  computer programs, 99-189
  data analysis, requirements, 43-97
  data automation, computerization, 99-189
  definition, 2
  emission inventory, 11-41
  engineering controls, 611-646
  exposure evaluation, abnormal pressure, 525-542
    biological agents, 543-565
    chemical agents, 319-357
    hot and cold environment, 447-464
    ionizing radiation, 359-404
    noise, 425-445
    nonionizing radiation, 405-424
    vibrations, 465-524
  extrapolation, toxicologic data, 567-594
  health surveillance programs, 595-609
  job safety and health laws, 681-718
  personal protection, 647-680
  practice, rationale, 1-9
  worker exposure measurement, 217-255
Informing worker, 606
Infrared radiation, 415-418
  biological effects, 416
  exposure control, 418
  exposure standards, 418
  measurements, 417
  physical characteristics, 415
  sources, 415
Infrasound, exposure limits for, 479

sensations due to, 479
Ingestion, see Routes of entry
Inhalable dusts, 233
Inhalation, see Routes of entry
Innovative devices, engineering control, 640
Instrumental methods, 204, 206, 209
  anodic stripping voltammetry, 206, 209
  atomic absorption spectrophotometry, 207, 209
  chromatography, 205
    gas, 207, 208, 209
    ion, 212
    liquid, 209
  electron spectroscopy, 209
  inductively coupled plasma emission, 208
  multiple substance, 206
  speciation by, 205
  spectroscopy, derivative, 212
  x-ray diffraction, 205, 209
  x-ray fluorescence, 206, 211
Investigations and inspections, 719
  constitutionality, warrentless inspections, 720
  inspection procedures, 720
    Accu-Namics, 720
    Chicago Bridge & Iron Company v. OSAHRC and Dunlop, 720
Ionizing radiation, bioassay techniques, 394-397
  external radiation exposure, 394
  internal radiation exposure, 395, 396, 397
    inhalation, 395
    puncture wounds, 397
  biological effects, 369-383
    electromagnetic radiation, neutrons, acute effects, 369
    intermediate effects, 369
    late effects, 372
    radionuclide radiation, 373, 374
      bone, effect on, 378
      lung, effect on, 376
      soft tissue, effect on, 378
  dosimetry, 365-369
    electromagnetic radiation, 365
    neutrons, 365
    particulate radiation, 366
      localized organ retention, 368
      particle distribution, 366
      respiratory tract deposition, 366
      respiratory tract retention, 367
  standards and guidelines, 385
    critical organ concept, 385

INDEX

dose-limiting recommendations, 386, 387
modifying dosage factors, 386
therapeutic measures, 387-394
  external radiation exposure, 387
    Lockport incident, 388, 389
  internal radiation exposure, 388
    chelation, 390
    pulmonary lavage, 391
Ionizing radiation energy sources, 397-399
  mining, uranium, 399
  mixed hazards, 398
  nuclear fission, 397
Ionizing radiation measurement, air monitoring, continuous, 364
  alpha particles, 364
  beta particles, 364
  electromagnetic, 365
  neutrons, 365
Ionizing radiation types, 360
  external, 360
    Compton scattering, 360
    gamma rays, 360
    neutron, 361
    pair production, 360
    photoelectric effect, 360
    x-rays, 360
  internal, 361
    alpha particles, 361
    beta particles, 361
Iron, hair, occurrence in, 304
  urine, analysis for, 293

Job energy requirements, 350
Job safety and health law, 681-718
  employees' rights and duties under, 706
  employer's duties under Occupational Safety and Health Act, 693
    general duty clause, 693
    specific duty clause, 695
      challenging validity of standards, 696
      economic and technological feasibility of standards, 697
      environmental impact of standards, 699
      process of promulgating standards, 695
      variance from standards, 700
  federal regulatory schemes other than OSHA, 684-690
    Atomic Energy Act of 1954 and other statutory sources of radiation control, 689
    Department of Agriculture, 690
    Department of Interior, 690
    Department of Labor, 689
    Environmental Protection Agency, 689
    Nuclear Regulatory Commission, 689
    Wash-Healy Public Contracts Act, 690
    Environmental Pesticide Act of 1972, 685
    Federal Aviation Act of 1958, 686
    Federal Consumer Product Safety Act, 688
    Federal Mine Safety and Health Act of 1977, 684
    Federal Noise Control Act of 1972, 686
    Federal Toxic Substances Control Act of 1976, 687
    Hazardous Materials Transportation Act, 686
    Hazardous Substances Act, 689
    Natural Gas Pipeline Safety Act of 1968, 686
    Outer Continental Shelf Lands Act, 690
    Railroad Safety Act of 1970, 685
  legislative history and background, 681
    agencies responsible for implementing and enforcing OSHA, 681
    ad hoc advisory committees, 682
    Department of Labor, 681
    National Advisory Committee on Occupational Safety and Health, 682
    National Institute for Occupational Safety and Health, 682
    Occupational Safety and Health Administration, 682
    Occupational Safety and Health Review Commission, 682
  occupational safety and health standards, 700-706
    asbestos, 702
    carcinogens, 703
    coke oven emissions, 704
    mine industry health standards, 705
    proposed standards, 704
    vinyl chloride, 703
  regulation of job safety and health by states, 690
Jute, 548

Ketones, biological monitoring, 345
Kinetosis, 471

Labels and warnings, 436, 702
Lead:
  administrative control, 336
  analysis for, blood, 302, 303
    urine, 293, 294

biological monitoring, 345
clinical cases, 332
hair, normal values, 308
  occurrence in, 304
maximal acceptable blood level, 303
urinary level versus exposure, 293
Legal considerations, health surveillance,
    exposure measurements, 219
  informing worker, 606
  wearing of respirators, certification of fitness, 606
Lint, 234
Lithium, in urine, 293
Los Alamos Scientific Laboratory, respirators, 659, 666
Low back pains, vibration and, 480, 482

Magnesium, urine, excretion, 293
Mandelic acid, ethyl benzene from, 286
  urine, normal values, 277
  styrene exposure, 284, 285
Manganese, hair, occurrence in, 304
  urine, analysis for, 293
Materials balance, 16-19
Maximum risk employee, 220
Mean ratio, proficiency testing, 197, 198
Measurement options, 237-242
Mechanism of toxicity, extrapolation related, 590
Medical examinations, hazard-oriented, 595
  justification, 600-603
    biochemical marker of cancer, 603
    cytologic examination, 602
    early detection of cancer, 601
    indicators of absorption, 600
    indicators of early effect, 600
    indicators of hypersusceptibility, 601
  medical surveillance, legally required, 598
    Federal Coal Mine Health and Safety Act of, 1969, 599
    OSHA, current, 598
      acrylonitrile, 598
      arsenic, 598
      asbestos, 598
      benzene, 598
      coke oven emissions, 598
      1,2,3-dibrom-s-chloropropane, 598
      suspected or proven carcinogens, 598
      vinyl chloride, 598
    proposed, 598
      asbestos, 598

beryllium, 598
lead, 598
medical surveillance recommended by NIOSH, 599
Mercury, administrative control, 336
  hair, normal values, 308
  occurrence in, 304
Metabolism, see Xentobiotics
Methanol, metabolism, 271
  pulmonary excretion, 296
Methods equivalence, 222
Methyl acetate, pulmonary excretion, 296
Methyl chloroform, skin absorption, 298
Methylene chloride, pulmonary excretion, 296, 302
  skin absorption, 298
Microsomal enzymes, inhibition, 267, 268
  oxidase, 266
  stimulation, 267, 268
Microspora faeni, 546
Mine Enforcement and Safety Administration, 324, 326, 656, 669, 672
Mixed atmospheres, airborne contaminants, 235, 241
  local effects, body organs, 235
  mist-vapor atmospheres, 231
Modes of entry, see Routes of entry
Molybdenum, hair, occurrence in, 304
Motion sickness, 471
Motivation, engineering control, 621
Mountain sickness, 535-536
  acute, 535
  chronic, 536
  pulmonary edema, 535
Multiphasic screening, 597
Mushroom growing, 555
Mutagens, 343
Mycobacterium marinum, 562
  sites of infection, 562
  skin tests, 562
  sources of exposure, 562

Nail analysis, measure of exposure, 312
National Advisory Committee on Occupational Safety and Health, 682
National Gas Pipeline Safety Act of 1968, 686
National Institute for Occupational Safety and Health:
  annual list of all known toxic substances, 682
  approved respirators, 669
  criteria documents, 100, 682

INDEX

gas masks, guidance in use, 657
hazard-oriented medical examinations, 598
medical surveillance guidelines for chemical hazards, 599
personal protective devices, 324
Proficiency Analytical Testing, 199
research of better respiratory protective devices, 661
research of better testing methods, 659
respirator approved system, 656
sampling strategy, 322
supplied-air respirators, 658
Neurological decompression sickness, 538
Neurological effects, vibration, 476, 483, 511
Noise, 425-445
  compliance evaluation, 444
    audio dosimeter, 444
    sound level meter, 444
  existing regulations, OSHA, 428
  exposure criteria, occupational, 426
    study data, 427
    test difficulties, 426
  legal aspects, 442
    apportionment, 443
    date of injury, 442
    hearing protection, 443
    presbycusis, 442
    waiting period, 442
  neighborhood noise, 440
  nonauditory effects, 426
    catecholamines, 426
    nausea, 426
    nervous reaction, 426
    startle reaction, 426
  nuisance noise, 439
  physiological effects on workers, individual reaction, 425
    audiometric testing, 425
    EPA, 425
    OSHA, 426
  present need, 444
  proposed regulation, OSHA, 429-438
    application and purpose, 429
    definition, 429
    hearing conservation, audiometric testing, 435
    information and warning signs, 436
    methods of compliance, engineering and administrative control, 434
    personal protection, 434
    monitoring, 431

    permissible exposure limits, 430
    records, audiometric tests, 436
      noise exposure measurements, 436
  speech communication, 440
Noise from exhaust systems, 638
Noise vibration and stress, 484
Nonionizing radiation, 405-424
  Hertzian (radio) waves, 418-423
  infrared radiation, 415-418
  ultraviolet radiation, 407-411
  visible light, 411-415
Nonliving biological agents, 545-557
  immunologically mediated effects, 545
    Arthus phenomenon, 546
    cell mediated immunity, 546
    diffuse fibrosis alveolitis, 546
    hypersensitivity pneumonitis, 546
    immune reaction, 545
  modes of action, 545
    histamine release, 545
    pharmacological effect, 545
  nature of materials, 545
    lipoproteins, 545
    mucopolysaccharides, 545
    split proteins, 545
Novel work shifts, 247
Nutrition, trace elements, 3

Occupational health limits, organizations recommending, 352
Occupational Safety and Health Administration:
  audiometric testing program, 426
  computer, automatic monitoring for vinyl chloride, 143
  contesting citations, 726
  criminal sanctions, 726
  employee duties and rights, 706
  engineering controls, definition, 611
  environmental impact of standards, 699
  feasibility, definition, 642
    standards, 697
  health standards, 321, 354, 700
  inspection procedures, 720
  labeling and warning signs, 436, 702
  legislative history, 681
  medical surveillance requirements, 598
  monitoring, legal requirements, 219
  noise, existing regulations, 428
    proposed rules, 429
  record keeping and reporting, 39, 99, 721
  respirator use, medical requirements, 676

respiratory protection, use of, 647
scope of coverage, 682
selection, use and care of respirators, 673, 677
specific duty clause, 695
violations, 723
Occupational Safety and Health Review Commission, 682
On-site consultation programs, 730
Ornithosis, control, 561
prevalence, 561
sources of exposure, 561
Outer Continental Shelf Lands Act, 690
Oxygen, hyperbaric, 525
transport, 528

Parathion, 340
Personal habits, drugs and stimulants, 4, 332
eating, 332
smoking, 4, 332
Personal protection, 434
Personal sampling, 238, 243
Perspiration, measurement of exposure, 312
Pharmacokinetics, toxicity extrapolation, 582
Phenol, analysis, interfering chemicals, 285
dietary sources, 278, 280
medicinal sources, 280
skin contact, 225
unsuitability for measuring benzene exposure, 278
urinary excretion half-life, 296
urine, normal concentration, 277, 281
Phenylglyoxalic acid, 277
Physical examinations, periodic, 596
preassignment, 596
preplacement, 596
specific purpose, 597
termination, 597
Plant debris, cotton dust, 233
Polynuclear aromatic hydrocarbons, 233-234
Potassium, hair, occurrence in, 304
Precision, analytical, 191, 192
Predicting human toxicity, 577
reliability in animals, 577
Presbycusis, 442
Pressure, abnormal, 349, 525-542
physics of, 526
Process changes, cost offset of, 617
Processes, evolution of, 616
Proficiency testing, analytical, 197, 199
Progenitors, index chemicals, 276
Promulgating standards, process of, 695

Protective clothing, 226
Pulmonary excretion, 295-302
Purchasing specifications, 639

Q fever, 561
control, 562
sources of exposure, 561

Railway Safety Act of 1970, 685
Raoult's law, material balance, 18
Rationale, industrial hygiene practice, 1-9
Raynaud's phenomenon, 492, 498, 500, 502, 503.
    See also Vibration syndrome, white fingers
Reactivity, exposure agents, 232
Recirculation, clean air, 633
Record keeping and reporting, 170-176
industrial hygiene, 322, 436, 721
legal requirements, 721
medical, 436, 603, 721
Reference materials, analytical, 192, 195
Research, animal studies, 327, 330
antagonism, 330
carcinogens, 328
clinical cases, 327
epidemiology, 327, 328
synergism, 4, 330
Residual, proficiency testing, 198, 199
Respirators, approved or accepted, 669
Bureau of Mines approval system, 651
chemical cartridge, 655
cleaning and sanitizing, 674
communication, 648
dust, fume, mist, 655
fatigue, 649
fitting, 648
gas masks, 653
immediately dangerous to life or health atmospheres, 665
inspection and maintenance, 674
maximum use limits, 667, 671
medical requirements, 676
NIOSH/MESA approval system, 656
oxygen deficiency, 665
pesticide, 658
program administration, 673
protection factors, 667
selection, 664
self-contained breathing apparatus, 652
storage, 675
supplied air, 655

INDEX 749

supplied air suites, 662
training, 675
use, 672
vision, 648
Respiratory protection, history, 647
  rationale for use, 647
  research, 659
Respiratory rate, pulmonary excretion, 298
Response time, instrumentation, 236
Routes of entry, ingestion, 226, 259, 272, 339, 544
  inhalation, 259, 338, 544
  skin absorption, 225, 226, 259, 264, 272, 340, 543

S. typhimurium T A 100, mutations in, induced by various aromatic amines, 267
Saliva, exposure, measure of, 312
Sample processing, 203
  analyte loss, 203
  ashing, 203, 204, 205
  methods, 203, 204, 205
Sampling, general, 193, 199
  analytical, 321
  automatic monitoring, 240
  bias, 191, 192
  breathing zone, 238
  calibration, 321
  chronological log, 324
  classification of methods, 197
  collection efficiency, 193, 202
  continuous, 240
  critical orifice, 201
  devices, 199
  equipment, 321
  errors, 191, 192
  experimental, 618
  filters, 199
  full period measurement, 251
  general air, 238, 243
  grab, 239, 247
  isokinetic, 240
  location, 69, 70
    compliance sampling, 70
    guidelines, general, 70
    methods, format, 196
      HIOSH/OSHA, 200
      personal, 200
      validation criteria, 193
  nitrogen dioxide, 202
  number, 250

options, 237-242
period, 236, 239, 247
phosphorus vapor, 202
problems, 213
processing, 204
quality control, 322
reactive liquids, 202
record keeping, 322
sequential, 240, 253
size, detect difference in data, 86-88, 90
  incidence data, 89
  quantitative measurements, 86
sorbents, 202, 231, 239
  activated charcoal, 201
  breakthrough capacity, 202
  coconut charcoal, 202
  disorption efficiency, 202, 213
  passive, 202
  porous polymers, 201
  silica gel, 201
  triethanolamine, 202
strategy, 218, 237, 242, 244, 245, 320
surveillance, 323
task oriented, 618, 619
time-weighted average, 249, 619
validation criteria, 193
vinyl chloride, 202
weather, 324
wipe, 340
Sanctions, safety and health law violations, 722
  citations, 722
  contesting citations and penalties, 726
  criminal sanctions, 726
    California Stevedore, 728
    Dan J. Sheehan, 727
  failure to abate violation, National Realty & Construction Company, 726
  penalties, 722
    De Minimis violation, 724
    nonserious violations, 723
    serious violations, *California Stevedore and Ballast,* 723
    willful and repeated violations, *Bethlehem Steel Corp. v. OSHARC,* 724
    Frank Irey Jr., Inc. v. OSHARC, 724
    George Hyman Construction Co., 725
Selenium, hair, occurrence in, 304
Sex related variations, 282, 459, 460
  biological monitoring, 282, 298
  blood volume, 298
  half-life urinary excretion, xentobiotics, 282

index chemicals, 282
xentobiotics, 282
Ships, vibration and motion in, 471, 485
Short term exposure limits, 237
Silica, old versus new evaluation methods, 330
Silicon, hair, occurrence in, 304
Sisal, 548
Size selection, 240
Skin absorption, 225, 226, 259, 264, 272
Skin protective devices, 662
Small industries, medical surveillance, 605
Solid sorbents, 201, 231, 239
Source sampling, 14-16
Spectroscopy, derivative, 212
Spine, whole-body vibration and, 480
Standard deviation, data analysis, 51
Standards, international, vibration, 487
    hand transmitted, 513
    whole-body, 486
Standards (OSHA):
  challenging validity of, 696
  economic and technological feasibility of, 697
    American Federation of Labor, Etc. v. Brennan, 697
    Atlantic & Gulf Stevedores, Inc., 699
    Brennan v. Butler Lime & Cement C., 699
    Brennan v. OSAHRC & HENDRIX (d/b/a Alsea Lumber Co.), 698
    Horne Plumbing and Heating Company v. OSAHRC, 699
    Industrial Union Department, AFL-CIO v. Hodgson, 698
    National Realty, 699
  emergency temporary standards, *Florida Peach Growers*, 696
  environmental impact, 699
  existing standards, 702-704
  interim standards, 695
  permanent standards, 696
  preemption, OSHA standards, 683
    Organized Migrants in Community Action Inc. v. Brennan, 684
    Southern Pacific Transportation Company, 683
    Southern Railway Company, 684
  promulgating standards, 695
  proposed standards, 704
  statutes other than OSHA, 683
    Atomic Energy Act of 1954 and other statutory sources of radiation control, 689
    Department of Agriculture, 690
    Department of Interior, 690
    Department of Labor, 689
    Environmental Protection Agency, 689
    Nuclear Regulatory Commission, 689
    Walsh-Healy Public Contracts Act, 690
    Environmental Pesticide Act of 1972, *Organized Migrants in Community Action*, 685
    Federal Aviation Act of 1958, 686
    Federal Consumer Product Safety Act, 688
    Federal Mine Safety and Health Act of 1977, 684
    Federal Noise Control Act of 1972, 686
    Federal Toxic Substances Control Act of 1976, 687
    Hazardous Materials Transportation Act, 686
    Hazardous Substances Act, 689
    Natural Gas Pipeline Safety Act of 1968, *Texas Eastern Transmission Corp.*, 686
    Outer Continental Shelf Lands Act, 690
    Railroad Safety Act of 1970, 685
  variances from, 700
State plans, safety and health laws, 690
Statistical analysis, exposure measurements, 59
  questions addressed to, 59
  test of significance, 60
    hypothesis test for compliance officer, 61
    hypothesis test for employer, 60
    power function, 62
    type I and II errors, 62
Statistical application, general, 43
Statistical design, methodology, 43-97
Statutes other than OSHA, see Standards (OSHA)
Stress, off-the-job, annoyance, 4
  diet, 3
  drugs, 4
  self medication, 4
  smoking, 4
  stimulants, 4
Stress effects, whole-body, vibration, 475, 477
Styrene, metabolism, 276
  pulmonary excretion, 296, 300, 301
  urinary excretion half-life, 296
Subchronic toxicity, 573
Subtilin, 556
Suits, personal protection, 662
Synergistic effect, 349

Talc, 338

INDEX

Target organs, 259
Tears, measure of exposure, 312
Temperature, abnormal, 349. *See also* Cold environment; Hot environment
Teratogens, 343
1,1,2,2-Tetrachloroethylene, pulmonary excretion, 296, 300
   skin absorption, 298
   urinary excretion, half-life, 292
Tetraethyl lead, biological monitoring, 345
   skin absorption, 340
Tetraethylthiuram, disulfide, effect of copper excretion, 272
Thallium, hair, occurrence in, 304
Therapeutic accident, injection exposure, 226
Thermopolyspora faeni, 545
Thesaurosis, 546
Threshold, response, 579
Threshold limit values, 7, 220, 236, 259, 262, 321, 334, 342, 349, 354
Time weighted average, 249
Titanium, hair, occurrence in, 304
Tobacco, 555
Toluene, metabolism, 276
   biological monitoring, 345
   pulmonary excretion, 296
   urinary excretion half-life, 292
Tools, vibration, 490, 496, 510
Toxicologic data extrapolation, 567-594
   acute toxicity, 572
      route of exposure, 573
      single versus repeated exposure, 572
      species differences, 573
   animal reliability, 577
   dose-response, 568
      cumulative response, 569
      differential safety factors, 571
      population distribution response, 569
   pharmacokinetics, extrapolation, toxicologic data, 582
   subchronic and chronic toxicity, 573
      dose-response data application, 575, 576
      inhalation exposures, 576
      length of exposures, 575
   threshold, response, 579
      carcinogenesis, teratogenesis, mutagenesis, 581
      dose-response curve, 579
   toxicity mechanism versus data extrapolation, 590
Toxic Substances Control Act (P L 94-469), 12, 331, 687
Trace elements, essential dietary elements, 3
   nutrition, 3
   toxicity, 3
Tractor, vibration effects, 480
Training, management, 8, 9
   workers, 8, 9, 326
   see also Education
Transmissibility, vibration, 474, 509
Traumatic vasospastic disease, vibration syndrome, 492
Trichloroacetic acid, delay, urinary appearance, 284
1,1,1-Trichloroethane, pulmonary excretion, 296
   urinary excretion half-life, 292
1,1,2-Trichloroethylene, pulmonary excretion, 296, 300
   skin absorption, 298
   urinary excretion half-life, 292

Ultraviolet radiation, 3, 407-411
   biological effects, 407
   exposure control, 411
   measurement, 410
   physical characteristics, 407
   sources, 407
   standards, 410
Urinalysis:
   creatinine correction, 275, 276
   disease, effect on, 274
   drugs, effect on, 274
   excretion rate, correction, 274
   metallic ions, analysis for, 293
   organic compounds, 290
   osmolality, correction, 274
   specific gravity, correction, 274, 275
   spot versus 24 hour sample, 289
   sulfate, correction, 274

Validation, analytical, sampling techniques, 193
   threshold limit values, 262
Vapor pressure, 231
Variance, 51, 236
Variance from standards, 700
Vehicle, vibration, 480
Ventilation:
   dilution, 35, 324
   energy conservation, 325
   evaluation, local exhaust, 335
   general room, 324
   local exhaust, 325

measurements, 335
Ventilation design, blast gates, 624
Vertical elutriator, 233, 234
Vibration, 465-524
  complexity of, 467
  factors affecting human response, 466
  frequency, 466
  hand-transmitted, 490
    energy absorption, 510
    international exposure standard, 486, 512
    morbidity from, 510
  intensity, affecting man, 468
  international exposure standards, 486, 487
  measurement of, 466, 468
  mechanical, definition, 465
  whole-body, 469, 472-474
    acute pathological effects, 479
    biodynamic effects, 473
    cardiopulmonary effects, 475
    direction of and human response, 469
    disequilibrium, 476
    effects, task performance, 478
    electroencephalograms in, 478
    etiology of disorders, 480
    international exposure standard, 486
    mitigating effects in man, 489
    neurophysiologic effects, 476
    occupational exposure to, 480
    physiologic response to, 474
    protection from, 485, 488
    resonance due to, 473
    sensation of, 476
    stress due to, 475, 477, 483, 485
Vibration syndrome, 490-517
  biodynamics of, 474, 509
  bone and joint changes, 506, 511
  climatic factors, 501, 507
  clinical tests for, 510
  etiology, 507
  factors influencing risks, 497, 516
  immunological manifestations, 508
  neurological signs, 504, 511
  pain, 498, 504
  prevalence in industry, 509
  prevention and treatment, 512
  white finger, 492
  severity, grading, 503
  see also Raynaud's phenomenon
  women, occupational exposure, 484
Vinyl chloride, pulmonary excretion, 296, 302
  standard, 703

Visible light, 411-415
  biological effects, 412
  exposure control, 415
  measurements, 414
  physical characteristics, 411
  sources, 411
  standards, 415

White finger, vibration, 492
Whole air samples, grab, 239
Women, occupational exposure, vibration, 484
Wood dust disease, 533
  cork, 554
  general features, 553
  maple bark disease, 554
  red cedar, 555
  sequiosis, 554
Wool sorter's disease, anthrax, 559
  anthrax pneumonia, 559
  disinfection, 560
  exposure source, 559
  prevalence, 559
  prevention, 559, 560
Worker exposure measurements, 217-255

Xentobiotics:
  accumulation, 263
  age, effects on, 280
  blood level variations, 302
  competition for enzymes, 270
  definition, 257
  elimination half-life, 266, 292
  excretion, 263
    hair, 304-310
    perspiration, 312
    pulmonary, 295, 296, 300
    milk, 310-312
    nails, 312
    urine, 265, 290, 293
  factors affecting, 265
  fetal, exposure to, 282
  metabolism, 263-267
  metabolism, variation, exposure routes, 286, 296-298
  saturation, metabolism, routes, 270
  storage, 263
Xylene, urinary excretion half-life, 292

Zinc, analysis, urine, 293
  hair, normal values, 308
  occurrence in, 304
Zoönoses, 560